TOWN AND COUNTRY PLANNING IN THE UK

Fourteenth edition

This extensively revised fourteenth edition of *Town and Country Planning in the UK* incorporates the major changes to planning introduced by the Planning and Compulsory Purchase Act 2004 and the government's mission to change the culture of planning. It provides a critical discussion of the system of planning – the institutions involved, the plans and other instruments that are used, the procedures for controlling development and land use change, and the mechanisms for implementing policy and proposals. It reviews current policy for sustainable development, housing and the Sustainable Communities Plan, the Barker Review, urban renewal and regeneration, the renaissance of city and town centres, the countryside, transport, and the heritage. Contemporary arrangements are explained with reference to their historical development, the influence of the European Union, the Labour government and changing social and economic demands for land use change.

Detailed consideration is given to

- the nature of planning and its historical evolution
- central, regional and local government, and the devolved administrations
- the EU and its environmental and regional policies
- mechanisms of controlling development
- policies for managing urban growth and delivering housing
- sustainable development principles for planning
- social and economic development of the countryside
- planning the natural environment, waste and pollution control
- conserving the heritage
- urban renaissance and regeneration
- community engagement in planning
- changes to the profession and education of planners.

Special attention is given to the objective of improving the coordination of government policies through the spatial planning approach. The many recent changes to the system are explained in detail – the new national policy statements and plans, regional spatial strategies and local development frameworks in England and other arrangements in Scotland, Wales and Northern Ireland; new forms of land use regulation; sustainability appraisal and strategic environmental assessment; community engagement and relations between planning and community strategies; partnership working; changes to planning gain; and new initiatives in urban and housing renewal.

Each chapter ends with notes on further reading and at the end of the book there are lists of official publications and an extensive bibliography, enhancing its reputation as the bible of British planning.

Barry Cullingworth was a Senior Research Fellow in the Department of Land Economy at the University of Cambridge, UK and Emeritus Professor of Urban Affairs and Public Policy at the University of Delaware, USA.

Vincent Nadin is Reader in the Centre for Environment and Planning at the University of the West of England, Bristol, UK; Visiting Research Fellow in the OTB Research Institute, Delft University of Technology, the Netherlands; and Visiting Research Fellow in the Institute for Environmental Planning, University of Hannover, Germany. He is editor of *Planning Practice and Research*.

TOWN AND COUNTRY PLANNING IN THE UK

Fourteenth edition

WITHDRAWN FROM STOCK

Barry Cullingworth and Vincent Nadin

Routledge
Taylor & Francis Group

LONDON AND NEW YORK

First published 1964
Fourteenth edition published 2006
by Routledge
2 Park Square, Milton Park, Abingdon, Oxon OX14 4RN

Simultaneously published in the USA and Canada
by Routledge
270 Madison Ave, New York, NY 10016

Routledge is an imprint of the Taylor & Francis Group, an informa business

Typeset in Garamond by Keystroke, Jacaranda Lodge, Wolverhampton
Printed and bound in Great Britain by Bell & Bain Ltd, Glasgow

British Library Cataloguing in Publication Data
A catalogue record for this book is available from the British Library

Library of Congress Cataloging in Publication Data
Cullingworth, J. B.
Town and country planning in the UK / Barry Cullingworth and
Vincent Nadin.–14th ed.
p. cm.
Includes bibliographical references and index.
1. City planning–Great Britain. 2. Regional planning–Great Britain.
I. Nadin, Vincent. II. Title.
HT169.G7C8 2006
307.1′216′0941–dc22   2005028475

ISBN10: 0–415–35809–4 (hbk)
ISBN10: 0–415–35810–8 (pbk)
ISBN10: 0–203–00425–6 (ebk)

ISBN13: 978–0–415–35809–5 (hbk)
ISBN13: 978–0–415–35810–1 (pbk)
ISBN13: 978–0–203–00425–8 (ebk)

Contents

Figures

Tables

Boxes

Preface

Since 1963, when the first edition of this book was drafted, it has become increasingly uncertain what should be included under the title of *Town and Country Planning*. At one time it could be largely defined by reference to a limited number of Acts of Parliament. Such a convenient benchmark no longer exists: planning policies are now far broader. Moreover, the importance of interrelationships with other spheres of policy, which has long been accepted, is now enshrined in 'the spatial planning approach'. Planners are encouraged to engage more effectively with other areas of government policy and action that have a spatial impact. It is therefore not easy (or even useful) to define the boundaries of town and country planning. Although the scope of the book has widened over successive editions to incorporate more of the issues with which planners are concerned, it cannot claim (as did the first edition) to provide 'an outline of town and country planning and the problems with which it is faced'. Such an enterprise would now take several volumes. Beyond basic statutory and administrative matters, selection of material is a personal matter, though we hope that other teachers and planners would agree with the choice.

The task of selection has been made more challenging by the considerable publication activity of government departments and agencies. The zeal with which civil servants have produced White and Green Papers, consultation papers, research reports, good practice guidance and much more over recent years is to be admired. With the flood of reports on Creating Sustainable Communities, and implementation of the 2004 reforms, the planners' bookshelves are now weighed down by a remarkable collection of material.

And these are not modest documents; numerous reports run to hundreds of pages (though much of the content is likely to be familiar). We look forward to some research on the impact of government publications, with analysis of exactly who reads all this material and with what effect.

In preparing this fourteenth edition, we have followed the pragmatic course of updating its predecessors – adding in some parts, deleting in others. It has been necessary, given the extensive changes since the last edition, to give more attention to some aspects of planning than others, for example, the new regime of strategies and plans at the regional and local levels and policies for growth. Throughout, the intention is to explain current policy and practice with reference to their historical development. The ambition to give some reference to all four nations of the UK under each topic has proved challenging and more work is needed on some subjects. So the outcome is not always satisfactory; too many compromises have had to be made, and too much has had to be omitted. But, like practising planners, the authors have had to operate within constraints which are externally determined.

Each chapter ends with a guide to further reading. They are intended to assist students who wish to follow up the discussion in the text, but they are only an introductory guide to some of the useful available material: they are in no way comprehensive. Though there may well be a need for an annotated bibliography of planning literature, this is not the place to provide it. The literature is now so vast that the selection of titles for recommendation is inevitably a personal (and, to some extent, an arbitrary) matter. However, it is not, we hope, idiosyncratic, though no doubt other teachers may prefer alternatives. The bibliography has been expanded, and the list of official publications has been trimmed, as some of the older material is now much less relevant for most readers.

Acknowledgement is made with sincere thanks to Betty Cullingworth for her tireless support; to Janet Askew for advice on the development control chapter; to the many people who have supplied information; and to the editors and staff at Routledge for their guidance and patience.

Barry Cullingworth
Vincent Nadin

Barry Cullingworth
1929–2005

Barry Cullingworth died in February 2005 just before this edition of *Town and Country Planning* was completed. He was particularly well known for this book but had a broad and distinguished academic record. As a researcher, consultant to government and prolific writer, he made an outstanding contribution to town and country planning and urban policy.

He was born in Nottingham and started his higher education by taking a degree in music at Trinity College, London. He switched to sociology and took a degree at the University of London. In 1955 he was appointed as a research assistant at Manchester University and subsequently held lecturing and research appointments at Durham and Glasgow Universities. He published his first book in 1960, *Housing Needs and Planning Policy*, followed in 1963 by *Housing in Transition*. In 1966 he set up the Centre for Urban and Regional Studies at the University of Birmingham and in 1972 moved back to Scotland to set up the Planning Exchange.

While at Birmingham and Glasgow, Cullingworth chaired numerous government inquiries into housing and the new towns, the most well known of which was on *Scotland's Older Houses*. The *Cullingworth Report*, as it is now known, revealed the parlous condition of private rented housing across the country and set the government on a path of radical reform. In later life he expressed disappointment with the relative lack of attention given to the quality and availability of affordable housing, especially in comparison to the priority given to protecting the countryside.

By the mid 1970s Cullingworth had published ten books, numerous official reports and undertaken consultancies at home and abroad, including reports for the OECD, WHO and United Nations. He was, therefore, the ideal candidate for appointment as Historian to the Cabinet Office to prepare the *Official History of Environmental Planning 1939–69*. With the late Gordon Cherry, he published the four volumes of the *History*, between 1975 and 1981. He explains in these volumes how 'a small group of visionaries in the civil service' reconstructed the government planning machinery intending 'to achieve a far greater degree of co-ordination and purposive action'. In many publications he was to advocate a positive role for planning as initiator of coordinated land use change.

In 1978, Cullingworth moved to North America, first as Chairman and Professor of Urban and Regional Planning at the University of Toronto and from 1983 as Unidel Professor of Urban Affairs and Public Policy at the University of Delaware. When he moved to Toronto this book was in its sixth edition and recognised as the 'leading review' in the field. He continued to publish in North America including *Urban and Regional Planning in Canada*, and *Planning in the USA*, now in its second edition.

Cullingworth returned to Britain in 1994, working in an ambassadorial role for the University of Delaware; taking on a visiting position at Cambridge's Department of Land Economy; and editing *British Planning: 50 Years of Urban and Regional Policy*. In recent years the writing of both the British and American textbooks has been shared with other authors. He was always an active partner, working energetically on the later editions until 2004. He was a generous co-writer too, with a willingness to update and change. His ability to digest vast quantities of information was matched only by his persistence in getting at the facts.

Cullingworth's publications reflect his energy, enthusiasm and commitment – and sheer capacity for work. They also owe something to the invaluable support of his wife Betty. He took a considered and meticulous approach to research and writing that lends authority to his publications. But he will be best remembered as an author who could draw out the significant from the routine and deliver his message in a meaningful and engaging style. He wrote with the intention of being understood and accessible.

His family remember him as a loving and funny man with a sense of mischief. He was, of course, usually surrounded by books, but it will be a surprise to many that he had a passion for DIY, finding time alongside the research and writing to work on renovations to the many houses the family moved into. He was an accomplished pianist too, with a passion for music.

Cullingworth's publications have guided many thousands of students and practitioners over more than forty years. Despite this success, he was unpretentious and modest. While making great efforts to be comprehensive in his research he would never claim that the findings were exhaustive. He preferred instead to say that he was pointing the reader to some useful material. He did much more than that. Many more students will continue to benefit from his writing.

Barry Cullingworth devoted his life to his work and family. He is survived by his wife Betty, and his children, Wendy, Jane and Peter.

Vincent Nadin

Acronyms and abbreviations

Acronyms and abbreviations are a major growth area in public policy. The following list includes all that are used in the text and others that readers will come across in the planning literature. No claim is made for comprehensiveness.

1990 Act	The Town and Country Planning Act 1990	ALBPO	Association of London Borough Planning Officers
1991 Act	The Planning and Compensation Act 1991	ALG	Association of London Government
2004 Act	The Planning and Compulsory Purchase Act 2004	ALNI	Association of Local Authorities in Northern Ireland
4Ps	Public Private Partnerships Programme	ALURE	alternative land use and rural economy
		AMA	Association of Metropolitan Authorities
AAI	area of archaeological importance	AMR	annual monitoring report
AAP	area action plan	ANPA	Association of National Park Authorities
ACBE	Advisory Committee on Business and the Environment	AONB	area of outstanding natural beauty
ACC	Association of County Councils	AOSP	areas of special protection for birds
ACCORD	assistance for coordinated rural development	APRS	Association for the Protection of Rural Scotland
ACO	Association of Conservation Officers	ARC	Action Resource Centre
ACOST	Advisory Council on Science and Technology	ASAC	area of special advertisement control
		ASNW	area of semi-natural woodland
ACRE	Action with Communities in Rural England	ASSI	area of special scientific interest (Northern Ireland)
ADAS	Agricultural Development and Advice Service	ATB	Agricultural Training Board
ADC	Association of District Councils	BACMI	British Aggregate Construction Materials Industries
AESOP	Association of European Schools of Planning	BANANA	Build Absolutely Nothing Anywhere Near Anything
AGR	advanced gas-cooled reactor	BATNEEC	best available techniques not entailing excessive cost
AIS	agricultural improvement scheme		
ALA	Association of London Authorities (now ALG)	BIC	Business in the Community
		BID	business improvement district

BNFL   British Nuclear Fuels Ltd
BPEO   best practicable environmental option
BPF   British Property Federation
bpk   billion passenger kilometres
BPM   best practicable means
BR   British Rail (now Network Rail)
BRE   Building Research Establishment
BRF   British Road Federation
BRO   Belfast Regeneration Office
BSI   British Standards Institution
BTA   British Tourist Authority
BUFT   British Upland Footpath Trust
BVA   best value authority
BVPI   best value performance indicators
BW   British Waterways
BWB   British Waterways Board

CA   Countryside Agency (formerly Countryside Commission)
CABE   Commission for Architecture and the Built Environment
Cadw   Not an acronym, but the Welsh name for the Welsh Historic Monuments Agency. The word means to keep, to preserve.
CAF   Coalfields Area Fund
CAP   Common Agricultural Policy
CAT   City Action Team
CBI   Confederation of British Industry
CC   Countryside Commission (now Countryside Agency)
CCS   Countryside Commission for Scotland (now Scottish Natural Heritage)
CCT   compulsory competitive tendering
CCW   Countryside Council for Wales
CDA   comprehensive development area
CDCR   Committee on the Development and the Conversion of the Regions (EU)
CDP   community development project
CEC   Commission of the European Communities (European Commission)
CEGB   Central Electricity Generating Board
CEMAT   Conférence Européene des Ministres Responsables de l'Aménagement du Territoire (European Conference of Ministers responsible for Regional Planning)
CEMR   Council of European Municipalities and Regions
CES   Centre for Environmental Studies
CFC   chlorofluorocarbon
CfIT   Commission for Integrated Transport
CHAC   Central Housing Advisory Committee
CHP   combined heat and power
CIA   commercial improvement area
CIEH   Chartered Institute of Environmental Health
CIPFA   Chartered Institute of Public Finance and Accountancy
CIS   community involvement scheme (Wales)
CIT   Commission for Integrated Transport
CITES   Convention on International Trade in Endangered Species
CLA   Country Land and Business Association
CLES   Centre for Local Economic Strategies
CLEUD   certificate of lawfulness of existing use or development
CLOPUD certificate of lawfulness of proposed use or development
CLP   central local partnership
CLRAE   Conference of Local and Regional Authorities of Europe (Council of Europe)
CMS   countryside management system
CNCC   Council for Nature Conservation and Countryside (Northern Ireland)
CNT   Commission for New Towns
CO   Cabinet Office
COBA   cost–benefit analysis
COE   Council of Europe
COI   Central Office of Information
COMARE Committee on Medical Aspects of Radiation in the Environment
COPA   Control of Pollution Act 1974
COR   Committee of the Regions (EU)
COREPER Council of Permanent Representatives
CORINE Community Information System on the State of the Environment (EU)

CoSIRA Council for Small Industries in Rural Areas
COSLA Convention of Scottish Local Authorities
COTER Commission for Territorial Cohesion (EU COR)
CPO compulsory purchase order
CPOS County Planning Officers' Society
CPRE Campaign to Protect Rural England (formerly Council for the Protection of Rural England)
CPRS Central Policy Review Staff
CPRW Campaign (formerly Council) for the Protection of Rural Wales
CRE Commission for Racial Equality
CROW Act Countryside and Rights of Way Act
CRP city-region plan (Scotland)
CRRAG Countryside Recreation Research Advisory Group
CS community strategy
CSD (1) Commission on Sustainable Development (UN)
CSD (2) Committee on Spatial Development (EU) (now CDCR)
CSERGE Centre for Social and Economic Research on the Global Environment
CSF community support framework
CSO Central Statistical Office
CSR Comprehensive Spending Review
CTRL Channel Tunnel Rail Link
CWI Controlled Waste Inspectorate

DAFS Department of Agriculture and Fisheries for Scotland
DATAR Délégation à l'aménagement du territoire et à l'action régionale (French national planning agency)
DBFO Design, build, finance, and operate (roads by the private sector)
DBRW Development Board for Rural Wales
DC (1) development corporation
DC (2) district council
DCA Department for Constitutional Affairs
DCAN development control advice note (NI)
DCC Docklands Consultative Committee

DCMS Department for Culture, Media and Sport
DDA Disability Discrimination Act 1995
DEA Department of Economic Affairs
DEFRA Department for Environment, Food and Rural Affairs
DETR Department of Environment, Transport and the Regions (1997–2000)
DEVE Committee on Development (EU COR)
DfEE Department for Education and Employment (now DfES)
DfES Department for Education and Skills (formerly DfEE)
DfID Department for International Development
DfT Department for Transport (formerly DoT)
DG Directorate General of the European Commission
DLG derelict land grant
DLGA derelict land grant advice note
DLR Docklands Light Railway
DLT development land tax
DNH Department of National Heritage
DoE Department of the Environment
DoENI Department of the Environment for Northern Ireland
DoT Department of Transport (now DfT)
DP development plan
DPD development plan document
DPM Deputy Prime Minister
DPOS District Planning Officers' Society
DRIVE dedicated road infrastructure for vehicle safety in Europe
DSD Department for Social Development (NI)
DTI Department of Trade and Industry
DTLR Department of Transport, Local Government and the Regions (2000–2)
DWI Drinking Water Inspectorate

EA environmental assessment
EAF environmental action fund

EAGGF European Agricultural Guidance and Guarantee Fund
EAZ education action zone
EBRD European Bank for Reconstruction and Development
EC European Community
ECMT European Conference of Ministers of Transport
ECOSOC Economic and Social Council (United Nations)
ECS Economic and Social Committee (EU)
ECSC European Coal and Steel Community
ECTP European Council of Town Planners
Ecu European currency unit (no longer in use)
EDU Equality and Diversity Unit (ODPM)
EEA (1) European Economic Area (EU plus Iceland, Liechtenstein, Norway and Switzerland)
EEA (2) European Environment Agency
EEC (1) European Economic Community
EEC (2) Energy Efficiency Commitment
EFS England Forestry Strategy
EFTA European Free Trade Association
EHCS English House Condition Survey
EIA environmental impact assessment
EIB European Investment Bank
EIF European Investment Fund
EIONET European Environment Information and Observation Network
EIP examination in public
EIS environmental impact statement
EMAS eco-management and audit scheme
EMP environmental management areas
EMU European Monetary Union
EN English Nature
EP English Partnerships
EPA educational priority area
EPA Environmental Protection Act 1990
EPC Economic Planning Council
ERCF Estates Renewal Challenge Fund
ERDF European Regional Development Fund
ERP electronic road pricing
ES environmental statement (UK)

ESA environmentally sensitive area
ESDP European Spatial Development Perspective
ESF European Social Fund
ESPON European Spatial Planning Observation Network
ESRC Economic and Social Research Council
ETB English Tourist Board
ETC English Tourism Council
ETLLD Scottish Executive Enterprise, Transport and Lifelong Learning Directorate
EU European Union
EUCC European Union for Coastal Conservation
EURATOM European Atomic Energy Community
EUETS EU Emissions Trading Scheme
EZ (1) employment zone
EZ (2) enterprise zone

FA Forestry Authority
FC Forestry Commission
FCGS Farm and Conservation Grant Scheme
FEOGA Fonds Européen d'Orientation et de Garantie Agricole (European Agricultural Guidance and Guarantee Fund)
FIFG Financial Instrument for Fisheries Guidance
FIG Financial Institutions Group
FMI financial management initiative
FoE Friends of the Earth
FOI Freedom of Information
FPS Fuel Poverty Strategy
FTA Freight Transport Association
FUA functional urban area (ESPON)
FWAG Farming and Wildlife Advisory Group
FWGS Farm Woodland Grant Scheme
FWPS Farm Woodland Premium Scheme

GATT General Agreement on Tariffs and Trade
GCR Geological Conservation Review
GDO General Development Order
GDP gross domestic product

GDPO General Development Procedure Order
GEAR Glasgow Eastern Area Renewal (Scheme)
GIA general improvement area
GIS geographic information systems
GLA Greater London Authority
GLC Greater London Council
GLDP Greater London Development Plan
GO government office
GO-East Government Office for Eastern Region
GO-EM Government Office for the East Midlands
GO-L Government Office for London
GO-NE Government Office for the North East
GO-NW Government Office for the North West
GOR Government Offices for the Regions
GO-SE Government Office for the South East
GO-SW Government Office for the South West
GO-WM Government Office for the West Midlands
GO-YH Government Office for Yorkshire and Humberside
GPDO General Permitted Development Order

HA Highways Agency
HAA housing action area
HAG housing association grant
HAP habitat action plan
HAT housing action trust
HAZ health action zone
HBF Home Builders' Federation
HBMC Historic Buildings and Monuments Commission
HC House of Commons
HCiS Housing Corporation in Scotland
HCLA hill livestock compensatory allowance
HERS Heritage Economic Regeneration Schemes (EH)
HHSRS housing, health and safety ratings system
HIA home improvement agency

HIDB Highlands and Islands Development Board (now HIE)
HIE Highlands and Islands Enterprise
HIP housing investment programme
HL House of Lords
HLCA hill livestock compensatory allowances
HLF Heritage Lottery Fund
HLW high-level waste
HMIP Her Majesty's Inspectorate of Pollution
HMIPI Her Majesty's Industrial Pollution Inspectorate (Scotland)
HMNII Her Majesty's Nuclear Installation Inspectorate
HMO (1) hedgerow management order
HMO (2) house in multiple occupation
HMSO Her Majesty's Stationery Office
HMT Her Majesty's Treasury
HO Home Office
HR human resources
HRF Housing Research Foundation
HSA Hazardous Substances Authority
HSE Health and Safety Executive
HWI Hazardous Waste Inspectorate

IACGEC Inter-Agency Committee on Global Environmental Change
IAEA International Atomic Energy Agency
IAPI Industrial Air Pollution Inspectorate
IAPs inner area programmes
IAURIF Institut d'aménagement du territoire et d'urbanisme de la région d'Ile de France
ICE Institution of Civil Engineers
ICNIRP International Commission on Non-Ionising Radiation Protection
ICOMOS International Council on Monuments and Sites
ICT information and communications technology
ICZM integrated coastal zone management
IDC industrial development certificate
IDeA Improvement and Development Agency

IEEP   Institute for European Environmental Policy
IEG   implementing electronic government
IIA   industrial improvement area
ILW   intermediate-level waste
IMPEL   EU Network for the Implementation and Enforcement of Environmental Law
INTERREG   European Community initiative for transnational spatial planning
IPA   integrated policy appraisal
IPC   integrated pollution control
IPCC   Intergovernmental Panel on Climate Change
IPPC   integrated pollution, prevention and control
IRD   integrated rural development (Peak District)
ISOCARP   International Society of City and Regional Planners
IUCN   World Conservation Union
IWA   Inland Waterways Association
IWAAC   Inland Waterways Amenity Advisory Committee

JNCC   Joint Nature Conservation Committee
JPL   *Journal of Planning and Environment Law*

LA21   Local Agenda 21 (UNCED)
LAAPC   local authority air pollution control
LATS   landfill allowance trading scheme
LAW   Land Authority for Wales
LAWDC   local authority waste disposal company
LBA   London Boroughs Association (now ALG)
LCO   landscape conservation order
LDC   local development company
LDD   local development document
LDDC   London Docklands Development Corporation
LDF   local development framework
LDO   local development order
LDP   local development plan (Wales)

LDS   local development scheme
LEAP   local environmental agency plan
LEC   local enterprise company (Scotland)
LEG-UP   local enterprise grants for urban projects (Scotland)
LETS   local employment and trading systems
LFA   less favoured area (agriculture)
LGA   Local Government Association
LGC   Local Government Commission for England
LGF   local government finance
LGMB   Local Government Management Board
LHS   local housing strategy (Scotland)
LLW   low-level waste
LNR   local nature reserve
LNYR   live near your work
LOTS   living over the shop
LPA   local planning authority
LPAC   London Planning Advisory Committee
LRT   Land Restoration Trust
LSC   Learning and Skills Council
LSP   local strategic partnership
LSPU   London Strategic Policy Unit
LT   London Transport
LTP   local transport plan
LTS   local transport strategy (Scotland)
LUCS   Land Use Change Statistics
LULU   locally unwanted land use
LWRA   London Waste Registration Authority

MAFF   Ministry of Agriculture, Fisheries and Food
MARS   Monuments at Risk Survey
MEA   Manual of Environmental Assessment (for trunk roads)
MEGA   metropolitan European growth area
MEHRA   marine environmental high risk areas
MEP   Member of the European Parliament
MHLG   Ministry of Housing and Local Government
MLGP   Ministry of Local Government and Planning
MMG   marine minerals guidance note

MMS   multi-modal study
MNR   marine nature reserve
MOA   Mobile Operators Association
MoD   Ministry of Defence
MPA   mineral planning authority
MPG   minerals policy guidance note
MPOS   Metropolitan Planning Officers' Society
MPS   minerals policy statement
MSC   Manpower Services Commission
MTAN   minerals technical advice note (Wales)
MTCP   Ministry of Town and Country Planning
MWMS   municipal waste management survey

NACRT   National Agricultural Centre Rural Trust
NAO   National Audit Office
NARIS   National Roads Information System
NATA   New Approach to Appraisal (roads)
NAW   National Assembly for Wales
NBN   National Biodiversity Network
NCB   National Coal Board
NCBOE   National Coal Board Opencast Executive
NCC   Nature Conservancy Council
NCCI   National Committee for Commonwealth Immigrants
NCCS   Nature Conservancy Council for Scotland (now Scottish Natural Heritage)
NCR   new commitment to regeneration
NCVO   National Council of Voluntary Organisations
NDC   New Deal for Communities
NDPB   non-departmental public body
NEC   noise exposure category
NEDC   National Economic Development Council
NEDO   National Economic Development Office
NERC   National Environment Research Council
NETCEN National Environmental Technology Centre
NFC   National Forest Company
NFFO   non-fossil fuel obligation
NGC   Northern Growth Corridor
NGO   non-governmental organisation
NHA   natural heritage area (Scotland)
NHMF   National Heritage Memorial Fund

NHS   National Health Service
NII   Nuclear Installations Inspectorate
NIMBY   not in my backyard
NIO   Northern Ireland Office
NIREX   Nuclear Industries Radioactive Waste Executive
NLUD   National Land Use Database
NNR   national nature reserve
NPA   national park authority
NPF   National Planning Forum
NPG   national planning guideline (Scotland)
NPPG   national planning policy guideline (Scotland)
NRA   National Rivers Authority (now Environment Agency)
NRF   Neighbourhood Renewal Fund
NRTF   national road traffic forecasts (GB)
NRU   Neighbourhood Renewal Unit
NSA (1) national scenic area (Scotland)
NSA (2) nitrate sensitive area
NTDC   new town development corporation
NUTS   nomenclature of territorial units for statistics: designates levels of regional subdivision in the EU
NVZ   nitrate vulnerable zone
NWDO   North West Development Office (NI)

ODPM   Office of the Deputy Prime Minister (since 2002)
OECD   Organisation for Economic Cooperation and Development
OeE   Office of the E-Envoy
OEEC   Organisation for European Economic Cooperation
OFLOT   Office of the National Lottery
OJ   *Official Journal of the European Communities*
ONS   Office for National Statistics
OPCS   Office of Population Censuses and Surveys (now part of ONS)
OPSR   Office of Public Services Reform
OS   Ordnance Survey

PAG (1) Planning Advisory Group
PAG (2) Property Advisory Group

PAN   planning advice note (Scotland)
PARSOL   Planning and Regulatory Services
          Online
PAT   policy action team
PDG   Planning Delivery Grant
PDL   previously developed land
PDO (1)   permitted development order
PDO (2)   potentially damaging operation (SSSI)
PDR   permitted development right
PEP   Political and Economic Planning
          (now PSI)
PFI   Private Finance Initiative
PGS   planning gain supplement
PI   Planning Inspectorate (also PINS)
PIC   Planning Inquiry Commission
PINS   Planning Inspectorate (also PI)
PIP   partnership investment programme
PIU   Performance and Innovation Unit
PLI   public local inquiry
POS   Planning Officers' Society
PPA   priority partnership area (Scotland)
PPC   Pollution, Prevention and Control Act
          2000
PPG   planning policy guidance note
PPP (1)   polluter pays principle
PPP (2)   public–private partnerships
PPS (1)   planning policy statement (previously
          PPG)
PPS (2)   planning policy statement (NI)
PRIDE   Programmes for Rural Initiatives and
          Developments (Scotland)
PSA (1)   Property Services Agency
PSA (2)   public service agreement
PSI   Policy Studies Institute
PSS   Planning Summer School
          (formerly TCPSS)
PTA   passenger transport authority
PTE   passenger transport executive
PTRC   Planning and Transport Research and
          Computation
PVC   polyvinyl chloride
PWR   pressurised water reactor

QUANGO   quasi-autonomous non-governmental
          organisation

RA   renewal area
RB   regional body
RAC   Royal Automobile Club
RAWP   regional aggregates working parties
RCAHMS   Royal Commission on the Ancient and
          Historical Monuments of Scotland
RCC   rural community council
RCEP   Royal Commission on Environmental
          Pollution
RCHME   Royal Commission on the Historical
          Monuments of England
RCI   Radiochemical Inspectorate
RCU (1)   Regional Coordination Unit (ODPM)
RCU (2)   Road Construction Unit
RDA (1)   regional development agency
RDA (2)   rural development area
RDC   Rural Development Commission
RDG   regional development grant
RDO   Regional Development Office (NI)
RDP   rural development programme
RDS   Regional Development Strategy
          Northern Ireland
REG   regional enterprise grant
RES (1)   race equality scheme
RES (2)   regional economic strategy
RHB   regional housing board
RHS   regional housing strategy
RIA   regulatory impact assessment
RIBA   Royal Institute of British Architects
RICS   Royal Institution of Chartered
          Surveyors
RIGS   regionally important
          geological/geomorphological sites
ROI   regional output indicator
RPB   regional planning body
RPG   regional planning guidance
RRAF   regional rural affairs forum
RSA (1)   regional selective assistance
RSA (2)   Regional Studies Association
RSDF   regional sustainable development
          framework
RSL   registered social landlord
RSPB   Royal Society for the Protection of
          Birds
RSS   regional spatial strategy

RTB (1) right to buy (public sector housing)
RTB (2) regional tourist board
RTPI Royal Town Planning Institute
RTS regional transport strategy
RWMAC Radioactive Waste Management Advisory Committee

SA sustainability appraisal
SAC special area of conservation (habitats)
SACTRA Standing Advisory Committee on Trunk Road Assessment
SAGA Sand and Gravel Association
SAP species action plan
SC standard charge
SCI statement of community involvement
SCLSERP Standing Conference on London and South East Regional Planning
SDA Scottish Development Agency (now Scottish Enterprise)
SDC Sustainable Development Commission
SDO special development order
SDP standard delivery plan (Scottish Housing)
SDS Spatial Development Strategy (London)
SDU Sustainable Development Unit
SE Scottish Executive
SEA (1) Single European Act 1987
SEA (2) strategic environmental assessment
SEDD Scottish Executive Development Department
SEEDA South East England Development Agency
SEEDS South East Economic Development Strategy
SEELLD Scottish Executive Enterprise and Lifelong Learning Department
SEERA South East England Regional Assembly
SEH Survey of English Housing
SEHD Scottish Executive Health Department
SEERAD Scottish Executive Environment and Rural Affairs Department
SEM Single European Market
SEPA Scottish Environment Protection Agency

SERC Science and Engineering Research Council
SERPLAN London and South East Regional Planning Conference
SEU Social Exclusion Unit
SFI Selective Finance for Investment in England
SHAC Scottish Housing Advisory Committee
SHG Social Housing Grant
SHQS Scottish Housing Quality Standard
SI statutory instrument
SIC social inclusion partnerships (Scotland)
SINC site of importance for nature conservation
SIP social inclusion partnership (Scotland)
SLF Scottish Landowners Federation
SME small and medium sized enterprises
SMR sites and monuments records (counties)
SNAP Shelter Neighbourhood Action Project
SNH Scottish Natural Heritage
SO Scottish Office
SOAEFD Scottish Office Agriculture, Environment and Fisheries Department
SODD Scottish Office Development Department
SOEnD Scottish Office Environment Department (now SOAEFD)
SOID Scottish Office Industry Department
SOIRU Scottish Office Inquiry Reporters Unit
SoS Secretary of State
SPA special protection area (for birds) (EU)
SPD (1) supplementary planning document
SPD (2) single programming document
SPG strategic planning guideline
SPP Scottish planning policy
SPZ simplified planning zone
SR Spending Review
SRA Strategic Rail Authority
SRB Single Regeneration Budget
SSHA Scottish Special Housing Association
SSSI site of special scientific interest
STB Scottish Tourist Board
SUD Committee on Spatial and Urban Development (EU)

SURF   Scottish Urban Regeneration Forum (Scotland)
SURI   small urban regeneration inititive (Scotland)

TAN   technical advice notes (Wales)
TCPA   Town and Country Planning Association
TCPSS   Town and Country Planning Summer School (now PSS)
TEC (1) training and enterprise council
TEC (2) Treaty establishing the European Community
TEN   Trans-European Network(s)
TEST   Transport and Environment Studies
TETN   Trans-European Transport Networks
TEU   Treaty on European Union
TfL   Transport for London
THORP   thermal oxide reprocessing plant
TPI   Targeted Programme of Improvements (DfT)
TPO   tree preservation order
TPPs   transport policies and programmes
TRL   Transport Research Laboratory
TSG   transport supplementary grant
TSO   The Stationery Office
TUC   Trades Union Congress

UCO   Use Classes Order
UDA   urban development area
UDC   urban development corporation
UDG   urban development grant
UDP   unitary development plan
UEI   urban exchange initiative
UKAEA   United Kingdom Atomic Energy Authority
UKBAP   UK Biodiversity Action Plan
UKBG   UK Biodiversity Group
UNCED   United Nations Conference on Environment and Development (Earth Summit, Rio, 1992)
UNCSD   United Nations Commission on Sustainable Development
UNCTAD United Nations Conference on Trade and Development

UNECE   United Nations Economic Commission for Europe
UNEP   United Nations Environment Programme
UNESCO United Nations Educational, Scientific and Cultural Organization
UP (1) Urban Programme
UP (2) urban partnerships (Scotland)
URA   Urban Regeneration Agency (known as EP)
URBAN   European Community initiative for urban regeneration
URC   urban regeneration company
UTF   Urban Task Force

VAT   value added tax
VFM   value for money
VISEGRAD four former communist countries: Poland, Czech Republic, Slovakia and Hungary
VOCS   volatile organic compounds

WAG   Welsh Assembly Government
WCA   waste collection authority
WCED   World Commission on Environment and Development
WDA (1) waste disposal authority
WDA (2) Welsh Development Agency
WDP   waste disposal plan
WES   wildlife enhancement scheme
WHO   World Health Organisation
WMEB   West Midlands Enterprise Board
WIP   waste implementation programme
WMO   World Meteorological Organization
WO   Welsh Office
WOAD   Welsh Office Agriculture Department
WQO   water quality objectives
WRA   waste regulation authority
WRAP   waste and resources action programme
WRAP   waste reduction always pays
WSP   Wales Spatial Plan
WTB   Welsh Tourist Board

WTO World Trade Organisation
WWF World Wide Fund for Nature
       (formerly World Wildlife Fund)
WWT Wildfowl and Wetlands Trust

YTSYouth Training Scheme

Encyclopedia refers to Malcolm Grant's *Encyclopedia of Planning Law and Practice*, London: Sweet and Maxwell, loose-leaf, regularly updated by supplements.

The nature of planning

If planning were judged by results, that is, by whether life followed the dictates of the plan, then planning has failed everywhere it has been tried. No one, it turns out, has the knowledge to predict sequences of actions and reactions across the realm of public policy, and no one has the power to compel obedience.

Wildavsky 1987: 21

Introduction

It is the purpose of this chapter to give a general introduction to the character of land use planning. Since this is so much a product of culture, it differs among countries. The understanding of one system is helped by comparing it with others and, for this reason, some international comparisons are introduced. The chapter presents a broad discussion of some basic features of the UK planning system, which is essentially a means for reconciling conflicting interests in land use. It is not a review of the theoretical literature on planning, though guidance on where this might be found is given in further reading at the end of the chapter.

Many of the arguments about planning revolve around the relationships between theory and practice. Planning theories (along with related theories on management, government, and other fields of human interaction) have often been based on abstract models based on notions of rationality, defined in normative terms. There are difficulties with the concept of rationality. Some of these stem from the fact that planning operates within an economic system that has a 'market rationality' which can differ from, and conflict with, the rationality which is espoused in some planning theories. But the crucial issue is that the concept of rationality cannot be divorced from objectives, ambitions and interests – as well as place

and time. These variables are the very stuff of planning: disputes and conflict arise, not because of irrationalities (though these may be present), but because different interests are rationally seeking different objectives. A brief discussion of these matters leads into issues such as those of incrementalism and implementation, both of which present their own rationales for behaviour and attitudes.

A notable feature of the UK system is the unusual extent to which it embraces discretion. This allows for flexibility in interpreting the public interest. It is in sharp contrast to other systems which, more typically, explicitly aim at reducing such uncertainty. The European and US systems, for example, eschew flexibility, and lay emphasis on the protection of property rights. Flexibility is highly regarded in the UK because it enables the planning system to meet diverse requirements and the constantly changing nature of the problems with which it attempts to deal. A major factor is the political climate – usually, but not always, reflecting the character of the party in power. The shifts in planning policy have been dramatic, and seem to be accelerating, though not on any clear pattern. Sometimes the movement is to a greater concern for market forces; at other times the opposite is dominant, though only within circumscribed limits and with surges in public and political support for and against development and conservation.

Conflict and disputes

Land use planning is a process concerned with the determination of land uses, the general objectives of which are set out in legislation or in some document of legal or accepted standing. The nature of this process will depend in part on the objectives which it is to serve. The broad objective of the UK system has been for many years to 'regulate the development and use of land in the public interest'. From 2004 a much wider purpose has been added: to contribute to the achievement of sustainable development.[1] Like all such policy statements, these have a very wide meaning. They are not, however, empty of content but establish the essential character of the UK planning system. Its significance is highlighted when it is compared with possible alternatives. These might be 'to encourage the development and use of land' or 'to facilitate the development of land by private persons and corporations'. Other alternatives include 'to plan the use of land to ensure that private property rights are protected and that the public interest is served' and (an example from Indiana) 'to guide the development of a consensus land use and circulation plan'. These scene-setting statements convey the overall philosophy or principles which at some times guide the planning system. They are important for that reason, and they are of direct concern in disputes about the validity or appropriateness of policies which are elaborated within their framework. They are called upon in support of arguments about specific policies.

Politics, conflict and dispute are at the centre of land use planning. Conflict arises because of the competing demands for the use of land, because of the externality effects that arise when the use of land changes, and because of the uneven distribution of costs and benefits which result from development. If there were no conflicts, there would be no need for planning. Indeed, planning might usefully be defined as the process by which government resolves disputes about land uses.

Alternatives arise at every level of planning – from the highest (supranational) level to the lowest (site) level. The planning system is the machinery by which these levels of choice are managed – from plan-making to development control. Though planning systems vary among countries, they can all be analysed in these terms. The processes involved encompass the determination of objectives, policy-making, consultation and participation, formal dispute resolution, development control, implementation and the evaluation of outcomes.

The explicit function of the processes is to ensure that the wide variety of interests at stake is considered and that outcomes are in the general public interest. In reality there are very many interests that might be served. Four groups or 'actors' stand out: politicians serving various levels of government, the development industry, landowners and 'the public'. The last is a highly diverse group which is attaining an increased role (not always meaningfully) by way of pressure group and public involvement. Another view of 'interests' going beyond actors who are present in person in the forums where the planning process is played out include, for example, unemployed people, homeless people, those searching for a first home, and even 'the environment'. Governments usually argue that a reasonable balance is being achieved between the different interests. Critics argue that intervention through land use planning serves to maintain the dominance of particular interests. Evaluations of planning suggest that those with a property interest are more influential and get more out of the planning system, but organised interest groups and even some individuals have had success in individual cases, so the outcomes are by no means certain.

One of the reasons for the increased importance attached to planning processes, and public involvement in them (apart from wider questions of democracy), lies in the belief that they are effective in reducing the scope for later conflict. The clearer and firmer the policy, and the wider its support, the less room there is for arguing about its application and implementation. Thus for the managers of the system efficiency is increased. But there are limits to this: there is no way that conflict can be planned away.

A central problem for the planning system is to devise a means for predicting likely future changes that may impact on the system. In fact, this is extremely difficult, and past attempts have demonstrated that there is a severe limit to prediction. This is one of the reasons why discretion has to be built into the processes: without this, it is difficult to take account of changing circumstances. A second, more immediate reason for discretion is the impossibility of devising a process which can be applied automatically to the enormous variety of circumstances that come to light when action is being taken. Plans and other policy documents provide a reference point for what has been agreed through the planning process, and against which proposals will be measured. Professional research and analysis together with opportunities for consultation, public participation and formal objection and adoption by political representatives give such documents legitimacy. But they cannot be blueprints. The implementation of a plan always differs from what is anticipated.

There are several reasons for this. For the individuals concerned, the actuality of implementation may appear different from the perceived promise of a plan. For the landowners involved, the market implications may prove to be unwelcome (whether or not market conditions have changed). There will generally be those whose objections at the plan-making stage were rejected in favour of alternatives, and who will naturally take advantage of any opportunity to repeat their objections at the stage of implementation: the passage of time and the changes that it has wrought provide that opportunity. Changed conditions may be so great that the plan is outdated or even counterproductive. Where this is the case, there is a clear need for a revised plan, but problems mount in the mean time and (since the process for elaborating a new plan has still to be completed) the areas of dispute multiply. In addition to these general issues, there is always scope for dispute on the detailed application of policy to individual cases: no plan can be so detailed as to be self-implementing. Finally, there are the cases where there is no policy, or where the policy is simply not relevant to the action that needs to be taken. For these and similar reasons, there has to be machinery for settling disputes concerning implementation.

Adjudication of disputes may be the responsibility of an administrative system (which is theoretically subservient to the political system), the courts, or an ad-hoc machine. The courts have a major role in countries where planning involves issues that are subject to constitutional safeguards (of property rights or due process), or where plans have the force of law. (Here it should be emphasised that plans in the UK are not part of the law but are made under the law. Typical practice in many other countries is for plans to be issued or enacted as law, which gives them, in theory, considerable force.) Where there are no such complications, as is typical in the UK, matters of dispute are more likely to be dealt with by an administrative appeal system. However, there is no hard and fast rule about this, and recent years have witnessed an increasing role for the UK courts. Nevertheless, there remains a huge difference in the role of the courts in the UK compared with the US and most European countries.

Planning, the market and the development process

In the early years of the less sophisticated postwar period, plans were drawn up in a vacuum which (ostensibly) ignored the manner in which the property market and development processes worked. Land was allocated to uses which seemed sensible in planning terms, but with little regard to the market. Indeed, market considerations were often explicitly expressed in terms which cast them as subservient to needs. Given the positive role which was envisaged for public development, this had some semblance of logic, but it rapidly disintegrated in the face of the realities of public finance and the incapacity of public authorities to take on the primary development role.

There is today a better (though very incomplete) understanding of how land and property markets work, and a greater appreciation of the need to take account of market trends (even if they have to be subjected to public control or influence). There is also a greater willingness on the part of both the public and the private sectors to pool their efforts and resources: the word 'partnership' is an important addition to the planning lexicon. Of course, this has not ushered in a new era of sweet harmonious cooperation: there are inherent conflicts of interest between (and within) the two sectors. The planning system provides an important mechanism for mediating among these conflicts.

There have been serious difficulties here (by no means resolved) which stem, in part, from a mutual ignorance between the planning and the development sectors. However, attitudes have changed somewhat as a result of a miscellany of forces, ranging from an increased concern on the part of local authorities to promote economic development, to the changed fortunes of the constituent parts of the development industry. Since there is no prospect of a future period of calm stability, attitudes will continue to change. This hardly promises a good base for a traditional type planning (as caricatured by the term 'end state planning'); rather, it promises an even greater role for flexibility and discretion.

A mention of some crucial features of the development process highlights the nature of the conflicts with which planning is concerned:

- Developers are concerned with investment and profits, particularly in a timeframe considerably shorter than is typical in the planning world where the preoccupation is with long-term land use.
- Developers need to act quickly in response to market opportunities and the cost of capital. Planners operate in a different timescale.
- Development is much easier on greenfield sites than on inner-city locations: to developers, projects are risky enough without being burdened with 'extraneous' problems. Problems which are 'extraneous' to developers can be central to the concerns and objectives of planners.
- Markets are very diverse: one location is not as good as another. They are, moreover, dynamic: timing may be the crucial factor in the feasibility of a development. Markets are frequently simply not understood by planners: their concern is more generally with the unfolding of a long-term plan. Any pressures they experience are more likely to be political rather than economic in origin.
- Developers are concerned with the particular; for planners, the particular is only one among many which add up to the general policy matters with which they are concerned.

Given such major differences between sectors, it is not surprising that their relationships can be difficult. The problem for planning is that full consideration for developers' concerns can quickly lead to ad-hoc responses which undermine planning policy. The very vagueness of policy statements and the high degree of discretion in the system increases the likelihood of this. The dilemma is inherent: there is no simple solution.

It is not surprising that the comprehensive planning philosophy which was dominant in the early postwar period is now discredited. It relates to a world which no longer exists. Thus, planning has moved from a preoccupation with grand plans to a concern for finding ways of reconciling the conflicting interests which are affected by development. This leads away from development control to negotiation and mediation. Paradoxically, this is happening at the same time that the central government is attempting to secure a greater degree of certainty through a plan-led system. Perhaps the circle will be squared if it is found that mediation leads to greater certainty? US experience shows how developers can work more easily under a negotiatory system than within a regulatory framework: they know how to operate it to their benefit.

Rationality and comprehensive planning

Central to planning is the concept of rationality. Since rationality requires all relevant matters to be taken into account, the use of the concept readily leads to a comprehensive conception of planning. Rationality also requires the determination of objectives (and therefore, though not always explicitly, of values), the definition of the problems to be solved, the formulation of alternative solutions to these problems, the evaluation of these alternatives, and the choice of the optimum policy. There are numerous conceptions of rationality but we are concerned here with the form of rationality that is derived from the scientific method of inquiry. The application of the idea of rational scientific method to policy-making gave rise to a particular form of planning and its critique from many different perspectives, has given rise to a wide range of other ideas about the nature of planning.

First, there are those who have criticised the simple notion of rationality noted above but have continued to maintain that the task for planning theorists is to elaborate the notion of planning as a set of procedures by which decisions are made (Davidoff and Reiner 1962; Faludi 1987). Second, there are those who reject the objectivity implied by the simple rational approach and instead focus on the role that planning plays in the distribution of resources among different interests in society. Part of this criticism has included the development of a body of thought variously described as social, community or equity planning, where planning is promoted as a tool that can redress inequalities, and work to benefit minority and disadvantaged groups (Gans 1991). Third, there are more fundamental criticisms from those who have used a neo-Marxist critique to draw attention to basic divisions of power in the political and economic structure of capitalist society (Paris 1982). This view asserts that irrespective of its explicit intentions, planning will inevitably 'serve those interests it seeks to regulate' (Ambrose 1986).

The persuasiveness of the concept of comprehensive rational planning is seen at every stage of the plan-making process, from the initial production of goals and objectives, to definitions of problems and proposed solutions. But all this is done in the context of the politics of the place and the time, and against the background of public opinion and the acceptability or otherwise of governmental action. Some important issues may be regarded not as problems capable of solution but as powerful economic trends which cannot be reversed. Others may be of a nature for which possible solutions are conceivable but untried, too costly, too administratively difficult, too uncertain, or even dangerous to the long-term future of the area. And, as will be apparent from later chapters, these acutely difficult problems (of urbanisation, congestion, inner-city decay, for example) have continually proved beyond the powers of governments to solve, at least in the short run, and the long run is unpredictable. Big differences of opinion exist among experts, politicians and electors on these matters. As a result, there are severe constraints operating on the planning process,

and there is little resembling a logical calm set of procedures informed by intellectual debate. Certainly, the process is far from scientific or rational.

Practitioners are quick to point out that planning involves deciding between opposing interests and objectives: personal gain versus sectional advantage or public benefit, short-term profit versus long-term gain, efficiency versus cheapness, to name but a few. It entails mediation among different groups, and compromise among the conflicting desires of individual interests. Above all, it necessitates the balancing of a range of individual and community concerns, costs and rights. It is essentially a political as distinct from a technical or legal process, though it embraces important elements of all.

The comprehensive rational process in its simplest form assumes that goals for planning can be identified and agreed by the political process and that the means to achieve those goals can be established and implemented through a technical process managed by professionals. These assumptions do not stand up in practice; it is exceedingly difficult to agree on anything more than the most general goals and the calculation of means is not in fact a separate process. For example, currently, one of the most difficult planning issues is concerned with reconciling the implications of the growth in traffic with traditional ideas about town centres and urban growth. The policy response has been as confused as the issues are complex and politically daunting. For example, a major focus has been on controlling the number of additional out-of-town shopping centres, and directing development to town centres and brownfield sites. Conflicts that arise here include the apparently irrepressible demand for car ownership and use, the traditional view that road space is free and should not incur any type of congestion charge, the desire of town centre businesses to maintain their custom and to avoid the risk of losing it because of tighter parking restrictions, the financial difficulties facing public transport, and so on. Any one of these issues on its own would be difficult enough: all of them together constitute a planning witches' brew. And, as so often happens, ends and means become intertwined in a hopelessly confusing way: protecting city centres, safeguarding inner city jobs, conserving the countryside, reducing pollution, facilitating ease of access for those without cars as well as the car traveller, providing greater choice for the shopper – which is the objective and which the solution? These issues are examined in later chapters; they are listed here to underline the essentially political nature of planning. Grand phrases about rational planning to 'coordinate land uses' crumble against the stark reality of the complex real world.

The concept of comprehensive planning in theory may be contrasted with the narrowly focused planning which takes place in practice. Each administrative agency takes its decisions within its particular sphere of interest, understanding, resources and competence. How can it be otherwise? The task of any agency is to undertake the task for which it is established, not to take on the complicating and possibly conflicting responsibilities of others (which, in any case would be resistant to a take-over). Thus, a conservation agency will take decisions of a very different character from an economic development agency: they have separate and potentially conflicting goals. The idea that there is some level of planning which can rise above the narrow sectionalism of individual agencies is inconceivable in terms of implementation: it also assumes that an overriding objective can be identified and articulated. But this should not prevent us from seeking coordination of the actions of different agencies and enterprises so as to achieve better outcomes. This would start with a shared understanding of the variety of goals that are being pursued. Goals are typically expressed in terms of the public interest; yet there are very many 'publics'. They have conflicting concerns and priorities which are represented by, or reflected in, different agencies of government. This simple point is worth emphasising at a time when planning is promoted as a means of sectoral policy integration and achieving 'joined-up government'.

Change takes place not only in physical terms but also socially, economically, institutionally and indeed in many other ways. The spatial restructuring is the most dramatic visually but, in terms of the quality of everyday life, other dimensions are of greater importance: income and income security, employment, health services and education, and also matters relating to race, handicap, age and gender. Each of these has its own brand of planning. It has been suggested at various points in planning history that there should be an overarching planning system which coordinates all of these. The promulgation of the idea of spatial planning from the mid 1990s is the latest variation. There is a danger that this can be conceived as an extreme form of comprehensive planning. It is inconceivable that there can be one single 'plan' or even process for planning across government. Those who are concerned with the limitations of purely physical planning need to bear in mind the weaknesses of the rational comprehensive approach. It is even less likely that the land use planning system could 'grow' from its place in the institutional hierarchy to become some sort of umbrella planning under which all the other activities of government (sectors) and their planning processes stand. Improving coordination and collaboration among government sectors (or free-standing policy silos) will not be so simple, but the spatial or territorial plan which spells out the geographical dimension of sector policies should play a part.

Even simple coordination among the various agencies of planning is difficult and, not surprisingly, rare. Planners have made more claims for comprehensiveness than other professionals; indeed, the search for this has been their distinguishing feature. The fact that they have neither the responsibilities nor the resources for such ambitious aims has not, in the past, prevented them from being articulated. A classic example is the Greater London Development Plan (GLDP), which was subject to a searching inquiry in 1970. The plan dealt with not only the land use issues which fell within the remit of the Greater London Council (GLC), but also a wide range of policy areas including employment, education, transport, health and income distribution. There was no doubt about the importance of these and similar issues, but they did not fall within the responsibility of the GLC; indeed, it had no way of exerting any influence in these fields (Centre for Environmental Studies 1970).

The experience of the Greater London Development Plan cast a shadow over the hopes of the more ambitious planners of the time, and it affected the attitude of central government to the definition of matters which were relevant to an official development plan. It took a long time for 'social needs and problems' to take their place in the official planning system. By the end of the twentieth century these achieved a central status in both planning thought and in the character of plans. Indeed, the publications of the Secretary of State have become voluminous. The fear of over-ambitious plan-making has receded: it is hoped that local planning authorities are now too experienced in the implementation of plans to seek the impossible. However, the range and scale of currently acknowledged problems is formidable: the apparent impossibility of tackling problems raised by affordable housing, 'social exclusion', widespread car ownership and such like could lead to a sclerosis and a recall for the older conceptions of land use planning. This is probably an exaggeration, but there is no question that the emerging comprehensive planning system will be presented with awesome challenges.

Incrementalism

The obvious failure of comprehensive planning to attain desired goals has led to a number of alternative models of decision-making processes. Many of these revolve around the problem of making planning effective in a world where values, attitudes, and aspirations differ, where market and political forces predominate, and where uncertainty prevails. Lindblom (1959) dismissed rational-comprehensive planning as an impractical ideal. In his view, it is necessary to accept the realities of the processes by which planning decisions are taken: for this he outlined a 'science of muddling through'. Essentially, this incremental approach replaces grand plans by a modest step-by-step approach which aims at realisable improvements on an existing situation. This is a method of 'successive limited comparisons' of circumscribed problems and actions to deal with them. Lindblom argues that this is what happens in the real world: rather than attempt major change to achieve lofty ends, planners are compelled by reality to limit themselves to acceptable modifications of the status quo. On this argument, it is impossible to take all relevant factors into account

or to separate means from ends. Rather than attempting to reform the world, the planner should be concerned with incremental practicable improvements. There has been much debate on Lindblom's ideas (a good selection is given in Faludi 1973); here it is necessary to make only two points.

First, incrementalism is theoretically different from opportunism: it is a rational and realistic approach to dealing with problems. It rejects a comprehensive analysis of all the available options, and concentrates on what appears practicable and sensible given the constraints of time and resources. The classic illustration of the infeasibility of its opposite is the zero-base budget which, instead of being based on a previous year's figures, rejects history, and questions the justification for every individual item. (The term comes from the base line of the new budget – zero.) As Wildavsky (1987) has demonstrated, this is a completely unmanageable approach: it overwhelms, frustrates and finally exhausts those who try it. (Of course, selected items may with benefit be isolated for such treatment; but that is a very different matter.)

Second, incrementalism is more of a practical necessity than a desirable model to be followed. All policies need thorough review at times – particularly those embedded in an established development plan. Without the occasional upheaval (and that is what zero-base budgeting or policy-making implies), policies can continue well after they have served their purpose: they may even have become counterproductive. Indeed, incrementalism can lead to disasters, wars often being dramatic illustrations of the point. The ease with which incrementalism continues does, in fact, make a break in the continuity difficult, and often both the political and the administrative systems are averse to change. Nevertheless, changes in direction are sometimes essential; events (particularly unexpected ones) may create the basis for a change, despite any fears about uncertain outcomes. The difficulties are well illustrated by the current heart-searching about transport policy. Here, a reversal of trends is necessary, and is increasingly being recognised; but how change shall be brought about, and what its character shall be, is highly problematic.

Implementation

The rational model of planning embraces the simplistic view that there is a logical progression through successive stages of 'planning', culminating in implementation. The beguiling logic does not translate into reality. On the contrary, it is highly misleading and dangerous to separate policy and implementation (or action) matters. In fact, sometimes policy emanates from ideas about implementation rather than the other way round. Thus a policy of slum clearance or redevelopment focuses on the clearly indicated types of action. The implementation becomes the policy, and the underlying purpose is left in doubt. If the objective is to improve the living conditions of those living in slum areas, there might be better ways of doing this such as rehabilitation, or area improvement through local citizen action. With such an approach, demolition might be merely an incidental element in the local programme. With clearance as a policy, however, there is a danger that quite different objectives might be served, such as commercial development, or provision for roads and car parking.

Unfortunately, the difficulties involved here are even greater than this suggests since clearly focused efforts are not enough. For instance, a policy of improving a low-income area by environmental improvements may be explicitly intended to benefit the existing inhab-itants, but the added attraction of the area may become reflected in higher rents and prices which could lead to gentrification, thus benefiting a very different group. Similarly, a policy of providing grants to industrialists to move to an area of unemployment may result in the substitution of capital for labour, or the influx of workers with skills not possessed by the local people. A policy of preserving historic buildings by prohibiting demolition or alteration may lead to accelerated deterioration as owners seek ways of circumventing the regulations (and, in the period before the prohibition comes into effect, a rash of demolitions – as with the 1928 art deco Firestone factory on the Great West Road, London, which was due to be awarded listed building status but was demolished on a bank holiday weekend in 1980). A policy of reducing urban congestion by controlling growth through the designation of green belts may result in 'leapfrogging' of development, increased commuting, and thus increased urban congestion. Examples could easily be multiplied (Derthick 1972; Hall 1980; Kingdon 1984).

To confuse matters further, arguments about such effects are often complicated by differing views on what the objectives of the policy really are. The green belt case is a particularly good illustration of the point, since defenders (and there are many) can slip from one objective to another with ease. If the green belt does not reduce urban congestion, it provides 'opportunities for access to the open countryside for the urban population' and 'opportunities for outdoor sport and outdoor recreation near urban areas'; if it does not do this, it does 'retain attractive landscapes and enhance landscapes, near to where people live'. Other objectives are to improve damaged and derelict land around towns; to secure nature conservation interest; and to retain land in agricultural, forestry and related uses. There is nothing unique in such a long list of miscellaneous policy objectives. It would be a very sad policy indeed that was unable to meet any objectives in such a list! (To quote a pithy observation by Wildavsky, 'objectives are kept vague and multiple to expand the range within which observed behaviour fits'; Wildavsky 1987: 35).

It should be added that sometimes policies have unintended good byproducts. Unfortunately, it is often difficult to relate cause and effect, but one example is the imposition by the US federal government of a 55 miles per hour (90 km/h) speed limit in 1974. This was introduced to reduce petrol consumption, but a welcome effect was a reduction in road accidents: this, for a time, became the basis of a powerful argument for retaining the speed limit after the fuel crisis had passed.

The points do not need labouring: the certainty which is required for the type of rational planning envisaged in some traditional theories is impossible. The underlying assumptions, relevance and political support can change dramatically; the outcomes of policy are difficult to predict, are frequently different from expectations, are hard to identify and to separate from all the other forces at work, and are rarely clear. Thus, not only is planning a hazardous exercise, with serious likelihood of failure, but also it is an exercise whose outcomes are remarkably difficult to evaluate, even when they are felt to be a resounding success.

It is perhaps unsurprising that most planners have neither the time nor remit to examine what went wrong with the last plan: they have moved on to the next one! It is, however, a matter of some surprise that there have been so few analyses of the (UK) planning scene to fill the vacuum. The wealth of US studies indicates how valuable this can be. Perhaps it is another cultural characteristic that there is little interest in learning why things go wrong?

The British planning system in comparative perspective

Since it is easier to understand one planning system by comparing it with another, it is worth exploring a little further the differences between the UK, US and other European systems. Three features are of particular interest: first, the extent to which a planning system operates within a framework of constitutionally protected rights; second, the degree to which a system embodies discretion; third, the importance of history and culture.

In many countries, the constitution limits governmental action in relation to land and property. The US Bill of Rights provides that 'no person shall . . . be deprived of life, liberty or property without due process of law; nor shall private property be taken without just compensation'. These words mean much more than is apparent to the casual (non-American) reader. Since land use regulations affect property rights, they are subject to constitutional challenge. They can be disputed not only on the basis of their effect on a particular property owner (i.e. *as applied*), but also in principle: a regulation can be challenged on the argument that, in itself, it violates the constitution (this is described in the legal jargon as being *facially* unconstitutional). Moreover, the constitution protects against arbitrary government actions, and this further limits what can be done in the name of land use planning. No such restraints exist in the UK system. Indeed, the UK does not have a codified constitution of the type common to most other countries (Yardley 1995).

Constitutions can influence the system in more subtle ways. In some European countries including Italy, the Netherlands and Spain, the constitution provides that all citizens have the right to a decent home. This may limit planning action, but may also influence policy priorities and provide legitimacy for intervention. In Finland and Portugal, landowners are granted the constitutional right to build on their land. This presents obvious difficulty in pursuing policies of restraining urban growth. Constitutions also often allocate powers to different tiers of government, which effectively ensures a minimum degree of autonomy for regional and local governments. Again, there is no such constitutional safeguard in the UK. As a result, the Thatcher government was able to abolish a whole tier of metropolitan local government in England and, in consequence, that part of the planning system that went with it. Such haughty action would be inconceivable in most countries. In the United States, for example, there is little to compare with the central power which is exercised by the national government in Britain. Plan-making and implementation are essentially local issues, even though the federal government has become active in highways, water and environmental matters and, in recent years, a number of states have become involved in land use planning. So local is the responsibility that even the decision on whether to operate land use controls is a local one, and many US local governments have only minimal systems. Similarly, in much of Europe, regional and local government would not tolerate the extent of central government supervision (they might say interference) in local planning matters. But there is a point where decisions have to be made at a higher level because opposition from local decision-makers might mean that some nationally or internationally important developments never happen or to coordinate developments that will affect more than one local territory. In these cases the subsidiarity principle is invoked. This is a relatively new import to government in Britain (where the assumption has been that the centre will decide) but simply means that decisions will be ceded up from the local level (community or individual) only if there is a demonstrated benefit or need. The European Courts have been kept busy mediating different understandings of benefit and need.

Lack of constitutional constraint allows for a wide degree of discretion in the UK planning system. In determining applications for planning permission, a local authority is guided by the development plan, but is not bound by it: other 'material considerations' are taken into account. In most of the rest of the world, plans become legally binding documents. Indeed, they are part of the law, and the act of giving a permit is no more than a certification that a proposal is in accordance with the plan. In practice, there are mechanisms that allow for variations from the provisions of a plan but, since these are by definition contrary to the law, they may entail lengthy procedures, and perhaps an amendment to the plan.

This discretion is further enlarged by the fact that the preparation of a local plan is carried out by the same local authority that implements it. This is so much a part of the tradition of British planning that no one comments on it. The American situation is different, with great emphasis being placed on the separation of powers. (Typically the plan is prepared by the legislative body – the local authority – but administered by a separate board.) The British system has the advantage of relating policy and administration (and easily accommodating policy changes) but, to American eyes, 'this institutional framework blurs the distinction between policy making and policy applying, and so enlarges the role of the administrator who has to decide a specific case' (Mandelker 1962: 4). The Human Rights Convention also focuses attention on the separation of powers since it provides for the right to appeal to an independent body against actions of government. While there is a limited right of appeal to the courts in the UK (which are independent) most appeals are heard by the government or its representatives. Changes to the planning system have already been made to meet the requirements of the Convention – and others will no doubt follow.

Above all, in comparing planning systems, there are fundamental differences in the philosophy that underpins them. Thus, put simply (and therefore rather exaggeratedly), American planning is largely a matter of anticipating trends, while in the UK there

is a conscious effort to bend them in publicly desirable directions. In France, *aménagement du territoire* (the term often incorrectly used as a translation for town and country planning) deals with the planning of the activities of different government sectors to meet common social and economic goals, while in the UK town and country planning is about the management of land use, albeit taking into account social and economic concerns.

Planning systems are rooted in the particular historical, legal, and physical conditions of individual countries and regions. In the UK, some of the many important factors which have shaped the system are the strong land preservation ethic, epitomised in the work of the Campaign to Protect Rural England (CPRE) (and its Scottish and Welsh counterparts) and, of longer standing, the husbandry of the landowning class. Added (but not unrelated to this), are the popular attitudes to the preservation of the countryside and the containment of urban sprawl which in turn is related to the early industrialisation of the UK; the small size of the country; the long history of parliamentary government; and the power of the civil service in central government and the professions in local government.

In comparison, land in the United States has historically been a replaceable commodity that could and should be parcelled out for individual control and development, and if one person saw fit to destroy the environment of his or her valley in pursuit of profit, well, why not? There was always another valley over the next hill. Thus the seller's concept of property rights in land came to include the right of owners to earn a profit from their land, and indeed to change the very essence of the land if this were necessary to obtain that profit.

In the Mediterranean countries of Greece, Italy, Portugal and Spain there has been a short history of democratic government, and planning regulation has enjoyed little general public support. Controlling land use has been much less a political priority than housing the population. In large parts of these countries rapid urbanisation has proceeded with little regard to regulations or plans. The historic cores of cities, meanwhile, have not until recently felt the scale of pressure for redevelopment which has been the norm in northern Europe.

However, in most countries, land for development is becoming more valuable, and the problem of coping with land use conflicts is of increasing importance. In Europe this has led to the growth of a conservationist ethic, with the restraint of urban growth being a top priority. In the USA, this has happened to a limited extent, particularly with environmentally valuable resources, but a major effect was in the opposite direction: to increase the attractiveness of land as a source of profit. Speculation has never been frowned upon in the USA. In many countries, land is regarded as different from commodities: it is something to be preserved and husbanded. In the USA, the dominant ethic regards land as a commodity, no different from any other. Though there is much rhetoric to the contrary, actions speak louder than words.

The contrast in the operation of planning in different countries is abundantly clear to anyone who travels.

Accommodating change

Having drawn the comparison, it is immediately necessary to qualify it: times and attitudes change, sometimes slowly, sometimes dramatically. The largest postwar change in the UK has been the move from 'positive planning' to a more market-conscious (and sometimes market-led) approach. The elements of this (which range from the abolition of development charges to the embracement of property-led urban regeneration) are discussed later, but it needs to be stressed that the extent of the change in planning attitudes toward market forces has been dramatic. The limits of the possible have been redefined in the light of experience and a recognition of the character of the forces at work in the modern world.

Governments are responsive to shifts in electoral opinion, particularly when changes can be made painlessly. The UK planning system provides a route by which change can be implemented, not only without pain, but also without much effort. Indeed, the ease with which it can accommodate change is quite remarkable. There has, for instance, been a see-saw

in the extent to which economic development, social needs and environmental concerns have had a high profile. In the 1980s, economic efficiency rose to prime place in the government's order of priorities. (This was the time when the planning system was attacked for its restrictive character: 'locking away jobs in filing cabinets'.) Environmental concerns later became salient: a result of a fascinating combination of conservative forces, ranging from green-belt voters keen to protect the belt, to a younger generation of protestors who had less to lose but saw more to protect. As already indicated, social considerations were for long regarded by central government as being outside the legitimate responsibility of the planning system (a curiously British myopia). Major arguments have raged between the centre and the localities on what is, and what is not, appropriate for inclusion in a development plan. After many years of pressing local authorities to exclude 'social factors', the central government made a curious about-turn in a planning policy guidance note (PPG) of 1992: 'authorities will wish to consider the relationship of planning policies and proposals to social needs and problems' (PPG 12, para. 5.48). More recent statements have gone much further along this road.

This flexibility (another aspect of the discretionary nature of the system) is a built-in feature. The statutory framework is essentially procedural; it is almost devoid of substantive content. Local authorities are given the duty to prepare development plans (a rare case where no discretion is allowed; unlike their US counterparts, a local authority is not free to decide not to have a plan). What goes into the plan, however, is very imprecise. More detailed requirements are, of course, spelled out in a variety of directions and advice from the central government. But that is the point: the content is added separately, and can be changed in line with what 'the Secretary of State may prescribe'.

Yet changes are often not easy to evaluate, even if only because the implementation of planning policy rests with local authorities and, despite much bandying of words, central government powers over them are limited. There are, of course, various control mechanisms and default powers, but these are cumbersome to use, and they carry political risks. Moreover, central government's understanding of how local government works, and its awareness of what happens in practice is even more circumscribed. These depths of ignorance have had surprisingly little academic light shed on them: few studies have been undertaken of the actual working of the planning machine. (Note the surprise which was expressed when the report on development control in North Cornwall (Lees 1993) revealed that the local councillors gave favourable consideration to the personal circumstances of local applicants for planning permission.)

Given such considerations, it can be difficult to chart (or even to be aware of) important changes. Legislative amendments and new policy statements are more apparent, but they may not be as important as trends which emerge over time. (For example, it *may* be that one of the most significant operational changes has been the way in which local authorities and housebuilders have evolved a system for negotiating housing land allocations; perhaps in time this model might be followed in other development sectors?) Moreover, major political statements and new laws typically follow rather than precede changes in attitudes and perceptions. The picture is also confused by grand claims for new approaches which seldom last far beyond an initial flurry. Much is obscured by political debate and the use of fuzzy jargon. Changes are more easily seen in retrospect than contemporaneously.

Planning questions

Planning is an imperative: only the form it takes is optional. At a minimum, some system is required to provide infrastructure – preferably in the right place at the right time. Something more than this is generally accepted to be necessary (and general acceptance is the bedrock of any form of effective planning). But there is no way of determining the extent to which a planning system *ought* to go in determining 'how much of what should go where and when'. The decision is a political one, even if it is taken by default (i.e. with no effective opposition). Hence, as stressed earlier, cultural influences are crucial though this does not mean that a planning system is hallowed or immune from review and radical change. It may be that the UK system has

reached precisely the stage at which this is required, though this is not the place to elaborate such an argument. It is, nevertheless, appropriate to point to some issues which need addressing if the planning system is to adapt to conditions which are very different from those which existed when it was introduced.

First, the UK planning system is highly effective in stopping development: it is much less effective in facilitating it. Comparative research on property markets in Europe (Williams and Wood 1994) underlines the lack of 'positive planning'. There are serious weaknesses in anticipating needs and allocating sufficient land for these to be met, in the assembling and acquisition of land (especially in inner cities) and integrating the planning of infrastructure with new development. Powers exist for such important planning actions, but they are underused since there is insufficient relationship between the (public) planning process and the (largely private) development process. In the 1947 Act, it was envisaged that 'positive' planning would be undertaken directly by the public sector. This proved infeasible and alternative mechanisms are underdeveloped.

Second, the most difficult issue facing any policy is defining the right questions. A mechanism is needed to facilitate this. It could be argued that current UK debates are focused on the wrong questions. Too many are concerned with the 'how' of planning policy rather than the 'why'. Why is the countryside to be protected? Why are city centres to be rehabilitated? Why are additional facilities for travel to be provided? There are many such questions, and though they do not have simple (or readily acceptable) answers, debate upon them would provide a firmer base for policy than exists at present. The debate would, however, raise a further level of policy questions. Thus, it might be asked where retail outlets should be located to maximise convenience, service and profitability (or whatever other criteria are to be employed), rather than posing the questions in terms of safeguarding existing patterns of development (particularly existing town centres). Instead of asking where the best locations are for housing an additional *x* million households, argument rages over protecting the countryside from housing development and concentrating new housing in urban areas. (There has been a major change in governmental attitudes since this was first written.)

Third, planning deals with a highly complex series of interrelated processes which are imperfectly understood. Though better understanding should be high on the research agenda, these processes will inevitably remain beyond the comprehension needed for fully competent land use planning. It follows that planning must proceed on the basis of either a high degree of ignorance, or belief in the efficacy of some overriding political or economic philosophy. In practical terms, this implies debating how far the planning process should ally itself to market forces (or socio-economic trends if that term is preferred).

These issues arise throughout this book. It should be evident from this introductory discussion that they are not easily resolved. Indeed, much planning effort is spent on wrestling with them. There seems no doubt that this will continue.

Further reading

Good starting points on the nature of planning are Allmendinger (2001) *Planning Theory*, Taylor (1998) *Urban Planning Theory since 1945*, the older but still relevant Chapter 2 of Healey *et al.* (1982) *Planning Theory: Prospects for the 1980s* and Ravetz (1986) *The Government of Space* (which contains a chapter on 'theoretical perspectives'). A useful collection of early articles is contained in Faludi (1973) *A Reader in Planning Theory*. (This includes the article by Lindblom referred to in the chapter, together with important articles by writers such as Davidoff, Etzioni, Friedmann and Meyerson.) Later collections contain more recent writings: Healey *et al.* (1982) *Planning Theory: Prospects for the 1980s*, Campbell and Fainstein (1996) *Readings in Planning Theory* and Fainstein and Campbell (1996) *Readings in Urban Theory*. A helpful analysis of 'Arguments for and against planning' is given by Klosterman (1985). For an insightful, succinct discussion of the constant flood of changes which besets planning, see Batty (1990) 'How can we best respond to changing fashions in urban and regional planning?'. Sillince (1986) *A Theory of Planning* gives useful summaries of rational comprehensive and incremental

models of procedural planning theory. A fuller account is provided by Alexander (1992) *Approaches to Planning*.

On politics and planning see Albrechts (2003) 'Reconstructing decision making'; for recent relations between political ideologies and planning see Tewdwr-Jones (2002) *The Planning Polity* and Allmendinger and Tewdwr-Jones (2000) 'New Labour, new planning'. On planning and citizenship see Neill (2004) *Urban Planning and Cultural Identity*. A particularly useful introduction to the analysis of policy issues is Kingdon (1984) *Agendas, Alternatives and Public Policies*. A clear and succinct account of policy processes is given in Ham and Hill (1993) *The Policy Process in the Modern Capitalist State*. There is a good range of readings in an accompanying volume edited by Hill (1993) *The Policy Process* and Tewdwr-Jones (1996) *British Planning Policy in Transition*. Hall (1980) *Great Planning Disasters* is required reading for all planners, as well as for non-planners who want to know why planning is so difficult. A more complex, but fascinating, study focused on the operation of US federal policy in one urban area is Pressman and Wildavsky (third edition, 1984) *Implementation*. Very interesting as well as revealing is Derthick (1972) *New Towns In-Town: Why a Federal Program Failed*. Such case studies are much more common in the USA than in the UK (a reflection of the cultural differences in the openness of government). Among the small number of British studies, see Muchnick (1970) *Urban Renewal in Liverpool*, Levin (1976) *Government and the Planning Process* (which focuses on the new and expanded towns), Healey (1983) *Local Plans in British Land Use Planning* and Blowers (1984) *Something in the Air: Corporate Power and the Environment*. Simmie (1993) *Planning at the Crossroads* summarises research findings on the impact of planning in the UK. A radical critique of the role of planning in society is Ambrose (1986) *Whatever Happened to Planning?* See also his *Urban Process and Power* (1994).

For a comparative study of 'certainty and discretion' in planning see Booth (1996) *Controlling Development*. Discretion is discussed at length (comparing the UK and the USA) in Cullingworth (1993) *The Political Culture of Planning*. A broader discussion of the two countries is given by Vogel (1986) *National Styles of Regulation*.

The challenge that Thatcherism and the market brought to ideas of planning has been addressed in many studies – notably Thornley (1993) *Urban Planning under Thatcherism: The Challenge of the Market*, Allmendinger and Thomas (1998) *Urban Planning and the British New Right* and Brindley *et al.* (1996) *Remaking Planning*.

Communication, negotiation and argumentation through planning have dominated discussions about planning theory during the 1990s. Early contributions are Forester (1982) 'Planning in the face of power' and other papers brought together in Forester's book (1989) *Planning in the Face of Power*. Later contributions include Innes (1995) 'Planning theory's emerging paradigm: communicative action and interactive practice', and Healey (1997) *Collaborative Planning*, (1998) 'Collaborative planning in a stakeholder society' and (1992c) 'Planning through debate: the communicative turn in planning theory'. Consequently, aspects of planning practice have been investigated using these ideas, for example, Healey (1993)

'The communicative work of development plans', Davoudi *et al.* (1997) 'Rhetoric and reality in British structure planning in Lancashire: 1993–95' and Tait and Campbell (2000) 'The politics of communication between planning officers and politicians: the exercise of power through discourse'. For a critique see Tewdwr-Jones and Allmendinger (1998) 'Deconstructing communicative rationality: a critique of Habermasian collaborative planning'.

For some amusing homespun philosophy on planning see Zucker (1999) *What Your Planning Professors Forgot to Tell You*.

Note

1 The 'purpose' was given by the Planning and Compulsory Purchase Act 2004 s. 39.

2 The evolution of town and country planning

The first assumption that we have made is that national planning is intended to be a reality and a permanent feature of the internal affairs of this country.

Uthwatt Report 1942

The planning system plays a crucial role in our national life – a vital tool in the process of change and renewal as well as conservation and environmental care, vital to our national prosperity. The planning system is at the heart of our shared national goals to raise productivity and ensure full employment, to encourage and foster strong vital communities, to help give everyone the opportunity of a decent home, and to achieve truly balanced and sustainable development and growth in every region and nation across the UK.

Gordon Brown, Chancellor 2003

The public health origins

Town and country planning as a task of government has developed from public health and housing policies. The nineteenth-century increase in population and, even more significant, the growth of towns led to public health problems which demanded a new role for government. Together with the growth of medical knowledge, the realisation that overcrowded insanitary urban areas resulted in an economic cost (which had to be borne at least in part by the local ratepayers) and the fear of social unrest, this new urban growth eventually resulted in an appreciation of the necessity for interfering with market forces and private property rights in the interest of social well-being.

The nineteenth-century public health legislation was directed at the creation of adequate sanitary conditions. Among the measures taken to achieve these were powers for local authorities to make and enforce building bylaws for controlling street widths, and the height, structure and layout of buildings. Limited and defective though these powers proved to be, they represented a marked advance in social control and paved the way for more imaginative measures. The physical impact of bylaw control on British towns is depressingly still very much in evidence; and it did not escape the attention of contemporary social reformers. In the words of Unwin:

much good work has been done. In the ample supply of pure water, in the drainage and removal of waste matter, in the paving, lighting and cleansing of streets, and in many other such ways, probably our towns are as well served as, or even better than, those elsewhere. Moreover, by means of our much abused bye-laws, the worst excesses of overcrowding have been restrained; a certain minimum standard of air-space, light and ventilation has been secured; while in the more modern parts of towns, a fairly high degree of sanitation, of immunity from fire, and general stability of construction have been maintained, the importance of which can hardly be exaggerated. We have, indeed, in all these matters laid a good foundation and have secured many of the necessary elements for a healthy condition of life; and yet the remarkable fact remains that there are growing up around our big towns vast districts, under these very bye-laws, which for dreariness and sheer ugliness it is difficult to match anywhere, and compared with which many of the old unhealthy slums are, from the point of view of picturesqueness and beauty, infinitely more attractive.

(Unwin 1909: 3)

It was on this point that public health and architecture met. The enlightened experiments at Saltaire (1853), Bournville (1878), Port Sunlight (1887) and elsewhere had provided object lessons. Ebenezer Howard and the garden city movement were now exerting considerable influence on contemporary thought. In the commentary to the 2003 republication of Howard's book, *To-morrow: A Peaceful Path to Real Reform* (originally published in 1898) it is described as 'almost without question the most important single work in the history of modern town planning'.[1] Howard's ideas about land reform and a 'socialist community', and his practical approach to the form of towns and how better towns could be created, were to influence many advocates for town planning.

The National Housing Reform Council (later the National Housing and Town Planning Council) was campaigning for the introduction of town planning. Even more significant was a similar demand from local government and professional associations such as the Association of Municipal Corporations, the Royal Institute of British Architects, the Surveyors' Institute and the Association of Municipal and County Engineers. As Ashworth has pointed out:

the support of many of these bodies was particularly important because it showed that the demand for town planning was arising not simply out of theoretical preoccupations but out of the everyday practical experience of local administration. The demand was coming in part from those who would be responsible for the execution of town planning if it were introduced.

(Ashworth 1954: 180)

The first Planning Act

The movement for the extension of sanitary policy into town planning was uniting diverse interests. These were nicely summarised by John Burns, President of the Local Government Board, when he introduced the first legislation bearing the term 'town planning' – the Housing, Town Planning, Etc. Act 1909:

The object of the bill is to provide a domestic condition for the people in which their physical health, their morals, their character and their whole social condition can be improved by what we hope to secure in this bill. The bill aims in broad outline at, and hopes to secure, the home healthy, the house beautiful, the town pleasant, the city dignified and the suburb salubrious.

The new powers provided by the Act were for the preparation of 'schemes' by local authorities for controlling the development of new housing areas. Though novel, these powers were logically a simple extension of existing ones. It is significant that this first legislative acceptance of town planning came in an Act dealing with health and housing. The gradual development and the accumulated experience of public health and housing measures facilitated a general acceptance of the principles of town planning.

Housing reform had gradually been conceived in terms of larger and larger units. Torrens' Act (Artisans and Labourers Dwellings Act, 1868) had made a beginning with individual houses; Cross's Act (Artisans and Labourers Dwellings Improvement Act, 1875) had introduced an element of town planning by concerning itself with the reconstruction of insanitary areas; the framing of bylaws in accordance with the Public Health Act of 1875 had accustomed local authorities to the imposition of at least a minimum of regulation on new building, and such a measure as the London Building Act of 1894 brought into the scope of public control the formation and widening of streets, the lines of buildings frontage, the extent of open space around buildings, and the height of

buildings. Town planning was therefore not altogether a leap in the dark, but could be represented
as a logical extension, in accordance with changing
aims and conditions, of earlier legislation concerned
with housing and public health.

(Ashworth 1954: 181)

The 'changing conditions' were predominantly the
rapid growth of suburban development: a factor which
increased in importance in the following decades.

> In fifteen years 500,000 acres of land have been
> abstracted from the agricultural domain for houses,
> factories, workshops and railways . . . If we go on in
> the next fifteen years abstracting another half a
> million from the agricultural domain, and we go on
> rearing in green fields slums, in many respects,
> considering their situation, more squalid than those
> which are found in Liverpool, London and Glasgow,
> posterity will blame us for not taking this matter
> in hand in a scientific spirit. Every two and a half
> years there is a County of London converted into
> urban life from rural conditions and agricultural
> land. It represents an enormous amount of building
> land which we have no right to allow to go unreg-
> ulated.
>
> (*Parliamentary Debates*, 12 May 1908)

The emphasis was entirely on raising the standards
of *new* development. The Act permitted local authorities (after obtaining the permission of the Local
Government Board) to prepare town planning schemes
with the general object of 'securing proper sanitary
conditions, amenity and convenience', but only for land
which was being developed or appeared likely to be
developed.

Strangely it was not at all clear what town planning
involved. It certainly did not include 'the remodelling
of the existing town, the replanning of badly planned
areas, the driving of new roads through old parts of a
town – all these are beyond the scope of the new
planning powers' (Aldridge 1915: 459). The Act itself
provided no definition: indeed, it merely listed nineteen 'matters to be dealt with by general provisions
prescribed by the Local Government Board'. The

restricted and vague nature of this first legislation was
associated in part with the lack of experience of the
problems involved.

Nevertheless, the cumbersome administrative procedure devised by the Local Government Board (in
order to give all interested parties 'full opportunity of
considering all the proposals at all stages') might well
have been intended to deter all but the most ardent of
local authorities. The land taxes threatened by the
Finance Act 1910, and then the First World War,
added to the difficulties. It can be the occasion of no
surprise that very few schemes were actually completed
under the 1909 Act.

Interwar legislation

The first revision of town planning legislation which
took place after the war (the Housing and Town
Planning Act 1919) did little in practice to broaden
the basis of town planning. The preparation of schemes
was made obligatory on all borough and urban districts
having a population of 20,000 or more, but the time
limit (January 1926) was first extended (by the
Housing Act 1923) and finally abolished (by the
Town and Country Planning Act 1932). Some of
the procedural difficulties were removed, but no
change in concept appeared. Despite lip-service to the
idea of town planning, the major advances made at
this time were in the field of housing rather than
planning.

It was the 1919 Act which began what Marion
Bowley (1945: 15) has called 'the series of experiments
in State intervention to increase the supply of working-
class houses'. The 1919 Act accepted the principle
of state subsidies for housing and thus began the
nationwide growth of council house estates. Equally
significant was the entirely new standard of working-
class housing provided: the three-bedroom house with
kitchen, bath and garden, built at the density recommended by the Tudor Walters Committee in 1918 of
not more than twelve houses to the acre. At these new
standards, development could generally take place
only on virgin land on the periphery of towns, and
municipal estates grew alongside the private suburbs:

'the basic social products of the twentieth century', as Asa Briggs (1952 vol. 2: 228) has termed them.

This suburbanisation was greatly accelerated by rapid developments in transportation – developments with which the young planning machine could not keep pace. The ideas of Howard (1898) and the garden city movement, of Geddes (1915) and of those who, like Warren and Davidge (1930), saw town planning not only as a technique for controlling the layout and design of residential areas, but also as part of a policy of national economic and social planning, were receiving increasing attention, but in practice town planning typically meant little more than an extension of the old public health and housing controls.

Various attempts were made to deal with the increasing difficulties. Of particular significance were the Town and Country Planning Act 1932, which extended planning powers to almost any type of land, whether built-up or undeveloped, and the Restriction of Ribbon Development Act 1935 which, as its name suggests, was designed to control the spread of development along major roads. But these and similar measures were inadequate. For instance under the 1932 Act, planning schemes took about three years to prepare and pass through all their stages. Final approval had to be given by Parliament, and schemes then had the force of law, as a result of which variations or amendments were not possible except by a repetition of the whole procedure. Interim development control operated during the time between the passing of a resolution to prepare a scheme and its date of operation (as approved by Parliament). This enabled, but did not require, developers to apply for planning permission. If they did not obtain planning permission, and the development was not in conformity with the scheme when approved, the planning authority could require the owner (without compensation) to remove or alter the development.

All too often, however, developers preferred to take a chance that no scheme would ever come into force, or that if it did no local authority would face pulling down existing buildings. The damage was therefore done before the planning authorities had a chance to intervene (Wood 1949: 45). Once a planning scheme was approved, on the other hand, the local authority

ceased to have any planning control over individual developments. The scheme was in fact a zoning plan: land was zoned for particular uses such as residential or industrial, though provision could be made for such controls as limiting the number of buildings and the space around them. In fact, so long as developers did not try to introduce a non-conforming use they were fairly safe. Furthermore, most schemes did little more than accept and ratify existing trends of development, since any attempt at a more radical solution would have involved the planning authority in compensation which they could not afford to pay. In most cases, the zones were so widely drawn as to place hardly more restriction on the developer than if there had been no scheme at all. Indeed, in the half of the country covered by draft planning schemes in 1937 there was sufficient land zoned for housing to accommodate 291 million people (Barlow Report 1940: para. 241).

A major weakness was, of course, the administrative structure itself. District and county borough councils were generally small and weak. They were unlikely to turn down proposals for development on locational grounds if compensation was involved or if they would thereby be deprived of rate income. The compensation paid either for planning restrictions or for compulsory acquisition, which had to be determined in relation to the most profitable use of the land, even if it was unlikely that the land would be so developed, and without regard to the fact that the prohibition of development on one site usually resulted in the development value (which had been purchased at high cost) shifting to another site. Consequently, in the words of the Uthwatt Committee,

> an examination of the town planning maps of some of our most important built-up areas reveals that in many cases they are little more than photographs of existing users and existing lay-outs, which, to avoid the necessity of paying compensation, become perpetuated by incorporation in a statutory scheme irrespective of their suitability or desirability.

These problems increased as the housing boom of the 1930s developed; 2,700,000 houses were built in England and Wales between 1930 and 1940. At the outbreak of the Second World War, one-third of all the houses in England and Wales had been built since 1918. The implications for urbanisation were obvious, particularly in the London area. Between 1919 and 1939 the population of Greater London rose by about 2 million, of which three-quarters of a million was natural increase, and over one and a quarter million was migration (Abercrombie 1945). This growth of the metropolis was a force which existing powers were incapable of halting, despite the large body of opinion favouring some degree of control.

The depressed areas

The crux of the matter was that the problem of London was closely allied to that of the declining areas of the North and of South Wales, and both were part of the much wider problem of industrial location. In the South East the insured employed population rose by 44 per cent between 1923 and 1934, but in the North East it fell by 5.5 per cent and in Wales by 26 per cent. In 1934, 8.6 per cent of insured workers in Greater London were unemployed, but in Workington the proportion was 36.3 per cent, in Gateshead 44.2 per cent, and in Jarrow 67.8 per cent. In the early stages of political action these two problems were divorced. For London, various advisory committees were set up and a series of reports issued.[2]

For the depressed areas, attention was first concentrated on encouraging migration, on training, and on schemes for establishing the unemployed in smallholdings. Increasing unemployment accompanied by rising public concern necessitated further action.[3] Special Commissioners were appointed for England and Wales, and for Scotland, with very wide powers for 'the initiation, organisation, prosecution and assistance of measures to facilitate the economic development and social improvement' of the special areas. The areas were defined in the Act and included the North East coast, West Cumberland, industrial South Wales, and the industrial area around Glasgow. The Commissioners' main task – the attraction of new industry – proved to be extraordinarily difficult, and in his second report, Sir Malcolm Stewart, the

Commissioner for England and Wales, concluded that 'there is little prospect of the special areas being assisted by the spontaneous action of industrialists now located outside these areas'. On the other hand, the attempt actively to attract new industry by the development of trading estates achieved considerable success, which at least warranted the comment of the Scottish Commissioner that there had been 'sufficient progress to dispel the fallacy that the areas are incapable of expanding their light industries'.

Nevertheless, there were still 300,000 unemployed in the special areas at the end of 1938, and although 123 factories had been opened between 1937 and 1938 in the special areas, 372 had been opened in the London area. Sir Malcolm Stewart concluded, in his third annual report, that 'the further expansion of industry should be controlled to secure a more evenly distributed production'. Such thinking might have been in harmony with the current increasing recognition of the need for national planning, but it called for political action of a character which would have been sensational. Furthermore, as Neville Chamberlain (then Chancellor of the Exchequer) pointed out, even if new factories were excluded from London it did not follow that they would forthwith spring up in South Wales or West Cumberland. The immediate answer of the government was to appoint the Royal Commission on the Distribution of the Industrial Population.

The Barlow Report

The Barlow Report is of significance not merely because it is an important historical landmark, but also because some of its major recommendations were for so long accepted as a basis for planning policy.

The terms of reference of the Commission were

to inquire into the causes which have influenced the present geographical distribution of the industrial population of Great Britain and the probable direction of any change in the distribution in the future; to consider what social, economic or strategic disadvantages arise from the concentration of industries or of the industrial population in large towns or in particular areas of the country; and to report what remedial measures if any should be taken in the national interest.

These very wide terms of reference represented, as the Commission pointed out, 'an important step forward in contemporary thinking' and, after reviewing the evidence, it concluded that

> the disadvantages in many, if not most of the great industrial concentrations, alike on the strategical, the social and the economic side, do constitute serious handicaps and even in some respects dangers to the nation's life and development, and we are of opinion that definite action should be taken by the government towards remedying them.

The advantages of more urban concentration at that time were clear: proximity to markets, reduction of transport costs and availability of a supply of suitable labour. But these, in the Commission's view, were accompanied by serious disadvantages such as heavy charges on account mainly of high site values, loss of time through street traffic congestion and the risk of adverse effects on efficiency due to long and fatiguing journeys to work. The Commission maintained that the development of garden cities, satellite towns and trading estates could make a useful contribution towards the solution of the problems of urban congestion.

The London area, of course, presented the largest problem, not simply because of its huge size, but also because 'the trend of migration to London and the Home Counties is on so large a scale and of so serious a character that it can hardly fail to increase in the future the disadvantages already shown to exist'. The problems of London were thus in part related to the problems of the depressed areas:

> It is not in the national interest, economically, socially or strategically, that a quarter, or even a larger, proportion of the population of Great Britain should be concentrated within twenty to thirty miles or so of Central London. On the other hand, a policy:

> (i) of balanced distribution of industry and the industrial population so far as possible throughout the different areas or regions in Great Britain;
>
> (ii) of appropriate diversification of industries in those areas or regions

would tend to make the best national use of the resources of the country, and at the same time would go far to secure for each region or area, through diversification of industry, and variety of employment, some safeguard against severe and persistent depression, such as attacks an area dependent mainly on one industry when that industry is struck by bad times.

Such policies could not be carried out by the existing administrative machinery: it was no part of statutory planning to check or to encourage a local or regional growth of population. Planning was essentially on a local basis; it did not, and was not intended to, influence the geographical distribution of the population as between one locality or another. The Commission unanimously agreed that the problems were national in character and required 'a central authority' to deal with them. They argued that the activities of this authority ought to be distinct from and extend beyond those of any existing government department. It should be responsible for formulating a plan for dispersal from congested urban areas – determining in which areas dispersal was desirable; whether and where dispersal could be effected by developing garden cities or garden suburbs, satellite towns, trading estates, or the expansion of existing small towns or regional centres. It should be given the right to inspect town planning schemes and 'to consider, where necessary, in cooperation with the government departments concerned, the modification or correlation of existing or future plans in the national interest'. It should study the location of industry throughout the country with a view to anticipating cases where depression might probably occur in the future and encouraging industrial or public development before a depression actually occurred.

Whatever form this central agency might take (a matter on which the Commission could not agree), it was essential that the government should adopt a much more positive role: control should be exercised over new factory building at least in London and the Home Counties, that dispersal from the larger conurbations should be facilitated, and that measures should be taken to anticipate regional economic depression.

The impact of war

The Barlow Report was published in January 1940, some four months after the start of the Second World War. The problem which precipitated the decision to set up the Barlow Commission, that of the depressed areas, rapidly disappeared. The unemployed of the depressed areas now became a powerful national asset. A considerable share of the new factories built to provide munitions or to replace bombed factories were located in these areas. By the end of 1940, 'an extraordinary scramble for factory space had developed'; and out of all this 'grew a wartime, an extempore, location of industry policy covering the country as a whole' (Meynell 1959). This emergency wartime policy, paralleled in other fields, such as hospitals, not only provided some 13 million square feet of munitions factory space in the depressed areas which could be adapted for civilian industry after the end of the war, but also provided experience in dispersing industry and in controlling industrial location which showed the practicability (under wartime conditions at least) of such policies. The Board of Trade became a central clearing-house of information on industrial sites. During the debates on the 1945 Distribution of Industry Bill, its spokesman stressed:

> We have collected a great deal of information regarding the relative advantage of different sites in different parts of the country, and of the facilities available there with regard to local supply, housing accommodation, transport facilities, electricity, gas, water, drainage and so on . . . We are now able to offer to industrialists a service of information regarding location which has never been available before.

Hence, although the Barlow Report 'lay inanimate in the iron lung of war',[4] it seemed that the conditions for the acceptance of its views on the control of industrial location were becoming very propitious: there is nothing better than successful experience for demonstrating the practicability of a policy.

The war thus provided a great stimulus to the extension of regional planning into the sphere of industrial location. This was not the only stimulus it provided: the destruction wrought by bombing transformed 'the rebuilding of Britain' from a socially desirable but somewhat visionary and vague ideal into a matter of practical and clear necessity. Nor was this all: the very fact that rebuilding would be taking a large scale provided an unprecedented opportunity for comprehensive planning of the bombed areas and a stimulus to overall town planning. In the Exeter Plan, Thomas Sharp (1947) urged that

> to rebuild the city on the old lines . . . would be a dreadful mistake. It would be an exact repetition of what happened in the rebuilding of London after the Fire – and the results, in regret at lost opportunity, will be the same. While, therefore, the arrangements for rebuilding to the new plan should proceed with all possible speed, some patience and discipline will be necessary if the new-built city is to be a city that is really renewed.
>
> (Sharp 1947: 10)

Lutyens and Abercrombie (1945) argued that in Hull,

> there is now both the opportunity and the necessity for an overhaul of the urban structure before undertaking this second refounding of the great Port on the Humber. Due consideration, however urgent the desire to get back to working conditions, must be given to every aspect of town existence.
>
> (Lutyens and Abercrombie 1945: 1)

This was the social climate of the war and early postwar years. There was an enthusiasm and a determination to undertake 'social reconstruction' (i.e. public sector intervention) on a scale hitherto considered utopian. The catalyst was, of course, the war itself.

At one and the same time war occasions a mass support for the way of life which is being fought for and a critical appraisal of the inadequacies of that way of life. Modern total warfare demands the unification of national effort and a breaking down of social barriers and differences. As Titmuss (1958: 85) noted, it 'presupposes and imposes a great increase in social discipline; moreover, this discipline is tolerable if, and only if, social inequalities are not intolerable'. On no occasion was this more true than in the Second World War. A new and better Britain was to be built. The feeling was one of intense optimism and confidence. Not only would the war be won, but also it would be followed by a similar campaign against the forces of want. That there was much that was inadequate, even intolerable, in prewar Britain had been generally accepted. What was new was the belief that the problems could be tackled in the same way as a military operation. What supreme confidence was evidenced by the setting up in 1941 of committees to consider postwar reconstruction problems: the Uthwatt Committee on Compensation and Betterment, the Scott Committee on Land Utilisation in Rural Areas, and the Beveridge Committee on Social Insurance and Allied Services. Perhaps it was Beveridge (1942) who most clearly summed up the spirit of the time, and the philosophy which was to underlie postwar social policy:

> The Plan for Social Security is put forward as part of a general programme of social policy. It is one part only of an attack upon five great evils: upon the physical Want with which it is directly concerned, upon Disease which often causes Want and brings many other troubles in its train, upon Ignorance which no democracy can afford among its citizens, upon Squalor which arises mainly through haphazard distribution of industry and population, and upon Idleness which destroys wealth and corrupts men, whether they are well fed or not, when they are idle.
>
> (Beveridge 1942: 170)

It was within this framework of a newly acquired confidence to tackle long-standing social and economic problems that postwar town and country planning policy was conceived. No longer was this to be restricted to town planning 'schemes' or regulatory measures. There was now to be the same breadth in official thinking as had permeated the Barlow Report. The attack on squalor was conceived as part of a comprehensive series of plans for social amelioration. To quote the 1944 White Paper *The Control of Land Use*, 'provision for the right use of land, in accordance with a considered policy, is an essential requirement of the government's programme of postwar reconstruction'.

The new planning system

The prewar system of planning was defective in several ways. It was optional on local authorities; planning powers were essentially regulatory and restrictive; such planning as was achieved was purely local in character; the central government had no effective powers of initiative, or of coordinating local plans; and the 'compensation bogey', with which local authorities had to cope without any Exchequer assistance, bedevilled the efforts of all who attempted to make the cumbersome planning machinery work.

By 1942, 73 per cent of the land in England and 36 per cent of the land in Wales had become subject to interim development control, but only 5 per cent of England and 1 per cent of Wales was actually subject to operative schemes (Uthwatt Report 1942: 9); there were several important towns and cities as well as some large country districts for which not even the preliminary stages of a planning scheme had been carried out. Administration was highly fragmented and was essentially a matter for the lower-tier authorities: in 1944 there were over 1,400 planning authorities. Some attempts to solve the problems to which this gave rise were made by the (voluntary) grouping of planning authorities in joint committees for formulating schemes over wide areas but, though an improvement, this was not sufficiently effective.

The new conception of town and country planning underlined the inadequacies. It was generally (and uncritically) accepted that the growth of the large cities should be restricted. Regional plans for London,

Lancashire, the Clyde Valley and South Wales all stressed the necessity of large-scale overspill to new and expanded towns. Government pronouncements echoed the enthusiasm which permeated these plans. Large cities were no longer to be allowed to continue their unchecked sprawl over the countryside. The explosive forces generated by the desire for better living and working conditions would no longer run riot. Suburban dormitories were a thing of the past. Overspill would be steered into new and expanded towns which could provide the conditions people wanted, without the disadvantages inherent in satellite suburban development. When the problems of reconstructing blitzed areas, redeveloping blighted areas, securing a 'proper distribution' of industry, developing national parks, and so on, are added to the list, there was a clear need for a new and more positive role for the central government, a transfer of powers from the smaller to the larger authorities, a considerable extension of these powers and, most difficult of all, a solution to the compensation-betterment problem.

The necessary machinery was provided in the main by the Town and Country Planning Acts, the Distribution of Industry Acts, the National Parks and Access to the Countryside Act, the New Towns Act and, later, the Town Development Acts.

The Town and Country Planning Act 1947 brought almost all development under control by making it subject to planning permission. Planning was to be no longer merely a regulative function. Development plans were to be prepared for every area in the country. These were to outline the way in which each area was to be developed or, where desirable, preserved. In accordance with the wider concepts of planning, powers were transferred from district councils (DCs) to county councils. The smallest planning units thereby became the counties and the county boroughs. Coordination of local plans was to be effected by the new Ministry of Town and Country Planning. Development rights in land and the associated development values were nationalised. All the owners were thus placed in the position of owning only the existing (1947) use rights and values in their land. Compensation for development rights was to be paid 'once and for all' out of a national fund, and developers were to pay a development charge amounting to 100 per cent of the increase in the value of land resulting from the development. The 'compensation bogey' was thus at last to be completely abolished: henceforth development would take place according to 'good planning principles'.

Responsibility for securing a 'proper distribution' of industry was given to the Board of Trade. New industrial projects (above a minimum size) would require the board's certification that the development would be consistent with the proper distribution of industry. More positively, the Board was given powers to attract industries to development areas by loans and grants, and by the erection of factories.

New towns were to be developed by ad-hoc development corporations financed by the Treasury. Somewhat later, new powers were provided for the planned expansion of towns by local authorities. The designation of national parks and 'areas of outstanding natural beauty' (AONBs) was entrusted to a new National Parks Commission, and local authorities were given wider powers for securing public access to the countryside. A Nature Conservancy was set up to provide scientific advice on the conservation and control of natural flora and fauna, and to establish and manage nature reserves. New powers were granted for preserving amenity, trees, historic buildings and ancient monuments. Later controls were introduced over river and air pollution, litter and noise. Indeed, there has been a steady flow of legislation, partly because of increased experience, partly because of changing political perspectives, but perhaps above all because of the changing social and economic climate within which town and country planning operates.

The ways in which the various parts of this web of policies operated, and the ways in which both the policies and the machinery have developed since 1947 are summarised in the following chapters. Here a brief overview sets the scene.

The early years of the new planning system

The early years of the new system were years of austerity. This was a truly regulatory era, with controls operating over an even wider range of matters than during the war. It had not been expected that there would be any surge in pressures for private development, but even if there were, it was envisaged that these would be subject to the new controls. Additionally, private building was regulated by a licensing system which was another brake on the private market. Building resources were channelled to local authorities, and (after an initial uncontrolled spurt of private house building) council house building became the major part of the housing programme.

The sluggish economy made it relatively easy to operate regulatory controls (since there was little to regulate), but it certainly was not favourable to 'positive planning'. It had been assumed that most of this positive planning would take the form of public investment, particularly by local authorities and new town development corporations. Housing, town centre renewal and other forms of 'comprehensive development' were seen as essentially public enterprises. This might have been practicable had resources been plentiful, but they were not, and both new building and redevelopment proceeded slowly. Thus, neither the public nor the private sectors made much progress in 'rebuilding Britain' (to use one of the slogans which had been popular at the end of the war).

The founders of the postwar planning system foresaw a modest economic growth, little population increase (except an anticipated short postwar 'baby boom'), little migration either internally or from abroad, a balance in economic activity among the regions, and a generally manageable administrative task in maintaining controls. Problems of social security and the initiation of a wide range of social services were at the forefront of attention: welfare for all rather than prosperity for a few was the aim. There was little expectation that incomes would rise, that car ownership would spread, and that economic growth would make it politically possible to declare (as Harold Macmillan later did) that 'you have never had it so good'. The plans

for the new towns were almost lavish in providing one garage for every four houses.

The making of plans went ahead at a steady pace, frequently in isolation from wider planning considerations, though the regional offices of the planning ministry made a valiant attempt at coordination; but even here progress was much slower than expected, and it soon became clear that comprehensive planning would have to be postponed for the sake of immediate development requirements.

For a time, the early economic and social assumptions seemed to be borne out but during the 1950s dramatic changes took place, some of which were the result of the release of pent-up demand which followed the return of the Conservative government in 1951 – a government which was wedded to a 'bonfire of controls'. One of the first acts of this government in the planning sphere was a symbolic one: a change in the name of the planning ministry – from 'local government and planning' to 'housing and local government'. This reflected the political primacy of housing and the lack of support for 'planning' (now viewed, with justification, as restrictive). The regional offices of the planning ministry were abolished, thus saving a small amount of public funds, but also dismantling the machinery for coordination. Though this machinery was modest in scope (and in resources), it was important because there was no other regional organisation to carry out this function.

The first change to the 1947 system came in 1953 when, instead of amending the development charge in the light of experience (as the Labour government had been about to do) it was abolished. At about the same time, all building licensing was scrapped. Private house building boomed, and curiously so did council house building, since the high building targets set by the Conservative government could be met only by an all-out effort by both private and public sectors. The birthrate (which – as expected – dropped steadily from 1948 to the mid 1950s) suddenly started a large and continuing rise.

The new towns programme went ahead at a slow pace, accompanied by a constant battle for resources which, so the Treasury argued, were just as urgently needed in the old towns. (The provision of 'amenities' was a particular focus of the arguments.) By contrast, public housing estates and private suburban developments mushroomed. Indeed, there was soon a concern that prewar patterns of urban growth were to be repeated. The conflict between town and country moved to centre stage. This was a more difficult matter for the Conservative government than the abolition of building controls, development charges and other restrictive measures. New policies were forged, foremost of which was the control over the urban fringes of the conurbations and other large cities where an acrimonious war was waged between conservative counties seeking to safeguard undeveloped land and the urban areas in great need of more land for their expanding house building programmes. On the side of the counties was the high priority attached to maintaining good quality land in agricultural production. On the side of the urban areas was a huge backlog of housing need. The war reached epic proportions in the Liverpool and Manchester areas where Cheshire fought bitterly 'to prevent Cheshire becoming another Lancashire'.

Similar arguments were used in the West Midlands, where a campaign for new towns (led by the Midland New Towns Society) was complicated by the government view that Birmingham was a rich area from which to move industry to the depressed areas. London, of course, had its ring of new towns, but these were inevitably slow in providing houses for needy Londoners, particularly since tenants were selected partly on the basis of their suitability for the jobs which had been attracted to the towns. The London County Council therefore, like its provincial counterparts, built houses for 'overspill' in what were then called 'out-county estates'. Similarly, Glasgow and Edinburgh built their 'peripheral estates'.

The pressures for development grew as households increased even more rapidly than population – a little understood phenomenon at the time (Cullingworth 1960a) – and as car ownership spread (the number of cars doubled in the 1950s and doubled again in the following decade). Increased mobility and suburban growth reinforced each other, and new road-building began to make its own contribution to the centrifugal forces.

Working in the opposite direction was the implacable opposition of the counties. They received a powerful new weapon when Duncan Sandys initiated the green belt circular of 1955. Green belts had no longer even to be green: their function was to halt urban development. Hope that all interests could be appeased was raised by the Town Development Acts (1952 in England, 1957 in Scotland). These provided a neat mechanism for housing urban 'overspill' and, at the same time, rejuvenating declining small towns and minimising the loss of agricultural land. But though a number of schemes were (slowly) successful, the local government machinery was generally not equal to such a major regional task.

It was this local machinery which was at the root of many of the difficulties. Few politicians wanted to embark on the unpopular task of reforming local government, and even those who appreciated the need for change could not agree on why it was wanted – whether to resolve the urban–rural conflict, to facilitate a more efficient delivery of services, or to provide a system of more effective political units. These and similar issues were grist to the academic mill, but a treacherous area for politicians. Perhaps the biggest surprise here was the decision to go ahead with the reorganisation of London government. The legislation was passed in 1963: this followed (in sequence but not in content) a wide-ranging inquiry. The surprise was not that the recommendations were altered by the political process, but that anything was done at all. One important factor in the politics of the situation was the desire to abolish the socialist London County Council (though ironically the hoped-for guarantee of a permanent Conservative GLC was dashed by the success of the peripheral districts in maintaining their independence).

One effect of the London reorganisation was that further changes elsewhere were taken very seriously. The writing was now on the wall for local government in the rest of the country, and campaigns and counter-campaigns proliferated. Three inquiries (for England, Scotland and Wales) were established by the Labour government which assumed office in 1964. These reported in 1969, but implementation fell to its successor Conservative government. For Scotland, the recommendations were generally accepted (with a two-tier system of regions and districts over most of the country). The city-region recommendations for England, however, were unacceptable, and a slimmer two-tier system was adopted. Wales was treated in the same way. The result south of the Border was that the boundaries for the urban–rural strife, though amended in detail, were basically unchanged in character. It would be only a matter of time before a further reorganisation would be seen to be necessary. What followed is discussed in the following chapter, but it is clear that the story is a continuing one.

More new towns

In the mean time, truly alarming population projections had appeared which transformed the planning horizon. The population at the end of the century had been projected in 1960 at 64 million; by 1965 the projection had increased to 75 million. At the same time, migration and household formation had added to the pressures for development and the need for an alternative to expanding suburbs and 'peripheral estates'. It seemed abundantly clear that a second generation of new towns was required.

Between 1961 and 1971, fourteen additional new towns were designated. Some, like Skelmersdale and Redditch, were 'traditional' in the sense that their purpose was to house people from the conurbations. Others, such as Livingston and Irvine, had the additional function of being growth points in a comprehensive regional programme for Central Scotland. One of the most striking characteristics of the last new towns to be designated was their huge size. In comparison with the Reith Committee's optimum of 30,000 to 50,000, Central Lancashire's 500,000 seemed massive. But size was not the only striking feature. Another was the fact that four of them were based on substantial existing towns – Northampton, Peterborough, Warrington and Central Lancashire (Preston-Leyland). Of course, town building had been going on for a long time in Britain, and all the best sites may have already been taken by what had become old towns. The time was bound to come when the only

places left for new towns were the sites of existing towns.

There were, however, other important factors. First, the older towns were in need of rejuvenation and a share in the limited capital investment programme. Second, there was the established argument that nothing succeeds like success; or, to be more precise, a major development with a population base of 80,000 to 130,000 or more had a flying start over one with a mere 5,000 to 10,000. A wide range of facilities was already available, and (it was hoped) could be readily expanded at the margin.

No sooner had all this been settled than the population projections were drastically revised downwards. It was too late to reverse the new new towns programme, though it was decided not to go ahead with Ipswich (and Stonehouse was killed by the opposition of Strathclyde because of its irrelevance to the problems of the rapidly declining economy of Clydeside). However, the reduced population growth prevented some problems becoming worse, though little respite was apparent at the time. Household formation continued apace, as did car ownership and migration. The resulting pressures on the South East were severe – and have remained so, with little resolution of the difficulties of 'land allocation'.

The rediscovery of poverty

While much political energy was spent on dealing with urban growth, even more intractable problems of urban decay forced their attention on government. Every generation, it seems, has to rediscover poverty for itself, and the postwar British realisation came in the late 1960s (Sinfield 1973). As usual, there were several strands: the reaction against inhuman slum clearance and high density redevelopment, the impact of these and of urban motorways on communities ('Get us out of this hell' cried the families living alongside the elevated M4: Goodman 1972), fear of racial unrest (inflamed by the speeches of Enoch Powell). These issues went far beyond even the most ambitious definition of 'planning', and they posed perplexing problems of the coordination of policies and programmes. Not surprisingly, the response was anything but coordi-nated, and programmes proliferated in confusion.

Housing policy was the clearest field of policy development. Slum clearance had been abruptly halted at the beginning of the war, when demolitions were running at the rate of 90,000 a year. It was resumed in the mid 1950s, and steadily rose towards its prewar peak. Both the scale of this clearance and its insensi-tivity to community concerns, as well as the inadequate character of some redevelopment schemes, led to an increasing demand for a reappraisal of the policy. Added force was given to this by the growing realisa-tion that demolition alone could not possibly cope with the huge amount of inadequate housing – and the continuing deterioration of basically sound housing. Rent control had played a part in this tide of decay, and halting steps were taken to ameliorate its worst effects, though not with much success.

More effective was the introduction of policies to improve, rehabilitate and renovate older housing: changing terminology reflected constant refinements of policy. Increasingly, it was realised that ad-hoc improvements to individual houses were of limited impact: area rehabilitation paid far higher dividends particularly in encouraging individual improvement efforts. A succession of area programmes have made a significant impact on some older urban neighbour-hoods, but a considerable problem remains; it is debatable whether the overall position is improving or deteriorating.

Housing policies have typically been aimed at the physical fabric of housing and the residential environ-ment. Their impact on people generally, and the poor in particular, has been less than housing reformers had hoped (the lessons of earlier times being ignored). This realisation followed a spate of social inquiries, of which *The Poor and the Poorest* by Brian Abel-Smith and Peter Townsend (published in 1965) was a landmark in raising public concern. A bewildering rush of programmes was promoted by the Home Office (including the urban programme in 1968, community development projects in 1969, and comprehensive community programmes in 1974), the Department of the Environment (DoE) (urban guidelines in 1972, area management trials in 1974, and 'the policy for the

inner cities' in 1977), the Department of Education (educational priority areas in 1968) and the Department of Health and Social Security (cycle of deprivation studies in 1973). This list is by no means complete, but it demonstrates the almost frantic search for effective policies in fields which had hitherto largely been left to local effort.

Despite all this, the problems of the 'inner cities' (a misnomer since some of the deprived areas were on the periphery of cities) grew apace. The most important factors were the rapid rate of deindustrialisation and the massive movement of people and jobs to outer areas and beyond. Unlike the interwar years, the problems were not restricted to the 'depressed areas': the South East, previously the source for moving employment to the North, was badly affected. In absolute (rather than percentage) terms, London suffered severely, losing three-quarters of a million manufacturing jobs between 1961 and 1984 (Hall 2002b).

There was initially little difference between the political parties here: both were searching for solutions which continued to evade them. Lessons from the USA indicated that more money alone was not sufficient, and academic writers pointed to the need for societal changes, but there were few politically helpful ideas around. Following a period in which the problems were seen in terms of social pathology, attention was increasingly directed to 'structural' issues, particularly of the local economy. In the 1980s, the Conservative government put its faith in releasing enterprise, though it was never clear how this would benefit the poor. New initiatives included urban development corporations, modelled on the new town development corporations but with a different private enterprise ethic. The London Docklands Development Corporation (LDDC) seemed almost determined to ignore, if not override, the community in which it was located, but this attitude eventually changed, and both the LDDC and later urban development corporations (UDCs) became more attuned to local needs and feelings. Indeed, later policies are characterised by an attempt to be much more sensitive to human needs, with an emphasis on bottom-up planning.

Land values

The issue of land values was addressed by both the two later Labour governments. In the 1964–70 administration, the Land Commission was established to buy development land at a price excluding a part of the development value and to levy a betterment charge on private sales. Its life and promise were cut short by the incoming Conservative government. Exactly the same happened with the community land scheme and the development land tax introduced by the 1974–9 Labour government. Thus, there were three postwar attempts to wrestle with the problem, and none was given an adequate chance to work. For a time, attention focused on land availability studies. These became a time-demanding ritual for planners, later transformed when increased household projections in the 1990s widened and intensified the debate.

The abandonment of attempts to solve 'the betterment problem' (which may no longer even be perceived as a problem) is more than a matter of land taxation or even equity. The so-called 'financial provisions' of the 1947 Act underpinned the whole system, and made positive planning a real possibility. Though it seems unlikely that the issue will return to the political agenda in the foreseeable future, it should not be forgotten that this vital piece of the planning machinery is missing. Planning is therefore essentially a servant of the market (in the sense that it comes into operation only when market operations are set in motion). This change, made in the 1950s, is far more fundamental than the high profile changes made under the Thatcher regime. Whether 'planning gain' can be made the basis of a new approach remains to be seen.

Entrepreneurial planning

The theme of the Conservative era which began in 1979 was a commitment to 'releasing enterprise'. This was translated into a miscellany of policies which had little in the way of a coherent underlying philosophy, but which could be characterised in terms of removing particular barriers which were identified as holding back initiative. The identified problems ranged from inner city landholding by public bodies (dealt with by requiring publicity of the vacant land which would thereby automatically trigger a market use); to the 'wasteful' and 'unnecessary' tier of metropolitan government in London and the provincial conurbations (simply abolished).

Many areas of public activity were privatised, large parts of government were hived off to executive agencies, and compulsory competitive tendering was imposed on local government. The emphasis on 'market orientation' and the concerted attack (regrettably the word is not an exaggeration) on local government had some strange results. More power was vested in central government and its agencies. Public participation was reduced. But, though the planning system was affected in tangible ways (Thornley 1993), in no sense was it dismantled, or even changed in any really significant way. True, it was bypassed (by urban development corporations); its procedures were modified (by government circulars, and changes in the General Development and Use Classes Orders); development plans were, for a time, downgraded, and threatened with severe curtailment; and simplified planning zones were introduced: a system in which 'simplification' meant less planning control, but might involve even more human resources in negotiation.

The list can be extended, but the rhetoric which preceded and accompanied the changes was harsher than the changes themselves. Moreover, the language of confrontation which the politicians employed disguised the fact that previous governments had done similar things, even if more *sotto voce*. The development corporation initiative, for example, was essentially the brainchild of a much earlier period and, indeed, as applied to redevelopment (as distinct from new town development) had for long been proposed by socialists as a means of assisting local authorities. Some of the early days of the UDC flagship – the London Docklands Development Corporation – were characterised by an excess of zeal, a lack of understanding of the way in which the administration of government is different from the administration of business (and an authoritarian style which was widely – and justifiably – criticised). Time, however, mellowed misplaced enthusiasm, and brought about a better understanding of the inherent slowness of democratic government. There was also a keener awareness of the need to pay attention to the 'social' issues of the locality as well as its physical regeneration.

More generally, an old lesson was relearned: it is extremely difficult for one level of government to impose its will on another unless it has some broad and powerful support from outside, as well as willing cooperation inside. (There is, however, the draconian alternative of simply abolishing a wayward layer of local government, as was done with the Greater London Council and the metropolitan county councils.)

An about-turn on structure plans illustrates the pragmatic nature (what some call the flexibility, and others the inconsistency) of the Conservative government's thinking. The initial decision to abolish them was one option for dealing with a problem which dates back to 1947: how to ensure that plans provide (without overwhelming detail) sufficient guidance for the land use planning of an area, while being adaptable to unforeseen changing circumstances. The option actually adopted was a 'streamlining' – not unlike earlier attempts. The 1965 Planning Advisory Group report had highlighted the problem: 'It has proved extremely difficult to keep these plans not only up to date but forward looking and responsive to the demands of change'.

Twenty years later, the 1985 White Paper, *Lifting the Burden*, was in a similar key: 'There is cause for concern that this process of plan review and up-dating is becoming too slow and cumbersome.' More effective structure plans require a framework of regional policy. In the 1990s, this began to be accepted and, following the election of the Blair government in 1997, regional policy moved to centre stage.

The environment

All governments operate with some degree of pragmatism: electoral politics force this upon them. So it was with the 1979–97 Conservative government. After many years of relegating environmental issues to a low level of concern, there was a sudden conversion to environmentalism in 1988. This was heralded in a remarkable speech by Margaret Thatcher in which she declared that Conservatives were the guardians and trustees of the earth. At base, this reflected a heightening of public concern for the environment which is partly local and partly global.

The action which followed looks impressive (though critics have been less impressed by the results). A 1990 White Paper *This Common Inheritance* spelled out the government's environmental strategy over a comprehensive range of policy areas (untypically this covered the whole of the UK). Environmental protection legislation was passed, 'integrated pollution control' is being implemented, 'green ministers' have been appointed to oversee the environmental implications of their departmental functions, and new environmental regulation agencies have been established. The latter follow a spate of organisational changes which remind one of the old saying: 'when in doubt reorganise'. But there are difficult issues here which, though including organisational matters, go much deeper. Questions about the protection of the environment underline a perhaps (to the layperson) surprising ignorance of the workings of ecosystems at the local, national and global levels. Additionally, new questions of ethics have come to the fore. Difficult problems of deciding among alternative courses of action are rendered ever more complex. Cost benefit analysis is of little help: indeed all forms of economic reasoning are being challenged. International pressures have played a role here as, of course, has the coming of age of the European Union (EU). This has added a new dimension to the politics of the environment (and much else as well).

Concern for historic preservation (now embraced in the term 'heritage') is of much longer standing. Though many historic buildings were destroyed during the war, the more effective stimulus to preservation came from the clearance, redevelopment, renewal and road-building policies which got under way in the 1950s and accelerated rapidly. As with housing, the emphasis has been mainly on individual historic structures, but a conservation area policy was ushered in by the Civic Amenities Act 1967, sponsored by a private member (Duncan Sandys), though with wide support. This proved a popular measure, and there are now over 9,000 of them. Indeed, there has been mounting concern that too many areas are designated, and too few resources applied to their upkeep and management.

The National Heritage Act 1983 bore a modern name that signified a new and wider appreciation of the historical legacy. A new executive agency, English Heritage (formally called the Historic Buildings and Monuments Commission for England) was established and took over many of the functions previously housed within the DoE. In Northern Ireland, Scotland and Wales, rather different administrative solutions were devised, as befits the distinctive character of these parts of the UK. Unfortunately, the new environmental and historical awareness was late in raising sufficient concern about transport to bring about any significant change from a preoccupation with catering for the car.

Road-building policies

Transport policy has traditionally been largely equated with road-building policy, and protests that alternatives need to be considered have been unavailing until recently. On a number of issues, however, the protests could not be ignored. One has already been mentioned: the brutal impact of urban motorways on the communities through which they passed. The outcry against this led to a reassessment of both the location of urban roads and their necessity. Compensation for 'distress' caused by new roads was increased as part of a policy labelled (in a 1972 White Paper) *Putting People First*. Closely related was a growing concern about the inadequacy of the road inquiry process, which resulted in a significant improvement of the provisions for public participation. These and other changes curbed but did not allay the concerns: indeed, they are still vocal. The turning point came in 1989 when new forecasts of huge increases in car ownership and use were published. It was widely considered to be impossible to satisfactorily accommodate the forecast amount of traffic. The results of a change in attitude were working their way through the political system before the era of 'integrated transport planning'. Traffic calming became part of the

contemporary vocabulary (and is now statutorily enshrined), road pricing moved on to the agenda for serious discussion (but little action) and road-building was slashed. This extraordinary reversal of the long-standing policy of building roads to meet the demand for them started under the Conservative government. It reflected that government's interpretation of public attitudes to road-building which nicely attuned with the political objective of reducing tax-related expenditure. In this area at least, a bankruptcy of political ideas (for which persuasive alternatives were sadly in short supply) led firmly into the doldrums.

The countryside

The countryside has always been dear to the hearts of conservatives, though support for the protection and enjoyment of the countryside has traditionally cut across party and class lines. Increasing concern for the rural landscape, growing use of the countryside for recreation (and investment) and huge changes in the fortunes of the agricultural industry have transformed the arena of debate on rural land use. At the end of the war, and for many years afterwards, the greatest importance was attached to the promotion of agriculture. There were, however, established movements for countryside conservation and recreation, some of which came together with the National Parks and Access to the Countryside Act 1949 (but a separate Nature Conservancy Council was also established, thus dividing the conservancy function).

The pressures for conservation and for recreation have varied over time, and the balance between them is inevitably an ongoing problem, particularly in areas of easy access (which now includes most of the country). Limited budgets held back incipient pressures in the early postwar years, but increasing real incomes and mobility led to mounting pressures which were acknowledged in the 1966 White Paper *Leisure in the Countryside* and the Countryside Act 1968. This replaced the National Parks Commission with a Countryside Commission, which was given wider powers and improved finance. At the same time, the powers of local authorities were expanded to include, for instance, the provision of country parks. Unlike national parks, these were not necessarily places of beauty, but were intended primarily for enjoyment. They were also seen as having the added advantage of taking some of the pressure off the national parks and similar areas where added protection was needed.

The 1972 reorganisation of local government was accompanied by a requirement that local authorities which were responsible for national parks should establish a separate committee and appoint a park planning officer. The modesty of this provision was clearly a compromise between concerns for local government and for the planning of national parks. It was a step forward, but an enduring case for ad-hoc park authorities continued. Local authorities had too many local interests to satisfy to give adequate resources for national parks – whose very name indicated their much wider role. The growth of pressures on the parks continued, and the administrative knot was finally cut when the Environment Act 1995 provided for the establishment of ad-hoc national park authorities for all the parks.

More widely, a long-standing debate continued on the divided organisational arrangements for nature conservation and amenity, and for scientific conservation and wildlife. In England, that separation continues (on the basis of arguments which are not easy to follow), but in Scotland and Wales the responsibilities are now vested in single bodies: Scottish Natural Heritage (SNH) and the Countryside Council for Wales (CCW). Of particular note was the first outcome of Scottish thinking on integrated countryside planning, which built upon the simple (but rarely used) notion that all countryside activities 'are based on use, in one way or another, of the natural heritage'. This thought has passed into the realm of 'ideas in good currency', and it is echoed in three highly coloured White Papers on the countryside, issued in 1995 and 1996.

The Blair government from 1997

The flood of proposals, discussion documents, consultation documents and legislation from the Blair government would justify a separate book, rather than a note towards the end of this chapter. However, important matters are discussed, or at least mentioned, at appropriate points later. Here a note is made of some of the outstanding features, in so far as they relate to town and country planning (generously defined).

It is on constitutional matters that the most dramatic changes have been made. Not only have devolution proposals been made, but also they have been passed into law, and both the Scottish Parliament and the National Assembly for Wales are operational. It is too early to comment on what the impacts on planning may be, though some preliminary indications are discussed in relevant chapters. Scotland in particular is engaged on some thorough-going reviews while Wales (where the advent of devolution was uncertain) has already produced a number of planning statements. Sadly, the Northern Ireland situation has proved too problematic for resolution and, at the time of writing, it is unclear whether the devolution plans will go ahead.[5]

One of the unknowns in these constitutional changes is their impact on England. There was already a consensus that regional planning needed more direction than it was getting through the regional planning guidance system. The Blair government rapidly made moves on two fronts. First, regional planning guidance was given a new lease of life, with a bottom-up involvement of local government and other 'regional stakeholders'. From 2004 this was taken further with a requirement for regional spatial strategies, with coverage of a wider range of issues including regeneration and transportation. The regional planning process has also 'gone public': examinations in public (EIPs) were held in the 1990s and are a requirement for the new strategies. Second, regional development agencies were established and led the production of regional economic strategies. Though these bodies are appointed by and responsible to the Secretary of State, regional assemblies are developing which will provide 'stakeholder' input, and (it is hoped) a much needed link between economic and land use planning. Things are happening rapidly on this front, and the outcome is by no means clear, but it seems that the regional planning dimensions are now becoming central to both land use and economic policies (even though the relationships between the two are difficult).

A possible resolution could have been in regionalism. The possibility of regional devolution is explicitly embraced in the Labour Party agenda, but the signs of a strengthening of regional consciousness have not proved sufficient to produce a regional tier of directly elected government. The position in London is more satisfactory for planning, with the elected mayor responsible for strategic planning, economic development traffic and other aspects of life in London.

The most problematic political issue in regional planning is the allocation of land for new housing. This was a very troublesome issue for the Conservative government, and it is proving no less so for its successor. A major commitment has been made to increase the proportion of brownfield sites, with an aspirational target of 60 per cent. Such targets have little rationale or credibility at the national level, but they concentrate effort and they also have political value. A revised PPG 3 on housing included a 'sequential' method for identifying housing sites. Another reform of housing land policy is imminent in the wake of the Barker Review which brought an economist's sharp (if narrow) analysis to bear on planning for housing. It is the government's hope that the new regional spatial strategy system will create the framework for agreement on housing figures without too much intervention by the Secretary of State. It is unclear whether this hope may be fulfilled. How far it will be possible to increase (and accelerate) the development of brownfield sites is equally unclear, although many of the recommendations of Lord Rogers' Urban Task Force Report of 1999 have now found their way into policy and practice.[6]

Devolution is not the only constitutional issue with which the Blair government has dealt. The European Convention on Human Rights has been incorporated into British law (which the previous government refused to do). However, though it passed the Human Rights Act (see 1997 White Paper *Rights Brought Home*), it has not yet accepted the need for a Human Rights Commission to advise and monitor the legislation.[7] The Act guarantees a number of basic rights and freedoms, including freedom from discrimination

and the right to the peaceful enjoyment of property. In effect, the Act marks an increase in the power of the courts over parliament. Judges will be looking beyond the letter of the law to its substance. There will be a greater role for judicial review, with a concern for the merits of a decision rather than the fairness of the way in which it was reached.[8]

The power of our domestic government is also curtailed by membership of the European Union. The direct impacts of membership on town and country planning are limited so far, though the indirect influence of Community competences in regional policy, environment, transport and other fields is important. Environmental policy, in particular, owes much to cooperation with other EU countries and Community legislation. The regional debate too, is now strongly influenced by Community policies. The 'spatial approach' now advocated by government has its origins in concerns about improving coordination of sectoral policies in the EU. Such ideas have been promoted before with corporate planning in the 1970s and 'joined-up government' in the 1990s. This time it is advocated both at national and EU level and with some resources to support it, but there is much learning to be done to avoid these aspirations being represented as a new form of comprehensive (but ineffective) planning.

European Community initiatives are tempting more planners to experiment with cross-national planning and exchange of experience, and government departments are looking to other countries for ideas for the *Modernising Planning* agenda. Increasing interdependence among the EU states may mean that transnational planning strategies (now commonplace elsewhere in Europe) will become accepted for the UK.

Finally, in this selective list of initiatives, mention must be made of the commitment to an integrated transport policy which has proved more elusive than expected. Rural policies have also presented serious difficulties as they have been beset by political controversy over hunting and the right to roam. A programme for 'Modernising Planning' is making more progress which, like many of the issues touched upon in this rapid survey, is discussed in relevant chapters.

Whither planning?

It is now well over half a century since the postwar planning system was put into place. Major changes have taken place during this time in society, the economy and the political scene – some of which have been touched upon in this rapid overview. In these shifting sands, 'town and country planning' has grown into (or been submerged by) a series of different policy areas which defy description, let alone coordination. Yet 'planning' is nothing if not a coordinative function, and the frenetic activity in reorganising machinery which has absorbed so much energy since the mid 1940s must, at some point, give way to substantive progress. The difficulty lies in determining the direction in which this lies.

One thing is clear: some of the most important underlying problems are well beyond any conceivable scope of 'planning'. For example, much urban change has been due to global forces which are currently beyond *any* political control. Multinationals and international finance were not in the standard vocabulary in the early postwar years. Planners find it easier to think in terms of 'need'. In recent years, they have been forced to recast some of their thinking in 'market' terms. But could they ever come to terms with the workings of the property investment market? As many studies have shown, 'the channelling of money to promote new urban development is determined not by need or demand, but by the relative profitability of alternative investments' (Bateman 1985: 32) – which may be in different sectors, such as industrial equities, or in quite different geographical locations. Much private sector development is now 'driven more by investment demand and suppliers' decisions than by final user demand – and even less by any sort of final user needs' (Edwards 1990: 175).

This widening gap between land use development and 'needs' throws considerable doubt on the adequacy of a planning system which is based on the assumption that land uses can be predicted and appropriate amounts of land 'allocated' for specific types of use. Overriding all other pressing considerations, of course, is the state of the economy. (It is little comfort that so many other countries share the same problem.) One result has been a strengthening of the 'partnership' philosophy which has gradually grown since the early 1980s. The term now means more than coordination of the efforts of different agencies: it implies that planning has to embrace the agents of the market, and adapt a regulatory system of planning to the need for negotiation. At the least, risks are shared.

The implications of all this are not clear. Although an obvious response may be to try harder to identify emerging trends, this is more difficult to do than ever before. Economic and social trends seem as unpredictable as the weather or the course of scientific inquiry. Comprehensive planning based on firm predictions of the future course of events is now clearly impossible. Incrementalism is the order of the day, and Burnham's famous aphorism ('make no little plans') has now been turned on its head: 'make no big plans'.

But planners have always strained for unattainable goals, whether they be frankly utopian or simply over-enthusiastic. Contemporary plans are more practicable in this regard than many earlier ones. The plans prepared at the end of the Second World War were often quite unrealistic in the assumptions that were made about the availability (and control) of resources – though that did not prevent them being very influential in moulding planning ideas. It remains to be seen whether the lesson has been learned – or whether some currently unpredictable change will transform the future. Be that as it may, there seems little doubt that in the perpetual planning conflict between flexibility and certainty, the former is the clear winner.

Further reading

Though the Barlow and Uthwatt reports are seldom read these days, they are well worth at least a perusal and another original source was republished in 2003: *To-morrow: A Peaceful Path to Real Reform* by Ebenezer Howard, with a commentary by Peter Hall, Dennis Hardy and Colin Ward – but it is expensive. Like other reports of the time (particularly Beveridge) they give an insight into the spirit of the times which produced the planning system. Hennessy (1992) narrates this wonderfully in

Never Again: Britain 1945–1951. A little-known but insightful essay is Titmuss (1958) 'War and social policy'. An excellent account of a longer period (1890–1994) is given by Ward (2004) *Planning and Urban Change*. Two of Peter Hall's books are also essential reading: *Cities of Tomorrow* (2002) and *Urban and Regional Planning* (2002). Ashworth (1954) *The Genesis of Modern British Town Planning* is a thorough account up to the passing of the 1947 Act. A clear exposition of the (original) 1947 Act is given by Wood (1949) – a civil servant who was heavily involved in drafting the legislation. Cherry (1996) *Town Planning in Britain since 1900: The Rise and Fall of the Planning Ideal* carries the story up to date while his *The Evolution of British Town Planning* (1974) incorporates a history of the planning profession and its Institute. A review of *British Planning: 50 Years of Urban and Regional Policy* is edited by Cullingworth (1999). A number of earlier writers are quoted in the text or in the endnotes, as are several of the wartime and postwar plans. LeGates (1998) has edited a useful selection of writings on *Early Urban Planning 1870–1940*.

Analyses and commentaries on the operation of the planning system rapidly become out of date. Among the books and articles published since the mid 1980s are Ambrose (1986) *Whatever Happened to Planning?*; Reade (1987) *British Town and Country Planning*; Healey *et al.* (1988) *Land Use Planning and the Mediation of Change*; Thornley (1993) *Urban Planning under Thatcherism: The Challenge of the Market*; Adams (1994) *Urban Planning and the Development Process*; Ambrose (1994) *Urban Process and Power*; Simmie (1994) *Planning London*; Allmendinger (1997) *Thatcherism and Planning*; Davies, H. W. E. (1998) 'Continuity and change: the evolution of the British Planning System 1947–97'; Taylor, N. (1998) *Urban Planning Theory since 1945*; Allmendinger and Chapman (1999) *Planning Beyond 2000*; Vigar *et al.* (2000) *Planning, Governance and Spatial Strategy in Britain*.

Notes

1 *To-morrow: A Peaceful Path to Real Reform* (1898) was republished as *Garden Cities of Tomorrow*. The 2003 republication is a facsimile of the original version with

a commentary by Peter Hall, Dennis Hardy and Colin Ward.

2 Royal Commission on the Local Government of Greater London (1921–3); the London and Home Counties Traffic Advisory Committee (1924); the Greater London Regional Planning Committee (1927); the Standing Conference on London Regional Planning (1937); as well as ad-hoc committees and inquiries, for example, on Greater London Drainage (1935) and a Highway Development Plan (Bressey Plan, 1938).

3 Government 'investigators' were appointed and, following their reports, the Depressed Areas Bill was introduced in November 1934, to pass (after the Lords had amended the title) as the Special Areas (Development and Improvement) Act.

4 The phrase was coined by Alix Meynell, a senior official in the Board of Trade (see Meynell 1959).

5 Following the Good Friday Agreement of April 1998, the UK and Irish governments passed legislation on referendums on the Agreement. These gave a clear endorsement by the electorates of both Northern Ireland and the Irish Republic. The agreement provides for devolution to an elected Assembly of legislative and executive powers for all matters which are currently the responsibility of the six Northern Ireland departments (thus including environmental and planning policies). Additionally, a North-South Ministerial Council will deal with matters of mutual interest.

6 *Towards an Urban Renaissance*. This is discussed in Chapter 6.

7 Such a body could also scrutinise proposed legislation, train lawyers, provide legal representation for test cases, and initiate its own cases. See Spencer and Bynoe (1998).

8 The Human Rights Act is currently thought unlikely to have much impact on the planning world. Nevertheless, there are areas where it *could* have significant impacts, particularly where discretion or personal liberty is involved. See Corner (1999) and Upton (1999). One author has speculated that the hearing and determination of local plan objections may well be in breach of the Convention's provisions relating to civil rights and the entitlement 'to a fair and a public hearing by an independent and impartial tribunal established by law'. See Kitson (1999) and also the succinct account by Johnston (1999b).

3   The agencies of planning

EUROPEAN GOVERNMENT

In view of the increasing Europeanisation of planning processes and networking within the United Kingdom, we consider it vital for actors in the process to develop the capacity for thinking in terms of EU space and spatial relationships, and to relate to non-British modes of planning thought.

(Tewdwr-Jones and Williams 2001: 162)

The growing influence of Europe

The impact of the European Union has been predominantly in the field of environmental controls but is now being felt more directly on mainstream planning practice and urban policy. The most striking and perhaps best known example of EU influence is environmental impact assessment, but other examples in cross-border and transnational spatial planning are emerging. Later chapters identify a range of agricultural, environmental, economic, and regional policies of the EU which are having an effect on parts of the British planning system. Chapter 4 includes a note on supranational and cross-border planning instruments and policies that have been introduced at the European level. Here, a brief and more general account is given of the main EU institutions and the parts of most importance to planning.

Britain in the EU

The UK was not an enthusiastic supporter of the postwar moves towards a federal Europe. Though it favoured intergovernmental cooperation through such

bodies as the Organisation for European Economic Cooperation (1948) and the Council of Europe (1949), it was opposed to the establishment of organisations which would facilitate functional cooperation alongside nation-states. It therefore did not join the European Coal and Steel Community (1952), nor was it a signatory to the 1955 Treaty of Rome which established the European Economic Community (EEC) and the European Atomic Energy Community. However, along with the other members of the Organisation for European Economic Cooperation, it formed the European Free Trade Association (EFTA) in 1960. Britain envisaged that EFTA would form the base for the development of stronger links with Europe. When it became clear that this was not viable, Britain applied for membership of the European Community. This was opposed by France and, since membership requires the unanimous approval of existing members, negotiations broke down. The opposition continued until a political change took place in France in 1969. Renewed negotiations led finally to membership at the beginning of 1973.

The Treaty of Accession provided for transitional arrangements for the implementation of the Treaty of Rome, which Britain agreed to accept in its entirety.

The objectives include the elimination of customs duties between member states and of restrictions on the free movement of goods; the free movement of people, services, and capital between member states; the adoption of common agricultural and transport policies; and the approximation of the laws of member states to the extent required for the proper functioning of the common market. These objectives are often referred to as the 'four freedoms': the free movement of goods, people, services and capital.

From May 2004 the number of member states increased from fifteen to twenty-five, with two more due to join in 2007.[1] The EU25 has a population of 454 million (EU15: 380 million) and a land area of almost 4 million square kilometres (EU15: 3.2 million square kilometres).[2] Compared to the USA, this is about 75 per cent more people living on under half the space – a population density for the EU of 114 people to each square kilometre (EU15: 117) compared with 27 people per square kilometre in the USA. Perhaps the key difference is diversity in language – in the EU there are twenty official languages (and many others that are not used for official purposes). The EU is also easily the world's largest trading bloc, having a share of exports more than three times its nearest rivals of the USA and Japan.

The organisational and political structure of the EU is complex and, like all such bodies, its actual workings are somewhat different from the formal organisation chart. Enlargement in 2004 prompted a major review of the treaties which govern the EU and a new European Constitution was proposed in 2004. The main institutions of the EU and the parts that are of particular interest to planning are shown in Figure 3.1. In brief, there is an elected Parliament which operates as an advisory body and for some matters as joint legislature with the Council of Ministers. The main legislature is the Council of Ministers which makes policy largely on the basis of proposals made by the executive, the European Commission. There is also a Court of Justice which adjudicates matters of legal interpretation and alleged violations of Community law. The distribution of competences between Parliament and the executive is very different from most national governments.

European Council

A summit of heads of state or government of the member states, together with the President of the European Commission, provides general political direction for the European Union, considers fundamental questions related to the 'constitution' and construction of the EU, and makes decisions on the most contentious issues (Dinan 1998). It is not the legislature: this is the function of the Council of the European Union. Decisions which require legislation have to go through the normal EU legislation process, but agreements and declarations reached in the European Council are binding on the EU institutions, and have been critical in shaping the evolution of the EU. The Presidency of the Council rotates on a six-monthly cycle.

Council of the European Union (Council of Ministers)

The main decision-making body of the EU is the Council of Ministers. This is the legislature of the Community, a task it shares for some matters with the European Parliament. Unlike most other legislatures it is indirectly elected – being composed of representatives elected in the member states – and it deliberates in private. These characteristics have given rise to the criticism of 'democratic deficit' in comparison to national legislatures and the European Parliament, which is directly elected and debates in public. But the characteristics reflect the fundamental nature of the EU as a pooling of national sovereignties and legislative powers, rather than a federal structure with a unitary legislature. This requires complex negotiation among the member states.

The Council meets in different compositions depending on the topic, with the relevant ministers representing each member state, as for example in meetings (councils) of ministers of transport, environment and agriculture. But there is only one Council of Ministers. There are many subcommittees and working groups with various functions and memberships. There is a Council meeting of some sort every working week, often lasting for three days, and 100,000 documents

THE EUROPEAN COUNCIL
Meeting of the Heads of State

Gives broad guidance and impetus to action

COREPER
Committee of Permanent Representatives – Civil servants from the member states who manage the work of the Council.

THE COUNCIL OF THE EU
Meetings of ministers (one from each member state). The Council meets in different configurations depending on the issue e.g.
The Environment Council
The Transport, Telecoms and Energy Council

The Presidency of the Council rotates every six months.

Legislature (on some matters shared with the European Parliament)

	January to June	July to December
2005	Luxembourg	United Kingdom
2006	Austria	Finland
2007	Germany	Portugal
2008	Slovenia	France
2009	Czech Republic	Sweden
2010	Spain	Belgium

INFORMAL MEETING OF MINISTERS OF SPATIAL PLANNING
Generally meets once during each Presidency.
It is not a formal council and has limited powers.

SUB-COMMITTEE ON SPATIAL AND URBAN DEVELOPMENT (SUD) of EU Regional Policy Committee (CDCR), previously the Committee on Spatial Development that produced the ESOP.

EUROPEAN COMMISSION
25 Commissioners, one from each country
36 Directorates (departments), including:

TRANSPORT AND ENERGY

ENVIRONMENT

REGIONAL POLICY
(includes spatial planning and Structural Funds)

Applies the Treaties by initiating legislation and implements policy as executive body and works in 20 official languages of the EU.

EUROPEAN PARLIAMENT
732 elected members (78 from UK)
20 standing committees, including:

Agriculture and Rural Development
Transport and Tourism
Environment, Public Health and Food Safety
Regional Development

Political driving force, supervising and questioning the Council and Commission. Joint power to adopt legislation with Council. Supervises appointment of Commission.

ECONOMIC AND SOCIAL COMMITTEE
317 nominated members from employers, workers and other interests (24 from UK)
6 sections, including:

Agriculture, Rural Development and the Environment
Transport, Energy, Infrastructure and the Information Society

A non-political body that is consulted and delivers opinions on proposed legislation.

COMMITTEE OF THE REGIONS
317 members representing regional and local government (24 from UK)
6 Commissions, including:

Commission for Territorial Cohesion Policy
Commission for Sustainable Development
Commission for Economic and Social Policy

Is consulted and delivers opinions where regional interests are involved.

THE COURT OF JUSTICE AND COURT OF FIRST INSTANCE
25 judges and 8 advocates general
25 judges, at least one from each country

Interpret the Treaties and apply judgments and penalties in cases of non-compliance.

Figure 3.1 *Institutions of the European Union and spatial planning*

are produced by the Council each year (Dinan 1998: 106). Representatives are usually senior ministers of national government, although regional ministers may also represent the country concerned, a point which may become more significant for the UK as devolution starts to bite. The criterion is that the representative must be authorised to commit the member state to the decisions made.

There is no formal Council of Ministers responsible for planning but, since 1991, there have been biannual informal meetings of ministers responsible for spatial planning. Under the Dutch Presidency in 2004 this was termed a meeting of ministers of territorial cohesion reflecting the terminology in the proposed Constitutional Treaty. A subcommittee on Spatial and Urban Development (SUD) of the Committee on the Development and the Conversion of the Regions (CDCR) is responsible for taking forward the territorial or spatial dimension of Community policy.[3] It consists of officials representing the planning ministries of member countries. The UK has been represented by the Office of the Deputy Prime Minister (ODPM), the Department of Trade and Industry (DTI) and the Scottish Office. It is the SUD and its predecessor, the Committee on Spatial Development, that has taken the most important action on European planning in preparing and taking forward the European Spatial Development Perspective, which is discussed in the next chapter.

The Council (in some cases in cooperation with the European Parliament) can make three main types of legislation. *Regulations* have direct effect and are binding throughout the EU. They require no additional implementing legislation in the member states and are used mostly for detailed matters of a financial nature or for the technical aspects of (for example) administering the Common Agricultural Policy (CAP). By contrast, *directives* provide framework legislation which, though equally binding, is implemented by national legislation. This leaves a degree of choice over the method of implementation to the member states. Environmental matters are typically dealt with in this way. The Council can also issue *decisions* which are binding on the member state, organisation, firm or individual to whom they are addressed. Finally, there are *common positions* or *actions*, *recommendations* and *opinions*, which have no binding force. The UK has 29 of 321 votes in the Council.

The work of the Council of Ministers is supported by officials in the Council of Permanent Representatives (COREPER). These are civil servants or permanent ambassadors to the EU of the member states. Indeed, it has been argued that COREPER is where the real decisions are made. It is the officials who conduct often very lengthy negotiations to reach agreement about measures among the member states before proposals are put before the ministers.

European Commission

The main work of the EU is undertaken by the executive of the Community, the European Commission. The Commission is a major driving force within the EU because it has the primary right to initiate legislation. It prepares proposals for decision by the Council, and oversees their implementation. (Only rarely can the Council of Ministers make a policy decision without a proposal from the Commission.) The Community's decision-making process is dominated by the search for consensus among the member states and this gives the Commission a crucially important role in mediation and conciliation. Of the same nature is the ethos of achieving compromise and of progressing in an incremental way. In promoting action at the EU level the key reference for the Commission is the European Treaties.[4]

Among the Commission's powers is that of dealing with infringements of Community law. If it finds that an infringement has occurred, it serves a formal notice on the state concerned requiring discontinuance or comments with a specified period (usually two months). If the matter is not resolved in this way, the Commission issues a reasoned opinion, requiring the state to comply by a given deadline. As a last resort, the Commission can refer a matter to the Court of Justice whose judgment is legally binding. Most matters are dealt with informally, but Britain has been subject to reasoned opinions on environmental matters (Haigh 1990: 153 and 160).

During the 1990s the influence of the Commission waned under fierce criticism of its perceived greed for power and the acquisition of national competencies. Its attempts at harmonisation of standards in the pursuit of the Single Market, though often well founded, have sometimes been inept, giving an impression of remoteness and arrogance, exacerbated by its poor control of Community funds. However, much of the popular criticism is misconceived. Thus, for example, to label the 17,000 officers of the Commission (including 3,000 translators and more than 3,000 scientists) as a massive bureaucracy is a gross exaggeration. (The ODPM has more than 15,000 staff.) Nevertheless, the media have harried the Commission on its interference in national affairs, and the cronyism of the Commissioners. Protecting national competencies in the face of expanding EU powers was a prime objective of the Thatcher administration, but other member states too have grown wary of the expanding competence of the Community. The European Parliament has taken the Commission to task on poor management. The Council took action during the 1990s to reduce unnecessary interference from the Commission, citing the principles of subsidiarity and proportionality. The effect has been dramatic, with a considerable fall in the amount of Community legislation and, less obviously, a weakening of its influence.

The culmination of mounting criticism came at the end of 1998, when the European Parliament, to which the Commission is accountable, threatened to sack all Commissioners. Although the proposal was defeated, the debate fuelled popular antagonism against the Commission and, in March 1999, the Commissioners resigned en bloc. A new Commission was approved by the European Parliament in September with major reforms to its organisational structure and procedures. A new Commission of twenty-five members was appointed in 2004. The main departments are each headed by a Director-General, but considerable influence over the work of the Directorate is exercised by the personal 'cabinet' of the Commissioner, and in particular by the chair (who is known as the *chef du cabinet*).

The departments with an interest in town and country planning or the broader concept of spatial planning

are Regional Policy (know as DG Regio) (whose main responsibility is for the Structural Funds), DG Environment, and DG Energy and Transport.

European Parliament

The European Parliament is a directly elected body consisting of 732 members who are elected every five years. Britain has 78 representatives, known as MEPs: Members of the European Parliament (down from 87 of 626 when there were 15 member states). The Single European Act and Treaty on European Union extended the powers of the Parliament, and the Amsterdam Treaty (which came into force in May 1999) has again increased its role in joint decision-making with the Council and its supervisory powers over the Commission. The Parliament is consulted on all major Community decisions, and it has powers in relation to the budget which it shares with the Council, and in approving the appointment of the Commission. The assent of Parliament is needed also for accession of new members and international agreements. However, it is important to note that the Parliament was established essentially as an advisory and supervisory body, while the Council of Ministers is the legislature. One reflection of the lack of legislative power is that the Parliament sits in plenary session for only three days each month and bizarrely continues to divide its sittings between two locations – Brussels and Strasbourg.

Parliament is organised along party political (not national) lines. The political groups have their own secretariats and are the 'prime determiners of tactics and voting patterns' (Nugent 1999: 130). Much of their work is carried out by standing committees and through questions to the Commission and Council. The Regional Development, Agriculture and Rural Development and Transport and Tourism Committees consider matters related to spatial development, including European regional planning policy and the common transport policy.

In 1995, the European Parliament established the office of the European Ombudsman charged with improving the quality of Community relations with the public. The Ombudsman can investigate complaints within all the Community institutions except for the Courts acting in their judicial role. Complaints can be made by anyone living in the European Union, and 1,372 were received in 1998. Almost 70 per cent of complaints are outside the mandate of the Ombudsman, and many of these are about the application of Community law within the member states. Three-quarters of admissible complaints were made against the European Commission, and the highest proportion, one-third, were related to access to information.

Committee of the Regions

The Committee of the Regions (COR) is the youngest European Institution, set up following the Treaty of European Union, and holding its first session in March 1994. It is intended to give a voice to the regions and local authorities in European Union debates and decision-making. It has 317 members representing the regions, including 24 from UK local authorities. (The UK representation is made up of 14 from England, 5 from Scotland, 3 from Wales, and 2 from Northern Ireland.) The COR has taken a particular interest in regional planning and in advocating wider use of the principle of subsidiarity, so as to strengthen the role of regional and local authorities.

The Treaty identifies particular areas where the COR has to be consulted by the Commission, including trans-European networks, economic and social cohesion, and structural fund regulations. It can also offer opinions in other areas that it thinks appropriate, typically when an issue has a specifically regional dimension. It has issued many opinions on planning, urban and environmental issues. A committee (confusingly known in the COR as a commission) has been established to deal exclusively with regional policy, spatial planning and urban issues which is known as the Commission for Territorial Cohesion (COTER) and another on Sustainable Development (DEVE).

European courts

There are two main European courts: the European Court of Justice and the Court of First Instance. The European Court of Justice has thirteen judges. It decides on the legality of decisions of the Council and the Commission, interprets Community law and ensures its consistent application and determines violations of treaties. Cases can be brought before it by member states, organisations of the Community, and private firms and individuals. Since 1989, the Court of First Instance has dealt with most actions involving private applicants. It is organised on a similar basis to the European Court of Justice. The Courts have played an important part in extending the competencies of the European Union by confirming that actions by the Community are legal under the treaties (Nadin and Shaw 1999), and by promoting harmonisation by ruling that certain actions are illegal (Nugent 1999: 263). These courts are quite separate from the European Court of Human Rights, discussed in the following section.

Council of Europe

The Council of Europe is not to be confused with the EU. It was set up in 1949 with ten member countries to promote awareness of a common European identity, to protect human rights and to standardise legal practices across Europe in order to achieve these aims. Since 1989, its main role has been to monitor human rights in the post-communist democracies, and to assist them in carrying out political, constitutional and legal reform. It now has forty-one member countries (including sixteen countries that were formerly part of the communist bloc).

It has a three-tier structure with a Council and Ministers, a Parliamentary Assembly, and a Congress of Local and Regional Authorities. With an annual budget of less than £100 million, it is much less powerful than the EU (which has an annual budget of over £50 billion), but nevertheless it has played an important part in maintaining and establishing democracy on the continent. It is best known for its

Convention on European Human Rights. Anyone who feels that their rights under the Convention have been breached may take a case to the European Court on Human Rights for a decision which will be binding on those states that have signed up to the Convention. The Convention is now incorporated into UK law and its impacts on planning are discussed in Chapter 12.

The Council has been active for many years in the field of regional planning and environment, and perhaps the most notable achievement is the Bern Convention on Conservation of Wildlife and Habitats. It has published conference and other reports on the implications of sustainability for regional planning, the representation of women in urban and regional planning, and many other topics. A conference of ministers of spatial and regional planning (CEMAT) has been meeting since 1970 and its most important contribution has been the European regional/spatial planning charter, known as the Torremolinos Charter. This was adopted in 1983 and committed the Council to producing a 'regional planning concept' for the whole of the European territory. It has taken some time but, as noted in the next chapter, CEMAT has also now published *Guiding Principles for Sustainable Spatial Development of the European Continent* (2000).

The Council was responsible for the European Campaign for Urban Renaissance (1980–2). This led to a programme of ad-hoc conferences, various reports and 'resolutions' on such matters as health in towns, the regeneration of industrial towns, and community development. In 1992, the Conference adopted *The European Urban Charter*. This 'draws together into a single composite text, a series of principles on good urban management at local level'. The 'principles' relate to a wide range of issues, including transport and mobility, environment and nature in towns, the physical form of cities, and urban security and crime prevention.

CENTRAL GOVERNMENT

We live in an age when most of the old dogmas that haunted governments in the past have been swept away. We know now that better government is about much more than whether public spending should go up or down, or whether organisations should be nationalised or privatised. Now that we are not hidebound by the old ways of government we can find new and better ones.

White Paper, *Modernising Government*, 1999

Modernising government

The quotation illustrates the style as well as the zeal of the Blair government in its attempt to change the nature of the governmental system. It is not, of course, the first government to enter office with such flourishes; nor is it unique in proclaiming innovations which are recognisably in line with secular social and political changes. But 'the third way' (Giddens 1998) is in marked contrast to at least the rhetoric of the long-living Conservative administration that was defeated in the election of 1997. Moreover, the years of office have witnessed a continuing torrent of measures to bring both policy and the machinery of implementation in line with the philosophy of the new government. A remarkable innovation designed to assist in the strategic planning of public expenditure is the Comprehensive Spending Review announced by the Chief Secretary of the Treasury in June 1997. Instead of adjusting departmental budgets at the margin, priorities are being attained by the use of zero-based budgeting:

Every department will scrutinise its spending plans in detail from a zero base, and ask, how does each item contribute to the Government's objective as set out in our manifesto? Why are we spending this money? Do we need to spend it? What is it achieving? How effective is it? How efficiently are we spending it? Its conclusions will inform a new set of public spending plans for the rest of this Parliament – a set that reflects our priorities.[5]

The outcome has been a significant shift in spending priorities towards education, health and capital expenditure in transport and housing, and away from defence, agriculture, the diplomatic service, and the legal system. The new arrangements represented 'the most ambitious re-engineering of the public expenditure system for several decades, shifting the emphasis away from annual negotiations and their emphasis on inputs, and towards objectives and outputs' (James 1999: 195).

This suggests that Whitehall departmentalism has become less rigid than previously, which may offer opportunities for increased coordination across the 'compartments' of government through planning. Certainly, Blair did not initially share Harold Wilson's experience in the early 1960s of the tardiness of the civil service in adapting to a Labour government after thirteen years of Conservative government. Indeed, he 'found a civil service almost startlingly keen to prove that they had not been politicised by eighteen years of Conservative rule'. However, Blair's drive for change has faced some problems with traditional departmentalism. He is reported as expressing frustration at civil servants 'defiantly defending their own departments: they are felt to oppose any structural changes to their fiefdom, particularly if it means ceding any territory'.[6] To combat this, he asked Lord Simon (formerly of BP) 'to introduce a revolution in civil service culture, including stripping out layers of management and imposing payment by results'. In the words of Michael White:

> Performance-related pay, targeted objectives for departments and individuals, more inter-departmental cooperation, fast-track promotion for bright young things, above all a shift from being preoccupied with policy and process to a new focus on outcome and delivery are what it's all about.[7]

Organisational responsibilities

A large number of governmental departments and agencies are involved in town and country planning. Those having the main responsibility for the planning Acts are the Scottish Executive Development Department (SEDD), the Transport, Planning and Environment Group of the Welsh Assembly, the Planning Service Executive Agency of the Department of the Environment for Northern Ireland (DoENI) and, for England, the Office of the Deputy Prime Minister. There are, of course, many planning functions that fall to departments responsible for agriculture, the countryside, the human heritage, national heritage, nature conservation, and trade and industry. Additionally, an increasing number of functions have been transferred from government departments to agencies and public bodies. Figure 3.2 shows the main institutional arrangements, and gives a flavour of their complexity.

Planning responsibilities have evolved over time and, though there have been numerous reorganisations, the machinery inevitably has a patchwork appearance. (As an example of the problems involved: in which department should questions of the rural economy be placed – the one concerned with agriculture, or natural resources, or economic development, or employment? Or should it form a separate department of its own?) The machinery is also unstable: changing perceptions, conditions, problems and objectives demand new policy responses which in turn can lead to organisational changes. For example, increased concern for environmental planning has resulted in the transfer of widespread environmental functions into a number of environment agencies. The agencies establish themselves, they extend their activities and the problems of cooperation and overlapping competences get more attention, leading to calls to unify and simplify the structure of agencies. Sometimes, different patterns emerge in different parts of the UK. Thus nature conservation and access to the countryside are the responsibility of one agency in Scotland (Scottish Natural Heritage) and in Wales (Countryside Council for Wales), but are divided between two in England (English Nature and the Countryside Agency).

Office of the Deputy Prime Minister

At the time of writing (2005) the central government planning department for England is the Office of the

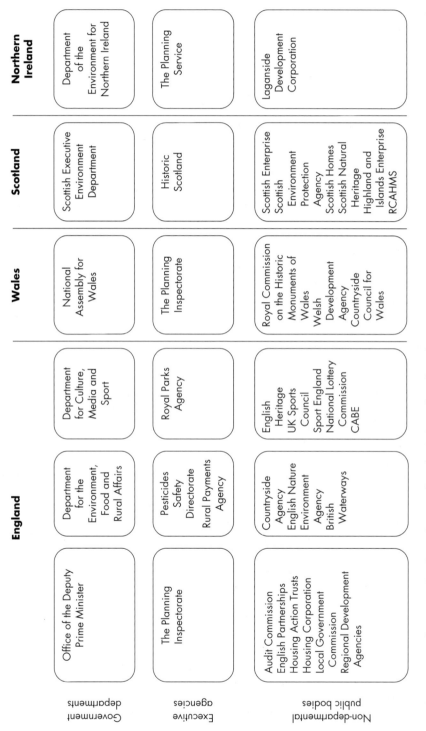

	England		Wales	Scotland	Northern Ireland
Government departments	Office of the Deputy Prime Minister	Department for the Environment, Food and Rural Affairs	Department for Culture, Media and Sport	National Assembly for Wales	Scottish Executive Environment Department	Department of the Environment for Northern Ireland
Executive agencies	The Planning Inspectorate	Pesticides Safety Directorate Rural Payments Agency	Royal Parks Agency	The Planning Inspectorate	Historic Scotland	The Planning Service
Non-departmental public bodies	Audit Commission English Partnerships Housing Action Trusts Housing Corporation Local Government Commission Regional Development Agencies	Countryside Agency English Nature Environment Agency British Waterways	English Heritage UK Sports Council Sport England National Lottery Commission CABE	Royal Commission on the Historic Monuments of Wales Welsh Development Agency Countryside Council for Wales	Scottish Enterprise Scottish Environment Protection Agency Scottish Homes Scottish Natural Heritage Highland and Islands Enterprise RCAHMS	Laganside Development Corporation

Figure 3.2 The organisation of central government for planning

Deputy Prime Minister, which was created in 2002. The ODPM is shown alongside other central government departments and agencies that have some relationship to the planning system in Figure 3.2. New governments and ministers are prone to reorganise the government machinery, and this has led to several changes to the name and location in the government system of the 'national' department for planning. So, in this book and other sources many references are made to the ODPM's predecessors when discussing the role of central government. From 2000 to May 2002 the department responsible for planning was the Department of Transport, Local Government and the Regions (DTLR); and prior to that from 1997 to 2000 it was the Department of Environment, Transport and the Regions (DETR). The big changes were the move of most environment functions to the new Department for Environment, Food and Rural Affairs (DEFRA) in 2000 and the creation of a separate Department for Transport (DfT) in 2002 (which recreated the separate department that existed before 1997).

The changing permutations of competences for planning, local government, environment and transport have a much longer history, as illustrated in Figure 3.3. Since 1942 there has been a separate ministry for land use planning with varying competence for other related policy fields. The Department of the Environment was formed in 1970 with the aim of providing more integration of planning, transport and some environmental policy but transport responsibilities were moved back to a separate Department of Transport in 1976. Transport was 'reintegrated' with planning by the Labour government in 1997, though this lasted only until 2002. The old Department of the Environment also had many heritage responsibilities, but these were moved to a separate Department of National Heritage (DNH) in 1992, which was superseded in 1997 by the Department for Culture, Media and Sport (DCMS). Other changes have included the gathering together of the pollution regulation functions within Her Majesty's Inspectorate of Pollution (HMIP), and later the establishment of Environmental Agencies for England and Wales, and for Scotland; these have taken over the functions of the HMIP, the National Rivers Authority and the waste

regulation functions of local government. It has been a restless time in Whitehall. More change is likely in 2005 with particular attention being given to the possible merger of the major environmental agencies in England and Wales.

The ODPM, as it is bureaucratically termed, has a wide range of responsibilities. The department brings together regional and local government (including the regional government offices) as well as the wider functions of housing, planning and regeneration, neighbourhood renewal and social exclusion. Also falling into the same complex of services are the responsibilities for implementing the range of new policies concerned with what in simpler days were collectively termed 'housing and planning'. Included in these is a much more explicit concern for poverty and social exclusion. Until 2004 it also took the lead on government devolution to the regional level, but this has now passed to the Department for Constitutional Affairs (DCA).

There has been substantial change in the objectives of the department since 2002, partly because of the changing organisation of responsibilities in government, especially the move of transport and environment functions to other departments; out go objectives relating to environmental improvement, the countryside, integrated transport and promoting elected regional government; in comes a much more specific objective to match the supply of housing to demand. The regional objective is directed more around economic performance and a less ambitious 'framework for regional governance'. The objectives are shown in Box 3.1 together with two of the seven public service agreements (PSAs) that related most closely to the planning system. Each department now has a small number of PSAs around which action and monitoring is organised.

The outcome of the reorganisation of competences in separate departments and a concentration of attention within them on their PSAs appears to have created a clearer separation of functions relating to planning with ODPM mostly dealing with urban planning, DEFRA addressing rural and environmental planning issues and DCMS heritage planning. This has happened at the same time that government has increased

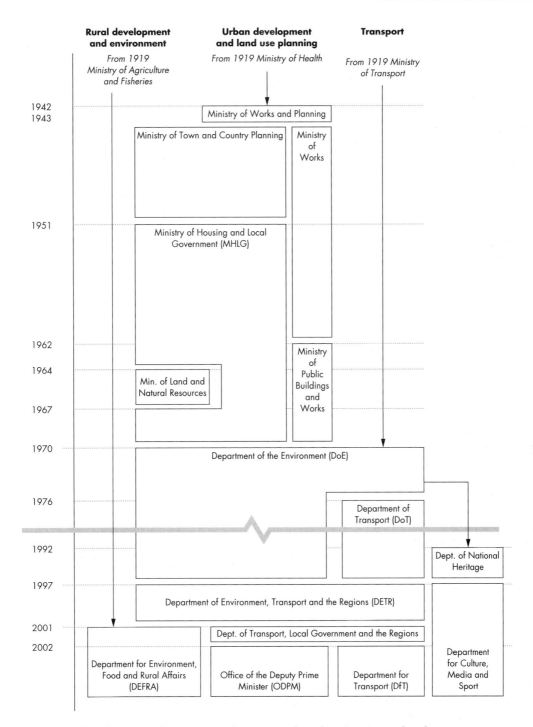

■ **Figure 3.3** *The changing departmental structure for planning in England*

BOX 3.1 THE ODPM'S STRATEGIC PRIORITIES AND PLANNING RESPONSIBILITIES

Strategic priorities

1 Delivering a better balance between housing supply and demand by supporting sustainable growth, reviving markets and tackling abandonment.
2 Ensuring people have decent places to live by improving the quality and sustainability of local environments and neighbourhoods, reviving brownfield land and improving the quality of housing.
3 Tackling disadvantage by reviving the most deprived neighbourhoods, reducing social exclusion and supporting society's most vulnerable groups.
4 Delivering better public services, by devolving decision-making to the most effective level – regional, local or neighbourhood:

   • Promoting high quality, customer-focused local services and ensuring adequate, stable resources are available to local government.
   • Clarifying the roles and functions of local government, its relationship with central and regional government and the arrangements for neighbourhood engagement, in the context of a shared strategy for local government.

5 Promoting the development of the English regions by improving their economic performance so that all are able to reach their full potential, and developing an effective framework for regional governance taking account of the public's view of what is best for their area.

2002 Public Service Agreements (PSAs) most relevant to planning

PSA 5: achieve a better balance between housing availability and the demand for housing in all English regions while protecting valuable countryside around our towns, cities and the green belt – and the sustainability of existing towns and cities . . .

PSA 6: all local planning authorities to complete local development frameworks by 2006 and to perform at or above best value targets for development control by 2006 . . . The Department to deal with called in cases and recovered appeals in accordance with statutory targets.

Source: ODPM Annual Report 2004

attention to coordinated action of departments around common objectives, especially the Sustainable Communities Plan. There is much less evidence of environmental sustainability in the ODPM's objectives and activities than delivering growth.

The ODPM manages about £56 billion of public money, of which £46 billion goes to local government as the main part of their funding from central government. The biggest expenditure (apart from local government) is the support to the regional development agencies (about £1.5 billion). ODPM spend on planning is about £178 million. The ODPM has increased its staffing with now about 4,400 posts, 900 of them in the government offices and 700 in the Planning Inspectorate. This administration costs about £333 million, £115 million of which is spent in the government offices and £33 million in the Planning Inspectorate. Much of the activity of the department is concentrated on the Sustainable Communities Plan and particularly the four growth areas in and around the South East (which are explained in Chapter 6). 'Delivery vehicles [*sic*] are being established in key development areas to plan and coordinate the growth programmes' (p. 31) which are overseen by a Cabinet Committee chaired by the Prime Minister.[8] The accent is on unlocking barriers to the delivery of houses, whether relating to infrastructure or recalcitrant planning authorities. For example, during 2003–4 the ODPM made £5 million available for the Bedford Western Bypass, but also, through the government offices, the government held face-to-face interviews with forty-four local authorities in the South East where there is an 'under-delivery' of housing. The reform of the planning system is also important but less prominent in spending since it does not require relatively big changes in funding.

ODPM has four executive agencies, notably the Planning Inspectorate (discussed on p. 49) and several non-departmental public bodies, for example, the Audit Commission, English Partnerships and the Housing Corporation.

Department for Culture, Media and Sport

The Department for Culture, Media and Sport was established in 1997, superseding the Department of National Heritage. It has a wide range of responsibilities, including the arts, sport and recreation, libraries, museums, broadcasting, film, press freedom and regulation, heritage and tourism. Its overall aim is 'to improve the quality of life for all through cultural and sporting activities, and to strengthen the creative industries'. There are now greatly enhanced resources

for these worthy objectives by way of the National Lottery. The areas of 'good causes' for which Lottery funds provide support are sport, the arts, heritage, charities, millennium projects, health, education and the environment (discussed further in Chapter 8). The department has important responsibilities for heritage planning, including listed buildings.

The DCMS is responsible for over forty executive and advisory non-departmental public bodies, including the British Library, the British Tourist Authority, the Millennium Commission, the National Heritage Memorial Fund and English Heritage (which is discussed in Chapter 8). It has also established close relationships with the Local Government Association (LGA) and is promoting local authority cultural strategies.

Department for Environment, Food and Rural Affairs

Other departments of government have special status in respect of town and country planning. From the 1940s particular status was afforded the Ministry of Agriculture, Fisheries and Food (MAFF), the functions of which are now part of the much wider Department for Environment, Food and Rural Affairs (DEFRA). An overriding concern of government after the war was the protection of agriculturally productive land. This secured a central place for the MAFF in land use decisions. It had to be consulted on important proposals, and the MAFF classification of agricultural land quality remains a potentially important consideration in development control (PPG 7: Annex B). The influence of the ministry waned somewhat in parallel with the decline of agriculture in the British economy, but agriculture still retains a special status. For example, if DEFRA has an unresolved objection to a development plan, the local planning authority must refer the dispute to the ODPM, and agricultural activity and development still have considerable exemption from planning control. DEFRA (like MAFF before it) also has to be consulted on any planning proposal which involves a significant loss of high quality agricultural land. Such objections have fallen considerably over recent years.[9] At the same time, it has assumed an increasing responsibility for countryside protection functions such as environmentally sensitive areas (discussed in Chapter 9).

Sustainable development is at the core of DEFRA's purpose, and in particular the formulation and implementation of the UK Sustainable Development Strategy. This means encouraging other departments and agencies to take practical note of the Strategy in their own plans and actions. The nature of the Sustainable Development Strategy is discussed in Chapter 7; it is promoted through a Sustainable Development Unit (SDU) that works in parallel with the Sustainable Development Commission (SDC) and coordinates a Research Network. DEFRA also estab-

lished a high-level Sustainable Development Task Force in 2003 including ministers from other departments and the devolved administrations. DEFRA plays an important role in the development of international policy on the environment, which also figures in its explicit objectives (shown in Box 3.2). DEFRA is responsible for over £5 billion of public spending and employs about 12,000 staff.

DEFRA is responsible for six executive agencies, but more important for planning are its three principal non-departmental public bodies, the Countryside Agency, English Nature and the Environment Agency. All are discussed in later chapters, but we give here a general explanation of executive agencies and non-departmental bodies.

BOX 3.2 DEFRA'S STRATEGIC PRIORITIES

DEFRA's aim

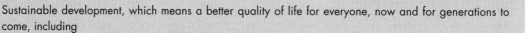

Sustainable development, which means a better quality of life for everyone, now and for generations to come, including

- a better environment at home and internationally and sustainable use of natural resources
- economic prosperity through sustainable farming, fishing, food, water and other industries that meet consumers' requirements
- thriving economies and communities in rural areas and a countryside for all to enjoy.

DEFRA's objectives

- to protect and improve the rural, urban, marine and global environment, and lead integration of these with other policies across government and internationally
- to enhance opportunity and tackle social exclusion in rural areas
- to promote a sustainable, competitive and safe food supply chain that meets consumers' requirements
- to promote sustainable, diverse, modern and adaptable farming through domestic and international actions
- to promote sustainable management and prudent use of natural resources domestically and internationally
- to protect the public's interest in relation to environmental impacts and health and ensure high standards of animal health and welfare.

Source: DEFRA Annual Report 2004

Executive agencies

The proliferation of new government agencies is confusing. Essentially it has taken two main forms: executive agencies (of which the highly successful Driver and Vehicle Licensing Executive was the forerunner) and non-departmental public bodies (exemplified by the Housing Corporation, the Local Government Commission for England, and the former new town development corporations).

Executive agencies remain part of their department, and their staffs are civil servants, but they have a wide degree of managerial freedom (set out in their individual 'framework' documents). They enjoy delegated responsibilities for financial, pay and personnel matters. They work within a framework of objectives, targets and resources agreed by ministers. They are accountable to ministers, but their chief executives are personally responsible for the day-to-day business of the agency. Ministers remain accountable to Parliament. If this sounds somewhat confusing, that is because it is. However, in principle, the stated intention is to increase accountability. A distinction is drawn between responsibility, which can be delegated, and accountability, which remains a matter for ministers – a contention which is the subject of considerable controversy. Examples of executive agencies are the Planning Inspectorate (discussed in the following section) and Historic Scotland.

By contrast, non-departmental public bodies (NDPB) are a type of quango (quasi-autonomous non-governmental organisation). These bodies (of which there are over 1,300) range enormously in function, size, and importance. They all play a role in the process of national government, but are not government departments or parts of a department. There are three types of NDPB: executive bodies such as the Countryside Agency, the Environment Agency and English Heritage; advisory bodies such as the Advisory Committee on Business and the Environment, the Radioactive Waste Management Committee and the Royal Commission on the Environment; and tribunals such as the lands tribunals, rent assessment committees and the Agricultural Land Tribunal.

Planning Inspectorate

The way in which the aims and objectives of executive agencies are cast is illustrated by the case of the Planning Inspectorate. This is a joint executive agency of the DETR and the Welsh Office. In Scotland, the equivalent is the Scottish Office Inquiry Reporters Unit (SOIRU). The major areas of work of the Inspectorate have fluctuated considerably since 1990. Determination of planning appeals has increased to 22,550 in 2003–4 from 12,619 in 1999–2000 in England (860 in Wales 2003–4); enforcement appeals are up to 3,376 in 2003–4 from 2,746 in 1999–2000; and 56 development plan inquiries were opened in 2003–4 compared with 36 in 1999–2000, but down on a high of 91 in 1995–6.[10] During the wave of plan-making in the mid 1990s local plan inquiries placed heavy demands on the Inspectorate, whereas from 2001 it has been the rapid increase in planning appeals. (The reasons for the increase in appeal work are examined in Chapter 5.)

Other responsibilities of the Planning Inspectorate include called-in applications, high hedges appeals, access appeals, highway inquiries and footpath orders under the Highways, Town and Country Planning, and Wildlife and Countryside Acts. Increasing resources are devoted to environmental matters such as inquiries under the Environmental Protection Act. The Inspectorate has an annual budget of £36 million (2003–4) and a total staff of around 780, of whom 300 are inspectors. The Inspectorate also employ 150 'fee-paid' (i.e. consultant) inspectors. In line with current ideas about governmental administration, it has performance targets which include deciding 80 per cent of written representation appeals within 16 weeks, and providing an inspector for local plan inquiries at the time the authority requested in 90 per cent of cases. The Inspectorate now also publishes its own journal.

The future of the Planning Inspectorate has been the subject of much debate during recent years. In 2000 it was subject to a major review by DETR (a review is required every five years), and in the same year the Environment, Transport and Regional Affairs Select Committee conducted an inquiry into *The Planning Inspectorate and Public Inquiries*. On the first review, the outcome was positive and it was decided to retain the Inspectorate as an executive agency. Attention then shifted to stage two of the DETR review on how performance could be enhanced. The report is generally positive: it notes that 'the Inspectorate is held in high regard by most who come into contact with it' and that there had been a dramatic improvement in efficiency over recent years. The main problems seemed to be in ensuring consistency and in the relatively small number of cases in which complaints arise where the committee found an apparently high-handed attitude to people querying decisions. (Since then complaints have risen dramatically as a result of the delays created by the rapid increase in appeals.) The report made other recommendations on the potential for more 'instant decisions'; the increasing need for inspectors to be able to provide specialist knowledge; and the difficulty of keeping up with incremental changes to government policy (or even knowing what government policy is), which is not so much an issue for the Inspectorate as for the government.

In the context of new legislation implementing the Human Rights Convention, there have been suggestions that the Inspectorate should be replaced by a system of environmental courts. The reason is that the Human Rights legislation requires that anyone whose rights have been affected should be 'entitled to a fair and public hearing . . . by an independent and impartial tribunal established by law'. Advocates of the environmental court argue that the Inspectorate does not meet this requirement because it is an executive agency of government and thus not independent.[11] This matter is taken up in the discussion of inquiries and hearings in Chapter 12, suffice to say here that the government (and the courts) have so far determined otherwise. The Select Committee came to the common-sense conclusion that if there is a need for more independence through a court system, the Inspectorate should be established as the first part of that, in effect: a court of first instance.

Central government planning functions

Relationships between central and local government vary significantly among various policy areas, 'reflecting, in part, the difference in weight and concern which the centre gives to items on its political agenda, and, in part, differences in the sets of actors involved in particular issues' (Goldsmith and Newton 1986: 103).

Under the Town and Country Planning Act 1943 (which preceded the legislation on the scope of the planning system), there was a duty of 'securing consistency and continuity in the framing of a national policy with respect to the use and development of land'. Though this is no longer an explicit statutory duty, the spirit lives on, and the Secretary of State has extensive formal powers. These, in effect, give the department the final say in all policy matters (subject, of course, to parliamentary control – though this is in practice very limited). For many matters, the Secretary of State is required or empowered to make regulations or orders. Though these are subject to varying levels of parliamentary scrutiny, many come into effect automatically. This delegated secondary legislation covers a wide field, including the Use Classes Orders (UCOs) and the General Development Orders (GDOs). These enable the Secretary of State to change the categories of development which require planning permission.

The formal powers over local authorities are wide-ranging. If a local authority fails to produce a 'satisfactory' plan, default powers can be used. The Secretary of State can require a local authority to make 'modifications' to a plan, or 'call in' a plan for 'determination'. Decisions of a local planning authority on applications for planning permission can, on appeal, be modified or revoked. Development proposals which the Secretary of State regards as being sufficiently important can be 'called in' for decision by the minister.

These powers are now frequently employed in the plan-making process, usually informally through the DETR regional offices.[12] In less interventionist times, they were reserved for cases where there was a deadlock between local and central government. This can amount almost to a game of bluff as, for instance, when a local authority wants to make a political protest, or to demonstrate to its electors that it is being forced by central government to follow a policy which is unpopular. Thus, opposition in Surrey to the M25 was so strong that the county omitted it from its structure plan. The Secretary of State made a direction requiring it to be included. Another battle arose over the Islington unitary development plan, where the Secretary of State took strong objection to the stringent controls which the borough proposed (inter alia) for its thirty-four conservation areas. The Secretary of State issued a direction requiring most of these to be changed. The Borough took the matter to court, which held that it had no power to intervene on the planning aspects of the case, and since the Secretary of State had not acted perversely or in conflict with his own policies, his action was quite legal. Judicial review cannot be used as an oblique appeal. It was therefore the responsibility of the Borough and the Secretary of State to resolve their differences to the satisfaction of the Secretary of State (*Journal of Planning and Environment Law* (JPL) 1995: 121–5).

A more recent case was a direction to Berkshire County Council to modify its proposed structure plan to increase the provision for new housing in the county (from 37,500 to 40,000) by the year 2006. As discussed in a later chapter, this is a common issue of friction between central government and a number of county councils, particularly in the South East. Perhaps the classic case of open political conflict was the North Southwark Local Plan which was formally called in by the Secretary of State. It was only the second plan to be called in and the first (and only plan) to be rejected entirely, because it opposed private investment and was hostile to the London Docklands Development Corporation (Read and Wood 1994: 11).

The interest of cases such as these lies in their exceptional nature. It is very rare for a local authority to engage in a pitched battle with central government. Equally, it is seldom that central government will feel compelled to use its reserve powers. It is perhaps noteworthy that these two cases arose in the politically charged areas of inner London between radical Labour authorities and a Conservative government that had become openly hostile to local government.

The North Cornwall case was handled in a way more consistent with tradition. The local authority was giving planning permissions for development in the open countryside contrary to national policies and the approved county structure plan. Pressure was brought to bear upon the district council by way of a special inquiry carried out by an independent professional planner (Lees 1993). Normally, informal pressures are sufficient: the threat of strong action by the Secretary of State is typically as good as – if not better than – the action itself. With the enhanced position of development plans in the so-called plan-led system, attention now focuses on the provisions of draft plans. Regional officials pore over the wording of local policies to ensure that they accurately reflect those established at the national level. To the outsider, this plan scrutiny can develop into a game of words, sometimes taking on the character of academic hair-splitting. For instance, at one time the word 'normally' was acceptable in policy, now it is not.

In spite of all this, it is not the function of the Secretary of State to decide detailed planning matters. In a ministerial statement, it was explained that:

It is the policy of the Secretary of State to be very selective about calling in planning applications. He will, in general, only take this step if planning issues of more than local importance are involved. Such cases may include, for example, those which, in his opinion,

- may conflict with national policies on important matters;
- could have significant effects beyond their immediate locality;
- give rise to substantial regional or national controversy;
- raise significant architectural or urban design issues; or
- may involve the interests of national security or of foreign Governments.

(*HC Debates* 16 June 1999, col. 138)

This echoes statements made by previous Secretaries of State: local planning decisions are normally the business of local planning authorities. The Secretary of State's function is to coordinate the work of individual local authorities and to ensure that their development plans and development control procedures are in harmony with broad planning policies. That this often involves rather closer relationships than might prima facie be supposed follows from the nature of the governmental processes. The line dividing policy from day-to-day administration is a fine one. Policy has to be translated into decisions on specific issues, and a series of decisions can amount to a change in policy. This is particularly important in the British planning system, where a large measure of administrative discretion is given to central and local government bodies. This is a distinctive feature of the planning system. There is little provision for external judicial review of local planning decisions (Scrase 1999; Keene 1999): instead, there is the system of appeals to the Secretary of State. The department in effect operates both in a quasi-judicial capacity and as a developer of policy.

The department's quasi-judicial role stems in part from the vagueness of planning policies. Even if policies are precisely worded, their application can raise problems. Since local authorities have such a wide area of discretion, and since the courts have only very limited powers of action, the department has to act as arbiter over what is fair and reasonable. This is not, however, simply a judicial process. A decision is not taken on the basis of legal rules as in a court of law: it involves the exercise of a wide discretion in the balance of public and private interest within the framework of planning policies.

Appeals to the Secretary of State against (for example) the refusal of planning permission are normally decided by the Planning Inspectorate. Inspectors represent or 'stand in the shoes' of the minister. Such decisions are the formal responsibility of the Secretary of State; there is no right of appeal except on a question of law. Inspectors also consider objections made to local development plans, and their binding decisions are put to the local authority.

Planning authorities, inspectors, and others are guided in their decisions and recommendations by government policy. Central government guidance on planning matters is issued by way of circulars and, since 1988, in policy guidance (as explained in Chapter 4). Since the introduction of planning guidance documents, circulars have been concerned mainly with the explanation and elaboration of statutory procedures. Policy guidance deals with government policy in substantive areas, ranging from green belts to outdoor advertising. Circulars and guidance are generally subject to some consultation with local authorities and other organisations prior to final publication, and they are often supported by research and sometimes prepared in draft by consultants, but the Secretary of State has the final word.

Circulars and the various forms of guidance are recognised as important sources of government policy and interpretation of the law, although they are not the authoritative interpretation of law (this is the role of the courts), nor are they generally legally binding. Indeed, advice can be conflicting, perhaps as a result of piecemeal revision at different times. Moreover, as is demonstrated repeatedly at public inquiries, differing interests can 'cherry pick' from the twenty-five policy statement and planning policy guidance notes to show how well their arguments meet the official guidance. Arguments for and against development in villages can be equally supported. While 'the overall strategy should be to allocate the maximum number of houses to existing larger urban areas' (PPG 13), the building of houses in villages can help to sustain the local services which are necessary for their economic survival (PPG 7). Nevertheless, circulars and guidance notes command a great deal of respect and form an important framework for development planning and development control.

Policy, of course, has to be translated into action. This presents inevitable problems: policy is general, action is specific. In applying policy to particular cases, interpretation is required; and often there has to be a balancing of conflicting considerations – of which many examples are given throughout this book. Policies can never be formulated in terms which allow clear application in all cases, since more than one 'policy' is frequently at issue. Even the most hallowed of policies has to be flouted on occasion: as witness developments in the green belts, in protected sites of natural or historic importance, and in national parks.

Such developments may be unusual (if only because they attract great opposition – of an increasingly strident nature), but they represent only the most obvious and the most public of the conflicts over land use.

Given the realities of land use controls, policies are usually couched in very general terms such as 'preserving amenity', 'sustaining the rural economy', 'enhancing the vitality of town centres' or 'restraining urban sprawl', and such like. This is a very different world from that of a zoning ordinance which is the principal instrument of development regulation in many countries. Such an ordinance may provide (for example) that a building shall be set back at least five metres from the road, have a rear yard of six metres or more, and side yards of at least two and a half metres. Zoning is intended to be clear and precise, and subject to virtually no 'interpretation'. Indeed, it was hoped that it would be virtually self-executing. Though these hopes failed to materialise, it is fundamentally different in approach from the British planning systems. Above all, the British systems embraces discretion and general planning principles rather than certainty for the landowner and developer.

It is important to recognise that discretion means much more than 'making exceptions in particular cases'. The system requires that all cases be considered on their merits within the framework of relevant policies. Local authorities cannot simply follow the letter of the policy: they must consider the character of a particular proposal and decide how policies should apply to it. But they cannot depart from a policy unless there are good and justifiable planning reasons for so doing. The same applies to the Secretary of State who is equally bound both by the formulated policies and the merits of particular cases. The courts will look into this carefully in cases which come before them and, though they will not question the merits of a policy, they will ensure that the Secretary of State abides by it. Thus, in a curious way, discretion is limited. All material considerations must be taken into account and justified. Arbitrary action is unacceptable as it is in the USA, which has written constitutional safeguards (Booth 1996; Purdue 1999).

DEVOLVED AND REGIONAL GOVERNMENT

The Union will be strengthened by recognising the claims of Scotland, Wales and the regions with strong identities of their own. The Government's devolution proposals, by meeting those aspirations, will not only safeguard but also enhance the Union.

White Paper, *Scotland's Parliament*, 1997

Devolution to Scotland and Wales

The campaign for devolution to Scotland and Wales failed at the end of the 1970s, but succeeded twenty years later. The aftermath of the earlier failure proved to be an important factor in the later success. The 1979 collapse of devolution led to the defeat of the Labour government and eighteen years of Conservative governments bent not on devolving power, but on centralising it. During this period, the strength of the movement for devolution increased, particularly in Scotland, where the Thatcher government displayed a marked insensitivity to Scottish feelings. As Vernon Bogdanor has put it:

The Thatcher Government's policies of competitive individualism were resented in both Scotland and Wales where they were seen as undermining traditional values of community solidarity; and policies such as privatisation and opting out from local authority control had little resonance there. But resented above all was the community charge, the poll tax. Only devolution, so it seemed, could protect Scotland and Wales against future outbursts of Thatcherism.

(Bogdanor 1999: 195–6)

Following the publication of White Papers (*Scotland's Parliament* and *A Voice for Wales*), the Scotland Act and the Government of Wales Act were passed in 1998. The very titles of the White Papers point to a major difference between them. Scotland has a Parliament with legislative powers over all matters not reserved to the UK Parliament. Wales has only executive functions, but it does have full powers in relation to subordinate legislation. The latter include environmental, housing, local government and planning functions. Thus Wales can change the provisions relating to the Use Classes Order, the General Development Order, the General Development Procedure Order, as well as the regulations concerning planning applications. The Assembly also gives its views to the departments preparing new legislation about special provisions for Wales. While much legislation is shared with England, there are often, and increasingly, special sections devoted to Wales.

The devolution to Scotland and Wales is of importance to England for a variety of reasons. One of these is its effect on the possible pressure for devolution to English regions. This might be fostered if Scotland and Wales were perceived to benefit economically from devolution at the expense of the poorer regions of England. Encouragement might lie in the new regional machinery being established in England. Support for regional government is stronger in the north but not sufficient to vote for regional government.[13]

Scottish Executive

Scotland has had a special position in the machinery of government since the 1707 Act of Union. It has maintained its independent legal and judicial systems, its Bar, its established Church (Presbyterian) and its heraldic authority (Lord Lyon King-at-Arms). The Scottish Office has a long history and, even before devolution, had a large degree of independence from Whitehall (though note that this is the UK government's Scottish department, not the Scottish Executive, which belongs to the Scottish government). Many years of responsibility for Scottish services, the relative geographical remoteness of Edinburgh (perhaps essentially psychological), the nature of the distribution of people and economic activity, the vast areas of open land, the close relationship between central and local administrators and politicians – such are the factors which gave Scottish administration a distinctive character.

The departments include: development (SEDD), enterprise, transport and lifelong learning (ETLLD) and environment and rural affairs (SEERAD). Following devolution, the ministers for these departments are members of the Scottish Parliament.

The Scottish Executive and the Scottish Parliament have begun to take an active role in the definition of distinctive planning policy for Scotland. A 1999 Consultation Paper *Land Use Planning under a Scottish Parliament* issued by the Scottish Office set out the potential:

> The form of any national planning policy guidance which emerges from the Scottish Executive could have significant implications for statutory development plans. A national plan would almost certainly be perceived as unduly centralist and excessively rigid. However, guidance produced by the Scottish Parliament and Executive, bringing together the various National Planning Policy Guidelines and incorporating spatial issues more explicitly, might be attractive. This could inform future development in Scotland and provide some degree of consistency in the pursuit of sustainable development. It could be a vehicle for high level coordination of the objectives of the major agencies as they relate to development and land use. It could also prove attractive for those areas where progress with structure plans has been slow.

This gives some idea of early thinking on the way in which the new machinery might work. Later in the document there is a more certain statement: 'there is a clear expectation that all national strategic policy guidance will be subject to scrutiny by the Scottish Parliament'.

National Assembly for Wales

In Wales, increasing responsibilities over a wide field have been gradually transferred from Whitehall to the (former) Welsh Office. This transfer has taken many years to achieve. Welsh affairs were dealt with by the Home Secretary until 1960, with many services being administered direct by the departments which served England. There has been a minister responsible for Wales since 1951, but it was not until 1964 that the (Labour) government established the Welsh Office and a Secretary of State for Wales (Bogdanor 1999: 157–62).

Following devolution, the National Assembly for Wales (NAW) took over responsibility for a wide range of functions from the Welsh Office and other government departments.[14] Relevant to the fields covered in this book are culture, economic development, environment, historic buildings, housing, local government, tourism, town and country planning, and transport. All these functions are now transferred to the Assembly. Particularly important are the powers of secondary or subordinate legislation. This is in contrast to the Scottish Parliament which has the wider powers of primary legislation.[15] However, in the field of town and country planning, the effective difference is smaller than might at first sight appear. This is because of the particular character of the British planning legislation. This provides only a very general framework for the substantive measures which are enacted in secondary legislation such as the Use Classes Order, the General Development Order, the General Development Procedure Order, and a host of statutory rules and regulations (Bosworth and Shellens 1999). The latter deal with such matters as advertisements, development plans, environmental impact assessment, inquiries procedures, and planning obligations. Additionally, of course, plans are the responsibility of the local authorities, now subject both to Welsh planning guidance,[16] and to approval by the Assembly.

The Assembly is both an executive and a deliberative body, and the executive is described as the Welsh Assembly Government (WAG). Planning is part of a Department for Environment, Planning and Countryside, which also has oversight of the Welsh built heritage agency (Cadw). However, the National Planning Strategy is prepared by a strategy section of the Strategy and Communications Department, reflecting the belief that it should cut across all the Assembly's activities and policies. Another department deals with Economic Development and Transport. The Assembly is also responsible for the Countryside Council for Wales, the Welsh Development Agency (incorporating the former Land Authority for Wales and the Development Board for Rural Wales) and the Welsh Tourist Board. Some of these are discussed in later chapters.

Northern Ireland Office

Government in Northern Ireland has a unique character and structure. National government performs, either directly or through agencies, virtually all governmental functions: local government has few responsibilities. Though there are twenty-six elected district councils, their powers are limited to matters such as building regulations, consumer protection, litter prevention, refuse collection and disposal, and street cleansing. The councils nominate representatives on the various statutory bodies responsible for regional services such as education, health and personal social services, and the fire service. They also have a consultative status in relation to a number of services including planning. All the major services, including countryside policies, heritage, pollution control, urban regeneration, transport, roads, and town and country planning are administered directly by the Northern Ireland Office. The DoENI is the responsible department for these. Housing is administered by the Northern Ireland Housing Executive, which was formed in 1971 to take control of the local authority housing stock. Other departments include Agriculture and Economic Development.

Given the tragic history of Northern Ireland, the Office's priority aims are significantly different from those of other parts of the UK: 'to create the conditions for a peaceful, stable and prosperous society in which the people of Northern Ireland may have the opportunity of exercising greater control over their own affairs'. Planning has an important role in this which is undertaken through an executive agency: the Planning Service Agency of the Northern Ireland Office. The general status of executive agencies is discussed below. The Agency's aim is 'to plan and manage development in ways which will contribute to a quality environment and seek to meet the economic and social aspirations of present and future generations (Trimbos 1997).

The consultative role of the district councils is regarded with great importance by the Agency, and it consults with them on a wider range of issues than is required by law. It is the government's intention that there should be a 'substantial democratic control of the planning process as soon as politically possible by cross-party agreements in the context of a comprehensive political settlement' (*NI Planning Service Agency Annual Report 1998–99*: 81).

It is also intended to reorganise the departments in the new administration. A new Department for the Environment will be responsible for planning control while a Department for Regional Development will be responsible for strategic planning. The Belfast Agreement gave a commitment to make rapid progress with a long-term regional strategy for consideration by the Assembly (once it has been established). This strategy will be a statutory document to which all NI and UK departments will require 'to have regard'.[17] At the time of writing, any statutory action is on hold as a result of the political impasse.

Towards regional government in England?

The institutions of government at the regional level in England are complex and potentially confusing. They are also evolving under the government commitment to regional devolution. It would not be an exaggeration to say that there has been a revolution (or very rapid evolution) of regional competences in England. This has been done in three ways creating three distinct regional institutions, as shown in Figure 3.4. First, the government has formalised and provided more resources to regional bodies (RBs), sometimes known as regional chambers or assemblies. These are comprised of nominated elected representatives from local government and other community and business interests. They are thus, indirectly elected bodies representing local interests. Second, regional development agencies (RDAs) were established with a specific remit to promote economic regeneration drawing together

funding from formerly national sources around a regional agenda. The RDAs work with local partners to develop their regional agenda but are accountable to national government. Third, central government has strengthened and integrated its presence at the regional level through government offices (GOs). This is central government operating at the regional level. While these are three distinct bodies, they share much of the same agenda for their regions. What gets done at the regional level relies on close cooperation and joint working among the three bodies, much of which is informal. While the regional body will prepare the regional spatial strategy, the regional development agency will be centrally involved in trying to ensure that it meets its own agenda, and the government office will effectively supervise the whole process. This sounds quite neat; in practice it is, understandably, a messy exercise, and more so given the number of other bodies that operate at the regional level. The following explanation necessarily concentrates on the formal powers and relationships.

The Labour government has made much progress on its commitments to regional devolution. The White Paper *Your Region, Your Choice* (p. 1) sets the scene:

> Experience in Scotland and Wales has shown how a tailored approach to economic regeneration can bring benefits: skills, jobs, prosperity. The Government is committed to revitalising the English regions. They contributed to establishing the UK as a great economic power as different regional strengths spurred our first industrial revolution. We must ensure that they can play their part in the knowledge-based economic revolution which is now taking place.

The 1997 Labour Manifesto made a firm commitment to elected regional government but only in those regions where there was a popular demand for it. Where regional government might be established the government said that a unitary system of local government would be expected (that is the complete loss of counties) and thus no increase in tiers of government. Regional government would be given powers over economic development and regeneration (including

GOVERNMENT OFFICES (e.g. GO-SW, GO-EM)
National government at the regional level; less departmental, more integrated and spatial orientation; coordinated by Regional Coordination Unit in ODPM

REGIONAL DEVELOPMENT AGENCIES (e.g. Advantage West Midlands)
To promote economic development in urban and rural areas, with regard to wider policies for sustainable development; report to DTI; current funding about £1 billion

REGIONAL BODIES
Voluntary assemblies made up of indirectly elected members from the constituent councils and government nominated representatives of business and civil society; Greater London Authority is the only directly elected body

Figure 3.4 *The organisation of regional government for planning*

control of the regional development agency), spatial development, housing, transport, skills and culture. The Secretary of State's powers to effectively approve and publish the regional strategies would have been devolved to regional government, but not call-in powers, which would stay at the centre. Legislative powers were not to be devolved, but would remain with the UK Parliament. Legislation was promised to allow for referenda in the regions on regional government but this was not for immediate action. In the mean time, regional chambers were to be developed to debate and formulate views about future policies. Again, there was

possibility for the arrangements to vary according to regional wishes, but the chambers were expected to be based on existing regional local authority organisations such as standing conferences and the like. As with these bodies, the regional chambers were to be local authority led, but they would include representatives from other regional stakeholders.[18] This proved an even more appealing idea than the government could have imagined, and all eight regions quickly established chambers based on previous voluntary cooperation arrangements, as shown in Table 3.1.

Table 3.1 *Regional government offices, regional bodies and regional development agencies in England*

Government office	Regional development agency	Regional body
East of England (GO-East)	East of England	East of England Regional Assembly
East Midlands (GO-EM)	East Midlands	East Midlands Regional Assembly
North East (GO-NE)	One North East	North East Assembly
North West (GO-NW)	North West	North West Regional Assembly
South East (GO-SE)	South East	South East of England Regional Assembly
South West (GO-SW)	South West RDA	South West Regional Assembly
West Midlands (GO-WM)	Advantage West Midlands	West Midlands Regional Assembly
Yorkshire and Humber (GO-YH)	Yorkshire Forward	Yorkshire and Humber Assembly
London (GO-L)		Greater London Authority

Note: The areas of all three types of organisation are now coterminous; the representative body for the regional chambers or assemblies is the English Regions Network.

Each regional body comprises 70 per cent local authority members and 30 per cent from the community (including higher education, the health service, parish councils and other 'stakeholders') and business (including the Confederation of British Industry (CBI)). The regional bodies' main jobs are to scrutinise the work of the RDA, and to lead preparation of the regional spatial strategy, the regional sustainability framework and the integrated regional strategy where it is being prepared. In order to support this work, which was previously done mostly by part-time and seconded staff, in 2001 the Deputy Prime Minister announced a £15 million funding over three years. In comparison with similar regional institutions overseas, the regional bodies have few competences and little resource, but their capacity to govern is expanding rapidly. There has been, for example, much progress on the provision of better information through 'regional observatories' which are often led by the regional bodies in cooperation with the RDAs and government offices.

While the regional chambers were established, lobbying continued on the proposal for directly elected regional government, especially in the three northern regions where there was greatest support, but it was not until 2004 that the first referendum was held. The negative result has put a hold on the development of regional government for some time.

Government Offices for the English Regions

The first set of initiatives to improve capacity at the regional level and for dealing with the regional dimensions of planning (apart from the voluntary associations of councils) was through establishing and then strengthening the Government Offices for the English Regions. These have built up a relationship with local government, and created a real governmental locus away from Whitehall. Set up in 1994, they have a range of functions within the remit of ten government departments (see Box 3.3). Their overall official role is to promote a coherent approach to competitiveness, sustainable economic development and regeneration.

Attention was initially focused on the publication and revision of regional planning guidance, generally prepared in draft by the regional conference of local authorities; but these have tended to be rather bland statements of general central government policies. The Blair government quickly adopted a different approach with its campaign to 'modernise' both local government and the planning system. The policy statement, *Modernising Planning*, pointed to the short-comings of the regional planning guidance (RPG), explained in more detail in Chapter 4. Crucially, it did not command the confidence or commitment of regional stakeholders, and needed a major overhaul. Fundamentally, this was seen to include a strong bottom-up approach:

> We propose a more inclusive process, involving the local authority conferences working with the Government Offices, business and other regional stakeholders, in producing drafts of the regional guidance itself. This would replace the current arrangement under which the regional planning conference merely provides 'advice' on the basis of which planning guidance is subsequently produced by the Government Office.
>
> (*Modernising Planning* 1998: para. 18)

The regional offices have strengthened considerably under the Labour government. The *Reaching Out Action Plan* of 2000 brought a more concerted effort on strengthening government at the regional level. The Action Plan had four objectives: to better coordinate area-based initiatives; to involve the government offices more in central policy-making (promoting the regional view); expanding the offices to include other government departments (the original contributors were Trade and Industry, Education and Science and Environment, Transport and the Regions (now ODPM); and to set up a Regional Coordination Unit (RCU) to act as a 'head office for the regional outposts'. The result is regional offices with a much more prominent profile and responsibilities. The government offices are not generally implementation organisations. Almost everything is done in partnership with other bodies that actually do the implementation – local

BOX 3.3 GOVERNMENT OFFICES FOR THE REGIONS

Sponsor departments

- Office of the Deputy Prime Minister
- Department for Education and Skills
- Department of Trade and Industry
- Department for Environment, Food and Rural Affairs
- Home Office
- Department for Culture, Media and Sport
- Department for Work and Pensions
- Department for Transport
- Department of Health
- Cabinet Office

Tasks

- Sponsor regional development agencies
- Carry out regulatory functions
- Provide a regional perspective informing central policy

Regional output indicators of particular relevance to planning (of twenty-two)

- Bus passenger journeys per 100,000 population
- Percentage of non-decent social housing
- Percentage of new homes built on previously developed land
- Percentage of household waste recycled or composted

government, the other regional organisations and agencies. Government offices facilitate and bring organisations together. They provide the necessary government support, which may often involve funding: in 2003–4 the offices allocated £9 billion of government money including managing much EU funding. The budgets are held by sponsoring departments but the GOs administer or influence its allocation. The RCU/GO network has an administrative budget of £99 million (2005–6 plans). It is also responsible for the delivery of more than forty public service agreements (PSAs) on behalf of the departments.[19] The Regional Coordination Unit operates from London and provides a communication channel between the offices and Whitehall departments, and monitors and supports the activities of the offices. A set of twenty-two regional output indicators (ROIs) have been agreed to help in comparing the performance of regions. They address the main policy themes including, for example, the number of VAT registrations which is an indicator of business start-up, and burglary offences per 1,000 households.[20]

Expansion of the government offices was envisaged as being a second-best solution: statutory planning at the regional level will have to await a democratically accountable statutory body to undertake it. There are strong indications that this might be emerging. In London the Greater London Authority (GLA) was created, but this is likely to be the only directly elected regional body in England for some time to come, despite the generally positive response to the government proposals.[21] A major stimulus has come from another policy area: regional development. In the absence of elected regional authorities, a new regional organisation has been established. Following the 1997 White Paper *Building Partnerships for Prosperity*, regional development agencies were set up in each of the eight regions outside London. (The London Development Agency has also been created and is responsible to the Mayor for London.)

Regional development agencies

Regional development agencies are, as their name indicates, agencies to promote economic development in their regions. The Regional Development Agency Act 1998 requires each agency to formulate and keep under review a strategy – or regional economic strategy (RES) – for implementing its statutory responsibilities to further economic development and regeneration, to promote business efficiency, to promote employment, to enhance the development and application of skills, and to contribute to the achievement of sustainable development. Given the traditional emphasis on urban areas (as well as the political prominence of rural concerns), the RDAs are specifically required to give equal attention to rural areas.[22] Indeed, funding is to be separately allocated or earmarked for rural projects.[23]

The RDAs are subject to any 'guidance' and directions issued by the Secretary of State. The first guidance was published in 1999 and dealt with regional strategies. Guidance has also been issued or announced on rural policy, sustainable development, regeneration policy, education and skills issues, competitiveness, inward investment, performance indicators, state aid rules, and equal opportunities.

An obvious problem arises on the relationship between the economic strategy led by the RDA and the regional spatial strategy led by the regional body. Much has been written on this, and the HC Select Committee report on regional development agencies includes a range of views.[24] Some have argued that RDAs should be required to work within the framework of a 'comprehensive overarching strategy' to be prepared and approved by the appropriate regional bodies. The then DETR rejected this on the grounds that the two strategies cover different issues, and that areas of mutual interest can be dealt with by constructive collaboration. What is not usually made explicit is the concern of central government that such an 'overarching' plan could be used to frustrate desirable economic development or housing provision. It is a nicely arguable question whether the requirements of regional land use planning should take priority over economic or housing 'requirements'. Those who see the protection of the countryside as an overriding policy objective will have no doubts on the answer to this, but central government takes a different view. The case of biotechnology clusters is a case in point, as is illustrated by the decisions on development proposals in the rural area south of Cambridge.[25] The RDA and others promoted the release of sites for the development of such clusters. One of the proposals was not accepted (40,000 square metres at the Welcome Trust Human Genome Project at Hinxton Hall) but another for 25,000 square metres was acceptable to central government in making the final decision, despite the location in an area of restraint. (It was the size of the development, not the location, that was unacceptable.) Two other proposals, both in the Cambridge Green Belt, were also accepted. Thus important economic developments can receive priority over countryside protection.

Nowhere has this been made more explicit than in the public examination on the regional planning guidance, now regional spatial strategy (RSS), for the South East. The Panel Report on this is scathing about the argument that economic and housing developments should be 'dampened down' in this region in an attempt to benefit other regions. The essence of this argument was that 'regional imbalances' should be tackled by preventing the 'economic magnetism' of the

THE AGENCIES OF PLANNING

'overheated' South East from 'draining away economic vitality and population from other UK regions'. The Panel castigated this view 'with its manifest overtones of postwar Barlow based industrial development policy'. Government policy was very different, with an emphasis on the economy being encouraged 'to go ahead at full speed on all engines'. The Panel Report continued:

> In our view it is high time that the ghost of Barlow (his report that is) be finally exorcised from regional strategy. Whereas in the 1940s and for some time thereafter, it may have been quite reasonable to consider the UK as the principal unit for economic planning, this is manifestly not the situation at the present time. Economic activity and investment discouraged from settling in the South East of England will not now find alternative landing places in the other UK regions; they are just as likely to go to other parts of the EU. The effect therefore of reducing development pressures by 'dampening down' the economy of the South East would have little or no beneficial effect on the economies of the other regions of the UK . . . The whole of the UK (and indeed the EU) has a vested interest in the economic success of the South East region as a core area for economic activity and a major source of capital and tax revenues. It is an engine of growth for the whole country . . . RPG needs to make it clear that there can be no question of doing anything but building on the success of the economy of the South East with a view to recovering its premier status in the EU and world league.[26]

Needless to say, this argument is not shared by all! Nevertheless, it now seems clear that the Blair government has made the decision to back the RDAs in maximising their individual growth potentials, irrespective of the impact on migration from the less favoured regions. In this, it is following in the steps of the previous Conservative government.[27]

The RDAs are financed with the funds of the government programmes they have taken over, such as urban regeneration, industrial land improvement, and rural development. In total they have a budget of around £800 million and will be expected to lever another £1 billion or so of private money for programmes such as urban regeneration and the redevelopment of derelict land. This may sound like big money but it is very small in relation to the size of the economies in which they intervene. Thus, they will only ever have marginal effects – though they may be very important marginal effects.

RDA staff also came mainly from the agencies subsumed, and perhaps some of the thinking with them. Some functions that might have been transferred to the RDAs have been retained by their departments: business support services, by the Department of Trade and Industry, and skills training by the Department for Education and Skills (DfES). But since 1997 the competences of the RDAs have expanded to include, for example, tourism. So there is the possibility of increasing integration of hitherto disparate policy areas at the regional level. The differences between regional and national priorities may be more significant. Conflicts between the centre and region are familiar to federal systems, and will no doubt come to the fore in the new system. Regional government means not only giving power to the regions, but also taking some power away from central and local government.

Greater London Authority

The abolition of the Greater London Council in 1986 left a gap in the machinery of government which was cumbersome, inefficient and indefensible. London became the only western capital city without an elected city government. Some functions carried out by the GLC were transferred in part to the London boroughs, but many were taken over by a range of joint bodies, committees, ad-hoc agencies and such like (including the London Planning Advisory Committee that prepared strategic planning guidance for the capital). The result was 'a degree of complexity that can be seen not so much as a "streamlining" as a return to the administrative tangle of the 19th century' (Wilson and Game 1998: 54). The election manifesto of the Labour Party promised a referendum to confirm popular demand for a strategic authority and mayor.

The referendum was held in 1998, and though only one-third of Londoners voted, there was an overall 72 per cent majority. The Greater London Authority Act provides for an elected Mayor and an elected Greater London Assembly. The Mayor is not a figurehead, but a highly influential leader. In the words of the 1998 White Paper *A Mayor and Assembly for London*, 'the Mayor will have a major role in improving the economic, social and environmental well-being of Londoners, and will be expected to do this by integrating key activities'. The main responsibilities include:

- The production of an integrated transport strategy for London (extending to transport issues for which the Mayor is not directly responsible) to be implemented by a new executive agency Transport for London (TfL) which will have responsibility for a wide range of services including London's bus and light rail services, the Croydon Tramlink, the Docklands Light Railway, Victoria Coach Station, taxis and minicabs, river services. It also acquires responsibility for a strategic London road network. Government funding is paid in a single block grant, and capital investment schemes within the budget available do not require central government approval.
- Preparation of strategic planning guidance for London in the form of a new Spatial Development Strategy (SDS). The content of this was for the Mayor to decide, but includes transport, economic development and regeneration, housing, retail development and town centres, leisure facilities, heritage, waste management, and guidance for particular parts of London such as the central area and the existing Thames Policy Area (there are also other strategies including transport). The unitary development plans of the Boroughs are required to be 'in general conformity' with the SDS. Development control remains with the Boroughs, but the Mayor is a statutory consultee for planning applications of strategic importance, and has defined powers of intervention, which are already being used for significant applications.
- The setting of an economic development and regeneration strategy for London. A London Development Agency has been appointed by, and responsible to, the Mayor.
- Improvement of London's environment, the development of an air quality management strategic plan, the production of a report every four years on the state of the environment in London.
- Appointment of half the members of a new independent Metropolitan Police Authority, and scrutiny of the policies of the authority.
- Overall responsibility for a new London Fire and Civil Defence Authority, and appointment of the majority of its members.
- Preparation of a strategy for the development of the culture, media and leisure sectors, appointments and nominations to the key cultural organisations.

Clearly this is a highly significant change to the government of London, providing an eloquent indication of the government's commitment to a more effective and democratic system of government. The position of Mayor is not an easy one, since it involves extensive and intensive negotiation with the London Boroughs and innumerable governmental bodies, as well as many professional and voluntary organisations. However, the arrangements were carefully thought through, and the benefits of a government for London are becoming apparent.

LOCAL GOVERNMENT

> The shift away from a system dominated by elected local governments is likely to be permanent given New Labour's enthusiasm for diverse service delivery; this demands new ways of thinking about local government and, indeed, new ways of being local government.
>
> (Stoker and Wilson 2004: 3)

Reorganising local government

Reorganising local government has become a tradition. Its functions and its structure have been subject to frequent change. The pace of this change has become almost frenetic since the early 1960s when it became apparent from the government's decision to reorganise London government that there were serious prospects for the reorganisation of local government elsewhere in the country. Since then analysis, debate, legislation, review, and further legislation have been ceaseless.

In summary, a uniform two-tier system of local government was established by legislation passed in 1963 for London, 1972 for Scotland, and 1973 for the rest of England and Wales. The two-tier system in London and the metropolitan counties was subject to drastic change by the Thatcher government in 1986 under the banner of 'streamlining the cities'. The upper tier (the Greater London Council and the metropolitan county councils of Greater Manchester, Merseyside, Tyne and Wear, West Midlands and West Yorkshire) were abolished, thus leaving a unitary system of local government in these areas.

Further reorganisation into unitary authorities took place in Scotland and Wales in 1996, and a number of unitary authorities were introduced in parts of non-metropolitan England between 1995 and 1997 (although much of the two-tier system remains). In Northern Ireland, a unitary system of local government was set up in 1973. Thus, while Northern Ireland, Wales and Scotland have a unitary system, England has a varied structure of local government, in which 115 areas have a unitary system and the remainder are two-tier, forming 34 counties and 238 districts.

English local government review

The anomalous structure of local government in England stems from the distinctive nature of the local government review that preceded it. A Local Government Commission was established to work within policy and procedural guidance published by the DoE. This 'guidance' proved to have greater power than the government expected: it had the effect of limiting the changes which could be made to the Commission's proposals. Moreover, because of the consultative way in which the Commission operated,

these proposals were significantly influenced by the views of articulate local interests. The government's initial proposals for change were set out prior to the establishment of the Commission in a consultation paper on *The Structure of Local Government in England*. This made the argument, widely accepted across the political spectrum, for a unitary structure of local government in the shires (the pattern which had been put in place in the metropolitan counties). It was argued that a single tier would reduce bureaucracy and costs, and improve coordination. It would clarify responsibility for services and, since taxpayers would be able to relate their local tax bills more clearly to local services, would provide for greater accountability. In the early stages of the Commission's review, the Environment Secretary, John Gummer, had stated unequivocally that the aim was to produce a unitary structure in England, with the two-tier system remaining in only exceptional circumstances. The way this was to be achieved was left open, with several possibilities: existing districts might become unitary authorities, two or more authorities could be merged into larger ones, and wholly new authorities might be created. The main criteria for judging the need for change was responsiveness to local needs and 'sense of identity', as well as the ubiquitous 'cost-effectiveness'.

During the two years of the Commission's review, district and county authorities sought to justify their existence through an expensive and sometimes bitter propaganda war. In fourteen cases this led to challenges to the Commission's recommendations in the courts. There was also a legal challenge by the Association of County Councils which successfully prevented the Secretary of State from modifying the guidance he had previously given to the Commission in an attempt to strengthen the case for unitary authorities. Despite the government's wish to see a unitary structure, the eventual undoubted winners were the counties. The Commission found little evidence that change would improve service provision. In the main, changes were limited to renewing unitary status for former county boroughs, and abolishing new and contrived counties created in the 1974 reorganisation.

After much debate, the Commission recommended only fifty new unitary authorities. These were mostly former county boroughs (unitary authorities before the 1974 reforms), although a significant number of 'special cases' were included on the basis of 'substantial local support for change'. The Commission explained the modest extent of its recommendations as due to the 'weight of evidence from national organisations pointing to the problems and risks associated with a breaking up of county wide services' – a view that was strongly supported by local opinion. However, these arguments failed to satisfy the many districts which were not proposed for unitary status and which had campaigned for this. More significantly, it did not satisfy the government, which was concerned to further increase the number of unitary authorities. Following these disagreements, the chairman of the Commission resigned, and the new chairman was given the remit to review again the case of twenty-one districts where

the government believed there was a strong case for unitary status. Further guidance was issued for this mini-review, stressing the potential benefits of unitary status particularly for areas needing economic regeneration (as in the Thames Gateway). It was argued that the 'single focus' of unitary local government would be more effective in promoting multi-agency programmes in these areas. This final review initially recommended unitary status for ten of the twenty-one districts, but this was reduced to eight after consultation. The new councils came into being in April 1997. See Box 3.4 for a summary of the type and number of local authorities.

The process of reorganisation in the shires has been the subject of considerable criticism and questions have been raised about whether it was worth while. Certainly, reorganisation seems to have been handled

BOX 3.4 LOCAL PLANNING AUTHORITY TYPES AND NAMES

Drawing a distinction between the type of local authority and the name it is given can help to reduce the confusion that arises about local government and the local planning authority. For example, the name 'city council' has only ceremonial significance and does not affect the competences of a council. Local authorities also describe themselves as boroughs but outside London this is just a hangover from structures long abolished. One unitary council, Rutland, calls itself a county council, recognising its historical county status.

England

In most parts of rural England there is a two-tier structure with both

- *county councils* (34) responsible for 'county matters': minerals and waste and also assistance to the regional planning body in preparation of the regional spatial strategy, e.g. Warwickshire County Council
- *district councils* (237) responsible for most local government planning functions except where there is a national park authority, e.g. Warwick District Council.

Many provincial cities and a few rural areas have a single-tier structure with

- *unitary councils* (46) responsible for all local government planning functions, e.g. Stoke-on-Trent City Council and Herefordshire Council.

In the six metropolitan areas of Greater Manchester, Tyne and Wear, Merseyside, South Yorkshire, West Midlands* and West Yorkshire (the metropolitan county councils were abolished but the metropolitan counties still exist) there is a unitary structure with

- *metropolitan district councils* (36) responsible for all local government planning functions, e.g. Birmingham City Council.

In London there is a single-tier structure with

- *London boroughs* (32) including the *Corporation of the City of London*, responsible for all local government planning functions but working within the strategic policy of the *Greater London Authority*, e.g. the London Borough of Wandsworth.

Wales

In the whole of Wales there is a single tier of local government with

- *unitary councils* (22) responsible for all local government planning functions except in the areas of the national parks.

Scotland

In the whole of Scotland there is a single tier of local government with

- *unitary councils* (32) with responsibility for all local government planning functions, except in the area of the national parks.

Northern Ireland

In the whole of Northern Ireland there is a single tier of local government with

- *district councils*, with limited responsibilities; the planning authority is the *Northern Ireland Planning Service*.

In England, Scotland and Wales there are *national parks* which are the planning authority for the area of the park, which may include parts of a number of local authorities, e.g. the Peak District National Park.

When *development corporations* are established they may become the planning authority or take on important planning functions, e.g. the Thurrock Urban Development Corporation will have DC powers over 'large scale and strategic developments'.

There are also about 10,200 *parish and community councils* in England, 1,000 community councils in Wales and 1,350 community councils in Scotland. Local authorities have a duty to consult them on planning matters. Note that nine parish or community councils have the ceremonial title city, e.g. Lichfield City Council, and some are known as town councils, e.g. Soham Town Council.

Notes: * There is both a metropolitan county called West Midlands and an administrative region called the West Midlands.

For further information see the Office for National Statistics website www.ons.gov.uk

much more expeditiously in Wales and Scotland.

Parish councils (or community or town councils) can play a role in the democratic process by providing an effective voice for local interests and concerns. Unlike their counterparts in Scotland and Wales they have statutory functions, though these are very restricted. Of particular importance (and widely used) is their right to be consulted on planning applications in their areas. They can also play a part in the consultation process for the preparation of development plans.[28]

Local government in Scotland

Even before devolution, the cultural history and physical conditions of Scotland have dictated that, to a degree, the administration of planning is distinctive. Changes to the law in Scotland require specific legislation, and the Scottish Office has for long had administrative discretion within which it could take account of the special circumstances which exist in parts of the country. Nevertheless, the broad thrust and impact of government policy have been much the same (Carmichael 1992).

In setting out to reorganise Scottish local government, the government was firmly committed to a single-tier structure, and the 1992 Scottish consultative paper provided options only on the number that were to be established. There were, of course, some political factors involved in this decision: the problem of conflicting interests within the Conservative Party was much less in Scotland since only a handful of the sixty-five Scottish local authorities were in Conservative control. The consultation document in Scotland was also more forthright about the role of local government reform in direct service provision. While the government confirmed its commitment to 'a strong and effective local authority sector', it also argued that local authorities no longer needed to 'maintain a comprehensive range of expertise within their own organisation', since 'much could be done by outside contractors.'

In reviewing the possible number and size of the proposed unitary authorities, a consultation paper provided four illustrations showing structures ranging from fifteen to fifty-one authorities. The choice between mainly small or mainly large authorities has important implications for the planning function, especially structure plans. Only the fifteen-authority option would have allowed for unitary authorities to prepare their own structure plans. Even then, special arrangements would have been needed for Glasgow to ensure effective strategic planning. The outcome of reorganisation in Scotland was thirty-two unitary councils, each of which has full planning powers for its area. The fragmentation of the strategic planning function across a larger number of authorities threatens a recognised strength of the Scottish system, and the need for special arrangements for strategic planning was acknowledged by the Scottish Office during the review. The country has been divided into seventeen structure planning areas, six of which require joint working between authorities. The plan framework is discussed further in Chapter 4.

Scottish legislation provides for the establishment of community councils where there is a demand for them, under schemes prepared by local authorities. As in England and Wales, their purpose is to represent the local community and 'to take such action in the interests of the community as appears to its members to be desirable and practicable'. A study of community councils concluded that 'in contemporary moves towards democratic renewal in local government, community councils are seen as having no special status or role by most local authorities, though some do accord them a distinctive role in consultation, and there is a wide variety throughout Scotland in their operations and effectiveness (Goodlad et al. 1999). Nevertheless, community councils may have a new role in the proposals of a working group of the Scottish Office and the Convention of Scottish Local Authorities (SO: Report of the Community Planning Working Group). (See Figure 3.5.)

Although there has been less than overwhelming support for community councils in Scotland, the debate on their future has been transformed into an enthusiastic promotion of the idea of 'community planning'. This is defined as any process through which a local authority comes together with other organisations to plan, provide for or promote the well-being of the communities they serve. The objectives are to improve levels of service and to increase the collective capacity of public sector agencies to tackle problems which require action from more than one agency. Though much cooperation between agencies already exists, there was a need for a more systematic approach which would provide an overarching strategy. The working group recommend that the Scottish Parliament should signify the importance of 'community planning' by providing a statutory basis for it.

Similarly, the 1999 McIntosh Report, on *Moving Forward: Local Government and the Scottish Parliament*, recommended both the retention of community councils (properly resourced) and the promotion of their role 'within the wider context of the area approach adopted by many councils, as a means of obtaining the fullest possible consultation at the local level'.

Scottish local government and the Scottish Parliament

The establishment of the Scottish Parliament raises a host of questions concerning local government, some of which have long been of importance (such as public apathy and mistrust: Carole Millar Research 1999), some of which arise because of devolution (particularly relationships between Parliament and local government), while others have arisen on the tide of reform which devolution has created (such as the electoral system). Whatever the reason, there is a major endeavour to improve the system of governance in Scotland.

The Commission on Local Government and the Scottish Parliament (the McIntosh Report) has recommended a number of wide-ranging proposals for reforming local government in the context of devolution. Its starting point is a declaration that

> relations between local government and the Parliament ought to be conducted on the basis of mutual respect and parity of esteem . . . Councils, like Parliament, are democratically elected and consequently have their own legitimacy as part of the whole system of governance. To play the role envisaged by the Commission local government

should take the initiative to respond to the challenges it now faces. It should review its procedures and renew itself. This involves citizen participation, not merely by way of consultation but also in decision-making; open transparent and intelligible methods of conducting business; a focus on the consumer; quality and cost-effectiveness in the delivery of services. Local government needs to develop new ways of working in partnership: it is uniquely placed to take an overview of local needs and to provide leadership in community planning.

It is on the basis of this type of thinking (and a long list of specific recommendations) that the Commission stressed that major changes lay ahead for Scottish local government. Among its recommendations are the ratification of a covenant between the Parliament and the thirty-two councils setting out their working relationship. This concept of a direct working relationship between local and central government is 'without parallel or precedent at Westminster', though it is in harmony with the European Charter of Local Self-Government, as well as the Hunt Report.[29]

Other proposals include a statutory power of general competence thus freeing local authorities from the limitations imposed by the constitutional position that they can carry out only those specific powers granted by legislation; further study of the ways in which local authorities may become financially more independent; and a review of local government elections, with the introduction of proportional representation in 2002.

This selection of recommendations gives some flavour of the extent to which Scottish local government has been under fundamental review.[30] In addition, a new ethical framework for local government in Scotland has been established.[31] This includes a review of aspects of the planning process, such as the training of members for the work of a planning committee, and the introduction of 'best practice' (see the discussion below and in Chapter 12). As with the somewhat more modest English ideas for modernising local government, these are big aspirations which can be met only by major changes in the culture of local government (Brooks 1999: 43).

■ Figure 3.5 *Planning authorities in the UK*

Note: Divisional planning offices are shown for Northern Ireland.

Non-metropolitan unitary councils

1	Darlington	17	Derby	32	Wokingham
2	Stockton-on-Tees	18	Nottingham	33	Slough
3	Hartlepool	19	Leicester	34	Windsor and Maidenhead
4	Middlesbrough	20	Rutland	35	Bracknell Forest
5	Redcar and Cleveland	21	Peterborough	36	Thurrock
6	York	22	Herefordshire	37	Southend
7	East Riding of Yorkshire	23	Milton Keynes	38	Gillingham and
8	Kingston upon Hull	24	Luton		Rochester upon Medway
9	North Lincolnshire	25	South Gloucestershire	39	Plymouth
10	North-East Lincolnshire	26	Bristol	40	Torbay
11	Blackpool	27	North-West Somerset	41	Poole
12	Blackburn	28	Bath and North-East	42	Bournemouth
13	Halton		Somerset	43	Southampton
14	Warrington	29	Thamesdown	44	Portsmouth
15	Stoke-on-Trent	30	Newbury	45	Isle of Wight
16	The Wrekin	31	Reading	46	Brighton and Hove

Metropolitan unitary district councils

Greater Manchester	Merseyside	Tyne & Wear	West Midlands	West Yorkshire
Bolton	Knowsley	Gateshead	Birmingham	Bradford
Bury	Liverpool	Newcastle upon	Coventry	Calderdale
Manchester	St Helens	Tyne	Dudley	Kirklees
Oldham	Sefton	North Tyneside	Sandwell	Leeds
Rochdale	Wirral	South Tyneside	Solihull	Wakefield
Salford		Sunderland	Walsall	
Stockport	South Yorkshire		Wolverhampton	
Tameside	Barnsley			
Trafford	Doncaster			
Wigan	Rotherham			
	Sheffield			

London borough councils

Barking and Dagenham	Ealing	Hillingdon	Newham
Barnet	Enfield	Hounslow	Redbridge
Bexley	Greenwich	Islington	Richmond upon Thames
Brent	Hackney	Kensington and Chelsea	Southwark
Bromley	Hammersmith and Fulham	Kingston upon Thames	Sutton
Camden	Haringey	Lambeth	Tower Hamlets
City of Westminster	Harrow	Lewisham	Waltham Forest
Croydon	Havering	Merton	Wandsworth

Wales

A	Flintshire	E	Vale of Glamorgan	I	Caerphilly
B	Wrexham	F	Rhondda, Cynon, Taff	J	Newport
C	Neath and Port Talbot	G	Cardiff	K	Torfaen
D	Bridgend	H	Merthyr Tydfil/Blaenau Gwent	L	Monmouthshire

Local government in Wales

The reorganisation of local government in Wales proceeded more quickly than in England. The review was carried out by the Welsh Office (rather than by an independent commission) and the country was considered as a whole (rather than by separate areas). After a two-year period of consultation, in 1993 a White Paper *Local Government in Wales: A Charter for the Future* was published, setting out detailed proposals. There was widespread agreement that the new structure should be unitary in character, and the debate was focused on the number and boundaries of the new local authorities.

The underlying thinking included a restoration of authorities which had been swept up in an earlier reorganisation (Cardiff, Swansea and Newport) and some of the traditional counties, such as Pembrokeshire and Anglesey. However, to fit into a unitary structure, the boundaries had to be stretched somewhat, and a number of counties had to be amalgamated. After consideration of proposals for thirteen, twenty and twenty-four unitary councils, the final outcome was twenty-two authorities. (Reflecting their history, these have varied formal names such as county borough, city and county council, but they all have the same functions.) They range in population from 66,000 in Cardiganshire to 318,000 in Cardiff.

In the White Paper, the unitary system was commended for its administrative simplicity, its roots in history, its familiarity and the relative ease with which residents could identify 'with their own communities and localities'. The intention was to create 'good local government which is close to the communities it serves'. The White Paper continued:

> Its aims are to establish authorities which, so far as possible, are based on that strong sense of community identity that is such an important feature of Welsh life; which are clearly accessible to local people; which can, by taking full advantage of the 'enabling' role of local government, operate in an efficient and responsive way; and which will work with each other, and with other agencies, to promote the well-being of those they serve.

These desirable objectives do not all work in the same direction, of course, and some compromise was inevitable. Some of the areas are very large. Powys, for example, has over 500,000 hectares: this is a very large area for local government. There is potential for the community councils to take on an increased role, but the Welsh Office has stressed that there is no intention of forming a second tier of local government.

Welsh local government and the Welsh Assembly

Preparations for devolution in Wales were far less advanced than they were in Scotland, where there

was a much firmer expectation that devolution would in fact take place. The first steps included the mounting of a consultation exercise on the establishment of a Partnership Council with local government. This Council was mandated by the Government of Wales Act. It consists of twenty-five members: ten from the Assembly, ten from the county and county boroughs, two from the community councils, and one each from the police authorities, the fire authorities, and the national park authorities (NAW, *The Partnership Council: Preparing the Ground*, 1999).

The Assembly has also produced a paper on the development of planning policy (*The Approach to Future Land Use Planning Policy*, 1999). Unlike the English and Scottish planning policy guidance publications, the Welsh published two guidance notes only. 'This enables the inter-relationships between policies to be clarified, and means that each revision has to be a full one across all policy topics, rather than piecemeal'. The guidance closely follows the English publications and reflects English research and policy development. The question of its Welsh distinctiveness has been debated, and the paper concludes:

> Planning policy for Wales should no longer track DETR priorities slavishly, nor should it diverge from GB policies unless this is for good reason. Both the process of developing planning policy, and its content, should be appropriate to Welsh circumstances, and be produced in a shorter time scale than hitherto.

It is stressed that in developing planning policies, there should be a partnership with local government, business and the voluntary sector. To facilitate this, a forum is to be established. This will include representatives from a wide range of organisations, and will have the remit 'to inform planning research and policy development. To ensure that Welsh needs are fully met, a Welsh Planning Research Programme is to be developed. A Research Scoping Study has been mounted to identify key research areas. These will include: speeding up the preparation of development plans; planning for rural areas; planning and integrated transport; improving local authority development control performance; waste planning; making planning more responsive to business; and locational policy for renewable energy (e.g. wind farms).

Local government in Northern Ireland

Local government in Northern Ireland was last reorganised in 1973, when thirty-eight authorities, made up of counties, county and municipal boroughs and urban districts, were replaced by a single tier of twenty-six district councils. Although this reduced the enormous variation in the size of districts (previously ranging from 2,000 to over 400,000) there is still a wide variation, from Moyle with a population of some 15,000 to Belfast City with a population approaching 300,000. Planning powers were centralised under the then Northern Ireland Ministry of Development. Since the demise of the power-sharing Northern Ireland Assembly in 1974, planning, like all public services, has been subject to 'direct rule' under the supervision of the Secretary of State for Northern Ireland. The preparation of plans and the control of development are functions of the Department of the Environment for Northern Ireland, which it exercises through the Planning Service (an executive agency). Local government is consulted only on the preparation of plans and development control matters.

The lack of accountability through local government (described as the 'democratic deficit') obviously needs to be seen in the light of the very special circumstances, though it has been judged to have operated with a 'considerable measure of success' (Hendry 1992: 84). Nevertheless, local councillors have been able to attack planning and to 'represent themselves as the champions of the local electorate against the imposed rule of central government' (Hendry 1989: 121). Even when the central bureaucracy has made determined efforts to open decision-making and involve local people, it has been accused of having ulterior motives (Blackman 1991a).

The promise of a 'lasting peace' in Northern Ireland during the ill-fated cease-fire brought with it ideas for reform which are still on the agenda. Several possible scenarios have been suggested, including the continuation of a central planning authority accountable to an elected Assembly, devolution to joint regional boards, and complete delegation of powers to the Districts (Royal Institution of Chartered Surveyors (RICS) 1994). The relatively weak position of local government over many years and the dearth of skills and experience will not be put right quickly. It is likely that any reform will be introduced incrementally.

Local strategic partnerships

The shift from local government as 'provider' to 'enabler' and the privatisation of much service provision since the 1980s have been accompanied by the dispersal of competences (responsibilities and powers) to many other agencies of government, voluntary bodies and the private sector. This is described as a shift from government to governance. With many actors involved, getting sensible coordination of policy and action is a considerable challenge. As explained in Chapter 10, 'partnership' or at least inter-organisational networking is the hallmark of much policy-making and implementation, especially in urban and rural regeneration. This has benefits, but the complexity of partnership working with large numbers of overlapping and often ad-hoc arrangements has become a problem. In recognition of this, the ODPM has promoted the notion of local strategic partnership (LSP) 'to provide a single overarching local coordination framework within which other, more specific local partnerships can operate'. A consultation document was issued in 2000,[32] noting the number of new partnership arrangements and promoting an additional strategic tier of partnership as the solution. This apparently contrary idea is in fact developing existing good practice, and is intended to lead to the reduction of other 'micro-partnerships' which will be subsumed under the new arrangements. One of the main objectives for government was to engage the mainstream programmes of other public sectors such as education, health, transport and crime prevention, together with the actions of community and private sectors in tackling urban regeneration, especially in areas of concentrated deprivation.

A local strategic partnership may cover any area and takes a lead role in the preparation of the community strategy, the powers of a local authority to promote or improve the economic, social or environmental well-being of their area (both provided by the Local Government Act 2000) and in coordinating neighbourhood renewal. LSPs are 'accredited' by the government offices. The creation of LSPs is mandatory for eighty-eight most deprived areas, and from 2002, spending under the Neighbourhood Renewal Fund is conditional on the local authority working with the LSPs. A national survey of local authorities found that most have set up LSPs.

The nature of the partnership has been left to the discretion of the local authority but must include the public, private, community and voluntary sectors. It must also cover all relevant sectors and levels of government. It is not surprising, therefore, that the partnerships are large. The Bristol LSP originally involved more than seventy organisations but proved unwieldy and frustrating in its processes for some partners, and was trimmed down to about twenty core partners for its second phase of work. Getting a representative membership that meets all demands is difficult. In his early review of LSPs, Bailey (2003) notes that complaints were made about only two community representatives in the Hackney LSP, while neighbouring Newham had included ten. The potential for variation was part of the original concept but highlights broader questions about representation and accountability – especially when statutory development plans are being asked to follow the work of the LSPs. His evaluation is charitable:

> It is too early to draw final conclusions about whether the development of LSPs represents a more advanced stage in the development of urban policy, whereby the previous *ad hoc* area-based initiatives are being required to work as a network with clearer organizational structure and greater strategic focus.
>
> (Bailey 2003: 456)

Moving in this direction will require fundamental changes in government and particularly central–local relations. Early findings from the national evaluation of LSPs are not hopeful. Lambert (drawing on Stewart 2002) concludes that 'the long standing silo culture of the UK government system is confirmed, driven by sector specific objectives and performance measures, and embedded in established policy and professional communities, legal frameworks and funding regimes' (Lambert 2004: 5). More is said about community strategies in Chapter 4.[33]

Managing planning at the local level

For town and country planning, the apparent and seemingly paradoxical outcome of change in the 1980s and 1990s has been a larger and stronger body of planners with strengthened statutory functions. The indirect effect of market deregulation, the increasing complexity of development issues, and the growing emphasis on environmental protection was bound to lead to a greater demand for planning skills (Healey 1989). The concept of an enabling local government also increased the need for strategic thinking and focused attention on the corporate planning function (Carter *et al.* 1991). The direct impact on the way in which the planning service is delivered has taken longer to come through. During the 1990s, planning was subject to only minimal change in comparison to other local authority services, and partly because of its statutory and regulatory functions was somewhat protected from the pressure for change. Compulsory competitive tendering (CCT) introduced by the Conservative government in the 1980s was not applied to planning.

Nevertheless, there has been strong pressure for change involving both sticks and carrots. The spread of auditing and value for money (VFM) has been given a new gloss by the Labour government with the concept of 'Best Value'.[34] These are not easy concepts to define for planning because of the difficulty of assessing quality in plans and planning decisions. The Audit Commission provides guidance for local authorities and district auditors on performance indicators for all services, including planning, but these have been criticised for the reliance on quantitative measures, the classic example being the proportion of applications decided within eight weeks.

Best Value requires that performance reviews are expected to look ahead over a five-year period, starting with areas of work where there are problems. The reviews must challenge why and how the service is being provided; invite comparison with others' performance across a range of relevant indicators; involve consultation with local taxpayers, service users and the wider business community in the setting of new performance targets; and embrace fair competition as a means of securing efficient and effective services. The reviews will produce new performance targets to be published in an annual local performance plan together with comparisons with other authorities (note the District Audit Service are to stop producing the annual local authority reports based on CIPFA (Chartered Institute of Public Finance and Accountancy) statistics); identification of forward targets for all services annually and in the longer term (at least five years); and commentary on how the targets will be achieved including proposed changes to procedures. Local performance plans are audited.

The Audit Commission's 1992 report *Building in Quality* addressed criticisms of the accent on efficiency rather than effectiveness of the planning system in performance review. It made a real attempt to introduce a wider assessment, recognising that there were many ancillary tasks in providing advice and negotiating with applicants and making 'complex professional and political judgments'. After consultation, the Audit Commission settled on six key 'best value performance indicators' (or BVPIs as they are inevitably called). As with the earlier version, these concentrate on matters of efficiency rather than on the effectiveness of the system, though the added breadth of performance review will be a significant improvement on previous practice. Because it is easier to measure the throughput of applications rather than the achievement of strategic objectives the indicators are mostly concerned with the development control function, and are discussed further in Chapter 5.

The 1998 White Paper *Modern Local Government: In Touch with the People* succinctly describes the duty of local authorities 'to deliver services to clear standards – covering both cost and quality – by the most effective, economic and efficient means available'. This is essentially a positive recasting of the enabling concept. (And is the norm in western Europe, with local government implementing its functions through a diverse range of agencies, often in partnership, but essentially seeing its role as prioritising community needs and acting as the focus for local political activity.) Best Value is also seen as an aid to local government 'to address the cross-cutting issues facing their citizens and communities, such as community safety or sustainable development, which are beyond the reach of a single service or service provider'. The very best performing councils are eligible for Beacon status normally for particular services. Applicant councils are chosen by an independent advisory panel, and are rewarded by being given wider discretion in the operation of the beacon service.

A new statutory duty has been placed on local authorities to promote the economic, social and environmental well-being of their areas. This is in line with the European Charter of Local Self-Government, which provides that local authorities can do anything to further the interest of their electors, unless prevented by statute. (The previous Conservative government refused to sign this charter, preferring to keep local authorities under central control: Jenkins 1995: 254–8.)

These are some of the elements of the local government modernisation programme of the Blair government. Others include the major recasting of the political structures of local authorities, with cabinet-style executives in place of the traditional committee system (White Paper, *Local Leadership, Local Choice*, 1999).

The ethical local authority

Local government has for long had a reputation for probity, particularly in planning, where foreign observers are quick to point out the obvious opportunities (nay, temptations) for corruption. That the temptations have not always been resisted is now well known. It was in the 1970s that the Poulson Scandal blew up: several local authority politicians and officials were found guilty of securing contracts for the architectural business of John Poulson. A number of well-known figures went to jail. It was an extreme case which shocked the local government world. It led to the setting up of a Royal Commission on Standards of Conduct in Public Life and to the introduction of a National Code of Local Government Conduct. There have been other cases of local government impropriety particularly in the planning arena, of which the most recent were in Warwick, Bassetlaw (Nottinghamshire) and Newark.[35] There is usually some discussion in the planning press about the misdoings of councillors and officials, as for example in Doncaster.[36]

Though only a small number of such cases have arisen, they are clearly unacceptable. In fact, the numbers involved were certainly fewer than the number of cases of Westminster 'sleaze' (a conveniently vague and all-embracing term) in the later years of the Conservative government.[37] This led to the appointment of a Committee on Standards in Public Life, under the chairmanship of Lord Nolan, which embarked upon a series of inquiries into various areas of public life. Its third report, devoted to local government, was published in 1997.[38]

The Nolan inquiry was concerned not to put local government on trial but to provide guidance on what standards of conduct should apply and how they could be maintained. The National Code was criticised for being inadequate, complicated and, in parts, inconsistent and even impenetrable. Moreover, in the words of Standards of Conduct in Local Government (para. 56), it represented 'something that is done to local authorities, rather than done with them'. Building upon the report, the government proposed a 'new ethical framework' to govern the conduct of elected members and also local government employees (who were not covered by the National Code). A Code of Conduct, based on a national model, is required of all local authorities, together with a Standards Committee to oversee ethical issues and to provide guidance on the code and its implementation. An independent Standards Board will have the responsibility of investigating alleged breaches of the local authority code.[39]

Planning was seen to require additional measures. Because of its complexity and the problems of dealing fairly and properly with planning law and its implementation, it was proposed that members of planning committees should be trained in the planning system.[40] There should also be a greater degree of openness in the planning process; this would, among other things, assist in dealing with the problems facing local authorities in granting permission for their own proposed developments, and 'the potential for planning permission being bought and sold'.

In coming to these conclusions, the Nolan Report noted that in 1947 'the need for postwar reconstruction was clear. Development enjoyed broad public support'. Things have now changed.

> Development is now a term which has a pejorative ring, and the planning system is seen by many people as a way of preventing major changes to cherished townscapes and landscapes. If the system does not achieve this (and it is a role which it was not originally designed to perform), then the result can be public disillusionment.
>
> (*Standards of Conduct in Local Government*, para. 277)

In Scotland, a 1998 consultation paper on the Nolan Report, *A New Ethical Framework for Local Government in Scotland*, broadly accepted its recommendations, but took issue with a number of them. It proposed a single code for all local governments (instead of a model code), and it favoured a national Standards Commission instead of local authority standards committees. It also argued that reasons should not be required for the granting of planning permission since such decisions are not subject to any appeal process, and it would not only add to the difficulties facing a planning committee but also put permissions at increased risk of legal challenge on purely technical grounds.

Further reading

European government

The institutions and policies of the EU are summarised in a series of free booklets, Europe on the Move, which are updated periodically (available from the UK Office of the European Commission at Jean Monnet House, 8 Storey's Gate, London, SW1) and a long list of free publications is available at www.cec.org.uk/sourcedi/catalog1.htm. For a summary of the history of the EU see Borchardt (1995) *European Integration: The Origins and Growth of the European Community*.

There are a great many general accounts of the making of the European Union, including the very comprehensive *Encyclopedia of the European Union* edited by Dinan (1998). For a more critical account see Chisholm (1995) *Britain on the Edge*; for a more theoretical view of European integration, see Nelsen and Stubb (2003) *The European Union* and Emerson (1998) *Redrawing the Map of Europe*. There are also a great number of texts on EU law and institutions, many of which are no more than reprints of official texts. Tillotson (2000) *European Union Law* and Craig and de Búrca (1999) *EU Law* are useful in their own right; these are also updated regularly and provide brief summaries of the history of the EU. The history of each policy area in which the Community has acted, including regional policy, environment and transport, is given in Moussis (1999) *Access to European Union: Law, Economics, Policies*. Williams (1996) *European Union Spatial Policy and Planning* gives an account of the European institutions from a planning perspective.

A chronological review of how Europe has influenced planning is given in Nadin (1999) 'British planning in its European context'. Two DETR research reports address the consequences of the development of European policies for planning in the UK: Nadin and Shaw (1999) *Subsidiarity and Proportionality in Spatial Planning Activities in the European Union* and Wilkinson *et al.* (1998) *The Impact of the EU on the UK Planning System*. See also Bishop *et al.* (2000) 'From spatial to local: the impact of the European Union on local authority planning in the UK' and Shaw *et al.* (2000) *Regional Planning and Development in Europe*. Further references on European spatial planning are given at the end of Chapter 4.

On European comparative planning systems, including the organisation of government, see the *EU Compendium of Spatial Planning Systems and Policies*, which comprises individual volumes describing the systems and policies of spatial planning in each member state, and Seaton and Nadin (2000) *A Comparison of Environmental Planning Systems Legislation in Selected Countries*.

Central government

An invaluable overview, though now dated, is given by Dynes and Walker (1995) *The Times Guide to the New British State: The Government Machine in the 1990s*. A more up-to-date account of *The British System of Government* is the deservedly popular account by Birch (1998). Jones and Kavanagh (1998) provide a good introduction to *British Politics Today*. On Northern Ireland see Connolly and Loughlin (1990) *Public Policy in Northern Ireland: Adoption or Adaptation?*

An up-to-date summary description of government departments and their functions is given in the annual Official Handbook *Britain*, prepared by the Central Office of Information. Greater detail is given in the annual *Whitaker's Almanack*. A principal source of information on the work of government departments is the *Departmental Annual Reports*. These are the *Government's Expenditure Plans* for the forthcoming three years and are sometimes referenced in this way. They are now available on the Internet.

An excellent insight into the operation of central government in exercising its controls over local government is given in Read and Wood (1994) 'Policy, law and practice'.

Devolved and regional government

On devolution the essential book is Bogdanor (1999) *Devolution in the United Kingdom*, which has an extensive bibliography. See also the excellent set of essays edited by Hazell (1999) *Constitutional Futures: A History of the Next Ten Years*, Connal and Scott (1999) 'The New Scottish Parliament: what will its impact be?', McCarthy and Newlands (1999) *Governing Scotland: Problems and Prospects – The Economic Impact of the Scottish Parliament* and Bosworth and Shellens (1999) 'How the Welsh Assembly will affect planning'. The government's position is given in *Your Region, Your Choice*, which has useful general information on the regions in many annexes. The two White Papers were *Scotland's Parliament* (Cm 3658) and *A Voice for Wales* (Cm 3718).

Mawson (1996) reviews 'The re-emergence of the regional agenda in the English regions: new patterns of urban and regional governance', while a review of the history of the two strands of regional planning (inter-regional economic and intra-regional land use) and the move towards a more integrated and comprehensive approach is well analysed in Roberts and Lloyd (1999) 'Institutional aspects of regional planning, management, and development: models and lessons from the English experience'. Roberts *et al.* (1999) contains an extensive discussion of *Metropolitan Planning in Britain*, with case studies of nine British metropolitan regions.

The progress of the RDAs and their strategies is being monitored by various academic centres. See, for example, Roberts and Lloyd (1998) *Developing Regional Potential*, Nathan *et al.* (1999) *Strategies for Success?* and Deas and Ward (2000) 'The song has ended but the melody lingers'. Charter 88's publications include Tomaney and Mitchell (1999) *Empowering the English Regions*. Peter Hall's essay on 'The regional dimension' (1999a) gives an overview of postwar regional economic policy, with a short comment on regional land use planning. Bradbury and Mawson (1997) *British Regionalism and Devolution: The Challenges of State Reform and European Integration* provide an informative analysis of the emergence of regionalism from a number of perspectives. Wannop (1995) *The Regional Imperative: Regional Planning and Governance in Britain, Europe and the United States* is an excellent account by a knowledgeable practitioner of the endeavours to plan on a regional scale, primarily in the UK, but also in Europe and the USA.

Local government

The principal textbooks which give a general introduction to local government structure and organisation are Chandler (1996) *Local Government Today*, and the third edition of Wilson and Game (2002) *Local Government in the United Kingdom*. For an overview of the politics of local government including the roles and relationships between councillors,

officers and political parties see Stoker (1991) *The Politics of Local Government*. On changing management approaches, see Stoker's volume of essays on *The New Management of British Local Governance* (1999); Stoker and Wilson (eds) (2004) *British Local Government into the 21st Century*; and Stewart (2003) *Modernising Local Government*. The government's agenda for local government is set out in ODPM (2004) *The Future of Local Government: Developing a 10 Year Vision*. For examination of the relationship between planning and changing government see Vigar *et al.* (2000) *Planning, Governance and Spatial Strategy in Britain*.

European comparisons are given in Hirsch (1994) *A Positive Role for Local Government: Lessons for Britain from Other Countries*. There are several reports of foreign experience and practice in local government prepared for the Commission on Local Government and the Scottish Parliament: Hughes *et al.* (1998) *The Constitutional Status of Local Government in Other Countries*, Hambleton (1998) *Local Government Political Management Arrangements: An International Perspective*, University of Edinburgh (1999) *Summary of Devolved Parliaments in the European Union* and Centre for Scottish Public Policy (1999) *Parliamentary Practices in Devolved Parliaments*.

On Scotland and Wales, see Boyne *et al.* (1995) *Local Government Reform: A Review of the Process in Scotland and Wales*; Midwinter (1995) *Local Government in Scotland*. On devolution to Scotland and Wales, Bogdanor (1999) is essential reading. On Northern Ireland, see Bannon *et al.* (1989) *Planning: The Irish Experience 1920–1988*, Hendry (1992) 'Plans and planning policy for Belfast', and a short article by Lipman (1999) 'Difficult decisions in a rural balancing act'.

Parish councils are the subject of a survey by the Public Sector Management Research Centre (1992) *Parish Councils in England*. A Welsh Office consultation paper was issued in 1992: *The Role of Community and Town Councils in Wales*. A particularly interesting document is the Scottish Office (1999) *Report of the Community Planning Working Group*.

The rate of change in local government under the Blair government requires a perusal of the relevant journals, such as *Local Government Chronicle* and *Municipal Review*. See also Hambleton (2000) 'Modernising political management in local government'. There has been a large number of official publications discussing and proposing changes in the operation of local government. Of particular importance are *Modern Local Government: In Touch with the People* (Cm 4014, 1998), *Local Leadership, Local Choice* (Cm 4298, 1999), *A Mayor and Assembly for London* (Cm 3897, 1998). For references on 'the ethical local authority' see Chapter 12.

Notes

1 The countries of the EU are (with the states which joined in 2004 marked with an asterisk): Austria, Belgium, *Cyprus, *Czech Republic, Denmark, *Estonia, Finland, France, Germany, Greece, *Hungary, Republic of Ireland, Italy, *Latvia, *Lithuania, Luxembourg, *Malta, The Netherlands, *Poland, Portugal, *Slovakia, *Slovenia, Spain, Sweden and the United Kingdom. Bulgaria and Romania are expected to join in 2007. Negotiations have been started with Turkey. Agreement was reached for Norway to join in 1972 and 1995, but on both occasions membership was rejected by a referendum; however, the Norwegian government has enacted legislation that requires the country to meet all EU law. Switzerland has also applied for accession, which was accepted by the EU but rejected in a referendum in 1992. Iceland is the only significant western European country which has not sought accession to the EU.

2 These figures are taken from Commission of the European Communities (CEC) (2004) *A New Partnership for Cohesion: Convergence, Competitiveness, Cooperation: Third Report on Economic and Social Cohesion*, Luxembourg: OOPEC; and CEC (2003) *Regions: Statistical Yearbook 2003*, Luxembourg: OOPEC. Statistics on the European Union are at http://europa.eu.int/comm/eurostat/.

3 The Committee on Spatial Development was not a formal committee of the Community, but was an informal intergovernmental arrangement.

4 The European Treaties lay the foundation for economic and political integration. The first was the European Coal and Steel Community (ECSC) of 1951, which was followed in 1957 by the European Atomic Energy Community (Euratom) and the Treaty establishing the European Economic Community (later to become the Treaty establishing the European Community: TEC), both signed in Rome. The latter is generally referred to as the 'Treaty of Rome', and set objectives for the creation of the Economic Community and established the basic institutions which would achieve it. The treaties have subsequently been amended by the Single European Act (SEA) 1987 which firmed the commitment to creating the Single Market; the Treaty on European Union (TEU) of 1992 known as the Maastricht Treaty, which considerably widened the areas of cooperation of the Union into foreign affairs, defence and justice; and the Amsterdam Treaty of 1997 (although not coming into force until 1999) which prepared for enlargement and made sustainable development an objective of the Union. During the process of preparing the Amsterdam treaties it was decided to consolidate the treaties, but in the event this was limited to a tidying up of the numbers of articles in the TEU and the TEC. The addition of numerous protocols (which have legal force) and declarations has increased the complexity and navigating the treaties is very much a job for experts.

5 Alistair Darling, HC Debates, 11 June 1997, col. 1144, quoted in James (1999: 194–5). See the 1998 White Paper, *Modern Public Services in Britain: Investing in Reform*, Cm 4011.

6 Jill Sherman (1999) 'Whitehall faces major revamp', *The Times* 12 July.

7 Michael White (1999) 'Whitehall warfare', *Guardian* 12 August.

8 This quotation and other examples here are taken from the ODPM Annual Report which summarises (but at length) the 'achievements' of the department. The relevant select committee scrutinises the annual departmental reports and their reports can provide useful complementary views. See also the Department's Business Plan available on the ODPM website.

9 Curiously, in 1999, the ministry made the first ever use of its powers under section 43(6) of the Planning

Act to veto the allocation of high grade farmland in East Yorkshire for development. This was overruled by the Secretary of State for the Environment. See *Planning* 26 February 1999, p. 1.

10 All figures refer to England only and are taken from the Planning Inspectorate (2000) *Statistical Report 1999–2000*, Bristol, PIEA and the Planning Inspectorate's Annual Reports and Accounts 2003–4 and 2002–3. The Inspectorate recovers full costs for development plan inquiries.

11 See the memoranda of the Environment Sub-committee on the Planning Inspectorate and Public Inquiries (2000) and the memorandum of evidence by Professor Malcolm Grant to the Select Committee Inquiry DETR (2000).

12   The Department undertakes the formidable task of monitoring all draft plans. In England this is done by the Regional Offices of the Department. For many years the Department has maintained a Planning Handbook which gives internal guidance to decision-making officers on procedural questions. This guides officers in monitoring development plans. The Handbook is now computerised and can be instantly updated. This is the primary tool for the coordination of Departmental action. It is not a public document. Officers are advised to check the wording of policies and proposals against current planning guidance. Realism is looked for, in particular whether sites said to be available can actually be developed. A most difficult task is to identify possible strategic implications of local policies. The Department also encourages informal approaches to regional offices by development plan teams in the course of plan preparation.

(Read and Wood 1994: 10)

13 See Chapter 2.

14 The Welsh Office has been renamed Wales Office, but the term is applicable only in Whitehall. In Wales, the reference is to the Assembly for Wales (Welsh Affairs Committee, *The Role of the Secretary of State for Wales*, 26 October 1999, HC 854, para. 36).

15 The decision to grant legislative powers to Scotland, but only executive to Wales 'had as much to do with political compromise and accident as with any rational argument' (Osmond 1977: 149).

16 Planning policy guidance in various forms have been issued by the Welsh Office for some years. See particularly *Unitary Development Plans* (1966) and *Planning Policy* (first revision 1999).

17 See Fourth Standing Committee on Delegated Legislation, 3 March 1999, *Draft Strategic Planning (Northern Ireland) Order 1999*.

18 See *Regional Chambers* (DETR 1999).

19 PSAs include regeneration of communities, tackling housing need and tackling countryside issues. Examples are provided in the RCO annual reports.

20 The indicators can be viewed and compared on the ROI website: http://intrago.go.regions.gsi.gov.uk/roi/website.

21 An encouraging insight is given by Kitchen (1999a).

22 In the inelegant language of the Act (section 4.2), 'A regional development agency's purposes apply as much in relation to the rural parts of its area as in relation to the non-rural parts of its area'.

23 The rural regeneration functions of the Rural Development Commission have been transferred to the RDAs. (Other functions have been merged in the new Countryside Agency.) An interesting history of the RDC is Rogers (1999).

24 HC Environment, Transport and Regional Affairs Committee, Environment Sub-Committee (1999) HC 232 (Session 1998–99), 3 vols. See e.g. memorandum by the TCPA (vol. 2, p. 45) and memorandum by DETR (vol. 2, p. 71).

25 How far these developments are to be judged as 'clusters' (or even one 'cluster') is problematic. There has been strong governmental backing for clusters, in the White Paper, *Our Competitive Future: Building the Knowledge Driven Economy* (Cm 4176, 1998) – generally referred to as 'the Competitiveness White Paper' – and the Sainsbury Report, *Biotechnology Clusters* (DTI 1999). There has been a surprisingly general enthusiasm for clusters, despite the vagueness of the concept. (Note their popularity in the RDA regional strategies.) For a critical view, see Perry (1999). For an informative

short account of the Cambridge proposals and the DETR decisions see Green (1999: 12).

26 Government Office for the South East (1999) *Regional Planning Guidance for the South East of England: Public Examination May–June 1999*, *Report of the Panel*, paras 4.8–4.10.

27 A good case in point is a Surrey plan of 1993 in which overriding importance was given to environmental and infrastructure policies. The Report of the Panel on the Examination in Public held that the plan did not 'adequately provide for industry and commerce' and concluded that the employment policies in the plan needed to be changed 'so as to be more responsive to the needs of business'. A similar stance was taken in relation to Hampshire, where policies were designed to control the rate of growth in order to safeguard the environment, character and heritage of the county. The Secretary of State backed the view of the Panel that Hampshire could not be regarded as an economic island. In strong terms he denounced the proposed policies as 'a recipe for economic decline' (Read and Wood 1994: 26).

28 For more information on parish and community councils see www.nalc.gov.uk.

29 The Hunt Report (1969) is the *Report of the Lords Select Committee on Relations between Central and Local Government*. The European Charter of Self-Government is reproduced in an appendix to the McIntosh Report (1999).

30 The Executive's response was issued in October 1999: *The Scottish Executive's Response to the Report of the Commission on Local Government and the Scottish Parliament* (www.scotland.gov.uk/library2/doc04). It is also noteworthy that a review of public transport includes a proposal for a Scottish Transport body (which is discussed in Chapter 11).

31 SO (1998) *A New Ethical Framework for Local Government in Scotland: Consultation Paper*. This stems from the Nolan Report, *Standards of Conduct in Local Government* (Cm 3702, 1997).

32 *Local Strategic Partnerships: Consultation Document* (DETR 2000). See also Carley *et al.* (2000b).

33 A national evaluation of local strategic partnerships is being undertaken for ODPM by a consortium of universities and consultants. Interim results were published in August 2005 and are available on the ODPM website: www.odpm.gov.uk.

34 The Local Government Act 1999 defines best value as 'securing continuous improvement in the exercise of all functions undertaken by the local authority, whether statutory or not, having regard to a combination of economy, efficiency and effectiveness'. The 'extremely uneven progress' in introducing Best Value into planning is reported by Thomas (1999). On the application of best value to parish councils, see DETR (1999) *The Application of Best Value to Town and Parish Councils*. See also Warwick Business School (1999).

35 See External Enquiry into Issues of Concern about the Administration of the Planning System in Warwick District Council (1994) and Report of an Independent Inquiry into Certain Planning Issues in Bassetlaw (1996). On the Newark case see *Planning* 29 October 1999: 2, and 5 November 1999: 15. Note also the North Cornwall case referred to on page 51.

36 See *Planning* 12 November 1999: 1.

37 Allen has commented that 'central agencies are often at least as incompetent, inefficient or corrupt as local bodies; local authorities are perennially in the news for alleged corruption and graft; one or two notorious cases can suffice to keep the whole concept of local government in disrepute' Allen (1990: 12). For a catalogue of cases of corruption in public administration, see Doig (1984).

38 The first report was on members of parliament, ministers and civil servants, and executive non-departmental public bodies (Cm 2850, 1995). This was followed by a report on further and higher education bodies, grant-maintained schools, training and enterprise councils, and housing associations (Cm 2170, 1996). The report on local government was published in 1997 (Cm 3702).

39 *Modern Local Government: In Touch with the People* (Cm 4014, 1998: Chapter 6), *Modernising Local Government: A New Ethical Framework* (1998) and *Local Leadership, Local Choice* (Cm 4298, 1999: Chapter 4).

40 The DETR has published a training syllabus prepared in conjunction with the LGA and the RTPI: *Training in Planning for Councillors* (DETR 1999).

4 The framework of plans

We want a system that is capable of reaching decisions that command public confidence and which is seen to be open and fair. A system that underpins our desire to improve productivity by being capable of reaching a proper balance between our desire for economic development and for thriving communities . . . The proposals in this consultation document are intended to help us produce such a system. It is time for fundamental change.

Planning Green Paper 2001: 2

The proposals show little sense of the purpose of planning, and make no contribution to the biggest and most urgent challenge facing the regulation and management of development: to deliver step change not just incremental improvement in climate change impacts.

Levett and Therivel 2002: 7

Introduction

This and the subsequent chapter discuss the two principal components of the formal system of town and country planning: the framework of plans (or policy instruments) and the development control process (or system of development regulation). In discussing the framework of plans, we are mostly concerned with the form and scope of policy instruments and the procedures by which they are created rather than their policy content, which is explained in later chapters (though these topics are not unrelated). The framework of plans is established by a huge library of statutes, rules, regulations, directions, policy statements, circulars, guidance and other official documents. However, it is important to appreciate at the outset that the formal system is one thing; the way in which matters work in practice may be very different. The informal planning system operates within the formal structure. It may continue with little modification even when major legislative changes are made; alternatively, there

may be significant changes in practice within a stable formal system. Political forces, professional attitudes and management styles will all affect the ways in which the system operates in practice.

It is also necessary to note that much development (in the everyday, rather than the legal, sense of that word) takes place without any help or hindrance from the planning system. Even where the development is clearly related to some action within the statutory framework for planning, the actual outcome is affected by 'extraneous' factors, and it may not be at all clear what effect planning has had on outcomes.

It is government policy to ensure that planning decisions are made with reference to an explicit and widely agreed framework of policies at national, regional and local levels, set out in plans and other policy instruments. This is described as plan-led development control. Reference to policy in plans reduces the amount of ad-hoc decision-making and the need for resolving conflicts around individual development proposals. Plans may help to improve the efficiency of decision-making and conflict mediation than decision-making on a project-by-project basis. Explicit policy statements can help to ensure accountability as the decision-makers are making their 'decision rules' or criteria explicit. Policy also provides a measure of certainty and coordination for the promotion of investment (Healey 1990).

A plan-led system requires a comprehensive and up-to-date hierarchy of national policy, regional strategies and local development plans. Where this is available for particular places and topics, it is a very important factor in decision-making, but it is always going to be difficult to establish and maintain a comprehensive policy framework. Rapid social and technological change can outpace planning policy and this has been an issue for national guidance on topics such as telecommunications and retail development. Until recently, the capacity and evidence base for formulating planning policy at the regional level has been limited, and at the local level there has been considerable variation in the performance of local planning authorities in preparing development plans and keeping them up to date. When there is a need to respond to change in planning policy, it can take considerable time to resolve conflicts. And even when a comprehensive policy framework exists there will always be important decisions that are not well informed by policy, or where policy has to be put to one side because there are other important material considerations to take into account.

As we shall see, there has been a constant flow of adjustments to the system of plan and policy-making since the mid 1960s, in order to address these problems and to try to make it more relevant and responsive to demands of the time. There have been periods of systematic reflection too, probing deeper into the operation of the system and leading to more extensive change, especially in 1968, 1991 and 2004, but the fundamental characteristics of the system have remained much the same. Each review of the framework has tackled similar questions.

- What framework of plans will ensure the accountability of decision-makers and safeguard the interests of those affected by planning, yet be expeditious and efficient in operation?

- How can the framework provide a measure of certainty and commitment, yet allow for flexibility to cope with changing circumstances, local conditions and new opportunities?
- What objectives should plans pursue, and how will these shape their form and content?
- Who should have influence in the planning process, and what should be the respective roles of central and local government and of local communities?

These perplexing questions have no easy answers which explains why the system is under almost constant appraisal and review. Acceptable answers rarely have stability since conditions and attitudes change over time. The biggest changes to the framework of plans in recent years have been the strengthening of the system at the regional level through the creation of regional spatial strategies, the reorganisation of plans at the local level into the local development frameworks, and the promotion of a broader scope for plans, or the 'spatial planning approach'. Figure 4.1 gives an overview of the various instruments used to express planning policy in the UK. These are the main instruments only. The figure notes the new instruments in England and Wales which are progressively replacing structure plans, local plans and unitary development plans. Changes are proposed also for Scotland. The discussion of this framework begins with the system at the EU level and works down through the national, regional and local levels.

SUPRANATIONAL PLANNING

The European Spatial Development Perspective, you might say, is a rather esoteric subject. Even those claiming the title of professional planner may not know too much about it. The same probably could have been said of the Treaty of Rome when it was signed by six European countries in 1957, an event which apparently went almost unreported in the British media. Ignorance can sometimes have serious consequences. In fact the ESDP is likely to have profound consequences for the lives of the

European Union

European Spatial Development Perspective
CoE Principles for Sustainable Development

Transnational spatial visions
NorVision (North Sea); Atlantic Area Spatial Strategy; Spatial Vision for North-West Europe

England	Scotland	Northern Ireland	Wales
Planning policy statements and guidance notes	Scottish planning policy Planning advice notes	Planning policy statements Development control advice notes	Planning Policy Wales Technical advice notes
Regional Spatial Strategies / Sub-regional strategies	National Planning Framework for Scotland	Regional Development Strategy for Northern Ireland 2025	Spatial Plan for Wales
Local development frameworks / Core strategy / Land allocations / Area plans	Structure plans	Area plans	Local development plans
	Local plans	Local plans	
		Subject plans	

Rows (left margin): **European** / **National/regional** / **Local**

Note: The local development framework in England and the local development plans in Wales are replacing structure plans, local plans and unitary development plans.

Figure 4.1 Overview of planning policy instruments in the UK

300 million people of the European Union and the many others soon to join it. Though they do not know it, it is already shaping the process of land-use planning at every level, from national to local. In the future, its influence can only grow.

Professor Sir Peter Hall in the Foreword to Faludi and Waterhout (2002)

The rationale for planning at the European scale

The EU is driven by the goals of economic competitiveness, social cohesion and sustainable development. These objectives have an obvious spatial dimension. The proposed Constitutional Treaty will also introduce the objective of 'territorial cohesion' if it is adopted by the member states and make the 'spatial dimension' of EU goals more explicit. The main obstacles to meeting these objectives are the great disparities in wealth, jobs, investment and access to services across Europe. Indeed, evidence suggests that, despite the actions of the EU, reducing disparities in the face of economic forces which tend to concentrate wealth and development, is a very difficult task (CEC 2004; OECD 1995).

The growing economic and social integration of European nations and regions in the context of globalisation is having a profound affect on spatial development patterns. Significant elements of economic activity together with political and cultural relations are effectively globalised and become independent of nation-states. The locational decisions of firms, and to some extent citizens, are now more likely to ignore regional and national boundaries. The extent and depth of globalisation is disputed but it is widely accepted that it has specific implications for changing patterns of spatial development. Of particular note in the European context are increased spatial concentration of economic activity and the central role of global and regional cities, intensified competition between cities across national boundaries, the depopulation of some rural and urban areas, the corresponding polarisation of economic prosperity, and the negative environmental consequences that result (Sassen 1995). Development in one country may have significant impacts in other countries, for example, flooding in the Netherlands may be affected by development in the same river catchment in Germany; prospective house buyers on one country may seek housing in another because of availability and affordability considerations. These effects are reinforced by Community policies especially in the fields of regional policy, transport, environment and agriculture, although their implications for spatial development are not always explicitly considered in the policy-making process. Spatial planning and state regulation in other spheres play a significant role in addressing these trends, by maximising the competitive position and growth potential of major urban areas while attempting to ensure that, at best, patterns of growth are sustainable and, at worst, the negative impacts are ameliorated (Dieleman and Hamnett 1994).

EU cohesion policy and regional policy

The EU pursues a 'cohesion policy' (or regional policy) through allocation of Structural Funds. The funding is to promote 'the harmonious, balanced and sustainable development of economic activities, and in particular the development of competitiveness and economic innovation'.[1] While not a planning instrument, cohesion policy is intended to have an effect on the distribution of development and investment. Allocation of the structural funds from 2000 is divided according to three 'objectives', the first two of which have a strong spatial dimension. Objective 1 is to assist designated regions lagging behind in development,[2] with less than 75 per cent of the Community average). Objective 2 is targeted at the economic and social conversion of designated areas facing structural difficulties. Objective 3 is to support the adaptation and modernisation of policies and systems of education, training and employment, and is available to all regions not designated as Objective 1. The Funds account for more than one-third of the total Community budget, and will amount to €195 billion (about £117 billion at 2004 exchange rates) in the programming period 2000–6; the UK will receive about €10 billion.[3]

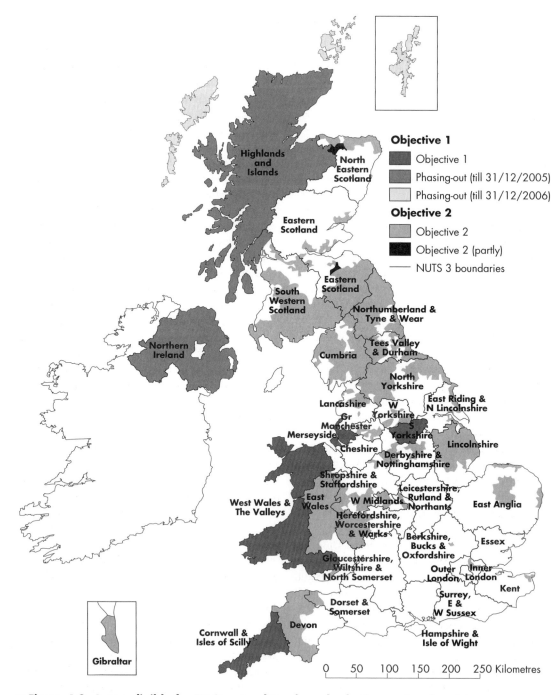

Figure 4.2 Areas eligible for EU Structural Funds and selective regional assistance

Source: European Commission Directorate General for Regional Policy

About 70 per cent of the structural funds go to Objective 1 areas, with 12 per cent to Objective 2 and 12 per cent to Objective 3; most of the rest are held in reserve or are allocated through Community initiatives. Thus some 82 per cent of Structural Funds are targeted on specific regions; this is perhaps not surprising since the main intention is to produce a better economic and social balance across the community. The areas of the UK which fall under these objectives are identified in Figure 4.2. The main beneficiaries in the UK are the Objective 1 regions, which are now Cornwall and the Isles of Scilly, Merseyside, South Yorkshire, and West Wales and the Valleys. The designations were effective for the seven years between 2000 and 2006.[4]

EU funding is also made available through Community initiatives which are used to tackle specific problems with a European dimension. In the 2000 to 2006 programme, there are four: INTERREG on transnational planning (€5 billion, €362 million for the UK), URBAN on urban regeneration (€700 million, €117 million for the UK), LEADER on rural development (€2 billion, €106 million for the UK) and EQUAL on employment and training (€2.8 billion, €376 million for the UK). The planning system has been involved in the implementation of all, but particularly INTERREG and URBAN.

For the programming period 2007–13 the European Commission has proposed a substantial revision to the way the funds are allocated. The new and old approaches are summarised in Table 4.1.[5] The proposed total allocation is €336.1 billion. The general thrust of the changes is simplification and a stronger focus on what are known as the Lisbon and Gothenburg objectives, that is for the EU to create 'the most successful and competitive knowledge based economy in the world' while 'protecting the environment and achieving more sustainable patterns of development'. The simplification entails nine separate objectives (or categories of funding support) to be reduced to three objectives: convergence, competitiveness and territorial cooperation. The Community Initiatives that have funded most spatial planning and urban activities (INTERREG and URBAN) will be 'fully integrated' into the new (mainstream) objective of territorial cooperation. The priorities go well beyond spatial

planning and address research and development (R&D), environmental measures, risk prevention and integrated water management, and access to transport and telecommunications. Cooperation will continue at three scales: cross-border (as between Northern Ireland and the Republic of Ireland), within large transnational zones (such as North West Europe or the North Sea) and within networks of regions, cities or other bodies addressing common issues across the whole of the EU. More attention will be given to urban areas (and areas with natural handicaps, e.g. mountainous zones) especially urban regeneration in medium sized towns (drawing on experience with URBAN). Member states will select regions for support under the competitiveness objective, thus ending the 'micro-zoning' of areas for Objective 2 support. Further devolution of implementation is planned; member states will have exclusive responsibility for managing interventions on the basis of a national strategic reference framework.

Irrespective of these changes, the accession of fifteen new member states in 2004 and the promise of two more before 2007 (see Chapter 3) means a quite different distribution of funding in the future. More resources will be concentrated in the poorer countries of central and eastern Europe, although intense political bargaining has ensured that considerable spending will continue in the west.

The UK government also has its own measures to promote the development of economically disadvantaged regions, although the level of funding has fallen considerably since the 1960s. (Note that these measures are known as regional policy and are to support the economic development of the regions. It should not be confused with regional planning.) Until 2004, regional selective assistance (RSA) was available across the UK from the Department of Trade and Industry and devolved administrations. The DTI has now devolved RSA in England for all but the largest projects, to the regional development agencies and it is now known as Selective Finance for Investment in England (SFI). The schemes (or products as they are known to the DTI) support projects to create or safeguard jobs and increase regional competitiveness, and most recently to improve a business's productivity. There were three tiers with different rates of assistance.

Table 4.1 EU cohesion policy 2000–6 and 2007–13

2000–6 €213 billion (15 member states) plus €44 billion for new members			2007–13 €336 billion (25 member states)		
Objective	Criterion	Funding (billion)	Objective	Criterion	Funding (billion)
Cohesion Fund	Member states with per capita GDP <90% of EU15 average	€19.7	Convergence	Member states with per capita GDP <90% of EU 25 average	€63 24%
Structural funds	Objective 1 Regions whose development is lagging behind	Regions with pc GDP<75% EU-25 average*	€151.0		Regions with pc GDP<75% EU-25 average*	€200 76%
	Objective 2 Areas facing economic restructuring	Micro-definition of regions	€22.2	Regional competitiveness and employment	Member states to propose regions**	€57.9 17.2%
	Objective 3 Modernising training and promoting employment	All regions not designated as Obj.1	€24.2			
Community initiatives	INTERREG URBAN EQUAL LEADER	Competitive application for projects	€5.3 €0.7 €3.1 €2.0	Territorial cooperation	Border regions and large transnational regions (to be reviewed)	€13.2 3.94%

Notes:
* €22.1 billion (8.38 per cent is set aside for the 'statistical effect' regions which will lose funding because they had <75 per cent average GDP for EU15 but have >75 per cent average GDP for EU25)
** Objective 1 regions 2000–6 not covered by convergence objective will receive 'phasing in' funding €9.58 billion (16.6 per cent)
Structural support for fisheries has moved from the structural funds.

The first two apply in 'assisted areas'. Tier 1 is designated according to the same criteria as Objective 1 Structural Funds and is thus conterminous with Objective 1.[6] Tier 2 regions were built up from ward level using other criteria and are much more concentrated than the Community Objective 2 regions, mostly applying to old industrial areas such as former Midlands coalfields. Grants generally range from 10 to 15 per cent of the project cost and the applicant must demonstrate that the project will not go ahead without the funding. Tier 3 areas are also designated where support is available to small and medium sized enterprises (SMEs) through Enterprise Grants. In Scotland and Wales RSA continues to apply in the same way for assisted areas, and the whole of Northern Ireland is designated an assisted area for much the same scheme, which is known there as Selective Financial Assistance.

EU competences in spatial planning

Spending on cohesion policy, while modest in comparison to aggregate public expenditure, has significant impacts on spatial development patterns through investment in infrastructure and changing locational decisions (Williams 1996). This is *de facto* spatial planning. However, coordination of this investment through explicit territorial planning strategy is another matter, which is why during the 1990s the European Commission started to take an interest in spatial planning with the strong support of some member states.

All the European institutions have now recognised the importance of spatial development that cuts across national borders. Member states are encouraged to work cooperatively on spatial planning in order to coordinate the spatial impacts of sectoral policies and promote more sustainable forms of development and economic competitiveness. It is argued that there are important cross-border and transnational dimensions to spatial planning which need to be taken up through appropriate institutions and instruments at jurisdictional levels above the nation-state. But there is a question about the legitimacy of such action, because, until recently, competence over spatial planning (in its various forms) has been considered to rest solely with the member state governments (or in some countries subnational governments). To what extent should there be a sharing of powers on spatial planning between the member states and the Community?

The European Treaties, which determine what the Community should do by specifying common objectives to be followed up by the Commission, do not assign competences for spatial planning to the Community. Nevertheless, as explained above, the core objectives of the EU have a spatial dimension. At the very least, many observers have thought that there needs to be action at the Community level on the spatial coordination of Community actions among sectors (transport, economic development, agriculture, etc.) and between jurisdictions (EU, national, regional and local) (CEC 1997). This is what has happened in practice; the European Commission has undertaken research on spatial development trends and planning systems, funded international cooperation on spatial planning and supported the development of EU policy on spatial planning through intergovernmental working of the member states.

In other sectors the Community has significant powers, for example, in transport the Community has pursued harmonisation of national transport policies and the development of Trans-European Transport Networks (TETN). In the environment field, the Community has produced Directives on Environmental Assessment, Air Quality, Waste, Birds and Habitats and Water, all with significant impacts on planning systems (and explained in relevant chapters of this book). The Common Agricultural Policy and other specific measures such as environmentally sensitive areas (ESAs) have been instrumental in spatial development changes in rural areas. The extent to which powers are shared with or transferred to the Community, depend on how member states apply the principle of subsidiarity (Nadin and Shaw 1999). In essence, subsidiarity means that competences should be located at the most appropriate level and that they should not be located at a higher jurisdictional level than is necessary. General concerns about the growing powers of the Community during the 1990s drew attention to the idea of subsidiarity and its use in controlling the extension of Community competences. Since then the making of Community legislation has slowed quite dramatically. Instead there is more emphasis on informal actions which come under less scrutiny and encouraging member states to cooperate through the 'intergovernmental process'. This has been the preferred route for action on spatial planning where action is led by the member states rather than the Commission (see Chapter 3).[7]

European Spatial Development Perspective

The first initiative on systematic planning at the European scale came from the German government, which prompted the establishment of a permanent Conference of European Ministers of Aménagement du territoire (CEMAT) through the Council of Europe in

1970. The early regional policy of the EU had little spatial content and instead focused on the need to support particular industrial and commercial sectors. Despite a resolution in the European Parliament to draw up a European scheme for spatial planning in 1983, progress has been slow. However, the French, German, Danish and Dutch governments continue to promote supranational planning studies, especially as regional policy funding was established and became such a large part of the EU budget. The studies introduced valuable spatial concepts at the European level such as the now infamous 'blue banana', used by Brunet (1989) to describe the area between South East England and Northern Italy with most cities over 200,000 population and where growth has been concentrated.

The Commission's first major contribution to the development of European supranational planning came with the publication of *Europe 2000: Outlook for the Development of the Community's Territory* (CEC 1991). This document was intended to provide a European reference for planners working on national or regional planning policies. It was effectively a geography text, raising awareness of European-wide spatial development issues. It adopted an approach which cut across country borders to identify seven transnational study areas having shared characteristics. (An eighth area was added with the inclusion of the eastern German *Länder*.) These eight regions, together with adjacent 'external impact areas', became the subject of extensive research studies. The initial findings for these study areas are reported in *Europe 2000+ Cooperation for European Territorial Development* (CEC 1994). As its title implies, this publication signalled a change in gear on supranational planning. It charts the trends in the physical development of the European territory and crucially makes strong assertions about the preferred development patterns for the future.

The main emphases of the *Europe 2000* studies are the need to control urban sprawl, a common feature of development predominantly, if not exclusively, in the southern European states, and the strengthening of small and medium size towns, especially where this can help in focusing the provision of services in rural areas. Underlying these ideas is the concept of a polycentric urban system, a balanced distribution of urban services

across the territory, with more emphasis on development at points along corridors joining the main centres. The rationale is that this will avoid both the congestion problems of very large conurbations and the decline in service provision in rural areas, but it is an idea which largely ignores the dominant centralising tendencies of the market. Other issues addressed were improved protection of areas of environmental importance, land abandonment in rural areas, the revitalisation of poor neighbourhoods, built heritage and the Trans-European Networks (TEN), which are explained briefly in Chapter 11.

In 1991 a start was made on moving from analysis to policy development and what was to become the European Spatial Development Perspective (ESDP) (CSD 1999). This is the most important initiative on spatial planning at this level, and a unique experiment in supranational planning. The ESDP is intended to promote 'coherence and complementarity' of the development strategies of the member states by coordinating the spatial aspects of EC sectoral policies. The task of achieving the necessary consensus among the then fifteen member states (and often autonomous regions within nations) is obviously a difficult one and, needless to say, the statements made are at a very general level. Nevertheless, the final document was endorsed by all governments at their meeting in Potsdam in May 1999. The main contents of the ESDP are shown in Box 4.1.

Because of uncertainty over Community competences on spatial planning, the ESDP was produced through an intergovernmental process with the lead responsibility changing from one member state to the next with changes to the Council Presidency (although in the later stages a Troika was established involving the current, previous and next presidencies to ensure some continuity). Given the scope for disagreement it is undoubtedly an achievement that it was produced at all, and despite its very general content, it has important messages, especially for those member states where planning is not well established. Critics have pointed out its strong orientation around economic growth 'as a precondition for balanced and sustainable development' (Richardson and Jensen 1999: 5) and thus it says much more about urban issues and the role

BOX 4.1 EUROPEAN SPATIAL DEVELOPMENT PERSPECTIVE (1999): A SUMMARY

Purpose

The main purposes of the ESDP are

- to raise awareness of the significance of spatial development trends for the objectives of the EU
- to provide a common reference framework to guide action on planning and development decisions
- to promote integration of policy across different sectors of activity, and coordination and complementarity of member state policies through a common strategy
- to provide a framework within which sustainable economic development can take place, and enable the EU to meet its international treaty obligations.

Development trends

The ESDP outlines a wide range of significant spatial development trends for Europe, including:

- little change in population, continuing urbanisation and urban sprawl
- intensification of agriculture, depopulation, and poor service provision in rural areas
- increasing waste and pollution, the ongoing loss of biodiversity
- increasing transport flows, congestion, missing links, bottlenecks, poor accessibility in peripheral areas.

Policy options

The ESDP has three general objectives: a more balanced system of towns and cities, parity of access to infrastructure, and prudent management of heritage. It promotes, for example,

- a balance of urban activity between and across regions by avoiding excessive concentration and the creation of alternative economic centres
- the sustainable development and control of physical expansion of cities
- more environmentally friendly access to services and the diversification of the rural economy
- special attention to the transport needs of land-locked and remote regions
- creation of buffer zones and completion of the network of protected areas
- innovative telecommunications services and applications.

Action and debate

Application of the ESDP will be through the re-orientation of member states' own strategic policies, and Community instruments and spending. The ESDP promotes a 'spatial planning approach': the cross-sectoral coordination of policy and action, including the spatial effects of EU actions through Cohesion Policy, the trans-European networks, the CAP, etc. A number of questions are raised:

- How can spatial planning help to coordinate sectoral policy?
- Is there a need for more spatial planning powers for the Commission?
- How can cooperation among member states be achieved? Is there a need for transnational planning instruments?

of cities than rural concerns. Practitioners also find it difficult to see how it might connect with local planning practice. Could it be otherwise? In its own sphere of operation, in the arena of intergovernmental working among member states and the Commission, it has certainly been influential. Faludi and Waterhout (2002: 164–5) draw attention to how 'the ESDP has been an important source of inspiration' for the Commission's developing interest in the ideas of territorial cohesion, and how this represents a reformulation 'of the spatial planning discourse as a Community concern'.

Spatial planning approach

Spatial planning activity at the European level has been important in drawing attention and practice towards 'the spatial planning approach'. A brief summary of where 'spatial planning' fits in relation to other Euro-planning concepts is given in Box 4.2. It should be emphasised that these words are used in various ways. Spatial planning has been used as a generic term to describe all planning systems (as in the *EU Compendium of Spatial Planning Systems and Policies*) and it is also a literal translation or close approximation to the name of planning systems in other countries, such as *Raumplannung* (Germany) and *ruimtelijke ordening* (the Netherlands). Other writers have for many years used the term spatial planning as a synonym for land use or physical planning. Here we are using the term spatial planning to highlight the difference with physical land use planning.

Physical planning describes government action to regulate development and land uses in pursuit of agreed

BOX 4.2 EURO-PLANNING JARGON

The planning system is loaded with jargon (in both senses) and the European dimension adds another layer, some of which is finding its way into planning in the UK. There can be no hard and fast rules – the meaning of words is inevitably ambiguous, especially so when working across languages and cultures. Meanings are also contested because they embody particular interests and concerns. So these definitions are offered as a starting point only.

Spatial development refers to the physical distribution of built and natural features and human activities across a territory and the qualities of those features and activities, for example, disparities in access to opportunities from one neighbourhood to another. A **territory** can be many things, for example, a neighbourhood, region or catchment area. Spatial development is also the process of change which results from a complex mix of market decisions and public intervention. All sectors of public policy have some impact on spatial development, although it is not always recognised. The spatial impacts of sectoral policy (such as transport, education and health) are known as **spatial policy**. For example, if a health authority decides to centralise its facilities in a smaller number of larger hospitals, there are implications for the distribution of access to those facilities.

Land use planning systems have relatively weak influence over spatial policy. Thus, the **spatial planning approach** concentrates on establishing better coordination on territorial impacts: *horizontally* across different sectors, *vertically* among different levels of jurisdiction, and *geographically* across administrative boundaries.

The **European Spatial Development Perspective** (CEC 1999) promotes the spatial planning approach to member states. It provides a reference point for spatial policy and recommends policy options to create more polycentric spatial development, improved urban–rural relationships, more parity in access to infrastructure and knowledge, and wise use of the natural and cultural heritage.

Polycentricity means 'many centres' and is promoted as a way to achieve more 'balanced' spatial development. It can operate at different scales from European to city-region (for example 'greater London' may have a polycentric structure at the scale of the South East but a monocentric character at the European scale). Polycentricity describes the functional complementarity of places (do they try to provide the same services?); their institutional integration (are we preparing different strategies for these places?) and political cooperation (is there scope for mutually beneficial cooperation?).

The proposed EU Constitutional Treaty will bring in an overarching objective for the Community of **territorial cohesion**. This makes more explicit the spatial dimension of the Community objective of a more fair (or equal) access for all citizens to services and opportunities (for example, housing, jobs and education) irrespective of their location. It will mean that sector decisions will have to pay more attention to the spatial policy impacts and act accordingly.

The Community's **INTERREG** Initiative has been the principal means of applying the ESDP through co-financing of spatial planning projects involving partners in different countries in three strands of action: **cross-border**, between geographically contiguous border regions, **transnational**, across large multinational spaces, and **inter-regional**, among non-contiguous regions. A number of the transnational programmes produced **transnational spatial visions** which drew together findings from individual projects and provide an agenda for future cooperation. The need to work across borders is justified with reference to **transnationality**, which means having an effect in more than one country. A railway line crossing national borders is obviously a transnational issue, but arguably, the concept might also embrace issues of common interest, such as rural out-migration. Demonstrating transnationality helps to meet the **subsidiarity** principle, that is, decisions should be ceded to higher jurisdictions only when there is demonstrable need or benefit to be gained.

Source: This is an edited version of Nadin and Dühr (2005)

objectives. This form of planning is one policy sector within government alongside policy sectors such as transport, agriculture, environmental protection and regional policy. Land use planning may incorporate mechanisms to coordinate other sector policies, but often these are weak. In the UK for example, it is only since the mid 1990s that transport has become more coordinated with land use planning strategies. *Spatial planning* in the European sense is more centrally concerned with the problem of coordination or integration of the spatial dimension of sectoral policies through a territorially based strategy. This flowed from the EU spatial planning work because it is through spatial policy (the spatial impacts of sectoral policies) that the EU has most effect on spatial development, since it has no role in land use planning. Those involved in developing a European dimension to planning draw attention to the contradictions among sectoral policies in particular places, perhaps for example promoting change in farming practices that undermine rural social policies, or economic investments that have damaging environmental consequences. The plan, a territorial strategy, is one way of ensuring more consistency and synergy among sector policies. In this view, land use planning is one of the sectors. However, we should not interpret spatial planning as a renewed effort at comprehensive rational planning; it is not. Rather it is suggesting that a territorial strategy can help to coordinate the actions of different sectors. In practice all planning systems in Europe tend to be land use systems with different degrees of sectoral coordination, but mostly weak with considerable sectoral compart-mentalisation (Nadin *et al.* 1997). This is not to say that sectoral coordination is not an aspiration of land use planning, but rather that it is not realised. At the European level the starting point is to provide planning that is a method of securing 'convergence and coordination between various sectoral policies' through a territorial development strategy (CEC 1999; see also Bastrup-Birk and Doucet, 1997). The need to improve both horizontal and vertical coordination of sectoral policies with a spatial impact is a common theme in the recent reform of a number of national planning systems (Seaton and Nadin 2000).

ESPON

Since the ESDP was 'adopted' by member states in 1999, attention has focused on the ESDP action pro-gramme, including the roles of the European spatial planning process in directing investments (especially through the EU Structural Funds) and in coordinating other Community actions. The most pressing demand was for a better understanding of spatial development patterns and trends at the transnational scale to provide a stronger evidence base for further policy develop-ment. To begin this task the member states with support from the Commission have established the

European Spatial Planning Observation Network (ESPON). More than twenty-five projects are being undertaken by more than a hundred research partners in transnational teams.

An interim report was published in 2004 based on sixteen interim project reports on spatial development trends and the impacts of policy and ESPON Briefing 1 followed later the same year. It reiterates earlier findings on the concentration of growth, which, with some exceptions, is concentrated in the 'Pentagon', an area bounded by London, Paris, Milan, Munich and Hamburg. The UK features prominently in the Pentagon but also is noted for population loss in Scotland. Between thirty-five and forty-five functional urban areas have been identified that could act as counterweights to the Pentagon 'if appropriate policies could be applied. This is particularly true for . . . Lyon, Marseille, Birmingham and Manchester'. These areas are described as metropolitan European growth areas (MEGAs). Although the analysis is incomplete and there are difficulties of generalisation (there are some-times greater differences within member states than between them) the report does not hesitate to make conclusions for policy. For example, because of the disparities identified in R&D and other innovation infrastructure, 'territorial cohesion requires strong innovation policies in favour of the less advanced countries and regions' (ESPON 2004a: 45). Findings from ESPON have already made important contribu-tions to the Third Cohesion report (CEC 2004) and the reform of Cohesion Policy described earlier.

INTERREG and spatial visions

While the member states and the Commission completed the ESDP, the Community initiatives INTERREG II and III have encouraged local authorities and other public bodies to take part in transnational planning projects. The 'III' denotes the third phase of the programme: INTERREG II ran from 1996 to 2000 and its successor INTERREG III from 2000 to 2006. The initiative is further broken down into 'strands' INTERREG IIIa is cross-border, IIIb is transnational and IIIc is inter-regional. It is the

IIIb strand, and its predecessor IIc: transnational, that have provided most support for cooperation on spatial planning. The original objective for INTERREG was to address the disadvantages faced by border areas (now predominantly part of the 'a' strand). The objectives of IIIb are to promote strategies for sustainable development, to foster transnational cooperation within a common planning framework, and to improve the impact of Community policies. A major stream of funding also promotes cooperative approaches to the problems of flood and drought.[8] Interreg IIc had a major impact in the UK in the sense of encouraging local authorities and other bodies to work cooperatively with partners in other countries on planning issues by co-financing projects.

The argument for these initiatives follows the logic of the ESDP – more integration of spatial planning policy between nations will contribute to a better balance of development between regions, and thus increase the social and economic cohesion and economic competitiveness of the Community. The funding available over the period 1995–9 amounted to 3.6 billion ecu (European currency units) of which 413 million ecu (about £248 million) was for IIc, a moderate amount in comparison with other Community initiatives. Total funding was increased to €5.3 billion for the period 2000–6, of which UK partners get about €362 million. Of this, 67 per cent goes to cross-border, 27 per cent to transnational and 6 per cent to inter-regional. As with most EU funding this is 'co-financing', which requires additional income to make up total project funding. The 'intervention rate' (the proportion of co-financing) is generally a maximum of 50 per cent in the UK, except for Objective 1 areas, where it may be higher. Funds are also top-sliced for the ESPON initiative and other coordinating activities. For INTERREG IIc, seven transnational regions were defined within which public bodies could bid for funding to support transnational spatial planning. With the inclusion of the new member states, the number of transnational regions grew to thirteen under INTERREG IIIb. The UK is involved in three transnational regions which were adjusted between the phases: the Atlantic Area, North West Europe and the North Sea Regions. The North West Europe

transnational zone covers the whole of the UK and overlaps with the other two regions. The idea of transnational regions was first mooted in the *Europe 2000* and arose from lobbying by clusters of administrative regions, particularly the Atlantic Area. The North West Europe region began as a Centre Capitals Region, but was then extended to cover the whole of the UK (Nadin and Shaw 1998b). The political lobbying that produces these regions undermines the rationale to some extent that these transnational regions share common spatial development problems.

To be eligible for funding a project has to involve at least two member states and have a general relevance for the rest of the EU. Examples of successful projects include collaboration on planning issues related to the high speed train network in North West Europe, sharing of experience and strategies on providing access to urban services in rural areas and increasing the vitality of small towns in the North Sea Region, and devising joint strategies for the conversion of fishing infrastructure to ecotourism in the Atlantic Area.[9] Three principles are important in the selection of projects. They must have a transnational dimension, there must be a strong multiplying effect giving added value, especially in improving the prospects for achieving planning strategies, and the projects must have the potential to influence other operational programmes so that they make a contribution to transnational planning strategies. In practice, many activities are accepted as 'spatial planning' which are not closely related to town and country planning in the UK such as support for innovation in industry, research and training. Under the INTERREG IIIb programme small sums have also been spent on infrastructure. Also, the transnationality criterion has been given a broad interpretation, and the level of actual joint working across national boundaries on many projects is quite low (Nadin and Brown 1999).

The strands of INTERREG are linked through a cross-cutting INTERACT programme, which seeks to improve their effectiveness and delivery and supports exchanges of experience, networking, and information dissemination. All the Interreg programmes were subject to mid-term evaluations in 2003 and INTERACT undertook a 'meta-evaluation' which was published in

2004 (all available on the INTERACT website). The review found that programmes were not spending money quickly enough, in some areas because of insufficient demand. It was not able to come to an overall conclusion about achievements, which sums up the initiative – it is very fragmented and bottom-up in terms of the creation of projects, and it relies very much on existing networks of contacts. Anecdotal evidence suggests that many projects arise from partners' concerns with their own problems (and finding funding to address them) rather than a desire to cooperate with other countries. There is no doubt that INTERREG has dramatically expanded the effort and experience on transnational planning in the UK. For those authorities and organisations with a history of transnational collaboration the initiative provides a welcome opportunity to contribute in a bottom-up way to the formulation of transnational planning policy and the ESDP. Where there is less experience, INTERREG has offered the potential for capacity building through education, discussion and the sharing of ideas. A UK evaluation by Zetter (2001) is supportive and provides case studies of positive project outcomes. The Commission and member states will be looking for practical and concrete results, but (as acknowledged by Zetter) to measure projects only by substantive outcomes would be to miss the point. A major benefit of INTERREG is to bring those engaged with similar planning problems in different countries together to share experience, to broaden horizons and to (slowly) develop common understandings and approaches.

An overriding theme of INTERREG is the need for programmes to be consistent with the aims of the ESDP and to contribute to the further elaboration of transnational planning strategies or frameworks. The operational programme measures and specific projects funded within them are intended to contribute to a broader transnational strategy or framework. Five such transnational frameworks, 'vision statements' or strategies were produced, including visions for the three INTERREG programmes that affect the UK. They are intended to provide a bridge between the ESDP and national and regional plans and to provide guidance for priorities for funding projects (Nadin 2002).

The transnational spatial vision is a new form of planning instrument, but there are examples of similar cooperation especially in the Benelux area where interactions across national borders are particularly strong. The first vision statement was the *Strategies around the Baltic Sea Region 2010* (VASAB) published in 1994. It predated transnational cooperation on INTERREG and provided some impetus for similar work elsewhere. The dramatic changes in the economic and social geography of the Baltic after the opening up of the former eastern Bloc states created a need for urgent collaboration across borders. The method of 'visioning' across the studies has been similar, involving the examination of the spatial impacts of sectoral policies, and investigating where further cooperation on spatial development might be needed or bring benefits (Nadin and Dühr 2005; Dühr *et al.* 2005). The Spatial Vision for North West Europe also provides a visualisation of the transnational region, illustrating in a very simple way some of the problems and possible solutions (Devereux and Guillemoteau 2001). But the vision statements are not in fact very visionary. They tend to be statements of 'what is' rather than 'what might be', though this is a potentially valuable exercise in identifying issues where the member states and regions might act. The Spatial Vision for North West Europe has certainly been cited in regional planning policy in the UK, notably in the north of England, where its comments on the need for alternative routes connecting to continental Europe have been welcomed.

Critical comment on the vision documents has noted that they are mostly descriptive, and there is also the question of ownership. They were prepared by a small group of national government representatives, planning academics and consultants only. The North West Europe Vision did include consultation in the member states, but again only with a narrow range of government interests, not generally involving politicians or the private sector, which inevitably means they will be less influential (Zonneveld 2005). It should be recognised that, like the ESDP, they represent first stage work in transnational cooperation. This work is being taken forward through more intensive investigations of the nature of transnational spatial development issues through studies under

INTERREG IIIB discussion with a wide circle of stakeholders. Visioning or long-term scenario building is also a feature of some national planning systems, though so far it has had little influence on the UK. The French national planning agency, DATAR, for example, has produced scenarios for France in Europe in its study *Aménager la France de 2020* (2002).

Convergence of European planning systems

Increasing attention to spatial planning policy at the European level, and cooperation across boundaries has drawn the attention of more practitioners to the differences in the way that planning operates across Europe. Each member state (and in some cases an individual region) has developed its system in response to local economic and physical development problems arising within particular cultural, legal and social contexts. There may be some degree of convergence of these systems as countries work together more.

In order to facilitate understanding about the way that spatial planning operates in different member states, and thus to promote more effective cross-border and transnational planning, DGXVI commissioned a *Compendium of EU Spatial Planning Systems and Policies*. This not only demonstrates the diversity in planning systems and policies (especially in their operation) but also notes similar trends as the different countries respond to the same macroeconomic forces. There is a distinct trend in much of Europe towards greater flexibility in the operation of regulation. New mechanisms are being introduced to establish more strategic planning frameworks and to allow for decisions which are contrary to the characteristic binding zoning plans.

Another common trend is the integration of spatial plans and sectoral spending programmes. The spatial plan is more widely recognised as the coordinative mechanism for sectoral policy and spending. New instruments are being introduced to tackle cross-border issues, and there is increased transnational cooperation between planners dealing with similar issues in different regions. In the UK this sort of cooperation

began in places like Kent that have strong connections with other countries (Kent County Council 1995) and some of the larger cities such as Birmingham that had a 'European agenda'. Now thanks mostly to INTERREG, such cooperation is very widespread among local authorities and other planning bodies.

As a postscript to this very brief review of developments in transnational planning, it should be mentioned that they have not been made without some resistance. The UK government in particular has been less than enthusiastic about 'universal spatial planning policies' and has instead emphasised the usefulness of exchanges of experience (although the Labour administration from 1997 has been more engaged). There is certainly room for debate on the very idea of a European dimension to spatial planning, on territorial cohesion, and on the many issues raised in the ESDP such as seeking dispersed but concentrated development, urban containment, and a focus on corridor development. There will be other important issues that have not yet been considered, such as the availability and price of land for housing development (and the Barker Review discussed in Chapter 6 made telling comparisons with other European countries on this issue). The proposals may not fully address powerful market forces, especially at a time when the private sector is taking a greater share of investment in virtually all member states. The limitations of the spatial planning systems in bringing about desired objectives of sustainable and balanced development across the community are recognised, and this may eventually lead the Commission to put more emphasis on other policy options for regulating and promoting development such as taxation measures or development incentives.

However, there is a general assumption in most member states that more cooperation on spatial planning is inevitable and that it is to be welcomed. This sentiment is shared by the new member states which, while not wishing to reinvent centralised state planning, do want to deal in a coordinated way with their massive problems of environmental degradation and economic decline.[10]

NATIONAL PLANS AND POLICIES

National spatial plans

There is no national land use planning in the UK in the sense that policies or plans are prepared for the whole country. So there is no UK national spatial plan or any specific national policy for planning across the UK, although other national (UK) policy documents such as the UK Strategy for Sustainable Development and the UK Biodiversity Action Plan have important guidance for planning. Even where there is a UK strategy, devolution has sometimes led to separate sub-strategies or programmes for England, Northern Ireland, Scotland and Wales. Even the use of the word 'national' has become difficult, but it is preferred to 'regional' when describing one of the four 'nations' of the UK. It may be surprising to readers, especially those from other countries, that despite the central-isation of UK government, a national (UK) spatial plan has not been prepared, indeed it was not on anyone's agenda until the mid 1990s. The UK government has since the 1960s eschewed the idea of any explicit form of national planning. Also, UK planning since the 1970s has paid little attention to the 'spatial' dimen-sion of planning, that is, shaping the overall structure of urban development, save for the general requirement for urban constraint through green belts. It has instead developed criteria based policies and concentrated on overall quantities of development, in particular the release of land for numbers of houses, as exemplified in the English regional planning guidance (discussed later). It was a combination of experiences with European spatial planning, recognition of the need for a more 'spatial planning approach', lessons from other countries, and the possibilities borne of devolution that gave rise to experiments with national spatial planning. The first was in Northern Ireland where there had been more interest in spatial regional planning than elsewhere in the UK.

The first regional plan in Northern Ireland was the Belfast Regional Plan published in 1964 (the Matthew Plan). This proposed the stopline, a system of radial motorways, and a major new town, Craigavon,

modelled on the English experience (Hendry 1989). Like its counterparts in England, the plan was over-taken by the effects of dramatic economic recession. The subsequent *Regional Physical Development Strategy 1975–85* sought to concentrate growth in the province to twenty-six key centres, but the depressing effects on other areas were widely challenged (Blackman 1985). A new rural planning policy published in 1978 took a much more relaxed approach to development in three-quarters of the rural territory, which led to extensive development of single houses in the countryside and ribbon development. This led to a reappraisal of the need for regional planning and the publication of *A Planning Strategy for Rural Northern Ireland* in 1993. This included both strategic objectives for the overall development of the territory and detailed development control policies, and could only have been the product of a system where central government sets the strategy, makes local plans and undertakes development control. The Strategy introduced new restrictions on devel-opment in the countryside while introducing the novel designation of 'dispersed rural communities'.

The Belfast Agreement of 1998 gave added impetus to the increasing activity on regional planning in Ulster. In the same year, the Secretary of State for Northern Ireland launched *Shaping our Future: Towards a Strategy for the Development of the Region*. Consultation in this draft document was followed by publication of the *Regional Development Strategy for Northern Ireland 2025* (RDS) in 2001. The strategy is very much in the mould of EU developments in spatial planning, placing Northern Ireland in its European and global context and seeking to integrate concerns about the physical development of its territory with social, economic and environmental objectives. It makes use of the 'hubs, corridors and gateways' concepts which are also familiar to European spatial planning and is consistent with the *National Spatial Strategy for Ireland 2002–20* published soon after in 2002. It describes itself as 'not a fixed blueprint or master plan. Rather, it is a framework, prepared in close consultation with the community, which defines a Vision for the Region and frames an agenda which will lead to its achievement' (p. 2). At 258 pages it is a very full framework (and it is accompanied by a *Family of Settlements Report* (2001) setting out housing need and requirements for each town). Policies are set out as strategic planning guidelines (SPGs) which give some long-term direction for local decisions.

The RDS was made under an Order,[11] which also spells out the status of the document; it requires Northern Ireland departments of government to

> have regard to the regional development strategy in exercising any functions in relation to development . . . In practice, this means that they should be in broad harmony with the strategic objectives and policies of the RDS. The RDS will also be material to decisions on individual planning applications and planning appeals.
>
> (p. 3)

The strategy was subject to strategic environmental assessment and an independent public examination following consultation which may give it more weight, although it remains to be seen what success it can have in its objective of concentrating development in a region where development in the countryside is the norm rather than the exception.

Until the 1980s, Scotland had a tier of regional reports which provided a corporate policy statement for the regions as well as a framework for the preparation of structure plans. There were few formal procedures governing their preparation, and they did not require central government approval, but were simply published with the Secretary of State's observations. They were much admired but, as the statutory development plan framework was put into place, they became regarded as redundant.[12] The Scottish Executive consulted on *A Review of Strategic Planning* in 2001. The proposal to prepare a national planning framework found wide support and in 2002 the Executive announced plans to prepare the framework with 'extensive stakeholder involvement'. The review concluded that a statutory 'national plan' for Scotland was not needed but that 'there are a limited number of subjects at the national level where the Scottish Ministers believe the planning system has an important role in delivering sustainable solutions' (p. 8). The potential to provide a spatial context for the Structural Fund spending was also noted. *The National Planning*

Framework for Scotland was published in 2004. It examines spatial development trends and the 'key drivers', identifies strategically important investments in infrastructure, and presents very general spatial scenarios for four regions within Scotland. Again the framework is 'not intended to be a prescriptive blueprint, but will be a material consideration in framing planning policy and making decisions on planning applications and appeals' (p. 1). It was subject to environmental appraisal and wide consultation. The Scottish Executive have given a commitment to update the Framework every four years as its publication is the start rather than the end of a process of debate on Scotland's 'long term spatial development'.

In Wales, a series of guideline documents was prepared as *Strategic Planning Guidance in Wales* by the Welsh Office. The documents were intended to 'consolidate and re-present the wide range of available strategic guidance material in a consistent and accessible form' and particularly to provide a framework for the preparation of structure plans. Given the changes to a unitary structure in Wales, the need for strategic policy became more pressing. The Welsh Assembly undertook consultation on a national spatial framework for planning in Wales in 2001 and a draft *Wales Spatial Plan* (WSP) was published in 2003. The consultation document called for a plan to 'support and influence the spatial expression of the policies and programmes of the Welsh Assembly', and a context for making development plans and major infrastructure projects at the national scale (p. 2). It is interesting to note that in the foreword to the draft Plan, Sue Essex, the minister responsible, noted that 'devolution has provided us with an opportunity to do things differently in Wales . . . adding a spatial perspective will enrich our understanding, challenge our thinking, sharpen our policy making, aid policy integration and improve service delivery'. One of the key objectives, as elsewhere in the UK, was to provide a 'distinctive approach' to planning. Also, in view of the broad ambitions of the Spatial Plan, responsibility was moved from the Planning Division to the corporate Strategy Division of the Assembly Government.

The final document *People, Places, Futures: The Wales Spatial Plan* was published at the end of 2004. Formal provision for the plan was made in s60 of the Planning and Compulsory Purchase Act, so it is a statutory document, indeed the Act states that 'there must be a spatial plan for Wales', and it has to be approved by the Assembly. The preparation process is not specified, though in keeping with current practice, it involved extensive consultation, including seminars and conferences across the country. Like the other national frameworks, it explicitly draws from the ESDP and transnational spatial visions, and explains the drivers for change in spatial development. It identifies large zones or sub-regions with an agenda of issues which cut across local authority boundaries. The differential performance of parts of Wales (and their potential) is important given the overall commitment to ensuring more balance and sustainable development. The Plan does not address the external connections as thoroughly as the Scottish example, though it does link parts of Wales into the metropolitan regions in England and their increasing influence. More emphasis is put on the need for inter-sectoral coordination within Wales as illustrated in the summary of the role of the Plan as

> making sure that decisions are taken with regard to their impact beyond the immediate sectoral or administrative boundaries; that there is coordination of investment and services through understanding the roles of and interactions between places; and that we place the core values of sustainable development in everything we do.
>
> (p. 4)

The complexity of the 'policy environment' in Wales is perhaps the best example of the 'strategy overload' that may be affecting all the countries to some degree. This presents the greatest challenge to these national spatial strategies. In Wales, there is the *Wales Spatial Plan, A National Plan for Wales* (2001), a *Strategic Agenda for the Welsh Assembly Government: A Better Country* (2003), *The National Economic Development Strategy: A Winning Wales* (2001), the *Rural Development Plan for Wales 2000–06* (2000) and a *Sustainable Development Scheme and Action Plan*, not to mention the programming documents for the Objective 1 area of West Wales, and national strategies for health, waste,

housing and more. As discussed later, the same problem is found in the English regions.

However, evaluation will need to bear in mind the experimental nature of the national plans. They each acknowledge the influence of European developments particularly the ESDP, and the opportunity that devolution offers, but they are still in search of the most appropriate form and content. They certainly provide an analysis of the existing spatial development situation, identifying the drivers of change such as demography, economic globalisation, European integration, and the knowledge economy, and their effect on different parts of the country. In looking beyond land use and connecting activity across sectors with a long-term horizon they are certainly much more like strategies than some of the other documents that come with this name. Their success will be determined by how well they are able to coordinate and influence activity in these other sectors.

The discussion above leaves one country in the UK now without a national plan. Some related work has been done by a consortium comprising the English Regions Network, RDAs, ODPM and DfT on *Regional Futures: England's Regions in 2030*. The study, led by Ove Arup and Partners and published in 2005, was not meant to be the first stage in the creation of a spatial plan for England, but rather 'to develop a "national perspective" on how England's regions (including London) relate to each other and to underlying forces in the economy, and how these relationships have been changing and will change in the future' (p. 1). It summarises the prosperity gap between the South East and regions in the North and Midlands and shows how the gap will widen in coming years. Its analysis, such as it is, starts with assumptions firmly rooted in current government policy: restraint in the South East would 'have very damaging consequences for the national economy. The effect would be much more to stifle growth through higher costs, inflation, off-shoring and lack of competitiveness to attract new business, rather than in redistributing activity to less prosperous regions' (p. 83). Moreover, it argues that the scope for government to be able to do anything about this, and especially the potential of spatial planning, is severely limited. Despite the eminent advisers who contributed to this project, the discussion lacks any imagination and its 'home truths' about the inevitability of further polarisation and weakness of public policy make for depressing reading for regional planners. Other equally relevant 'home truths' are not given much attention: stringent restraint in the South East has been a fact of life since the early 1950s; the size and growth rates of cities are not good indicators of their performance and quality of life of citizens; the environmental impacts of the South East mega city are spread across the globe; and the consequences for infrastructure (both overuse and underuse) will be enormous.

A more positive approach is promised from the Town and Country Planning Association (TCPA), which has begun a national debate on *A Vision for England's Future* with a view to joining up existing regional strategies. A report is promised in 2006. This report will have to address the very different situation in England to those of its neighbours: England is extremely complex in spatial development terms, having both a global 'command centre' in London and relatively remote rural areas. Parts of England share more in common with their neighbours in Scotland and Wales than they do with the South East, and parts of London have much more connection with other countries than the rest of England. This initiative follows earlier lobbying for a UK spatial planning framework after a wide-ranging study found that, in principle, it was both desirable and feasible (Wong *et al.* 2000). The authors argue that the main purpose of such a framework should be issued by the Cabinet Office and would seek to join up and fill in the gaps between existing strategies and monitor spatial policy targets. Its main task would be to cut across the compartmentalised sectoral thinking of government. Further discussion of this idea is not needed since it was not accepted, nor is likely to be in the future. But there is a pressing need for a forum where national spatial development issues can be discussed and their consequences understood, both among levels of government but especially among policy sectors. Experience in other countries suggests that 'national spatial frameworks' are mostly in fact analysis of spatial development trends rather than plans for intervention, but that may be no bad thing. A national spatial development report would be a good start to more joined-up spatial policy, in the same way that the German federal government prepares a statement periodically on the state of spatial development of the country.

National policy statements

The UK planning systems have paid much more attention to the provision of generic policy on planning issues through planning policy statements of various kinds. All four systems make use of policy statements at this level.

In Scotland, the creation of national policy was prompted by the need to deal with the unprecedented problems posed by North Sea oil and gas. Since these were considered to be of national importance, the Scottish Office decided to issue guidelines for use by local authorities, especially in relation to coastal development (Gillett 1983). *North Sea Oil and Gas Coastal Planning Guidelines* was published in 1974, and *national planning guidelines* (NPGs) on other topics soon followed. These guidelines were intended to fill a gap between relatively inflexible policy expressed in circulars, and general advice that could be ignored (Raemaekers 1995). They had formal status, but they did not tie ministers or local authorities to particular solutions. They did not go so far as to constitute a 'national plan', and they were not intended to be comprehensive, but they were locationally specific.

The benefits of this system soon became apparent (Diamond 1979). National guidance enabled local authorities to explain the way in which their plans took account of national policies; a higher degree of coordination was possible among the various branches of central government, and national interests in which the Secretary of State needed to be involved could be readily separated from local matters (Nuffield Foundation 1986; Rowan-Robinson and Lloyd 1991). Nevertheless, as the series expanded to cover more topics and non-locationally specific guidance, questions arose about the precise status of NPGs and their overlap with other policy statements (Rowan-Robinson and Lloyd 1991; Planning Exchange 1989).

In response to these concerns, the Scottish Office introduced in 1993 a rationalised structure for national policy and advice through a series of *national planning policy guidelines* (NPPGs) together with continued use of *planning advice notes* (PANs) and circulars. The role of NPPGs is 'to provide statements of government policy on nationally important land use and other planning matters, supported where appropriate by a locational framework'. NPPGs were broader in scope and provided more comprehensive coverage of topics of national concern. The Scottish Executives Review of Strategic Planning concluded in 2002 that the NPPG should be renamed *Scottish planning policy* (SPP) on the grounds that each is a statement of Executive Policy rather than broader guidance. A programme of review and updating was begun to make the documents more concise and the policy content more explicit: the current list therefore has both forms (and they are listed in Official Publications). PANs have continued with more titles being published to disseminate good practice and to provide advice to all involved in the planning system.

A review of experience of Scottish national planning policy (Raemaekers *et al.* 1994) concluded that it was a success and called for greater breadth in topic coverage.[13] A further review was conducted in 1999 (Land Use Consultants 1999) again with generally positive findings. Many detailed recommendations have been made for improving the form and content of national policy and guidance, especially on consultation which is now firmly embedded in the process.

In England and Wales, national guidance did not arrive until 1988, although its form and content has provided a model for more recent changes to the Scottish system. National planning policy in England is now expressed in *planning policy statements* (PPSs) which are steadily replacing the former *planning policy guidance notes* (PPGs). There are also *minerals planning guidance notes* (MPGs) and *marine minerals guidance notes* (MMGs). As in Scotland, circulars are used mainly for elaboration of procedural matters. National planning policy in England has certainly clarified the national criteria for decision-making but it has tended to be more general than its equivalent in Scotland, broader in scope, and not at all location specific.

National policy has a considerable impact on planning practice. An evaluation of the effectiveness of the PPGs concluded that they had 'assisted greatly in ensuring a more consistent approach to the formulation of development plan policies and the determination of planning applications and appeals' (Land Use Consultants 1995a: 47). This is because they are important material considerations in development control and have a determining influence on the content of development plans. Conformity between national policy and plans is ensured through regional office scrutiny of development plans, but the study also found that most professional planners have a high regard for national policy statements and welcome the order and consistency that they bring. Councillors are generally more sceptical because national policy constrains their discretion to respond according to their interpretation of local needs.

Having successfully introduced more systematic planning policy at national level, the government's problem is now to maintain consistency and clarity in the series. There have been many calls for more statements on new topics, while at the same time some concerns have been voiced over perceived contradictions between one statement and another, and between the series and other government policy statements. One example (from the Land Use Consultants 1995a study) was concern over the different explanations of the term 'sustainable development' in government policy. The need for some rationalisation and 'weeding' of national policy – to draw this out from more general guidance – was accepted as part of the modernising planning agenda, while at the same time providing more specific policy on some topics such as major infrastructure developments so as to reduce the effort needed to debate national policy at inquiries on development projects of national significance.[14] This would require some aspects of national policy guidance to be much more locationally specific (currently, only minerals guidance mentions locations), though the new regional spatial strategies may meet this requirement to some extent.

In Wales, the publication of national guidance was, until 1995, mostly shared with England through joint publications. On occasion, separate advice was thought necessary to reflect distinctive Welsh conditions (for

THE FRAMEWORK OF PLANS

example on land for housing and on strategic planning), but these followed the English version quite closely. By 1995, the Welsh Office had decided to go its own way, and published two draft planning policy guidance notes intended to replace policy statements shared with England. This was published first in 1996 in two documents and was subsequently revised by the National Assembly into one. The current version is *Planning Policy for Wales* (2002) and, though published as one document, deals with a wide range of matters. Concerns have been raised about the dilution that the reduction of the amount of policy suggests, but it is a much more concise statement than those in England and Scotland. *Technical advice notes* (TANs) have also been published, with a minerals technical advice note (MTAN) filling in much of the detail. The *Spatial Plan for Wales* addresses questions about strategic direction.

Northern Ireland lagged behind in the production of 'national' guidance until the mid 1990s. Planning officers in the Province made use of policy statements in other parts of the UK to keep in touch with policy developments (Royal Society for the Protection of Birds (RSPB) 1993). Since 1995 national planning policy statements have been published by the Planning Service for Northern Ireland. The statements are similar in form to those in the rest of the UK but reflect the special planning and political circumstances in Northern Ireland – not least the centralisation of planning activity in the Planning Service. It may be the strong competence of central government is responsible for the definitive nature of some policies, for example that 'there is no justifiable need for any new regional out-of-town shopping centres' (PPS1 para. 35). The Planning Service also publishes *development control advice notes* (DCANs) providing more detailed guidance on good practice. An explicit policy framework from national government is especially important in Northern Ireland because it is national government which is making almost all decisions. The PPSs must conform to the Regional Development Strategy, discussed earlier.

In whatever form, national planning policy issued by national government and the devolved administrations carries considerable weight and thus the content of policy statements is discussed in some detail in subsequent chapters. But though local planning authorities are required to have regard to national policy, they are not bound by it. Indeed, other material considerations may be of greater importance in particular cases, and planning authorities may wish to take a different line, so long as they can give adequate reasons. Moreover, the advice in one statement may contradict another, perhaps as a result of piecemeal revisions at different times. Nevertheless, national policy commands a great deal of respect and is closely followed in development control and development planning, and is quoted profusely in the decision-making process, especially at inquiries. Some national policy is still to be found in circulars and also, from time to time, in ministerial statements. Major changes in policy are often published in White Papers. All of these documents can be regarded as material considerations in planning, and thus central government has an array of instruments in which national policy can be expressed. Indeed the result can be confusing, even for professionals including planning inspectors. The reviews underway may go some way to addressing this problem, but it will always exist. Other developments in this area are the more strategic spatial planning frameworks discussed above and possibly also more spatially specific policy on a few issues such as infrastructure. A critical task is engaging other sector policy, the policy-makers and operators such that there is wider ownership of national spatial policy among government departments, agencies, and service providers.

Regional spatial strategies in England

In parallel with developments at the national level, regional planning policy has also been gradually strengthened since the mid 1980s, and this has involved three major steps. The first began with the 1986 consultation paper on *The Future of Development Plans*, which recognised the value of voluntary cooperation among local authorities in some regions such as East Anglia and the West Midlands in producing a general regional overview to guide the production of development plans. Official encouragement was given in this Consultation Paper to the formation of other regional groupings, though no precise procedures were suggested. Regional planning conferences of local authorities were invited to prepare draft regional planning guidance looking ahead twenty years or more. The Secretary of State would then consider and publish final guidance. The 1989 White Paper recommended the involvement of business organisations and other bodies as well as local authorities in the preparation of guidance. Conservation and agricultural interests were later added to this list. Strategic guidance was also produced in the metropolitan counties but this was gradually merged with the regional guidance. The full set of regional planning guidance took some time to prepare but was welcomed by all sides. Early efforts came in for considerable criticism, being described as little more than a detailing of national guidance and restatement of current policies (Minay 1992), though Roberts (1996) gave a more positive appraisal. A later review by Baker (1998) noted the increasing specificity of regional guidance through the use of sub-regional divisions, first signs of attempts to integrate a wider range of sectoral policy interests (particularly transport and economic development) and growing institutional capacity for planning at the regional level. The most contentious task was in defining and allocating regional housing targets, which led to considerable central government amendments to regional guidance, especially in the South.

From 1997, the new Labour government's commitment to regionalisation provided the right context for a second step to be made on strengthening regional planning.[15] In February 1998 a ministerial statement was published on *Modernising Planning* together with a consultation paper on *The Future of Regional Planning Guidance*. One of the main themes of the ministerial statement was the need to strengthen strategic planning capabilities at the regional and sub-regional levels. The consultation paper accepted the validity of criticisms of regional planning guidance in not providing a real strategic direction for the regions and not having the confidence of regional stakeholders. Proposals were made for extending the scope and

specificity beyond land use, by making the process for its production more inclusive and transparent; and by giving the regional bodies the main competence for its production. The proposals stopped short of recommending that regional guidance should become a statutory document on the basis that this would require primary legislation. Local authorities were encouraged to take the principles of the consultation document forward prior to the publication of formal guidance. The renewed emphasis on the regional level was reflected in the publication of *Planning Policy Guidance Note 11: Regional Planning*, published in 2000.

EU thinking on spatial planning, explained above, had a considerable influence on the new approach to regional planning. The RPG was to provide a *regional spatial strategy* for the region with a planning horizon of fifteen to twenty years, covering new housing, the environment, transport, infrastructure, economic development, agriculture, minerals and waste treatment and disposal. It was also expected to provide a strategic context for the preparation of local transport plans and regional economic strategies (PPG 11 (2000) para. 1.03). There was a strong emphasis in the new arrangements on the need for more concise and regionally specific guidance not repeating national policies and with more attention to how planning could help to deliver regional policies, including those of the regional economic strategies.

The third step followed soon after. The 2001 *Planning Green Paper* (discussed in more detail later) repeated many earlier criticisms: regional planning guidance was still too long and too detailed, it was poorly integrated with other regional strategies (especially the regional economic strategies), there was duplication of effort at the regional and county level and difficult decisions were being avoided, not least the provision of sufficient housing land in the South East. All this needs to be seen in terms of the weak institutional arrangements at the regional level and the very limited resources available for the preparation of regional guidance, After decades of neglect, the regional planning function was weak and 'dependent upon local government officers coming together and carrying out the necessary studies on the back of their mainstream jobs' (Kitchen 1999: 12).[16] The Green

Paper addressed these issues with a proposal, later implemented in the 2004 Act, to replace regional guidance with regional spatial strategies. The new strategies would be more focused on regional level strategic issues; they would be used as a tool to integrate strategies at the regional level; and there would be sub-regional strategies that would replace structure plans (see p. 109). The transitional arrangements give structure plans a life of only three years (to 2007), though if there are good reasons for keeping some structure plan policies longer, then a case can be made to government. Otherwise they will cease to have effect after the three years, which may be some incentive to make rapid progress on the RSSs.

The earlier changes had stopped short of giving the regional guidance statutory status but the Planning Act 2004, following proposals in the Green Paper, has made the regional spatial strategy a statutory planning instrument. Furthermore, at a stroke it recast all the existing regional planning guidance notes as regional spatial strategies so as to give them this status. This means that the regional spatial strategy becomes part of the statutory development plan, local development documents must be in conformity with the strategy and development control decisions must be made in accordance with it (see Chapter 5). The statutory duty in s. 39 of the 2004 Act requiring plans to contribute to sustainable development also applies. The new status for the regional strategy is in effect replacing the status that structure plans enjoy. In practice this may not make much difference in the short term. Although fierce arguments about meeting regional targets for housing development are almost routine in the South East, regional policy still contains little else to argue with. In the longer run, however, there is potential for the regional spatial strategies to make much more of an impression. Some regions are certainly making use of the broader and stronger remit to recast regional strategy so as to address more vigorously the objectives of economic competitiveness and sustainable development.

The renaming of regional planning guidance notes as regional strategies is confusing. At least one RPG, for the East Midlands region, has been published at the time of its approval with the name of regional spatial strategy because it was after the Act. But all RPGs have now been given the name 'spatial strategy' though they were published as (and the documents say they are) regional planning guidance. The more recent regional guidance or their revisions have taken into account the changing government view about their scope and purpose, others have not, and none has yet been approved through the new procedure.

There are many other questions about form, scope and procedure which are answered in summary by *PPS 11: Regional Spatial Strategies* published in 2004. The main components of the strategy are set out in Figure 4.3. This shows that the *regional transport strategy* (RTS) is an integral component of the RSS, though separate advice is given on its preparation. Thus local transport plans must also be in conformity with the RSS. The strategy must contain a vision statement, a spatial strategy with a key in diagrammatic form, and an implementation plan. Sub-regional strategies may be prepared for parts of the region and these will throw up some of the most difficult issues cutting across local authority boundaries, especially where city and rural authorities meet. Indeed the rationale given for replacing structure plans with the regional strategy was the need for policy that cuts across county boundaries in addressing functional sub-regions such as travel to work or housing market areas. Exceptionally, government policy accepts that additional non-statutory strategic planning policy may need to be prepared for functional regions that cross over regional boundaries. The Thames Gateway is one such case, involving three regions. The responsibilities for the preparation of RSS are in some ways now clearer than for RPG, and also to some extent stronger. Regional planning bodies are now charged with the preparation of RSS, but they still lack the authority of directly elected regional government. As a result, the government offices are heavily involved in the process (bringing the force of central government to bear), and the regional development agencies also get heavily involved because of their influence and interest in implementation. County planning authorities are also involved in partnership with the regional body in preparing the sub-regional elements of the RSS. Along with the national parks and unitary districts, they can take on

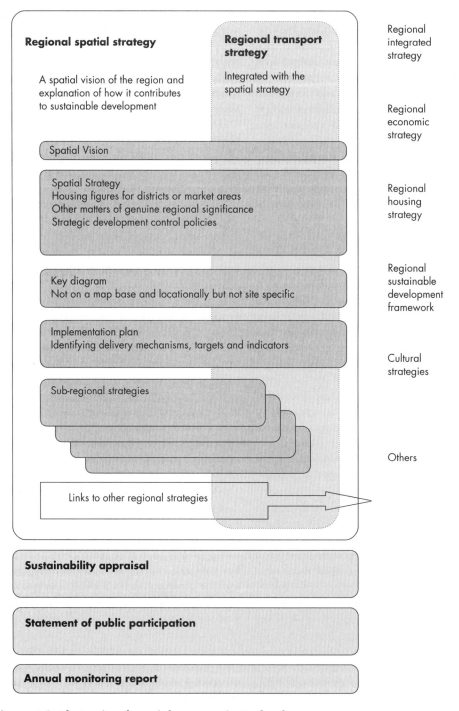

Figure 4.3 *The regional spatial strategy in England*

work on the regional strategy in the preparation of sub-regional strategies and in other work such as monitoring and analysis. Arrangements for London were made separately and earlier, so that the *London Plan: Spatial Development Strategy for Greater London* was the first 'regional' strategy to be prepared and adopted. In this case, the Greater London Authority is the responsible body for both preparation and 'approval'.

The increased status of regional planning policy together with demands to improve community involvement have required introduction of even more testing procedures for its preparation and approval. The procedures introduced by the 2004 Act and accompanying Regulations are summarised in Figure 4.4. In summary, the regional planning body prepares a draft strategy in close collaboration with the government office and regional development agency, which is submitted to the Secretary of State, who then appoints a panel to hold an examination in public. The panel reports to the Secretary of State, who then amends the strategy and consults further before issuing the final strategy.

The first rounds of RPG in the 1980s were subject to some delay in approval by government, partly because of unresolved contentious issues. The West Midlands region experimented with a public examination for the RPG in a similar way to that conducted at the time for structure plans. The government took up this innovation and made the examination in public a requirement for all RPG in the changes made in 2000. The general procedure remains the same after 2004 for the RSS. There is even more emphasis on project planning and adhering to a quite difficult 'indicative timetable' set out by central government. Figure 4.5 shows the emerging structure of regional spatial strategies along with the national frameworks already in place elsewhere in the UK. Because every RPG is now an RSS, all the new strategies are revisions, but in some cases they will need to be substantial revisions to meet the new broader objectives set by government, especially in providing sub-regional guidance. The RSS is 'tested' at an examination before an independent panel with invited participants. The panel's report forms part of the input to the Secretary of State's consideration of the guidance before a draft

RPG is published. After experiencing the examination of the South East RPG, Crow (2000) concluded that it can make a positive contribution – although the final decision still rests with the Secretary of State.

Two other innovations carried forward into the 2004 reforms are the requirement for sustainability appraisal (considered further in Chapter 7) and the identification of clear targets and performance indicators.[17] The DETR commissioned research on both matters which provides the basis for good practice guidance for the RPBs.[18] An ECOTEC study on targets and indicators highlights the haphazard proliferation of targets and indicators (and might usefully be replicated for development plans). The study found little systematic consideration of indicators or targets in relation to the policy objectives but they mostly represent general aspirations for the region. Lack of data here is a problem; indeed, the whole evidence base for regional strategies (and not just RSS) is weak and the ODPM plans to commission research on this in 2005. The newly formed regional observatories are making some progress on providing more regionally relevant data. The requirement for an annual monitoring report on RSS will also demand better data and information.

The central purpose of a regional spatial strategy is still to provide a regional framework for the preparation of local authority development plans. However, the RSS must contribute to the joining up of policy at the regional level. As the PPS puts it:

> The government's policy on spatial planning goes beyond traditional land use planning to bring together and integrate policies for the use and development of land with other policies and programmes which influence the nature of places and how they function . . . In line with this, RSS policies should draw out the links with related policy initiatives and programmes to deliver the desired spatial change.
>
> (para. 1.6)

This is rather a woolly statement and sets a difficult task for which the regional planning bodies and RSS are not yet well equipped. The new approach recognises the need to involve regional stakeholders more fully

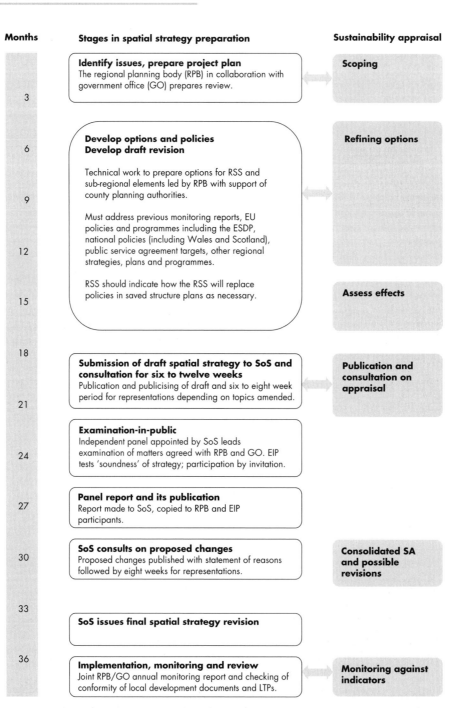

Figure 4.4 *The procedure and indicative timetable for revising regional spatial strategies*

Source: ODPM (2004) Planning Policy Statement 11, *Regional Spatial Strategies*

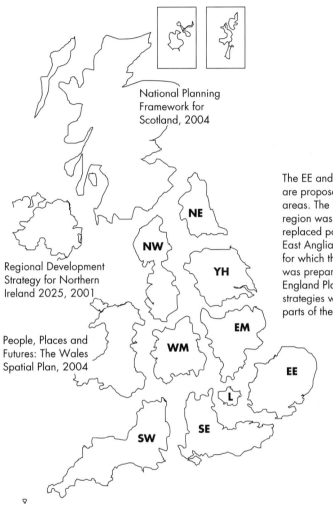

National Planning
Framework for
Scotland, 2004

The EE and SE areas of RSS
are proposed rather than actual
areas. The East of England
region was created in 2000 and
replaced parts of the regions of
East Anglia and the South East
for which the current RPG/RSS
was prepared. The East of
England Plan will replace these
strategies where they cover
parts of the East of England

Regional Development
Strategy for Northern
Ireland 2025, 2001

People, Places and
Futures: The Wales
Spatial Plan, 2004

NE
NW
YH
EM
WM
EE
L
SE
SW

NE	The Regional Spatial Plan for the North East, 2002
NW	Regional Spatial Strategy for the North West, 2005 Draft
YH	Regional Spatial Strategy for Yorkshire and Humber, 2005 Draft
EM	Regional Spatial Strategy for the East Midlands, 2005 Draft
WM	The West Midlands Spatial Strategy, 2005 Draft
EE	The East of England Plan (RSS), 2005 Draft
SE	The South East Plan: Regional Spatial Plan for the South East, 2005 Draft
SW	Regional Spatial Plan for the South West, 2005 Draft
L	The London Plan: Spatial Development Strategy for Greater London, 2004

The EM RSS also covers the whole of the Peak District National Park including the
parts in other regions, and the SE RSS includes the whole area of the New Forest National Park.

Figure 4.5 *National and regional spatial planning instruments in the UK*

in the process, notably the regional development agencies, other public bodies in such sectors as health and education, the Environment Agency, business and commercial organisations including transport operators and utility companies, and voluntary bodies. The statutory status of the RSS may give more incentive to these other interests, some of whom have hitherto showed little interest in regional spatial planning, to become more involved in the process. Nevertheless, the expectations that an enhanced regional planning policy can command the commitment of a whole range of regional and national actors seems ambitious in the face of the dominance of a few corporate stakeholders (Vigar and Healey 1999). There is a need for a system which obliges all the key regional interests to work cooperatively on strategic spatial planning process and to sign up to its conclusions. While the new arrangements move in this direction, they are still firmly rooted in the land use planning sector, with little incentive for some interests to get involved. However, practice varies considerably across the country, and where there is a common enemy or crisis (as in the North) there is likely to be more cooperative working. For much of the country, especially on the fringes of the metropolitan areas and big cities, conflict and not cooperation is the norm. In these places the more difficult decisions are sometimes swept under the carpet; certainly it is more difficult to come up with radical solutions. Much of the revised regional planning procedure is intended to enable (or force) regional planning bodies to make these difficult decisions and particularly to plan for substantial growth in the south. The challenge will be for the regions to come up with meaningful strategies in the face of a process that requires much compromise, and this will need convincing analysis and arguments and perhaps helping hands from the government offices and development agencies.

The relationship between the RPG and RDA strategies will be critical. In this approach much emphasis is given in government policy on cooperative working between the regional bodies to ensure complementarity of the strategies. The regional economic strategies will need to operate within spatial strategy and, in turn, the spatial strategy will need to reflect the economic strategy's analysis of the regional economy so as to devise a strategy that can support objectives such as the promotion of the knowledge economy and improving productivity in the regions. Kitchen asks: 'What will happen when push comes to shove, as at some time in most regions it will? Will the RPG with its environmental and sustainability appraisals have sufficient teeth to make a real difference to what RDAs actually do?' (Kitchen 1999: 130). The relationships between the RPBs and RDAs is to be monitored by the government offices, and there is no doubt that there will need to be considerable work to avoid inconsistency and contradiction.

DEVELOPMENT PLANS

Establishing development plans 1947–68

The main instrument of land use control in Britain until 1947 was the planning scheme. This was, in effect, development control by zoning. As discussed in Chapter 2, zoning was replaced in 1947 by a markedly different system which attempts to strike a distinctive balance between flexibility and commitment. The approach is, in many important ways, the same in 2005 as it was in the 1950s. It is fundamentally a discretionary system in which decisions on particular development proposals are made as they arise, against the policy background of a generalised plan. The 1947 Act defined a development plan as 'a plan indicating the manner in which a local planning authority propose that land in their area should be used'.

Unlike the prewar operative scheme, the development plan did not of itself imply that permission would be granted for particular developments simply because they appeared to be in conformity with the plan. Though developers were able to find out from the plan where particular uses were likely to be permitted, their specific proposals had to be considered by the local planning authority. When considering applications, the authority was expressly directed to 'have regard to the provisions of the development plan', but the plan was not binding and, indeed, authorities were instructed to have regard not only to the development plan but also to 'any other material considerations'. Furthermore, in granting permission to develop, local authorities could impose 'such conditions as they think fit'.

However, though the local planning authorities had considerable latitude in deciding whether to approve applications, it was intended that the planning objectives for their areas should be clearly set out in development plans. The development plan consisted of a report of survey, providing background to the plan but having no statutory effect, a written statement, providing a short summary of the main proposals but no explanation or argument to support them, and detailed maps at various scales. The maps indicated development proposals for a twenty-year period and the intended pattern of land use, together with a programme of the stages by which the proposed development would be realised. The plans were approved by the minister (with or without modifications) following a local public inquiry. Initially, a three-year target was set for submission of the plans, but only twenty-two authorities met this, and it was not until the early 1960s that they were all approved.

By this time, the requirement to review plans on a five-yearly cycle had brought forward amendments, many taking the form of more detailed plans for particular areas. These had to follow the same process of inquiry and ministerial approval as the original plans, and many authorities were still engaged on the first review in the mid 1960s. Furthermore, although the system of development control guided by development plans operated fairly well without significant change for two decades, 1947 style plans did not prove flexible in the face of the very different conditions of the 1960s. The statutory requirement for determining and mapping land use led inexorably towards greater detail and precision in the plans and more cumbersome procedures. The quality of planning suffered, and delays were beginning to bring the system into disrepute. As a result, public acceptability, which is the basic foundation of any planning system, was jeopardised.

It was within this context that the Planning Advisory Group (PAG) was set up in May 1964 to review the broad structure of the planning system and,

in particular, development plans. In its report, published in 1965, PAG proposed a further fundamental change to the planning system, one which would distinguish between strategic issues and detailed tactical issues. Only plans dealing with the former would be submitted for ministerial approval: the latter would be for local decisions within the framework of the approved policy. Legislative effect to the PAG proposals was given in 1968 (for England and Wales) and 1969 (for Scotland), creating a two-tier system of structure plans and local plans.

Structure plans and local plans since 1968

The essential features of the 1968 system are still in place, though there have been some major and many minor amendments. *Structure plans* were introduced in 1968 to provide a strategic tier of development plan and, until 1985, were prepared for the whole of England by county councils (and the two former national park boards). They are no longer prepared but most will be extant (in currency as part of the statutory development plan) until 2007, so some explanation of the history is needed. They were originally subject to the Secretary of State's approval, but since 1992 have been adopted by the planning authority itself. They consist of a written statement of policies and proposals, a key diagram setting out the broad land use policies (but not detailed land allocations) for the area and an explanatory memorandum in which the authority summarises its justification for the policies.

The detailed arrangements for structure planning have been amended on numerous occasions and central government has vacillated on their legitimate scope and content. The initial conception was that they should be wide-ranging but the government narrowed the range of competence of structure plans over the years, only to widen it again in 1999. The content of plans is discussed further below. The functions, last set out in the now superseded PPG 12 (1999), are 'to state in broad terms the general policies and proposals of strategic importance for the development and use of land in the area, taking into account national and regional policies' (para. 3.7). The structure plan should indicate the scale of provision including figures for housing and other land uses, 'and the broad location of major growth areas and preferred locations for specific types of major development . . . [and] the general location of individual major and strategic developments likely to have a significant effect on the plan; and . . . broad areas of restraint on development' (para. 3.8).

The structure plan makes use of a key diagram rather than a map, thus avoiding the identification of particular parcels of land. This limits debate to the general questions of *strategic location* rather than the use of specific sites. General land use policies can thus be determined before detailed land use allocations are made, albeit not always to the liking of those affected by later more detailed plans. In practice, over much of the history of structure planning, counties have formulated their 'policy and general proposals' in greater detail than anticipated by government, including quite detailed land allocations and development control policies in some cases. An argument in favour of more detail was that few local plans were being produced, but more detail also gave the county council more control over the implementation of policy.

Local plans provide detailed guidance on land use. They too are to be replaced under the new system introduced in 2004, but many will be extant for some time, so explanation is needed. They consist of a written statement, a proposals map and other appropriate illustrations. The written statement sets out the policies for the control of development, including the allocation of land for specific purposes. The proposals map must be on an ordnance survey base, thus showing the effects of the plan to precise and identifiable boundaries. Under the 1968 system, there were three types of local plan: general plans (referred to as 'district plans' before 1982), action area plans and subject plans. *General local plans* were prepared 'where the strategic policies in the structure plan need to be developed in more detail'.[19] *Action area local plans* dealt with areas intended for comprehensive development and *subject plans* dealt with specific planning issues over an extensive area, typically minerals and green belt, but many others such as caravans and pig farming.

Local plans have never been subject to approval by the Secretary of State, but are adopted by the local planning authority (although the Secretary of State does have rarely used powers to call in plans and to require modifications). The original rationale for this was that a local plan would be prepared within the framework of a structure plan, and since structure plans would be approved by the Secretary of State, local authorities could safely be left to the detailed elaboration of local plans. This went to the very kernel of the philosophy underlying the 1968 legislation, namely that central government should be concerned only with strategic issues, and that local matters should be the clear responsibility of local authorities.

This division of plan-making functions was predicated on the creation of unitary planning authorities responsible for preparing both the structure and local plan. But the Local Government Act 1972 established two main types of local authority in England and Wales, and divided planning functions between them. The two levels of local government do not share the same views about planning policy across much of Britain, which exacerbated conflict in the system. For some this has always been a fundamental weakness of the system leading to calls for the abolition of structure plans, but for others it has been a useful separation of powers, with the conflict usefully exposing critical issues in planning. Two mechanisms were introduced to promote effective cooperation in the planning field and to minimise delay, dispute and duplication – the development plan scheme (later the local plan scheme) and the certificate of conformity.[20]

Evaluation of the 1968 development plans

There has probably never been a time when development plans, of whatever vintage, did not have their critics – and many of the criticisms have never changed. A decade after the start of the new system, Bruton (1980: 135) summarised the problems as 'delay and lack of flexibility; an over-concentration on detail; [and] ambiguity in regard to wider policy issues'. The same is probably true nowadays.

Plans were very slow in coming forward to statutory approval and adoption.[21] The first structure plan cycle took fourteen years to complete and many took more than two years to get approval once prepared.[22] They were long, complex and contained policies thought to be not relevant to planning, for example, costs of waste collection, the development of cooperatives, standards of roads maintenance and even 'nuclear-free zones'. These delays held back the adoption of local plans; indeed, the first local plan was not adopted until 1975. However, the rate of deposit and adoption increased sharply in the 1980s and, by March 1987, 495 local plans had been adopted in England and Wales (Coon 1988). Unfortunately, many of the plans were out of date by the time their processing, which took an average of five years, was complete. (Again, this problem is still as relevant in 2005.) One result in the 1970s and 1980s was the proliferation of non-statutory planning documents ('informal policy') which outnumbered statutory plans by about ten to one (Bruton and Nicholson 1985). They took many forms, from single issue or area policy notes to comprehensive but informal plans, but much of it was correctly described as 'bottom drawer policy' which had been subject to little consultation. There is a need for some policy or guidance to be supplementary to the statutory plan and not subject to the same statutory procedures but much informal policy at this time was prepared in this way to avoid public scrutiny and/or the formal procedures.

Structure planning was not coming up to the expectations of the PAG report. Though it undoubtedly provided a forum for debate about strategy, it did not provide the firm lead that was promised. The uncertainties and complications of structure planning in practice carried over to local planning and contributed, in some areas, to a professional culture that was at best indifferent to statutory plans (Shelton 1991). There were more positive attitudes in other areas. Where the stakes involved in development applications were high, as in London and counties such as Hertfordshire (where full statutory plan cover was completed during the 1980s), statutory plan-making was vigorously pursued. Also, despite turbulent economic conditions, the plans proved to be reasonably robust and effective in implementing policy and

defending council decisions at appeal. This variation in practice was identified in research at the time: Healey *et al.* (1988) concluded that plans had proved to be effective in guiding and supporting decisions, and in providing a framework for the protection of land. They were particularly useful in shaping private sector decisions, especially in the urban fringe. Conversely, the difficulty encountered in controlling public sector investment in housing, economic development, inner city policy and infrastructure provision, was shown to be an impediment to effective implementation of strategy (Carter *et al.* 1991).

Davies *et al.* (1986a, 1986b, 1986c) concluded that plans might play only a small part in guiding development control decisions overall, but were much more important when a case went to appeal – what they termed the 'pinch points' of the system. They suggested that this reflected the system's chief virtue: its ability to enable a sensitive response to local conditions. It was recommended that the government should encourage local authorities to provide better written policy cover, to reduce its complexity, to use statutory plans, and to facilitate more speedy adoption.[23] A big contribution to the failure to produce plans has been the confused attitude of central government. The status of statutory plans reached a low point in 1985 when the White Paper, *Lifting the Burden*, denigrated both structure and local plans, and criticised the procedures for preparing plans as 'too slow and cumbersome'. More flexibility was also called for – somewhat at odds with previous advice which had sought to reduce administrative discretion in the system by a planning framework which offered more certainty, clarity, and consistency to private sector investors (Healey 1986).

Unitary development plans in England from 1985

The Thatcher government's precipitate decision to abolish the GLC and the MCCs forced hasty changes to the planning system in these areas. This was simple in the extreme: London boroughs and the metropolitan districts became 'unitary' planning authorities, and a new *unitary development plan* (UDP) was proposed to be prepared for each authority. In precisely those parts of the country where there is a particular need for a strategic approach to planning, the structure planning tier was lost. It was replaced by a tier of *strategic guidance* produced on a cooperative basis by the constituent districts but issued by the Secretary of State. A joint planning committee was established for this purpose in Greater London – the London Planning Advisory Committee.[24] Subject and action plans also disappeared in favour of the new 'unitary' approach to plan-making.

Unitary development plans will be replaced by the new local development frameworks but remain until such plans are put into place. They are in two parts: Part I is the structure plan element and has the characteristics of the structure plan described above. Part II is the local plan element with a written statement of the authority's policies and proposals, a map showing these proposals on an Ordnance Survey base and a reasoned justification of the policies. The parts of a UDP are presented as one document, and they have a ten year horizon. The UDP is adopted by the district council, and is not subject to the approval of the Secretary of State (although reserve powers of central intervention have been maintained).

There was a good deal of initial scepticism about these new arrangements, though they are more closely allied to the 1965 thinking of PAG than the system that was then put into place. There were particular concerns about the future of strategic thinking in the metropolitan areas, difficulties of cooperation between districts, and problems of participation and coping with the statutory right to objection in plans which embrace such large areas (Nadin and Wood 1988). For the districts themselves, many of these worries have proved unfounded. It has been possible to accommodate policy and political differences among districts, but this has been very much on a lowest common denominator level (Hill 1991; G. Williams *et al.* 1992). Serious concerns have been voiced about the extent to which the public, interest groups and even some professionals can engage effectively in the process. There have also been considerable delays in some metropolitan districts, where very detailed UDPs were produced in particularly contentious circumstances

generating great conflict and many thousands of objections.

The future of development plans

In 1985, prompted by the concern for 'freeing' enterprise from unnecessary restraints, the White Paper *Lifting the Burden* announced that there were to be changes in the development plan framework. The following year a consultation paper was published proposing the abolition of structure plans in England and Wales (but not in Scotland, where they had 'not in general given rise to the same problems as have been experienced south of the Border') and their replacement by statements of county planning policies on a limited range of issues (to be specified by the Secretary of State), more policy at regional and sub-regional level, and the introduction of single-tier district development plans covering the whole of each district. These proposals were a response to growing dissatisfaction about the making of many ad-hoc and apparently inconsistent decisions by both the Secretary of State and local authorities. The lobby for change included both the development and the conservation lobbies, who had a common demand for more certainty in the system and a reduction in the growing number of speculative applications. There was also some dissatisfaction among government supporters about decisions taken centrally which went against local (often Conservative) opinion. Local authorities were concerned that more of their decisions were being overruled, and complained at the lack of clarity in central policy. By comparison, matters looked better in Scotland and in the emerging system in the metropolitan counties.

Many of the 500 responses to the consultation paper argued very strongly against the proposed abolition of structure plans and (for the time being) they were saved. Instead PPG 12 was published in 1988 urging local authorities to extend statutory plan coverage, normally by district wide plans, and to replace non-statutory policy which it described as 'insufficient and weak'. Strategic green belt boundaries were singled out as requiring further specification in detailed local plans (and some were still not completed in 2004). In return, the government offered an enhanced status for plans. The same proposals were published in the 1989 White Paper *The Future of Development Plans*, with the addition of a mandatory provision for all counties to prepare minerals development plans. County councils were urged to press ahead with the revision and updating of structure plans and to cooperate on the elaboration of regional guidance. The counties, for their part, were to ensure that plans were less bulky and concentrated on strategic issues. Shortly afterwards the government announced proposals to end the requirement for the Secretary of State to approve all structure plans and alterations in favour of adoption by the local authority. During debate on the Planning and Compensation Bill that followed, provisions were added to further increase the status of the statutory plans in development control.

The plan-led system 1991

The Planning and Compensation Act 1991 made four major changes to the planning framework. The first made the plan the primary consideration in development control. In commending the amendment (formerly section 54A of the 1990 Act), Sir George Young coined a phrase in saying that 'the approach shall leave no doubt about the importance of *the plan-led system*'. Perhaps more significantly, the adoption of *district wide local plans* was made mandatory, and the requirement for central approval of structure plans was abolished (central government has retained its powers of intervention). Small area local plans and subject plans were abandoned except for minerals and waste, which henceforth would be prepared for the whole of the authority's area.

In the first part of the 1990s, it seemed that the framework of local planning policy in England and Wales was to become more coherent. At that time most local authorities had little coverage of statutory plans, but a mix of interlinked subject and small area-based policy documents and informal plans. The 1991 Act offered a much clearer system. Those needing to know about planning policy would make reference to the structure plan, the district wide local plan and the minerals and waste plans, with more chance that they would exist. In the event, the prospect of an orderly structure of development plans in England was dashed by local government reorganisation in the mid 1990s.

Reorganisation of local government in England between 1994 and 1997 is explained in Chapter 3. It affected the two-tier system of counties and districts beyond the metropolitan areas by introducing unitary district councils. Some counties were abolished completely to be replaced by unitary councils. Other unitary councils were established in counties, mostly for the provincial cities, creating an island unitary authority within the remaining two-tier structure. Where the two-tier system remained, the planning framework was not affected: counties prepared the structure plan and waste and minerals plans (or one plan for both topics) and districts prepared the district wide local plan. Where new unitary authorities were created, they take on the county as well as district functions. Most prepared joint structure plans with the neighbouring county councils. The joint arrangements for structure planning are summarised in Table 4.2. The exceptions are Halton, Warrington, Herefordshire, the Isle of Wight and Thurrock, which prepared unitary development plans (as in the metropolitan districts). All the other unitary authorities prepared their own district wide local plans. The metropolitan districts were unaffected by local government reorganisation and continued with their unitary development plans. The content of structure plans, unitary development plans and local plans will all remain in currency until it is replaced or reviewed by development plan documents prepared under new arrangements established in 2004.

Planning Green Paper 2001

The Planning Green Paper *Planning: Delivering a Fundamental Change* published at the end of 2001 marked the formal start of substantial change to the planning system in England which will continue for some time as the new arrangements are put into place. The New Labour agenda for modernising and

Table 4.2 *Structure plan areas in England*

Previous structure plan authority	New arrangements
Avon County Council	Joint structure plan: Bristol UA, North Somerset UA (formerly Woodspring), Bath and NE Somerset UA (formerly Wansdyke and Bath), South Gloucestershire UA (formerly Northavon and Kingswood)
Bedfordshire County Council	Joint structure plan: Luton, Bedfordshire County Council
Berkshire County Council	Joint structure plan: Bracknell Forest UA, Newbury UA, Reading UA, Slough UA, Windsor and Maidenhead UA, Wokingham UA
Buckinghamshire County Council	Joint structure plan: Milton Keynes UA, Buckinghamshire County Council
Cambridgeshire County Council	Joint structure plan: Peterborough UA, Cambridgeshire County Council
Cheshire County Council	UDPs for the UAs and a structure plan for the remainder Halton UA, Warrington UA, Cheshire County Council
Cleveland County Council	Joint structure plan also with Darlington UA, Middlesbrough UA, Hartlepool UA, Redcar and Cleveland UA (formerly Langbaurgh-on-Tees), Stockton-on-Tees UA
Cornwall	Structure plan
Cumbria	Joint structure plan: Cumbria CC, Lake District NPA
Derbyshire County Council	Joint structure plan: Derby City UA, Derbyshire County Council
Devon County Council	Joint structure plan: Plymouth UA, Torbay UA, Devon County Council UA
Dorset County Council	Joint structure plan: Bournemouth UA, Poole UA, Dorset County Council
Durham County Council	Darlington UA: joint structure plan with former Cleveland LAs Structure plan for Durham County Council
East Sussex County Council	Joint structure plan: Brighton and Hove UA, East Sussex County Council
Essex County Council	UDP for Thurrock Joint structure plan: Southend UA, Essex County Council
Hampshire County Council	Joint structure plan: Portsmouth UA, Southampton UA, Hampshire County Council
Hereford and Worcester	Structure plan for Worcestershire County Council UDP for Herefordshire
Humberside County Council	Joint structure plan: Kingston upon Hull UA, East Riding UA (formerly East Yorks, Beverley, Holderness and part of Boothferry) Joint structure plan: North East Lincolnshire UA (formerly Cleethorpes and Great Grimsby), North Lincolnshire UA (formerly Glandford, Scunthorpe and part of Boothferry)
Isle of Wight	UDP: Isle of Wight UA

Kent County Council	Joint structure plan: Medway Towns UA (formerly Rochester and Gillingham), Kent County Council
Lancashire County Council	Joint structure plan: Blackburn with Darwen UA, Blackpool UA, Lancashire County Council
Leicestershire	Joint structure plan: Leicester City UA Rutland UA, Leicestershire County Council
North Yorkshire County Council	Joint structure plan: York UA, North Yorkshire County Council, Yorkshire Dales NPA
Northamptonshire	Structure plan
Nottinghamshire County Council	Joint structure plan: Nottingham City UA, Nottinghamshire County Council
Oxfordshire	Structure plan
Shropshire County Council	Joint structure plan: The Wrekin UA, Shropshire County Council
Somerset	Joint structure plan: Exmoor NPA, Somerset County Council
Staffordshire County Council	Joint structure plan: Stoke on Trent City, Staffordshire County Council
Suffolk	Structure plan
Surrey	Structure plan
Warwickshire	Structure plan
West Sussex	Structure plan
Wiltshire County Council	Joint structure plan: Swindon UA, Wiltshire County Council

Notes:
For clarity the term unitary authority (UA) is used here rather than district council. The Peak District National Park and Lake District National Park are also structure plan authorities. From April 1997 all national parks became the sole planning authority for their area.

Under the 2004 changes structure plans are being replaced by regional spatial strategies and are expected to expire in 2007 when their 'transitional arrangements' three-year life is ended and the RSS is approved.

'joining-up' government in pursuit of priority outcomes was central to the reform process, alongside oft repeated criticisms that planning is a brake on economic growth and that it does not do enough to protect the environment or promote social cohesion. Wider discourse on 'spatial planning' at the EU level provided inspiration for the direction of change (see p. 90). Recommendations on reform varied but there was a measure of agreement that planning had become marginalised in government decision making at a time when there is an urgent need for more effective coordination of the impacts of disparate strategies, policies and actions for particular places. A CBI-TUC Investment Group, for example, had drawn attention to the failure of the planning system 'to deliver the level of speed, certainty and responsiveness that businesses need to make successful investment decisions in a modern economy' (p. 31). The Group emphasised problems from inconsistencies among policies at national, regional and local levels, the complexity of national policy guidance, delays in the adoption of plans and decisions on 'major commercial applications' and delays of many years on major infrastructure projects (p. 50). These interests are concerned with the role of planning in economic productivity, particularly in facilitating and encouraging investment. From this perspective they were not impressed with the performance of the planning system. Others, such as the RSPB and LGA highlighted weaknesses in the coordination of the spatial impacts of all policies.

The 2001 Green Paper was the ODPM's response. (HM Treasury later launched another, and in some ways more significant initiative, the *Barker Review of Housing Supply* (2004), which is discussed in Chapter 6.) The Green Paper relates only to England. The Welsh Assembly published a separate consultation paper, and Scotland is in the process of considering further reforms (both are discussed in later sections). Four 'daughter documents' were published at the same time alongside the Green Paper with proposals for change on procedures for dealing with major infrastructure projects, planning obligations (planning gain), compulsory purchase and the Use Classes Order.[25] The discussion of the Green Paper is placed here because the most significant changes are in relation to the framework of plans. Other proposals of the Green Paper and its aftermath are examined in the relevant chapters.

The Green Paper starts with a positive note on the role of planning in society and the commitment of the government to the system, followed with a brief review of what was wrong with the existing system. The thrust of the government thinking (as presented by ODPM) is very much in line with the CBI–TUC conclusions, the flavour of which is given in the foreword by the then minister, Stephen Byers:

> Good planning can have a huge beneficial effect on the way we live our lives. It must have a vision for how physical development can improve a community. But some fifty years after it was first put in place, the planning system is showing its age. What was once an innovative emphasis on consultation has now become a set of inflexible, legalistic and bureaucratic procedures. A system that was intended to promote development now blocks it. Business complains that the speed of decision is undermining productivity and competitiveness. People feel they are not sufficiently involved in decisions that affect their lives.

This sentiment was widely shared and not only by the business communities. The reality of performance on development plans was inescapable. By 2002, 13 per cent of relevant authorities in England had not adopted a local or unitary development plan and 214 plans were 'out of date'. It took twenty-one county authorities more than five years to revise structure plans after the approval of new regional guidance.[26] It may be surprising, therefore, that the fundamental principles of the system were not challenged; all parties agree on the importance of the planning system in guiding and regulating development. But there was agreement that this needs to be done more effectively and that this required substantial changes in the way the task is done. The Green Paper proposed redesigning the tools and redistributing competences, a lot of filling of gaps, and many smaller changes. The analysis of the problem underlying the proposals was that the system was too complex and difficult to understand, with unclear rules about planning permission, delays, unpredictability and poor attention to the customer and good service standards. Understandably, it put much more emphasis on weak community engagement and legalistic procedures than the recommendations from business interests, indeed much of the publicity surrounding the Green Paper was almost wholly couched in terms of 'putting the community first'. Commendably, the government also acknowledged and later acted on the very difficult resource and skills base for those who had to provide the service, perhaps encouraged by the business community which specifically noted the need for action on 'enhancing local authorities' capabilities and resources'.[27]

The analysis is better on symptoms than causes. The main failures of the development plan system were obvious, if complex, but not well explained in the Green Paper. The main problems were the failure to tackle issues lying at the boundaries between authorities and between policy sectors; the 'weight' of local planning policy arising from too much attention to comprehensive coverage and not enough concentration of effort where change was anticipated or needed; an unwillingness of politicians and communities to accept new development and make difficult decisions; the questionable qualities of the eventual outcomes of new development from the planning process; and poor management of the system. Important issues are addressed to some extent in the daughter papers, for example, the costs and benefits arising from new development, the provision of necessary infrastructure where growth takes place and the ownership of plans – corporately within local authorities and more broadly among the public sector. Sustainable development is barely mentioned, though it was later to figure prominently in the changes and government rhetoric about planning reform. These and other problems were, and are, worthy of more consideration before embarking on change, especially the conundrum that some planning authorities seem to be able to do the job very well, while others with similar characteristics struggle. The Green Paper and other statements generalise far too readily and in doing so make unwarranted criticism of planning authorities and planners that are more successful. But government believed that urgent action on the whole system was needed to address poor performance of some.

The main proposals for the plan framework (which by and large were carried through) were

- the abolition of structure and local plans to be replaced at the strategic level with statutory regional and sub-regional spatial strategies (discussed above)
- at the local level, new local development frameworks offering more flexibility in the production and presentation of planning policy
- a more strategic and selective approach to local planning with more focused general planning policies and more detailed action plans where needed
- increased requirements for participation and consultation in the interests of 'engaging the community' more effectively
- business planning zones which would 'lift regulation' where it is not necessary.

It was anticipated that carrying through these proposals would cause an upheaval of regional and local development planning and, indeed, it has. It is fair, therefore, to ask if it was necessary or desired by those who work with the system. The overwhelming weight of opinion supported reform. Over 15,000 responses

were made to the Green Paper and interests from the development industry, community, environment, government and business interests all endorsed the need for change (Smith & Williamson Management Consultants 2002). However, there was even more overwhelming objection to the government's specific proposals: 88 per cent of respondents did not want to see local plans replaced by local development frameworks (96 per cent of members of the public) and 90 per cent disagreed with the proposal to abolish structure plans and replace them with regional strategies. (Only local government and the elite of the professional planning institution showed any enthusiasm.) In evidence given to the 2002 Transport, Local Government and the Regions inquiry into the reform of planning, David Lock, representing the Town and Country Planning Association, described the problem of producing development plans thus:

> It has been terribly slow, very expensive and unsatisfactory in many ways but we have been through that great loop and the Committee should know that amendments, revisions, updating of local plans are now happening very quickly. In other words, many years of investment are now yielding results.
> (HC 476 III 24/04: para. 519)

He pointed to other more simple solutions that were tried and tested elsewhere. Statutory timetables (used in the production of local transport plans) would prevent councils stalling the process at the time of an election, change of power, or indecision; and changes in policy guidance would be sufficient to prevent councils worrying too much 'as to whether this settlement boundary goes through the greenhouse of number 27 or does not' (para. 529). The failure of local authorities to work cooperatively across boundaries could be tackled through a package of incentives and sanctions. Lock explained how these relatively minor adjustments could 'get us closer to a faster, fairer and speedier system than chucking the whole lot out and spending several years constructing a brand-new one which is not yet designed' (para. 529).

Despite such reasoning it is no surprise that the government went for the more radical solution while presenting it as a means to engage the community more effectively (which was supported by 80 per cent of respondents).[28] Figure 4.6 gives an overview of the framework for expressing planning policy in England before and after 2004. This shows that all levels of planning policy have been affected though the biggest changes are at the county and district levels (the impacts on national and regional policy are explained above). The role of the structure plan is replaced both by the sub-regional element of the regional spatial strategy and the more strategic elements of new development plan documents at the local. The role of the local plan will be taken on by a portfolio of development plan documents prepared at the district level. Ironically, the impasse created by moving from one system to another has given some local authorities a breathing space as they abandon their current programme of adoption of contentious plans while expressing much enthusiasm about making progress on the adoption of the new type of framework. But the incremental approach was unlikely to be enough to satisfy ministers, officials and the planning elite who need to demonstrate to other government interests that they were really tackling the issue. Perhaps also it would not have provided sufficient impetus for changing attitudes about planning that is so clearly needed. It would be wrong to underplay this important dimension of the reform. New instruments and procedures are one thing, but the real challenge is to replace the tired and negative corporate mind-set of the planning profession. The Green Paper was part of a much broader campaign to change the very culture of planning.

Local development framework 2004

The concerns of the majority who opposed the abolition of structure and local plans have largely been borne out. The outcome is on the face of it a very complex revision of the tools of local development planning. There are opportunities to make this a much more user-friendly system for the consumer in the medium term, but a lot of activity on plan-making and review has temporarily been diverted to learning about the nature of the new tools and project planning. The government has sought to help with a whole catalogue of documents providing 'good practice guidance' (though there has been little practice on the new system so far). There is no doubt a need for guidance on specific issues, particularly given the battery of new acronyms and terms, though the flood of documentation (some of which is quite repetitive) may not have helped to sell the new system. The puzzle is that many local authorities seem to want yet more guidance, while the new system should give more flexibility, enable innovation and the creation of more locally relevant solutions. There are other questions about why the various components of local planning had to be expressed in such a con-voluted way, and why the government policy needs to adopt such a prescriptive 'painting by numbers' approach, especially given the very different conditions under which plans are produced in different parts of the country. The reform process began with the intention of reducing and sharpening the focus of government guidance and the requirements for detailed local authority 'project planning' on plan-making, closely supervised by the regional government offices. Government has taken the need to monitor and evaluate progress very seriously and parallel research projects are underway that will provide feedback to those who are operating the system. The stream of publication also provides a plentiful supply of information for students about what the government think the new system is about and how it should be operated. *PPS 12 Local Development Frameworks* and its sister document *Creating Local Development Frameworks: A Companion Guide to PPS12* are the main sources, though the system that has been created (and in places the way it is explained) provides much opportunity for confusion. For example:

> The local development framework will be comprised of local development documents which include development plan documents, that are part of the statutory development plan and supplementary planning documents which expand policies set out in a development plan document or provide additional detail.
>
> (para. 1.4)

The local development framework is not a plan, but a folder or portfolio that contains all the local development documents of the local authority together with other related information. Local development framework is a non-statutory term not defined in the Act, and is only used to describe the full portfolio of the local development documents. These documents are statements of the local planning authority's planning policy. So, for the visitor to the planning office, there should be one source for all information on the local authority's policies – which if it comes off will be a major step forward. But local authorities will need to give some attention to how this is presented to those who need to know about planning. We are all used to the Windows user interface which hides the complexity of what lies behind the computer screen. A similar approach is needed for the local development framework. In the mean time, those who need to know about planning face a potentially confusing array of documents.

There are two types of local development document: *development plan documents* and *supplementary planning documents*. Both are statements of planning policy but they have different status. It is important to distinguish first the development plan documents because these contribute to the *development plan*, which has special status in decision-making. The development plan is the starting point for planning decisions and decisions 'should be made in accordance with the plan unless material considerations indicate otherwise' (section 38(6) of the 2004 Act).[29] In all cases the development plan will comprise a core strategy, site specific allocations and a proposals map. The local authority may also prepare action area plans and other development plan documents, and it is likely that all development plans will have some optional components. As shown in Figure 4.7 the development plan documents in the local development framework combine with other documents at the regional level to make up the complete development plan: the regional spatial strategy (or the London Plan) and its component regional transport strategy and sub-regional spatial strategies, and minerals and waste development plan documents prepared by county councils where they exist. The development plan also comprises saved policies from

structure and local plans that will be replaced in due course by the new development plan documents; indeed it is inevitable that we will be using structure and local plans in some places for a considerable time to come. Policies in structure and local plans are saved for three years initially, and the Secretary of State makes special provision for them to continue beyond that date. Local planning authorities will need to seek this special provision if they think it is required.

Some development plan documents are mandatory, others can be prepared at the discretion of the council. The mandatory components are the core strategy, site specific allocations and the proposals map. The *core strategy* comprises a long-term spatial vision and should express the broad policies that are needed to achieve that vision and provide a monitoring and implementation framework to measure progress towards it. The word strategic is important. It means that the core strategy should look over a long-term horizon (in the government's view at least ten years), it should take a comprehensive view showing how the planning strategy fits in with other strategies in its own and neighbouring areas, and that it should identify broad locations for policies but not individual sites. It should generally be the first document to be prepared, but there is great variety in the way that local planning authorities can bring forward components of the local development framework as explained below. The core strategy is not so much an innovation for many authorities that had already presented a vision statement at the start of their local plans together with a diagrammatic representation of the district – a key diagram can also be used in the new system.

Site specific allocations are exactly what they say and will be designated also on the proposals map. Other policies setting out the criteria for planning permission on such sites will be needed too, and may be contained in the same or a different document. *Area action plans* are now back in the system (the first local plans adopted in the 1970s were action plans). They may be prepared for areas where there is a need, either because of significant change or the need for careful conservation. They are not mandatory, but if they are prepared will become parts of the development plan. Examples of where they might be used are given in PPS 12: to

Before 2004

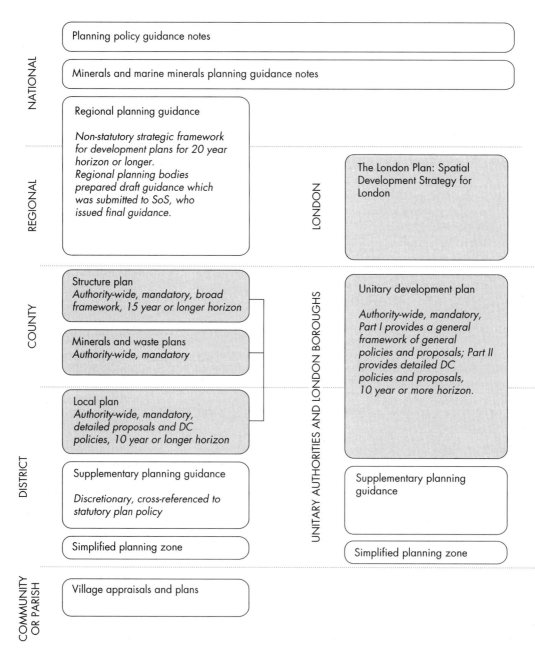

Figure 4.6 *The planning policy framework in England before and after 2004*

After 2004

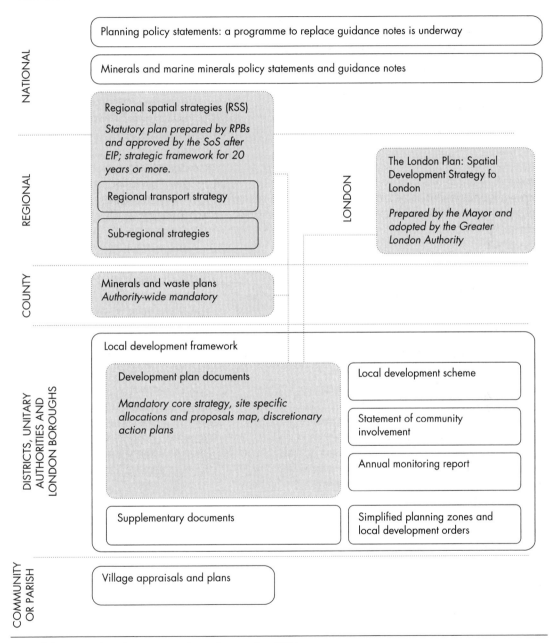

NATIONAL

Planning policy statements: a programme to replace guidance notes is underway

Minerals and marine minerals policy statements and guidance notes

REGIONAL

Regional spatial strategies (RSS)

Statutory plan prepared by RPBs and approved by the SoS after EIP; strategic framework for 20 years or more.

Regional transport strategy

Sub-regional strategies

LONDON

The London Plan: Spatial Development Strategy fo London

Prepared by the Mayor and adopted by the Greater London Authority

COUNTY

Minerals and waste plans
Authority-wide mandatory

DISTRICTS, UNITARY AUTHORITIES AND LONDON BOROUGHS

Local development framework

Development plan documents

Mandatory core strategy, site specific allocations and proposals map, discretionary action plans

Local development scheme

Statement of community involvement

Annual monitoring report

Supplementary documents

Simplified planning zones and local development orders

COMMUNITY OR PARISH

Village appraisals and plans

Regional planning guidance became regional spatial strategy from 2004
Policies in structure plans, local plans and UDPs are 'saved' until 2007 but remain extant for longer
Shaded area is the 'development plan'
National parks and the Broads Authority also prepared their own development plans from 1997

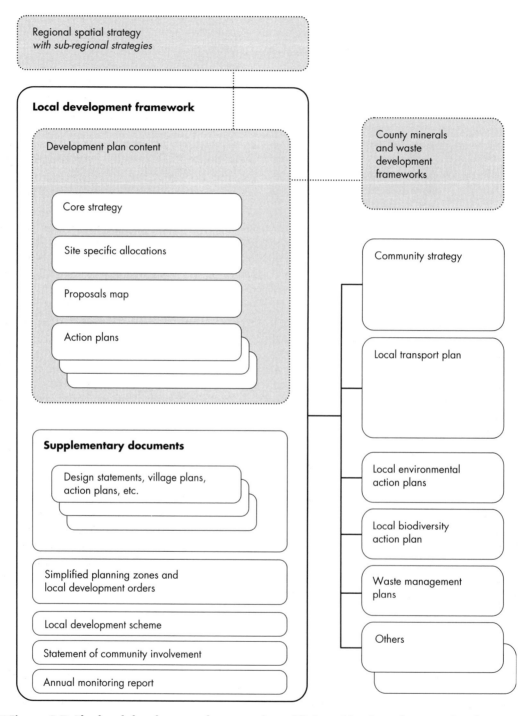

Figure 4.7 The local development framework and links with other plans (England)

deliver planned growth areas, to protect areas sensitive to change, and in areas where there is a need to resolve conflicting objectives, all sound plausible. The idea of action plans in areas to stimulate regeneration and for area-based regeneration initiatives is less plausible. Although the idea apparently builds on the importance of 'delivery' of change through planning, previous extensive studies and experience in the big cities suggest that statutory plans are not the best way to facilitate change in areas of low market demand or for that matter in town centres where there is lots of rapid change. The *proposals map* will be as before, expressed on a map where it is possible to identify the precise boundaries relating to policies and proposals. The map will be updated each time it is affected by the adoption of a new document and it will also include for information designations which are not designated through the development plan process, such as conservation areas. The philosophy again is to provide a complete picture in one source.

As well as development plan documents, the framework *must* include a local development scheme,[30] a statement of community involvement and annual monitoring reports. It may also include supplementary planning documents (and in all cases there will be some supplementary documents), local development orders (discussed in Chapter 5) and simplified planning zones where they exist (see zoning instruments on p. 140). Documents from the required sustainability appraisal also need to be made available. Some local authorities have proposed joint working on the preparation of the local development framework or action plans. This is encouraged where there may be benefits in preparing a single cross-boundary document, or simply to share resources and expertise in the preparation of separate documents.

All local authorities will have a local development framework, although what they contain will vary considerably, especially in the early years of this new system. The transitional arrangements allow local planning authorities to identify those parts of their existing plans that they wish to save and all the content is automatically saved for three years from 2004 or for plans still in the former process, from the date of adoption. Policies in structure and local plans can be saved beyond the three-year period if the authority or regional body can convince the regional office (acting for the Secretary of State) that they meet the general requirements of the new system and that there is no good reason to replace them. The *local development scheme* prepared by all local planning authorities sets out what parts of plans, if any, the authority wish to save and the arrangements for the replacement of other policies by new development documents. Similarly, regional bodies must identify how they intend to replace policies in structure plans with revisions to the regional strategy. If a draft structure plan, local plan, replacement plan or alterations had reached the 'first deposit' stage under the previous procedure, the authority were able to continue through to adoption if it wished. Where the inquiry inspector or panel had been appointed, the old procedure applied with inspectors' recommendations and local authority modifications to the plan. Where an inspector or panel had not been appointed then transitional arrangements come into play, allowing for the inspector's report to be binding on the authority and requiring no modifications stage (see below for further explanation). If the draft plan had not reached the first deposit stage then that process ceased and the proposals would have to be taken through the new process from scratch.

While the 2004 changes have given the development plan system a thorough shake up, the same elements of policy that were provided for in structure and local plans remain, but in a different guise. Strategic policy, indicating general locations but not sites, and providing for 'agreement in principle' on development and protection will be prepared through the regional (and sub-regional) strategy process, and also in the core strategy of the local development framework. Development control policies, land allocations and other policies and proposals that are designated on a map will be prepared in the development plan document process that will be presented together with other relevant policy and information in the local development framework. There should be less general policy for the whole authority area and more specific policy and proposals on areas that are subject to change or special pressure. There is more opportunity to develop individual policy instruments (documents) that are designed in form, content and procedure to be 'fit for the task'. In this way local authority and government resources and expertise should be more concentrated on priority topics. There might also be better connection to strategies and plans in other sectors, though this 'spatial planning' aspect of the reform is less in evidence in the guidance. The changing scope and content of plans is discussed in a later section.

Development plans in Northern Ireland

The formal change to a discretionary system of development plans and control did not come to Northern Ireland until 1972. Prior to this, the system was much the same as for the rest of Britain before 1947, with local authorities able to prepare planning schemes. Practice was similar also in that very little progress was made on the preparation and approval of such schemes, and a system of interim development control operated. The 1972 Order introduced the development plan, with similar status to those in the rest of the UK. There are three types of development plan (area, local and subject plans) which are produced and adopted by the DoENI (shown in Figure 4.8). Area plans which can cover the whole or a substantial part of one or more district council areas are the main reference for development control, and include both strategic and detailed policies.

The provisions of the 1991 Act amended by the 2004 Act, which make the development plan the first and primary consideration in development control, were not implemented in Northern Ireland, although in 1999 the NI Office consulted on proposals to make the plan the primary consideration. This was in response to the House of Commons Northern Ireland Affairs Committee's 1996 report on *The Planning System in Northern Ireland*. The Committee expressed serious concerns about the lack of a clear strategy for the Province as a whole (which is now met by the creation of the *Regional Planning Strategy*) and the inadequacy of the development plans system. Northern Ireland is not affected by changes in local government. Area and local plans will continue to be prepared by the six

divisional offices of the DoENI. In 2005 consultation began on fundamental reforms to the planning system in Northern Ireland which follow change made in England and Wales by the 2004 Act.

Development plans in Scotland

The Scottish system differs in several significant ways from that in England and Wales, but the two-tier system of development plans and the procedures for the adoption and approval were broadly similar until 1996. Some differences can be attributed to the particular geographical characteristics of Scotland; others may legitimately be attributed to a desire to avoid some of the difficulties of the English system. Because of the different administrative structure and larger planning areas in Scotland, there is a slightly different emphasis in the functions of structure plans which are to indicate policies and proposals concerning the scale and general location of new development, and to provide a regional policy framework for accommodating development (PAN 37: 7). Progress on the approval of structure plans was a significant problem, with an average of seventeen months needed for Secretary of State approval.

Changes to the development plan system itself have followed closely those introduced south of the Border. For example, the procedure for making alterations to structure plans and local plans has been made simpler. Also, certain adjustments have been made to the division of planning responsibilities between regions and districts.

The 1991 Act brought some of the same changes made in England and Wales to Scotland, notably the enhanced status of development plans in development control and insertion of section 18A into the 1972 Act, with the same effect as s. 54A (now s. 38(6) of the 1990 Act) in England and Wales, calls for more succinct statements of policy and the emphasis on 'physical land use development' (PAN 37). However, in a number of ways the Scottish development plan system remains distinctive. The structure plan still has to be approved by the Secretary of State. The survey still plays a part in the approval, and must be put on deposit and

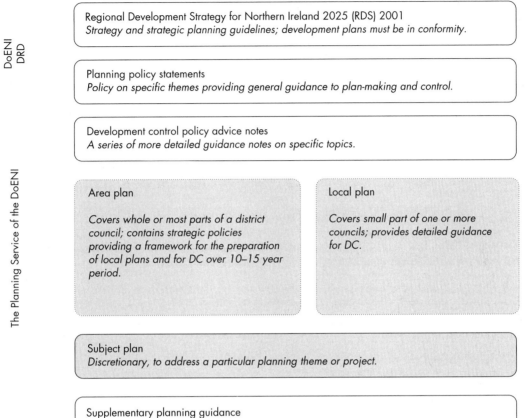

DoENI
DRD

The Planning Service of the DoENI

Regional Development Strategy for Northern Ireland 2025 (RDS) 2001
Strategy and strategic planning guidelines; development plans must be in conformity.

Planning policy statements
Policy on specific themes providing general guidance to plan-making and control.

Development control policy advice notes
A series of more detailed guidance notes on specific topics.

Area plan

Covers whole or most parts of a district council; contains strategic policies providing a framework for the preparation of local plans and for DC over 10–15 year period.

Local plan

Covers small part of one or more councils; provides detailed guidance for DC.

Subject plan
Discretionary, to address a particular planning theme or project.

Supplementary planning guidance

■ **Figure 4.8** *The planning policy framework in Northern Ireland*

accompany the deposited plan in the submission. Local government reorganisation created unitary authorities in Scotland in 1996 and although the two-tier system of structure and local plans was retained, joint working is now necessary for the production of some structure plans, and the arrangements are summarised in Figure 4.9 and Table 4.3. The Scottish Office designated seventeen structure plan areas, six of which cover more than one unitary authority, while local planning continued unchanged in the new unitary districts.

In 2001 the Scottish Office launched a review of strategic planning in Scotland which suggested that there was no real need for a second higher tier of development plans for much of Scotland, especially with the proposal for a national planning framework.

This met with some agreement and it is now proposed to amend legislation to require structure plans only for the four major city regions (Glasgow, Edinburgh, Aberdeen and Dundee) and unitary plans for the rest of Scotland. This has now been incorporated in the wider 'modernising planning agenda' in Scotland. A 2004 consultation paper *Making Development Plans Deliver* followed a 2003 paper on *Options for Change* and explained that the new *city-region plans* (CRPs) would be more selective and strategic but continue to be part of the statutory development plan and to require ministerial approval. The same paper made proposals for changes to local plans, and introduced a new local development plan reflecting a specific Scottish agenda, but echoing the thrust of change in England and calling

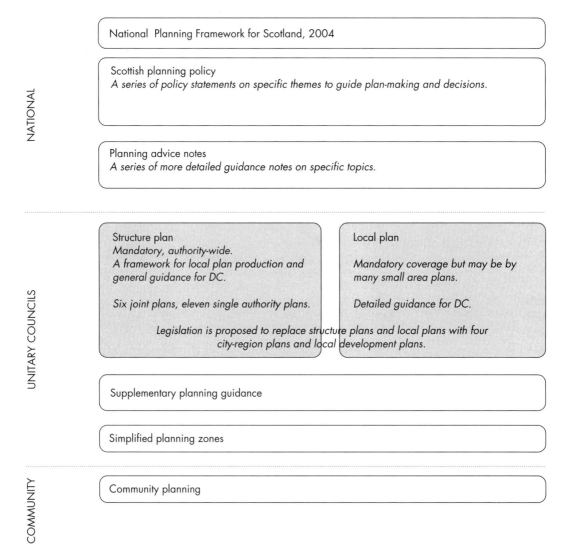

Figure 4.9 *The planning policy framework in Scotland*

for 'more urgency and confidence in the process with a greater focus on content and outcomes' (para.1). The review of progress presented illustrates the need for change, with seven out of ten local plans being more than five years without review and 20 per cent adopted more than twenty years before. The experience of delivering development plans in Scotland is similar to that in England with local government reorganisation delaying the production of plans (Hillier Parker *et al.* 1998). It was not until 1989 that full structure plan cover was achieved. However, progress on local plans has been better overall than in England and Wales, mainly because of the earlier introduction of a mandatory requirement for full cover. The current proposals are likely to be taken up in new Planning Bill, though some will be introduced through the

Table 4.3 *Structure plan areas in Scotland*

Single authority structure plan areas	Joint structure plan areas	Proposed city-region plan areas
Argyle and Bute	Aberdeen City Aberdeenshire	Aberdeen and Aberdeenshire councils
Borders	East Ayrshire North Ayrshire South Ayrshire	
Dumfries and Galloway	Angus Dundee	Angus, Dundee City, Fife and Perth and Kinross Councils
Falkirk	Stirling Clackmananshire	
Fife	East Lothian Edinburgh Midlothian West Lothian	City of Edinburgh, East Lothian, Fife, Midlothian, Scottish Borders and West Lothian Councils
Highland	Dumbarton and Clydebank East Dumbartonshire East Renfrewshire Glasgow City Inverclyde North Lanarkshire Renfrewshire	East Dunbartonshire, East Renfrewshire, Glasgow City, Inverclyde, North Lanarkshire, Renfrewshire, South Lanarkshire and West Dunbartonshire Councils
Moray Orkney Islands Perthshire and Kinross Western Isles Shetland Islands		

review of Scottish planning policy. The agenda includes widening the ownership of plans through more effective involvement of other interests and building more confidence in plans by ensuring fewer departures from the plan in decisions avoiding over complexity. Care will be needed not to confuse changes in Scotland with those in England since there are numerous subtle and not so subtle differences. An *action plan* is proposed as a requirement for each new local development plan but it will not be a development plan (as in England) but a 'schedule for delivery' setting out agreements among those responsible for delivery. Other changes bring the systems closer together. A general area-wide policy framework is envisaged (and most councils are moving towards this anyway) but retaining provision for detailed policy for specific smaller geographical areas.

Development plans in Wales

In Wales, the system of development plans was virtually the same as that for England until 1996. One important variation was that the responsibility for waste rested with the districts (rather than counties) and thus waste policies were included in local plans rather than separate county-wide subject plans.

Local government reorganisation created unitary councils in 1996, and the plan framework was amended to require each authority (including the national parks) to prepare a unitary development plan. The form of the Welsh UDP is similar to the UDPs in England and has a Part I and Part II (discussed above). Provisions were made for joint UDP preparation (though this was always unlikely given the very large area of Welsh local authorities) while the organisation of Part II of the plan could be organised around smaller areas. Arrangements for the transition to the new framework allowed planning authorities to seek approval from the Secretary of State to continue through to the adoption of plans already in preparation (see Figure 4.10). With changes to local authority boundaries, local plans (including some yet to be adopted) may cover only part of an

authority's area or be split between two. There was barely time to establish this system before the 2004 Act brought further changes. Section 6 of the Act made separate provisions for development plans in Wales (their implementation in detail in regulations and orders made by the Assembly) and made the Wales Spatial Plan a statutory requirement. The Assembly had previously canvassed views on a Welsh approach to reform of development plans through *Planning: Delivering for Wales* in 2002. They were later to consult on more detailed proposals in *Delivering Better Plans for Wales* in 2004.[31]

The 2004 Act retains a unitary structure of development plans in Wales but the UDP is replaced by the *local development plan* (LDP) and the two-part structure is abandoned. The same broad changes are made in Wales as in England with an intention to create simpler and more focused general policies for the whole of the authority's area linked to an overall vision

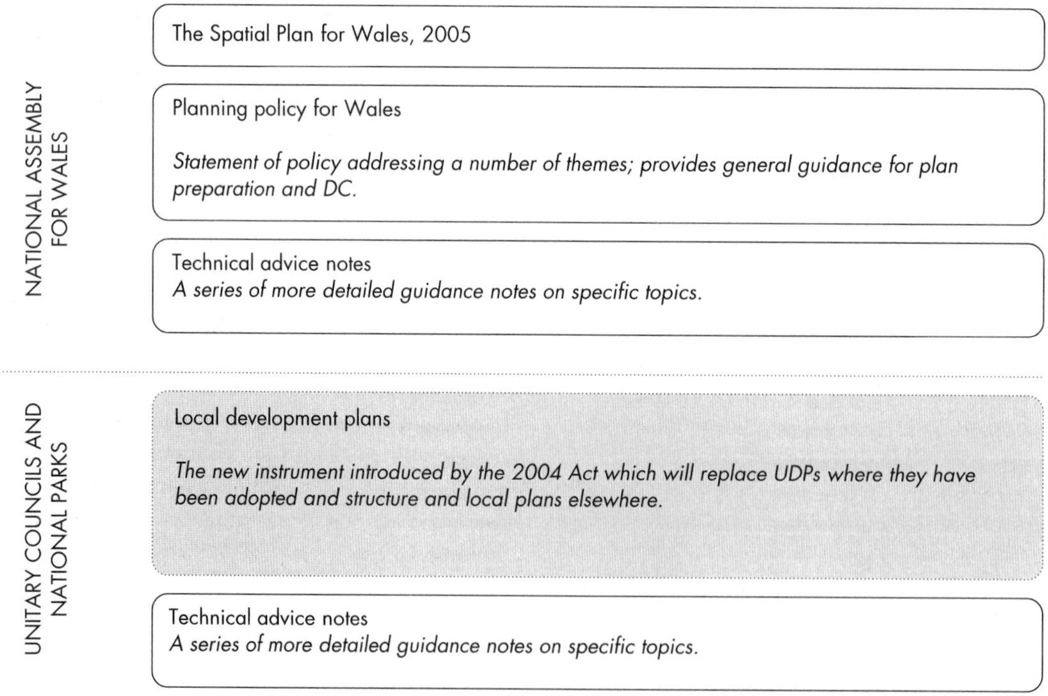

NATIONAL ASSEMBLY FOR WALES

The Spatial Plan for Wales, 2005

Planning policy for Wales

Statement of policy addressing a number of themes; provides general guidance for plan preparation and DC.

Technical advice notes
A series of more detailed guidance notes on specific topics.

UNITARY COUNCILS AND NATIONAL PARKS

Local development plans

The new instrument introduced by the 2004 Act which will replace UDPs where they have been adopted and structure and local plans elsewhere.

Technical advice notes
A series of more detailed guidance notes on specific topics.

Figure 4.10 *The planning policy framework in Wales*

for the area, to provide detailed planning policy only where it is needed, to address interdependencies with other plans and programmes and across administrative boundaries and to introduce apparently simpler procedures for the adoption of plans (discussed below). However, the detailed arrangements are neater, with the LDP as a single document, which will make for an interesting comparison in future as Wales seeks to achieve the same objectives through a much simpler arrangement. Improved community engagement is encouraged through a requirement for each authority to prepare a community involvement scheme (CIS) and a programme of local development plan preparation must be agreed with the Assembly. Formal annual monitoring is also required. Because the system has only 'recently' (in planning chronology) been changed in Wales, few UDPs have been adopted, but where they are in place, they will remain in force until they are replaced by local development plans. Many authorities will not have adopted a UDP and will continue to use structure and local plans as the extant development plan policy, while they adjust their UDP preparation to the revised approach. A number of options are open to authorities in the path they take from UDP to local development plan preparation depending on their progress so far, but all need to move immediately to 'LDP principles' which means, in short, cutting the length and complexity of draft plans, demonstrating effective community engagement (and preparing a CIS) and ensuring that plans are subject to the sustainability appraisal process. The 2004 Assembly consultation on Better Development Plans included a draft policy statement, *Local Development Plans Wales*, which were due to be published in final form in 2005.

The content of plans

The 1947 legislation was largely concerned with land use: 'a development plan means a plan indicating the manner in which a local planning authority propose that land in their area should be used'. The 1968 Act signalled a major shift in focus: emphasis was laid on major economic and social forces and on broad policies

or strategies for large areas. It was held that land use planning could not be undertaken satisfactorily in isolation from the social and economic objectives which it served. Thus the plans were to encompass such matters as the distribution of population and employment, housing, education and leisure.

This broader concept of planning did not survive, and by 1980 central government had moved back to a predominantly land use approach. This radical departure from the ideas of 1968, and the contraction of the scope of structure plans has been widely documented (Cross and Bristow 1983; Healey 1986). Central government also intervened to significantly restrict plan content. Thornley (1991: 124) provides a useful summary of what he describes as the 'attack on structure plans'.[32] During the 1990s departmental advice about plan content became increasingly specific and restrictive. The impact was that, while local plans embraced wide-ranging social and economic objectives, their proposals nevertheless are 'primarily about land allocation' (Healey 1983: 189). Moreover, while local plans vary substantially in form, and 'appear local in orientation and specific to particular areas and issues', there is considerable consistency in scope and content. Consistency arises from the need for central government support for policy and the use of planning inspectors and government offices to strip out policies thought not to be the concern of the land use plan and if necessary, powers of direction or intervention in the adoption process. Another reason is the professional training and reproduction of a particular culture in the planning profession which had become firmly rooted in land use and physical concerns and the regulation of development.

Criticisms of the weaknesses of environmental planning, the sustainable development agenda, the failure to deliver on regional targets and the obvious need for more joined up policy on spatial development led to an about-turn during the late 1990s and government began to promote a wider scope for development plans. The 1999 revision of PPG 12 provided some clarification of the government's position on plan content, as shown in Box 4.3. There now appears to be no succinct list of the matters to be considered as within the scope of local development planning, although there is much

BOX 4.3 SCOPE AND CONTENT OF DEVELOPMENT PLANS IN ENGLAND

The 2004 Act requires local planning authorities 'to keep under review':

(a) the principal physical, economic, social and environmental characteristics of the area of the authority
(b) the principal purposes for which land is used in the area
(c) the size, composition and distribution of the population of the area
(d) the communications, transport system and traffic of the area
(e) any other considerations which may be expected to affect those matters
(f) such other matters as may be prescribed . . . (s. 13) (see also endnote 33).

PPS 11, *Regional Spatial Strategies* Annex A gives a list of topics that should be taken into account in the preparation of regional spatial strategies. (It is proposed to keep this up to date with reviews appearing on the government website.) The main topics are

> sectoral policies of the EU that have a spatial impact; and the EU spatial planning documents; air quality and links to land use planning, biodiversity and nature conservation, climate change, the coast, culture, economic development, energy, green belt, local health improvement, delivery of new housing, affordable housing, the government's Communities Plan, minerals, retail and leisure, rural development and the countryside, soil, transport (roads, rail, freight, ports, aviation, cycling and walking), waste management, and water quality and resources.

The now superseded Planning Policy Guidance Note 12 provided further information on the 'other issues that may be addressed in plans, either as land use policies or as considerations which influence policies in the plan'.

- environmental considerations: energy, air quality, water quality, noise and light pollution, biodiversity, habitats, landscape quality, the character and vitality of town centres, tree and hedgerow protection and planting, revitalisation of urban areas, conservation of the built and archaeological heritage, coastal protection, flood prevention, land drainage, groundwater resources, environmental impacts of waste and minerals operations, unstable land
- economic growth and employment: revitalisation and broadening of the local economy and employment opportunities, encouraging industrial and commercial development, types of economic development; and generally to take account of the needs of businesses while ensuring that proposals are realistic
- social progress: impact of planning policies on different groups, social exclusion, affordable housing, crime prevention, sport, leisure and informal recreation, provision for schools and higher education, places of worship, prisons and other community facilities, accommodation for gypsies; but 'to limit the plan content to social considerations that are relevant to land use policies'.

reference to having regard to the plans and programmes prepared in other sectors. To get the complete picture the reader would need to wade through the planning policy statements, many of which have much to say on what plans should and should not do.[33] However, the 2004 Act and associated policy documents have three core and related messages about the scope of plans – to ensure that development is sustainable, to deliver for the economy and to adopt the spatial planning approach. These objectives are not altogether compatible of course. Sustainable development has become the statutory objective for the planning system by its inclusion in the 2004 Act – the relevant section is given in Box 4.4. This may have more symbolic than practical importance since it leaves the definition of sustainable development to other government policy which was already well established. Some of the important dimensions of development that may

contribute to sustainable development still lie outside the planning system, although progress is being made, most recently with a widening of the scope of plans in the field of energy (discussed in Chapter 7). Nevertheless, there are still few levers within and outside the system to ensure that the development which planning is expected to deliver is more sustainable.

On the positive side, the spatial planning approach suggests a considerable widening of the scope of plans in seeking not only to 'have regard to' but also to influence strategies and investment in other sectors. Box 4.5 gives an indication of the government's view of the spatial planning approach (see also the discussion in the first section of this chapter). Success will depend very much on the planning authority's skills in building networks, establishing collaborative relationships and planning processes with other sectors rather than

---

BOX 4.4 SPATIAL STRATEGY CONTENT OF THE DEVELOPMENT PLAN

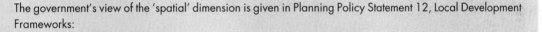

The government's view of the 'spatial' dimension is given in Planning Policy Statement 12, Local Development Frameworks:

> Local planning authorities should adopt a spatial planning approach to local development frameworks to ensure the most efficient use of land by balancing competing demands within the context of sustainable development. Spatial planning goes beyond traditional land use planning to bring together and integrate policies for the development and use of land with other policies and programmes which influence the nature of places and how they function. This will include policies which can impact on land use, for example, by influencing the demands on or needs for development, but which are not capable of being delivered solely or mainly through the granting of planning permission and may be delivered through other means.
>
> (para. 1.8)

The other strategies and plans that local development documents should take account of should include:

> The community strategy and strategies for education, health, social inclusion, waste, biodiversity, recycling and environmental protection. Local development documents should be prepared taking into account urban and rural regeneration strategies, local and regional economic and housing strategies, community development and local transport plans.
>
> (para. 1.9)

BOX 4.5 THE SUSTAINABLE DEVELOPMENT OBJECTIVE OF PLANNING IN ENGLAND AND WALES

The 2004 Act (s. 39) requires

bodies responsible for a regional spatial strategy, local development documents or in Wales, the Wales Spatial Plan or local development plan, to 'exercise that function with the objective of contributing to the achievement of sustainable development'. In doing this they must have regard to national policies and advice contained in guidance issued by the secretary of state or National Assembly for Wales.

any formal powers of control and influence. And this in turn depends crucially on the objectives that are set in those sectors (at national level) which are often narrow and driven by targets. Building capacity to engage with the wider agenda for spatial planning will also be critical. But the change of attitude on the scope of planning and the wider government 'joining-up process' across all sectors do provide opportunities, though these will have to be taken up in the context of the 'congested state' with its proliferation of agencies, special initiatives and collaborative structures (Sullivan and Skelcher 2002).

Of the collaborative relationships government gives priority to the community strategy. Government policy is that the local development framework should be the spatial expression of the community strategy and that collaborative working should be established with the local strategic partnership that prepares it, so that the development documents can help to deliver its policies for the area (paras 1.9 to 1.11 of PPS 12). A study on the relationship between local development frameworks and community strategies (Entec 2003) concluded that a better relationship could improve understanding of community needs and aspirations in the development plan process, give a more integrated approach to future development and help to join up the approach to community planning. But it also acknowledged the variability in the quality of community strategies 'and the great difference in purpose and procedure of community strategies and development plans'. The community strategy is prepared

through voluntary cooperation and although there is a statutory requirement for their production in each authority in England and Wales, there is no statutory procedure to be followed. The community strategy is intended to 'allow communities to articulate their aspirations, needs and priorities; coordinate the actions of the local authority, and the public, private, voluntary and community organisations that operate locally; and focus and shape existing and future activity' (Entec 2003: 4). The report acknowledges that 'some planners tend to regard LSPs as ephemeral, highly self-selecting and unrepresentative partnerships of the powerful' (p. 19). Lambert (2004) makes a similar point

Where LSPs were newly established strategy preparation tended to be a tentative and drawn out process. The product is seen by some partners as too broad brush and aspirational, or as a regurgitation of existing plans and strategies, and the key objectives can have something of 'motherhood and apple pie' flavour.

(Lambert 2004: 4)

These are early days for both local development frameworks and community strategies and both the Entec (2003) and Lambert (2004) reviews note that the community strategies are likely to become more important and detailed corporate strategic documents in the future, at least in some authorities. The test for an effective relationship will come when the local derived aspirations in the community strategy come face to face with local development documents that are strongly influenced by central departmental policies and priorities.

A consequence of the increasing attention that many organisations now give to plans in the light of the 'plan led system' has been the production of 'model policies'.[34] Numerous national and local organisations such as the Environment Agency, English Nature and Friends of the Earth (1994a) have suggested policy wording that can be taken 'off the shelf' rather than written anew for each local plan. General advice published by the Planning Officers' Society (POS) in 1997 was updated in 2004 to accommodate the reformed system. *Policies for Spatial Plans* provides both general advice about how to write policy, what is admissible content, and a bank of generic policy statements that can be tailored to local circumstances. Other advice is given in individual planning policy and good practice guidance, notably in *Planning for Sustainable Development: Towards Better Practice* (DETR 1998). The guidance from the centre (and experience) strongly suggests that plans will need to be more succinct and selective (a message that has been repeated many times), but many interest groups want to see their policies in the plan and suggest detail that goes well beyond what is appropriate, though which may be supported by local politicians. This explains why, despite many successive attempts to streamline development plans and other planning policy documents, they have tended to grow in size and complexity and be subject to delay and uncertainty. The question is whether the new system can really generate a simpler, more focused plan while also widening its scope and influence in integrating and helping to realise other strategies.

Statutory procedures and management of the plan process

While the apparent outcome of the 2004 reform in England is a more complicated set of policy documents at the local level, the procedures for adopting these documents are now less complex. The general procedure for the preparation of development plan documents in England is illustrated in Figure 4.11 and it should be

remembered that while practice is similar across the UK, devolution is resulting in a more varied picture.

The procedure for the adoption and approval of plans provides 'safeguards' to ensure the accountability of government and the consideration of many interests in the planning process. It also upholds the rights of private property interests to have their say when proposals affect their interests. This is particularly important in the UK, where there is no constitutional safeguard of private property or other rights (other than that provided by the European Convention on Human Rights) and where there is wide administrative discretion in decision-making. There is no appeal to the courts on the policy content of plans, although they may be used to ensure that statutory procedures are followed. The Secretary of State is the final arbiter on the content of plans, either through direct approval of documents in the case of regional spatial strategies (explained above) or through powers of 'examination' of the content of local development plan documents and if necessary intervention and direction in the process. The procedures also provide for increased involvement of other organisations and the public in policy formulation. The process of open discussion and formal adoption lends authority and standing to plans. See Box 4.6.

In the following discussion, the focus is on the key safeguards, the main criticisms of the procedure, and recent amendments. The knotty questions about the extent to which the public and other objectors are effectively able to make use of the safeguards and how this influences plan content are dealt with in Chapter 12. The main safeguards in plan preparation and adoption are

- the opportunity for all interests to be consulted in the formative stages of plan preparation
- the need for authorities to consider conformity between plans and regional and national guidance
- the right to make representations to both strategic and local development plan documents (which may be objections or indications of support)
- to have representations to local development documents considered, and if desired, heard before an independent inspector

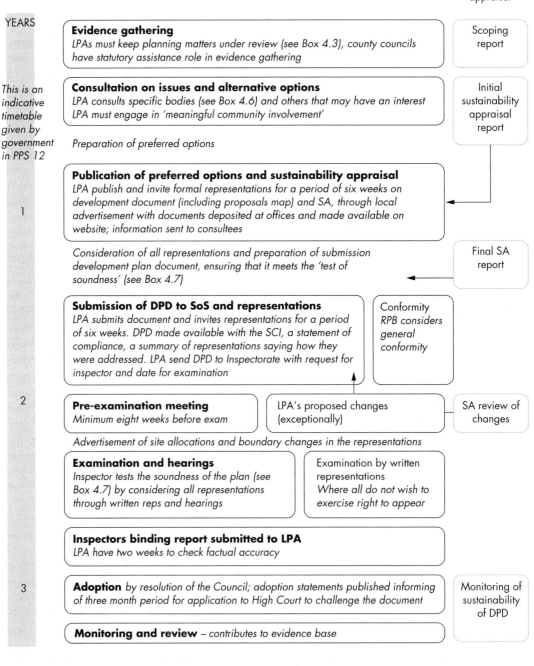

Sustainability
appraisal

YEARS

Evidence gathering
LPAs must keep planning matters under review (see Box 4.3), county councils
have statutory assistance role in evidence gathering

Scoping
report

*This is an
indicative
timetable
given by
government
in PPS 12*

Consultation on issues and alternative options
LPA consults specific bodies (see Box 4.6) and others that may have an interest
LPA must engage in 'meaningful community involvement'

Initial
sustainability
appraisal
report

Preparation of preferred options

Publication of preferred options and sustainability appraisal
LPA publish and invite formal representations for a period of six weeks on
development document (including proposals map) and SA, through local
advertisement with documents deposited at offices and made available on
website; information sent to consultees

1

*Consideration of all representations and preparation of submission
development plan document, ensuring that it meets the 'test of
soundness' (see Box 4.7)*

Final SA
report

Submission of DPD to SoS and representations
LPA submits document and invites representations for a period
of six weeks. DPD made available with the SCI, a statement of
compliance, a summary of representations saying how they
were addressed. LPA send DPD to Inspectorate with request for
inspector and date for examination

Conformity
RPB considers
general
conformity

2

Pre-examination meeting
Minimum eight weeks before exam

LPA's proposed changes
(exceptionally)

SA review of
changes

Advertisement of site allocations and boundary changes in the representations

Examination and hearings
Inspector tests the soundness of the plan (see
Box 4.7) by considering all representations
through written reps and hearings

Examination by written
representations
*Where all do not wish to
exercise right to appear*

Inspectors binding report submitted to LPA
LPA have two weeks to check factual accuracy

3

Adoption by resolution of the Council; adoption statements published informing
of three month period for application to High Court to challenge the document

Monitoring of
sustainability
of DPD

Monitoring and review – contributes to evidence base

Figure 4.11 *The procedure for the adoption of local development frameworks*

BOX 4.6 DEVELOPMENT PLAN CONSULTATION BODIES

The list contains only those bodies that *must* be consulted as part of the local development document or regional spatial strategy process. Longer lists of bodies that also may need to be consulted are given in PPS 11 and PPS 12.

- government office for the region
- Scottish Executive (for RSS bordering Scotland)
- regional planning body (for LDDs)
- county councils (for the RSS)
- Mayor of London (for London Boroughs)
- authorities whose area is in or adjoins the area of the planning authority
- town or parish councils
- Countryside Agency
- Environment Agency
- Highways Agency
- Historic Buildings and Monuments Commission for England
- English Nature
- Environment Agency (RSS)
- Strategic Rail Authority
- a regional development agency whose area is in or adjoins the area of the planning authority
- electronic communications code network operators (who are able to make use of the permitted development rights) and owners of electronic communications apparatus
- strategic health authorities
- gas suppliers
- sewerage undertakers
- water undertakers
- electricity companies (RSS)

- the overarching right of the Secretary of State to intervene and to direct changes; and
- a limited right to challenge the plan in the courts on procedural matters.

The central focus of the formal adoption procedure is the hearing – now described by government as an 'independent examination'. For a regional spatial strategy the hearing is heard before an appointed panel; for a local development plan it is heard before an inspector (or team of inspectors) of the Planning Inspectorate.

For many years the examination for local plans at the local level was known as a public local inquiry (PLI) and structure plans were heard at an examination in public. While the names have changed, the procedures are much the same as before but incorporate many of the lessons learned over many years about how best to organise such hearings so that it meets the expectations of those making representations while not adding undue delay to the process. The development plan document examination is in the form of a public inquiry. An 'independent' inspector hears representations, which are mostly objections and counterproposals. There is a statutory right to be heard, although this tends to be exercised most by those who are better organised and resourced. The 'quasi-judicial' nature of the examination tends to make it an adversarial debate with questioning of evidence, but the Inspectorate has done much work to make the process less adversarial while maintaining the formality needed to ensure that all can participate. The inspector is independent only of the local authority, not government. He or she 'stands in the shoes' of the Secretary of State, and one of the main jobs of the inspector is to ensure that the plan is in accordance with national and regional policy. Anyone can object to a development plan, and the planning authority has a duty to consider all objections. The examination into the regional spatial strategy follows the procedure for the examination in public and is a 'probing discussion' of selected matters which the Secretary of State needs to consider as part of the approval process. There is no right to be heard, but contributions are by invitation. (The forms of hearing are discussed further in Chapter 12.)

The formal procedures for plan approval and adoption have come under constant scrutiny as one of the contributors to the continual failure of many authorities to prepare plans in good time and keeping them up to date. This was especially true during the 1990s when there was strong support from central government for the preparation and effective use of plans. During the 1990s the plan-making process took about five and a half years on average (Steel *et al.* 1995), and in excess of four years in Scotland (Hillier Parker *et al.* 1998: 11). These averages mask great variation extending from three to ten years. In England the time taken to adopt plans has tended to get longer because the enhanced status of plans and the district wide format has led to an increase in the number of objections. However, overall productivity in the system may have improved because the increase in time taken to prepare and adopt is much less than the extra work that is entailed. The greatest proportion of time taken in the process is still in the preparation of a draft plan prior to deposit – a problem confirmed by research on structure plans (Baker and Roberts 1999).

Overall, the generalisations made here need to be treated with care. Some planning authorities are able to cope very well with the procedures and produce plans in good time and keep them up to date.[35] In contrast, the performance of some planning authorities has been abysmal, with very slow progress and little information about programmes. There have been periodic calls from ministers since the 1970s for improved performance.

In the 1980s numerous changes were made, though with little effect. They included reducing the requirements for public participation, allowing the local plan to be adopted in advance of a structure plan review, powers for the Secretary of State to request modifications and providing more opportunity for objections after the inquiry.[36] After the 1991 Act, new targets were set for complete cover of plans by 1996. It quickly became apparent that the targets were not going to be met. In 1994, the DoE made more proposals for further amendments to the procedure and in 1996 revised the Code of Conduct for inquiries and EIPs, and regulations governing the preparation of plans. Local authorities were encouraged to remove excessive detail from plans, undertake more effective consultation early in the process and give more emphasis on dealing with objections in writing. The Inspectorate simplified its reporting to authorities.

This (much abridged) story of changes to the procedure is given to emphasise that the problem of delay in the process is not new and that repeated attempts to do something about it by changing the procedures have had little effect in the face of increasing numbers of objections to plans. By 2001 two authorities had not even reached the deposit stage with their local plan, eleven more had not reached the inquiry stage, and thirty-four more had not adopted the plan after the inquiry. More than 200 plans had reached or passed their end date.

Many ideas have come forward for tackling the problem. The TCPA, local government associations and Planning Officers' Society all published recommendations. They all focus on change to the formal procedure while less attention was given to the form and content of plans and management of the process which dominated explanations for problems in research findings. The recommendations included abolition of hearings and making the inspector's recommendations binding on the local authority (Roberts, T. 1998; Royal Town Planning Institute (RTPI) 1999b). In Scotland, a 1998 Hillier Parker report on development planning came to similar conclusions arguing that the report of the public inquiry should be binding and that there should be a national timetable for plan production. Not everyone agreed. Research on the efficiency and effectiveness of local plan inquiries (Steel *et al.* 1995) suggested that making the inspector's report binding would be difficult to implement (and probably counterproductive) and pointed instead to management weaknesses (in terms of both officer and member involvement), and the importance of the form and content of the plan. Various studies have pointed to lack of political will and/or professional expertise, failures in management and confusing advice from central government as explanations for poor performance.[37] We should also recognise that the plan-making process nowadays is far more complex and contentious in most places than it was in the 1970s when the system was established. This is because of the increasing participation of interests who recognise the plan's significance for later development control decisions and outcomes. Objections to plans in the 1990s were typically counted in thousands, while few plans would have been subject to this level of objection twenty years earlier. One example (among many) is the case of the East Lothian Local Plan where in 1998–9 the Scottish Inquiry Reporters Unit engaged three reporters on the inquiry, who simultaneously addressed eight associated planning appeals. Central government also became more active in scrutinising plans prior to the inquiry, checking plans for consistency with central government policy and regional guidance. This resulted in departmental objections to plans and frequent and sometimes lengthy requests for changes.

The Inspectorate has gathered evidence from many hundreds of planning inquiries and its views and advice have been widely circulated.[38] The main problems have been failure to plan ahead from the start; poor attitudes to objectors who are seen as a nuisance; shifting priorities during the plan-making process, which reflects a lack of commitment of officers and members; unnecessary conflicts with the regional offices and DETR; provoking conflict through unnecessary plan content that seeks to cover all contingencies; too much prescription and detail including policies not capable of implementation or monitoring; and failure to identify costs of these problems for senior managers and politicians. Again, it should be emphasised that many successful authorities do not exhibit these problems. Also external factors have played a part: fluctuating attitudes of central government to plan-making, problems with the two-tier system, political conflicts and the Conservative government's negative attitude towards planning and local government.

In 1996 the DoE began a more fundamental review of the development plan process. A 1997 consultation paper set out the options for speeding delivery of plans and a further paper in 1998, under the *Modernising Planning* initiative, made specific proposals for *Improving Arrangements for the Delivery of Local Plans and Unitary Development Plans*. Revised guidance (PPG 12) and Regulations were published in 1999 and more research was undertaken on the structure plan process (Baker and Roberts 1999). The 1999 version of PPG 12 rejected calls to make the inspectors' recommendations binding or to limit the rights to object and appear at the inquiry and concentrated on a new procedure for pre-deposit consultation, drawing out objections at an early stage in the process, reducing the length and detail of plans, and improving local authority management through a publicly adopted timetable for plan production. A two-stage deposit for local plans and UDPs was introduced. The first *initial deposit* was to allow local authorities to gather objections early on in the process to be followed by a period of negotiation and revision of the plan. A second *revised deposit* stage would then gather objections only on any of the changes made. A statutory requirement was introduced for consultation with consultees prior to deposit so as to reinforce the importance of getting as much information as possible at the start of the process. The post-inquiry modifications stage was also retained with the possibility of further objection if those affected had not been able to object in the first two rounds of deposit.

These changes did not prove effective but produced a very complex process with authorities producing many changes to plans between deposit stages, and confusing objectors, consultees and, sometimes, themselves. Certainly, inspectors were given a very difficult task in sorting objections to different versions of the plan, and local authorities were faced with often difficult choices about modification. The plan-making process became unfit for its purpose. The record of plan-making in much of the country is poor, coverage of up-to-date plans is patchy, the outcomes do not reflect the costs of the service and monitoring and review are weak. The situation is an embarrassment for central government and local authorities alike. The *Planning Concordat* drawn up between the LGA and DETR in 1999 recognised that 'under-performance of some local authorities . . . could undermine the plan-led system'. Ministerial meetings were held with planning officers in the regions during 1999.

Inevitably, a more radical revision and simplification of the procedure came forward as part of the 2004 reforms. The current procedure is shown in Figure 4.11. After many years of considering and rejecting this solution, inspectors' reports on local development documents are now binding on the authority. Interestingly the planning profession has long since advocated this, though it does raise fundamental questions about where policy is being made. There is no doubt that the binding nature of the inspector's report will reduce the time in the very difficult adoption process after the inquiry. Other consequent amendments have been made to try to ensure that the inspector and objectors have complete information about options, possible land allocations and the boundaries of plan designations. The government rather inelegantly call this 'front-loading', which means engaging effectively at an early stage in the process with consultees (who now have a statutory duty to respond to requests for information) and ensuring that there is early and meaningful consultation with local communities and interests. The local authority should seek to expose all options as early as possible and to engage in wide participation on them. The changes to the 'document production' part of the procedure take it to something like earlier variants with two six-week stages when representations can be made. The first is a consultation stage on 'options' and the second an

opportunity to object (or support) the local authority's 'preferred options'. In PPS 12 this second stage is described as 'participation'. Students of community involvement may be confused by this, since it is not participation in the usual sense of the term but an opportunity to make formal representations in writing which will be taken forward to an 'independent examination'. The opportunity for changes which previously had been widened has now been reduced, and only 'exceptionally' should local authorities propose changes after submission to the Secretary of State and the second 'deposit' period. This change is understandable and should make the examination easier to organise and manage, but it does mean that the authority will have to have a sound plan prior to submission and to move quickly on the process so that changing circumstances are not allowed to undermine the validity of the plan.

The role of the inspector at the examination also changes. Previously the inspector considered representations to the plan and was formally tied to making recommendations to the authority only on matters raised by representations. Henceforth the inspector will have a wider role in 'testing the soundness' of the plan as a whole (see Box 4.7). The inspector will do this by examining the representations but will not be tied to them. The authority must indicate how the document meets the test of soundness and it will be assumed that it meets this test unless it can be shown through the examination that it does not. Objectors will need to explain also how their counterproposals help to meet or affect the test of soundness and the authority will have to publicise counterproposals so that others have a chance to consider these options prior to the examination. The inspector will have a big job to do, but the other changes to the form and content of documents should mean that the volume of work will be more manageable. The Planning Inspectorate's *Annual Report* for 2004–5 highlighted the workload implications following submission of large numbers of development plan documents (DPDs). They estimate that 554 will be submitted in the year 2007–8 (Planning Inspectorate 2005: 26).

Statutory procedures have also been introduced for the adoption of supplementary planning documents

BOX 4.7 TEST OF SOUNDNESS OF DEVELOPMENT PLAN DOCUMENTS

There is a presumption that the development plan document is sound unless shown to be otherwise through the examination process.

A development plan document will be sound if it meets the following tests:

Procedural

i.   it has been prepared in accordance with the local development scheme;
ii. it has been prepared in compliance with the statement of community involvement;
iii. the plan and its policies have been subjected to sustainability appraisal;

Conformity

iv. it is a spatial plan which is consistent with national planning policy and in general conformity with the regional spatial strategy for the region or, in London, the spatial development strategy and it has properly had regard to any other relevant plans, policies and strategies relating to the area or to adjoining areas;
v.   it has had regard to the authority's community strategy;

Coherence, consistency and effectiveness

vi.   the strategies/policies/allocations in the plan are coherent and consistent within and between development plan documents prepared by the authority and by neighbouring authorities, where cross boundary issues are relevant;
vii. the strategies/policies/allocations represent the most appropriate in all the circumstances, having considered the relevant alternatives, and they are founded on a robust and credible evidence base;
viii. there are clear mechanisms for implementation and monitoring; and
ix.   the plan is reasonably flexible to enable it to deal with changing circumstances.

Source: ODPM (2004) PPS 12 *Local Development Frameworks* (para. 2.24)

(SPD). The local development scheme will set out proposals for the preparation of supplementary documents and the review of existing supplementary planning guidance. The procedure requires local planning authorities to consult on a draft document, to deposit the document for comments for between four and six weeks, and to consider representations before publishing the final version. The requirements

for 'soundness' and sustainability appraisal apply to supplementary documents though they are not subject to an examination. A supplementary planning document is a document providing additional information on policies expressed in development plan documents as in the case of design guides. It is important that they do not contain free-standing policies that will provide the basis for considering planning applications because this procedure does not provide an opportunity for interested persons to object to them (Upton 2005).

The management failures are also being addressed through procedural changes. Each local planning authority (including counties for the minerals waste framework) must prepare a *local development scheme*, which is a project plan for the preparation of the portfolio of documents in the local development framework. The first round of schemes was submitted in March 2005. This is a statutory requirement and consultation is required with the government office and Planning Inspectorate so that the feasibility of the scheme can be checked. When complete it is submitted to the Secretary of State, who has a four-week opportunity to intervene, otherwise the scheme comes into effect automatically. Another influence on management of the process is the responsibility of local authorities to ensure that they are meeting the requirements of Best Value (this is discussed further in Chapter 5). Under Best Value, authorities should demonstrate that they are delivering quality services and judge their performance in development planning against performance indicators and targets, some set at national level and others by the authority itself (the national best value performance indicators are listed in Chapter 3). The authority must evaluate and benchmark its performance in comparison to other similar authorities. One option to address management issues is for consultants to be used for assistance in preparing plans but very few such jobs have been put out to tender.

Despite the constant amendment to the plan-making process, and except for losing the principle that the authority should be responsible for the final modifications to the plan, the basic procedure has proved resilient. This is perhaps because it nicely balances the concerns of local authorities (who typically call for fewer procedural requirements in order to speed the process) and the concerns of objectors of all kinds (who naturally desire more influence in the local planning process). The problems and costs of the formal procedures have been a major factor explaining a distinct lack of enthusiasm of some planning authorities for statutory plan-making. The new system has the potential to make a significant improvement on

previous practice, but many of the same problems will remain.

Zoning instruments

There are two examples of attempts to reintroduce the zoning approach in the UK planning system, and a third is in preparation. They are enterprise zones, simplified planning zones and business planning zones. All reflect economic rather than land use planning objectives. Zoning is a fundamental break from the UK system of planning in that the grant of planning permission is effectively made in advance of the proposal coming forward.

A major plank in the Conservative government's response to economic recession in the early 1980s was the proposed reduction in the 'burden' of regulation on business and enterprise.[39] In enterprise zones (EZs), amendments to the planning regime were part of a much wider range of advantages offered, including exemption from rates on industrial and commercial property. The enterprise zone scheme had the effect of granting planning permissions in advance for such developments as the scheme specifies, and it was up to the planning authority to determine what planning concessions were offered. The scheme was simply proposed by the local authority and approved by the Secretary of State. Thirty-two enterprise zones were designated.[40] The enterprise zone initiative was closely monitored, and findings show in some cases a dramatic increase in development activity.[41] Overall, however, the liberalisation of land use planning controls made only a minor contribution to any success. A considerable amount of negotiation (whether it be termed 'planning' or not) still had to take place, both between the developers and local authorities and also between developers and other agencies.

Whatever the research on enterprise zones might have concluded, the government was so enamoured of the idea that it introduced a new type of simplified planning zone (SPZ) based upon it. The general notion of zoning as an alternative to the development plan had been rejected, but the DoE did see a limited role for zoning in particular locations where greater certainty, and some flexibility in the detail of development proposals, would contribute to economic development objectives. An SPZ is the local equivalent to a development order made by the Secretary of State. It replaces the normal discretionary planning system with advance permission for specified types of development.

Two broad types of scheme are possible: the *specific scheme*, which lists certain uses to be permitted, and the *general scheme*, which gives a wide permission but excludes certain uses. Conditions can be made in advance, and certain matters can be reserved for detailed consideration through the normal planning process. SPZs cannot be adopted in national parks, the Broads, AONBs, sites of special scientific interest (SSSIs), approved green belts, conservation areas and other protected areas. SPZs were particularly promoted for older industrial sites (especially those in single ownership) where there is a need to promote regeneration (Lloyd 1992).

The introduction of the SPZ provisions has excited very limited interest, and slow progress initially has now stopped completely. Such interest as there was tended to come from authorities with experience of (or failure to obtain) enterprise zones: these authorities had fewer fears about the loss of normal development control powers over the quality of development (Arup Economic Consultants 1991). In all important respects the procedures for adoption of SPZs were identical to those of local plan preparation and adoption. The prospect of taking a scheme through these lengthy procedures was daunting, and it rapidly became clear that they were (in the words of the research report) 'undoubtedly cumbersome'.

As with development plans the procedures have been amended to try to encourage more interest but with little effect (Blackhall 1993, 1994). Some of the reasons are perhaps obvious. There is little difference between the allocation of land in a development plan and an SPZ: both indicate the type of development that is acceptable. Moreover, the extra 'certainty' provided by an SPZ designation is to some extent illusory since formal relationships are replaced by informal discussions. Additionally, decisions on the fulfilment of conditions and negotiations on reserved matters may still be needed. There is only a very small number of

zones, and these operate in a narrow range of circumstances.

Overall, as Allmendinger (1996a) argues, the SPZ concept largely failed because it lacked clear and consistent objectives. It sought to offer deregulation and more certainty for developers but, in fact, led to greater uncertainty when put into practice. But, above all, the recession at the end of the 1980s undermined property-led development on which the idea rested. The reintroduction of zoning into the British planning system through SPZs has so far been unsuccessful, but the provisions for zones are still on the statute and amendments were made by the 2004 Act to bring the procedures for their adoption into line with the reformed system.

Government also consulted about a new instrument, the business planning zone, in the 2001 Green Paper. The idea was the same as the SPZ; the local planning authority would be able to designate such zones 'where no planning permission would be necessary for development, if it is in accordance with tightly defined parameters'. The object of this proposal is 'fast moving businesses such as leading edge technology companies' (p. 38). It was proposed that such zones would be limited to 'low-impact' businesses which do not create great demands for travel, housing or infrastructure. The Green Paper further suggested that each region should have at least one zone for the promotion of technology companies. (Interestingly, the DTI requires each regional development agency to develop a science park within its region.) A study on *Planning for Economic Development* (ECOTEC and Roger Tym and Partners 2004) found that there was some confusion about the purpose of this proposal and that those who might be interested thought other mechanisms could address the issues better. What is the problem that business planning zones address? The need for planning permission is not the main problem for the location of business, but the availability of the right sort of property may be. The study found that other tools would be more appropriate: positive planning policies, financial resources, legal powers, and initiatives to promote development. Nevertheless, HM Treasury has promised to bring forward this proposal.

Further reading

Supranational planning

The literature on European spatial planning is growing rapidly. For short papers reviewing the subject from a UK perspective see the special edition of *Town and Country Planning*, edited by Nadin and Dühr (March 2005). There is also a collection of papers in a special edition of *Town Planning Review* 76(1) (2005) on territorial cohesion. The reader edited by Faludi (2002) provides a broader perspective: Faludi provides an overview of the ESDP process and other contributions are by Martin and Robert on the history; Drevet on enlargement; Doucet on North West Europe; Nadin on visioning; and Zetter on the future. For a complete account of the making of the ESDP see Faludi and Waterhout (2002). Other material includes the ESDP itself (see DG Regio Website). For interpretation and critique see Bengs and Böhme (1998) *The Progress of European Spatial Planning*, Böhme and Bengs (1999) *From Trends to Visions: The European Spatial Development Perspective*, the special edition of *Built Environment* 23(4) (1997), especially the article by Bastrup-Birk and Doucet, and Faludi (2000) 'The European Spatial Development Perspective: what next?' The European Council of Town Planners (ECTP) 2003 conference report on the ESDP by Mark Tewdwr-Jones is available on the ECTP website.

On the impact of the ESDP in the UK see Shaw and Sykes (2003); Tewdwr-Jones and Williams (2001); Nadin (1999) 'British Planning in its European context', Wilkinson *et al.* (1998) *The Impact of the EU on the UK Planning System*; and Bishop *et al.* (2000). The influence of the 'spatial' approach is considered by Harris and Hooper (2004). There is limited material on INTERREG, especially evaluation: see articles by Samson, Jordan, Mills and Millar in the March 2005 edition of *Town and Country Planning*; also Nadin and Shaw (1998b) 'Transnational spatial planning in Europe: the role of Interreg IIc in the UK'. On ESPON see Bengs (2002) *Facing ESPON*, Davoudi (2005) 'The ESPON: past, present and future' and Gestel and Faludi (2005) 'Towards a European territorial cohesion assessment network'.

There is increasing interest in the different planning systems in Europe. The European Commission's *Compendium of Spatial Planning Systems and Policies* (Nadin *et al.* 1997) is the most comprehensive source and Seaton and Nadin (2000, 2002) provide updates on a smaller number of countries. See also Healey *et al.* (1994) *Trends in Development Plan Making in European Planning* and the updated ISOCARP *International Manual of Planning Practice* (Lyddon and Dal Cin 1996). The CEMAT of the Council of Europe are developing a web-based collection of planning system summaries (see below).

National plans and policies

A collection of papers exploring the subject was published in *Town Planning Review* 70(3) (1999). The national and regional spatial strategies themselves are key references (and are listed at the end of the book). Beyond that, there are several critical reviews of national and regional planning policy including Allmendinger (2003) 'Integrating planning in a devolved Scotland', Land Use Consultants (1995a) *Effectiveness of Planning Policy Guidance Notes*, Roberts (1996) 'Regional planning guidance in England and Wales', Quinn (1996) 'Central government planning policy', Kitchen (1999a) 'Consultation on government policy initiatives' and Baker (1998) 'Planning for the English regions'. On Scottish guidance, see Land Use Consultants (1999) *Review of National Planning Policy Guidelines* and on Wales see Alden and Offord (1996) 'Regional planning guidance'. The standard textbook on regional planning is by Glasson (2000) *An Introduction to Regional Planning*. Wannop (1995) *The Regional Imperative* provides a review with international comparisons. The resurgence of interest in strategic planning is examined by Roberts and Lloyd (1999) 'Institutional aspects of regional planning, management and development'. For an international comparison see Alterman (2002) *National-level Planning in Democratic Countries*.

Development plans

A good starting point for investigation of the policy framework in England is *PPS 1: Delivering Sustainable Development* (2004), *PPS 11: Regional Spatial Strategies* (2004) and *PPS 12: Local Development Frameworks* (2004) (and their equivalents in the other countries of the UK).

There has been much publication of good practice guidance, most recently ODPM (2005) *Local Development Framework Monitoring* and the Planning Officers' Society *Policies for Spatial Plans*. Baker and Roberts (1999) *Examination of the Operation and Effectiveness of the Structure Planning Process* provides a comprehensive review of structure planning practice. The RTPI has published a guide to good practice and a research report both entitled *Fitness for Purpose: Quality in Development Plans* (2001a, 2001b). The Planning Officers' Society *Learning and Dissemination Project* presents findings from evaluation of implementation of the LDFs system in ten local authorities (see www.planningofficers.org.uk) and findings from ODPM's *Spatial Planning in Practice* project are available on the ODPM website. A document on *Definitions of the Local Development Documents* was made available on the ODPM planning website in 2005.

There is more material on local development planning, although much of it concentrates on procedures. There are few recent relevant textbooks; see Adams (1994) *Urban Planning and the Development Process*. Tewdwr-Jones (1996) *British Planning Policy in Transition: Planning in the 1990s* gives a wide-ranging set of papers. The standard texts are now ageing. Their explanations for success and failure are still pertinent, but the details of practice less so: Healey (1983) *Local Plans in British Land Use Planning*, Bruton and Nicholson (1987) *Local Planning in Practice* and Fudge *et al.* (1983) *Speed, Economy and Effectiveness in Local Plan Preparation and Adoption*. Other notable sources on local planning include Wenban-Smith (2002) 'A better future for development plans', Poxon (2000) 'Solving the development plan puzzle in Britain', Hull (1998) 'Spatial planning', Hull and Vigar (1998) 'The changing role of the development plan in managing spatial change', Vigar and Healey (1999) 'Territorial integration and "plan-led" planning', Healey (1994) 'Development plans', Healey (1990) 'Places, people and politics', Healey (1986) 'The role of development plans in the British planning system', and several chapters of case studies in Greed (1996a) *Implementing Town Planning*.

An important text summarising a major evaluation of the impact of development plans is Healey *et al.* (1988) *Land Use Planning and the Mediation of Urban Change*. See also MacGregor and Ross (1995) 'Master or servant?'. Kitchen (1997) *People, Politics, Policies and Plans* gives a view of plan-making from inside a local authority.

Up-to-date summaries of procedures are given in the latest editions of Telling and Duxbury's (2002) *Planning Law and Procedure* and Moore (2002) *A Practical Approach to Planning Law*. Steel *et al.* (1995) *The Efficiency and Effectiveness of Local Plan Inquiries* examines the procedures in practice, as does the RTPI study by Cardiff University and Buchanan Partnership (1997) *Slimmer and Swifter: A Critical Examination of District Wide Local Plans and UDPs*. See also articles by Upton (2005) on supplementary planning documents and editorial in *Journal of Planning and Environment Law* (January 2005) on examinations. For Northern Ireland, see Dowling (1995) *Northern Ireland Planning Law* and for Scotland, see Collar (1999) *Greens Concise Scots Law: Planning*.

The value of enterprise zones is considered by PA Cambridge Economic Consultants (1995) *Final Evaluation of Enterprise Zones*, while the number of publications on SPZs outnumbers the zones. See Blackhall (1993) *The Performance of Simplified Planning Zones*, Lloyd (1992) 'Simplified planning zones, land development, and planning policy in Scotland' and Allmendinger (1996b) 'Twilight zones'. Official guidance is given in PPG 6, PAN 31 and TAN (W) 3, all entitled *Simplified Planning Zones*.

Notes

1 Council Regulation (EC) no. 1260/1999 *Official Journal L* 161 21.6.99. For a review of the differences between the 2000–6 and 1995–9 programmes see the European Commission paper *Reform of the Structural Funds 2000–2006 Comparative Analysis*, June 1999.
2 Areas are designated according to the nomenclature of territorial units for statistics (NUTS). In England NUTS 1 equates to the standard regions, NUTS 2 to groups of counties and NUTS 3 to individual counties or groups of local authorities. The definition of the NUTS regions is very contentious as it can determine whether the area is eligible for Community assistance

(Casellas and Galley 1999). The DTI has favoured the use of smaller statistical areas (wards) for the definition of assisted areas Tier 2 in the UK.

3 The Funds are the European Regional Development Fund (ERDF), the European Social Fund (ESF), the Guidance section of the European Agricultural Guidance and Guarantee Fund (EAGGF) and the Financial Instrument for Fisheries Guidance (FIFG). Objective 1 areas make use of all four; Objective 2 areas are eligible for ERDF and ESF and Objective 3 makes use of the ESF only. There is in addition 18 billion Euro Cohesion Fund available for structural assistance to Greece, Ireland, Portugal and Spain, and 21 billion Euro earmarked for the accession countries. The comparative figure for the five year programming period to 1999 is €155 billion, of which the UK received €13 billion (about £7.8 billion).

4 A transitional assistance mechanism has been established until 2005 to soften the blow of the loss of Community funding for areas previously designated as Objective 1: the Highlands and Islands and Northern Ireland. Northern Ireland also benefits from a special PEACE programme under the Community Initiatives which is worth €500 million (€100 million of which is to be spent in the Republic of Ireland).

5 There is a short explanatory factsheet at http://europa. eu.int/comm/regional_policy, under 'sources of information'. The European Commission adopted the proposals on 14 July 2004, decisions by the Council and Parliament were expected in 2005 and the detailed regulations will be published in 2006.

6 Previously, the areas designated for UK and EU funding support were different; the government justified this because of the programmes' differing objectives. In 1998 the European Commission published guidelines on regional aid to promote comparable and transparent systems across the EU Official Journal 98/C 74/06 and the DTI had to review the boundaries of Tier 1 on the same criteria as Objective 1. More details on SFI are available at www.dti.gov.uk or from the regional development agencies.

7 The EU is empowered to 'adopt measures concerning town and country planning, land use with the exception of waste management . . . and management of natural resources' (Article 175) but it is agreed that this would support action in these fields only in so far as they help to achieve measures related to environmental protection, and then only when agreed by unanimity (Nadin and Shaw 1999: 33; Bastrup-Birk and Doucet 1997).

8 INTERREG IIc was launched in 1996 (OJ 96/C200/07) and builds on previous cross-border cooperation programmes through INTERREG I and IIa. It is funded under the ERDF Regulation EEC 4254/88. Competence for the EU to provide funding in this way comes from Article 161 (former Article 130d) under Title XVII Economic and Social Cohesion, and the Structural fund regulations under this Article give more detailed justification. Article 10 funding was also used to promote transnational planning in some areas. A parallel programme TERRA has been launched to promote transnational cooperation on spatial planning in areas which are 'vulnerable'.

9 Details of projects funded through INTERREG IIc are available on the programme websites which are accessible via http://www.interregiicuk.org.uk/

10 These arguments have been developed and consolidated over time in a succession of documents. Those not mentioned elsewhere in this chapter include the *European Regional and Spatial Planning Charter* (CoE 1983), *Guiding Principles for Sustainable Spatial Development* (CEMAT 1999) and the *Fifth Environmental Action Programme on the Environment: Towards Sustainability* (CEC 1992) (and the sixth to be published in 2001).

11 Article 3, The Strategic Planning (Northern Ireland) Order 1999.

12 Raemaekers and colleagues' (1994) report on national planning guidance in Scotland recommended a tier of regional planning guidance for Scotland, and a return to regional reports for the city regions in the Central Belt to provide greater coordination in joint structure plan preparation in the wake of local government reorganisation, but this was not taken up.

13 The Scottish Executive has also taken a very positive attitude to the dissemination of national planning guidance. Copies are freely distributed and many are available for reference on the Scottish Executive website www.scotland.gov.uk/.

14 The DETR consultation paper on *Planning for Major Projects* draws attention to the problem of long inquiries debating policy in the absence of more specific national guidance on significant developments such as airports.

15 The Town and Country Planning Association has for many years been at the forefront of the campaign for better strategic planning, see for example Hall, D. (1991) and TCPA (1993).

16 The ECOTEC *Scoping Study: RPG Targets and Indicators* (1999) provides some interesting findings on the capacity of the regional bodies to undertake regional planning, and has some positive conclusions too, for example, recent considerable improvements in the information base in some regions.

17 Some examples of targets were given in the superseded PPG 11: traffic reduction and modal split; development of town centre versus out of town floorspace; rural accessibility to services; enhancement of bio-diversity and the use of recycled materials. The research found current practice only sets quantified housing targets.

18 The research projects were the *Scoping Study: RPG Targets and Indicators* undertaken by ECOTEC (1999) and *Proposals for a Good Practice Guide on Sustainability Appraisal of Regional Planning Guidance* undertaken by Baker Associates (1999).

19 Unlike the structure plan, which was prepared by all relevant authorities for the whole of their area, local authorities were advised that local plans would not be needed in all areas, for example where there was little pressure for development and no need to stimulate growth. This discretion was used: a small number of authorities prepared a single plan for the whole area, others prepared one or more plans for parts of their area, while others prepared none at all.

20 The *development plan scheme* was later replaced by the *local plan scheme*, which is now the *local development scheme*, it set out the agreed programme for the preparation and amendment of local plans. With the introduction of a mandatory requirement to produce district wide local plans in 1992 such schemes were made redundant. Now there are a number of development plan documents, it is required again. The 1986 Act required local authorities to keep a register of development plan policies and the new regulations require a similar index of information in respect of the development plan. The *certificate of conformity* was given by the structure planning authority. Before 1992 the lack of a certificate would delay the local plan process, but today any disputes about conformity are taken to the inquiry as an objection.

21 By 1977, only seventeen of the necessary eighty-nine structure plans for England and Wales had been submitted, and seven approved. By 1980, of seventy-nine English structure plans which were expected, sixty-four had been submitted and thirty-eight approved.

22 Over the years 1981 to 1985 the time taken from the submission of structure plans to their final approval averaged twenty-eight months. Many of the written statements and explanatory memoranda were very lengthy: in the first round, several contained more than 100,000 words.

23 Further support has been lent to these arguments by subsequent research. Rydin *et al.* (1990) and Collins and McConnell (1988) argued not only for recognition of the value of plans, and for a stronger development plan framework, but also for flexibility for local variation in form and content.

24 Strategic guidance was published for all the metropolitan regions, but except for London (where it has a statutory character), and the Thames Gateway, it has been incorporated into RPG, now RSS.

25 The 'daughter papers' were *New Parliamentary Procedures for Processing Major Infrastructure Projects*, *Reforming Planning Obligations: A Consultation Paper*, *Compulsory Purchase and Compensation: the Government's Proposals for Change*, *Possible Changes to the Use Classes Order and Temporary Uses Provisions* (all DTLR, December 2001).

26 ODPM Press Release: Planning Bill puts Community First, 4 December 2002. Performance on development control and appeals was no better.

27 This quote is taken from the 2001 CBI Planning Brief, *Planning for Productivity: A Ten Point Action Plan*. It is a moot point whether the government would have accepted arguments about resources and skills (made often by local government) had the business community not also recognised this problem.

28 While local authorities were generally more positive about the introduction of the LDF, a survey revealed that they are very sceptical about the possibility of preparing the LDF in three years while also improving community involvement (71 per cent thought it would not be very or at all easy to achieve).

29 Also, the 2004 Act establishes the principle that where policies conflict, the conflict will be resolved in favour of the most recent plan to be adopted.

30 In 2005 all authorities preparing LDFs submitted local development schemes to their respective regional offices for scrutiny, and to get them all in is certainly an achievement. The regional offices are collecting and collating very extensive information about progress and characteristics of the proposed LDFs some of which should be made more widely available.

31 The Assembly published *Revised Initial Guidance Notes on the Implications for Development Plans in Wales of the Planning and Compulsory Purchase Act* in 2003 and reissued them in 2004. They provide a very useful summary of the transition to the new system and are available on the Assembly website. The provisions of the Act relating to development plans in Wales are not in force at the time of writing but that is expected during 2005.

32 In the 1980 Manchester Structure Plan, for example, the Secretary of State 'deleted more than 40 per cent of the policies, and a further 20 per cent were substantially modified'. Thornley argues that such actions reflected the government's intention to allow market forces to operate at the cost of social and other wider objectives.

33 The 2004 Town and Country Planning (Local Development) (England) Regulations says that county councils

> shall keep under review . . . (a) the principal physical, economic, social and environmental characteristic of the authority; (b) the size, composition and distribution of the population of the area; (c) the communications, transport system and traffic of the area; (d) any other considerations which may be expected to affect those matters.

34 The idea of model policies for local plans was proposed in the early 1970s (Fudge *et al.* 1983) but received little support.

35 Some of the authorities with the best records of plan production are in the areas where the planning process is under most pressure, as in the home counties around London.

36 The Local Government and Planning Act 1980 had previously introduced an expedited procedure which, in certain circumstances, allowed the local plan to be adopted in advance of a structure plan review. The Housing and Planning Act 1986 gave powers to the Secretary of State to request modifications to plans (in addition to the seldom used powers to call in). The 1991 Act abandoned the need for the six-week consultation period prior to the deposit for objections, and an extra opportunity has been provided for objections after the inquiry where the planning authority does not accept the inspector's recommendations.

37 For example, in the early 1990s government called on local authorities to include all policies that might be used to reject planning applications in the statutory development plan, which contributed to the production of compendia of all possible development control policies, complex designations and, in turn, many objections.

38 Most of this advice is written up in the Inspectorate's guidance to local authorities. This account also incorporates comments made during the delivery of University of the West of England short courses by the former head of local plan inquiries at the Inspectorate, David John. See also the RTPI (2001a) guide to good practice listed under further reading.

39 Sir Geoffrey Howe credited the notion to Peter Hall, who in turn identified the origins of the concept in a 1969 article (Banham *et al.* 1969). Hall, P. (1991) has reviewed the ways in which this notion was transposed and 'sanitised' into the enterprise zone initiative in Britain.

40 The remaining enterprise zones are the East Midlands, Dearne Valley, North East Derbyshire, and Tyne Riverside.

41 The Corby EZ, for example, was virtually fully committed after seven years with 5,600 jobs and 3.15 million sq. ft of new floorspace (PA Cambridge Economic Consultants, 1987).

5 The control of development

The extent of vilification to which development control has been subject in Britain over the past 25 years suggests that it may be a rather more important process than its detractors allow. The importance of the process has, at one level, to do with matters of substance. Questions of land-use and urban form affect profoundly the welfare and enjoyment of life of those who live in urbanized societies like ours. Decisions taken in the course of development control have a long-term impact. At another level, however, the development control process serves as a focus for a whole range of questions about how we govern ourselves and on whom we confer power to take decisions on our behalf.

Booth 1996: 1–2

The scope of control

Most forms of development (as statutorily defined) are subject to the prior approval of the local planning authority, though certain categories are excluded from control because they are thought to be trivial or beneficial. Those seeking approval must submit application for planning permission to the local planning authority (LPA). Legislation gives considerable discretion to the local planning authority in granting permission, although decisions must be made in accordance with the development plan unless material considerations indicate otherwise. Few matters are excluded as potential material considerations. Since 1989 more significance has been given to the development plan in decision-making, such that it is the first and primary point of reference. Nevertheless, the planning authority can approve a proposal that does not accord with the provisions of the plan, and in practice it will not provide definitive guidance on many applications for development. The development plan may also be out of date or superseded by other guidance or considerations. The discretion of decision-makers at the time the application is made for approval is a

hallmark of planning in the UK. In most other countries, the decision is effectively made with the adoption of the plan.

Planning decisions are made subject to conditions or refused and in both cases there is a right of appeal. If the action of the LPA is thought to be *ultra vires* (beyond their legal powers), there is also a right of recourse to the courts. Furthermore, some major infrastructure projects and other applications of more than local importance may be 'called in' for decision by central government. Development control necessarily involves measures for enforcement. This is provided by procedures which require anyone who carries out development without permission or in breach of conditions to 'undo' the development, or stop the carrying out or continuation of development which is in breach of planning control.

These and other elements of control are discussed in the following sections. The general process of decision-making on planning applications is illustrated in Box 5.1. Readers should beware that this is only a general guide and reference should be made to the reading listed at the end of this chapter on matters of detail.

BOX 5.1 CHANGES MADE TO DEVELOPMENT CONTROL BY THE 2004 ACT AND AMENDMENTS TO THE GDPO

The Planning and Compulsory Purchase Act 2004 enables introduction of a number of provisions intended to speed the development control process and to focus the system more centrally on the achievement of government objectives. The ODPM began consultation on changes to the GDPO and the provisions in the 2004 Act in 2003. In November 2004 the ODPM began further consultation on revised proposals.*

The Planning and Compulsory Purchase Act 2004 introduces many new provisions in its own right and inserts sections into the 'principal act', the Town and Country Planning Act 1990. Many of the changes simply enable government to make changes and will require secondary legislation before they come into force.

- **Local development orders** will allow a local planning authority to extend permitted development rights within all or part of its area and for specific development or a class of development (other changes to the A classes in the Use Classes Order have been made in 2005 under existing powers).
- Local planning authorities are given powers to decline to determine **repeat applications** similar to one previously refused by the authority within two years (previously the authority was only allowed to decline to determine applications dismissed on appeal or after call-in). They will also be able to decline to deal with identical applications, the process known as **twin-tracking**.
- Outline applications will need to be accompanied by a design statement giving more detail of building heights, access and landscaping.
- The **duration of a full planning consent** is reduced from five to three years, though planning authorities retain the power to vary the duration; development initially approved through outline consent must begin within two years from the date of final approval of reserved matters.
- A duty is introduced on **statutory consultees** to respond to consultation within twenty-one days, and to submit a report to the Secretary of State on their performance against the deadline (the duty is generally only in relation to consultations under the GDPO). The duty also applies in relation to pre-application consultations.
- Regional planning bodies replace county councils as statutory consultees on applications that may affect implementation of the **regional spatial strategy**.
- New arrangements are introduced for consideration of **major infrastructure projects** (see relevant section in this chapter). Where such applications are called in by the Secretary of State, the applicant will need to submit an **economic impact report**.
- Temporary stop notices will be able to require an immediate stop to activities.
- New provisions for **simplified planning zones** are introduced (see Chapter 4).
- Provisions will allow for a new **planning tariff** which will replace much negotiation on planning obligations; this is planned for introduction in 2006.
- The Secretary of State is enabled to make regulations requiring fees in relation to call-in and recovered appeals and to set timetables for their consideration.

Note: * ODPM (2004) *Changes to the Development Control System: Consultation Paper*. Further consultation was planned for 2005. See also ODPM Circular 08/2005 *Guidance on Changes to the Development Control System*.

The control of development involves very strong powers that determine rights to use land and property with all that entails, not least the financial costs and benefits. At the centre of this system is the definition of *development*, particularly since the term has a legal connotation far wider than in ordinary language.

Definition of development

In brief, development is 'the carrying out of building, engineering, mining or other operations in, on, over or under land, or the making of any material change in the use of any buildings or other land'. There are many legal niceties attendant upon this definition with which it is fortunately not necessary to deal in the present outline. Some account of the breadth of the definition is, nevertheless, needed. 'Building operations', for instance, include rebuilding, structural alterations of or additions to buildings, some categories of demolition and, somewhat curiously, 'other operations normally undertaken by a person carrying on business as a builder'; however, maintenance, improvement and alteration works which affect only the interior of the building or which do not materially affect the external appearance of the building are specifically excluded. An exception to this general rule was made in 2004 when provisions were made to allow for interior alterations 'which have the effect of increasing the gross floor space of a building' to be brought within the definition of development. This was particularly to enable control of the creation of mezzanine floors in large shops.[1]

The second half of the definition introduces a quite different concept: development here means not a physical operation, but a change in the *use* of a piece of land or a building. To constitute 'development', the change has to be *material*, that is, substantial, a concept which it is clearly difficult to define, and which, indeed, is not defined in the legislation. A change in *kind* (for example from a house to a shop) is material, but a change in *degree* is material only if the change is substantial. For instance, the fact that lodgers are taken privately in a family dwelling house does not of itself constitute a material change so long as the main use of the house

remains that of a private residence. On the other hand, the change from a private residence with lodgers to a declared guest house, boarding house or private hotel would be material. Difficulties arise with changes of use involving part of a building, with ancillary uses, and with the distinction between a material change of use and a mere interruption.

This is by no means the end of the matter, but enough has been stated to show the breadth of the definition of development and the technical complexities to which it can give rise. Reference must, nevertheless, be made to one further matter. Experience has shown that complicated definitions are necessary if adequate development control is to be achieved, but the same tortuous technique can be used to exclude matters over which control is not necessary. First, there are certain matters which are specifically declared not to constitute development (for example, internal alterations to buildings, works of road maintenance, or improvement carried out by a local highway authority within the boundaries of a road). Second, there are others which, though possibly constituting development, are declared not to require planning permission.

There is provision for the Secretary of State to make a *General Permitted Development Order* (GPDO) specifying 'permitted development rights' for matters that constitute development but do not require permission because it is effectively granted by the Order. (This is not to be confused with the *General Development Procedure Order* (GDPO) which sets out the many procedures to be followed in the development control process.) The *Use Classes Order* (UCO) specifies groups of uses within which a change of use does not constitute development and is therefore permissible. Also, the Secretary of State can make *special development orders* (SDOs) granting planning permission for specific locations or categories of development; from 2005 local planning authorities can make *local development orders* (LDOs) to remove or add local permitted development rights (see Figure 5.1). Planning permission may also effectively be granted in advance through adoption of simplified planning zones and business planning zones (discussed p. 141).

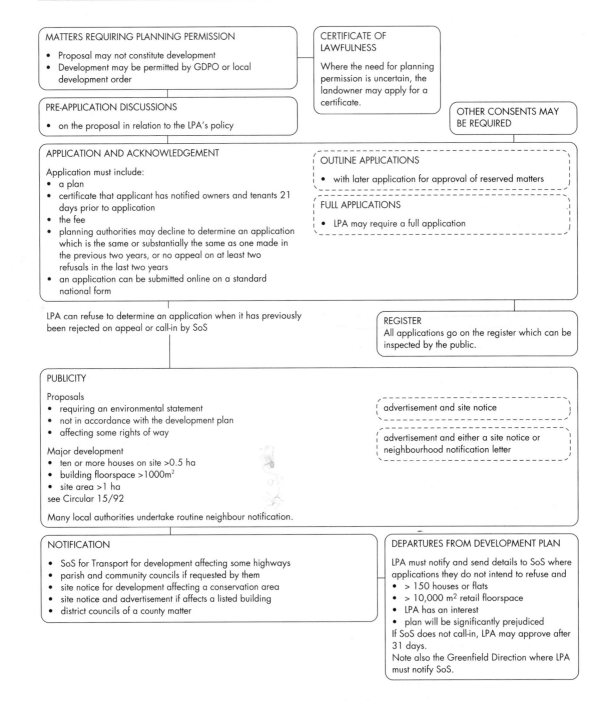

MATTERS REQUIRING PLANNING PERMISSION

- Proposal may not constitute development
- Development may be permitted by GDPO or local development order

CERTIFICATE OF LAWFULNESS

Where the need for planning permission is uncertain, the landowner may apply for a certificate.

PRE-APPLICATION DISCUSSIONS

- on the proposal in relation to the LPA's policy

OTHER CONSENTS MAY BE REQUIRED

APPLICATION AND ACKNOWLEDGEMENT

Application must include:
- a plan
- certificate that applicant has notified owners and tenants 21 days prior to application
- the fee
- planning authorities may decline to determine an application which is the same or substantially the same as one made in the previous two years, or no appeal on at least two refusals in the last two years
- an application can be submitted online on a standard national form

OUTLINE APPLICATIONS

- with later application for approval of reserved matters

FULL APPLICATIONS

- LPA may require a full application

LPA can refuse to determine an application when it has previously been rejected on appeal or call-in by SoS

REGISTER

All applications go on the register which can be inspected by the public.

PUBLICITY

Proposals
- requiring an environmental statement
- not in accordance with the development plan
- affecting some rights of way

Major development
- ten or more houses on site >0.5 ha
- building floorspace >1000m²
- site area >1 ha
see Circular 15/92

Many local authorities undertake routine neighbour notification.

advertisement and site notice

advertisement and either a site notice or neighbourhood notification letter

NOTIFICATION

- SoS for Transport for development affecting some highways
- parish and community councils if requested by them
- site notice for development affecting a conservation area
- site notice and advertisement if affects a listed building
- district councils of a county matter

DEPARTURES FROM DEVELOPMENT PLAN

LPA must notify and send details to SoS where applications they do not intend to refuse and
- > 150 houses or flats
- > 10,000 m² retail floorspace
- LPA has an interest
- plan will be significantly prejudiced
If SoS does not call-in, LPA may approve after 31 days.
Note also the Greenfield Direction where LPA must notify SoS.

Figure 5.1 *The planning application process in England*

CONSULTATION

In various circumstances consultation is required with:
- neighbouring authorities
- Health and Safety Executive
- Sports Council
- highway authority
- The Coal Authority
- Environment Agency (NRA)
- HMBC (English Heritage)
- The Theatre Trust
- waste regulation authorities
- regional planning bodies
- Secretaries of State for Transport and National Heritage

LPA will have long list of local consultees.

There is formally 14 days to respond.

See article 10 of GDPO. Statutory consultees are required by 5.24 PCP Act and the GDPO to respond within 21 days. Other legislation also specifies time limits for certain consultees.

PREPARATION OF REPORT

Planning officers will prepare a report on the application, undertake discussions and negotiations with the applicant and other interests, consider consultation returns and policy context, undertake site visits and request further information or changes to the proposal.
Reports are considered by planning committees and sometimes area committees or parish councils.
Many minor decisions are delegated to planning officers.

DECISION

Application is determined in accordance with the development plan, unless material considerations indicate otherwise, decision will be made within eight weeks.

REFUSED

LPA must give clear and precise reason with reference to development plan policies.

GRANTED

Development to be begun within a specified period or three or two years from approval of reserved matters.

APPEAL

Made to SoS within six months – must include:
- original application
- plans and correspondence
- notices

Determined by Inspectorate by
- written representations
- informal hearing (no cross-examination)
- public local inquiry (inquiry procedure rules apply)

SoS 'recovers' some
- appeals for own decision e.g. if >150 houses or of significant controversy, or for other reasons

DECISION

Inspector makes most decisions but may report to SoS if 'recovered'.

CHALLENGE

Appellant can seek 'statutory review' in the High Court within six weeks on the grounds that decision
- not within powers of the Act
- procedural requirements not met.

Decision may only be to quash or uphold previous decision.

For a more comprehensive explanation see Moore (2005).
For variations in Scotland see McAllister and McMaster (1994) and Collar (1994).
For variations in Northern Ireland see Dowling (1995).
Recent changes to development control introduced by the Planning and Compulsory Purchase Act 2004 are explained in Circular 8/2005.

The Use Classes Order and the General Permitted Development Order

The Use Classes Order groups all land uses into classes. Table 5.1 shows the use classes in different parts of the UK. Changes within each class do not constitute development and therefore do not need planning permission. Thus, class A1 covers shops used for all or any of a list of ten purposes, including the retail sale of goods (other than hot food); the sale of sandwiches or other cold food for consumption off the premises; for hairdressing; for the direction of funerals; and for the display of goods for sale. Class A3 covers 'use for the sale of food or drink for consumption on the premises or of hot food for consumption off the premises'. As a result of these classes, a shop can be changed from a hairdresser to funeral parlour or a sweet shop (or vice versa), but it cannot be changed (unless planning permission is obtained) to a restaurant or take-away, which is in a different class. The classes, it should be stressed, refer only to changes of use, not to any building work, and the Order gives no freedom to change from one class to another. Whether such a change constitutes development depends on whether the change is 'material'.[2]

The General Permitted Development Order gives the developer a little more freedom by listing classes of 'permitted development' – or, to be more precise, it gives advance general permission for certain classes of development, typically of a minor character.[3] If a proposed development falls within these classes, no application for planning permission is necessary: the GPDO itself constitutes the permission. The Order includes minor alterations to residential buildings, and the erection of certain agricultural buildings (other than dwelling houses). A summary of permitted development rights is given in Box 5.2. The GPDO also permits certain changes of use within the UCO, such as a change from an A3 use (the Food and Drink class) to an A1 use (shop), but not – because of the possible environmental implications – the other way round. While the use changes allowed by the UCO are all 'bilateral' (any change of use within a class is reversible without constituting development), the GPDO builds upon this structure by specifying a number of 'unilateral' changes *between* classes for which permission is not required. The rationale here is that the permitted changes generally constitute an environmental improvement. The rights given by the GPDO can be withdrawn by Article 4 directions and conditions on planning permissions (discussed below). At the time of writing, the Office of the Deputy Prime Minister expected to consult on changes to the GPDO during 2005.

The cynic may be forgiven perhaps for commenting that the freedom given by the UCO and the GPDO is so hedged by restrictions, and frequently so difficult to comprehend, that it would be safer to assume that any operation constitutes development and requires planning permission (though it may be noted with relief that painting is not normally subject to control, unless it is 'for purpose of advertisement, announcement or direction'). The legislators have been helpful here. Application can be made to the LPA for a *certificate of lawfulness of proposed use or development* (CLOPUD). This enables a developer to ascertain whether or not planning permission is required.

The Orders are modified from time to time, usually with the intention of lifting the burden of regulation. Intentions are fine, but once the rules are exposed for discussion, the result is often more rules not fewer. Deregulation is difficult here because changes of use can have dramatic effects on amenity, traffic generation and the quality of places.[4] New uses too have to be accommodated in the system, and some fall out of fashion. In England, the UCO refers anachronistically to 'dance halls' but does not adequately recognise the large city pubs and clubs that are associated with 'binge drinking'. Pubs were classified under A2: food and drink alongside restaurants and thus large city bars could be developed in any property with the A2 use despite their more extensive and difficult impacts and association with antisocial behaviour. Pubs could be converted into fast food restaurants within the A3 use class, and often were.

In 2001 the Department of Transport, Local Government and the Regions commissioned a review of the UCO (and Part 4 of the GDPO which deals with temporary uses) (Baker Associates 2001). The research

Table 5.1 Summary and comparison of the Use Classes Orders

England and Wales (Town and Country Planning (Use Classes) Order 1987) (SI No. 764) (as amended)			Scotland (The Town and Country Planning (Use Classes) (Scotland) Order 1989) (SI N. 147) (as amended)			Northern Ireland (Planning (Use Classes) Order (NI) 1989) (as amended)		
Class	Use	Development permitted by the GDPO (which may be subject to limitations)	Class	Use	Development permitted by the Permitted Development Order	Class	Use	Development permitted
A1	Shops	from A3; A2 if premises have a display window at ground floor level, or for the display or sale of motor vehicles	1	Shops	from sale and display of motor vehicles	1	Shops	from a betting office or from food or drink
A2	Financial and professional	to A1 from A3	2	Financial, professional and other services	to 1	2	Financial, professional and other services	from a betting office or from food or drink; to 1 if it has a display window at ground floor level
A3	Food and drink	to A1 and A2	3	Food and drink	to 1 and 2			
A4* A5*	Pubs and bars Takeaways	to A1, A2 and A3 to A1, A2 and A3						
B1	Business	to B8 (max 235 m^2) from B2 and B8 (max 235 m^2)	4	Business	to 11 (max 235 m^2)	3	Business	
						4	Light industrial	to 11
B2	General industrial	to B1 and B8 (max 235 m^2)	5	General industrial	to 4 and 11 (max 235 m^2)	5	General industrial	to 4 or 11 (max 235 m^2)
			7–10	Special industry groups		7–10	Special industrial groups	
B8	Storage or distribution	to B1 (max 235 m^2) from B1 or B2 (max 235 m^2)	11	Storage or distribution	to 4	11	Storage and distribution	to 4 (max 235 m^2)

Table 5.1 continued

England and Wales (Town and Country Planning (Use Classes) Order 1987) (SI No. 764) (as amended)			Scotland (The Town and Country Planning (Use Classes) (Scotland) Order 1989) (SI N. 147) (as amended)			Northern Ireland (Planning (Use Classes) Order (NI) 1989) (as amended)		
Class	Use	Development permitted by the GDPO (which may be subject to limitations)	Class	Use	Development permitted by the Permitted Development Order	Class	Use	Development permitted
C1	Hotels, boarding houses and guest houses		12	Hotels and hostels (not including public houses)		12	Guest houses and hostels	to 14
C2	Residential institutions		13	Residential institutions		13	Residential institutions	to 14
C3	Dwelling houses		14	Houses		14	Dwelling houses	
D1	Non-residential institutions		15	Non-residential institutions		15	Non-residential institutions	
D2	Assembly and leisure		16	Assembly and leisure		16	Assembly and leisure	

Notes: The subdivision of residential dwellings into two or more separate dwellings is a change of use
* A4 and A5 classes were introduced in England and Wales in 2005

BOX 5.2 SUMMARY OF PERMITTED DEVELOPMENT RIGHTS IN ENGLAND

Permitted development rights are granted by the General Permitted Development Order 1995. The GPDO grants planning permission for certain minor forms of development which are listed in Schedule 2. The permissions can be withdrawn by Article 4 directions or conditions attached to planning permissions. The application of the Order is complex and this is only a brief summary.

- Development within the cartilage of a dwelling house, limited to 10 per cent of the cubic content of terraced houses and 15 per cent of detached houses, and an overall maximum of 115 cubic metres.
- Minor operations such as painting and erection of walls and fences but not over 2 metres in height
- Temporary buildings and uses in connection with construction, and temporary mineral exploration works
- Caravan sites for seasonal and agricultural work
- Agricultural and forestry buildings and operations (although the local planning authority must be notified in certain circumstances)
- Extension of industrial and warehouse development up to 25 per cent of the cubic content of the original building
- Repairs to private driveways and services provided by statutory undertakers and local authorities (including sewerage, drainage, postboxes), maintenance and improvement works to highways by the highway authority
- Limited development by the local authority such as bus shelters and street furniture
- Certain telecommunications apparatus not exceeding 15 metres in height, and closed circuit television cameras, subject to limitations.
- Restoration of historic buildings and monuments
- Limited demolition works

Permitted development rights are similar in other jurisdictions in the UK. At the time of writing recommendations had been made for many (mostly small) amendments to the GPDO, including, for example, abandoning the cubic content measure but instead using plot ratio and proximity to cartilage only.

found that the main concerns were with the food and drink class, and especially noise from bars. The main recommendations were to combine A1 (shops) with A2 (financial and professional services) in one 'mixed retail' class. This class was also recommended to include food and drink premises and pubs and bars if they fell below a threshold of 100 square metres. Above that threshold these uses would have their own classes. The government subsequently consulted on a range of options for change in 2002. The outcomes announced in 2005 are new classes A4 (pubs and bars) and A5 (takeaways),[5] but this will include all premises that fall within that category, irrespective of size; the threshold idea was rejected. The changes also put Internet cafes in Class A1 and retail warehouse clubs and nightclubs become *sui generis*.

Concerns about the complexity and difficulty of the interpretation of the GDPO have also brought it under review by ODPM. The research commissioned from Nathaniel Lichfield in 2003 provides a blow-by-blow account of the operation of the GPDO with separate sections on the many categories of permitted development. The research did not fully address telecommunications and temporary uses though they both figure as among that type of development giving most problems (they are being addressed in other ways). The main issues raised by consultees were the inconsistencies and difficulty of interpretation of the GPDO, the adverse impacts which arise from inadequate control particularly in sensitive areas and the failure of the system of permitted development overall to contribute to government policy, not least, achieving more sustainable development.

Consultation is expected in 2005 on revisions to the GPDO and although the details are not known, the Lichfield research suggests that there will be more tightening than relaxation of controls. Among the many recommendations are the much needed changes to define work by statutory undertakers as development and permitted development for such work to be conditional on reinstatement to the original standard, and to allow removal of redundant buildings provided under permitted development rights.

Withdrawal of permitted development rights

The development rights that are permitted by the GPDO can be withdrawn by a Direction made under Article 4 of the Order (and hence are known as *Article 4 Directions*). The effect of such a direction is not to prohibit development, but to require that a planning application is made for development proposals in a particular location. The direction can apply either to a particular area (such as a conservation area) or, unusually, to a particular type of development (such as caravan sites) throughout a local authority area. Article 4 Directions should be made only in exceptional circumstances and where there is 'a real and specific threat'.[6]

The most common use of an Article 4 Direction is in areas where special protection is considered desirable, as with a dwelling house in a rural area of exceptional beauty, a national park or a conservation area. Without the direction, an extension of the house would be permitted up to the limits specified in the

GPDO. The majority of Article 4 Directions in fact relate to 'householder' rights in conservation areas. They are also used in national parks and other designated areas to control temporary uses of land (such as camping and caravanning) which would otherwise be permitted (Tym *et al.* 1995a).

Since the Article 4 Direction involves taking away a legal right, compensation may be payable and the Lichfield (2003) research mentioned previously recommends that this right be removed. The report also generally advocates a different approach: removing permitted development rights most often removed by Article 4 Directions and then allowing local authorities to bring in permitted development through local development orders.

Local development orders

Provisions for the introduction of local development orders were made by amendments to the 1990 Act introduced in the 2004 Act. The purpose of LDOs is to allow local planning authorities to extend 'national' permitted development rights for all or part of their area for specific developments or general classes of development. It is another idea intended to speed the planning system and provide more certainty for business, although one that has been received with considerable scepticism, including opposition from 'the Planning Inspectorate, CPRE, the Civic Trust and the Law Society' (Land Use Consultants and Wilbraham and Co. 2003: 12). The main concerns were the potential fragmentation of the system with local planning authorities making different requirements and thereby creating confusion and, given that standards will vary among local authorities, undermining confidence in the system. The Home Builders' Federation, among others, supported the idea in principle on the grounds that it may speed the system by removing many smaller applications and others where there were agreements about their implementation.

The effect of an LDO is to grant planning permission 'in advance' so as to speed the implementation that has been adopted in the development plan. Like the simplified planning zone, it borrows from the approach to development regulation in continental countries where the 'regulation plan' determines the grant of a permit and removes the planning authority's discretion once the order is made. Also, like the early days of SPZs, LDOs have been suggested as means to facilitate house building. The links to the government's concerns about the rate of house building (see Chapter 6) are obvious. As explained in Chapter 4, SPZs were never popular. Developers, investors and planners alike found it a difficult concept in a system built on negotiation and compromise at the time a proposal comes forward.

However, the provisions for LDOs now set out in s. 61A of the 1990 Act appear to be very flexible.[7] Permission could be granted for a specific form of development on one site, or to any development of a particular type within the authority's area, and it can specifically exclude any type of development or location. No separate hearing is required in the procedure and the order is adopted by resolution of the local planning authority, although, as always, with provision for government to intervene if needed. Planning Policy Statement 12: Local Development Frameworks suggests that an LDO would best be prepared in conjunction with the plan. This would allow the authority to consult on both at the same time, although given the likely controversial and uncertain nature of the latter this proposal is not likely to speed the process.

A consultation exercise on how LDOs might be used and drafted was conducted at the end of 2003 (Land Use Consultants and Wilbraham and Co. 2003). It concludes with a draft guidance note and includes examples of current practice which might lend themselves to the LDO approach. In Northampton, for example, English Partnerships have been working very closely with the local planning authority and other interests on a detailed masterplanning of the Upton urban expansion area, including detailed design codes and site specific development briefs. Where this extensive preparatory work has led to broad agreement, it may be that an LDO would speed the planning part of the process.

Special development orders

While the GPDO is applicable generally, special development orders relate to particular areas or particular types of development. SDOs (like other orders) are subject to parliamentary debate and annulment by resolution of either House. They have provided an opportunity for testing opinion on controversial proposals, such as the reprocessing of nuclear fuels at Windscale, but most of the nineteen SDOs made in England and Wales were to facilitate the operation of urban development corporations. In these cases the order granted permission for development that was proposed by the corporations and approved by the Secretary of State. The use of the SDO procedure raises considerable controversy since it involves a high degree of central involvement in local planning decisions. One very contentious case was the granting of permission for over a million square feet of offices and homes at the eastern end of Vauxhall Bridge. At that time the DoE said that 'the purpose of making fuller use of SDOs would not be to make any general relaxation in development control, but to stimulate planned development in acceptable locations, and speed up the planning process' (Thornley 1993: 163). In practice, central government has not made use of the orders in recent years and has instead opted for other means to shape major decisions. In 1999, SDOs became subject to the provisions for environmental assessment.

Planning application process

All planning authorities provide guides on the planning application process and readers should make reference to them for the finer points. For many minor applications it is a straightforward process, but in some cases it can become very complex and time consuming. Figure 5.1 gives an overview of the process in England, and it is much the same elsewhere. Many applications will begin with pre-application discussions with the local authority. The 2004 Act introduces provisions that will allow planning authorities to charge for this service. It is especially important for the local authority to ensure that the application is complete and meets its requirements so that there is minimum delay in processing.

These more routine aspects of the control process, the planning application form and procedures for its acknowledgement and registration came under close scrutiny in 2003 as part of the government's drive to improve efficiency. In 2004, Arup with Nick Davies Associates reported (at great length) on a wide-ranging study for the ODPM of how local planning authorities deal with the receipt of applications. The outcomes include recommendations for a standard application form and lots of guidance for local authorities intended to provide more certainty for applicants and consistency among authorities in, for example, starting the clock during the application registration on the eight-week determination target. That this basic evaluation of the application process had not been made years before is perhaps testament to the need for a shake-up of the system. Many planning authorities have made great progress in service provision, and the best practice needs to become more widespread. But all authorities have come some distance since the 1970s when (from personal experience) the very first job done on receipt of an application was to apply for an extension of time to consider it.[8]

On receipt of the application and fee the authority will acknowledge and begin publicity, notification and consultation procedures, all of which will vary depending on the type of application. A review of the arrangements for publicising planning applications in England (Arup 2004) provides a summary of arrangements and practices in planning authorities together with many recommendations for improvement. This is considered in Chapter 12, with comparisons of practice elsewhere in the UK and alongside other forms of public involvement in planning.

As shown in Figure 5.1 the local planning authority will consult with many bodies, some of which are statutory consultees, meaning that they have to be consulted by law. Research on statutory and non-statutory consultation found some confusion among planning authorities about who should be consulted for what purpose. Not all consultees have had the capacity and/or been willing to cooperate effectively in this process, and so from 2005 there is a statutory duty to

respond to consultation within twenty-one days. Circular 8/2003 also enables a planning authority to forgo consultation with certain statutory consultees if it believes that the development is subject to standing advice issued by the relevant consultees.

Many planning applications will also require other consents from the authority and other agencies, notably building regulations approval. Since the mid 1990s attention has been given to the idea of creating a 'one stop shop' approach providing a more user friendly service for those who will be seeking more complicated consents.[9] In 2004 ODPM reported on a more fundamental approach to the *Unification of Consent Regimes* (Halcrow Group 2004). The review concentrated on potential unification of planning, listed building and conservation area consents but in a mammoth report (about 350 pages) provides a Cook's tour of other regimes, including enforcement, building regulations, hazardous substances consents and more. The report identifies many benefits of the existing, largely separate, regimes, but also many potential benefits from unification. The critical issue is whether one procedure could incorporate a number of regimes without diluting the effect of any one, especially heritage. The scope for a measure of unification seems obvious (and some local authorities already integrate the procedures).

On the basis of consultation returns, the relevance of national and local policies, previous decisions and a site visit, the planning officer will make a report to the planning committee with a recommendation on the decision to be made.[10] This report along with the committee agenda and minutes and consultation returns are public documents. The applicant may be able to make a presentation to committee but this is at the discretion of the authority (see Chapter 12). Decision notices are sent to the applicant, who can appeal against refusal or conditions imposed. Amendments to the GDPO in 2000 and 2003 (applying to England) introduced requirements for decisions to include an explanation of the reasons for any grant of planning permission and a summary of the policies and proposals that were relevant to the decision.[11]

Most applications in many authorities will be decided by the planning officer under delegated powers, subject to meeting criteria such as the application being in accordance with development plan policies and below certain thresholds. In 2004 planning authorities delegated 87 per cent of applications on average, although three authorities delegated less than 10 per cent.[12] When elected members consider applications they may not always agree with officers and there are some celebrated cases where members have decided applications against the advice of their officers such as Ceredigion and North Cornwall.[13] This is one category of cases that has led to planning authorities' decisions being subject to judicial review (discussed in Chapter 3).

More complex applications will require negotiations between the applicant or agent and the authority. The officer will be seeking to ensure that the application meets policy and will be working from past experience of committee decisions. The discretionary nature of the British planning system allows for negotiation prior to the final decision. In theory this offers scope to ensure that the final development is closer to meeting the needs of all parties, so long as officers and applicant recognise the benefits of negotiation to achieve better outcomes (Claydon 1998). In practice it appears that local authority officers are less prepared to make good use of the opportunity of negotiation than developers (Claydon and Smith 1997).

In complicated cases it is sometimes convenient for an applicant or the LPA (or both) to deal with an application in outline. *Outline planning permission* gives the applicant permission in principle to carry out development subject to *reserved matters*, which will be decided at a later stage. This device enables a developer to proceed with the preparation of detailed plans with the security that they will not be opposed in principle. In a few cases there will need to be an environmental impact assessment – the procedures for which are described in Chapter 7.

The development plan in the determination of planning applications

Crucial to the development control process is the concept of *material considerations*. These are exactly what the term suggests: considerations that are material to the taking of a development control decision. The primary consideration is the development plan.[14] Plans have always been important considerations in development control, but during the 1970s and 1980s many local planning authorities did not have adopted statutory local plans and even if they did, they were not always given the weight they deserved in decision-making by both local and central government. Thus, the status of the plan became ambiguous.

In 1989 the government began the move to a plan-led system and asked all planning authorities to ensure that they had an adopted up-to-date local plan in place. In 1991 statutory force was lent to the role of the plan in decision-making through insertion of s. 54A into the 1990 Act.[15] This may sound strange to those new to planning, and it may be appropriate to ask if there can be any other sort of planning system. The implications of s. 54A 1990 Act (which is now superseded by s. 38(6) of the 2004 Act) have been the subject of much debate. In the light of experience and court rulings the meaning has been clarified in various revisions of policy guidance for England, Wales and Scotland. The current guidance is given in Box 5.3.

Section 54A certainly had a major impact on the planning system. There is much more emphasis on the preparation of statutory plans to ensure that there is an adequate framework of policy against which to test applications. The 'presumption in favour of development' dating back to the beginnings of planning control (Harrison, M. 1992) has effectively been changed to a presumption in favour of the development plan, or more accurately in the words of Malcolm Grant,

> it is if anything, a presumption *in favour* of development that accords with the plan; and a presumption *against* development that does not. In each case, the development plan is the starting point, and its provisions prevail until material considerations indicate otherwise.
>
> (*Encyclopedia* P54A.07, emphasis in original)

But in most cases other material considerations will also play a part in the decision and this has always been

BOX 5.3 THE PLAN-LED SYSTEM

The government is committed to a plan-led system of development control.
The Planning and Compulsory Purchase Act 2004 s. 38(6) says:

> If regard is to be had to the development plan for the purposes of any determination to be made under the planning Acts the determination must be made in accordance with the plan unless material considerations indicate otherwise

Planning Policy Statement 1: *Delivering Sustainable Development* says:

> [The] plan-led system, and the certainty and predictability it aims to provide, is central to planning and plays the key role in integrating sustainable development objectives. Where the development plan contains relevant policies, applications for planning permission should be determined in line with the plan, unless material considerations indicate otherwise.

The Planning System: General Principles adds:*

> If the Development Plan contains material policies or proposals and there are no other material considerations, the application should be determined in accordance with the Development Plan. Where there are other material considerations, the Development Plan should be the starting point, and other material considerations should be taken into account in reaching a decision. One such consideration will be whether the plan policies are relevant and up to date.

Scottish Planning Policy 1 *The Planning System* (2002) makes similar statements referring to ss. 25 and 37(2) of the Town and Country Planning (Scotland) Act 1997, and usefully sets out the approach to decision-making on planning applications set out by a House of Lords judgement of 1998:**

- identify any provisions of the development plan which are relevant to the decision;
- interpret them carefully, looking at the aims and objectives of the plan as well as detailed working of policies;
- consider whether or not the proposal accords with the development plan;
- identify and consider relevant material considerations, for or against the proposals; and
- assess whether these considerations warrant a departure from the development plan.

The weight to be attached to any relevant material consideration is for the judgement of the decision-maker (para. 47).

Notes: * This document was published 'alongside' Planning Policy Statement 1, and is an updated version of content in the previous Planning Policy Guidance Note 1: *General Policy and Principles* (1997). It is, therefore, government policy, even though, for some reason, it could not be published in PPS 1. Note that this quotation also gives a good example of the government's affectation with capital letters, as in Development Plan!
** *City of Edinburgh v the Secretary of State for Scotland* 1998 SLT120.

the case. Whether or not other material considerations outweigh the development plan is a matter of judgement for the decision-makers.[16] Even with much more comprehensive plan coverage, many issues raised by planning applications will not be addressed in policy, and there is a limit to which governments at any level can, or wish to, commit policies to paper. The more this is done, the more inflexible will planning become, the less will it be able to adapt to changing circumstances, the greater is the likelihood of conflict between policies, and the more confusing the situation will be. The discretionary 'hallmark' of the British development control system mentioned at the start of this chapter, is also, in comparison to systems elsewhere, a great advantage. But this only applies so long as there are effective safeguards to ensure that discretion is exercised in the proper way.

Legal niceties aside, how do planning authorities actually decide planning applications? On this central question, findings from research in the 1980s on the role of development plans are probably still most enlightening. Davies *et al.* (1986b) found that many considerations were not covered by plans, and policies were typically expressed in general ways and needed 'translation' into operational terms for each application. Supplementary guidance and other planning documents including design guides, development briefs, informal local plans and 'policy frameworks' were important. With some caveats, notably the more comprehensive nature of many plans in the 1990s, these findings are probably no less valid today. The same authors found in a study of appeals that 'Inspectors nearly always dismissed appeals, and supported the local authority, on proposals for which there was relevant cover in the development plan'. On the other hand, they 'more often allowed appeals which turned on practical appeal considerations lacking firm local policy coverage, but in which national policies were invoked in favour of the appellant'. The message here is that development control and appeal decisions tend to abide by policy where it exists. And these findings predate the so called 'plan-led system'. It suggests that greater coverage of statutory plans has been more important than the statutory requirement to make it the starting point for decision-making.

Other material considerations

Since planning is concerned with the use of land, purely personal considerations are not generally material (though they might become so in a finely balanced case). The courts have held that a very wide range of matters can be material.[17]

The list of possible considerations begins with the siting and appearance of the proposed buildings; the suitability of the site and its accessibility; relationship to traffic and infrastructure provision; landscaping and the impact on neighbouring land and property. But many other matters may be relevant: environmental impacts; the historical and aesthetic nature of the site; economic and social benefits of the development; considerations of energy and 'sustainable development'; impact on small businesses; previous appeal decisions; the loss of an existing use; whether the development is likely to be carried out; and in a few cases financial considerations, including the personal circumstances of occupiers. Whether or not any of these considerations is material depends on the circumstances of each case. Very few considerations have been held by the courts to be immaterial but include the absence of provision for planning gain; and to make lawful something that is unlawful (Moore 2000: 206; see also Thomas, K. 1997).

Planning policy statements (and their equivalents) and circulars are important material considerations. Although they have no formal statutory force (they are not legally binding), the local planning authority must have regard to them. Where the local authority does not follow national guidance it must give 'clear and convincing reasons'.[18] Changes to national policy that postdate the development plan are particularly important, as in the case of the revisions to PPG 3, *Housing*, made in 2000. Government policy is helpful where there are special requirements. For example, for out-of-town shopping centres, it is explicitly advised (in PPG 6, para. 1.16) that 'key considerations should be applied' including the likely impact of the development on the vitality and viability of existing town centres, their accessibility by a choice of means of transport, and their 'likely affect on overall travel patterns and car use'. But for many topics guidance can be found to justify alternative positions. And the courts have found that it may be expressed in policy, 'previous decisions, written parliamentary answers, and even after dinner speeches' and conference speeches (Read and Wood 1994: 13).

Two considerations warrant further discussion: the design and appearance of development, and amenity.

Good design

Much of the built heritage is worth preserving because it is well designed. It is therefore of more than contemporary concern that new buildings should be well designed. Nevertheless, the extent to which 'good' design can be fostered by the planning system (or any other system) is problematic.

Good design is an elusive quality which cannot easily be defined. In Holford's (1953) words, 'design cannot be taught by correspondence; words are inadequate, and being inadequate may then become misleading, or even dangerous. For the competent designer a handbook on design is unnecessary, and for the incompetent it is almost useless as a medium of instruction.' Yet local authorities have to pass judgement on the design merits of thousands of planning proposals each year, and there is continuous pressure from professional bodies for higher design standards to be imposed.

There is a long and inconclusive history to design control (well set out by John Punter, in various publications from 1985). A 1959 statement by the Ministry of Housing and Local Government (MHLG) stressed that it was impossible to lay down rules to define good design. Developers were recommended to seek the advice of an architect (presumably a good one!). The policy should be to avoid stifling initiative or experiment in design, but 'shoddy or badly proportioned or out of place designs' should be rejected – with clear reasons being given.

The reader is referred to Punter's work for the fascinating details of the continuing story, recounting the personal achievements of Duncan Sandys, particularly in founding the Civic Trust in 1957, and later in promoting the Civic Amenities Act; the high buildings controversy ('sunlight equals health'); the problem of

protecting views of St Paul's Cathedral; the arguments over the Shell Tower (which prompted the quip that the best view of the Shell Tower was to be obtained from its roof); the publication of Worskett's *The Character of Towns* (1969); the unpublished Matthew-Skillington Report on *Promotion of High Standards of Architectural Design* which led to the appointment of a Chief Architect in the Property Services Agency; the property boom and a spate of books bearing titles such as *The Rape of Britain* (Amery and Cruikshank 1975) and *The Sack of Bath* (Fergusson 1973); the *Design Guide for Residential Areas* (Essex County Council 1973) – 'the most influential local planning authority publication ever'; the attempt (in 1978) to prevent the building of the National Westminster Tower; and the unprincely attack in 1984 by the Prince of Wales on the 'monstrous carbuncle' of the proposed extension to the National Gallery.[19]

In his case study of office development control in Reading, Punter (1985) demonstrates the interesting point that it is only since the late 1970s that the local authority 'have begun to influence the full aesthetic impact of office buildings, though they have controlled height, floorspace and functional considerations since 1947'. Moreover:

> Aesthetic considerations do not operate in a vacuum: they are merely one set of considerations among many in deciding whether a development gets planning permission. In the case of office development, despite its visual impact, the control of floorspace and the provision of associated facilities and land uses have been higher order goals in Reading . . . Aesthetic considerations are inevitably the first to be sacrificed in the cause of 'speed and efficiency' in decision-making, by clients, developers, architects and planners.
>
> (Punter 1985)

The Conservative administration of 1979 started off with a strong bias against design controls with DoE Circular 22/80. Michael Heseltine was responsible, saying that 'far too many of those involved in the system – whether the planning officer or the amateur on the planning committee – have tried to impose their standards quite unnecessarily on what individuals want to do'.[20] The 1992 version of PPG 1 included an annex on design control (based on a draft prepared jointly by the Royal Institute of British Architects (RIBA) and RTPI) and tried to square the circle by advising that 'planning authorities should reject obviously poor designs' but they 'should not impose their taste on applicants for planning permission simply because they believe it to be superior'. The 1997 version of PPG 1 followed in similar vein, but with the overall balance in favour of intervention. More is made of the role of the development plan and supplementary planning guidance (now SPD) (if subject to public consultation) in justifying control 'to promote or reinforce local distinctiveness', but 'local planning authorities should not concern themselves with matters of detailed design, except where such matters have a significant effect on the character or quality of an area' (para. 18).

Meanwhile, public, private and voluntary bodies have led numerous campaigns for improved standards of design, notably John Gummer's (then Secretary of State) *Urban Design Campaign*,[21] CPRE's *Local Attraction* campaign, the DETR and Commission for Architecture and the Built Environment's *By Design: Urban Design in the Planning System – Towards Better Practice*, and others from the (then) Countryside Commission, the Royal Fine Art Commission, Common Ground and English Partnerships. Some very attractive publications have emerged as a result of this interest. Whether there has been a parallel emergence of better designed buildings as yet is an open question.

Many of the more difficult aesthetic decisions are made by inspectors. Durrant's (2000) explanation of the reasoning that an inspector makes in cases of dispute over quality of design reveals the very subjective nature of the task: in his case including an example of allowing a twenty-storey 'glass mountain' adjacent to a grade 1 listed parish church on the south bank of the Thames at Battersea. Durrant argues that the reasoning process has two principal components: the context (both aesthetic and functional) and the scale of buildings, but at the appeal stage the options available to the inspector's decision are really only yes or no.

The design qualities of the most 'significant' devel-

opments come under particular scrutiny through the Design Review Committee of the Commission for Architecture and the Built Environment (CABE 2004), and the Design Commission for Wales. CABE scrutinises about 500 projects a year, 100 of which are discussed in the committee. Significant for CABE means that they are prominent or they may affect an important site, or are out of the ordinary. But CABE does not try to replicate the job of the local planning authority in testing designs against national and local policy and design guidance. Rather CABE, in this and other activities, seeks to change the development process overall, so that improvements can be made to the quality of proposals. Recent government statements are therefore welcome in that they are directed equally to developers:

> We are not going to beat about the bush. When applying for planning permission, house builders will have to demonstrate to local planning authorities how they have taken the need for good design into account. The point is that good urban design is not just about aesthetics . It concerns the quality of life people experience. For example, it can help prevent crime and the fear of crime. It can help create a sense of community. It is not trite to say that good urban design helps make good places and satisfied people. It will help us put land to better use, because wasting land in the towns means more land lost in the countryside.
>
> (The Minister for Housing and Planning, Nick Raynsford, in a speech to the House Builders Federation, 27 January 2000)

This statement also reflects a widening of attention in design considerations from aesthetics to social inclusion. This is also taken up by the successor to PPG 1, *Planning Policy Statement 1: Delivering Sustainable Development*.

> Design which is inappropriate in its context, or which fails to take the opportunities available for improving the character and quality of an area and the way it functions, should not be accepted . . . High quality and inclusive design should create

well-mixed and integrated developments which avoid segregation and have well-planned public spaces that bring people together and provide opportunities for physical activity and recreation.

(paras 34 and 35)[22]

Planning Policy Guidance Note 3: *Housing* (2000) (due for replacement with PPS 3 in 2005) continues in the same way, with guidance on design and layout; it can be used to make the best use of sites, especially previously developed land. Regeneration policy has also promoted the way urban design 'can position development in the market, change perceptions of place and create value' (p. 9). This quotation is taken from the English Partnerships and Housing Corporation joint *Urban Design Compendium* prepared by Llewellyn-Davies (2000), following recommendations from the Urban Task Force. Among a list of constraints that prevent making quality places the norm, the *Compendium* picks out 'reactive planning and development control approaches and mind-sets, applying quantitative standards (zoning, density, car parking, privacy distances, etc.) rather than providing qualitative advice and judgements' (p. 12). This is not to say that planning is not a consideration, but rather how it is applied.

Carmona's (1998, 1999) survey of residential design guidance shows that most authorities are making efforts to improve the quality of design, although practice 'remains varied in the extreme'. About half of all planning authorities have at least three forms of design guidance often linked in hierarchical fashion from strategic through local to site specific. But

together, the evidence illustrates a strong belief in the value of pre-conceived prescription as the basis for controlling residential design, but tremendous variety – and therefore inconsistency – in the chosen approaches used to prescribe that design.

(Carmona 1999: 36)

It should be noted that good design is a necessary but not sufficient condition in achieving social and economic aspirations. The *Design Improvement Controlled Experiment* began in 1989 with radical improvement to the design of local authority housing estates, including replacing open courtyards with more private gardens and removing overhead walkways. The improvements were welcome in themselves, but evaluation by Price Waterhouse published in 1997 found that the social and sustainability objectives were not met. Indeed, this sort of investment performed less well than Estate Action.[23]

Nevertheless, for Punter, 'design issues occupy a more important position in contemporary planning practice today than at any stage over the last 50 years' (1999: 151). Design is an important consideration in planning decisions, and not just the aesthetic qualities of buildings, but also for social and economic goals.

Amenity

'Amenity' is one of the key concepts in British town and country planning, yet nowhere in the legislation is it defined. The legislation merely states that 'if it appears to a local planning authority that it is expedient in the interests of amenity', it may take certain action, in relation, for example, to unsightly neglected waste land or to the preservation of trees. It is also one of the factors that may need to be taken into account in controlling advertisements and in determining whether a discontinuance order should be made. It is a term widely used in planning refusals and appeals: indeed the phrase 'injurious to the interests of amenity' has become part of the stock-in-trade jargon of the planning world. But like the proverbial elephant, amenity is easier to recognise than to define, and there is considerable scope for disagreement on the degree and importance of amenities: which amenities should be preserved, in what way they should be preserved, and how much expense (public or private) is justified.

The problem is relatively straightforward in so far as trees are concerned. It is much more acute, for example, in connection with electricity pylons, yet the electricity generating and supply companies are specifically charged not only with maintaining an efficient and coordinated supply of electricity but also with the preservation of amenity. Here the question is not merely one of sensitivity but also of the additional cost of preserving amenities by placing cables underground.

Apart from problems of cost, there is the problem of determining how much control the public will accept. Poor architecture, ill-conceived schemes, mock-Tudor frontages may upset the planning officer, but how much regulation of this type of 'amenity-injury' will be publicly acceptable? And how far can negative controls succeed in raising public standards? Here emphasis has been laid on design bulletins, design awards and such ventures as those of the Civic Trust, a body whose object is 'to promote beauty and fight ugliness in town, village and countryside'. Nevertheless, planning authorities have power not only to prevent developments which would clash with amenity (for example, the siting of a repair garage in a residential area) but also to reject badly designed developments which are not intrinsically harmful.

Conditional permissions

A local planning authority can grant planning permission subject to conditions, and almost all permissions are conditional. This can be a very useful way of permitting development which would otherwise be undesirable. Many conditions are simple, requiring for example, that materials to be used are agreed with the local authority before development starts. But there are many more complex permutations. Thus a service garage may be approved in a residential area on condition that the hours of business are limited. Residential development may be permitted on condition that landscape works are carried in accordance with submitted plans and before the houses are occupied.

The power to impose conditions is a very wide one. The legislation allows planning authorities to grant permission subject to 'such conditions as they think fit'. However, this does not mean 'as they please'. The conditions must be appropriate from a planning point of view: 'the planning authority are not at liberty to use their power for an ulterior object, however desirable that object may seem to them to be in the public interest. If they mistake or misuse their powers, however *bona fide*, the court can interfere by declaration and injunction' (*Pyx Granite Co Ltd v Minister of Housing and Local Government* 1981).

DoE Circular 11/95, *The Use of Conditions in Planning Permissions*,[24] stresses that

> If used properly, conditions can enhance the quality of development and enable many development proposals to proceed where it would otherwise have been necessary to refuse planning permission. The objectives of planning, however, are best served when that power is exercised in such a way that conditions are clearly seen to be fair, reasonable and practicable. (para. 2)

As might be expected, there is considerable debate on the meaning of these terms. Circular 11/95 elaborates specifically on the meaning of six tests: conditions should be necessary, relevant to planning, relevant to the development to be permitted, enforceable, precise and reasonable in all other respects.[25] Numerous court judgments provide guidance on how the tests should be applied. To meet the test of being necessary, the local authority should ask if permission would be refused if the condition were not imposed. Relevance to planning and to the development may be difficult to judge. While planning conditions should not be used where they duplicate other controls such as those of pollution control, they may be needed if the other method of regulation does not secure planning objectives. At one time, development may have been permitted subject to means of access for people with disabilities being agreed but this is now covered by the other legislation, the Disability Discrimination Act. Conditions should not be imposed on one site to seek to improve conditions on a neighbouring site, for example, where existing car parking is insufficient. But it may be appropriate to impose conditions to address problems elsewhere as a result of the new development, for example, increasing congestion on another part of the site. And it is possible to impose conditions on the use of land not under the control of the applicant. The enforceability test requires that the local planning authority should be able to monitor and detect whether the applicant is complying with it. Enforceability is also closely related to precision in drafting of the condition. Both the authority and the applicant need to be able to understand exactly what is required by a condition. An Appendix to the Circular gives numerous examples of how conditions should be drafted so as to avoid vagueness and ensure clarity.

The reasonableness test requires that the condition is not unduly restrictive. In particular it should not nullify the benefit of the permission. A condition may also be unreasonable if it is not within the powers of the applicant to implement it, for example where it relates to land in the ownership of a third party. A striking example of a condition which was quite unreasonable was dealt with in the *Newbury* case. There the district council gave permission for the use of two former aircraft hangers for storage, subject to the condition that they be demolished after a period of ten years. The House of Lords held that since there was no connection between the proposed use and the condition, it was *ultra vires*. In granting permission for development at Aberdeen Airport the planning authority sought to impose a number of conditions to minimise the impact on the local area. One condition restricted the direction of take off and landing of aircraft, but this was found to be both unreasonable and unnecessary since the Civil Aviation Authority (and not the airport) controls flight paths (McAllister and McMaster 1994: 136–7).

Up to 1968, conditions were also imposed to give a time limit within which development had to take place. The 1968 Act, however, made all planning permissions subject to a condition that development is commenced within five years, which in 2004 was reduced to three years. If the work is not begun within this time limit, the permission lapses, and it need not be renewed if the circumstances have changed.

The purpose of this provision is to prevent the accumulation of unused permissions and to discourage the speculative land hoarder. Accumulated unused permissions could constitute a difficult problem for some planning authorities: they create uncertainty and could make an authority reluctant to grant further permissions, which might result in, for example, too great a strain on public services.[26] The provision relates, however, only to the beginning of development, and this has in the past been deemed to include digging a trench or putting a peg in the ground.[27]

As well as the imposition of conditions the local authority may also reach agreement with the developer about planning obligations or 'planning gain', where the developer pays for related works without which planning permission could not be granted. Chapter 6 explains planning agreements and proposed changes. They typically cover the provision of infrastructure such as traffic management and access, public open spaces and other improvements as 'compensation' for loss through development, a proportion of affordable housing in residential schemes and even commuted payments to support public transport serving the development. The government emphasise that planning conditions should be used in preference to planning obligations, but for planning authorities the obligations are more important in larger scale developments. Here it should be noted that planning conditions should not duplicate obligations, and permission cannot be granted on condition that an obligation is entered into, although 'it is possible to see conditions as a prelude to obligations being entered into, so as to enable the application to be determined, but preventing implementation of the permission until such time that alternative arrangements, i.e. a s. 106 obligation have been put in place' (Chesman 2004; Kirby 2004). However, while delaying the decision on the application until agreement has been reached is preferred, Kirby notes that this approach has not yet been tested in the courts. While a developer can appeal against planning conditions, there is no such possibility for obligations which are entered into 'voluntarily'.

Fees for planning applications

Fees for planning applications were introduced in 1980. This represented a break with planning traditions, which had held (at least implicitly) that development control is of general communal benefit and directly analogous to other forms of public control for which no charges are made to individuals. The Thatcher administration had a very different view. The 1980 Bill provided additionally for fees for appeals but this was dropped in the face of widespread objections from both sides of the House. The 2004 Act enabled regulations to be made on fees for pre-application discussions and fees for call-in and appeals recovered by the Secretary of State. Previous attempts by authorities to charge fees for pre-application discussions were halted by a decision of the House of Lords.

The fee structure is subject to change over time, and a detailed schedule is therefore not appropriate; the last setting of fees was in 2002, with another review due to be put into place in 2005.[28] The government's ultimate aim is to recover the full administrative costs of dealing with planning applications, indeed some local authorities already effectively do this. A review of planning costs and fees in 2003 (Arup Economics and Planning 2003) took forward a commitment in the 1999 Green Paper and considered the potential from the changes introduced by the 2004 Act which allow for fees to be levied for any function of the local planning authority. The Act also allows for charges, which suggests the possibility of charging for the actual cost of a service rather than a predetermined fee. The report estimates the total fee income for planning at £174 million, which would suggest a 15–35 per cent increase to effect a full recovery of costs across all authorities. Cost recovery is weakest for the largest applications. This is one issue that is addressed in the 2005 review of fees. Initial proposals for new fee rates in 2004 were revised following consultation returns suggesting more significant increases were called for.

Planning appeals

An unsuccessful planning applicant can appeal to the Secretary of State, and a large number in fact do so. Appeals are allowed on the refusal of planning permission, against conditions attached to a permission, where a planning authority has failed to give a decision within the prescribed period, on enforcement notices and other matters as discussed below. Appeals decided during 1998–9 (England and Wales) numbered 12,877 of which about one-third were allowed. Figure 5.2 illustrates trends in the number of appeals. About half of the recent rapid increase in appeals is largely put down to the decision in September 2003 to reduce the

time limit for making an appeal from six to three months. The sharp rise in appeal numbers and the resulting backlog in dealing with appeals led to a government U-turn in January 2005 when the time limit was put back to six months. By 2004 the Planning Inspectorate was failing to meet any of its performance targets for appeals and appellants would typically wait one year for the appeal to be dealt with. While returning the system to a six months' deadline the ODPM also extended the time local planning authorities have for making decisions on major applications from eight to thirteen weeks before the applicant can appeal against non-determinations.

Arup conducted an investigation into the increasing number of appeals and the increasing popularity of the hearings procedure (see p. 169) for the ODPM, reporting in 2004 (Tunnel et al. 2004). The authors found that the rate of appeals against applications had been constant until 2001, when a number of factors acting together led to a sharp increase. In explanation, the report points to radical changes in policy especially in housing, telecommunications and parking, local authorities' desire to meet best value performance indicators (see p. 186) and resource constraints in planning leading to less opportunity for pre-application discussions. Their judgement is that there will be a 'settling down' in the policy environment as new approaches become more widely understood and accepted, and the resource issues would be addressed, and therefore concluded that refusal rates are likely to return to their historical norm.

Although the appeal is made to the Secretary of State, the vast majority are considered by inspectors 'standing in the Secretary of State's shoes'. The same applies in the other countries of the UK, but the Welsh Assembly is so far unique in establishing a cross-party Planning Decision Committee, with four members to make the final decision on important appeals and called-in applications (see p. 170).[29] Until 1969 the ministry responsible for planning dealt with all appeals.[30] In view of increasing delay in reaching decisions and the huge administrative burden, the Planning Act 1968 introduced a system whereby decisions on certain classes of appeal were 'transferred' to professional planning inspectors who had previously

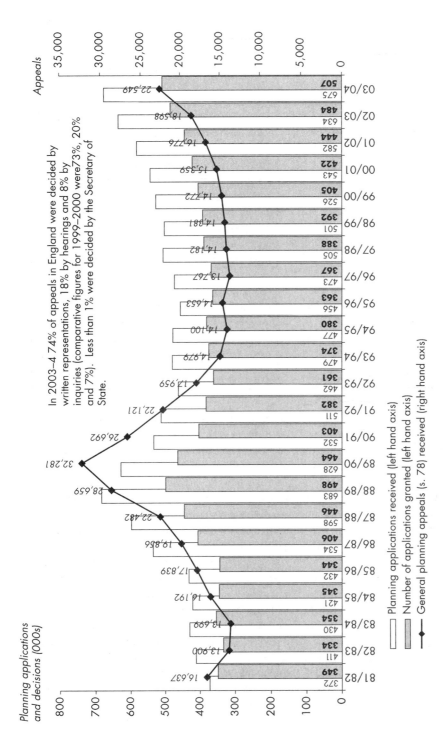

Figure 5.2 Planning applications, appeals and decisions in England 1981 to 2004

Planning applications and decisions (000s)

Appeals

In 2003–4 74% of appeals in England were decided by written representations, 18% by hearings and 8% by inquiries (comparative figures for 1999–2000 were73%, 20% and 7%). Less than 1% were decided by the Secretary of State.

☐ Planning applications received (left hand axis)
☐ Number of applications granted (left hand axis)
◆ General planning appeals (s. 78) received (right hand axis)

Sources: DETR/ODPM Statistical Releases and Planning Inspectorate Annual Statistical Reports
These figures exclude county matters, and other types of application. The appeals figures exclude enforcement and other appeals.
Scottish planning authorities received 48,751 planning applications in 2003–4 (up from 40,000 in 1998–9) and 805 appeals were lodged.
Welsh authorities received about 36,742 planning applications in 2003–4.
The Northern Ireland Planning Service received about 18,000 in 2003–4.

only made recommendations to the minister. Over time the range of decisions transferred to inspectors has been extended such that virtually all are now decided by the Planning Inspectorate. Matters of major importance may be 'recovered' for determination by the Secretary of State. In fact, less than 1 per cent of all appeals are recovered, although it can be argued that the significance is much greater than the figure suggests. Even where decisions are recovered, it is the senior civil servants in the department rather than the minister who make most decisions.[31]

Wide powers are available to the Secretary of State and inspectors. These include the reversal of a local authority's decision or the addition, deletion or modification of conditions. The conditions can be made more onerous or, in an extreme case the Secretary of State may even go to the extent of refusing planning permission altogether if it is decided that the local authority should not have granted it with the conditions imposed.

Before reaching any decision, the inspector or Secretary of State needs to consider the evidence and this can be done in three ways: by inquiry, hearing or written representation. Most appeals are considered by written representation with 73.1 per cent of all planning appeals in England in 1998–9, while hearings account for 19.1 per cent and inquiries 7.8 per cent. The procedures are governed by the rules of natural justice and by inquiry procedure rules, which have been updated in England.[32] Both the applicant and planning authority have the right to demand a full inquiry if they so wish, but the emphasis over recent years has been to get as many appeals as possible heard by the other two less expensive and time-consuming methods. The efficiency of procedures leading up to and during inquiries has been strongly criticised (Graves *et al.* 1996; O'Neill 1999).

Inquiries are *adversarial* debates conducted through the presentation and questioning (cross-examination) of evidence. The proceedings are managed by inspectors but advocates, often barristers, play a dominant role in the proceedings, thus lending a courtroom atmosphere. Such an approach has benefits in safeguarding the principles of *open, impartial and fair* consideration of the issues. Nevertheless, it is widely acknowledged

as unnecessary for certain less complex appeals, especially where one party is not professionally represented. Thus, the hearing procedure has been created; this proceeds in an *inquisitorial* way with the inspector playing an active role in structuring a round table discussion and asking questions, but with no formal cross-examination. But the most popular and straightforward procedure is through 'written reps'.

Over the years the mechanisms for considering appeals have been streamlined. The substantial increase in the number of appeals in the late 1980s led to reviews of the process; the first in 1985 introduced rules to govern the written representation procedure in a similar way to the rules for inquiries, which were also strengthened. Further minor changes were made in 1992 and further substantial revisions in 2000, aimed at speeding up the process, providing statutory rules for the hearings process and reducing the time allowed for submission of statements.[33]

The rules govern the exchange of information among the parties to the appeal and set a timetable for this to happen. The latest amendments are intended to ensure that the appeal processes follow the predetermined timetable more often than is the case now. Sanctions have been strengthened such that evidence may be disregarded unless it is submitted on time, or in some cases to impose costs on the guilty party (except for written representations). There is a stronger emphasis on the appellant and local authority agreeing the matters in dispute beforehand, keeping evidence concise,[34] and inspectors are encouraged to take more control over unnecessary cross-examination. Evaluations of more informal methods of holding appeals are generally positive (Stubbs 1999, 2000).

The appeals procedure is a microcosm of the whole planning system. It is where the system and its policies are challenged and where the most contentious and difficult issues are addressed. It is a 'pinch point' of the system, and at the time of writing might be described as being 'in crisis'. Is it acceptable for developers to wait a year for the appeal process to effectively start?

The state of the appeals process is critically important for the system as a whole, both in terms of planning policy and how the system should be operated. Although each appeal is considered on its own merits, the cumulative effect is to operationalise policy. It is here that the sometimes vague, sometimes contradictory, messages in government policy must be resolved. The wider effect of appeal decisions may be difficult to assess but clearly they have a very real influence on other decisions made by planning authorities, and are a route for the imposition of central government policy on local authorities. Inspectors pay particular attention to national policy which is the determining factor in many appeals (Rydin *et al.* 1990; Wood *et al.* 1998).[35]

On the operation of the system, discussions on the appeal system give a pointer to the government's overall philosophy on decision-making in planning as explained by Shepley (1999: 403). The approach is one which not only supports the fundamental principles of openness, fairness and impartiality but also recognises the need to make decisions more quickly, more cheaply, and earlier in the development process. Experiments have been conducted on 'alternative dispute resolution' in planning through mediation. The intention is that the 'win-lose' style of deciding planning appeals may be replaced with a process that seeks a solution which is acceptable to both parties. Pilot studies involved the Planning Inspectorate providing trained mediators where local authorities suggested cases that might benefit. Evaluation of the pilot found that there is a role for mediation generally for householder applications involving disputes over design or layout. But further use of mediation would bring about only a modest reduction in appeal cases, and some incentives would be needed to establish more use (Welbank *et al.* 2000). Perhaps the most important finding was that the current planning application regime was unsuitable for householder applications and an alternative is needed. In 2004 the ODPM began a review of householder consents with a view to simplify and speed up the application process. It will take into account findings from other research projects including the Lichfield (2003) review of permitted development rights discussed earlier.[36]

Another idea under consideration is the 'environmental court' which would 'extend public access to environmental justice' but also involve far reaching changes to the system – making for example the Planning Inspectorate part of the machinery of the courts rather than government departments.[37] Irrespective of the merits of environmental courts it is likely that there will have to be some changes to the appeal and enforcement procedures in the light of the Human Rights Act 1998.

Call-in of planning applications

The power to 'call in' a planning application for decision by the Secretary of State is quite separate from that of determining an appeal against an adverse decision of a local planning authority. The power is not circumscribed: the Secretary of State may call in *any* application. During the year 1998–9, 119 applications in England (of 503,000) were called in. The Scottish Executive called in about 27 applications each year. There are no statutory criteria or restrictions, and no prescribed procedures for handling representations from the public, although if either the applicant or the local planning authority so desire, the Secretary of State must hold a hearing or public inquiry. Answers to a House of Commons written question confirm that call-in will be 'very selective' and only be taken 'if planning issues of more than local importance are involved. Such cases may include, for example, those which, in the opinion of the Secretary of State may conflict with national policies on important matters, could have significant effects beyond their immediate locality, give rise to substantial regional or national controversy, raise significant architectural or urban design issues or may involve the interests of national security or of foreign governments' (*Hansard* 16 June 1999 col. 138).

To assist the Secretaries of State in making these decisions, all applications for development involving a substantial departure from the provisions of a development plan which the planning authority intends to permit must be sent to the Secretary of State together with a statement of the reasons why it wishes to grant the permission. This procedure enables the Secretary of State to decide whether the development is sufficiently important to warrant its being called in.

The Secretary of State also makes directions requiring local authorities to consult with him or her on certain types of application, so that consideration can be given to the use of call-in powers. For example, directions have been made requiring local authorities to consult the Secretary of State on leisure, retail and office uses over 5,000 square metres and which do not conform to the development plan (reduced in 1999 from 10,000 square metres). There is also now a requirement to consult where the authority intends to grant permission for development of playing fields against the advice of the Sports Council. In 2003, the Residential Density Direction (London and the South East) informed the relevant authorities that they must consult the Deputy Prime Minister before giving planning permission on planning applications for housing that proposed a density of below thirty per hectare on sites of one hectare or more.[38] The Density Direction now applies to all growth areas including the South East, South West, East of England and Northamptonshire. In 2005 the government announced plans to consult on a 'Greenbelt Direction' which will require councils to refer applications for 'inappropriate development' to the government.

Further powers are available for directions that prevent a planning authority granting permission for a particular application or a class of application, again this may be used to give the Secretary of State time to consider if the application should be called in.

Certain types of development tend to invite central government involvement. In the light of the government's commitments to increasing the delivery of new housing but in a sustainable way, new settlements and other very large housing developments figure prominently; so do applications involving the green belt, large-scale minerals proposals and development affecting buildings of national significance are most common. Mineral workings often raise problems of more than local importance, and the national need for particular minerals has to be balanced against planning issues. It is argued that such matters cannot be adequately considered by local planning authorities (who will invariably face massive local opposition) and such cases involve technical considerations requiring expert opinion of a character more easily available to central government. A large proportion of applications for permission to work minerals have been called in. Furthermore, there is a general direction calling in all applications for the winning and working of ironstone in certain counties where there are large-scale ironstone workings.

On important questions of design, CABE has, in its terms of reference, the power 'to call the attention of any of our departments of state . . . to any project or development which [it considers] may appear to affect amenities of a national or public character'.[39] Inevitably the Secretary of State has the job of balancing local concerns with national policies and priorities.

Variations in Northern Ireland and Scotland

Development control operates in a similar way across the whole of the UK, although it is established by separate law in Northern Ireland and Scotland. Legislation is generally made for England and Wales together, although it increasingly makes specific provision for Wales as required by the National Assembly for Wales. Planning policy is separately made for all four.

The comparison of use classes orders illustrated in Table 5.1 shows how minor variations build up to reflect 'local' conditions, concerns and priorities. These small variations (and some very big ones) can be found right across the planning systems.

The most important difference is that in Northern Ireland development control is operated by the Planning Service, an executive agency of the Department of Environment for Northern Ireland, which operates through six divisional planning offices. Local authorities in Northern Ireland have only a consultative role and planning applications are made to the department (Trimbos 1997). The Planning Service makes recommendations to the local district councils, which can request the Planning Service to reconsider. It may do so, but if there is no agreement, the matter is referred to the Chief Executive's Office and a decision is made by the Management Board (senior civil servants).

Appeals in Northern Ireland are heard by the Planning Appeals Commission; this is an independent body whose members are appointed by the Secretary of State for Northern Ireland. The Commission also hears inquiries into major planning applications and development plans. Another variation is in neighbour notification, where Northern Ireland has had a more thorough system. This is guided by a non-statutory notification scheme requiring, for example, advertisement of all applications.

Enforcement of planning control

If the machinery of planning control is to be effective, some means of enforcement is essential. Under the prewar system of interim development control, there were no such effective means. A developer could go ahead without applying for planning permission, or could even ignore a refusal of permission. The developer took the risk of being compelled to 'undo' the development (for example, demolish a newly built house) when, and if, the planning scheme was approved, but this was a risk that was usually worth taking. And if the development was inexpensive and lucrative (for example, a petrol station) the risk was virtually no deterrent at all. This flaw in the prewar system was remedied by the strengthening of enforcement provisions.

These are required not only for the obvious purpose of implementing planning policy, but also to ensure that there is continuing public support for, and confidence in, the planning system. To quote PPG 18 (1991):

> The integrity of the development control process depends on the planning authority's readiness to take effective action when it is essential. Public acceptance of the development control process is quickly undermined if unauthorised development, which is unacceptable on planning merits, is allowed to proceed without any apparent attempt by the LPA to intervene before serious harm results from it.
>
> (para. 4)

Enforcement provisions were radically changed by the Planning and Compensation Act 1991 following a comprehensive review by Robert Carnwath, QC, published in 1989. Current provisions are summarised in DoE Circular 10/97.[40] The 1991 Act provided a range of tools in addition to the long standing provision for enforcement notices.

Development undertaken without permission is not an offence in itself, but ignoring an *enforcement notice* or *stop notice* is an offence, and there is a maximum fine following conviction of £20,000. (In determining the amount of the fine, the court is required to 'have regard to any financial benefit which has accrued'.) There is a right of appeal against an enforcement notice. An appeal also contains a deemed application for development for which a fee is payable to the planning authority. Appeals can be made on several grounds, for example, that permission ought to be granted, that permission has been granted (e.g. by the GPDO), and that no permission is required. There is also a limited right of appeal on a point of law to the High Court. New procedures for enforcement appeals came into effect from December 2002 and brought them into line with changes made to the planning appeals procedure in 2000, for example, in the use of hearings rather than inquiries, simultaneous submission of evidence and new stricter timetables.

Enforcement can be a lengthy process. For example, South Hams District Council issued an enforcement notice in January 1990 for the removal of a house built without consent. In 1993 the owner was fined £300 for breaching the enforcement notice. In 1995 he was jailed for three months for contempt of a court order requiring demolition. He had demolished only the upper storey and grassed over the lower half.[41]

Where it is uncertain whether planning permission is required, a LPA has power to issue a *planning contravention notice*. This enables it to obtain information about a suspected breach of planning control and to seek the cooperation of the person thought to be in breach. If agreement is not forthcoming (whether or not a contravention notice is served) an enforcement notice may be issued, but only 'if it is expedient' to do so 'having regard to the provisions of the development plan and to any other material considerations'. In short, the local authority must be satisfied that enforcement is necessary in the interests of good planning.

In view of the government's commitment to fostering business enterprise (discussed further below), planning authorities are advised in PPG 18 to consider the financial impact on small businesses of conforming with planning requirements. 'Nevertheless, effective action is likely to be the only appropriate remedy if the business activity is causing irreparable harm.' Development 'in breach of planning control' (development carried out without planning permission or without compliance with a planning condition) might be undertaken in good faith, or ignorance. In such a case, application can be made for retrospective permission. It is unlikely that a local authority would grant unconditional permission for a development against which it had served a planning contravention notice, but it might be willing to give conditional approval.

The 1991 Act also introduced a *breach of condition notice* as a remedy for contravention of a planning condition. This is a simple procedure against which there is no appeal, though there may be some legal complexities that will prevent its widespread use (Cocks 1991). Further, there is a new provision enabling a local planning authority to seek an injunction in the High Court or County Court to restrain 'any actual or apprehended breach of planning control'. In Scotland, the provision is for an interdict by the Court of Session or the Sheriff.

Where there is an urgent need to stop activities that are being carried on in breach of planning control, a LPA can serve a *stop notice*. From 2004 the stop notice can be served as soon as building works start or unauthorised use begins, and there is no way to delay its effect. This is an attempt to prevent delays in the other enforcement procedures (and advantage being taken of these delays) resulting in the local authority being faced with a *fait accompli*. Development carried out in contravention of a stop notice constitutes an offence. The introduction of the temporary stop notice with immediate effect (introduced by amendment to the Planning and Compulsory Purchase Bill) has been described as a 'power to close down a business for up to twenty-eight days without any liability for compensation' with a warning that judicial review may lead to a finding of incompatibility with the Convention on Human Rights (Kinloch 2005).

Robinson draws attention to one underused power associated with planning enforcement, that enables planning authorities to exercise a 'more flexible means of maintaining visual amenity without any unauthorised development having occurred' (p. 2). It is s. 215 of the 1990 Act: land adversely affecting the amenity of neighbourhoods. Robinson argues that in the relatively small number of cases where it is used, 's. 215 has been very successful in ensuring that land is tidied and restored to its former condition'.

The provisions for enforcement are complex and there are many difficulties in their operation. The DETR consultation paper *Modernising Planning: Improving Enforcement Appeal Procedures* (1999) made numerous recommendations including requirements for a list of relevant development plan policies and time limits for notification and representations to be made. Further proposals for change came forward from the ODPM in 2002 in a *Review of Planning Enforcement*. The response by the Planning Officers' Society gives a frank assessment of the state of the system and what needs to be changed. At the root of problems, the POS believe, is the lack of resources; 'the excessive protection of those against whom enforcement action is taken; and protracted regulatory procedures' (p. 1). The position can be exacerbated by the lowly esteem in which the enforcement system (and those who staff it) are often held. Several commentators have termed enforcement 'the weakest link in the planning chain'. The POS recommended, among other things, the early stop notice (explained above) and a change in culture in the magistrates' courts such that fines are imposed which act as a stronger deterrent. Magistrates have generally imposed small fines that bear no relation to the potential financial benefits that arise from the unauthorised development or use; the POS cite the example of advertisement hoardings.

A study in Scotland found great variation in the use of enforcement powers. For example, one authority had served 156 planning contravention notices between 1992 and 1996 while another had served none (Edinburgh College of Art and Brodies 1997).

While planning authorities are mostly happy with enforcement powers, they are sometimes reluctant to use them.[42] Fortunately, the majority of alleged contraventions of planning control are dealt with satisfactorily and without any recourse to legal action, but the minority have a disproportionate effect on the credibility of the planning process as a whole.

Revocation, modification and discontinuance

The powers of development control possessed by local authorities go considerably further than the granting or withholding of planning permission. They can interfere with existing uses and revoke a permission already given, even if the development has actually been carried out.

A *revocation order* or *modification order* is made when the development has not been undertaken (or before a change of use has taken place). The local authority must 'have regard to the development plan and to any other material considerations', and an opposed order has to be confirmed by the Secretary of State. Compensation is payable for abortive expenditure and any loss or damage due to the order. Such orders are rarely made.

Quite distinct from these powers is the much wider power to make a *discontinuance order*. This power is expressed in extremely wide language: an order can be made 'if it appears to a local planning authority that it is expedient in the interests of the proper planning of their area (including the interests of amenity)'. Again confirmation by the Secretary of State is required, and compensation is payable for depreciation, disturbance and expenses incurred in carrying out the works in compliance with the order. An Order will be confirmed only if the case is a strong one. Indeed, cases have established the principle that a stronger case is needed to justify action to bring about the discontinuance of a use than would be needed to warrant a refusal of permission in the first instance.

British planning legislation does not assume that existing non-conforming uses must disappear if planning policy is to be made effective. This may be an avowed policy, but the planning Acts explicitly permit the continuance of existing uses.

Purchase and blight notices

A planning refusal does not of itself confer any right to compensation. On the other hand, revocations of planning permission or interference with existing uses do rank for compensation, since they involve a taking away of a legal right. In cases where, as a result of a planning decision, land becomes 'incapable of reasonably beneficial use' the owner can serve a *purchase notice* upon the local authority requiring it to buy the property. In all cases, ministerial confirmation is required. The circumstances in which a purchase notice can be served include:

* refusal or conditional grant of planning permission
* revocation or modification of planning permission
* discontinuance of use.

In considering whether the land has any *beneficial use*, 'relevant factors are the physical state of the land, its size, shape and surroundings, and the general patterns of land-uses in the area; a use of relatively low value may be regarded as reasonably beneficial if such a use is common for similar land in the vicinity' (DoE Circular 13/83).

A purchase notice is not intended to apply in a case where an owner is simply prevented from realising the full potential value of the land. This would imply the acceptance in principle of paying compensation for virtually all refusals and conditional permissions. It is only if the existing and permitted uses of the land are so seriously affected as to render the land incapable of reasonably beneficial use that the owner can take advantage of the purchase notice procedure.

There are circumstances, other than the threat of public acquisition, in which planning controls so affect the value of the land to the owner that some means of reducing the hardship is clearly desirable. For example, the allocation of land in a development plan for a school or for a road will probably reduce the value of houses on the land or even make them completely unsaleable.

In such cases, the affected owner can serve a *blight notice* on the local authority requiring the purchase of the property at an 'unblighted' price. These provisions are restricted to owner occupiers of houses and small businesses who can show that they have made reasonable attempts to sell their property but have found it impossible to do so except at a substantially depreciated price because of certain defined planning actions. These include land designated for compulsory purchase, or allocated or defined by a development plan for any functions of a government department, local authority or statutory undertaker, and land on which the Secretary of State has given written notice of his or her intention to provide a trunk road or a *special road* (i.e. a motorway).

The subject of planning blight takes us into the much broader area of the law relating to compensation. This is an extremely complex field, and only an indication of three major provisions can be attempted here.

First, there is a statutory right to compensation for a fall in the value of property arising from the use of highways, aerodromes and other public works which have immunity from actions for *nuisance*. The depreciation has to be caused by physical factors such as noise, fumes, dust and vibration, and the compensation is payable by the authority responsible for the works. Second, there is a range of powers under the heading 'mitigation of injurious effect of public works'. Examples include sound insulation; the purchase of owner occupied property which is severely affected by construction work or by the use of a new or improved highway; the erection of physical barriers (such as walls, screens or mounds of earth) on or alongside roads to reduce the effects of traffic noise on people living nearby; the planting of trees and the grassing of areas; and the development or redevelopment of land for the specific purpose of improving the surroundings of a highway 'in a manner desirable by reason of its construction, improvement, existence or use'. Third, provision is made for *home loss payments* as a mark of recognition of the special hardship created by compulsory dispossession of one's home. Since the payments are for this purpose, they are quite separate from, and are not dependent upon, any right to compensation or the *disturbance payment* which is described below.

Logically, they apply to tenants as well as to owner occupiers, and are given for all displacements whether by compulsory purchase or any action under the Housing Acts. These provisions were slightly extended in the 1991 Planning and Compensation Act.

Additionally, there is a general entitlement to a *disturbance payment* for persons who are not entitled to compensation. Local authorities have a duty 'to secure the provision of suitable alternative accommodation where this is not otherwise available on reasonable terms, for any person displaced from residential accommodation' by acquisition, redevelopment, demolition, closing orders and so on.

Development by the Crown, government departments and statutory undertakers

Part 7 of the 2004 Act brings an end to Crown immunity from planning control (or strictly speaking, will do when it comes into force during 2006). The change was announced as far back as 1992. This long-standing, and for many, frustrating anomaly was implemented by insertion in the 1990 Act of s. 292A(1) which states simply 'This Act binds the Crown'. (Much more detail is also added, mostly to do with procedures when national security is at issue.)

Before 2005, because the Crown is generally not bound by statute, development by government departments did not require planning permission. However, since 1950, there have been special arrangements for consultations. Increased public and professional concern about the inadequacy of these led to revised, but still non-statutory, arrangements culminating in DoE Circular 18/84. This said that, before proceeding with development, government departments will consult planning authorities when the proposed development is one for which specific planning permission would, in normal circumstances, be required. In effect, local authorities should treat notification of a development proposal from government departments in the same way as any other application. Where the local authority is against the development the matter is referred to the Secretary of State.

Development by private persons on 'Crown land' (that is, land in which there is an interest belonging to Her Majesty or government department) has required planning permission in the normal way, although there are limitations on the ability of the planning authority to enforce in these cases.

Development undertaken by statutory undertakers is subject to planning control but it is also subject to special planning procedures. Where a development requires the authorisation of a government department (as do developments involving compulsory purchase orders, work requiring loan sanction, and developments on which government grants are paid) the authorisation is usually accompanied by *deemed planning permission*. Much of the regular development of statutory undertakers and local authorities (for example, road works, laying of underground mains and cables) is *permitted development* under the GPDO. Statutory undertakers wishing to carry out development which is neither permitted development nor authorised by a government department have to apply for planning permission to the local planning authority in the normal way, but special provisions apply to *operational land*. The original justification for this special position of statutory undertakers was that they are under an obligation to provide services to the public and could not, like a private firm in planning difficulties, go elsewhere.

Development by local authorities

Until 1992, planning authorities were also deemed to have permission for any development which they themselves undertook in their area, as long as it accorded with the provisions of the development plan; otherwise they had to advertise their proposals and invite objections. The only requirement was for the local authority to grant itself permission by resolution. These 'self-donated' planning permissions were problematic: although local authorities are guardians of the local public interest, they can face a conflict of interest in dealing with their own proposals for development. Pragmatic consideration of the merits of a case involving their own role as developers can easily distort a planning judgment. Examples include attempts by authorities to dispose of surplus school playing fields with the benefit of permission for development, and competing applications for superstore development when one of the sites is owned by the authority itself. The local authorities' position was not helped by judgments against them that found many irregularities in the necessary procedures (Moore 2000: 311).

Because of these difficulties, new regulations were issued in 1992 which require planning authorities to make planning applications in the same way as other applicants, and generally follow the same procedures including publicity and consultation. There must be safeguards to ensure that decisions are not made by members or officers who are involved in the management of the land or property, and the planning permission cannot pass to subsequent land and property owners. Where other interests propose development on local authority owned land they must apply for permission in the normal way. The new procedures did not go as far as some had hoped and criticism continues, and inevitably so since the accusation of bias is always possible while local authorities are able to grant themselves planning permission. The Scottish Local Government Ombudsman has for long complained about 'the ease with which planning authorities breach their own plans, particularly considering the time, effort, and consultation which goes into them'. One solution would be for the Secretary of State to play a role in all applications in which the local authority has an interest (as proposed by the Nolan Committee on Standards of Conduct in Local Government).

Control of advertisements

The need to control advertisements has long been accepted. Indeed, the first Advertisements Regulation Act 1907 antedated by two years the first Planning Act. But, even when amended and extended (in 1925 and 1932), the control was quite inadequate. Not only were the powers permissive, but also they were limited. For instance, under the 1932 Act, the right of appeal (on the ground that an advertisement did not injure the amenities of the area) was to the Magistrates Court –

hardly an appropriate body for such a purpose. The 1947 Act set out to remedy the deficiencies. There are, however, particular difficulties in establishing a legal code for the control of advertisements. Advertisements may range in size from a small window notice to a massive hoarding, in the form of a poster, a balloon or even lasers; they vary in purpose from a bus stop sign to a demand to buy a certain make of detergent; they could be situated alongside a cathedral, in a busy shopping street, or in a particularly beautiful rural setting; they might be pleasant or obnoxious to look at; they might be temporary or permanent, and so on. The task of devising a code which takes all the relevant factors into account and, at the same time, achieves a balance between the conflicting interests of legitimate advertising and 'amenity' presents real problems. Advertisers themselves frequently complain that decisions in apparently similar cases have not been consistent with each other. The official departmental view is that no case is exactly like another, and hard and fast rules cannot be applied: each case has to be considered on its individual merits in the light of the tests of amenity and – the other factor to be taken into account – public safety.[43]

The control of advertisements is exercised by regulations,[44] which are explained in PPG 19: *Outdoor Advertising Control*. The Secretary of State has very wide powers of making regulations 'in the interests of amenity or public safety'. The question of public safety is rather simpler than that of amenity, though there is ample scope for disagreement: the relevant issue is whether an advertisement is likely to cause danger to road users, and also to 'any person who may use any road, railway, waterway (including coastal waters), docks, harbour or airfield'. In particular, account has to be taken of the likelihood of whether an advertisement 'is likely to obscure, or hinder the ready interpretation of, any road traffic sign, railway signal, or aid to navigation by water or air'. Amenity includes 'the general characteristics of the locality, including the presence of any feature of historic, architectural, cultural or similar interest'. The definition of an advertisement is not quite as complicated as that of development, but it is very wide:

Advertisement means any word, letter, model, sign, placard, board, notice, awning, blind, device or representation, whether illuminated or not, in the nature of, and employed wholly or partly for the purposes of, advertisement, announcement or direction and . . . includes any hoarding or similar structure used, or designed or adapted for use, and anything else principally used, or designed or adapted principally for use, for the display of advertisements.

(1990 Act s. 336(1))

It is helpfully added that the definition excludes anything 'employed as a memorial or as a railway signal'. Various classes of advertisement are currently excepted from all control, although the classes are currently under review. They are advertisements displayed on a balloon, on enclosed land, within a building and on or in a vehicle. Also excepted are traffic signs, election signs and national flags. As one might expect, there are some interesting refinements of these categories, which can be ignored for present purposes (though we might note, in passing, that a vehicle must be kept moving or, to use the more exact legal language, must be normally employed as a moving vehicle).With these exceptions, no advertisements may be displayed without *consent*. However, certain categories of advertisement can be displayed without *express consent*; so long as the local authority takes no action, they are *deemed* to have received consent. These include bus-stop signs and timetables, hotel and inn signs, professional or business plates, 'To Let' and 'For Sale' signs, election notices, statutory advertisements and traffic signs.

It needs to be stressed that amenity and public safety are the only two criteria for control. The content or subject of an advertisement is not relevant, and a local authority cannot refuse express consent on grounds of morality, offensiveness or taste. Thus an advertisement which contained the words 'Chish and Fips' was considered by the Secretary of State, on appeal, to be questionable on grounds of taste, but not detrimental to amenity: the appeal was allowed (*JPL* 1959: 736).

If an advertisement displayed with deemed consent becomes unsafe, unsightly or in any way 'a substantial injury to the amenity of the locality or a danger to members of the public', the LPA can serve a *discontinuance order*. There is the normal right of appeal to the Secretary of State. Advertisements displayed with express consent can be subject to revocation or modification, again with the normal rights of appeal.

Complex though this may seem, it is not all that there is to advertisement control. In some areas, for example, conservation areas, national parks or areas of outstanding natural beauty, it may be desirable to prohibit virtually all advertisements of the poster type and seriously restrict other advertisements including those normally displayed by the ordinary trader. Accordingly, local planning authorities have power to define *areas of special advertisement control* (ASAC) where very strict controls are operated. Within such areas, the general rule is that no advertisement may be displayed; such advertisements as are given express consent are considered as exceptions to this general rule.

These special controls originated primarily from the need to deal with the legacy of advertising hoardings which were such a familiar sight in the 1930s. It is now argued that they are obsolete, and can be replaced by simpler controls. Added justification is given to this argument by the fact that, in 1995, nearly half of the area of England and Wales had been defined by local planning authorities as being within areas of special control. Consultation papers in 1996 and 1999 have argued that many orders were out of date since they no longer corresponded to the current limits of the built environment,[45] while the system was either obscure or widely misunderstood by the public. The last paper proposes to limit ASACs to national parks, AONBs, conservation areas, SSSIs and the Broads. Other changes are proposed to update the regulations, to close loopholes and to reflect developments in the advertising industry, for example, in relation to bringing balloon advertisements under control, to add the flying of the European flag as an exemption from control, and to bring lasers into the meaning of advertisement. Particular attention is being given to fly-posting following research on the subject (Arup Economics and Planning 1999).

Control of mineral working

The reconciliation of economic and amenity interests in mineral working is an obvious matter for the mineral planning authorities (MPAs). It would, however, be misleading to give the impression that the function of planning authorities is simply to fight a continual battle for the preservation of amenity. Planning is concerned with competing pressures on land and with the resolution of conflicting demands. Amenity is only one of the factors to be taken into account. The general policy framework for minerals is set out in MPG 1. It is interesting to compare the current policy, as set out in the 1996 MPG with that of the earlier (1988) version (both of which are illustrated in Box 5.4). The 1996 version places a significantly greater emphasis on conservation and environmental considerations.[46]

Planning powers provide for the making of the essential survey of resources and potentialities, the allocation of land in development plans, and the control (by means of planning permission) of mineral workings. The MPA has to assess the amount of land required for mineral working, and this requires an assessment of the future demand likely to be made on production in their area. Obviously, this involves extensive and continuing consultation with mineral operators. All MPAs are now required to prepare *minerals plans* (which may be produced jointly with their waste plan).

Powers to control mineral workings stem from the definition of development, which includes 'the carrying out of . . . mining . . . operations in, on, over or under land'. However, a special form of control is necessary to deal with the unique nature of mineral operations. Unlike other types of development, mining operations are not the means by which a new use comes into being, they are a continuing end in themselves, often for a very long time. They do not adapt land for a desired end use: on the contrary, they are essentially harmful and may make land unfit for any later use. They also have unusual location characteristics: they have to be mined where they exist. For these reasons, the normal planning controls are replaced by a unique set of regulations.

BOX 5.4 OBJECTIVES OF SUSTAINABLE DEVELOPMENT FOR MINERALS PLANNING

The objectives are

(i) to conserve minerals as far as possible, while ensuring an adequate supply to meet needs;

(ii) to ensure that the environmental impacts caused by mineral operations and the transport of minerals are kept, as far as possible, to an acceptable minimum;

(iii) to minimise production of waste and to encourage efficient use of materials, including appropriate use of high quality materials, and recycling of wastes;

(iv) to encourage sensitive working practices during minerals extraction and to preserve or enhance the overall quality of the environment once extraction has ceased;

(v) to protect areas of designated landscape or nature conservation from development, other than in exceptional circumstances where it has been demonstrated that development is in the public interest; and

(vi) to prevent the unnecessary sterilisation of mineral resources.

Previous statement of objectives (1988)

(a) to ensure that the needs of society for minerals are satisfied with due regard to the protection of the environment;

(b) to ensure that any environmental damage or loss of amenity caused by mineral operations and ancillary operations is kept at an acceptable level;

(c) to ensure that land taken for mineral operations is reclaimed at the earliest opportunity and is capable of an acceptable use after working has come to an end;

(d) to prevent the unnecessary sterilisation of mineral resources.

Source: MPG 1 (1996, 1988)

Two major features of the minerals control system are that it takes into account the fact that mineral operations can continue for a long period of time, and that measures are needed to restore that land when operations cease. It is, therefore, necessary for MPAs to have the power to review and modify permissions and to require restoration. Under current legislation, MPAs have a duty to review all mineral sites in their areas. This includes those which were 'grandfathered' in by the 1947 Act. These old sites, of which there may be around a thousand in England and Wales, often lack adequate records. They present the particular problem that they can include large unworked extensions which are covered by the permission; if worked these could have serious adverse effects on the environment. The provisions relating to these sites are even more complicated than those pertaining to the generality of mineral operations, and they have been significantly altered by the 1995 Environment Act. Details are set out in MPG 14.

Policies for restoration (and what the Act quaintly calls 'aftercare') have become progressively more stringent, mainly in response to what the Stevens Report (1976) referred to as a great change in standards and attitudes to mineral exploitation. A lengthy guidance note (MPG 7) fully explains restoration policies and options. In view of the ongoing nature of mineral operations, particular importance is attached to schemes of progressive restoration which are phased in with the gradual working out of the site. (A very effective policy is to make new working dependent upon satisfactory restoration of used sites.) A good idea of the current policy is gained from the following quotation from MPG 7:

> The overall standards of reclamation have continued to improve over recent years, and with the development and implementation of appropriate reclamation techniques, there is potential for land to be restored to a high standard suitable for a variety of uses. Consistent and diligent application of the appropriate techniques will ensure that a wide range of sites are restored to appropriate standards. This may lead to the release of some areas of land which would not otherwise be made available for mineral working, for example, the best and most versatile agricultural land. Conversely where there is serious doubt whether satisfactory reclamation can be achieved at a particular site, then there must also be a doubt whether permission for mineral working should be given.
>
> (MPG 7 1996: para. 3)

The extraction of minerals is one of the most obvious examples of a 'locally unwanted land use' (LULU) and one that has a disproportionate effect on particular locations (Blowers and Leroy 1994). But minerals extraction may also bring economic benefits especially in more remote rural locations. Minerals development control seeks to reconcile these conflicting interests, and reviews of minerals planning guidance have taken much more account of the need for sustainable development (see Box 5.4). Nevertheless, a major limitation of the control of minerals exploitation is the emphasis on finding suitable locations albeit in the interests of mitigating environmental impacts.[47] Much less attention is given to managing the demand for these resources, a question which is taken up in Chapter 7.

Major infrastructure projects

Decision-making on major infrastructure projects is very difficult and the process tends to be long-winded. This has been exacerbated in the UK by limited national policies or strategies concerning investment in roads, bridges, airports and the like. From time to time, this problem reaches public attention, or rather, the inquiry part of the process does, most recently through the inquiry into Terminal 5 at Heathrow. The inquiry sat for 525 days, heard 700 witnesses and received 6,000 documents, and got a panning from the press as a model of government red tape. It is understandable, therefore, that the government should seek to improve on performance in dealing with major projects.

Approval for major infrastructure projects such as railways, light rail systems and bridges can be given in different ways. Before 1992, most projects were approved through private members bills or the hybrid bill procedure and Act of Parliament.[48] The Channel Tunnel Rail Link was approved in this way in two years. The procedure involves select committees in each House hearing petitioners' requests for amendments to the scheme. Following the Act planning permission is still required from local planning authorities for detailed 'reserved matters'.

The Transport and Works Act 1992 provided a new procedure for railway, tramway and inland waterway schemes whereby the Secretary of State is able to make *works orders* which among other things will normally include deemed planning permission. Orders are subject to objection and public inquiry if necessary, but they are not normally debated in Parliament. The Secretary of State makes the decision taking into account objections and the inquiry report. For works of national significance the works must still be approved by Parliament. A special development order can also be used to grant planning permission for projects but is seldom used (see p. 157). Special provisions also exist for approval of port infrastructure, trunk roads and nuclear power stations. Some projects still come forward through the planning process and call-in or recovered appeals by the Secretary of State.

The ODPM consulted on New Parliamentary Procedures for Processing major Infrastructure Projects in 2001. The proposals included more up-to-date statements of government policy on infrastructure, an improved regional policy framework (now coming forward in regional spatial strategies), a procedure to allow Parliament to give approval to the project in principle, improved inquiry procedures and changes to compulsory purchase and compensation provisions (see Chapter 6). Progress has been made on all, except the new parliamentary procedure. Changes have focused on the inquiry procedure which plays a part in Transport and Works Act approvals as well as normal planning applications. Circular 2/2002 explains the new procedures for handling major infrastructure projects, which will be reviewed after five years. The changes bring major infrastructure inquiries into line with practice for other plan and appeal inquiries and include provision for round table discussion, stricter timetabling and provisions for inspectors to limit cross-examination. Because of the complexity of large projects, the role of independent technical adviser has been introduced to summarise and make recommendations on scientific issues beyond the competence of the inspector. The inspector will also be able to appoint mediators to facilitate reaching agreement among the parties (if only agreement on what they disagree about).[49]

Caravans

During the 1950s, the housing shortage led to a boom in unauthorised caravan sites. The controversy and litigation to which this gave rise led to the introduction of special controls over caravan sites (by Part I of the Caravan Sites and Control of Development Act 1960).[50] The Act gave local authorities new powers to control caravan sites, including a requirement that all caravan sites had to be licensed before they could start operating (thus partly closing loopholes in the planning and public health legislation). These controls over caravan sites operate in addition to the normal planning system: thus both planning permission and a licence have to be obtained. Most of the Act dealt with control, but

local authorities were given wide powers to provide caravan sites.

Holiday caravans are subject to the same planning and licensing controls as residential caravans. To ensure that a site is used only for holidays (and not for 'residential purposes'), planning permission can include a condition limiting the use of a site to the holiday season. Conditions may also be imposed to require the caravans to be removed at the end of each season or to require a number of pitches on a site to be reserved for touring caravans.

One group of caravanners is particularly unpopular: gypsies or, to give them their less romantic description, 'persons of nomadic life, whatever their race or origin' (but excluding 'members of an organised group of travelling showmen, or persons engaged in travelling circuses, travelling together as such'). The basic problem is that no one wants gypsies around: 'all too often the settled community is concerned chiefly to persuade, or even force, the gypsy families to move on'. The Caravan Sites Act 1968 gave local authorities in England and Wales (but not in Scotland) the duty to provide adequate sites for gypsies 'residing in or resorting to' their areas. In 1979 100 per cent grants were made available for capital works on sites.[51] But the problems persisted; indeed, they got worse with increases in the estimated number of gypsy caravans, although many of these are in fact 'new age travellers'. 'The public visibility of gypsies has grown, while the tolerance of the settled community to them has declined' (Home 1993).[52]

Consultation on the Reform of the Caravan Sites Act 1968 in 1992 heralded a marked shift in policy. The new regime abolished (or privatised) the obligation of local authorities to provide gypsy sites. For a time central government grants for gypsy caravan sites were also phased out, although they are now reinstated. Local authorities should 'continue to indicate the regard they have had to meeting gypsies' accommodation needs', with 'broad strategic policies' in structure plans and detailed policies in local plans (DoE Circular 1/94). However, gypsy sites will not be appropriate in green belts or other protected areas which had been previously allowed for by Circular 28/77.

Significantly, the legislation implementing the new policy is not of a planning character: it is the Criminal Justice and Public Order Act 1994 (explained in Circular 1/94). In addition to repealing the obligations imposed on local authorities by the Caravan Sites Act, it provides stronger powers to remove 'unauthorised persons', though the DoE Circular espouses a policy of toleration towards gypsies on unauthorised sites. Gypsies have pursued their cases through planning inquiries, the courts and to the European Court of Human Rights. While some claims have been held to be admissible, they have generally been unsuccessful in their 'human rights arguments'. Local authorities have had more success in challenging decisions where inspectors have granted permission for sites on appeal. The high court and the European Court of Human Rights have determined that they are not well placed to challenge decisions made by inspectors, who are better placed to strike a balance in the light of all the evidence (Maurici 2004).

The abolition of 'the privileged position of gypsies' did little to address this sorry saga, as a research report for ODPM on the provision and condition of gypsy sites amply illustrates (Niner 2002, 2004). This was the first major investigation since 1977 and was followed by a House of Commons Select Committee Report *Gypsy and Traveller Sites* in 2004. In the same year the ODPM consulted on a proposed Circular, *Planning for Gypsy and Traveller Sites*, taking forward recommendations from both reports.

Telecommunications

One area of development control work that has expanded rapidly and with some controversy is telecommunications. The expansion of masts to service mobile phone networks has been a particular concern, at first because of their visual impact, but more recently because of potential health effects. In England, PPG 8, *Telecommunications* (revised 2001) and Circular 4/99 are the main sources for policy on telecommunications in England. Similar guidance applies in the other parts of the UK (but see p. 183). Given the speed at which this technology is advancing it will be no surprise that,

since 1995, policy guidance has been under constant review.

There are more than 50 million mobile phones in the UK and more than a quarter of a million people employed in this sector directly. In 2004 there were 40,000 'masts', predicted to rise to 88,000 by 2007.[53] In addition a new communications network for the police will require an extra 3,500 masts. Government policy is strongly behind the expansion of telecommunications, and this presents many challenges for the planning system, not least in its effect on spatial development patterns (see, for example, Graham and Marvin 1999). But the erection of masts and related equipment has been the main talking point so far, and government policy is clearly influenced by the economic argument for efficient communications networks and fostering competition among rival networks. There is a tension between the central government's policy for expanding mobile telecommunications which has brought in considerable revenue (led by the DTI), and its local implementation in the face of concerted opposition in some places from the public. It should be said also that 'the public' faces two ways on this, with people wanting the convenience of mobile communications while not always accepting the infrastructure that provides it.

In England, masts under 15 metres in height are permitted development (with some exceptions),[54] but there is a *prior approval procedure* which gives the planning authority fifty-six days to say whether it wishes to approve details of the siting and appearance of the development. The local authority consults in the same way as for planning permission, and it may refuse or add conditions to the approval. If the authority fails to notify the applicant within the fifty-six day period the application is deemed granted.

Local development frameworks should include general policies for telecommunications related development and may allocate sites for large masts. The planning authority should also encourage different operators to share facilities, though competition among the networks limits their willingness to cooperate. There is also an obligation on the developer to site the mast so that it has least effect on the external appearance of buildings. Where this is not followed, the planning authority may serve a breach of condition notice on the basis that a condition of the permitted development right has not been complied with. Masts over 15 metres in height require planning consent.

The main proposed change to planning policy relates to health considerations (which are barely mentioned in the 1992 PPG 8). The Independent Expert Group on Mobile Phones conducted an assessment of the health effects and concluded that 'there is no general risk' (quoted in the 2000 consultation paper). Nevertheless, the Group recommended a precautionary approach and removal of permitted development rights. This was not accepted by government in England, (although the period for prior approval was extended to fifty-six days from forty-two). Permitted development rights were removed in Scotland. Revised planning guidance accepts that the perception of risk can be a material consideration in determining applications for telecommunications apparatus.

Though the public consultation requirements are similar in the prior approval process, there is little confidence with this system among local interests, and there is considerable variation in practice across the country. The All Parliamentary Mobile Group (APMobile) on *Mobile Phone Masts* (Askew 2004a, 2004b) which examined the issues and presented recommendations on the planning aspects of phone mast development, gives examples, including Basingstoke and Deane's 'telecommunications inquiry', where all interests took part in the preparation of supplementary guidance for the district. The industry, through its representative body the Mobile Operators Association, points to the *Code of Best Practice on Mobile Phone Network Development* (2002) produced jointly with central and local government. It incorporates 'ten commitments' which, if implemented in a uniform way, would address many of the interest group's concerns. However, anecdotal evidence suggests that this is far from the case, with some operators constructing masts even before permission is granted. The APMobile Report repeated the recommendation of the Stewart Report that permitted development rights be revoked for the erection of all base stations. It also addressed the consultation issue, suggesting that more attention could be paid to the 'pre-rollout stage' when operators

are planning the spatial distribution of masts, and the inclusion of a 'telecommunications plan' in the local development framework.

Interestingly the APMobile Report also highlighted the inconsistencies across the four jurisdictions in the UK, a question which might be raised for many other planning topics. It recommended a comparative review of law and practice and some collaboration to ensure consistent best practice. Mention has already been made that Scotland revoked permitted development rights in 2001,[55] and operators are also required to notify the local authority when new antennae are installed on existing masts, which addresses concerns about intensification of antennae, and possible risks, once a mast is approved. An evaluation of the effects of these changes on operators in Scotland is due.

ODPM consulted again in 2005. This is one of those cases where difficult decisions call for more research. At the speed that the technology is advancing, the issue may no longer be a problem (for central government) when new policy is adopted. In the mean time, the Minister for Planning, Keith Hill, made a parliamentary statement (another APMobile recommendation) to clarify that the government expects proper consultation and calling for more joint working on local strategies for telecommunications development.

Efficiency and resourcing of development control

There has been a succession of attempts on the part of central government to 'streamline the planning process' and to make it more 'efficient', though the reasons have varied. (More explanation is given in Chapter 3.) In the early 1970s, the concern was with the enormous increase in planning applications and planning appeals which, of course, stemmed from the property boom of the period. By 1981, government concern was with the economic costs of control, with cutting public expenditure and with 'freeing' private initiative from unnecessary bureaucratic controls. During the 1990s the emphasis has been on speeding and raising standards of the 'planning service' so as to achieve better efficiency and value for money. In 2001 the Green Paper *Planning: Delivering a Fundamental Change*, continued in a similar way, although with a different rhetoric and, unarguably, more determination to effect change through new legislation and a comprehensive review of policy and practice. The Green Paper and implementation of change through the 2004 Act are discussed more fully in Chapter 4. The concern here is with efficiency in development control. A brief review of the history of tackling delay is needed.

A good starting point is the analysis of an inquiry chaired by George Dobry QC in 1975. In the 1970s there was lengthening delay in the processing of planning applications during a property boom. Dobry was quick to point out that 'not all delay is unacceptable: it is the price we must pay for the democratic planning of the environment'. He also took account of increasing pressure for public consultation and participation in the planning process; and the 'dissatisfaction on the part of applicants because they often do not understand why particular decisions have been made', and general concerns that the system was not doing enough to protect good environment or promote high quality development.

Dobry, like his successors, had the difficult task of reconciling apparently irreconcilable objectives: to expedite planning procedures while at the same time facilitating greater public participation and devising a system which would produce better environmental results. His solutions attempted to provide more speed for developers, more participation for the public *and* better quality development and conservation. His solution was to divide applications into minor and major, such that minor applications could be dealt with more expeditiously through a simpler process (see the 2005 review of householder applications on p. 158) though with the opportunity for some participation and with a safety channel to allow them to be transferred to the major category if this should prove appropriate.

Dobry's scheme was a heroic attempt to improve the planning control system to everyone's satisfaction (Jowell 1975). Inevitably, therefore, it disappointed everyone, not least his overriding concern for expediting procedures forced him to compress 'simple' applications into an impracticable timescale. In the mean time the boom had collapsed and a new Labour government had other concerns. The government rejected all Dobry's major recommendations for changes in the system, though it was stressed that their objectives could typically be achieved if local authorities adopted efficient working methods. Dobry's view that 'it is not so much the system which is wrong but the way in which it is used' was endorsed, and his *Final Report* was commended 'to students of our planning system as an invaluable compendium of information about the working of the existing development control process, and to local authorities and developers as a source of advice on the best way to operate within it'.

Following the next change of power in 1979, the incoming Conservative government quickly picked up the theme of planning delay and lost no time in preparing a revised development control policy. A draft circular created alarm among the planning profession, partly because of its substantive proposals but also because of its abrasive style. 'The Most Savage Attack Yet' expostulated *Municipal Engineering*, while *Planner News* remonstrated that the results of the circular 'could be disastrous'. The revised circular, as published (22/80), was written with a gentler touch, but much of the message was very similar. The emphasis was on securing a 'speeding up of the system' and ensuring that 'development is only prevented or restricted when this serves a clear planning purpose and the economic effects have been taken into account'. It was at this point that the infamous target eight-week period for deciding on planning applications was instigated, with regular publication of comparative performance figures. Quarterly figures have been published since 1979, and are used by both the government and the development industry to bolster criticisms of the system.

The policy 'to simplify the system and improve its efficiency' (to use the words of the 1985 White Paper, *Lifting the Burden*) continued with revised circulars, new White Papers, and the introduction of planning mechanisms which reduced or bypassed local government control such as simplified planning zones and urban development corporations. However, towards the end of the 1980s, a greater emphasis on 'quality' emerged as environmental awareness and concern increased. A change in direction was signalled by the 1992 Audit Commission report on development control, significantly entitled *Building in Quality*. Though the major emphasis was still on the process of planning control rather than its outcome, there was a very clear recognition of the importance of the latter. The report noted that there had been a preoccupation with the speed of processing planning applications 'ignoring the mix of applications, the variety of development control functions, and the quality of outcomes'. But there had been no 'shared and explicit' concept of quality and added value. What that 'added value' may be is dependent upon the authority's overall objectives: 'in areas under heavy development pressure or in rural areas, environmental, traffic, or ecological considerations may be paramount'; in Wales, 'the impact of the development on the Welsh language can be a consideration'.

The effect of *Building in Quality* was been to redress the balance somewhat from the emphasis on lifting the burden of regulation, but the importance of meeting the eight-week target (and for major applications, thirteen-week target) remains. There was a considerable overall improvement in performance in the first part of the 1990s: from 46 per cent of applications decided within eight weeks in 1989–90 to 65 per cent in 1993–4. Performance remained much the same during the 1990s but has improved from 1999. In 2004, the rate was 77 per cent, an improvement of 5 per cent over the previous year, suggesting that concerted efforts from 2002 (explained below) are beginning to take effect. After thirteen weeks authorities had made 90 per cent of decisions. In Scotland, 66 per cent of applications were decided in eight weeks in 2003 and in Wales 63 per cent.

A review of progress on *Building in Quality* in 1999 points to the value of increasing delegation of decision-making to officers for those authorities whose performance has improved. The reduced number of applications from the peak in 1988–9 (illustrated in Figure 5.2) and the increasing coverage of local development plan policy are also significant factors. However, recent improvements have been made in the face of a sharp increase in the number of applications and appeals from 2000, and there is still great variation

in performance, all of which makes for difficulties in generalising about performance. The varying conclusions of two reports published in 2002 give a very good impression of the complexity of the problem. The 2002 Audit Commission Report on *Development Control and Planning* identified 'intractable barriers' to improvement:

> these include resource limitations, competing priorities within local government and the inherited complexities of the planning system. But the slow pace of change is also symptomatic of a wider malaise: there has been a reluctance to accept the need for improvement in many cases. In the area of customer service, for example, there is evidence that planning has failed to keep pace with improvements in other council services.
>
> (p. 7)

This is not an encouraging report for planning managers. Of the planning services inspected by the Audit Commission, almost 60 per cent were rated fair or poor, and 40 per cent of those judged unlikely to improve. In 2002, more than 90 per cent of councils were failing to meet the 80 per cent target for deciding planning applications in eight weeks, and 13 per cent had no statutory local plan in place. (There have been improvements since publication.) The report concludes that 'existing good practice advice is not being applied consistently within the planning service'. It recommends that authorities address five key areas: focus on what matters to local people with more responsive dialogue with those affected by planning decisions; assessing the value-added of development control and improving enforcement; enhancing customer care; reducing delay in development control and ensuring that resources are used to best effect (p. 15). Many planning managers would find the criticisms partial, and the recommendations self-evident. The problems from their perspective are resources and meeting quality objectives as well as quantitative ones, a position which has been backed up by government.

The second report paints a quite different picture of authorities struggling to meet rapidly increasing demands with declining resources. *Resourcing of*

Planning Authorities, a report by Arup with the Bailey consultancy for the ODPM, notes 'the overwhelming finding is that resources have declined significantly over the past five years and performance has generally worsened, albeit in different functions in different authorities' (p. 13).[56] This very full report points to many other factors that contribute to poor apparent efficiency in planning including the shortage of investment in information and communications technology (ICT), the particular problems for authorities with lots of heritage of valuable natural environments and diversion of resources at times to the development plan process. Officers interviewed also criticised the Best Value scheme (see Chapter 3) for diverting resources. The report sticks by the performance measure of percentage of applications determined within a specified time, and notes the difficulty of measuring quality, except by the proportion of appeals. The report confirms that good decision-making tends to reduce the number of appeals and therefore the overall costs of the planning service to the country. The report also addresses the acute staffing problems, including high turnover and problems in recruitment arising from declining popularity of planning courses, low salaries, competition with the private sector, and above all, the poor image of planning.

In the light of their findings, and taking a 'highly conservative assumption', the researchers concluded that

> to achieve the equivalent of 1996/97 levels of gross expenditure would therefore require increases on the 2001/02 levels of gross expenditure of 37 per cent for unitary authorities and district authorities and 23 per cent for county authorities . . . The increase would equate to between four and five additional staff, on average, per authority.
>
> (p. 16)

The report was prepared in time to feed into the Green Paper modernising planning process and was important in bringing forward the Planning Delivery Grant (PDG: explained below). But despite general problems the research was unable to explain the great variation in performance among authorities using the same level of resources. It is difficult to explain the relationship between resources and performance; many other factors come into play. In 2003 ninety planning authorities with particularly poor records falling below the minimum thresholds for development control were selected as best value standards authorities. The best value performance indicator 109 requires an authority to make decisions on 60 per cent of major applications in thirteen weeks, 65 per cent of minor applications in eight weeks and 80 per cent of other applications in eight weeks (see Box 5.5). The authorities were set individual targets and required to submit additional returns on performance. A review of progress with these authorities (Addison and Associates 2004) showed that the speed of the DC process had improved (thirty-four authorities met or nearly met the standard) but others starting from a very low base were still well below the standard at the end of the first year.

Planning Delivery Grant

In the face of a rapid increase in applications and workload, very variable performance in dealing with planning applications, and difficulties in staff recruitment and retention, the ODPM introduced a system of financial incentives for local planning authorities – the Planning Delivery Grant. The grant aims to improve performance and resources for local planning authorities (and latterly other planning bodies). Perhaps counter-intuitively the grant is targeted at authorities that perform well. In the first year of operation (2002–3) planning authorities received between £75,000 and £475,000 with a mean of £129,000, with the nine authorities that had always met government targets getting the most money and 152 authorities receiving the 'basic' £75,000.

The figures come from the first evaluation of the Planning Delivery Grant by Addison and Associates published by ODPM in 2004. It concluded that the PDG had provided a demonstrable incentive, considerably raising the profile of the planning function within local authorities and focusing attention on the effectiveness of the planning service. The PDG is not ring-fenced, but most money is spent within the

BOX 5.5 BEST VALUE PERFORMANCE INDICATORS FOR PLANNING

Strategic objectives

- percentage of new homes built on previously developed land
- cost and efficiency
- planning cost per head of population.

Service delivery outcomes

- the number of advertised departures from the statutory plan approved by the authority as a percentage of total permissions granted
- percentage of applications determined within eight weeks (excluding applications involving environmental assessment)
- average time taken to determine all applications.

Quality

- percentage of applicants and those commenting on planning applications satisfied with the service received (based on a list of questions specified by the DETR)
- score against a check list of planning practice – for development control these include

  - providing for pre-applications discussions
  - having a published charter setting targets for stages of the process
  - having fewer than 40 per cent of appeals overturning the Council's original decision
  - delegation of 70 per cent or more decisions
  - no costs or Ombudsmen reports finding against the authority
  - provision of a one-stop service
  - equal access to the planning service for all groups.

In 2004 ODPM consulted on additional indicators: the percentage of appeals allowed against the authority's decision to refuse and a quality of service checklist, deletion of planning cost per head of population, and the proportion of decisions delegated to officers. The Planning Officers' Society agreed to the first two but not the last deletion.

planning service, with 46 per cent going to the retention and recruitment of staff and 22 per cent on ICT. Training too was important, in particular to bring non-planning graduates up to speed in the system. The PDG has also generated co-financing from local authority budgets for particular projects, although this is offset by spending (about half the PDG) on projects which would have happened anyway, that is, the low 'additionality' factor. But overall the PDG has been a success and so has been continued.

The PDG was increased in 2004–5 to £130 million with twenty-four planning authorities receiving £700,000 or more. In this phase, the criteria for funding were widened to include plan-making performance and housing delivery in high demand areas. Authorities were also penalised for poor performance in appeals with a 10 per cent reduction in PDG. In 2005–6 the criteria will be widened further in spending a planned £169 million to take into account housing need in low demand areas and the quality of the e-planning service. There is also now top-slicing for the regional bodies' preparation of spatial strategies, the Greater London Authority's planning work, the Planning Inspectorate, and a new mid-career distanced learning course in spatial planning from the University of the West of England.

The introduction of Best Value, the new initiative to improve performance in the delivery of local government services across the board, is explained in Chapter 4. Best Value seeks to marry the need for increased efficiency with recognition of the importance of maintaining and improving quality, and has established a wide framework of performance indicators and targets, some dictated 'nationally' and others defined by the local authority itself. Box 5.5 shows the performance indicators for planning; most relate to the development control function and specific reference is made to the quality of the service. Local authorities prepared five-year plans for improving performance under Best Value, to be reviewed annually. Authorities are required to supplement the national indicators with their own local indicators in making comparisons with other authorities. Good practice in implementing Best Value has been prepared by the Planning Officers' Society with support from government.

Although planning authorities have long made the case for assessing quality as well as efficiency there is no doubt that more than a few will continue to struggle to meet the criteria. But dramatic improvements are possible. North Wiltshire was commended in the 2004 Planning Awards for rising from the eighth slowest English planning authority to one of the top 10 per cent, following extensive management changes involving both officers and elected members. Other authorities are finding innovative ways to speed and provide more certainty in the process. Developers are taking on more of the responsibility for consultation on their development proposals including consultations with key interests and the public. Birmingham City Council, for example, has reached agreements in the form of a 'concordat' with major developers such that the City will expedite the decision-making process subject to the developer having conducted consultation and other matters prior to the application being submitted (Carmona *et al.* 2003). Other performance improvements are anticipated from the e-government initiative, including online planning application processes and consultation, which is considered in Chapter 12.

Further reading

Legal texts

The law and procedure of development control is explained fully in several textbooks. For England and Wales, see Thomas, K. (1997) *Development Control: Principles and Practice*, Moore (2002) *A Practical Approach to Planning Law* and Duxbury (2002) *Planning Law and Procedure*; for Northern Ireland: Dowling (1995) *Northern Ireland Planning Law*; for Scotland: Collar (1994) *Greens Concise Scots Law: Planning* and McAllister and McMaster (1999) *Scottish Planning Law*. Grant (1997) *The Encyclopedia of Planning Law and Practice* provides excellent commentary on planning legislation and policy. The Scottish Executive provide a separate PAN on *Development Control*.

Use Classes and Development Orders

The Nathaniel Lichfield (2003) Study for ODPM considers each part of the GPDO in detail. See also Edinburgh College of Art *et al.* (1997) *Research on the General Permitted Development Order and Related Mechanisms* and also BDP Planning and Leighton Berwin (1998) *The Use of Permitted Development Rights by Statutory Undertakers*. On Article 4 Directions, see Tym *et al.* (1995a) *The Use of Article 4 Directions by Local Planning Authorities* and Larkham and Chapman (1996) 'Article 4 Directions and development control'. The standard legal text is Grant (1996) *Permitted Development*. Halcrow Group's (2004) report on *Unification of Consent Regimes* is very informative in placing the planning and listed building regimes alongside others.

The development plan as a consideration

The role of plans in appeal decisions is considered in Bingham (2001) 'Policy utilisation in planning control'. The impact of the introduction of s. 54A now s. 38(6) of the 2004 Act (s. 25 of the Scottish Act) is reviewed by Gatenby and Williams (1996) 'Interpreting planning law', and this early assessment is still very relevant. See also their earlier article (1992) 'Section 54A: the legal and practical implications'. Other sources include MacGregor and Ross (1995) 'Master or servant?', Purdue (1991) 'Green belts and the presumption in favour of development', Harrison, M. (1992) 'A presumption in favour of planning permission?' and Herbert-Young (1995) 'Reflections on section 54A and plan-led decision-making'.

Other material considerations

The ODPM statement *The Planning System: General Principles*, published alongside PPS 1, presents a succinct statement concerning material considerations. See also SPP 1, *The Planning System* (and PAN 40 as above) for Scotland, and PPS 1 for Northern Ireland. A categorisation of considerations, drawing on work by Lyn Davies, is summarised in Thomas, K. (1997) *Development Control: Principles and Practice*. See also HC Welsh Affairs Committee (1993) *Rural Housing*, where the issue is examined in depth.

Design

There are two very good starting points for considering the role of design as a consideration in planning: the DETR and CABE (2000) *By Design: Urban Design in the Planning System – Towards Better Practice* includes checklists of design considerations and a list of other references, and Carmona's two-part article in *Planning Practice and Research* (1998, 1999) 'Residential design policy and guidance: prevalence, hierarchy and currency'. CABE has produced many publications – see its website www.cabe.org.uk and *Protecting Design Quality in Planning* (2003). Another starting point is Taylor (1999) 'The elements of townscape and the art of urban design'. Punter's work is notable in this field. See Punter (1990) *Design Control in Bristol, 1940–1990* and (1986–97) 'A history of aesthetic control: the control of the external appearance of development in England and Wales'. See also Smith Morris (1997) *British Town Planning and Urban Design*.

There is a huge library on particular design issues. See, for example, Barton *et al.* (2002) *Shaping Neighbourhoods* and Barton *et al.* (1995) *Sustainable Settlements*; FPD Savills Research (2003) *The Value of Housing Design and Layout*; Llewelyn-Davies (1998a) *Planning and Development Briefs: A Guide to Better Practice*; DoE (1994) *Quality in Town and Country* (Discussion Paper) and (1995) *Quality in Town and Country: Urban Design Campaign*; National Audit Office (1994) *Environmental Factors in Road Planning and Design*; NIDoE (1994) *A Design Guide for Rural Northern Ireland*; Scottish Office (1994) *Fitting New Housing Development into the Landscape* (PAN 44); Bishop (1994) 'Planning for better rural design'; Owen (1991) *Planning Settlements Naturally*. There are numerous guides to better design in the built environment including English Partnership's *Urban Design Compendium* (2000); the Scottish Office PAN 59, *Encouraging Higher Standards of Design*, DETR and CABE (2000) *By Design: Urban Design in the Planning System – Towards Better Practice* and (2000) *Training for Urban Design*. Tall buildings are among current concerns, see the Greater London Authority's *Interim Strategic Planning Guidance on Tall Buildings* (2001).

Amenity

Despite its significance there have been few studies of amenity. For a discussion of statutory provisions see Sheail (1992) 'The amenity clause', and also an unusual historical study of the development of the notion of amenity in Millichap (1995a) 'Law, myth and community'.

Appeals

The *Planning Inspectorate Journal* has provided numerous perspectives on appeals. The former Chief Planning Inspector gave an account of 'Decision-making and the role of the Inspectorate' (Shepley 1999). An example of the analysis of appeals data is given by Wood *et al.* (1998) 'The character of countryside recreation and leisure appeals'.

Enforcement

Two standard legal texts are Millichap (1995b) *The Effective Enforcement of Planning Controls* and Bourne (1992) *Enforcement of Planning Control*, still relevant although they are superseded by recent changes. The DETR published *Enforcing Planning Control: A Good Practice Guide* and see also PPG 18. The operation of the procedures in Scotland has been investigated by Edinburgh College of Art *et al.* (1997) *Research on the General Permitted Development Order and Related Mechanisms*; see also 'The stunning powers of environmental inspectors' by Upton and Harwood (1996), which provides a striking contrast to the planning enforcement system. The DETR consultation paper *Improving Enforcement Appeal Procedures* (1999) reviews the finer points of implementing the 'new' system.

Advertisements

The regulations are explained in DoE Circular 19/92; policy guidance is given in PPG 19 and TAN (W) 7. The fullest exposition of the law of advertisement control is given in Mynors (1992) *Planning Control and the Display of Advertisements*, and a useful update is given in the 1999 DETR consultation paper on *Outdoor Advertisement Control*. For Scotland see SOEnD Circular 31/92.

Minerals

The minerals planning guidance notes provide a rich source of information, in particular MPG 1: on general considerations and MPG 2 on how development control of minerals is undertaken. A bibliography on reclamation for various uses is given in MPG 7; on aggregates see MPG 6 (revised 1994) (DoE). The DETR sponsor extensive research on minerals and the reader is directed to the *Planning and Minerals Research Newsletter*. An excellent review of mineral resource planning and sustainability is given by Owens and Colwell (1996) *Rocks and Hard Places*. Minerals guidance is being reviewed and further consultation papers are expected.

Caravans and gypsies

The central text is the research by Niner (2002) at the Centre for Urban and Regional Studies, Birmingham University, *The Provision and Condition of Local Authority Gypsy and Traveller Sites in the English Countryside*. Another account is given by Morris (1998) 'Gypsies and the planning system' see also Gentleman (1993) *Counting Travellers in Scotland*. The main official reference is DoE Circular 1/94 *Gypsy Sites and Planning* although this was due to be replaced in 2005.

Efficiency in development control

The Audit Commission (1992) report, *Building in Quality: A Study of Development Control*, and the subsequent (1998) *Building in Quality: A Review of Progress on Development Control* and (2002) *Development Control and Planning* are the main sources. The various responsible government departments publish quarterly figures on the development control performance. The Planning Officers' Society has made its good practice guidance on Best Value and planning available via the web at www.planningofficerssociety. org.uk. Further details on Best Value are available at the ODPM and Audit Commission websites. See also the DETR good practice guide: *The One Stop Approach to Development Consents* and the report of the Scottish Executive (1999) *Targets Working Group on Planning Services* which includes an analysis of what factors delay planning permissions.

Notes

1 The provision is intended to enable controls over large increases in floor space by using basements or building mezzanine floors in existing large retail stores. The Secretary of State issued a consultation document in March 2005 suggesting that the threshold should be 200 square metres.

2 Changes of use in sites of special scientific interest may also require approval from English Nature although allowed by the UCO.

3 Until 1995, the General Development Order contained both permitted development rights and procedural matters (relating to planning applications). In 1995 these were separated (following the Scottish model introduced in 1992). There is therefore now a General Permitted Development Order and a General Development Procedure Order. See Circular 9/95 *General Development Order Consolidation*. Though these new orders are predominantly consolidations, they contain a number of changes.

4 See, for example, Bell (1992) on problems arising from changes of use.

5 Statutory Instrument 2005 no. 84 The Town and Country Planning (Use Classes) (Amendment) (England) Order 2005.

6 Circular 9/95, *General Development Order Consolidation* 1995 para. 1. The Secretary of State approves Article 4 Directions except those that are specifically related to dwelling houses in conservation areas. PPG 15 explains the application of Article 4 Directions in conservation areas in England.

7 The General Development Procedure Order will be amended in 2005 including provisions for local development orders.

8 I am referring here to practice at Nottinghamshire County Council in the early 1970s (until 1974 counties dealt with all planning applications). This is not to disparage this and other councils who did a great deal of good work, but to illustrate the great difference in attitude to service provision which was then commonplace in local government.

9 In 1998 the DETR published a good practice guide, *The One-Stop Shop Approach to Planning Consents*. There is also increasing interest in comparisons of practice in the UK with other countries; on development control see GMA Planning *et al.* (1993) *Integrated Planning and the Granting of Permits in the EC*.

10 The changes in local authority management structures mean that many decisions are now made by a cabinet rather than committees, but in the case of planning and other regulatory activities local authorities must retain the committee decision-making procedure (although some decisions will be delegated). Some large authorities will divide up the committee into smaller local area committees.

11 The Planning Officers' Society has published a Practice Note on Reasons for the Grant of Planning Permission (see www.planningofficers.org.uk).

12 These and other development control figures are from the quarterly returns on planning application statistics provided by the ODPM.

13 Essex (1996) reviews these two cases and the general issue of relationships between officers and members in decision-making. The issue is also taken up in Chapter 12.

14 During the transition phase following the 2004 Act, the development plan in England may mean the unitary development plan or the structure and local plan including minerals and waste plans (and even old subject plans) depending on the area in question. As the new plan system is put into place it will comprise the regional spatial strategy and the development plan documents in the local development framework (see Chapter 4 for definition of development plans elsewhere in the UK).

15 Exactly the same provision is made in the Scottish legislation as s. 25 of the 1997 Act (formerly s. 18A of the 1972 Act). No such provision has been made for Northern Ireland.

16 In his judgment in the case of *The City of Edinburgh v. the Secretary of State for Scotland*, Lord Hope said:

> it requires to be emphasised however, that the matter is nevertheless still one of judgment, and that this judgment is to be exercised by the decision taker. The development plan does not, even with the benefit of section 18A, have absolute authority. The planning authority . . . is at liberty

to depart from the development plan if material considerations indicate otherwise.

(*Encyclopedia* P54A.05/2)

17 In the oft-quoted Stringer case, it was stated that 'any consideration which relates to the use and development of land is capable of being a planning consideration' (*Stringer v. Minister of Housing and Local Government* 1971). Whether a particular consideration falling within that broad class in any given case is material will depend on the circumstances. In another important case (Newbury), the House of Lords formulated a threefold 'planning test': to be valid a planning decision had to (i) have a planning purpose; (ii) relate to the permitted development; and (iii) be reasonable (*Newbury District Council v Secretary of State for the Environment* 1981).

18 The government's statement *The Planning System: General Principles* (2005), like its predecessors, notes that the government's statements of planning policy

> cannot make irrelevant any matter which is a material consideration in a particular case. But, where such statements indicate the weight that should be given to relevant considerations, decision-makers must have proper regard to them. If they elect not to follow relevant statements of the Government's planning policy they must give clear and convincing reasons (*EC Grandsen and Co Ltd v. SSE and Gillingham BC* 1985). Emerging planning policies, in the form of draft Departmental Circulars and policy guidance, can be regarded as material considerations, depending on the context. Their very existence may indicate that a relevant policy is under review and the circumstances which have led to that review may need to be taken into account.

(paras 13 and 14)

19 The Prince of Wales followed up his criticisms with *A Vision of Britain* (1989), 'a personal view of architecture' spelling out, with telling illustrations, how 'we can do better'.

20 The role of regulation of design was taken forward in Circular 31/85 which emphasised that 'a large proportion of planning appeals involve detailed design matters' and that 'far too many planning applications are delayed because the planning authority seeks to impose detailed design alterations'.

21 Chapman and Larkham (1999) note the poor level of commentary on and failure to disseminate lessons from this initiative in an article generally sceptical of its wider impact in the face of lack of interest after a change in government.

22 PPS 1 goes on to say (in the traditional way):

> design policies should avoid unnecessary prescription or detail . . . local planning authorities should not attempt to impose architectural styles or particular tastes and they should not stifle innovation, originality or initiative . . . it is, however, proper to seek to promote or reinforce local distinctiveness.

23 Price Waterhouse (1997) *The Design Improvement Controlled Experiment: Evaluation of the Impact, Costs and Benefits of Estate Re-modelling*, London: DETR.

24 Circular 11/95 does not deal with conditions in respect of minerals or waste, which are dealt with in the minerals planning guidance notes and PPG 23, *Planning and Pollution Control*.

25 In Scotland guidance is given in Circular 4/1998, and in Wales it is Welsh Office Circular 35/95, both entitled *The Use Conditions in Planning Permissions*.

26 PPG 3, *Housing* (2000) notes that it is common practice to renew planning permissions, but encourages local planning authorities to review permissions in the light of current planning policy and if necessary not renew permissions or impose new conditions (para. 40).

27 But the trench-digger may be brought up against a further provision: the serving of a *completion notice*. Such a notice states that the planning permission lapses after the expiration of a specified period (of not less than one year). Any work carried out after then becomes liable to enforcement procedures.

28 The fees are amended on a regular basis. For illustration, at the time of writing, the fee in England and Wales for residential development is £190 per dwelling (up to a maximum of £9,500 or £4,750

for outline applications) and £95 for extensions to dwellings. For commercial and industrial buildings it varies according to gross floor space created: £35 for 40 square metres, and for larger developments £190 for each 75 square metres up to a maximum of £9,500 (The Town and Country Planning (Fees for Applications and Deemed Applications) (Amendment) Regulations 1997 Statutory Instrument 1997 no. 37). Fees in Scotland are set out in The Town and Country Planning (Fees for Applications and Deemed Applications) (Scotland) Amendment Regulations 2000, SI 2000 no. 150. The Planning Portal provides a 'fee calculator'.

29 In England appeals are made to the Deputy Prime Minister and in Wales to the Welsh Assembly. In both cases the Planning Inspectorate Executive Agency considers and makes decisions on most (the nature of the Agency is explained in Chapter 3). In Scotland the Inquiry Reporters Unit considers appeals representing the Scottish Minister for Planning. In Northern Ireland, the Planning Appeals Commission has the same role as the Planning Inspectorate.

30 The Franks Committee on Administrative Tribunals and Inquiries argued that it was not satisfactory 'that a government department should be occupied with appeal work of this volume, particularly as many of the appeals relate to minor and purely local matters, in which little or no departmental policy entered' (Franks Report 1957: 85).

31 Cases may be recovered by the Secretary of State where they involve substantial development (over 150 houses or retail development over 100,000 square feet), significant proposed development in the green belt, major mineral planning appeals, where other government departments have an interest, or where there is major controversy over the development.

32 Circular 5/2000 explains the procedures and gives references to the inquiry, hearing and written representation rules. At the time of writing Scotland and Wales had not reviewed their appeal procedures although devolution is likely to make them more distinctive to the specific needs of these countries.

33 See the DETR consultation papers *Modernising Planning: Improving Planning Appeal Procedures* (1998) and *Modernising Planning: The Recovery of Costs of Public Local Inquiries Held into Planning Matters* (1998).

34 The Inspectorate has agreed and published *Better Presentation of Evidence in Chief* with the Local Government Planning and Environment Bar Association (2000). See also the RTPI Practice Advice Note no. 9 *Development Control: Handling Appeals* (1995) although this does not take account of the new procedures in England.

35 The difficulties of the interpretation of aggregate appeals data (since each decision is made on its merits) have been a subject of continuing debate. For example, see Brotherton (1993).

36 A website has been set up to disseminate the work of the review: www.odpm.gov.uk/householderconsents.

37 See Department of Land Economy, Cambridge University (1999) *Environmental Court Project*.

38 Circular 1/2002 set out the requirements of the Direction and this was superseded by Circular ODPM 01/2005, which added the areas outside the south east.

39 Until September 1999 this role was undertaken by the Royal Fine Arts Commission, which requested intervention by the Secretary of State on numerous occasions but not always successfully. Another important body is the Urban Design Alliance, which comprises professional bodies who seek to improve the quality of life through urban design.

40 The Circular should be read in conjunction with Planning Policy Guidance Note 18: *Enforcement* and the DETR *Enforcing Planning Control: Good Practice Guide for Local Planning Authorities*. In Scotland the key references are Circular 4/1999, and Planning Advice Note 54, *Enforcement*. See also RTPI Practice Advice Note no. 6 (1999) and *Enforcement of Planning Control* (1996).

41 *Independent*, 9 December 1995: 9.

42 The study found that stop notices are not used because of the fear of compensation payments; that breach of condition notices may be difficult to employ because conditions are not worded with sufficient specificity; and that there was some frustration at the difficulty of employing the 'ultimate sanction' through the courts.

43 The government has suggested that a third test should be added – the policies of the development plan.

44 In England the Town and Country Planning (Control of Advertisements) Regulations 1992 (SI 666) as amended in 1994 (SI 2351) and 1999 (SI 1810).

45 DoE consultation paper, *Outdoor Advertisement Control: Areas of Special Control of Advertisements* (1996) and DETR consultation paper *Modernising Planning: Outdoor Advertisement Control* (1999).

46 The first revision of MPG 1 was published in 1994. The second and sixth points in the 1996 list were added in 1996 strengthening policy on both preventing negative environmental impacts and ensuring that mineral resources are kept available.

47 The DETR has funded a series of research projects on the environmental impacts of minerals exploitation that inform national policy on development control. The most recent reports are Arup Environmental and Ove Arup and Partners (1995) *The Environmental Effects of Dust from Surface Mineral Workings*, Vibrock Ltd (1998) *The Environmental Effects of Production Blasting from Surface Mineral Workings*, ENTEC UK Ltd (1998) *The Environmental Effects of Traffic Associated with Mineral Workings*, and University of Newcastle upon Tyne (1999) *Do Particulates from Opencast Coal Mining Impair Children's Respiratory Health?*

48 An Annex to the ODPM's consultation paper on New Parliamentary Procedures for Processing Major Infrastructure Projects gives a useful summary of the ways in which major infrastructure projects may be approved. The examples here are drawn from the paper.

49 A new Code of Practice on the Dissemination of Information about major infrastructure projects was also published in 2002.

50 This legislation has remained as a separate code and is not consolidated in the Town and Country Planning Act 1990. The Caravan Sites Act 1968, which deals mainly with the protection from eviction of caravan dwellers and gypsies, is similarly separate.

51 The report by John Cripps (1977) *Accommodation for Gypsies: A Report on the Working of the Caravan Sites Act 1968* was instrumental in the changes. Circular 28/77 clearly conveyed the government policy of the time to give gypsies special protection in the planning system: it even accepted the necessity of establishing gypsy sites in the protected areas such as green belts and AONBs.

52 The number of gypsies has been estimated at '9,000 families in 13,500 caravans, 9,000 of which are parked on legal sites' (excluding new age travellers) (Morris 1998).

53 These figures are taken from the 2004 All Parliamentary Mobile Group Report on Mobile Phone Masts (Askew 2004b).

54 Permitted development applies to ground based masts and those installed on buildings or other structures, and a public call box. Some masts or antennae may be so small that they do not constitute development – for example television aerials have been treated as outside the definition of development (despite their sometimes significant impact on the external appearance of buildings). The exceptions from permitted development for masts under 15 metres include proposed masts on listed buildings, scheduled ancient monuments and where the planning authority have made an Article 4 Direction withdrawing permitted development rights.

55 See Circular 5/2001 The Town and Country Planning (General Permitted Development) (Scotland) Amendment (no. 2) Order, *Development by Telecommunications Code System Operators*.

56 The study found that the typical cost of dealing with a householder application is about £200 whereas the fee at that time was £95 (so even if the system operates efficiently it does so making a loss) and that the ideal ratio is about one member of staff 'for every 150 to 200 applications, plus support services'.

6 Land policies

It is clear that under a system of well-conceived planning, the resolution of competing claims and the allocation of land for the various requirements must proceed on the basis of selecting the most suitable land for the purpose, irrespective of the existing values which may attach to the individual parcels of land.

Uthwatt Report 1942

The UK planning system is underpinned by an extraordinary feat of nationalisation which was passed without the revolution that might have been expected in many other countries. It was the nationalisation of the right to develop land. Instead of any outcry, or even any political opposition, the issues were considered to be of a technical nature that could be pondered upon by a selected body of wise men. The Committee was required to 'make an objective analysis of the subject of compensation and recovery of betterment in respect of public control of the use of land . . . and to advise on possible means of stabilizing the value of land required for development or retirement'. The terms of reference were radical, though they decided not to recommend the obvious solution of land nationalisation. This chapter considers how successive governments have tried to deal with the problem, from land development taxes to planning gain supplements.

Uthwatt Report

Effective planning necessarily controls, limits or even completely destroys the market value of particular pieces of land. Is the owner to be compensated for this loss in value? If so, how is the compensation to be calculated? And is any 'balancing' payment to be

extracted from owners whose land appreciates in value as a result of planning measures? This problem of compensation and betterment faced the Uthwatt Committee. It arises fundamentally 'from the existing legal position with regard to the use of land, which attempts largely to preserve, in a highly developed economy, the purely individualistic approach to land ownership'. This 'individualistic approach', however, has been increasingly modified during the past hundred years. The rights of ownership were restricted in the interests of public health: owners had (by law) to ensure, for example, that their properties were in good sanitary condition, that new buildings conformed to certain building standards, that streets were of a minimum width, and so on. It was accepted that these restrictions were necessary in the interests of the community (*salus populi est suprema lex*) and that private owners should be compelled to comply with them even at cost to themselves.

All these restrictions, whether carrying a right to compensation or not, are imposed in the public interest, and the essence of the compensation problem as regards the imposition of restrictions appears to be this – at what point does the public interest become such that a private individual ought to be compelled to comply, at his own cost, with a restriction or requirement designed to secure that public interest? The history of the imposition of obligations without compensation has been to push that point progressively further on and to add to the list of requirements considered to be essential to the well-being of the community.

(*Uthwatt Report*, para. 33)

But clearly there is a point beyond which restrictions cannot reasonably be imposed on the grounds of good neighbourliness without payment of compensation – and 'general consideration of regional or national policy requires so great a restriction on the landowner's use of his land as to amount to a taking away from him of a proprietary interest in the land'. This, however, is not the end of the matter. Planning sets out to achieve a selection of the most suitable pieces of land for particular uses. Some land will therefore be zoned for a use which is profitable for the owner, whereas other land will be zoned for a use having a low, or even nil, private value. It is this difficulty of *development value* which raises the compensation problem in its most acute form. The expectations (or hopes) of owners extend over a far larger area than is likely to be developed. This *potential* development value is therefore speculative, but until the individual owners are proved to be wrong in their assessments (and how can this be done?) all owners of land having a potential value can make a case for compensation on the assumption that their particular pieces of land would in fact be chosen for development if planning restrictions were not imposed. Yet this *floating value* might never have settled on their land, and obviously the aggregate of the values claimed by the individual owners is likely to be greatly in excess of a total valuation of all pieces of land.

Furthermore, the public control of land use necessarily involves the shifting of land values from certain pieces of land to other pieces: the value of some land is decreased, while that of other land is increased. Planning controls, so it was argued, do not destroy land values: in the words of the Uthwatt Committee, 'neither the total demand for development nor its average annual rate is materially affected, if at all, by planning ordinances'. Nevertheless, the owner of the land on which development is prohibited will claim compensation for the full potential development of his

land, irrespective of the fact that the value may shift to another site.

In theory, it is logical to balance the compensation paid to aggrieved owners by collecting a betterment charge on owners who benefit from planning controls (Hagman and Misczynski 1978), but previous experience with the collection of betterment had not been encouraging.[1] The Uthwatt Committee concluded that the solution to these problems lay in changing the system of land ownership under which land had a development value dependent upon the prospects of its profitable use. They maintained that no new code for the assessment of compensation or the collection of betterment would be adequate if this 'individualistic' system remained. The system itself had inherent 'contradictions provoking a conflict between private and public interest and hindering the proper operation of the planning machinery'. A new system was needed which would avoid these contradictions and which so unified existing rights in land as to 'enable shifts of value to operate within the same ownership'. The Uthwatt Committee's solution was the nationalisation of development rights in undeveloped land.

The 1947 Act

Essentially, this is what the Town and Country Planning Act 1947 did: development rights and their associated values were nationalised. No development was to take place without permission from the local planning authority. If permission were refused, no compensation would be paid (except in a limited range of special cases). If permission were granted, any resulting increase in land value was to be subject to a development charge. The view was taken that 'owners who lose development value as a result of the passing of the Bill are not on that account entitled to compensation'. This cut through the insoluble problem posed in previous attempts to collect betterment values created by public action. Betterment had been conceived as any increase in the value of land arising from central or local government action. The 1947 Act went further: all betterment was created by the community, and it was unreal and undesirable (as well as virtually impossible) to distinguish between values created, for example, by particular planning schemes, and those due to other factors such as the general activities of the community or the general level of prosperity.

If rigorous logic had been followed, no payment at all would have been made for the transfer of development value to the state but this, as the Uthwatt Committee had pointed out, would have resulted in considerable hardship in individual cases. A £300 million fund was therefore established for making 'payments' (as distinct from 'compensation') to owners who could successfully claim that their land had some development value on the appointed day – the day on which the provisions of the Bill which prevented landowners from realising development values came into force. Considerable discussion took place during the passage of the Bill through Parliament on the sum fixed for the payments, and it was strongly opposed on the ground that it was too small. The truth of the matter was that, in the absence of relevant reliable information, any global sum had to be determined in a somewhat arbitrary way, but in any case it was not intended that everybody should be paid the full value of their claims. Landowners would submit claims to a centralised agency, the Central Land Board, for *loss of development value*, that is, the difference between the *unrestricted value* (the market value without the restrictions introduced by the Act) and the *existing use value* (the value subject to these restrictions). When all the claims had been received and examined, the £300 million would be divided between claimants at whatever proportion of their 1948 value the total would allow. (In the event, the estimate of £300 million was not as far out as critics feared: the total of all claims finally amounted to £380 million.)

These provisions, of which only the barest summary has been given here, were very complex and, together with the inevitable uncertainty as to when compensation would be paid and how much it should be, resulted in a general feeling of uncertainty and discontent which did not augur well for the scheme. The principles, however, were clear. To recapitulate, all development rights and values were vested in the state: no development could take place without permission from the local planning authority and then only on payment of a betterment charge to the Central Land Board. The nationalisation of development rights was effected by the 'promised' payments in lieu of compensation. As a result, landowners owned only the existing use rights of their land and it thus followed, first, that if permission to develop was refused no compensation was payable, and, second, that the price paid to public authorities for the compulsory acquisition of land would be equal to the existing use value, that is, its value excluding any allowance for future development.

The scheme did not work as smoothly as was expected. Land changed hands at prices which included the full development value. This was largely due to the severe restrictions which were imposed on building. Building licences were very scarce, and developers who were able to obtain them were willing to pay a high price for land upon which to build. The Labour government was in the process of reviewing the scheme when it lost office.

The 1954 scheme: the dual land market

The Conservative government which took office in 1951 was intent on raising the level of construction activity and particularly the rate of private house building. Although, within the limits of building activity set by the Labour government, it is unlikely that the development charge procedure seriously affected the supply of land, it is probable that the Conservative government's plans for private building would have been jeopardised by it. This was one factor which led the new government to consider repealing development charges. There is no doubt that these charges were unpopular, particularly since they were payable in cash and in full, whereas payments on the claims on the £300 million fund were deferred and uncertain in amount. Given the political and technical problems involved, it was decided that the best solution was the complete abolition of development charges. However, to safeguard the public purse, acquisitions of land by public authorities were to remain at the existing use value.

The effect of the complicated network of legislation which now (in 1954) operated was basically to create two values for land according to whether it was sold in the open market or acquired by a public authority. This was an untenable position and as land prices increased, due partly to planning controls, the gap between existing use and market values widened, particularly in suburban areas near green belt land. The greater the amount of planning control, the greater did the gap become. Thus, owners who were forced to sell their land to public authorities considered themselves to be very badly treated in comparison with those who were able to sell at the enhanced prices resulting in part from planning restrictions on other sites. The inherent uncertainties of future public acquisitions – no plan can be so definite and inflexible as to determine which sites will (or might) be needed in the future for public purposes – made this distinction appear arbitrary and unjust. The abolition of the development charge served to increase the inequity. The contradictions and anomalies in the 1954 scheme were obvious. It was only a matter of time before public opinion demanded further amending legislation.

The 1959 Act: the return to market value

Opposition to this state of affairs increased with the growth of private pressures for development following the abolition of building licences. Eventually the government was forced to take action. The resulting legislation (the Town and Country Planning Act 1959) restored *fair market price* as the basis of compensation for compulsory acquisition. Owners now obtained (in theory at least) the same price for their land irrespective of whether they sold it to a private individual or to a public authority.

These provisions thus removed a source of grievance, but they did nothing towards solving the fundamental problems of compensation and betterment, and the result proved extremely costly to public authorities. If this had been a reflection of basic principles of justice there could have been little cause for complaint but, in fact, an examination of the position shows clearly that this was not the case.

In the first place, the 1959 Act (like previous legislation) accepted the principle that development rights should be vested in the state. This followed from the fact that no compensation was payable for the loss of development value in cases where planning permission was refused. But if development rights belong to the state, surely so should the associated development values? Consider, for example, the case of two owners of agricultural land on the periphery of a town, both of whom applied for planning permission to develop for housing purposes – the first being given permission and the second refused on the ground that the site in question was to form part of a green belt. The former benefited from the full market value of the site in residential use, whereas the latter could benefit only from its existing use value. No question of compensation arose since the development rights already belonged to the state, but the first owner had these given back without payment. There was an obvious injustice here which could have eventually led to a demand that the 'penalised' owner should be compensated.

Second, as has already been stressed, the comprehensive nature of the planning system has a marked effect on values. The use for which planning permission has been, or will be, given is a very important factor in the determination of value. Furthermore, the value of a given site is increased not only by the development permitted on that site, but also by the development not permitted on other sites. In the example given above, for instance, the value of the site for which planning permission for housing development was given might be increased by virtue of the fact that it was refused on the second site.

Land Commission 1967–71

Mounting criticism of the inadequacy of the 1959 Act led to a number of proposals for a tax on betterment. The Labour government which was returned to power in 1964 introduced the Land Commission Act, which provided for a new levy and had two main objectives: 'to secure that the right land is available at the right time for the implementation of national, regional and local plans' and 'to secure that a substantial part of the development value created by the community returns to the community and that the burden of the cost of land for essential purposes is reduced' (White Paper, *The Land Commission*, 1965).

To enable these two objectives to be achieved, a Land Commission was established. The Commission could buy land either by agreement or compulsorily, and it was given very wide powers for this purpose. The second objective was met by the introduction of a betterment levy on development value. This was necessary not only to secure that a substantial part of the development 'returned to the community', but also to prevent a two-price system as had existed under the 1954 Act. The levy was deducted from the price paid by the Commission on its own purchases and was paid by owners when they sold land privately. Landowners thus theoretically received the same amount for their land whether they sold it privately, to the Land Commission, or to another public authority.

The levy differed from the development charge of the 1947 Act in important ways. Most significantly, it did not take all the development value. Though the Act did not specify what the rate was to be, it was made clear that the initial rate of 40 per cent would be increased to 45 per cent and then to 50 per cent 'at reasonably short intervals'. (It never was.) The Land Commission's first task was to assess the availability of, and demand for, land for house building, particularly in the areas of greatest pressure. In its first annual report, it pointed to the difficulties in some areas, particularly in the South East and the West Midlands, where the available land was limited to only a few years' supply. Most of this land could not, in fact, be made available for early development. Much of it was in small parcels; some was not suitable for development at all because of physical difficulties; of the remainder, a great deal was already in the hands of builders. Thus there was little that could be acquired and developed immediately by those other builders who had an urgent need for land. All this highlighted the need for more land to be allocated by planning authorities for development.

The Land Commission had to work within the framework of the planning system, and was subject to the same planning control as private developers. The intention was that the Commission would work harmoniously with local planning authorities and form an important addition to the planning machinery. As the Commission pointed out, despite the sophistication of the British planning system, it was designed to control land use rather than to promote the development of land. The Commission's role was to ensure that land allocated for development was in fact developed, by channelling it to those who would develop it. It could use its powers of compulsory acquisition to amalgamate land which was in separate ownerships and acquire land whose owners could not be traced. It could purchase land from owners who refused to sell for development or from builders who wished to retain it for future development.

In its early reports, it expressed the hope that they would fulfil their role in acting 'as a spur to those local planning authorities whose plans have not kept pace with the demand for various kinds of development'. However, this was not to be so. Although it became increasingly active, it is not easy to appraise what success the Land Commission achieved. It was only beginning to get into its stride in 1970 when a new government was returned which was pledged to its abolition on the grounds that it 'had no place in a free society'. This pledge was fulfilled in 1971 and thus the Land Commission went the same way as its predecessor, the Central Land Board.

The Conservative years 1970–4

Land prices were rising during the late 1960s (with an increase of 55 per cent between 1967 and 1970), but the early 1970s witnessed a veritable price explosion. Using 1967 as a base (100), prices rose to 287 in 1972 and 458 in 1973. Average plot prices rose from 908 in 1970 to 2,676 in 1973 (DoE, *Housing and Construction Statistics 1969–1979*). Not surprisingly, considerable pressure was put on the Conservative government to take some action to cope with the problem, though it was neither clear nor agreed what the basic problem was (Hallett 1977: 135). The favourite explanation, however, was 'speculative hoarding', and it was this which became the target for government action (in addition to a series of measures designed to speed up the release and development of land). A White Paper, *Widening the Choice: The Next Steps in Housing*, set out proposals for a *land hoarding charge*. This was to be levied 'for failure to complete development within a specified period from the grant of planning permission'. After this 'completion period' (of four years from the granting of outline planning permission or three years in the case of full planning permission), the charge was to be imposed at an annual rate of 30 per cent of the capital value of the land. The scheme was clearly a long-term one and, to deal with the urgent problem ('urgent' in political if not in any other terms), a *development gains tax* and a *first letting tax* were introduced.

The development gains tax provided for gains from land sales by individuals to be treated, not as capital gains, but as income (and thus subject to high marginal rates). The first letting tax, as its name implies, was a tax levied on the first letting of shops, offices, or industrial premises. In concept, it was an equivalent to the capital gains tax which would have been levied had the building been sold. Both taxes came into operation at the time when the land and property boom turned into a slump. Indeed, it has been suggested that they contributed to it (Hallett 1977: 137).

Community Land Scheme

The Labour government which was returned to power in March 1974 lost little time in producing its proposals for a new scheme for collecting betterment. The objectives of this were 'to enable the community to control the development of land in accordance with its needs and priorities' and 'to restore to the community the increase in value of land arising from its efforts'. The keynote was 'positive planning', which was to be achieved by public ownership of development land. In England and Scotland, the agency for purchasing development land was to be local government (thus avoiding the inter-agency conflict which arose between local authorities and the Land Commission). In Wales, however, with its smaller local authorities, an ad-hoc agency, the Land Authority for Wales was

to be created, now part of the Welsh Development Agency (Morgan and Henderson 1997).

In order 'to restore to the community the increase in value of land arising from its efforts', it was proposed that 'the ultimate basis on which the community will buy all land will be current use value'. Sale of the land to developers, on the other hand, would be at market value. Thus, all development value would accrue to the community. Provisionally, however, development values were to be recouped by a development land tax. The ensuing legislation came in two parts: the Community Land Act 1975 provided wide powers for compulsory land acquisition, while the Development Land Tax Act 1976 provided for the taxation of development values. Thus the twin purposes of 'positive planning' and of 'returning development values to the community' were to be served.

The Community Land Scheme was complex, and became increasingly so as regulations, directions and circulars followed the passing of the two Acts. The intention was for it to be phased in gradually, thus enabling programmes to be developed in line with available resources of finance, staffing and expertise. The scheme, like its two predecessors, had little chance to prove itself before the return of a new government. The economic climate of the first two years of its operation could hardly have been worse, and the consequent public expenditure crisis resulted in a central control which limited it severely (Grant 1979; Emms 1980).[2]

Thus three attempts to solve the compensation and betterment problem failed, though the problems to which they were directed are still very much with us. Moreover, as the following discussion shows, there are still attempts to secure the recoupment of betterment.

Planning agreements and obligations

The failure of comprehensive schemes for the collection of betterment was one of a number of factors which, in the early 1980s, stimulated an already established trend for increasing the levying of charges on developers. Other influences included a general move from a regulatory to a negotiation style of development control, increased delays in the planning system, and the financial difficulties of local authorities in providing infrastructure (Jowell 1977a; Sheaf Report 1972).

Planning authorities have had power to make 'agreements' (with the approval of the Secretary of State) since 1932, but it was not until the property boom of the early 1970s that they became widely used – or, as some argue, 'abused'. The term *planning gain* is popularly used, but with two different meanings. The term can denote the provision of facilities which are an integral part of a development, but it can also mean 'benefits' which have little or no relationship to the development, and which the local authority requires as the price of planning permission. There has been very extensive debate on this issue, and the list of relevant publications is very long. Unfortunately, neither publications nor statutory changes and ministerial exhortations have done much to settle the arguments. The extremes range from the Property Advisory Group's (1981) categorical statement that planning gain has no place in the planning control system, to Mather's (1988) proposal that planning gain should be formalised by allowing local authorities to sell or auction planning consents. Essentially, the issue is the extent to which local authorities can legitimately require developers to shoulder the wider costs of development: the needed infrastructure, schools and other local services.

The extremes are easy to identify: the cost of access to a development is clearly acceptable, while financial contributions to the cost of running a central library are not. But, of course, most items fall well within these extremes. The general view, supported by a number of studies, was that the majority were legitimate (Byrne, S. 1989; Eve 1992; Rowan-Robinson and Durman 1992a). These studies effectively demolish the argument that there was widespread extortion by way of planning gain, though the range of infrastructure and community facilities secured by planning authorities through planning obligations has steadily widened, and not always in accordance with the guidance (see Box 6.1 and Box 6.2). Many agreements deal with occupancy conditions (for example, restrictions required for sheltered housing, agricultural dwellings and social housing). Other agreements secure provision of infrastructure and facilities that have been necessitated by a development (particularly local roads) and in environmental improvement (such as landscaping). Only a very small number of agreements are concerned with wider planning objectives. In Scotland, research led to conclusions that 'most agreements are useful adjuncts to the development control process; abuse of power does not present a problem; and for the most part, the benefits secured by agreements have been related to the development proposed: where they have not, the benefits have been of a relatively minor order' (Rowan-Robinson and Durman 1992a: 73).

The statutory provisions relating to agreements were amended by the Planning and Compensation Act 1991. Agreements were replaced by 'obligations' and can now be unilateral – not involving any 'agreement' between a local authority and a developer at all. This provision allows a developer to make an agreement to provide the necessary off-site works even if the local authority is not prepared to be a party to the agreement (a *unilateral undertaking*). DoE Circular 7/91 also confirmed that local authorities could negotiate with developers for the provision of social housing through planning obligations. This represented a major extension of the arena of planning agreements which takes it well beyond the provision of facilities required by the proposed development, and into the territory of a tax on development to pay for the delivery of public services. In the words of the then minister (Sir George Young) planning gain 'would provide facilities that the public would never have afforded'. But by this time there was some support for planning obligations since they allowed the development industry to free up restraints, so long as the costs were offset by the potential profits to be made; a major change in opinion since the Property Advisory Group (1981) declared the pursuit of planning gain to be unacceptable (RICS 1991; Rowan-Robinson and Durman 1992a). For profitable developments such as major retail stores, substantial payments can be made on the promise of future profits which are safeguarded by the planning system which would effectively 'protect' the development from further competition. By the 1990s a fundamental change in the roles of the private and public sectors in land development had become accepted. It has become virtually unanimously accepted that the

BOX 6.1 EXAMPLES OF FACILITIES SECURED BY PLANNING AUTHORITIES THROUGH PLANNING AGREEMENTS

Residential developments

Direct consequences of development

- Offsite highways
- Parking
- Landscaping
- Open space
- Sports facilities
- Community centres
- Schools
- Health services
- Public transport facilities
- Waste and recycling facilities
- Emergency services
- Childcare facilities

Affordable housing

- Social rented housing
- Key worker housing
- Sheltered housing

Contributions to community needs

- Construction, training and recruitment initiatives
- Town centre improvement
- Public art
- Countryside managements
- Contributions to cultural plans, theatres, museums, etc.

Commercial developments

- Offsite highways
- Parking
- Landscape
- Open space
- Public transport
- Green transport plans

- Housing via mixed use policies

- Training and recruitment initiatives
- Town centre improvement
- Public art

Source: GVA Grimley *et al.* (2004: 16)

public sector is financially unable to meet the associated costs of development, and developers are willing to shoulder them as part of the development value created through planning permission (they may be passed on to landowners or users). Planning gain has become the accepted way of dealing with the state recouping some development value. But there is also virtual unanimity about the manifest deficiencies of the system of planning obligations, 'variously described as opaque, slow, unfair, complex and reactive' (ODPM 2004).

Diagnosis of the problem has been much easier than prescribing solutions for planning obligations. For Grant, 'planning gain is a relatively simple and straightforward phenomenon that we have managed to

BOX 6.2 PLANNING GAIN: THE PAIGNTON ZOO CASE

The Paignton Zoo case is a revealing case of the extent to which planning benefits are acceptable as legitimate. A proposed development included a 65,000 sq. ft retail store, parking spaces for 600 cars, a petrol station, and the refurbishment of the zoo.

The proposals clearly raised major issues of policy including those set out in PPG 6 *Town Centres and Retail Development* and PPG 21 *Tourism*. There were several conflicting considerations, including the likely effect of the retail development on the town centre, and the precarious economic position of the zoo (which was 'likely to close unless it receives a capital injection of the size that only this proposal is likely to provide, thereby causing a loss to the local economy of approximately 6 million per annum and a significant loss of jobs'). The Secretary of State decided that these and other benefits to tourism and the local economy (together with highway improvements) more than outweighed any harm which might be done to the vitality of the town centre, and he therefore granted planning permission. In the words of the decision, 'the harm likely to arise from the proposals is less clear cut than the effects that would result from the decline and possible closure of the zoo; the balance of advantage lies in favour of allowing the proposal; the zoo's leading role in the local economy places it in a virtually unique position'. However, the Secretary of State stressed that the decision 'should not be regarded as a precedent for other businesses seeking to achieve financial stability'.

Source: JPL (1995: 657)

convert into a complex and secretive transaction' (2003: 1). He identifies four strands in the story of what has gone wrong: the use of unconventional legal mechanisms rather than conditions; the negotiated form of agreements rather than fixed charges; the false notion that contributions were to meet planning needs arising from the development and not a tax on development value; and the presentation of firm government requirements for affordable housing as a voluntary negotiated contribution. In seeking a solution the government has had to tread carefully avoiding slipping into a situation where changes might be contrived as, on the one hand, the selling of planning permissions, or on the other hand, a development value tax. Of particular concern is the problem not envisaged when the system was established, that it would be in the commercial interests of the applicant to offer obligations that are not strictly necessary to resolve concerns that make the development 'unacceptable in planning terms'.

The current system of planning obligations (known also, from the part of the 1990 Act that provides for them, as *section 106 agreements*) is set out in Circular 1/97, but a new circular was proposed in 2005. The general approach to obligations is shown in Box 6.3. This is a holding measure, making urgently needed adjustments to the current system, while further, more substantial changes are proposed in the longer term. The recent history of formulating reform of obligations is complicated. The ODPM consulted on *Planning Obligations: Delivering a Fundamental Change* in 2002 as part of the wider agenda for reform of the planning system. A rather late consultation took place at the end of 2003 on *A New Approach to Planning Obligations* so as to allow provisions to allow for changes to planning obligations to be inserted into the Planning and Compulsory Purchase Act 2004, and in particular to allow for a 'standard (planning) charge'. In January 2004 the ODPM provided a statement on its response to the consultation returns in *Contributing to Sustainable Communities: a New Approach to Planning Obligations*. There had been some consternation at the

BOX 6.3 PLANNING OBLIGATIONS: GENERAL POLICY

Proposed revision of Circular 1/97 Planning Obligations (extract)

5 The Secretary of State's policy requires, among other factors, that planning obligations are only sought where they meet all of the following tests:

A planning obligation must be:
(i)   necessary to make the proposed development acceptable in planning terms;
(ii) relevant to planning;
(iii) directly related to the proposed development;
(iv) fairly and reasonably related in scale and kind to the proposed development;
(v)   reasonable in all other respects.

6 The use of planning obligations must be governed by the fundamental principle that planning permission may not be bought or sold. It is therefore not legitimate for unacceptable development to be permitted because of benefits or inducements offered by a developer which are not necessary to make the development acceptable in planning terms (see 5(i)).
7 Similarly, planning obligations should never be used as a means of securing for the local community a share in the profits of development, i.e. as a means of securing a 'betterment levy'.

This revision of Circular 1/97 repeats the same tests, but simplifies and clarifies them and gives more emphasis to the necessity test.

hasty consultation, but the statement explained that 'it would be quite wrong for the Government not to take the rare opportunity that the Planning Bill represents to make decisive reform in this area' (p. 5). In fact the necessity of changes to the planning obligations system had been well known for many years with many studies on the subject, as illustrated in the Further Reading on the subject at the end of this chapter. In any case, the provisions in the Act only allow for subordinate legislation (regulations) to be made to introduce changes; they leave considerable scope for alternative solutions, and the government were going to need this flexibility. In the mean time, the Barker *Review of Housing Supply* was published (see p. 221) with a specific recommendation for a *planning gain supplement* (PGS) which was quickly followed by commitment by the Chancellor in the 2003 Budget Report to consider carefully with a decision by the end of 2005. (It would appear that this recommendation was made in isolation from the ODPM's longer standing work on reform of the system.) A new shorter Ministerial Statement on *Planning Obligations in England* in June 2004 explained the next steps: ODPM went ahead with its revision of Circular 1/97 and published a revision for consultation at the end of 2004 together with proposals for further good practice guidance; the proposals for the planning charge are to be considered 'on a timetable consistent with that for decisions on the PGS'. That last cryptic message does not give any confidence about joined up government thinking on this issue, especially given the complexity of the proposals under discussion. In contrast, the proposed revisions to Circular 1/97 are unequivocal about the purpose of planning obligations:

> It is the government's view that s106 is not the right mechanism with which to achieve the successful capture of development gain. We are therefore proposing in the revised Circular that s106 should continue to be an impact mitigation or positive planning measure linked to planning necessity and that it should not be used for tax-like purposes such as the capture of land value increases for purposes not directly necessary for development to proceed . . . [The Circular] is seeking to discourage the offering by developers of facilities that are not required by the development, in order to make clear that planning permission is not being bought or sold.
>
> (ODPM (2004) *Draft Revised Circular on Planning Obligations*)

More research was undertaken on the potential further use of *standard charges* (SCs) (GVA Grimley *et al.* 2004). Some local authorities were found to be already using 'contributions frameworks' as a way of providing consistency in their approach to negotiating obligations. The researchers concluded that a national system of standard charges is feasible, but recognised 'substantial difficulties in making a system work fairly and simply alongside the planning system's presumption that, where possible, community facilities and affordable housing should be provided on site'. It is likely to be some time before charges are implemented.

In the meantime local authorities have implemented a form of standard charge through s. 106 agreements. The best known is the Milton Keynes 'roof tax' where developers are paying a standard charge of £18,500 per dwelling on a development of 1,400 houses. The authority say that only half of the funding created will go to pay for typical s. 106 type facilities. The rest will go to strategic projects including 'highways improvements and medical services' (*Planning* 1629: 22 July 2005: 2). This suggests that the 'roof tax' lies somewhere between planning obligations and a development land tax, and although apparently contrary to what has already been said about the government's view it has been endorsed by the ODPM.

A formal extension to obligations to recoup development land value through introduction of a planning gain supplement may prove more difficult. Certainly the surveying profession (Johnson and Hart 2005) could hardly be more critical of the idea, concluding that the planning gain supplement is 'based on a misunderstanding of how land is valued, how planning gains arise, and how the property market operates' (para. 5.2). This RICS report suggests that the supplement has advantages over previous attempts at taxing new development since it will tax site values and sales prices rather then 'development gain', but also points out serious flaws which would make it unworkable (para. 2.26–28). One major problem it claims is that the 'tax point' is when permission is granted but (as numerous studies have shown) this does not necessarily guarantee development proceeds and the gain made. The report is more positive about standard charges or planning tariffs to replace obligations. A similar system has been in apparently successful operation in Ireland since the reform of the Irish planning system in 2000. In that case variations on tariffs promote other planning objectives such as increased densities in certain locations.

Planning, affordable housing and housing supply

The planning system provides the government with an alternative means to deliver affordable housing alongside the traditional approach where housing associations use Social Housing Grant (SHG) and this method is becoming more important. Research sponsored by the Joseph Rowntree Foundation has found that the use of planning agreements became more important between 1999 and 2003 in a context of overall falls in the provision of affordable housing, although with expectations of rising numbers over future years following more consistent use of s. 106 (Monk *et al.* 2005). About half of all affordable homes are now provided by planning through s. 106 planning obligations and 'exceptions sites'. But the stance of central government on the role of planning obligations in relation to affordable housing has been a curious one for some time. On the one hand 'planning conditions and agreements cannot normally be used to impose restrictions on tenure, price or ownership', but 'they can properly be used to restrict the occupation of property to people falling within particular categories of need'. Both statements are from Circular 7/91 on *Planning and Affordable Housing* which was an early attempt to wrestle with this politically difficult issue. This was replaced by Circular 9/98 (with the same title) which repeats the warning against tenure conditions, and defines affordable housing in these terms:

> The terms 'affordable housing' or 'affordable homes' are used in this Circular to encompass both low-cost market and subsidised housing (irrespective of tenure, ownership whether exclusive or shared or financial arrangements) that will be available to people who cannot afford to rent or buy houses generally available on the open market.

This interpretation was criticised as quite inadequate on three grounds. First, it leads to the provision of small houses for sale at full market prices. Second, housing may be less expensive than other housing in a development, but not 'affordable' to local people. Third, on resale, houses are sold at full market prices, thus losing the benefit of any discount and also control over future occupants (Chartered Institute of Housing *et al.* 1999). Current policy now defines affordable housing as 'non-market housing, which can include social-rented housing and intermediate housing'; intermediate means housing 'at prices or rents above those of social rent but below market prices or rents'.

Though the Circular 'uses the language of voluntary provision and relies upon developers' contributions being secured through negotiation, neither the purpose nor the effect of its requirements is voluntary' (Grant 1999a: 71). Developers are expected to provide affordable housing on developments above a certain size; in PPG 3 the policy was twenty-five dwellings, or more than one hectare, except in Inner London, where the requirement related to fifteen dwellings or half a hectare; the proposed 2005 revision says not normally above fifteen dwellings or half a hectare, and lower where need cannot be met on larger sites alone. Where

a developer is unwilling to accept such a condition planning permission may be refused. This policy is seen as a means of catering for a range of housing needs and of encouraging the development of 'mixed and balanced communities in order to avoid areas of social exclusion'. It is to be noted that this policy has no specific legislative provision and, though this does not make it illegal, a developer has little chance of successfully opposing it. An appeal is hardly like to succeed when the principle is set out as ministerial policy. But developers have argued that reducing the threshold may affect the viability of some sites, and the wish to deliver development is at the root of some local authorities' unwillingness to be more strict on quotas of affordable housing.[3]

Even more curious is the policy of 'exceptional release' of land, outside the provisions of the development plan, for 'local needs' housing. This is an explicit 'use of the planning system to subsidise the provision of low cost housing through containment of land value'.[4] Until 2005 exception sites were only that – they were sites not allocated in the development plan (and therefore they had no development value above their existing, usually agricultural, value). Now local authorities can allocate small sites in the development plan (explained below). Exception sites are typically small parcels of land adjacent to a village and provided at existing use or even donated free by landowners. More than 900 sites were granted through the exceptions route in 2001–2.

The extent to which authorities can achieve planning benefits depends, of course, on their bargaining power, which in turn may be related to current (and local) economic conditions. The situation varies over time and by region. In some circumstances, 'getting a developer to build anything is, in our eyes, a planning gain' (quoted in Jowell 1977a: 428); in others, the local pressures for development are so strong that local authorities can secure considerable benefits, perhaps asking as much as 50 per cent social housing on sites with high development value. The London situation, however, is unique, with housing costs at record levels, and acute pressures on affordable housing.[5] Figures from Shelter show that a first buy house in 2004 was 60 per cent less affordable than it was in 1994. Average first time buyer prices have increased from £48,000 to £134,000. First time buyers in London are now paying on average 23 per cent of their income as mortgage interest payments, whereas in 1994 it was 13 per cent. In contrast, average social housing rents have stayed fairly constant at about 13 per cent and 15 per cent of average male earnings or £51 per week in the South East and £41 in the North East.[6] The problem will worsen as the number of households in London continues to increase (by 600,000 between 1996 and 2021 according to the latest household projections) and new provision remains low, though action is being taken as explained in the later section on the *Sustainable Communities Plan*. The affordability issue has brought planning to a wide audience, with stories such as Dave Gilmour of Pink Floyd selling a London mansion for £4 million and donating the proceeds to Shelter for a housing centre for key workers and homeless people.[7] (See also Box 6.4 for a more usual case.) Local communities have also sought protection from better-off incomers. A (perhaps surprisingly) little used policy is to reserve new housing for local people. This is particularly appropriate in areas where there is great competition for housing on the part of commuters or holiday home buyers. One example is where new houses are restricted to locals in the North York Moors National Park.

A review of *Delivering Affordable Housing through Planning Policy* (Entec 2002) is critical of general practice, found 'few examples of good practice' and much variability in policy. Local authorities do not have a clear definition of what constitutes 'affordability' in their locality, they tend to develop policy for administrative areas rather than housing market areas, they may be unwilling to set low thresholds, especially in rural areas, and developers, understandably, are organizing applications so as to avoid liability for providing a quota of affordable homes.

An argument has been played out for many years about the value of designating sites specifically for social or affordable housing, this might stigmatise particular locations and reproduce concentrations of less well-off households. Until 2005, local authorities were restricted to indicating types and densities of housing, and PPG 3 said 'it would be inappropriate for policies to identify particular sites and allocate them for affordable housing'. An interim PPG 3 *Housing Update* published early in 2005 reversed this advice, at least for small sites, providing an additional tool for planning to assist in the provision of affordable housing. The new guidance says that development plan documents may 'allocate sites solely for affordable housing' (p. 2). The Update also gives much more emphasis to the requirement for local planning

BOX 6.4 LACK OF LOW COST HOUSING LEADS TO DISMISSAL OF APPEAL

McCarthy and Stone proposed to develop a brownfield site in Stockport with two blocks of sheltered flats for elderly owner occupiers. The development would have met a market demand and would have improved the character and appearance of a derelict site. However, it was rejected on appeal because the scheme did not provide any low cost housing. The inspector said that he considered that the failure to make provision for an element of low cost housing on what is a suitable site would be so harmful as to amount to a compelling planning objection. He also maintained that the provision of affordable housing would not render the development cost of the flats uneconomic. Moreover, there was no 'convincing evidence that the development's success would be jeopardised because of any incompatibility between affordable housing and sheltered housing for the elderly'.

Source: Planning (17 September 1999: 9)

authorities 'to make sufficient land available either within or adjoining existing rural communities to enable . . . local requirements to be met' (p. 1).

These were among proposals for revising PPG 3 and Circular 6/98 in the 2003 consultation document *Influencing the Size, Type and Affordability of Housing* and were brought forward while discussion continued on the main changes. A second set of proposals for revising PPG 3 were published early in 2005. The new emphasis of policy is reflected in the revised title of the proposed PPS 3, *Planning for Mixed Communities*. It requires more from local planning authorities in the analysis of housing needs through a local housing assessment, and more attention to enabling a variety of housing types that is more likely to result in mixed communities. It is accepted that the notion of affordability is going to vary across the country and thus local authorities are expected to make an assessment of local affordability taking into account availability and incomes, and working across boundaries on actual housing market areas. There is also more attention to the role of the regional strategy in identifying local housing markets and providing a framework for local action on housing alongside the regional housing strategy.

Compulsory purchase

Local authorities and other bodies have long held considerable powers of compulsory acquisition of land, which have been amended incrementally over more than 150 years such that, according to one judge dealing with the subject, 'UK law on the subject is bogged down in complexity and obscurity'.[8] The 'enabling powers' for compulsory purchase are widely spread among public bodies; we concentrate here on clarifying some of the basic points on compulsory purchase orders (CPOs) made under planning powers.

The compulsory purchase regime has been under review for some time, and continues to be. The reasons were spelled out in an *Interim Report* of the review,[9] noting the perception that the process 'is slow in operation, inefficient, and not always fair to those whose property is acquired' (p. 7). As a result, compulsory purchase is now less often used than local authorities would like to, and is therefore not the aid to urban regeneration that it might be (Freilich 1999). The Urban Task Force (1999f) recognised the problems of land assembly and Adams *et al.* (2002) systematically review the numerous types of 'ownership constraints' that can impede progress on urban regeneration including 'ransom strips' where owners (including the public sector) make unreasonable demands for critical parcels of land needed to release larger sites; multiple ownerships (the most significant barrier) where owners of one site need to purchase adjacent land to complete the development; or just 'owner apathy' where the land is retained but for no specific purpose. In sixty-four out of eighty sites examined in their research, ownership constraints 'disrupted plans to use, market, develop or purchase the land' (p. 214). Drawing on Lichfield and Darin-Drabkin (1980), they call for a more imaginative public land policy: 'although compulsory purchase has dominated British thinking on so much of developmental land policy, despite its evident faults and limited potential to deliver rapid regeneration, other countries have more varied forms of land assembly, including land readjustment or re-parcelling.'[10]

For the time being at least, land policy will continue to focus on compulsory acquisition powers. They have been used widely for land assembly, and the redevelopment of many towns from the 1950s to the 1990s would have been impossible without them, though many of the results are hardly an advertisement for increasing their use now. Ward (2004) explains how expansion of compulsory purchase powers in the 1940s eased land assembly powers for comprehensive redevelopment of town and city centres. The close and then mostly private relations (partnerships) between the planning authority and developers would not be tolerated nowadays. There is also more emphasis now on protecting the rights of existing land and property owners, especially through the Human Rights Act (Redman 1999).

The 2001 *Planning* Green Paper included proposals for compulsory purchase which were taken forward in *Compulsory Purchase Powers, Procedures and Compensation: the Way Forward*. Circular 02/03 was published clarifying CPO powers and procedures and was superseded by Circular 06/04 *Compulsory Purchase and the Crichel Down Rules*,[11] after substantial changes were introduced by the Planning and Compulsory Purchase Act 2004. The 2004 Act amended s. 226 of the Town and Country Planning Act 1990, which provides powers for compulsory acquisition for planning purposes. Local authorities, joint planning boards and national park authorities can acquire land for the purpose of development, redevelopment or improvement if they think that

- the acquisition will facilitate the carrying out of development, redevelopment or improvement on, or in relation to, that land; and
- the development, redevelopment or improvement is likely to contribute to the promotion or improvement of the economic, social or environmental well-being of their areas.[12]

The revision is similar to the preceding purpose, but removes the purpose 'in the interests of the proper planning of an area and provides a more meaningful statement under the second point. The government had recognised in its 2002 consultation the need for a clarification of the powers to enable authorities to use them for 'a full range of planning and regeneration purposes'. The powers can be exercised not only for the authorities' own development but also to facilitate private development and for disposal to a private developer. Indeed, the government has made it clear that these 'planning purposes' powers (which could be of particular importance in bringing land on to the market) were generally to be used to assist the private sector. But

> a compulsory purchase order should only be made where there is a compelling case in the public interest . . . [and it must] sufficiently justify interfering with the human rights of those with an interest in the land affected . . . The more comprehensive the justification which the acquiring authority can present, the stronger its case is likely to be. But each case has to be considered on its own merits.
>
> (Circular 6/04, paras 17 and 18)

The authority will need to show that there are not likely to be 'impediments' to implementation of proposals for the land, including the grant of planning permissions (and this applies to all types of compulsory purchase, not just those made on planning grounds). On this, the development plan is an important consideration and the 2004 Act requires the scheme to be in accordance with the development plan unless material considerations indicate otherwise. So there is some discretion as for consideration of planning applications. However, where the scheme is not in accordance with the development plan, the government expect that the authority will have prepared supplementary guidance covering the proposals. It will be particularly important that the authority has consulted on this, so as to provide an opportunity for those affected by the compulsory acquisition to comment on the planning proposals and for the authority to consider how it affects their rights.

Compensation is payable 'on the principle that the owner should be paid neither less nor more than his loss'. So compensation is what the land might fetch on the open market (the open market value) together with a payment for 'severance'. The 2004 changes also introduced a component of compensation for 'disturbance and other losses not directly based on the value of the land', in the interests of making the procedures operate more fairly. The procedure operates in two stages, the making of the order and then its confirmation by the Secretary of State, and the provisions for objections and consideration of objections are similar to those for dealing with planning appeals involving an inquiry or, if all parties agree, written representations.

Brownfield, vacant and derelict land

It was a major objective of the postwar planning system to ensure that land required for development would become available – if necessary by the use of compulsory purchase powers. As previous discussion has shown, things did not work out like this despite three attempts (in 1947, 1967 and 1975). Except in special cases, such as new towns and comprehensive development areas, there has been little use of compulsory purchase powers. Thus the land 'allocations' in plans remained just that – allocations on paper. There is no necessary relationship between the allocation of land and its *availability*. It is therefore not surprising that there has been considerable controversy over the extent to which allocated land is in fact available for development. In Hooper's (1980) words: 'The planning system and the house building industry operate not only with a different definition, but with a different conception, of land availability – the former based on public control over land use, the latter on market orientation to the ownership of land'. However, land availability studies were the centre-pin of the planning system until they were supplanted by urban capacity studies.

This new system, introduced by the revised PPG 3 *Housing* (2000) represents a major change in policy. It places emphasis on the reuse of land in urban areas, and generally 'the compact city'. The town or city is the favoured location for new development in view of its assumed 'sustainability'. This is interpreted in various ways: it is held that urban locations reduce traffic (and emissions) and help to safeguard the countryside; they provide accessibility to goods and services, and allow new energy-saving technologies such as combined heat and power systems; and they provide a more lively and interactive social milieu.[13] The policy has widespread popular support, particularly in terms of 'saving' the countryside. Eloquent of this is the Select Committee's forthright declaration that 'the only way that the government's proposals for urban regeneration and for greater use of recycled land can be achieved are by restricting the amount of greenfield land brought forward'. This has been backed up by a new *Greenfield Housing Direction* (2000) which requires local authorities to consult the Secretary of State on planning applications for major housing developments of more than 5 hectares or 150 dwellings (discussed in Chapter 5).

The same commitment is evident in changes in government policy, above all the commitment to brownfield development, or in the government's terms: 'to maximising the re-use of *previously developed land*' (PDL) and the conversions of buildings for housing in

order both to promote regeneration and to minimise the amount of greenfield land being taken for development. This policy permeates the revised PPG3 in which the policy is spelled out in some detail. Potential sites should be assessed against a number of criteria such as the availability and net cost of previously developed sites, their location and accessibility by public transport, the capacity of the infrastructure and services such as schools and hospitals, and the potential for developing and sustaining local services, and physical constraints on development.

A sequential approach to the phasing of sites is introduced under which greenfield sites should not be developed for housing until the following options have been considered:

- using previously developed sites within urban areas
- exploiting fully the potential for the better use and conversion of existing dwellings and non-residential properties
- increasing densities of development in existing centres
- releasing land held for alternatives uses, such as employment
- identifying areas where, through land assembly, area-wide redevelopment can be promoted.

The urban capacity studies (mentioned earlier) are promoted to aid in this exercise, which should take account of the National Land Use Database and examine the implications of policies for increasing densities, reducing car parking, and reviewing the potential over-allocation of land for employment as the principal means for determining the location of potential housing sites.[14] Regional spatial strategies should also make use of capacity studies in proposing land recycling targets and allocating them among planning authorities.

The government's target is for 60 per cent of new housing to be provided on previously-developed land or through conversions. The Urban Task Force promoted this target and devised its own estimates for the various types of recyclable land and also of the number of dwellings that are likely to be accommodated on this land under current policies. At the time it was argued that these estimates are heavily influenced by wishful thinking, particularly since they are almost three times the estimate of the National Land Use Database. They are shown in Table 6.1. On a range of assumptions, the Task Force (p. 305) reached an 'attainable target' of 62.2 per cent. This proportion was regarded by some (such as the TCPA) as over-ambitious, and by others (such as the CPRE) as too low.[15] It was also argued that rates might decline, because increases in using recycled land for housing arose from the use of vacant brownfield sites which had been relatively easy to deal with (Llewelyn-Davies 1996). Certainly, the actual figures calculated by the Task Force are debatable; nevertheless it maintains that they show that 'over a significant period, the cumulative effect of a consistent and continued policy commitment could be considerable'. In one sense therefore, their detailed calculations are of less import than the message they tried to convey, captured by the title of the report: *Towards an Urban Renaissance*.

The actual figure for 1998 was 55 per cent or 53 per cent excluding conversions, and by 2003 this had reached 67 per cent (see Table 6.2). The proportion varies considerably among regions, from 53 per cent in the East Midlands to 93 per cent in London. The comparable figures for 1996 were 54, 37 and 82 per cent. At the county level the highest rates in 2003 were in Merseyside with 95, Surrey 92 and Berkshire 89 per cent; the lowest were in Humberside 26 per cent and Cornwall and the Isles of Scilly 27 per cent. The National Land Use Database (NLUD) 2003 Survey found that about 66,000 hectares of brownfield land could be readily available for development. The various categories of previously developed land, either vacant or suitable for redevelopment, are show in Table 6.3.

The reliability of these figures has been questioned and further research undertaken on the discrepancies between the Land Use Change Statistics (LUCS), produced by the Ordnance Survey, and the returns made by local authorities for the Land Use Database (Roger Tym and Partners 2004). As an illustration of the difficulties, there were two quite different figures for the amount of derelict land: 34,500 hectares according to the Derelict Land Survey, and 17,300 hectares recorded in the National Land Use Database. Findings from the Roger Tym study (2004: 6) confirm the problem and provide a sober warning to those who are enthusiastic about quantitative targets:

> To a large extent, LUCS and local authority statistics for any one year do not cover the same collections of sites and dwellings. This is partly due to timing differences; there are also many permanent errors of omission and miscounting, which mean that the two sources do not cover the same sites and dwellings at all. Two data sets which aim to measure the same thing may diverge more than they overlap in the sites and dwellings they cover in any single year.

The English Partnerships (2003) *Towards a National Brownfield Strategy* noted that *regional brownfield action*

Table 6.1 *Estimated number of houses likely to be built on previously developed land, England 1996–2021*

Existing stock of land	
Vacant previously developed	173,729
Derelict land/buildings	152,000
Existing stock of buildings	101,000
Projected windfall and other sources (1996–2021)	1,526,000
Total	1,952,729

Source: Urban Task Force (1999f: 305). Updated by Government Statistical Bulletin 500, *National Land Use Database*, which gives a much less optimistic forecast for the reuse of existing vacant buildings (101,800 rather than 247,000). The period 1996–2021 is that used for the current household projections.

Table 6.2 *Percentage of new dwellings built on previously developed land by region 1996–2004*

	1996	1997	1998	1999	2000	2001	2002	2003	2004	Future target
North East	53	46	50	40	47	45	56	53	60	60
North West	65	58	62	60	67	70	72	72	77	70
Yorkshire and the Humber	51	49	51	50	57	55	63	64	72	60
East Midlands	37	37	35	37	43	48	54	53	52	60
West Midlands	50	56	54	54	55	60	67	69	71	65
East of England	53	53	54	58	53	58	58	59	60	n/a
London	82	89	92	89	89	90	91	94	94	n/a
South East	56	54	56	61	62	65	66	63	71	n/a
South West	35	34	37	39	44	48	49	58	57	50
England	54	53	55	56	58	60	64	65	68	

Source: ODPM (2004) *Land Use Change in England: Residential Development to 2004: Update July 2005*. Future targets are as given in regional spatial strategies, some were not finalised for this Bulletin.

Table 6.3 *Previously developed land unused or available for development in England 2003*

	All previously developed land	Suitable for housing development	
	Area (ha)	Area (ha)	Number of homes
Vacant and derelict land and buildings			
Vacant land	14,610	5,820	189,700
Vacant buildings	4,550	2,670	121,600
Derelict land and buildings	20,550	6,580	167,300
All vacant and derelict land	39,710	15,070	478,600
Currently in use			
Allocated in development plan or with planning permission for any use	17,580	9,790	296,500
No allocation or planning permission but known redevelopment potential	8,470	4,630	174,700
All currently in use	26,050	14,410	471,200
All land types	65,760	29,480	949,800

Source: ODPM (2004) *Previously Developed Land That May be Available* (Table 1, p. 7)

plans were to be prepared by the regional development agencies to speed up the delivery of development on previously developed land. It seems that the other parts of the public sector are major culprits in holding back development and most effective use of brownfield land. The Development Director of English Partnerships said in 2003:

> far too often, the first that English Partnerships and RDA or local authority learns that a surplus public sector property is becoming available for redevelopment is when they see an advert in the paper with 28 days to make an offer.[16]

The Brownfield Strategy seeks to join up public action a little better, and an Assets Register detailing surplus public sector land and property is in preparation, but the government will also need to make more progress on getting the objectives of many public sector organisations into conformity with government's sustainability and planning objectives. And that is only the start of the story.

Once identified and acquired, brownfield land will almost certainly require restoration or other improvements such as access or removal of old uses. Policies on the reclamation of derelict and contaminated land predate the debate on the increased use of 'brownfield' by many years. Much of the early land restoration work was done to remove eyesores and potential dangers caused by spoil heaps and other waste, and to return the land to agriculture or forestry, or to make it available for public open space (known in the jargon as a 'soft end use'). The current preoccupation with getting housing and other 'hard end uses' on such sites has accelerated the programmes of remediation. As noted above, extensive tracts of land that was once useful and productive has become waste land, particularly in the inner cities and in mining areas. It is unsightly, unwanted and, at worst, derelict and dangerous. The planning system is not designed to deal with such land easily: its essential characteristic is to allocate land between competing uses. Where there are no pressures for development, there is a severe limit to what can be done, especially when the amount of waste land is large, as it is in older industrial and coalfield areas.

Major efforts have been made to deal with the problems. Between 1988 and 1993, some 9,500 hectares of derelict land were reclaimed, but a large amount of new dereliction is continually being created as older primary industries close and the total amount has remained high. Though the amount of derelict land in England decreased by 2 per cent between 1988 and 1993, the 1993 Derelict Land Survey showed a total of 39,600 hectares of derelict land at the latter date. Unfortunately, this was the last year that the survey was undertaken. The latest figure, from the National Land Use Database in 2003, is 20,550 hectares in England (17,000 hectares in 1999) and 14,610 hectares of vacant land (16,000 hectares in 1999).[17] This is an area about twice the size of the city of Glasgow.

A range of policy instruments to deal with derelict land had been developed. Some of these have been part of broader policies in relation to urban regeneration (through urban development corporations, enterprise zones and the Urban Programme).

Vacant land is conceptually different from derelict land, though the two categories can overlap.[18] Research since the early 1990s showed that though vacancy may be a transient feature of the environment, two-thirds of a sample of sites had been vacant for more than twelve years. The research pointed to the barriers to further use of the land, which successive policy initiatives have subsequently sought to address. About two-thirds of sites remained vacant because of institutional factors, owners' intentions or poor demand. As the evaluation study explained:

> Many sites remain vacant for non-physical reasons. Some are delayed by the legitimate workings of the planning system, and by legal and other institutional difficulties. Existing policy instruments can do little to overcome these difficulties. Others are delayed by owners', particularly private sector owners', intentions that they should remain vacant for various (largely obscure) reasons.
>
> (Whitbread *et al.* 1991, para. 3.147)

An earlier report suggested a long list of reasons why vacant land is not put to temporary uses: expenditure by the owner in meeting fire, safety and insurance requirements, in providing access, and in site clearance; temporary tenants tend to be unreliable and to cause environmental problems; demand from temporary users is deficient and uncertain, and often provides landowners with a very low financial return; there are often problems in securing vacant possession; landowners may be unaware of the potential of temporary uses; or they may think that keeping sites vacant preserves existing use rights, or puts pressure on local authorities to grant planning consent for development (Cameron *et al.* 1988). Much of this land is in private ownership and, at the time, it was thought that short of compulsory acquisition, there was little that 'policy' could do to speed up the reuse of the land.[19]

One specific category of vacant land is redundant military land. The 'peace dividend' following the demise of communism in Russia and certain European states has generated significant amounts of land in the Ministry of Defence (MoD) Estate, as well as closed US Air Force bases in the UK.[20] Much of this is located in rural areas or economically depressed urban areas (including former naval dockyards). Disposal of MoD land is subject to the Crichel Down rules requiring surplus land to be offered back to the original owners at current market values. Such values will take into account the condition of the site, which may be contaminated. Many sites also contain listed buildings, monuments, and environmental and landscape designations that will affect reuse. The MoD is required to maximise income from disposals, which has often led to proposals for new housing, irrespective of location.

Contaminated land

There is no clear line between vacant, derelict and contaminated land (or neglected, underused, waste and despoiled land). It is all previously developed land. The terms are used in different ways, sometimes for different purposes, sometimes with the same or similar meanings (and new terms arise from time to time, such as brownfield and recycled land). Contaminated land is particularly difficult to define, though the term is commonly used to imply the existence of a hazard to public health. The Environment Act 1995 introduced

a statutory definition which incorporates this long-standing idea (shown in Box 6.5). Though there is an overlap with 'derelict' land, there are important differences. A chemical waste tip may be both derelict and contaminated; a disused chalk quarry may be derelict but not contaminated; an active chemical factory may be contaminated but not derelict. Previously developed and vacant land may not be contaminated. It is the additional health danger which is the characteristic feature of contaminated land, and this also implies a severe degree of pollution and, typically, an increased difficulty in abating it. However, the health risk arises only in relation to the use to which the land is to be put. A piece of land may pose no risk if used for one purpose, but a severe risk if it is used for another. The site of an oil refinery may be contaminated, but that is of no consequence if no other use is intended (and assuming that there are no effects beyond the site). 'A scrap yard contaminated by metal traces would constitute a hazard for subsequent agricultural use, but the contamination would be of no account in the construction of an office block'.[21]

Partly because of a characteristically pragmatic approach, there has never been an attempt to quantify the amount of contaminated land in Britain. Instead of identifying contaminated land and then determining appropriate policies for dealing with it, the British approach has been to regard contamination as a general concept which is given substance only in relation to particular sites and particular end uses. The nature of policy flows from this: policy is to ensure that the quality of land is fit for the purpose to which it is being or will be used. There is no requirement for land to be brought up to a minimum quality standard regardless of use, unless that land poses a threat to the public health or the environment. The House of Commons Environment Committee considered this approach to be inadequate since (in its judgement) there is land which is so contaminated that it is 'a threat to health and the environment both on site and in the surrounding area'. The Committee also recommended that local authorities should be given a duty 'to seek out and compile registers of contaminated land'.

There was a remarkably swift response to this: the Environment Protection Bill was amended to provide

BOX 6.5 LAND DEFINITIONS

Brownfield land or previously developed land (PDL)

Previously developed land is that which is or was occupied by a permanent structure and associated fixed surface infrastructure [in] built-up and rural settings. The definition includes defence buildings and land used for mineral extraction and waste disposal where provision for restoration has not been made through development control procedures. The definition excludes land and buildings that are currently in use for agricultural or forestry purposes, and land in built-up areas which has not been developed previously (e.g. parks) and land that was previously developed but where the remains have blended into the landscape in the process of time (to the extent that it can reasonably be considered as part of the natural surroundings).

Source: Paraphrased from PPG 3 (2000) and used by NLUD

Greenfield land

Any land outside the above definition.

Vacant land

Land that was previously developed and is now vacant which could be developed without treatment (see below for definition of treatment). Land previously used for mineral extraction or waste disposal which has been or is being restored for agriculture, forestry, woodland or other open countryside use is excluded.

Vacant buildings

Unoccupied for one year or more, that are structurally sound and in a reasonable state of repair (i.e. capable of being occupied in their present state). Includes buildings that have been declared redundant or where re-letting for their former use is not expected. Includes single residential dwellings where they could reasonably be developed or converted into ten or more dwellings.

Derelict land

Land so damaged by previous industrial or other development that it is incapable of beneficial use without treatment, [which may include] demolition, clearing of fixed structures or foundations and levelling. Includes abandoned and unoccupied buildings . . . in an advanced state of disrepair . . . Excludes land . . . which has been or is being restored for agriculture, forestry, woodland or other open countryside use [and] land damaged by a previous development where the remains of any structure or activity have blended into the landscape in the process of time.

Source: NLUD Data Specification (www.nlud.org.uk)

Contaminated land

Statutory definition

Any land which appears to the local authority . . . to be in such a condition, by reason of substances in, on, or under the land that:

- significant harm is being caused or where there is a significant possibility of such harm being caused, or
- pollution of controlled waters is being, or is likely to be caused.

Source: Environmental Protection Act 1990 (quoted in PPS23 Annex 2, para. 2)

Definition for planning policy

Where the actual or suspected presence of substances in, on or under the land may cause risks to people, property, human activities or the environment, regardless of whether or not the land meets the statutory definition.

Source: PPS 23 Annex 2 (paras 2.5 and 2.13)

such a duty. The implementation of this, however, rapidly ran into severe difficulties, and the initial proposals had to be drastically changed.[22] The problem underlying all this is that it is relatively simple to register land that is *possibly* contaminated, but extremely laborious and costly to identify land that is in fact contaminated. Even at the low rate of £15,000 per hectare, it would cost around £600 million merely to investigate the 40,000 hectares of land identified in the 1988 Derelict Land Survey (Thompson 1992: 22). To cover all relevant land would cost many times this amount, and would take many years to complete.[23]

It was because of difficulties such as these that the government was eventually forced to abandon the scheme as originally envisaged. The difficulties of changing from the traditional British reactive approach to a genuinely proactive approach are manifest (Harrison, A. 1992: 809). However, a renewed attempt was made in 2000. The revised provisions are set out in the long and complex Circular 02/2000.[24] It will take some time for the implications of the new regime to be understood.

Suffice it to say at this stage that they will create a regime to enforce remediation on certain contaminated sites where there is a serious degree of health or environmental risk and where there is a justification for requiring compulsory remediation. Compulsory remediation is a drastic remedy, and these powers are likely to be used only in limited circumstances where voluntary remediation, e.g. in the course of redevelopment (or otherwise) is unlikely to occur. It is therefore a legislative supplement to the planning and development control process which is likely to continue to govern the overwhelming majority of remediation of contaminated sites.

(Winter 1998: 10)

Local authorities are required to prepare and implement a strategy for identifying land falling within the statutory definition (Box 6.5) 'contaminated' and to require its remediation. This means inspection of sites, identification of responsibilities for remediation and monitoring – which is generally a job for the environmental health profession. The Environment Agency will also monitor implementation and provide advice on specific problems, especially those related to water pollution. The planning system has to address contaminated land when making plans (at both regional and local levels) and when receiving proposals for development on land that may be subject to contamination. Given the complexities, Annex 2 to PPS 23 has provided a broader definition for planners of 'land affected by contamination'. In order to avoid blighting land only sites that are identified as contaminated land, and where the local authority is taking action, will be listed on registers.

In 1994 English Partnerships came into operation to implement 'a new approach to vacant land' which includes unused, under-used or ineffectively used urban land, land which is contaminated, derelict, neglected or unsightly, or land which is likely to be affected by subsidence. The work of English Partnerships is discussed in Chapter 10. Most of the work of English Partnerships is geared towards economic benefits of land restoration. To complement this activity, it cooperated in 2003 with Groundwork, the Forestry Commission and Environment Agency to create an independent *Land Restoration Trust*. The Trust is modelled on the National Trust and will work with local partners, including communities, to restore derelict land to create new green amenities for the benefit of local people. This is a return to consideration for 'soft end uses' and will be welcomed, because many sites officially defined as derelict become important green spaces and informal recreation places, such as the Kirkstall Valley in Leeds.

Increasing densities

Alongside brownfield development there has been much advocacy of increasing the density of development in cities, especially around transport nodes. This compact city idea is promoted because it provides opportunities for more effective public transport and increased cycling and walking, sharing of resources, including local energy production; it may reduce development of greenfield sites, and it may provide more social interaction through local provision of services. This theory may be difficult to realise in practice: urban life is often characterised by traffic congestion, poor environmental quality and 'town cramming' (Williams, K. 1999: 169). At what point do higher urban densities give rise to cramming? There is, of course, no mathematical answer to the question, though there is an abundant literature on the issue. Aspects of design are often of greater significance, as are even more elusive elements of 'character'. But most

important is the very *richness of cities*, so well captured in the 1999 report with this title by Worpole and Greenhalgh. This richness is created in complex ways; increased density alone is not sufficient to deliver it, although it may be a necessary condition. The potential benefits of more sustainable and 'liveable cities' are only part of the equation; the government is also exercised by the need to deliver more housing, especially in the South East, while not encroaching on the green belts. This is such an important issue for government (mainly because of the need to feed the economy) that the Secretary of State needs to be informed on applications for development falling below thirty homes per hectare (explained in Chapter 5). The Strategic Plan for London has also been at the forefront of promoting increased densities, both in the City and in the new growth areas.

What is conspicuously missing from much of the debate is the question of the acceptability of increased densities or 'urban intensification'. It is not easy to measure this in any straightforward way since the term is capable of varying interpretations, but a good proxy is provided by the findings of the authoritative DoE sponsored *Housing Attitudes Survey* (Hedges and Clemens 1994). This showed 'central urban dwellers' to be much less satisfied than those in the suburbs, and these again less than those in rural areas. This finding is reinforced by the analysis of population density 'which shows a marked inverse relationship between satisfaction and density'.[25] The survey also showed a clear preference for houses rather than flats. This can hardly be surprising since this has been a consistent finding of housing research, but the issue has gained prominence in view of the very large increase in one-person households shown in the household projections (who make up three-quarters of the total increase). Although it may seem reasonable to assume that many of these will want small dwellings, possibly in flats, the evidence is that the greater part of the demand is for houses with gardens (Hooper *et al.* 1998). 'A preference for a flat starts at 11 per cent, falls to 1 per cent as the family grows, and then climbs to 31 per cent among single older people' (Hedges and Clemens 1994). The 1996 White Paper, *Household Growth: Where Shall We Live?*, concluded that, despite the increase in small households, 'there is little evidence of any increase in demand for smaller housing units; there has, moreover, been a decline in one bedroom houses and flats completed in the last ten years, and a growth in the number of larger houses (four bedrooms)'.

Even more persuasive is the fact of long-term decentralisation from the cities. This has eased the traditional problems of cities, though it has proved difficult to attune policies to the problems which remain. Movement out of the cities has been a dominant feature of demographic and economic geography for a century. (However, it should be stressed that the arithmetic of this is usually expressed in net terms, ignoring the fact that people are moving *into* as well as out of urban areas.)

Much of the debate on the urban renaissance is couched in terms of redevelopment of the inner city, ignoring the problems and opportunities of the suburbs (where the majority of people live and where much development activity has been concentrated during the 1990s). The suburbs do not typically need large-scale redevelopment plans but, as a Civic Trust study shows, they can be in need of careful improvement to arrest decline and to enable them 'to play a more positive and sustainable role within city regions' (Gwilliam *et al.* 1998).

There are also a variety of measures that can improve both suburban and inner city environments while, at the same time, providing additional housing. Policies in relation to empty properties can clearly make a modest but useful contribution to both, as the work of the Empty Homes Agency (Plank 1998) demonstrates.[26] The LOTS (living over the shop) scheme was less successful largely because of the lengthy and often difficult negotiations required with the owners of the shops! It does, however, have potential when included as an element of wider-based regeneration schemes.[27] In addition to housing issues, a relatively neglected matter is that of the geography of jobs. Though there are no figures on this for recent years, employment in the 1980s showed an employment exodus from urban areas. Patterns of commuting have become more complex, and there is now suburb-to-suburb and even city-to-suburb commuting. As more housing is provided in the cities (often involving the replacement of places of employment), will reverse commuting grow? Does this matter?

The brownfield target and increasing densities assist in promoting better use of redundant urban land and they deflect constant criticisms of greenfield development but attention also needs to be paid to providing effective planning tools for the assembly and delivery of urban land, and they should not be seen as a replacement for a fuller analysis of the sustainability of urban development.

Household projections

Demographic analysis and forecasting are crucial to any method of determining housing needs and land requirements. Projections of households are made on a periodic basis by the Government Statistical Service. Until quite recently, these were widely accepted as a basis for policy. The national figures are used by the central department to determine regional and county housing requirements. Concern about these projections grew in the 1980s, particularly in the South East 'where years of continuous housing development have generated a militant resistance to what are seen as excessive impositions of yet further housing development' (Breheny 1997). The publication of the household projections for the period 1991–2016, published in 1995, gave rise to an even more vociferous and wider debate, which was kept informed by a series of CPRE publications that strenuously put forward the case both for protecting the countryside against housing development and for disparaging the methodology used in the official household projections (see Box 6.6).[28] The press also took up the popular outcry and led a 'greenfield campaign'. There were sufficient legitimate grounds of criticism in their arguments for them to be credible, particularly among those who were convinced of their conclusions. A DETR research project reinforced some of these criticisms; for example that the projections 'extrapolate forward past trends in a technically complex way, but take limited account of the underlying causal processes or relationships that might affect the rates at which households form' (Bramley *et al.* 1997).

BOX 6.6 HOUSEHOLD PROJECTIONS FOR ENGLAND TO 2016

The number of households in England is projected to grow from 20.2 million in 1996 to about 24.0 million in 2021, an increase of 3.8 million or about 150,000 households per year. Slightly more than three-quarters of the projected increase in the number of households can be attributed to changes in the size and age structure of the adult population.

The South East, East of England and the South West are all projected to have around a quarter more households in 2021 than in 1996. For London and East Midlands growth is around a fifth, and in other areas projected growth is significantly lower. The North East has the lowest projected growth of just 8 per cent.

If international migration increased or decreased by 40 thousand per annum over the projected period, this could mean a projected change at national level at 2021 of over 0.4 million households. Similarly, if real interest rates throughout the period were one percentage point higher or lower, the projected number of households in 2021 could change by 0.2 million.

Source: DETR (1999) *Projections of Households in England to 2021* (selected passages from pp. 5–6)

The official household projections are now more than a technical input to the planning system: they are matters of widespread controversy. Particularly attractive to critics are two points. First, that no projections are wholly satisfactory, and second, that housing supply does have *some* effect on household formation. This latter point is popularly viewed in terms which are similar to the now accepted argument that new roads generate traffic. Thus the shortcomings of 'predict and provide' which were seen to be valid in relation to roads were translated into housing terms: more houses lead to more households in the same manner as more roads lead to more traffic. This appealing (and greatly exaggerated) argument has been taken over in the government's redesign of the arrangements for determining house-building needs at regional and local authority level (DETR 1998, *Planning for the Communities of the Future*). In place of 'predict and provide' there is now 'plan, monitor and manage'. This is a neat piece of political semantics, which appears to mean that both the assessment of housing requirements and its distribution within the region should be kept under review, and if there are signs of either under- or over-provision both RPG and development plans should be reviewed accordingly.[29] The 2003 study on

Delivering Planning Policy for Housing found that local authorities were making changes in line with government policy, in that the household projections were being used alongside other factors to assess the housing land requirement.[30]

More fundamentally, the arguments about household predictions reflect a widespread opposition to change. This is a compound of a desire to maintain existing amenities, fears of increased traffic and congestion, and the traditionally strong countryside preservation ethic. They do not take sufficient note of the qualitative aspects of housing development, the various components of change and the forces which create them, and their relationship with jobs, services and other aspects of quality of life (Daniels 2001). The battles over the latest regional planning guidance amply illustrate this. They are dressed up in emotive and vague slogans which confuse the issues. Thus the draft RPG for the South East states that 'the countryside should be more strongly protected from inappropriate development',[31] but, as the Panel report on the public examination pointed out, though this 'sounds incontrovertible at first blush, the use of the term inappropriate without qualification begs the question of what is inappropriate'. The report continues:

All too often we found that it simply meant any form of urban expansion, particularly for house building. While it must be an objective to minimise the loss of countryside to urban expansion, we do not consider that this one objective should dominate all others. It should not result in denying the opportunity of a decent home for all who desire one in the region, nor should it stand in the way of economic success, nor – and we see this as a particular danger – should it compromise real urban renaissance by providing an excuse for town cramming . . . If urban concentration is forced upon towns for reason of preserving countryside and without due balance of the other elements of urban renaissance, then the cities and towns will simply become worse places to live in, and the pressures on the countryside will be unnecessarily increased.[32]

This critique of the anti-development stance taken by the London and South East Regional Planning Conference (SERPLAN) and the local authorities in the South East permeates the Panel report, which describes the process as 'short-term incremental decisions of planning to meet need as and when it arises'. (The Draft RPG suggested a baseline housing provision of 862,000 dwellings between 1991 and 2016; the panel proposed a figure of 1.1 million.) In the panel's view this is 'the antithesis of a plan-led system'. The essence of planning lies in taking a view of what is likely to happen in the future and planning to meet it. It continues: the approach

will serve only to perpetuate planning by appeal resulting on the ground in disjointed increments of added on development in apparently random locations with little coherence to the established structures of towns nor genuine opportunities for their development to be accompanied by planning extension of public transport and other infrastructure. This is not a sustainable way to meet development needs, and it is hardly surprising that it attracts so much opposition from local people when it occurs.[33]

The SERPLAN strategy typifies much of current planning for development needs. It reflects public opinion, and presents a major problem for central government. It is difficult to see how responsible planning at the regional level can be squared with planning which is responsive to public opinion. An additional problem is the effect of migration and travel across regional boundaries, especially in the South East, where the 'growth areas' are on borders between this and other regions. The East Midlands, for example, has been simultaneously experiencing development pressures from the South East, West Midlands, Greater Manchester and South Yorkshire. Much of the opposition to the household projections is concentrated in the South. This is, at least in part, due to the fact that migration from the North to this region has markedly increased housing demand in this part of England. Indeed, 'the speed of migration appears to have significantly increased with the upturn in the economy since 1993'.[34] The government's concern for intra-regional policy is not matched by its action on inter-regional issues.[35] Indeed, the essential remit of the regional development agencies is the fostering of regional economic development, and accordingly this is being fostered in the South East as in the other regions.[36]

There is also a marked movement out of the urban areas which has created a repopulation of small towns in the countryside. As Peter Hall has pointed out,

already by the 1980s, the map of population change was the exact reverse of the equivalent map of the 1890s: the counties and the districts that were then suffering the biggest population losses have become the areas with the biggest gains.

(Hall and Ward 1998: 106)

What is interesting about this centrifugal movement is what Champion and Atkins (1996) have termed 'the counterurbanisation cascade'. 'At the beginning of the 1990s, migration within Britain was producing a clear redistribution of population down the settlement hierarchy from larger metropolitan areas to medium-sized and smaller cities and towns and more rural areas' (Champion and Atkins 1996: 26).[37] However, the 2000 Urban White Paper notes that during the late 1990s there were indications of a slowing of population decline in the metropolitan areas and some were even growing, particularly London.

Housing land supply: the *Barker Review*

The *Barker Review* (2004) (which names both the author and the title) is a wide-ranging and impressive study commissioned by the government because of concern about the long-term upward trend in real house prices in the UK and its effect on the wider economy.[38] Three points should be made at the outset: first, the review was commissioned by the Treasury in association with the ODPM; second, it is a much more fundamental investigation of the problem and possible solutions than the typical research reports sponsored by ODPM;[39] and third, the report has had a very substantial impact on policy and action, supporting pre-existing policies for housing development in the South East (see p. 280), and promoting fresh thinking about the relationship between planning and the economic health of the country. It has given a much needed 'shot in the arm' to debates about housing land and planning. Almost certainly, this and the Communities Plan (discussed on p. 226), will be remembered long after the ODPM's *Planning Green Paper* is forgotten. It will be remembered by some for its persuasive argument and by others because it swung planning back to being driven by the market. It will certainly have some impact because of the weight of its principal sponsor, HM Treasury. Naturally, therefore, it concentrates on the economics rather than the more difficult politics of housing supply. Nevertheless, it should be required reading for planning students and practitioners.

The argument presented is that high house price inflation (2.4 per cent in the UK compared with a European average of 1.1 per cent) creates problems of affordability and, because of the volatility of the housing market, has exacerbated problems of macro-economic instability and had an adverse effect on economic growth. This is not a current 'crisis' but a long standing trend which explains why consumers see housing as an investment and hold expectations that price inflation will continue. New house-building (134,000 a year in England) is significantly lower than the rate of new-household formation (179,000), it is half what it was in the mid 1970s, and there is a shortfall of 39,000 new dwellings per year. Even the government's current targets set out in RPG/RSS are not being achieved, with a shortfall of 15,000 per year between 1996 and 2001; high demand (in the housing boom years) does not seem to affect the rate of completions; and there is an increasing rate of refusal of large housing applications. On the economy, the report argues that higher rates of house-building would:

- help to reduce volatility in house prices, thereby improving macroeconomic stability and supporting growth
- improve flexibility and performance of the UK economy via greater labour mobility
- bring greater access to housing for many households, avoiding unwelcome distributional effects, and the ill-effects of poor housing.

The government has already recognised that 'doing nothing is not an option' (the reference here is to the Sustainable Communities Plan discussed later). Barker's views of what could be done range widely over the role of the Housing Corporation, the failure of the development industry to innovate, and the impacts of taxation. Deficiency in the administrative machinery for assessing housing needs and influencing the market are noted, particularly at the regional level where three bodies deal with aspects of housing more or less independently: *regional bodies* (assemblies) and *regional spatial strategies* make broad allocations of housing land, *regional housing boards* advise on the demand for and funding of social and other 'sub-market' (non-market) housing, through the *regional housing strategy*, and the *regional development agencies' economic strategies* support regeneration which is closely linked to housing demand and supply. It hardly needs to be said that 'they often use a different evidence base and operate over different timescales'. However, the Review found that the single most important barrier to the delivery of housing (and thus part remedy to house price inflation) is availability of land through the planning system.

In order to reduce house price inflation to levels experienced in other European countries, Barker recommended that the delivery of new housing would need to almost double, from 150,000 to 295,000 homes per year, though there are numerous cautionary notes about the uncertainty of the calculations (since repeated by other commentators).[40] For this to happen, a more 'effective' planning system is needed. This would be

- a system that responds to market signals;
- decision-making procedures that take full account of the wider cost and benefits of housing development, including environmental and amenity costs;
- appropriate incentives for development at the local level;
- clear and timely mechanisms to provide the necessary infrastructure and services to support development and deliver sustainable communities; and
- sufficient resources to enable effective decision-making.

(p. 32)

The main specific recommendations are given in Box 6.7. There is a sense of déjà vu about much of the list. All of the recommendations have been voiced in various forms before. The most interesting discussion is outside the specific recommendations; for example the possibility of moving to a binding system of local plans (where the Review gets well beyond its expertise) as tried before with simplified planning zones, and techniques to introduce market price signals into the system, also tried before by some authorities. Inevitably, the Review came back to the 'old chestnut' of taxing development land value, with recommendations for a planning gain supplement. The difference here (as noted above) is that the recommendations are being made to the Treasury as well as the ODPM. The Chancellor responded to this report in the 2004 Budget Statement and decisions were promised before the end of 2005. The ODPM has already acted on a number of the points.

The detailed recommendations should not divert attention from the main themes of the Review, which are to ensure that more land is allocated for housing and associated development, to give more weight to the economic considerations, costs and benefits of decisions on housing land and to seek ways of distancing the decisions on housing land from the political process. These have not gone unnoticed among the main protagonists and present the main areas of dispute; indeed the arguments are at opposite poles, partly because Barker (because of the terms of reference) seriously underestimates the local political nature of housing and of the planning system; the reasons why many communities and their representatives are against further housing growth; and the environmental consequences of growth. A good summary of the alternative views is provided by Neil Sinden of CPRE:

> Boosting house building at the levels proposed by Kate Barker would result in an unnecessary environmental disaster . . . there is no solid evidence of undersupply of new homes in the UK and no evidence that a massive increase in house building would solve the problem of the lack of homes people on lower incomes can afford.[41]

The CPRE (and others) point to the potential to tackle the housing crisis through more brownfield development, promoting more even demand across the country, and demand side measures (reducing people's willingness to pay more and expectations of house price increases) on the basis that there is a surplus of dwellings over households in all regions in England. They highlight that the decline in house building since the late 1970s is explained by the drastic reduction in building of social housing and support a substantial boost in spending on affordable housing. Of equal concern to opponents is the potential 'relaxation' of planning controls over house building, and democratic control over planning. The suggestion that housing market indicators should trigger the automatic release of housing land undermines the role of planning in taking account of the full range of considerations, including the environmental interest. Certainly, Barker gave insufficient attention to the question of how the housing crisis could be tackled while also contributing to sustainable development (mentioned in the terms

BOX 6.7 THE *BARKER REVIEW*: MAIN RECOMMENDATIONS FOR PLANNING

For the planning system

- Regional planning bodies to set targets to improve affordability.
- The merger of regional planning and housing bodies.
- New techniques so that housing assessments reflect a full consideration of the economic, social and environmental costs and benefits of housing.
- More 'realistic' allocations of housing land with more flexibility to bring forward land for development and to meet historic shortfalls.
- Allocation of a 'buffer' of land allocation to facilitate quick responses to changes in demand. The preservation of the principle of containing urban sprawl through green belt designation, but with possibility to change green belts where forcing development elsewhere would create perverse environment impacts.
- Alternative routes to gaining permission limiting the involvement of elected members to matters of principle.
- A stronger national policy statement on housing preventing restrictions on housing development without compelling evidence they are needed.
- Increasing permitted development rights.
- Use of local development orders and increase in fees to provide more resources for planning.
- Dedicated project teams for larger-scale developments involving other public sector stakeholders.
- Additional access for planners to planning and legal expertise and resources.

Other relevant recommendations

- More coordination of and resources for infrastructure that would bring housing forward.
- More use of 'area-based special purpose vehicles' (take planning control and development promotion from local authorities).
- Measures to 'share in windfall development gains accruing to landowners so that increase in land values can benefit the community more' (a tax on development value) – a planning gain supplement (PGS). This would be a charge on the developer at the time the planning permission is granted, which it is thought would be met by the landowner, rather than the house buyer.

of reference) and particularly within environmental capacities.

Notwithstanding the weaknesses of the Review, it is difficult for planning to sidestep the criticisms both expressed and implied. Planning has failed to deliver even its own estimates of land needed for housing. A pointed 'case study' of York and Harrogate is used to illustrate the costs and benefits of housing restraint

(p. 38 of the Analysis Report). Both are attractive areas that suffer from a lack of development sites and both are constrained by green belt designation. House price growth in York has exceeded 12 per cent as 'many of the new dwellings in these areas are bought by newcomers, some no doubt moving from London or the south east, having cashed in their capital gains'. The result is that public sector workers and others in the tourism business are priced out of the market, creating labour shortages, longer commuting distances, affordability problems and forcing developers into targeting other land and property (such as pubs, shops and businesses) for lucrative housing development. The illustration in the report is too kind, however, and does not give the full story. It failed to mention that the local authority had yet to adopt a statutory local plan some thirty-six years after the system was created and fifteen years since the government called on all authorities to adopt local plans with haste. Procrastination over the plan, of course, happens because of resistance to new development and unwillingness and inability of the system to make difficult decisions, and that includes the failure of the Regional Office and ODPM to sort things out with the local authority. With examples like this it is fair to suggest that planning has got off lightly in the criticism. York is only one of many authorities that have failed to deliver housing land. It can be argued forcefully that this is the democratic process in operation, and (as noted above) the abysmal quality of much previous (sub)urban development is a good deterrent to any support for further growth. Local 'NIMBYs' (not in my back yard) are acting 'rationally' in opposing new growth. Weak support from government in providing physical and social infrastructure should also be noted. But in the absence of a properly argued case in a local plan, the defence is weak. And the costs of not acting are most heavily felt by those who are most in need of housing and jobs (see Table 6.4). In the case of York, the potential of one of the few locations in the north of England that might succeed in the knowledge economy may not be met.

Accommodating growth: new settlements

The conclusion of the new towns programmes, coupled with increasing concern with the 'land for housing' problem, naturally prompted debate on additional new towns. The TCPA had traditionally maintained that this should be a major plank in regional policy but, during the 1980s, against the background of a buoyant housing market, proposals came from the private sector for private enterprise new towns that would fill the gap left by the completion of the existing new towns. The best known of these came from the now disbanded Consortium Developments, which proposed a ring of new villages around the South East which would form 'balanced communities' developed to high standards of design.

> Consortium Developments Ltd, by working on a relatively large scale, can negotiate a keen price that allows investment in a quality product: a high quality infrastructure in the paving and road surfaces, high quality landscaping, sensitive design of public spaces, variety in both form and tenure of housing provision, and a wide range of supporting facilities.
>
> (Roche 1986: 312)

Table 6.4 Population of the UK 1981–2001 and projected 2011–26 (million)

	1981	1991	2001	2011 projected	2021 projected	2026 projected
England	46.80	48.5	49.5	51.6	54.0	55.0
Wales	2.80	2.90	2.9	3.00	3.1	3.1
Scotland	5.20	5.10	5.1	5.0	5.0	4.9
N. Ireland	1.50	1.60	1.70	1.70	1.8	1.8
UK	56.4	57.4	59.1	61.4	63.8	64.9

Source: Annual Abstract of Statistics 2005 (Table 5.1)

These were words in the direct tradition of the new towns movement, but their spokespersons now had to contend with a sophisticated planning machine. Proposals for Foxley Wood in Hampshire, Stone Bassett in Oxfordshire, Westmere in Cambridgeshire, and Tillingham Hall in Essex were all rejected on appeal. As Hebbert (1992: 178) comments, their experience 'demonstrated that even the presence of the most radical free enterprise British government of recent times is no guarantor of profitable large scale private developments in green field sites'. However, they have not been completely ruled out: the 1992 version of PPG 3 notes (with no conscious irony) that there had now been 'considerable experience of planning proposals' (*sic*) for new settlements which had 'almost invariably been deeply controversial'. It advised that future proposals should be contemplated only in cases where they represented a clear expression of local preference supported by local planning authorities. Politically, the importance attached to 'local choice' effectively meant that any proposal for a new settlement was likely to be killed, though a study was commissioned of 'alternative development patterns for new settlements'.[42]

The analysis of the differing types of development did not go very far in demonstrating the superiority of any one type of development over another.[43] This is not surprising: general issues of urban form are of limited practical value since the real problems are not general but site specific. The advantages and disadvantages of particular development forms vary according to the features of alternative sites and their location in a specific sub-region (and, with larger development, perhaps the wider region as well). They will vary also according to the size, character and purpose of the development, its transport links and potentialities, and its present and future relationships with the surrounding areas. Additionally, there are issues of finance, administration, politics and such like that can prove of decisive importance. All these (and no doubt other) factors combine to make generalisations highly problematic, and thus any major development proposal requires thorough and lengthy study and negotiation. Given the high sensitivity to development almost anywhere (no doubt increased by

the generally poor quality of design) it is not surprising that proposers of new settlements have had a very tough time making any progress at all. (An American environmental acronym points to the problem: BANANA – Build Absolutely Nothing Anywhere Near Anything.)

In the mean time, development has proceeded (or did not proceed) on particular sites for which builders sought planning permission. This non-planning approach was checked by a process that contained one or more of the elements of strong local opposition, public inquiries and ministerial decisions. Somewhere buried in this process was a vestige of planning policy, but it was a hit-or-miss affair. Certainly, it was a far cry from positive planning or ensuring that the right development went ahead at the right place at the right time. But, of course, the basic dominant political philosophy was not only unsympathetic to 'planning policies': it held that market mechanisms were superior. And so little progress was made in fashioning the planning system to the needs of the time.

Much consideration has been given to more public sector led new settlements since then, but the planning system has not been able to deliver. It would involve huge public expenditure, particularly on infrastructure, and strong opposition from vested interests. As a result there was little change in the attitude to new settlements on the part of the Blair government, though the revised PPG 3 (2000) was arguably slightly more positive:

> The Government is not against new settlements and believes that in the right location and with the right concept, they can make a contribution to meeting the need for housing. However, the cost of developing a new community from scratch, including the full range of new services and infrastructure, means that they will not always be a viable solution. New settlements will not be acceptable if their principal function is as a dormitory of an existing settlement. New settlements, whether large-scale additions to existing settlements or completely new, may under certain circumstances prove to be a sustainable development where

- they are large enough to support a range of local services, including schools, shops and employment;
- they exploit existing or proposed public transport by locating in a good quality public transport corridor;
- they can make use of previously used land; and
- there is no more sustainable alternative.

Proposals for 'larger new settlements' have to be brought forward through the new regional guidance machinery (discussed in Chapter 3). It is warned that proposals for new settlements will be controversial and all schemes will need to be agreed between the tiers of plan-making authorities. That this warning is fully justified is illustrated in the declaration of the Sane Planning in the South East protest group:

> The Sane Planning in the South East protest group have presented to the Secretary of State a declaration to mark the tenth anniversary of the protest against Foxley Wood (when an effigy of the then Secretary of State was burned). The group maintains that new settlements still have no place in the South East.
> (*Planning*, 30 July 1999)

Planned extensions to existing urban areas were more favoured although they often ran into the need to amend green belt boundaries which have to be reviewed where possibilities for development within urban areas are limited. However, the Panel Report on the South East Draft Regional Planning Guidance proposed that 'areas of plan-led expansion should be designated in Ashford, the Milton Keynes / Bedford / Northampton triangle, the Crawley / Gatwick area, and an area close to Stansted'.[44] Though the report does not suggest any special mechanism for the last three of these, for Ashford it comments that 'substantial town expansion, possibly up to a population of 150,000, should be assisted by action under new town legislation'. The intensity of government action to drive forward these growth areas in the South East was not anticipated. The equivalent of two provincial cities will be built around London in the next 10 years.

Sustainable communities and growth areas

The Communities Plan, *Sustainable Communities: Building for the Future*, was launched by the Deputy Prime Minister in 2003. It is not so much a plan, as a consolidated list of government activity: spending, investments, changes to institutional and administrative arrangements, proposals for new legislation and policy, and lots more. Much of it had been presented in previous plans and programmes; some items were very significant, others relatively minor. The ODPM collected most things that might be associated with 'communities' into the basket. Box 6.8 gives an indication of the content and Table 6.5 shows how resources are allocated. The result was that the main thrust was lost, or perhaps that was the intention. The message has become clearer with publication of subsequent progress reports. For some the Communities Plan is a brave attempt to harness disparate government activities across different sectors around the task of creating 'successful, thriving and inclusive communities, urban and rural, across England' (p. 2). For others it is a hastily contrived and largely unsustainable attempt to tackle the mounting contradictions of allowing the market to lead continued growth in and around London.

The two most significant components for planning are 'sustainable growth', that is, increasing the delivery of housing and other development in four growth areas identified in the Regional Planning Guidance/Spatial Strategy for the South East, and tackling 'low demand and abandonment', especially in nine of the most affected areas in the north of England. (The objectives in relation to 'decent homes' are examined in Chapter 10.) The broad thrust is to ensure that housing numbers are delivered according to plan, that the plan includes increased numbers of homes, and that a larger proportion of homes are affordable. RPG targets for housing in the growth areas were proposed to be increased by 200,000, and up to a total of 900,000 homes by 2031. The four growth areas are shown in Figure 6.1.

The criticisms of the Plan have been fierce. The House of Commons Select Committee on

BOX 6.8 *SUSTAINABLE COMMUNITIES* (2003): MAIN ACTION POINTS

Sustainable Communities: Building for the Future (2003)

Decent homes decent places (see Chapter 10)

- All social housing and 130,000 private sector homes to meet decent homes standard by 2010
- Improving neighbourhood environments through the liveability programme and fund (£201 million), new Best Value performance indicators, and neighbourhood warden schemes
- Business improvement districts
- Support CABE to drive up design standards and launch CABE Space.

Low demand and abandonment

- Nine pathfinder projects and a Market Renewal Fund to execute plans (£500 million)
- Improve compulsory purchase system
- Gap-funding to pump prime development where costs exceed returns
- Continued support for the Coalfields Programme and Regeneration Trust.

A step change in housing supply

- Incentives and sanctions to ensure that targets for new homes in the South East are delivered
- More and targeted resources for affordable housing, especially in the growth areas (total of £1 billion per year for three years)
- A new role for English Partnerships in land assembly for housing
- Continued support for key worker housing (£1 billion over three years)
- New Homelessness Directorate in ODPM and £260 million fund.

Land, countryside and rural communities

- Maintain 60 per cent brownfield target, and new national brownfield strategy
- All local authorities to do an urban capacity study
- Create the Land Restoration Trust
- Regional targets to maintain or increase green belt land
- Increased targets for rural affordable housing.

Sustainable growth

- To provide for major increased growth in the four growth areas identified in RPG for London and the South East
- Translate growth proposals into RPG/RSS and provide £446 million investment for Thames Gateway and £164 million for other growth areas
- Review of DfT transport plans in relation to the four growth areas and specific transport improvements for Thames Gateway
- Specific implementation arrangements, partnerships, URCs and UDA/UDCs
- Cabinet Committee chaired by Prime Minister to consider plans for Thames Gateway

Reforming the delivery

- Legislation to reform the planning system and funding for regional assemblies to improve regional planning
- Planning Delivery Grant of £350 million over three years
- Business improvement districts

Sustainable Communities: People, Places and Prosperity: A Five Year Plan (2005)

- National framework for 'neighbourhood arrangements', devolving powers for limited local services and environmental issues, and local charters setting out arrangements and expectations
- The *Cleaner, Safer, Greener Communities Programme*, with 'How To' guides on town centres, streets and parks
- Local area agreements rationalising funding streams, piloted in twenty-one areas
- Merger of regional planning and housing bodies
- Recognition of need for regional growth outside South East
- City-region action plans
- More decentralisation of government department activity to the regional offices
- Support for inter-regional growth strategies

Note: The 2005 plan repeats some of the content of the 2003 statement. There are numerous other action points. Many were already provided for by the Urban White Paper and other commitments.

Environmental Audit (which has the specific remit to examine the impact of government policy on sustainable development) found that the Plan was a positive change 'in the way the government approaches growth and regeneration', but said 'we deplore the absence of any reference to environmental protection, or the need to respect environmental limits'. They called for a definition of 'sustainable community' (which has now been provided),[45] on the grounds that ODPM seemed not to understand the concept:

> ODPM seems to have taken the approach to sustainability and the SCP that by simply calling 'sustainable' and by mentioning the environment occasionally . . . the Plan is inherently and obviously fully compatible with the principles of sustainable development. This is clearly not the case. (para. 47)

On the use of the word 'sustainable' the Report is damning, describing efforts to direct the Plan towards sustainable development as 'a window dressing exercise'. Of equal concern to the committee was the admission during presentation of evidence that the additional growth was to accommodate incomers drawn to the area for economic reasons; and that DEFRA was not properly consulted in the early stages. These comments draw on evidence presented to the committee by, for example, Friends of the Earth (FoE) who described the Plan as 'a piecemeal approach to a housing crisis and . . . it makes political judgments about growth and where it will take place before any effective assessment of environmental limits has been made' (para. 49). That FoE should say this is not surprising, but the Plan was clearly a fragmented set of initiatives. Although the growth areas had been previously identified, the determination of government to push so strongly for rapid delivery of housing,

Table 6.5 Sustainable communities' resources (£m)

	2002–3	2003–4	2004–5	2005–6	Total 2003–4 to 2005–6
Housing: London, East and South East	995	1,573	1,558	1,605	4,736
Housing: other regions	719	852	892	914	2,658
Arm's length housing management organisations	59	323	851	820	1,994
Transitional funding for housing finance reforms	500	175	140	65	380
Disabled facilities grants	97	99	99	99	297
Homelessness	90	93	83	83	259
Other housing programmes	501	466	394	355	1,215
Market Renewal Pathfinders	25	60	150	290	500
Thames Gateway	0	40	198	208	446
Other growth areas	0	40	58	66	164
Local environment/liveability	13	41	79	81	201
Regional development agencies	1,322	1,521	1,551	1,607	4,679
European regional development fund	210	229	229	229	687
English Partnerships	145	163	179	179	521
Other urban programmes	21	35	30	29	94
Planning, including Delivery Grant	27	73	153	194	420
Neighbourhood Renewal Fund	300	400	450	525	1,375
New Deal for Communities	350	265	287	298	850
New Ventures Fund	77	99	99	94	292
Total	5,451	6,547	7,480	7,741	21,768

Note: The resources include existing and new commitments.

without a more considered review of the environmental costs and benefits through, for example, a sustainability appraisal, was unacceptable to many. Even the government's own adviser was rather flippant in accepting the economic case for growth in the South East.[46] The South East Region Office of the Environment Agency published a press release saying 'the development of 800,000 new homes in the South East could set off an environmental time bomb' (para. 65).[47]

The alternative view is to be found in the government's presentation of the Plan. As explained above, the Barker Review was to back up the need for more concerted action in the light of failure to deliver housing. HM Treasury's interest in facilitating growth in the South East together with a government view that regional policy directing growth to other regions is largely ineffective were important influences in

bringing forward the Communities Plan for accelerated growth in the South East. There may appear to be contradictions with the government's objective of devolving responsibility so that regions can decide best how to invest resources, but HM Treasury has decided that this is an issue of national significance for central action, thus resulting in the Prime Minister chairing the committee overseeing progress in the Thames Gateway.

The first progress report on the Communities Plan (published as early as July 2003) *Making it Happen: Thames Gateway and the Growth Areas* said more about the new bodies being created to implement the proposals (delivery vehicles, as the ODPM calls them) which are the new urban development corporations and regeneration companies discussed elsewhere in this book. It also added the completion of the Channel

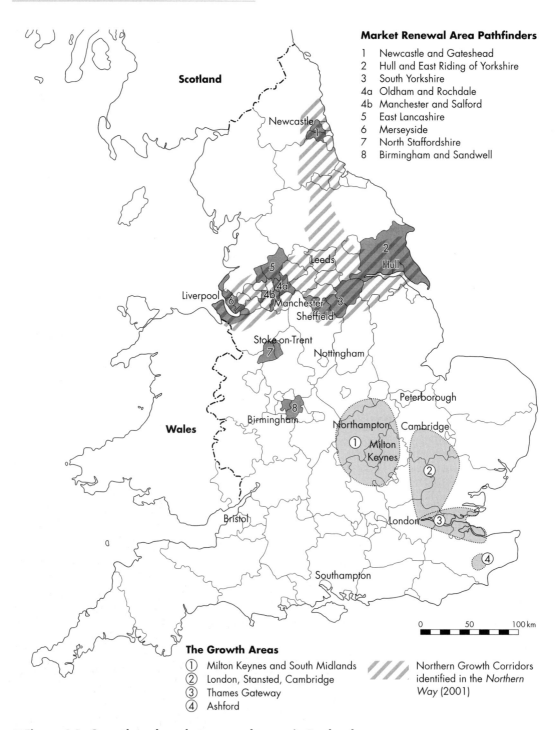

Market Renewal Area Pathfinders

1   Newcastle and Gateshead
2   Hull and East Riding of Yorkshire
3   South Yorkshire
4a   Oldham and Rochdale
4b   Manchester and Salford
5   East Lancashire
6   Merseyside
7   North Staffordshire
8   Birmingham and Sandwell

The Growth Areas

①   Milton Keynes and South Midlands
②   London, Stansted, Cambridge
③   Thames Gateway
④   Ashford

Northern Growth Corridors identified in the *Northern Way* (2001)

Figure 6.1   *Growth and market renewal areas in England*

Tunnel Rail Link as a 'sustainable communities' action! It predicted that by 2016 120,000 new homes would be built in the Thames Gateway (40,000 more than the original planning targets) and 133,000 in the Milton Keynes–South Midlands area (44,000 more than the planning target). The revised targets will have to be confirmed in the regional spatial strategies, and at the time of writing that is a contentious issue, despite increasing government funding for infrastructure. The cost of infrastructure to support the Communities Plan was estimated at £40 billion to £50 billion over twenty years. The South East and East of England Assemblies, those affected by the growth proposals, have both signalled opposition to the plans unless accompanied by very substantial infrastructure investment to improve its sustainability. The 2004 Progress Report, *Making it Happen: The Northern Way*, concentrates on progress in the North and Midlands, with snapshots of progress in each of the regions with many examples of local regeneration activity

Early in 2005 the agenda was represented in *Sustainable Communities: People, Places and Prosperity: A Five Year Plan*, which again was accompanied by regional profiles.[48] It says more about changes to planning brought forward through the 2004 Act and particularly the requirements for more and earlier community involvement in drawing up plans. There is much discussion of other new powers for the 'community', the 'neighbourhood' and parish councils, but it also spells out the need for decisions to be made at higher levels, whether by the local authority, regional body, inter-regional grouping (see below) or at the national level. The South East growth areas (where most investment is concentrated) are strangely absent, save for a brief summary of progress. Instead this 'plan' gives attention to the need for more balanced regional development, including the inter-regional strategies, EU regional policy funding (neither of which are ODPM initiatives) and more attention to city regional planning through action plans. There is mention of the work of the core cities group and of city-region action plans. This is connected to the growing appreciation in government that prosperous neighbourhoods require prosperous regional economies. A definition of sustainable communities is given too (four pages), but it

does not go as far as to say that growth should be within environmental capacities, rather communities should be 'environmentally sensitive', through for example, minimising waste and protecting the environment.

The sustainable communities documents leave an impression of much activity, and perhaps, real change in the way government values and understands towns and cities and the neighbourhoods that make them up. But with their eclectic and inconsistent mix of initiatives large and small, policies and proposals new and old, monitoring statements, wishful thinking and spin, they present a chaotic picture of policy-making and delivery. They are not nearly as useful as they might be in presenting the threads of government policy. And where is the plan?

The Northern Way

The government has an objective, expressed through a public service agreement target, 'to make sustainable improvements in the economic performance of all English regions and over the long term reduce the persistent gap in growth rates between the regions'. This will be a difficult target to achieve, first because the gap between London, the South East and the rest of England is widening, and second, because the further massive investment in the South East will ensure that this continues. In terms of aggregate economic gain for the UK this may not be a problem, and the PSA may be a more symbolic than real expression of aspirations. But it is a problem for the well-being of the North of England, and other 'peripheral regions', and for the sustainable development of the UK territory as a whole. Even economically it may be that the country is losing out through unexploited potential of people and infrastructure elsewhere in the country.

For this reason, and at the invitation of the government, key regional stakeholders in the three regions of the North, led by the RDAs have collaborated in the creation of an inter-regional strategy *Moving Forward: The Northern Way* (2004) and an action plan (2005).[49] These are preliminary statements of intent and a search for ways forward, rather than a complete plan of action, but they pick up strands of thinking about spatial development at the EU level discussed in Chapter 4. The UK has very little experience of planning at this scale, though emerging national plans for Northern Ireland, Scotland and Wales, together with the strategy for the Thames Gateway, are showing the way. The argument is that there is benefit in inter-regional working to avoid damaging competition, to pool resources, to develop complementary strategies, and above all to get all parts of government at all levels working in concert around a common plan.

Growth corridors and city-regions spanning local authorities and regions figure prominently. *The Northern Way* has identified eight city regions where growth has been concentrated over recent years and three growth corridors around transport axes, where action may be concentrated.[50] There are now also similar initiatives for the Midlands Way and South West Way – though in the latter case it might have been more usefully defined to incorporate parts of South Wales. In comparison with plans for the South East, there is little on the table for regions in the North and Midlands. Compare the government's contribution so far, a £100 million Northern Way Growth Fund, with the costs of infrastructure for the South East noted above. And the proposals are swimming against the tide to say the least. Harding *et al.* (2004: 5) have calculated that current growth rates in the Northern Way regions would have to increase by 3 per cent 'to prevent any further widening of the gap' and increase by 15 per cent to reduce the gap in prosperity by half by 2014.

Market renewal areas

Earlier parts of this chapter have stressed the policy problem related to delivering more housing, but this sits uncomfortably with a surplus of certain kinds of housing in some locations, particularly in the North, Midlands and parts of Scotland. It is difficult to generalise about the problem, except that to some extent it is all connected with the failure (or 'restructuring') of regional economies. Bramley and Pawson (2002: 396) offer three categories of explanation – regional demographic change linked to job loss;

changes in preferences and behaviour and the declining popularity of social housing; and the process by which particular 'neighbourhoods are stigmatised by reputations for poverty, crime and other problems'. Migration has not, however, been at a level to bring about any general collapse of housing demand in the north of England, but it has led to increased 'departures and higher vacancies in local authority housing and has produced local surpluses in the least popular localities' (Holmans and Simpson 1999). The role of the planning system has not been insignificant:

> In our view, there has been a tendency to release too much land (both greenfield and brownfield) for new private housing development in some of the sub-regions of England affected by generic low demand . . . The problem is a consequence of plans being based on outdated or optimistic population projections, plus an emphasis on housing-led regeneration, plus a degree of competition between districts to attract such development and the associated population (typically working family households).
>
> (Bramley and Pawson 2002: 410)

It should also be noted that much 'release' of housing land has barely been shaped by detailed local plans or assessments of local needs and demand (which may not have existed or been up-to-date). Where new suburban homes have been built around affected towns and cities they tend to be relatively cheap and coupled with available mortgages provide a strong pull factor from existing estates. Changes have been made to strengthen the option of de-allocating land as explained in PPG/PPS 3, although this has always been a (difficult) option.

The government has responded on numerous fronts, including, in the worst hit areas, the designation of nine housing market renewal pathfinder areas in 2003.[51] The government is targeting £500 million for implementation of 'strategic action plans' and the actions are also supported in places by the RDAs, English Partnerships, urban regeneration company activity and other initiatives. The approach varies depending on the particular causes of low demand and tenure affected – although social housing is generally a greater problem. Action often involves selling on social housing into the private market and selective redevelopment in partnership with house builders, which also leads to more mixed tenure neighbourhoods, though comprehensive clearance has also been needed where the housing is completely redundant.[52]

Green belts

The policy of maintaining an adequate supply of land for housing can be difficult to reconcile with policies relating to green belts and the safeguarding of agricultural land. Although 1987 saw a major policy shift on the latter (which is discussed later), green belts have, for a variety of reasons, remained a strong policy issue for both central and local government, well supported by public opinion.

Green belt policy formally emerged in 1955 although the idea first gained currency in the 1920s and designations were made around London in 1938 (Ward 2004). The campaigners for green belts had expressed considerable concern about the implications for urban growth of the expanded house building programme. Unusually, the policy can be identified with a particular minister – Duncan Sandys (who later made another contribution to planning with the promotion of the Civic Trust and the Civic Amenities Act). Sandys' personal commitment involved disagreement with his senior civil servants, who advised that it would arouse opposition from the urban local authorities and private developers who would be forced to seek sites beyond the green belt. Experience with the Town Development Act (which provided for negotiated schemes of 'overspill' from congested urban areas to towns wishing to expand) did not suggest that it would be easy to find sufficient sites. Sandys, however, was adamant, and a circular was issued asking local planning authorities to consider the formal designation of clearly defined green belts wherever this was desirable in order to check the physical growth of a large built-up area, to prevent neighbouring towns from merging into one another, or to preserve the special character of a town.

The policy had widespread (and long-lasting) appeal to county councils, who now had another weapon in their armoury to fight expansionist urban authorities, but also more widely. One planning officer commented that 'probably no planning circular and all that it implies has ever been so popular with the public. The idea has caught on and is supported by people of all shades of interest'. Another noted that

> the very expression *green belt* sounds like something an ordinary man may find it worthwhile to be interested in who may find no appeal whatever in 'the distribution of industrial population' or 'decentralisation' . . . Green belt has a natural faculty for engendering support.
>
> (Elson 1986: 269)

The green belt also formed a tangible focal point for what is now called the environmental lobby. However, initially, its biggest support came from the planning profession which in those days still saw planning in terms of tidy spatial ordering of land uses. Desmond Heap, in his 1955 presidential address to the (then) Town Planning Institute, went so far as to declare that the preservation of green belts was 'the very *raison d'être* of town and country planning'. Their popularity, however, has not made it any easier to reconcile conservation and development. The green belt policy commands even wider support today than it did in the 1950s. Elson concluded his 1986 study with a discussion of why this is so:

> It acts to foster rather than hinder the material and non-material interests of most groups involved in the planning process, although it may be to the short term tactical advantage of some not to recognise the fact. To *central government* it assists in the essential tasks on interest mediation and compromise which planning policy-making represents . . . To *local government* it delivers a desirable mix of policy control with discretion. To *local residents* of the outer city it remains their best form of protection against rapid change. To the *inner city local authority* it offers at least the promise of retaining some economic activities that would otherwise leave the area; and to the *inner city resident* it offers the

prospect, as well as often the reality, of countryside recreation and relaxation. To the *agriculturist* it offers a basic form of protection against urban influences, and for the *minerals industry* it retains accessible, cheap, and exploitable natural resources. *Industrial developers* and *house builders* complain bitterly about the rate at which land is fed into the development pipeline, yet at the same time are dependent on planning to provide a degree of certainty and support for profitable investment. Planning may be an attempt to reconcile the irreconcilable, but green belt is one of the most successful all-purpose tools invented with which to try.

(Elson 1986: 264)

The latest policy statement on green belts in England (the revised PPG 2 of 1995) confirms the validity and permanence of the green belts policy which now covers over one and a half a million hectares (13 per cent) of England. (The main elements of green belt policy are shown in Box 6.9.) The general location of the green belts is shown in Figure 6.2. Table 6.6 illustrates the growth in the area of green belt land in England and Table 6.7 gives the area of green belt in Scotland which is discussed separately later. Elson *et al.*'s (1993) study, which was undertaken at a time when the earlier (1988) PPG 2 was operative, concluded that the green belts had been successful in checking unrestricted sprawl and in preventing towns from merging. Green belt boundary alterations in development plans had affected less than 0.3 per cent of green belts in the areas studied over an eight-year period. Most planning approvals in green belts had been for small-scale changes which had no significant effect on the open rural appearance of green belts. The appeal system had strongly upheld green belt policy. The situation has remained much the same, despite some high profile large releases of land from green belt. About 13 per cent of England's

---

BOX 6.9 GREEN BELT POLICY IN ENGLAND

Purposes

- to check the unrestricted sprawl of large built-up areas
- to prevent neighbouring towns from merging into one another
- to assist in safeguarding the countryside from encroachment
- to preserve the setting and special character of historic towns
- to assist in urban regeneration, by encouraging the recycling of derelict and other urban land.

Use of Land in Green Belts

- to provide opportunities for access to the open countryside for the urban population
- to provide opportunities for outdoor sport and outdoor recreation near urban areas
- to retain attractive landscapes, and enhance landscapes near to where people live
- to improve damaged and derelict land around towns
- to secure nature conservation interest
- to retain land in agricultural, forestry and related uses.

Source: DoE (1995) PPG 2 *Green Belts*

Figure 6.2 Green belts in the UK

Table 6.6 *Green belts, England 1997 and 2003 (ha)*

By green belt *	1997	By region**	1997	2003
Tyne and Wear	52,500	North East	63,410	66,330
York	25,400	North West	255,760	260,610
South and West Yorkshire	252,800	Yorkshire and Humber	261,350	262,640
North West	251,700	East Midlands	79,710	79,520
Stoke-on-Trent	44,100	West Midlands	269,170	269,140
Nottingham and Derby	62,000	East Anglia	26,690	26,690
Burton and Swadlincote	700	London and South East	600,320	600,470
West Midlands	230,400	South West	105,900	106,180
Cambridge	26,700			
Gloucester and Cheltenham	7,000			
Oxford	35,100			
London	512,900			
Avon	68,500			
SW Hampshire and SE Dorset	82,300			
Totals	1,652,300		1,652,300	1,671,600

Notes:
*DETR Information Bulletin 1183, December 1999, *Green Belt Statistics: England 1997*: North West includes Greater Manchester, Merseyside, Cheshire and Lancashire; London excludes metropolitan open land; SW Hampshire and SE Dorset includes the New Forest area. The 1997 green belt statistics for England cannot be compared with earlier figures since they are based on a new and more accurate method.

** ODPM Statistical Bulletin Local Planning Authority Green Belt Statistics: England 2003. A new method has been adopted using digitised proposals maps from development plans. Only those plans that have changed will produce updated figures.
The New Forest area of green belt was redesignated as national park in 2005, but that is not reflected in these figures.

Table 6.7 *Green belts, Scotland 1999 (ha)*

Aberdeen	23,039
Ayr / Prestwick & Troon	3,024
Clackmannan	981
Edinburgh	15,869
Falkirk / Grangemouth	3,803
Glasgow	109,917
Total Scotland	156,633

Source: Figures supted by the Scottish Executive Development Department

land area is green belt, and in 2000 4,710 dwellings were built in the green belt taking up 430 hectares of land (0.03 per cent of land designated as green belt). The proportion developed on previously developed land was 60 per cent which increased to 68 per cent in 2003. Note the area remains designated as green belt until it is changed in a development plan.

The relationship between green belt restraint and the preservation of the special character of historic towns was much more difficult to evaluate. Though the idea had 'a well-established pedigree', and though the green belt boundaries were particularly tight, there was little evidence to connect policy and outcomes. It was difficult also to assess how far green belts had assisted in urban regeneration. Though the green belts did 'focus development interest on sites in urban areas', local authorities tended to regard the creation of jobs as more important than any land development objective *per se*. Indeed, urban regeneration was often seen as requiring the selective release of employment sites in the green belt. The supply of adequate sites within urban areas was not sufficient for development needs (though it might be increased by an expanded programme of land reclamation). Moreover, refusal to allow development on the periphery of an urban area could lead to leapfrogging beyond the green belt, or development by the intensification of uses in towns located within the green belt. A note is made of the suggestion that the inner city will rarely be a substitute location for uses seeking planning permission on the urban fringe:

> The housing market potential in the two locations is quite different (in terms of the size and price range of houses which may be marketed for example), and many of those developing other uses require the better accessibility (normally by private car) which a peripheral or outer location affords.
>
> (Elson *et al.* 1993: para. 2.37)

Further research on the Oxfordshire settlement strategy, which concentrates development in selected country towns beyond the Oxford green belt, found that new residents in three of these towns (Bicester, Didcot and Witney) exhibit 'high travel distances, high levels of car use, little use of public transport, and almost 90 per cent of employed residents travelling to work outside the town'. By contrast, a new housing development on the edge of Oxford has far less car travel since the public transport system provides a better alternative. The DoE laconically comments that 'these conclusions suggest that local authorities will need to consider carefully the regional dimension of location planning, and the transport policies applied in individual settlements' (DoE 1995, *Reducing the Need to Travel through Land Use and Transport Planning*).

In Scotland, green belts have been established around Aberdeen, Ayr/Prestwick, Edinburgh, Falkirk/Grangemouth, and Glasgow. Interestingly, the Dundee green belt has been replaced by a general countryside policy (Regional Studies Association 1990: 22). Scottish green belts have had somewhat wider purposes than those in England: these include maintaining the identity of towns by establishing a clear definition of their physical boundaries and preventing coalescence; providing for countryside recreation and institutional uses of various kinds; and maintaining the landscape setting of towns. There is a greater emphasis on the environmental functions of the green belts, and recreation is included as a primary objective. The title of the Scottish Circular (24/85) is significant: *Development in the Countryside and Green Belts* underlines the links between general countryside policies and green belts. 'As a result, a much more integrated approach to the planning of green belt and non-green belt areas is achieved in Scotland.' The Regional Studies Association study commends the Scottish approach, arguing that 'green belts have become an outmoded and largely irrelevant mechanism for handling the complexity of future change in the city's countryside'. This may have something to do with the variability by which green belt policy in Scotland is devised and implemented, as revealed in a *Review of Green Belt Policy in Scotland* (Bramley *et al.* 2004b). The Review found that the original purposes of green belt remained important, and recommended that the role in protecting landscapes and green environments should be recognised more explicitly. Surprisingly the Review also found that green belt was being used as 'a strategic land reserve', which is the opposite of its purpose, and there was much redesignation and development of former green belt land. Recommendations are made to clarify green belt functions and to ensure that they serve a strategic approach to the developing settlement form. Another idea for a two-tier green belt is more difficult to follow. A first tier, described as 'green heritage areas', would be effectively permanent, whereas a second tier, described as 'urban fringe greenspace', would last the life of development plans.

Green belts are the first article of the British planning creed. They are hallowed by use, popular support, and fears of what would happen if they were 'weakened'. Fierce arguments are raged by a wide range of groups from national bodies such as the CPRE to local green belt residents. There are, however, other issues which until the 1990s did not attract the same concern, such as the costs imposed by green belts, and the inadequacy of a planning policy which lays such a great emphasis on protection and a lesser emphasis on instruments for meeting development needs. In some circles the green belt is thought of as a barrier to more sustainable settlement patterns. On this line of argument, green belts should be part of a more comprehensive land use and transport policy. The position is exemplified by the Town and Country Planning Association's *Policy Statement: Green Belts* (2002) which calls for a 'reappraisal of the roles, purposes and extent of the green belts'. While wishing to maintain containment policies, the Association argues that

> green belt policies restrict the scope for considering forms of urban growth that conform better to sustainable development principles . . . [and that] a key option for further expansion of some urban areas should be the development of planned extensions well related to public transport corridors . . . such extensions can directly conflict with green belts.

This is convincing so far, but they propose adoption of 'a flexible approach', and that the regional and structure planning policies (to become RSS policies) should make provision for this flexibility. Such language was never going to appeal to those who recognise that the success of green belt designations is largely down to their inflexibility and simplicity. There is, in any case,

an opportunity to revise green belt boundaries (and it happens) even though their main characteristic is permanence. If regional local authorities were to come up with convincing arguments in well reasoned regional and sub-regional plans for strategic developments to which they all agree, and which demonstrate unequivocally their superior benefits in terms of sustainability, then there might be a case for a fundamental change to a boundary. So far they have, in general, not been willing, able or capable to do this. Indeed, in at least one place, York, detailed green belt boundaries still need to be designated. General claims for flexibility that might create more sustainable settlement patterns are not enough. The government (in England), for one, is unmoved by these calls and has strengthened its general policy on green belts (as part of the package on Sustainable Communities discussed earlier), while no doubt it will agree to adjustments where they are proven to be the only or best alternative.

In this connection the more recent articulation of green belt policy in Wales (where there are only draft proposals for green belts) is noteworthy (Tewdwr-Jones 1997). There has been increasing pressure from environmentalists for the establishment of green belts around the main urban areas which are under development pressure, and the 1999 *Planning Guidance for Wales* set out guidelines separately for Wales for the first time. The 2002 *Draft Revision of Planning Policy for Wales* says more about the potential of 'green wedges' which must be reviewed in the development plan process whereas the green belt should be more permanent. The policy echoes the English PPG, and says all local authorities which are subject to pressures for development must consider their use, though they 'must justify the need for such areas, demonstrate why normal planning and development control policies, including green barrier/green wedge policies would not be adequate' (para. 2.6.6). As in England, the debate on Welsh green belts largely ignores the issue of managing the countryside within green belts though both specifically refer to opportunities for access and for outdoor sport and outdoor recreation. These do not figure significantly in the public debate: the overwhelming concern is with preventing development.[53] A proposal for the first green belt in Wales, between

Newport and Cardiff, was included in the deposit UDP for Newport in 1999, with objections considered at the inquiry in 2005.

Town centres and shopping

There has been a turnaround in retail development since the mid 1990s. In 2000, the amount of new retail floorspace developed in town centres exceeded the amount in out-of-town shopping centres and retail parks for the first time since the 1990s. That position has strengthened subsequently. Out-of-town shopping centres were blamed for weakening or even killing off traditional town and district centres, for increasing car travel (and its accompanying pollution) and for decreasing accessibility to services for those without cars. They are very popular in themselves, and on a market test are successful, but concern about the decline of the centres of smaller towns led to initiatives to promote *Vital and Viable Town Centres* as to the introduction of much more restrictive planning policies on retail development. These are now set out in Planning Policy Statement 6, *Planning for Town Centres* (2005).[54] Since 1994, policies on transport in town centres and retailing have become increasingly strict (Truelove 1999: 207), a trend which has been strongly supported by the House of Commons Environment Committee and more recently the 'ODPM committee'.[55] Planning guidance for town centres is to promote 'their vitality and viability' by:

- planning for the growth and development of existing centres; and
- promoting and enhancing existing centres, by focusing development in such centres and encouraging a wide range of services in a good environment, accessible to all.

(PPS 6 para. 1)

The emphasis on planning for growth was added in the 2005 addition, in recognition of the reactive and restrictive position of the previous policy. Although the PPS has been successful, it explains 'it is not the purpose of the planning system to restrict competition,

preserve existing commercial interests or to prevent innovation' (para. 1.7). But, of course, the impact of policy may have exactly these effects, since new competitors are seldom allowed. There have been concerns that the 'town centres first policy' was leading to constraint on businesses, therefore the need for local authorities to be positive in finding ways to provide opportunities for the market in and around existing town centres and to extend the centres as well as refusing developments elsewhere. The Statement identifies a number of other objectives for policy on town centres: to enhance consumer choice, especially for socially excluded groups; to support the improvement of productivity in sectors using town centres like retail and leisure; to improve accessibility by a choice of means of transport and promote less car use; to promote social inclusion by ensuring that everyone has access to town centre services and facilities; to encourage investment in deprived areas and create economic growth, employment and improvements to the physical environment; and to deliver more sustainable patterns of development, by high-density, mixed-use development.

There were rumours that HM Treasury has been behind the emphasis on productivity and the need to plan and release land to allow town centre uses to expand but these were roundly rejected by the Deputy Prime Minister. But the general thrust of government policy around issues of both economic competitiveness and social exclusion are much more apparent in this than earlier town centre and retail policy. The latter was more concentrated on issues of town centre decline, transport and sustainable development. In assessing proposed site allocations and proposals for development, local planning authorities must:

- identify the appropriate scale of development;
- apply the sequential approach to site selection;
- assess the impact of development on existing centres; and
- ensure that locations are accessible and well served by a choice of means of transport.

(PPS 6 para. 2.28)

Perhaps the most significant innovation has been the

sequential approach which in considering development proposals gives first priority to town centre sites, followed by edge-of-centre sites, district and local centres, and then out-of-centre sites, giving preference to those that are accessible by public transport. The authority must give particular attention to the town centre, and the possible extension of town centre uses, before considering out-of-centre locations. The same message is given to developers who are told they 'will need to be flexible and innovative, and should explore fully the possibility of fitting development onto more central sites' (para. 2.25). In general developers have been able to work with the retail and town centre policy, despite dire warnings about the effect on shopping and the development industry. This is the conclusion of an ODPM review of the 1996 version of PPG 6 (Hillier Parker and Cardiff University 2004) the main findings of which were that planning authorities were doing little to positively plan and bring forward sites that might be usefully located to meet the policy. Also, while developers had obviously adjusted their strategies to fit into the policy and new regional shopping centre developments had ended in favour of alternative locations (which is abundantly obvious to anyone who lives in a city) it was too soon to be sure about the specific impacts of the policy. There is a tendency for concentration of comparison shopping in the top fifty town and city centres; the bulky goods and DIY sectors have not found the policy an 'impediment' to freestanding locations; convenience stores such as Sainsbury and Tesco have been able to expand with store extensions (as well as the new smaller formats), and have found authorities in the North more willing to accommodate new stores. The retail sector continues to innovate too; the latest manifestation is the forecourt store alongside the petrol station which has as yet received little attention. In sum the researchers conclude that PPG 6 has changed retailers', developers' and investors' perceptions. It has brought about a modest shift in activity towards established town centres, and has facilitated and encouraged innovation by retailers and developers, within town centres (Hillier Parker and Cardiff University 2004: 13).

The same report points to continuing uncertainties over definitions, such as 'out-of-town', and particularly the 'need' for new retail development. Students of planning appeals will know what arguments can rage over such an apparently simple statement. First, how is 'edge-of-town' to be defined? A proposal by Sainsbury to build a store on a redundant site in Brighton was rejected on appeal on grounds which included its location being not genuinely on the edge of town. It was 145 metres from the primary shopping area (*Planning* 16 October 1998). Second, what evidence is there that proximity to the town centre will have the benefits that are claimed? A study for the DoE suggested (on the rather small sample of two case studies) that 'in terms of linked trips, edge-of-centre stores do not necessarily generate significantly higher degrees of linkage with town/district centres than out-of-centre stores'.[56] There are many issues of this kind which make 'town centres' and retailing more complex than they may seem; the latest 'retail concept' is the factory outlet centre, the numbers of which increased from one in 1990 to forty-one built or approved by 2001 (Walker, G. 2001). The government has helped with standard definitions of some of the key terms, but these are contested and much depends, as the PPS accepts, on the particular characteristics of the centre and the development. 'Need' is one of the more difficult concepts for retailing. A ministerial statement on the subject in 1999 produced the following:

> the requirement to demonstrate 'need' should not be regarded as being fulfilled simply by showing that there is capacity (in physical terms) or demand (in terms of available expenditure within the proposal's catchment area) for the proposed development. Whilst the existence of capacity or demand may form part of the demonstration of need, the significance in any particular case of the factors which may show need will be a matter for the decision-taker.

This requires several readings before one realises that it means (in David Lock's words): 'you must prove "need", but the Government will not tell you whether you have succeeded until you have succeeded' (Lock 1999). As Sainsbury's representative pointed out, the one thing that is clear, however, is that the minister has certainly provided the lawyers with potentially rich pickings when arguing about how need is defined and by whom (Williams, H. 1999). The uncertainties about definitions are exposed by developers and retailers seeking an advantage in the planning application, appeal and inquiry processes, and increasingly, through the courts. In 1999 and 2003 the government accepted the need for further clarification after legal challenges did not go their way, and ministerial statements were issued to try to provide a more unambiguous interpretation. They are named after the ministers in charge at the time, the Caborn and McNulty Statements. The new PPS 6 has incorporated these reviews.

These issues have been set out here at some length, not because they are exceptional, but because they are the very stuff of planning arguments. As with the debates on 'vital and viable' town centres, these can mask secular social and economic trends such as changes in retail trading patterns and distribution, changes in trading laws (as with the relaxation of the Sunday trading laws which affect small retailers much more than the superstores), changes in branch banking, and (still unclear) the effect of the Internet on buying patterns.

Scottish land reform

There are numerous policies relating to land, but rarely is there anything which might be termed a 'land policy'. Scotland presents a fascinating exception. Following a very long history of attempts to reform the Scottish feudal land system, the Scottish Parliament was embarking on 'an integrated programme of action and legislation' over a four- or five-year period from 1999. This was summarised in the very first White Paper to be published by the Scottish Executive (*Land Reform: Proposals for Legislation*, 1999).

The Scottish system is extraordinarily complex, and only a short indication of its character can be given here.[57] Essentially, feudal land ownership is a hierarchical system in which land rights derive from the highest authority, theoretically God, but in practice the Crown. The Crown is known as the Paramount Superior, and all other landowners are known as Vassals of the Crown. The relationship need not be direct, however, and a vassal can convey land (to a new vassal), retaining interests which are set out in the title deed. There is no limit to the number of times this 'feuing' can take place. Each superior can reserve rights and impose additional 'burdens' (such as a restriction on building on the land or carrying on a business). Nearly all privately owned land in Scotland is held under feudal tenure and the survival of such characteristically feudal elements as superiorities and feu duties is indicative of the extraordinary archaic and complex nature of Scotland's current system of land ownership.

Previous reforms have attempted to simplify this system, but have not tackled the more political issue of land ownership. It is claimed that Scotland has the most concentrated pattern of private landownership in the world: '343 landowners own over half of the entire privately owned rural land in the country'. In the Highlands and Islands, half of all the private land (about 1.5 million hectares) is owned by fewer than a hundred landowners.[58]

Proposals for reform were issued by a Land Reform Policy Group appointed by the government, and a White Paper *Land Reform: Proposals for Legislation* was issued in 1999. This first instalment of reform is limited in its scope to giving 'community bodies the right to buy rural land which is to be sold', and to creating a right of 'responsible access to land'. The latter is outlined in Chapter 9. Here a brief summary is given of the proposals relating to the former.

The intention is to create new opportunities for 'community ownership'. This is to be done by providing for the registration of community bodies (set up for the purpose and incorporated) who are interested in acquiring land when it comes to be sold in their area. Registered bodies will have the right to buy such land (whether privately or publicly owned). The price will be assessed by a government-appointed valuer, with disputes being settled by the Lands Tribunal for Scotland. A minimum percentage of those aged 18 or over and who live and/or work on the land in question must support the proposed purchase. To deter evasion, Scottish ministers will be able to exercise a new compulsory purchase power where this is in the public interest.

The proposals have been characterised by Wightman as 'based on a flawed, shallow and partial analysis of the problem [and revealing] a timidity and poverty of imagination when it comes to tackling landed power' (*Guardian* 30 August 1999). Others might argue that it is perhaps early days to judge. Some progress has already been made (in advance of general legislation) in the Highlands and Islands where a Community Land Unit has been established and is operating schemes of both technical and financial assistance in its region.[59]

Further reading

Land values and prices

Hall and Ward (1998) *Sociable Cities: The Legacy of Ebenezer Howard* provides a historical overview of the land question; Bramley *et al.* (1995) *Planning, the Market and Private House-building* also covers much of this material. There has been surprisingly little study of the operation of the various experiments in capturing land values for the public benefit. A long and detailed account of the legislative history is given in Cullingworth (1980) *Land Values, Compensation and Betterment, vol 4: Environmental Planning 1939–1969*. More digestible accounts are provided by McKay and Cox (1979) *The Politics of Urban Change* (Chapter 3) and Cox (1984) *Adversary Politics and Land*.

The effect of planning on the land market has been the subject of a long-standing debate both in theoretical terms, as in Evans (1983) 'The determination of the price of land' and in the context of British planning as in the same author's (1988) *No Room! No Room! The Costs of the British Town and Country Planning System* and (1991) 'Rabbit hutches on postage stamps'. Less tendentious are the studies commissioned by government, for example, Monk *et al.* (1996) 'Land-use planning, land supply, and house prices' and Bramley and Watkins (1996) *Steering the Housing Market: New Building and the Changing Planning System*.

Planning gain

By contrast there has been a large number of studies of planning agreements, planning obligations and planning gain. Selected titles are Rowan-Robinson and Young (1989) *Planning by Agreement in Scotland*; Eve (1992) *Use of Planning Agreements*; and Bunnell (1995) 'Planning gain in theory and practice'. A comprehensive study is Healey *et al.* (1995b) *Negotiating Development: Rationales and Practice for Development Obligations and Planning Gain*. The December 1997 special issue of *Urban Studies* is devoted to 'Developer contributions: the bargaining process', and includes articles on the USA, Canada and the Netherlands. Of particular interest are Ennis (1997) 'Infrastructure provision, the negotiating process and the planner's role', and Claydon and Smith (1997) 'Negotiating planning gains through the British development control system'. DoE Circular 1/97 *Planning Obligations* and its 2005 replacement are important, together with the numerous consultation documents given in the text. See also Cornford (1998) 'The control of planning gain' and Planning and Environment Law Reform Working Group's (1999) 'Planning obligations'. A succinct overview is given in Wenban-Smith and Pearce (1998) *Planning Gains: Negotiating with Planning Authorities*, and critical review in Grant (2003) 'Planning gain'. The importance of the increased cost of infrastructure in negotiations for contributions from developers is considered in Marvin and Guy (1997) 'Infrastructure provision, development processes and the co-production of environmental value'. See also the GVA Grimley 2004 report on *Developing a Methodology to Capture Land Value Uplift Around Transport Facilities*. The report by Johnson and Hart (2005) *The Barker Review of Housing*, for the RICS contains a summary of government attempts at taxing development gains as well as an assessment of proposals for a planning gain supplement and tariffs.

Planning and affordable housing

PPG 3/PPS 3 (a revision was expected in 2005) is the first source for government policy, together with the ODPM 2003 consultation document on *Influencing the Size, Type and Affordability of Housing*. The use of planning powers to require the provision of affordable housing has attracted much debate. See, for example, Kirkwood and Edwards (1993) 'Affordable housing policy: desirable but unlawful?'; Barlow *et al.* (1994b) *Planning for Affordable Housing*; Elson *et al.* (1996) *Green Belts and Affordable Housing: Can We Have Both?*; 'Planning mechanisms to secure affordable housing' in Joseph Rowntree Foundation (1994) *Inquiry into Planning for Housing*; and Gallent (2000) 'Planning and affordable housing'. See government sponsored research by Entec (2002) *Delivering Affordable Housing through Planning Policy* and by Environmental Resources Management (2003) on *Improving the Delivery of Affordable Housing in London and South East*. The Shelter report by Holmans *et al.* (1998) *How Many Houses Will We Need?* is an assessment of the need for affordable housing in England. Specific issues are considered by Farthing and Ashley (2002) 'Negotiations and the delivery of affordable housing through the English planning system', Gallent *et al.* (2002) 'Delivering affordable housing through planning', Gallent *et al.* (2004) 'Second homes' and Morrison (2003) 'Assessing the need for key-worker housing'. On the strange 'exceptions policy' (the exceptional release of land for local needs housing), see Annex A to PPG 3 (or its replacement) and Circular 6/98 *Planning and Affordable Housing*; and Gallent and Bell (2000) 'Planning exceptions in rural England'.

Land availability

See *The Supply of Land for Housing: Changing Local Authority Mechanisms*; Bramley *et al.* (1995) *Planning, the Market and Private House-building* and Bramley and Watkins (1996) *Steering the Housing Market: New Building and the Changing Planning System*. UK Round Table on Sustainable Development (1997) *Housing and Urban Capacity* contains a review of studies. See also Llewelyn-Davies (1997) *Sustainable Residential Quality: New Approaches to Urban Living* and Lord Rogers' (1999) Task Force report *Toward an Urban Renaissance*. A useful collection of essays is edited by Jenks *et al.* (1996) *The Compact City: A Sustainable Urban Form?*

Brownfield, vacant, derelict and contaminated land

Llewelyn-Davies (1996) *The Re-use of Brownfield Land for Housing* deals with the difficulty of the remaining brownfield sites and the need for substantial government subsidies. The difficulties are illustrated in a short report by the Civic Trust (1999) *Brownfield Housing: 12 Years On*. See also Alker *et al.* (2000) 'The definition of brownfield' and Bibby and Shepherd (1999) 'Refocusing national brownfield housing targets'. The research on vacant land includes: Cameron *et al.* (1988) *Vacant Urban Land: A Literature Review* and Whitbread *et al.* (1991) *Tackling Vacant Land: an Evaluation of Policy Instruments*. The latter provides a review of previous research. For a broader overview of urban land policies, see Chubb (1988) *Urban Land Markets in the United Kingdom*.

Derelict land is dealt with in a number of DoE reports including *Assessment of the Effectiveness of Derelict Land in Reclaiming Land for Development* (1994); *Derelict Land Prevention and the Planning System* (1995); Evans, C. (1998) *Derelict Land and Brownfield Regeneration* on the legal aspects.

Policy on contaminated land is succinctly set out in PPG 23 *Planning and Pollution Control* and Circular 2/2000. A commentary on this is given by Graham (1996) 'Contaminated land investigations: how will they work under PPG 23?'

On redundant military land, see Bateman and Riley (1987) *The Geography of Defence*, National Audit Office (1992) *Ministry of Defence: Management and Control of Army Training Land*, Farrington (1995) 'Military land in Britain after the cold war', Fuller Peiser and Reading University (1999) *Development of the Redundant Defence Estate* and Fyson (1999) 'Iron out defence land policy to get the full benefits'.

Household projections

The most accessible discussions of household projections are given in Breheny and Hall (1996) *The People: Where Will They Go?* More technical is Bramley *et al.* (1997)

The Economic Determinants of Household Formation: A Literature Review. See also Allinson (1999) 'The 4.4 million households: do we really need them anyway?' For rural areas see Rural Development Commission (1998) *Household Growth in Rural Areas: The Household Projections and Policy Implications.*

New settlements

An account of the long-standing British campaign for new settlements is discussed in Ward (1992) *The Garden City: Past, Present and Future* and by Hardy (1991a) *From Garden Cities to New Towns* and (1991b) *From New Towns to Green Politics* (a two-volume history of the TCPA). A volume in the official history *Environmental Planning 1939–1969* by Cullingworth (1979) provides a detailed dead-pan record of government policy over this thirty-year period. Breheny *et al.* (1993) *Alternative Development Patterns: New Settlements* gives a view during that phase, and an analysis of current issues is Hall and Ward (1998) *Sociable Cities: The Legacy of Ebenezer Howard.*

Green belts

An up-to-date review of green belt policy in Scotland has been published by Bramley *et al.* (2004b) and a detailed *Strategic Sustainability Assessment of the Notts-Derby Green Belt* was undertaken by Baker Associates (1999) which addresses some difficult questions. Two major earlier publications on green belts are Elson (1986) *Green Belts: Conflict Mediation in the Urban Fringe*, and the report of a study for DoE by Elson *et al.* (1993) *The Effectiveness of Green Belts.* Broader in scope is the classic study by Peter Hall *et al.* (1973) *The Containment of Urban England.* On green belts in Scotland, see Regional Studies Association (1990) *Beyond Green Belts* and Pacione (1991) 'Development pressure and the production of the built environment in the urban fringe'. On Welsh policy in relation to green belts, see *Planning Guidance (Wales): Planning Policy First Revision* (1999) and Tewdwr-Jones (1997) 'Green belts or green wedges for Wales?' A short critical appraisal of green belt policy is Cherry (1992) 'Green belt and the emergent city'. For a review of green wedges, green buffers, strategic gaps and the like, see Lyle and Hill (2003) and the ODPM (1999) research report *Strategic Gap and Green Wedge Policies in Structure Plans.*

Town centres and shopping

PPS 6 and its equivalents should be the starting point, then for an alternative view see the Retail Forum: Newsletter of the Retail Planning Forum, available at www.nrpf.org/. The equivalent Planning Policy Statement for Northern Ireland is PPS 5 *Retailing and Town Centres* (1996) and for Scotland it is NPPG 8 *Town Centres and Retailing* (1998). Good reviews of the issues involved arising with out-of-town shopping centres include Hillier Parker and Cardiff University (2004) *Policy Evaluation of the Effectiveness of PPG6*, BDP Planning and Oxford Institute of Retail Management (1994) *The Effects of Major Out-of-Town Retail Developments* and CB Hillier Parker and Savell Bird Axon (1998) *The Impact of Large Foodstores on Market Towns and District Centres.* See also Sparks (1998) *Town Centre Uses in Scotland*, URBED (1994) *Vital and Viable Town Centres: Meeting the Challenge*, HC Environment Committee (1997) *Shopping Centres*, Ravenscroft (2000) 'The vitality and viability of town centres' and National Retail Planning Forum (1999) *A Bibliography of Retail Planning.* An analysis of the changing economics of superstore development is Wrigley (1998) 'Understanding store development programmes in post-property-crisis UK food retailing'. For a discussion of retail parks, see Guy (1998) 'High Street retailing in off-centre retail parks', and 'Alternative-use valuation, open A1 planning consent, and the development of retail parks'. More generally, see Guy (1994) *The Retail Development Process* which is still the only main text on retail planning.

Scottish land reform

The main book used in the text is Callander (1998) *How Scotland is Owned.* See also Callander (1987) *A Pattern of Landownership in Scotland.* Another author in this field is Wightman (1996) *Who Owns Scotland?* and (1999) *Scotland: Land and Power.* See also Ogilvie (1997) *Birthright in Land* and McCrone (1997) *Land, Democracy and Culture in Scotland.*

Notes

1 The principle had been first established in an Act of 1662 which authorised the levying of a capital sum or an annual rent in respect of the 'melioration' of properties following street widenings in London. There were similar provisions in Acts providing for the rebuilding of London after the Great Fire. The principle was revived and extended in the planning Acts of 1909 and 1932. These allowed a local authority to claim first 50 per cent, and then (in the later Act) 75 per cent, of the amount by which any property increased in value as the result of the operation of a planning scheme. In fact, these provisions were largely ineffective since it proved extremely difficult to determine with any certainty which properties had increased in value as a result of a scheme or, where there was a reasonable degree of certainty, how much of the increase in value was directly attributable to the scheme and how much to other factors. The Uthwatt Committee noted that there were only three cases in which betterment had actually been paid under the planning Acts.

2 Previous editions of this book give a more detailed account of the Community Land Scheme.

3 See *Planning* 4 February 2005: 'Developers fear for viability'.

4 *Planning Policy and Social Housing* (RTPI 1992: 5). Grant (1999a) discusses this policy explicitly as a form of betterment recoupment. He adds that

> the tenuous link drawn in the circular between private and affordable housing is demonstrated by the government's willingness for the obligation to be commuted to a financial contribution by the developer towards the provision of affordable housing elsewhere in the local authority's area.

5 See Chartered Institute of Housing *et al.* (1999), National Housing Federation (1999), Whitehead *et al.* (1999) and Environmental Resources Management (2003).

6 These figures are from Shelter's affordability index: more details at www.england.shelter.org; and from Wilcox (2004).

7 *Independent* 10 January 2002.

8 Lord Nicholls in *Waters v Welsh Development Agency*, quoted in Parry (2005).

9 *Fundamental Review of the Laws and Procedures Relating to Compulsory Purchase and Compensation: Interim Report* (London: DETR, 1999). See also the report of the 1999 Symposium on Compulsory Purchase on the ODPM Website.

10 In another article, the same research team make a case for 'urban partnership zones' as an alternative to compulsory acquisition in some cases (Adams *et al.* 2001).

11 The Crichel Down rules refer to the arrangements by which surplus government land that was originally acquired by or under threat of compulsory purchase is offered back first to previous owners, their successors or to sitting tenants; see Part 2 of Circular 6/04.

12 This is a paraphrase of s. 99 of the 2004 Act amending s. 226 of the 1990 Act. An annex to Circular 6/04 provides a summary of the changes made by the 2004 Act.

13 The claimed benefits of living in compact cities vary greatly. Arguments in favour include Jacobs (1961), Elkin *et al.* (1991), Sherlock (1991), ECOTEC (1993b) and various official publications on sustainable development. Arguments suggesting that the benefits are illusory, infeasible or overstated include Breheny (1997), P. Hall (1999c) and K. Williams (1999).

14 A review by Llewelyn-Davies revealed that few local authorities had undertaken such studies. Where they had been carried out, they seriously underestimated the amount of land available for housing. They recommended that studies should (1) be based on original site work, (2) include a significant physical design element, and (3) not be constrained by existing policies and standards. None of the studies reviewed met these criteria (UK Round Table on Sustainable Development 1997).

15 Evidence to the HC Select Committee on *Housing PPG 3*, HC 490-I. See also the Friends of the Earth report by Rudin (1998).

16 Reported in *Outlook*, the magazine of English Partnerships, Spring 2003: 5.

17 The figures are not completely comparable (the earlier

ones were recognised as possibly underestimates). The 1999 figures are from the Government Statistical Service Information Bulletin 500 (20 May 1999) and the later ones from NLUD findings, available at www.nlud.org.uk.

18 For Scotland, see *Scottish Vacant and Derelict Land Survey 1998* (Edinburgh: The Stationery Office, 1999).

19 Policies can be founded on myths as well as on adequate understanding of problems. So it was with the land registers established by the Local Government, Planning and Land Act 1980. The myth was that one of the major causes of urban dereliction was the hoarding of land by public authorities. By requiring local authorities and other public bodies to 'register' their land, it was expected that it would find its way into the development process. In fact, with the reality being much more complicated than the perception, the registers were of little effect. (See the evaluation undertaken for the DoE by Whitbread *et al.* 1991.)

20 This section draws on the DETR study by Fuller Peiser and University of Reading (1999) which notes that the Ministry of Defence is the second largest estate in single ownership in the UK with about 226,000 hectares of land. (Only the Forestry Commission has more land.)

21 This, and the following quotations, are from the HC Environment Committee report on *Contaminated Land* (1990).

22 The crux of the problem lay in the concept of 'contamination'. Instead of referring to land that is contaminated, the Act relates to 'land which is being or has been put to any use which may cause that land to become contaminated with noxious substances'. This very inclusive definition was made particularly onerous in the initial draft regulations because of the very large number of contaminative uses which were specified. There was strong criticism that the registers would create widespread blight and, in an attempt to pacify objectors, the number of specified uses was greatly reduced.

23 Another objection to the initial regulations was that they prohibited the deregistration of sites. This was defended on two grounds: one is that factual information on the site's history (which cannot by

definition change) will be necessary when any future change of use is proposed. The other is that contamination from the site may have migrated to adjacent sites; owners, regulatory authorities and developers are expected to use registers to identify such sources of contamination.

24 See also the DETR web pages on contaminated land at www.environment.detr.gov.uk/landliability/index. htm which provides a summary of the current regime.

25 Hedges and Clemens (1994) Tables 6.17 and 7.17 and commentary pp. 132 and 158. Breheny (1997) discusses these and other relevant issues. See also Todorovic and Wellington (2000).

26 There is also the issue of empty properties owned by government departments: see DETR (1999) *Revised Guidance on Securing the Better Use of Empty Homes*. From April 2000, council tax is payable at the rate of 50 per cent on dwellings that have been vacant for a year or more.

27 See DETR (1997) *Evaluation of Flats over Shops*, London Planning Advisory Committee (1998b) and Urban Task Force (1999a: 253–4).

28 See, for example, Council for the Protection of Rural England (1994c), Bramley and Watkins (1995), Bramley (1996a) and Green Balance (1999).

29 DETR (1999) *Planning Policy Guidance Note 11: Regional Planning Public Consultation Draft*, para. 5.4. See also HC Select Committee on the Environment, Transport and Regional Affairs, Tenth Report (session 1997/98) *Housing*, vol. 1, para. 2.11, and *The Government's Response* (Cm 4080), paras 127–36. Stephen Crow, in his evidence to the HC Environment Subcommittee, argued that both the expressions 'predict and provide' and 'plan, monitor and manage' were 'slogans which can mean all things to all men'.

30 Interestingly, in his evidence to the Select Committee, the Deputy Prime Minister gave his view that it did not: 'I do not think that planning, predict and provide is contradictory to planning, monitoring and managing; one is a process and the other one is how you achieve it' (HC Select Committee op. cit. para. 210). For the CPRE view on this see Wenban-Smith (1999) and its *Sprawl Patrol* campaign and briefing sheet *Plan, Monitor and Manage* (details at www.cpre.

org.uk). The Select Committee's report on *Housing PPG 3* (1999) criticised the draft PPG for its lack of clear and specific guidance (op. cit. para. 14).

31 SERPLAN *A Sustainable Development Strategy for the South East* (SERP 500, 1998). Accompanying documents are listed in Appendix 2 of *RPG for the South East, Public Examination: Report of the Panel*.

32 *RPG for the South East, Public Examination: Report of the Panel*, paras 4.54 and 4.63.

33 Ibid., para. 4.67.

34 Evidence of Professor Tony Crook and Dr Christine Whitehead to the Select Committee, op. cit. p. 74.

35 It is curious and unfortunate that 'there is no mechanism in England whereby the desirability of inter-regional migration can be debated' (Breheny and Hall 1996).

36 See the discussion and references on RDAs in Chapter 3.

37 The major factor, of course, is the changing pattern of employment. Though there has been a general loss of manufacturing jobs, the loss has been most dramatic in the conurbations. These losses have not been offset by a corresponding growth in alternative employment in the affected areas. The expansion in service jobs has been located almost entirely in towns and rural areas where there are attractive and cheap sites. See Turok and Edge (1999), Turok and Webster (1998) and Rowthorn (1999).

38 An *Interim Report: Analysis* was published in December 2003; the *Final Report: Recommendations* was published in March 2004. The government published its initial response with the 2004 Budget Paper.

39 By this we mean that ODPM has commissioned much research which is evaluation of existing policies with the effect of building in many assumptions and constraining findings; there is also a tendency for ODPM research to use questionnaire surveys of local authorities and other interests, and case studies of 'good practice', though this is an impression rather than empirical finding. A Planning Research Network has been created which may stimulate more fundamental research on the operation of planning.

40 Despite the uncertainties, the reports invaluably bring together a wide range of data, information and explanation – to make some telling points, for example, at the current new house replacement rates a home built now would have to last 1,200 years (p. 47).

41 CPRE Press Release 12 March 2004. See also the analysis undertaken on behalf of CPRE by Europe Economics (2004); Wenban-Smith (2004); and other CPRE publications available at www.cpre.org.uk.

42 Breheny *et al.* (1993). Much of this is of a technical nature, comparing the costs and benefits of different forms of development. This is a difficult and complex matter, since so much depends on site-specific issues. The authors neatly point up the difficulties by stressing that their analysis is 'intended to focus discussion rather than present a definitive assessment'. But central government is urged to come off the fence, and to give a clear statement on the management of urban growth. It is unequivocally stated that 'unless much tougher containment policies are introduced – at the very time when concerns are being expressed over urban intensification – it is inevitable that significant greenfield/village development will take place in the UK'.

43 It did, however, carefully avoid making the mass of assumptions which flawed an earlier study by the National Institute of Economic and Social Research (Stone 1973; see also Cullingworth 1979: 473).

44 See also Hall and Ward (1998), particularly Chapter 9 on 'Sustainable social cities of tomorrow'.

45 First Report in 2005 was *House Building: A Sustainable Future*. The ODPM definition of sustainable community is on the ODPM website under sustainable communities.

46 This section draws on evidence given to the Environmental Audit Committee Report. The Communities Plan and growth areas do not require a sustainability appraisal under the legislation, but the Committee recommended that this should be undertaken. DEFRA commissioned a *Study on the Environmental Impacts of Increasing Housing Supply* from Entec (2004). The government expert is Sir John Egan who contradicted the explanation in the plan for the need for housing on the grounds of increasing households in the UK and said it was in response to the 'urgent need to find housing of high quality for the best people in the world who want to come here' (para. 57).

47 A more considered review of the connection of the sustainable communities plan and sustainable development has been made by Anne Power on behalf of the Sustainable Development Commission. It notes that the Plan is 'essentially a "top down" programme, with little to encourage community involvement or ownership of the proposals, possibly for fear of opposition to its overall purpose' (p. 3).

48 It is a sister document to *Sustainable Communities: Homes for All – A Five Year Plan*, explained in Chapter 10.

49 These and other documents are available at thenorthernway.co.uk.

50 The city regions are Central Lancashire, Liverpool, Manchester, Sheffield, Leeds, Hull and the Humber Ports. The three development corridors centre on the M62 and the northern parts of the A1 and M6.

51 The areas are Birmingham and Sandwell, East Lancashire, Humberside, Manchester and Salford, Merseyside, Newcastle and Gateshead, North Staffordshire, Oldham and Rochdale and South Yorkshire.

52 Bramley and Pawson (2002) suggest that this is a very expensive option, with costs for clearance per private sector unit of between £25,000 and £30,000.

53 But see the study commissioned by the Sports Council for Wales (Elson 1991). Tewdwr-Jones (1997) suggests that the alternative policy of green wedges in areas of possible development pressures could provide a flexible way of meeting both current recreation needs and future development needs.

54 The original planning guidance was Planning Policy Guidance Note 6: *Town Centres and Retail Development* (1993, revised 1996); a number of ministerial policy statements have supplemented this document, as explained in PPS 6.

55 Two inquiries by the HC Environment Committee (1994) *Shopping Centres and their Future* (and the *Government Response* 1995), and *Shopping Centres* (1997) were important. The *Government Response* to this was published later in the same year. In July 2004, the HoC Housing, Planning, Local Government and the Regions Committee reported on Draft Planning Policy Statement 6, though it has little to say.

56 CB Hillier Parker and Savell Bird Avon (1998: para. 10.12). See also the series of reports on the employment impact of out-of-town superstores published by the National Retail Planning Forum.

57 This account leans heavily on Callander (1998), from which extensive quotations are taken.

58 Wightman, A. (1999) 'A land (un)divided: land reform proposals for Scotland fall far short of what is needed for the redistribution of power', *Guardian* 30 August; see also Wightman (1996, 1999).

59 Highlands and Islands Enterprise Community Land Unit *Action Framework 1998–2001*. In its first year the Unit's achievements included financial assistance to Abriachan Forest Trust towards the purchase of 50 hectares of woodland on the side of Loch Ness, and assistance to some twenty smaller community land initiatives.

7 Planning, the environment and sustainable development

In the last few decades, much has been achieved in reversing the environmental damage of previous centuries. Few people, for example, would have foreseen, even fifty years ago, that a river like the Don, despoiled by the filth of two centuries of industrial intensification and decline, would flow clean enough to support thriving fish populations by the dawn of the new Millennium. Few probably even spared a thought for whether such a turn-around in environmental fortunes might be desirable, let alone achievable.

Sir John Harman, Chairman of the Environment Agency for England and Wales
in the Foreword of *Creating an Environmental Vision Consultation* Draft, 2000

The environment

In one sense, all town and country planning is concerned with the environment, but the reverse is not true, and it is difficult to decide where to draw the boundaries. The difficulty is increased by the rate of organisational change over recent years including the shifting of responsibilities from local government to ad-hoc bodies, and by the flood of new legislation, prompted in part by the EU. Further complications arise because of the increased concern for the environment and the rise of sustainable development as a political goal.

The implications for 'town and country planning' are still working themselves out, and not always easily as some of the implications touch at the heart of the planning system. Thus, it has been a long-standing feature of planning control that permission is given unless there are good reasons for refusal. It is for the local planning authority to demonstrate (to the Secretary of State if necessary) that an application should be refused. With 'environmental' procedures, however, the onus shifts somewhat: the developer's proposals have to be demonstrably acceptable, and permission can be refused if they are not. Though official

pronouncements and advice are coy in acknowledging this, it is clear that environmental factors can be decisive in a planning decision and that applicants may even be required to discuss the merits of alternative sites. In the words of Annex 1 to PPS 23, environmental statements must include 'an outline of the main alternatives studied by the applicant and the reasons for his/her choice . . .' (para. 1.39).

Local authorities have specific powers in relation to some environmental issues such as certain aspects of pollution, waste, and noise, but they are not environmental planning authorities. As planning authorities they must also pay attention to pollution issues in deciding planning applications, as discussed below. The roles of pollution and planning regulation are set out in Box 7.1. Other specific 'pollution control regimes' exist for this purpose, but there is no clear dividing line. A related issue here is that of *sustainability* – a concept around which much environmental policy revolves (see Box 7.2).

BOX 7.1 PLANNING AND POLLUTION CONTROL

The planning and pollution control systems are separate but complementary. Pollution control is concerned with preventing pollution through the use of measures to prohibit or limit the release of substances to the environment from different sources to the lowest practicable level. It also ensures that ambient air and water quality meet standards that guard against impacts to the environment and human health. The planning system controls the development and use of land in the public interest. It plays an important role in determining the location of development which may give rise to pollution, either directly or from traffic generated, and in ensuring that other developments are, as far as possible, not affected by major existing, or potential sources of pollution. The planning system should focus on whether the development itself is an acceptable use of the land, and the impacts of those uses, rather than the control of processes or emissions themselves. Planning authorities should work on the assumption that the relevant pollution control regime will be properly applied and enforced. They should act to complement but not seek to duplicate it.

Source: ODPM (2004) PPS 23 *Planning and Pollution Control* (para. 10)

Sustainability

Words cast a spell which can, at one and the same time, command respect and create great confusion. No word illustrates this better than the ubiquitous 'sustainability'. There is a view that the word has been so badly abused and misused that it has lost any useful meaning; it now serves to obscure rather than reveal the real issues. General public awareness and understanding of the concept remains low.[1] That there is a broad political consensus on the importance of the general idea of sustainability is surely an indicator of how widely it can be interpreted. Thus sustainability and sustainable development are not capable of precise scientific definition.[2] They are instead social and political constructs used as a call to action but with little in the way of practical guidance (O'Riordan 1985; Baker *et al.* 1997).[3] Indeed the ambiguity inherent in the terms can be seen as a positive as it presents an opportunity for local political debate on sustainability issues among competing positions. Debate around the sustainability concept ensures that some of the key conflicts and contradictions in public policy (and planning practice) are at least exposed and perhaps addressed (Meyerson and Rydin 1996).

But acceptance of the political, vague and uncertain meaning of the sustainability concept is not an excuse for inaction (any more so than the contested nature of the term democracy is an excuse not to improve our democratic processes). Many academics, environmental groups and government officials are devoting earnest effort to establishing what sustainability means – or what it should mean, for public policy. There are, without question, important implications for town and country planning arising from the fundamental principles of sustainability – but the nature of these principles can be confusing because of the great variety of definitions. One famous poetic rendering is by Chief Seattle: 'We do not inherit the world from our ancestors: we borrow it from our children'. This encapsulates the essential idea, which is more prosaically expressed in the well-known formulation of the Brundtland Report (1987): 'Sustainable development is development that meets the needs of the present without compromising the ability of future generations to meet their own needs'.

Shiva (1992: 192) has pointed to two very different uses of the concept. One ('the real meaning') relates to the primacy of nature: 'sustaining nature implies the integrity of nature's processes, cycles and rhythms'. This is to be contrasted with 'market sustainability',

BOX 7.2 DEFINITIONS OF SUSTAINABILITY

Sustainable development: development that meets the needs of the present without compromising the ability of future generations to meet their own needs. It contains within it two key concepts:

- the concept of 'needs', in particular the essential needs of the world's poor, to which overriding priority should be given
- the idea of limitations imposed by the state of technology and social organisation on the environment's ability to meet present and future needs.

Source: Brundtland (1987) *Our Common Future*

To promote development that enhances the natural and built environment in ways that are compatible with

- the requirement to conserve the stock of natural assets, wherever possible offsetting any avoidable reduction by a compensating increase so that the total is left undiminished
- the need to avoid damaging the regenerative capacity of the world's ecosystems
- the need to achieve greater social equality
- the avoidance of the imposition of added costs or risks on future generations.

Source: Blowers (1993) *Planning for a Sustainable Environment: Report to the TCPA*

Sustainability means making sure that substitute resources are made available as non-renewable resources become physically scarce, and it means ensuring that the environmental impacts of using those resources are kept within the Earth's carrying capacity to assimilate those impacts.

Source: Pearce (1993) *Blueprint 3: Measuring Sustainable Development*

Sustainable development is not simply about creating wealth and protecting the environment. It is also about caring for people and their quality of life. It is about ensuring that the quality of life of future generations will be as good as, or better than, it is for us.

Source: Environment Agency (2000) *Creating and Environmental Vision*

which is concerned with conserving resources for development purposes, and, if they become depleted, finding substitutes. On this latter approach, sustainability is convertible into substitutability and hence a cash nexus. The distinction is given eloquent expression in the words of a Native American elder who, in epitomising the non-convertibility of money into life, said: 'Only when you have felled the last tree, caught the last fish, and polluted the last river, will you realize that you can't eat money' (Shiva 1992: 193).[4]

This distinction between fundamental (or strong) definitions of sustainability and superficial (or weak) definitions has been made in numerous ways.[5] Owens (1994b) explains that the strong definition places fixed and inviolable constraints on economic activity, whereas the weak definition simply gives environmental capacities greater weight in the decision process. Broadly speaking, the first formulation challenges whether it is right to continue to meet various demands and needs if this cannot be accomplished without reducing current levels and quality of environmental stock. Thus demand management of resource use should be the central policy response. Needless to say, it is generally the weaker formulations that actually dominate. The policy response at this level has been described as *ecological or environmental modernisation* (Jacobs 1999). Here there is an emphasis on meeting sustainability through securing greater eco-efficiency through reducing waste, conserving energy, and reducing pollution, while, to put it crudely the economy continues to function as before. The objective is to influence market forces rather than regulate or replace them, for example, with devices such as environmental designations or financial mechanisms, though the limitations in providing adequate rewards to the market are well understood (Milton 1991). A key role of planning here is in finding appropriate locations to meet resources demands where environmental costs are lower or where the trade off of environmental loss against economic gains is more acceptable. Directions for the planning in the ecological modernisation frame, according to Davoudi (2001: 90) are to 'facilitate economic processes while making them benign' and 'focus on centrally formulated, non-spatial, apolitical regulatory criteria'. She also presents the alternative conception of sustainability, the risk society, which sees the current mode of production as irreconcilable with maintaining the state of the environment and eco-systems. In this framework planning would 'defend the environment against risks associated with economic processes, and focus on strategic and holistic approaches to place-making'.[6] In practice, Cowell and Owens (1998) have shown how the planning system mediates the questions of demand management and spatial location in a case study of aggregates planning – though the general argument can be applied more widely.

Those who advocate that sustainability is familiar in the history of planning (Hall, D. *et al.* 1993) are in effect presenting the weak interpretation: the planning system's traditional role has been to deal with the locational issues so as to reduce environmental damage and achieve some sort of 'balance', or more correctly, 'trade-off' between new urban development and environmental protection. But strong (or even moderate) interpretations of sustainability raise questions about

the capability of the planning system to deal with the structural questions of the relationship between social justice – the distribution of costs and benefits – economic demands and environmental capacities. This is not to say that the spatial or territorial questions are unimportant – they are – but that additional dimensions should also be considered, not least in demand management. Changes to the planning system such as in relation to meeting housing land requirements (discussed in Chapter 6) hint at changes in this direction. They also reflect growing consensus about the fundamental and very challenging principles which should govern public policy for sustainability.

In this respect it should be noted that the UK approach has traditionally differed from that in other European countries, particularly Germany. An important difference in principle (differences in practice may be less marked) is that of 'anticipation' as distinct from reaction. Whereas the UK has taken the view that environmental problems should be defined in terms of their measurable impacts, other countries have gone beyond this, and anticipated problems before the degree of environmental damage can be ascertained: this is related to the *precautionary principle*.

In Germany the concept of *Vorsorgeprinzip* is applied meaning broadly the principle of 'prevention' or 'anticipation' (but this fails to capture its full meaning). The German word connotes a 'notion of good husbandry which represents what one might also call best practice'. Möltke (1988) comments that '*Vorsorgeprinzip* is more than just prevention as an efficient means to an end but rather prevention as an end of itself.' The aim is, therefore, to establish pollution control policy, not merely as a means of reducing economic or social cost but also as a means of preserving wider ecosystems. Typically, the European approach involves the avoidance of 'excessive cost'. This, of course, is no easier to define than concepts such as 'reasonable cost', but it is clearly intended to be more demanding. Shed of its more philosophical overtones, the issue is fundamentally 'whether to protect environmental systems before science can determine whether damage will result, or whether to apply controls only with respect to a known likelihood of environmental disturbance' (O'Riordan and Weale 1989: 290). The government now sanction the use of the precautionary principle in planning (indeed they are 'committed' to using it) and say it should be invoked when

- there is good reason to believe that harmful effects may occur to human, animal or plant health, or to the environment; and
- the level of scientific uncertainty about the consequences or likelihood of the risk is such that best available scientific advice cannot assess the risk with sufficient confidence to inform decision-making.

(PPS 23, para. 6)

The principles for sustainability for territorial development and land use planning have been explained in many different ways but are summarised in the EU (1996) *European Sustainable Cities Report*, Blowers (1993), Selman (1999) and others. They are

- to develop within *environmental capacities* and apply the *precautionary principle* where these are uncertain
- to protect and enhance the stock of *natural capital* ensuring that it is passed on in good condition to future generations (*intergenerational equity* and *futurity*)
- to ensure that most human benefit is obtained from economic activity, and that there is a fair distribution of the benefit from the use of resources (*intra-generational equity*)
- not to export the costs of economic growth and environmental quality to other places (however distant) and promote *local self-sufficiency*
- to close *resource loops* through reuse and recycling and the active management of resource flows
- to ensure that the costs of environmental damage are borne by those who cause them (*the polluter pays principle*)
- to ensure active *involvement* of local communities in decisions that affect them.

The shortlist has been developed into a more comprehensive framework of sustainability principles as shown in Table 7.1. This conceptualisation of sustainability was developed specifically for spatial planning with the aim of assisting in transposing the very general notions of sustainability into planning and development practice and also for appraising existing planning policies and actions.[7] The conclusions from the research indicate how sustainability has been 'operationalised' or put into practice by identifying which sustainability principles are actually used in plan and decision-making and how (see Table 7.2).

Assessment of the take-up of sustainability principles into aspects of town and country planning are now coming forward (some references are given in the further reading). In sum, there has been only partial and fragmented conversion of the principles into planning policies and actions. Policies tend to follow well-worn formulae or 'checklists' and are seldom ambitious in addressing the strong definition of sustainability through for example, demand management. Where there is a stronger position on sustainable development, the planning response tends to be understood in relatively narrow terms, predominantly the organisation of land uses and transport links, and because of institutional fragmentation there has been difficulty in coordinating impacts in fields such as energy, waste air, noise and water. Policy compartmentalisation and departmentalism are strong barriers to effective coordinated approaches to sustainable development. The positive results derive largely from linkages between the planning process and Local Agenda 21 and the application of environmental appraisal (discussed on p. 280).

Owens (1994b) suggested there was a lot of 'sustainability rhetoric' but in practice 'business as usual'. Counsell (1998) reported that translation of sustainability principles into operational policies in structure plans was still 'proving difficult' though there is great variation in performance – perhaps as much related to local short-term self-interest as concerns about long-term intergenerational equity. In the mean time the stock of advice to planning authorities about how to incorporate sustainability into plans and decisions has increased sharply.[8] But aspirations still outstrip achievements. Even the most ambitious experimental projects such as the government's *Millennium Villages* have 'not yet delivered the order of magnitude of improvement needed to demonstrate true sustainability' (Llewelyn-Davies 2000: 3). The explanation is

Table 7.1 *Sustainability principles for spatial planning*

Principles	Criteria
Overarching	
Futurity and intergenerational equity	Precautionary principle (no irreversible decisions)
	Include cumulative and long-term impacts in decision-making
Intersocietal equity	Commitment to equity at local, national and international levels
	Ensure commitment to equity so environmental impacts and the costs of protecting the environment do not unfairly burden any one geographic or socio-economic sector
Local and regional self-sufficiency	Reducing externality effects so that environmental impacts and costs do not unfairly burden any one geographic group or socio-economic sector
	Using close in preference to distant resources
Risk prevention and reduction	Natural disasters
	Human-made disasters
Environmental	
Maintain the capacity of natural systems	Absolute protection of critical natural capital
	Defence of improvement of soil quality and stability
	Defence and improvement of key habitats and biodiversity
	Respecting absorption and assimilation capacities of natural systems
	Efficient use of renewable resources
Minimise resource consumption	Minimum depletion of renewable resources
	Minimum depletion of non-renewable resources
	Energy efficiency
	Minimisation of waste, recycling and re-use
Environmental quality	Reduction of pollution emissions; protection of air and water quality and minimisation of noise
	Protection and enhancement of environmental amenity and aesthetics
	Protection of natural and cultural heritage
Economic and societal	
Protect and develop the economic system	Encourage and develop connections between environmental quality and economic vitality
	Satisfy and protect basic needs (shelter, food, clean water etc.)
	Provide entrepreneurial and employment opportunities
Develop the human social system (education, democracy, human rights)	Protect basic human rights
	Ensure health and safety
	Improve local living conditions
	Satisfy the economic and living standards to which people aspire
Develop the capacity of the political system	Ensure transparent decision-making processes
	Develop open, inclusive and participatory governance
	Apply subsidiarity and ensure that competences are exercised at the most appropriate level

Table 7.2 *Main events in the growth of the sustainable development agenda*

1972	UN Conference on the Human Environment, Stockholm
1973	First EC Action Programme on the Environment
1985	First EC Directive on Environmental Assessment
1987	World Commission on Environment and Development: Brundtland Report, *Our Common Future*
1990	*This Common Inheritance: Britain's Environmental Strategy*
1992	UN Conference on Environment and Development (UNCED or the Earth Summit), Rio and creation of the UN Commission on Sustainable Development (CSD)
	Agenda 21: a comprehensive world-wide programme for sustainable development in the twenty-first century
	Climate Change Convention: international agreement to establish a framework for reducing risks of global warming by limiting 'greenhouse gases'
	Biodiversity Convention: international agreement to protect diversity of species and habitats
	Statement of Forest Principles for management, conservation and sustainable development of the world's forests
1994	*Sustainable Development: The UK Strategy*
1996	UN Habitat II Conference, Istanbul
	EU Expert Group on the Urban Environment Report on European Sustainable Cities
	The Aalborg Charter on Local Agenda 21 and the setting up of the European Sustainable Cities and Towns Campaign
1997	Earth Summit +5, five year review and adoption of Programme for the Further Implementation of Agenda 21 by UN General Assembly
	EU Amsterdam Treaty incorporates sustainable development as a fundamental objective of the EU
1998	Consultation on draft revised UK Strategy on Sustainable Development *Opportunities for Change* and supplementary strategies on business, tourism, biodiversity, forests, construction and sustainability indicators
	EU Communication on *Sustainable Urban Development: A Framework for Action*
1999	*A Better Quality of Life: A Strategy for Sustainable Development for the United Kingdom* (and additional special papers)
	Down to Earth: A Scottish Perspective on Sustainable Development
2000	EU Global Assessment of the Fifth Action Programme on the Environment
2001	EU Sixth Framework Programme on the Environment
	OECD Analytical Paper on Sustainable Development
2001	UN Habitat III Conference
2002	World Summit on Sustainable Development, Johannesburg
2005	*Securing the Future: Delivering UK Sustainability Strategy* (DEFRA)

of course complex and the references noted here point to many factors, but planners will often cite the contradictory and unhelpful nature of national policy and actions (especially outside the planning system) and the limited scope of planning. Recent changes to housing policy suggest that significant efforts are being made to provide a stronger framework, but considerable ambiguity remains at the national level. The Sustainable Communities programme is notable in the controversy it has engendered about just how sustainable it is (a question discussed in Chapter 6). One facet of the plan, the Millennium Communities Programme, is of particular interest since it has the central task of providing demonstration projects to promote more sustainable development in mainstream housing development, and this is discussed in Chapter 10.

Agenda 21 in the UK

The UK has made a very positive response to the commitments of Agenda 21. The 1992 Rio Earth Summit gave a major impetus to the elaboration of 'sustainable' policies. Agreement was given to Agenda 21, a comprehensive world-wide programme for sustainable development in the twenty-first century. In formulating this programme, major emphasis was placed on a very wide degree of participation. In the UK this is organised at central and local government levels.

Two years after the Rio Summit the government published *This Common Inheritance: Britain's Environmental Strategy* which was followed by annual monitoring reports. In 1994 this was effectively replaced by *Sustainable Development: The UK Strategy*. (This was the first national sustainability strategy arising from the Rio Declaration to be published.) The 1997 Labour administration undertook to revise the strategy and published numerous consultation documents during 1998. The revised strategy, *A Better Quality of Life*, was published in May 1999. Scotland published its own sustainable strategy consultation document *Down to Earth* in the same year. Wales published its Sustainability Strategy in 2004 and the Northern Ireland Executive planned to publish a strategy in 2005.

The 1999 Strategy promoted four main objectives: social progress (the main addition from the previous strategy), protection of the environment, prudent use of natural resources and maintenance of high levels of economic growth (see Box 7.3). The strategy identified 147 sustainable development indicators including 15 headline indicators and made a fairly frank assessment of the baseline position and trends for each.[9] The range of indicators, including for example, levels of crime, give a clear indication of the very broad definition that the government has given to sustainable development.[10] The sustainable development indicators have been given some bite by their incorporation into the government's Annual Spending Review whereby each department prepares a sustainable development report. The indicators are also embedded into some of the Public Service Agreement targets which define objectives for government to pursue. (But this all assumes that indicators are true measures of sustainable development.) The devolved administrations have developed their own indicators; the Scottish Executive has a list of 24 indicators; the Welsh Assembly uses 12 indicators.

The indicators for England are summarised in Table 7.3, together with the assessments of baseline performance made in 1999 as published in *A Better Quality of Life*, and performance between 1999 and 2005 as published in the last annual review of 1999 *Strategy in Achieving a Better Quality of Life*.[11] The strategy found favour with many interests, not least because all areas of public policy are given some prominence in the objectives of sustainable development. There is a strong theoretical argument for a holistic perspective that recognises the part that must be played by all sectors of government in achieving social, economic and environmental sustainability objectives. But clarity of purpose is sorely compromised, especially in comparison with approaches elsewhere, and there is little doubt that the economic imperative still holds sway. The strategy is in the ecological modernisation approach with a concentration on increasing economic growth but to be achieved while reducing pollution and the use of natural resources. Thus some indicators for UK sustainability have more than a passing resemblance to the OECD's indicators for economic competitiveness.

Levett (2000) describes the list of indicators as 'a towering achievement' especially in their breadth but notes that many are concerned with inputs as proxies for ends or measuring actual progress towards greater sustainability – as for example in measuring the existence of Agenda 21 strategies rather than their impacts. Such criticisms of indicators are well known. Selection is intensely political because the indicators are in effect the definition of sustainability, and they may reveal great shortcomings. Above all, as the strategy itself accepts, increasing eco-efficiency will not be able to keep pace with 'business as usual economic growth'. As Levett (2000) explains, 'eco-efficiency may have a useful contribution to make, but it is fanciful to the point of irresponsibility to expect it to be the main means of reconciling economic and environmental aims'. Thus ecological modernisation is not a

BOX 7.3 PRINCIPLES OF SUSTAINABLE DEVELOPMENT FOR THE UK GOVERNMENT

The 1999 UK Strategy for Sustainable Development identified four central aims:

- social progress which recognises the needs of everyone
- effective protection of the environment
- prudent use of natural resources
- maintenance of high and stable levels of economic growth and employment.

The framework goal of the 2005 Strategy, *Securing the Future* (p. 16) is as follows:

The goal of sustainable development is to enable all people throughout the world to satisfy their basic needs and enjoy a better quality of life, without compromising the quality of life of future generations. For the UK Government and the Devolved Administrations, that goal will be pursued in an integrated way through a sustainable, innovative and productive economy that delivers high levels of employment; and a just society that promotes social inclusion, sustainable communities and personal wellbeing. This will be done in ways that protect and enhance the physical and natural environment, and use resources and energy as efficiently as possible. Government must promote a clear understanding of, and commitment to, sustainable development so that all people can contribute to the overall goal through their individual decisions. Similar objectives will inform all our international endeavours, with the UK actively promoting multilateral and sustainable solutions to today's most pressing environmental, economic and social problems. There is a clear obligation on more prosperous nations both to put their own house in order, and to support other countries in the transition towards a more equitable and sustainable world.

The 2005 Strategy also sets out five principles that will form the basis for policy:

- living within environmental limits: respecting the limits of the planet's environment, resources and biodiversity – to improve our environment and ensure that the natural resources needed for life are unimpaired and remain so for future generations
- ensuring a strong, healthy and just society: meeting the diverse needs of all people in existing and future communities, promoting personal wellbeing, social cohesion and inclusion, and creating equal opportunity for all
- achieving a sustainable economy: building a strong, stable and sustainable economy which provides prosperity and opportunities for all, and in which environmental and social costs fall on those who impose them (polluter pays), and efficient resource use is incentivised
- promoting good governance: actively promoting effective, participative systems of governance in all levels of society – engaging people's creativity, energy, and diversity
- using sound science responsibly: ensuring policy is developed and implemented on the basis of strong scientific evidence, while taking into account scientific uncertainty (through the precautionary principle) as well as public attitudes and values.

Table 7.3 *The UK's strategic objectives and headline indicators for sustainable development*

Headline indicator	Baseline assessment	Overall performance 1990–8	1999–2005
Maintaining high and stable levels of economic growth and employment			
H1 Total output of the economy (GDP and GDP per head)	Between 1970 and 1998 the output of the economy has grown 86 per cent in real terms	+	+
H2 Total and social investment as a percentage of GDP	Total investment has declined from 20 per cent of GDP in 1970 to 17 per cent of GDP in 1998 – and business has invested consistently less per head than other G7 countries	–	0
H3 Proportion of people of working age who are in work	In May/July 1999 the employment rate was 74 per cent of those of working age, about the same as 1970 though it has increased for women and decreased for men	0	+
Social progress which recognises the needs of everyone			
H4 Success in tackling poverty and social exclusion (children in low income households, adultswithout qualifications and inworkless households, elderly in fuel poverty)	Little change since 1990 with 19 per cent of children in low income households; 17 per cent of working age people with no qualifications; 13 per cent people in workless households and about 60 per cent of single elderly households in fuel poverty	–	+
H5 Qualifications at age 19	The proportion of 19 year olds with NVQ level 2 (or 5 GCSEs grade C or above) was 45 per cent in 1984 and 74 per cent in 1999.	+	+
H6 Expected years of healthy life	Life expectancy has increased (74 years for men and 79 years for women in 1995) but more years are spent in poor health	0	0
H7 Homes judged unfit to live in	Improvement from 8.8 per cent unfit homes in 1986 to 7.2 per cent in 1996 (1.5 million homes)	0	+
H8 Level of crime	Recorded crime of all types has increased substantially since 1970; burglary and theft from cars has decreased since 1993 but violent crime continues to rise	+ and –	+
Effective protection of the environment			
H9 Emissions of greenhouse gases	UK emissions of greenhouse gases fell by 9 per cent between 1990 and 1997 mainly because of the switch from coal to gas and nuclear power electricity generation; transport emissions are becoming more significant	+	+
H10 Days when air pollution is moderate or higher	The average number of days recorded as moderate or higher per recording site fell from 60 days in 1993 to 25 days in 1998	+	–

H11 Road traffic	Over the last 20 years, the amount of car mileage per head has increased by 65 per cent, road traffic is now eight times that in 1950, (car traffic fourteen times) and it is forecast to grow by a third over the next 20 years	0	+ and −
H12 Rivers of good or fair quality	Nearly 95 per cent of the river network is of good or fair quality. River lengths that are of good chemical quality rose from 48 per cent in 1990 to 59 per cent in 1998	+	+
H13 Populations of wild birds	Populations of some farmland and woodland birds have fallen by more than half since the mid 1970s, though populations of others, including open water birds, have been fairly stable	−	0
H14 New homes built on previously developed land	The proportion of new homes built on previously developed land has been much the same since 1989 (though it increased from 1985) and in 1997 was 55 per cent	0	+

Prudent use of natural resources

| H15 Waste arisings and management | Household waste has increased by 26 per cent from 1983/84 to 1997/98 and now stands at between 170 and 210m tonnes, 60 per cent of which is disposed of by landfill | − | − and 0 |

Sources: *Quality of Life Counts: Indicators for a Strategy for Sustainable Development for the United Kingdom: A Baseline Assessment,* Government Statistical Service, 1999 and 2004 *Achieving a Better Quality of Life: Review of Progress Towards Sustainable Development, Government Annual Report 2003.* Note that the first performance column above is from the baseline assessment from 1990 to 1998 not change from 1990 to 2005 as given in the last Annual Report. In the context of devolution the Northern Ireland Executive, Scottish Executive and Welsh Assembly are responsible for elaborating on these national goals and indicators.

long-term solution. Nevertheless, improvement in the sustainability indicators is fast becoming an end in itself, while the political significance and impact of the strategy has been questioned, especially in relation to public awareness.[12] Consequently, considerable effort has gone into publicising the sustainable development goals of government, although often stressing the economic growth elements. Indeed, Davoudi (2001) points out that the Foreword to the 1999 *Sustainable Development Strategy* by Tony Blair barely mentions the environment.

The Sustainable Development Commission has been at the forefront of monitoring the government's commitment to sustainability and its conclusions make the title of its 2004 report on progress, *Shows Promise But Must Try Harder,* which was based on an independent report on the headline indicators by Levett-Therivel

Sustainability Consultants (2004). The Commission challenged the government 'to create a new Strategy that is unified and much more strongly driven by a fundamental overarching commitment to sustainability at all levels and in all parts of Government; it should be a core part of the programme of all Departments, led from the centre' (p. 4). Twenty challenges are made in all, many to do with the government setting an effective lead in its own departmental activities. The report calls for new indicators especially on measuring economic progress beyond output and employment; a more fundamental approach to transport, especially to 'tackle head-on the failure of many parts of the transport sector to bear their full environmental costs' (p. 5); and the use of price signals through taxation to ensure that consumers understand better the sustainability impacts of their behaviour.

A new Strategy was already on the agenda and after wide consultation on a draft revision *Taking It On*, it was published in 2005 as *Securing the Future: Delivering UK Sustainability Strategy*. It may be noted here that although widely publicised, only 900 written responses were made to the consultation, which is a fraction of the responses made to the 2001 Green Paper on the planning system and is an indicator of public awareness and concern about the issue perhaps. The main development in the new strategy is a revised and more integrated discussion about the nature of sustainable development. The Strategy itself reports that government departments cherry picked from the four principles set out in the 1999 version. Along with this, and in line with the challenge set by the Sustainable Development Commission, there is more said about how government departments, including the devolved administrations' share ownership and responsibility for its application. The four principles are expanded with a stronger statement on respecting environmental limits and working within the capacity of the environment to absorb development; a revised approach to 'a sustainable economy . . . in which environmental and social costs fall on those who impose them' (p. 16). Climate changes figures much more prominently,[13] and the planning system is identified as a 'key lever' for making necessary changes to help meet targets for slowing the growth of greenhouse gases and energy use.

The contribution of planning to sustainable development is conveniently summarised in the Strategy on one whole page of the document (p. 116), which refers to policies already in place (in some cases for a considerable time) such as the brownfield land targets, sequential test and others discussed elsewhere in this book. The centrepiece of this explanation is s. 39 of the Planning and Compulsory Purchase Act 2004 which (though rather distorted by the legal construction) has the effect of requiring those operating the planning system at both regional and local levels 'to exercise the function with the objective of contributing to the achievement of sustainable development'. For further advice on what this means the Act points to national policies and advice contained in guidance issued by the Secretary of State. Thus the 2005 Sustainable Development Strategy is required reading.

Regional sustainable development frameworks

In February 2000 the DETR published guidance for the preparation of regional sustainable development frameworks (RSDFs) with a requirement for them to be in place by the end of 2000 (although this has proved to be optimistic).[14] The regional bodies have been responsible for adopting the frameworks in coordination with the sustainable development work of the regional development agencies and regional planning guidance, now regional spatial strategies. The agencies had previously been issued with guidance on incorporating the principles of sustainable development into their economic strategies and some have set up extensive sustainability issue networks or round tables.

The frameworks are non-statutory guidance but it is widely recognised that the regional and sub-regional levels are crucial for many sustainable issues such as waste, water management, renewable energy, agriculture, tourism and urban–rural interdependencies (McLaren *et al.* 1998). Progress on regional sustainable development frameworks was underway in some regions, not least because of concerns that neither the regional economic strategies nor regional guidance has fully addressed Agenda 21. To counter this, the frameworks propose a long-term and high-level vision and establish regional indicators and targets. The objective is to join up resource considerations and they should certainly provide a common context for the preparation of both RDA strategies and RPG. Evaluation of the regional sustainable development frameworks concludes that they have not been a strong influence on activity or policy-making in the regions (CAG Consultants and Oxford Brookes University 2002). The government has promised yet more advice.

Local Agenda 21

At the local level, Local Agenda 21 calls for each local authority to prepare and adopt a local sustainable development strategy. These local efforts have been aided by the work and publications of the Improvement and Development Agency (IDeA, formerly the Local Government Management Board or LGMB).[15] A major feature of the consultation programme at the local level is that it involves much more than the term 'consultation' often means. Groups have been established in local areas to debate the meaning of sustainability and to determine how progress towards it can be achieved and assessed, following the principle of 'you can only manage what you can measure'. These local endeavours are designed to produce policies and indicators which are locally appropriate. The research has underlined the importance of this local 'ownership'. There is a positive and a negative aspect to this. Positively, 'Agenda 21 is as much concerned with the *process* of sustainable development – participative, empowering, consensus-seeking, and democratic – as it is with *content*' and 'social processes of securing agreement on and commitment to sustainability aims are indispensable' even where the requirements for sustainability are determined externally (LGMB 1995a). Also, sustainable development strategies draw together many actors into an inclusive network, but 'this, paradoxically, is potentially its greatest weakness, as excessive inclusivity may lead to a lack of clear purpose, direction and commitment' (Selman and Wragg 1999).

In short, the changes in attitudes and behaviour which will be required by policies of sustainability will come about only if they are acceptable. The negative side to this is the widespread distrust of both local and central government which research has uncovered (Macnaghten *et al.* 1995). Agenda 21 emphasises equality and economic, social and political rights. Among the top concerns are poverty, unemployment and deterioration in the quality of life and the health of local communities. These are reflected to some extent in the local sustainability indicators chosen.[16] But similar to practice at the national level, the indicators generally reflect the data that is routinely collected and readily available, and there is limited opportunity for comparison from one authority to another (Cartwright 2000). The process can 'easily become cosmetic and bogged down in group dynamics and inertia' (Scott 1999). In addition, although almost all local authorities have prepared a Local Agenda 21 Strategy, their commitment has varied considerably (Cartwright 1997).

Local Agenda 21 has certainly contributed to the growing awareness of environmental and sustainability issues in local politics, but the sum of evaluations (and a review of examples of strategies) suggests that they have succeeded simply in presenting the agenda, with limited impact on mainstream policy. The question now is how the Local Agenda 21 process proceeds. The likely direction is integration with community planning and Best Value (LGA and IDeA 1997; Hams 2000; Christie 2000).

Environmental politics and institutions

Environmental politics has become an energetic force on the British scene since the early 1970s and this is reflected in the growth of environmentally related government units and agencies, advisory panels and interest groups. Its rise has been prompted by a miscellany of matters, with the most significant first step being prompted by the oil crises of the 1970s which prompted a new look at resource depletion. Fear of environmental disasters has also played a part, and these seemed more credible after international catastrophes such as Seveso, Bhopal and Chernobyl, and in the UK at Windscale and Flixborough. The impact of development on natural resources has become clearer with the swing from widespread drought in the late 1990s to even more devastating floods in 2000. Radical campaigners, especially the anti-road tree-dwellers, have also played their part. Thus the environment has become part of the political coinage, and the parties vie with each other in producing convincing statements not only of their concern but also of their workable programmes of action.

Curiously, part of the growth of environmental consciousness was due initially to the lack of government concern. The environment was rarely the subject of political battles. Yet England has been a world pioneer on a number of environmental issues. The Alkali Inspectorate, which was established in 1863, was the world's first environmental agency. Some of the earliest voluntary organisations had their origin in England: for example, the Commons, Open Spaces and Footpaths

Preservation Society in 1865, and the National Trust in 1895 – an organisation that (with over 2 million members) has grown to be the largest conservation organisation in Europe. The Town and Country Planning Act 1947 introduced a remarkably comprehensive land use planning system (even though, in the circumstances of the time, much of rural land use was purposely omitted). Legislation on clean air has a long history, with its major landmark being the 1956 Act, passed following the killer smog of 1952. The UK also had the first cabinet-level environment department (the Department of the Environment was established in 1970) though its name was, for many years, more impressive than its achievements. Yet these historical events stand as lonely peaks in an otherwise flat plain: until the 1980s, the environment was not a salient political issue (McCormick 1991; Robinson 1992). Part of the reason for this has been the idiosyncratic nature of British pollution control: instead of the formal, legalistic, and adversarial styles common elsewhere. Britain has traditionally operated a system of comfortable negotiation between government technicians and industry. This curiously informal and secretive system avoids confrontation and legalistic procedures (McAuslan 1991).

All recent administrations have had strong advocates of the environmental cause at Cabinet level but the topic has not quite made it to the premier division issues in the UK as it has in some other countries. Thatcher, for example, was initially averse to environmental concerns, which she viewed as a brake on enterprise. Her administrations followed traditional British practice in responding 'pragmatically and flexibly, even opportunistically, when environmental issues have threatened to become too contentious' (Lowe and Flynn 1989: 273).[17] Blair was in government for more than three years before making a speech on environment policy, although in 1997 he made a call for all local authorities to complete Agenda 21 strategies by the end of 2000 and by 2003 with the UK Presidency of the G8 and EU in sight, he adopted the climate change cause. Accusations of hyperbole followed his headline-grabbing claim that climate change is a 'challenge so far-reaching in its impact and irreversible in its destructive power, that it alters radi-

cally human existence'.[18] Apart from that he has said little about the need for fundamental changes and regulation to achieve more environmental sustainability, but rather continued with an explicit commitment to the tradition of voluntary agreements with business (Warburton 2000).

While, global summits aside, prime ministers have not prioritised the environment, certain ministers, parliamentary select committees, agencies, advisory bodies and interest groups have continued to raise the profile of environmental issues. Parliamentary committees are often regarded as ineffectual, but they have been of great value to environmental groups by providing a new public platform and a route for exerting pressure on Parliament. In particular, the reports of the Environment Select Committee and the Royal Commission on Enviromental Pollution (RCEP) have become a respected source of alternative wisdom and relatively accessible information.

In 2000 the government established the Sustainable Development Commission,[19] which subsumed the UK Round Table on Sustainable Development and the British Government Panel on Sustainable Development. Its purpose, like its predecessors, is to review the extent to which sustainable development is being achieved, identify trends in unsustainability and deepen understanding of the concept. The two previous organisations made a considerable contribution to government policy with annual and ad-hoc reports. The Royal Commission on Environmental Pollution has also been an important advocate of improved environmental policy through such reports as *Transport and the Environment* (1994) and *Energy: The Changing Climate* (1999). It conducted a review of *Environmental Planning* and published its report in 2002,[20] pointing to the lack of integration of environmental considerations in planning arising from complex and overlapping legislation and functions. It called for substantial reform; many of the issues it raised are being addressed by the 2004 Act and other measures. It was particularly eager to see planning legislation amended 'to include both a statement of its general purpose and a set of criteria to be taken into account in decision-making' (p. 108). The discussion above shows that it succeeded on the first with the statutory purpose of sustainable development, though it had warned that 'the drafting needs to provide sufficient flexibility, but avoid such blandness or vagueness as to have little longer-term effect' (p. 108).

Another feature of British environmental politics is the active character of some of the important interest groups. Some of these are not merely interest groups: they own and manage extensive areas of land, and they fulfil a range of executive responsibilities. The National Trust and the Royal Society for the Protection of Birds, for instance, own and manage large areas of protected land. Such bodies are also characteristically charities and therefore debarred from overt political activity. Lobbying is thus not only well mannered but also discrete. The emphasis may be more on education than propaganda, though the distinction can be a fine one.

Governments may try to outflank environmental groups, but increasingly they cannot ignore them, particularly with their new access to power via the EU. Some thirty British groups, together with eighty from other countries, are members of the European Environmental Bureau which gives them access to the European Commission and the Council of Ministers (Deimann 1994). The British groups have been able to make good use of their experience in lobbying. According to Lowe and Flynn (1989: 272), they 'have adapted more easily than many of their counterparts to the successive rounds of consultation and detailed redrafting of directives and regulations that characterise Community decision-making'.

Impact of the EU

There can be no doubt that the EU has had a major impact on British environmental policy. Indeed, it is not much of an exaggeration to say that much of the government's policy has been dictated by its directives (Milton 1991: 11; Wilkinson *et al.* 1998). This is so despite the fact that the Treaty of Rome imposed no environmental obligations on member states, and the Community initially had no environmental competences. Indeed, Article 2 of the Treaty provided that sustained rather than sustainable growth was

the aim: 'a continuous and balanced expansion'. The international scene changed in the late 1960s and early 1970s, with a significant influence being the UN Conference on the Human Environment which was held in Stockholm in 1972. In the same year, the EC determined that economic expansion should not be 'an end in itself', and that 'special attention will be paid to protection of the environment' (Robins 1991: 7).

In 1973, the first EC *Action Programme on the Environment* was agreed, covering the period 1973–6. Further programmes followed: the fifth covers the period 1993–2000 and a sixth was published in 2002. The Single European Act 1987 gave added legitimacy by including environmental goals in the Treaty and, significantly, added the important provision that 'environmental protection requirements shall be a component of the Community's other policies' (Haigh 1990: 11). Since then the European Environment Agency (EEA) has been established with headquarters in Copenhagen providing a monitoring service for the European institutions.[21]

The environmental action programmes have had increasing impact on policy and practice in member states. They are 'forward planning' documents for emerging policies to be implemented by the EU and followed by national, regional and local governments. While they have no binding status, many of the proposals result in directives and other action. The Fifth Action Programme has brought a more comprehensive and long-term approach. The overriding aim of the programme was to ensure that all EU policies have an explicit environmental dimension. It stressed the potential of spatial planning instruments. EU documents are not noted for their brevity, and the programme documents are far too wordy to reproduce, but the following gives some flavour of their character. It also illustrates the importance attached to spatial planning instruments:

> The community will further encourage activities at local and regional level on issues vital to attain sustainable development, in particular to territorial approaches addressing the urban environment, the rural environment, coastal and island zones, cultural heritage and nature conservation areas. To this

purpose, particular attention will be given to: further promoting the potential of spatial planning as an instrument to facilitate sustainable development.

(EC 1992: *Towards Sustainability*)

But these are only objectives; they need to be transposed into agreed Community law and action. The great majority of EU environmental laws are in the form of directives (see Chapter 3) which give member states some freedom to choose the manner in which they are transposed into national law. It is unusual for directives to be transposed into national legislation by the due date – which is typically two months after adoption by the Council of Ministers. Nevertheless, they must be implemented 'in a way which fully meets the requirements of clarity and certainty in legal situations'. States cannot rely on administrative practices carried out under existing legislation (Wägenbaur 1991). Moreover, if a directive is not implemented by national law, it is possible for legal action to be taken by private parties to seek enforcement. The use of Community legislation has tended to give way in some areas to more general agreements and guidelines. (Chapters 3 and 4 consider the institutions and spatial planning actions of the EU.)

The 1987 Amsterdam Treaty incorporated sustainable development as a fundamental objective of the EU and since then there have been commitments to ensure environmental appraisal of all Community policies and actions. In 1999 the Commission undertook an evaluation of the Fifth Action Programme and reported in the *Global Assessment* which recognised that while some environmental improvements have been made 'less progress has been made overall in changing economic and societal trends which are harmful to the environment'. The report notes that economic growth 'simply outweighs the improvements attained by stricter environmental controls'. A Sixth EU Environmental Action Programme followed in 2002: *Environment 2010: Our Future, Our Choice*.[22] The proposals range over such matters as environmental taxation, improving the implementation of existing initiatives, completing the European network of habitats through *Natura 2000*, and preventing urban sprawl, especially along coasts. Two of the proposals

in the agenda for action are of particular interest here. The first is the commitment to encourage better land use planning and management decisions while ensuring that 'environmental issues are properly integrated into planning decisions'. The Commission was given the task of following this up through preparation of a communication on environment and planning. In this case the lead is taken by the Environment Directorate, whereas other activity on INTERREG and the ESPON (explained in Chapter 4) is led by the Regional Policy Directorate. Suffice to say that the European Commission is no better joined up than the UK government. The proposed communication is likely to concentrate on promoting changes to existing Community legislation and actions, rather than introducing new measures, although they are not ruled out. A scoping study undertaken by consultants ECA and Scott Wilson (2002) was based mostly on interviews with government representatives, academics and environmental non-governmental organisations (NGOs). The interviewees agreed that the EU could say more about the integration of environment in planning, not least to address the commonplace trade-off of environmental concerns for apparent economic and social gains, but voiced concerns about the Commission intervening in any other way in spatial planning in the member states. This ambivalent response may be one reason why further progress has not been made on the Communication.

More attention has been given to the second main commitment in the Sixth Environment Action Programme, the *Thematic Strategy on the Urban Environment*, one of seven thematic strategies to be prepared. The Commission began a wide-ranging consultation on the Urban Environment Strategy in 2003 and published a communication, *Towards a Thematic Strategy for the Urban Environment* in 2004. This set out the overall aim to improve the 'environmental performance and quality of urban areas'. Four themes have been identified:

- *sustainable urban management:* concerning, for example, the adoption of explicit environmental targets, actions and monitoring by local authorities in an integrated urban management system

- *sustainable construction:* concerning, for example, the minimisation of resource inputs to construction, recycling of construction materials, and maximising energy efficiency in new construction
- *sustainable urban design:* concerning, for example, the appropriate physical form of urban areas for more sustainability, redesigning and retro-fitting existing urban areas and building on brownfield land
- *sustainable transport:* including, for example, the types of measures to promote more sustainable mobility and tools for evaluating the impacts of transport measures.

New EU laws are being considered which will require cities with a population of more than 100,000 to prepare environmental management and sustainable transport plans. Needless to say, the UK government is against these proposals (which may not add much to what is already done for big cities) but local government is more positive, on the basis 'that the proposed legislation would give badly-needed weight to environmental work' (Atkinson and Mills 2005: 107). The final strategy was published early in 2006 and was issued as a joint decision of the European Council and Parliament. It could be of help to local authorities (subject to the commitment of resources to the task) but it is a tall order to deliver

> tools which oblige and enable local authorities and their partners to apply all the policy instruments at their disposal (land use and other plans, environmental permitting and inspection, existing EMAS [eco-management and audit scheme] work, procurement, fiscal measures, and so on) in a mutually supportive way to achieve measurable improvement in the urban environment.
>
> (Atkinson and Mills 2005: 108)

Environment agencies

The Environment Act 1995 provided for the establishment of an Environment Agency for England and Wales and an equivalent Scottish Environment Protection Agency. The idea of such an agency had been resisted by the government for a number of years, and the change of heart was primarily in response to demands from industry for a one-stop shop for environmental regulation.[23] Another factor in the debate was the importance of having an agency that was able to negotiate from a position of strength with the EU.

Against this background, the Environment Agency has taken over the responsibilities of bodies which had been established by a reorganisation only a few years earlier. In England and Wales, these were the National Rivers Authority, Her Majesty's Inspectorate of Pollution, and the local waste regulation authorities. In Scotland, they were the river purification authorities, HM Industrial Pollution Inspectorate, and the waste regulation and local air pollution responsibilities of the district and islands councils. In Northern Ireland, the DoENI has all the responsibilities for environmental protection except waste disposal which lies with the local authorities.

The agencies are non-departmental public bodies; the management has a large degree of freedom within the framework of ministerial guidance and its management framework. The framework is based on the government's overall strategy for sustainable development explained above. It is therefore important that they take an integrated approach to their responsibilities: this, indeed, is its essential raison d'être. Sustainable development is also leading the agencies to reflect on the traditional reactive and regulatory approach and to add a 'more forceful dimension'. Part of this for the Environment Agency includes a commitment to creating a single regulatory system that covers 'the environmental impact of processes and their resource use, products and their effects and their impact on land use'. And it foresees an increasing role for local authorities and development agencies such that land use planning and development control are 'more closely aligned to environmental risks and steps necessary to avoid them' (2000: 38–9).[24]

Establishing more integration and prevention rather than regulation is a considerable challenge. The functions of the agencies are already very wide including industrial pollution, aspects of waste including radioactive substances, water resources and quality, the implementation of a number of EU Directives (and in Scotland local air pollution control). Further links with land use planning are also anticipated. The long-term strategies of the environment agencies adopt a thematic approach (shown in Box 7.4). The list illustrates the breadth of their portfolios. Planning comes into contact with the Agency on a number of its key concerns, for example, air quality, waste and flooding which are discussed separately below.

BOX 7.4 GOALS OF THE ENVIRONMENT AGENCIES

England

- People will know that they live in a healthier environment, richer in wildlife and natural diversity.
- Wildlife will thrive in urban and rural areas. Habitats will improve for the benefit of all species. Everyone will understand the importance of safeguarding biodiversity.
- The emission of chemical pollutants into the atmosphere will decline greatly and will be below the level at which they can do significant harm.
- Our rivers, lakes and waters will be far cleaner. They will sustain diverse and healthy ecosystems, water sports and recreation.
- Our land and soils will be exposed far less to pollutants. They will support a wide range of uses including production of healthy, nutritious food and other crops without damaging wildlife.
- Industry and businesses will value the assets of a rich and diverse natural environment. In the process, they will reap the benefits of sustainable business practices, improve competitiveness and value and secure trust in the wider community.
- All organisations and individuals will minimise the waste they produce. They will reuse and recycle materials far more intensively and use energy and materials more efficiently.
- Drastic cuts will be made in the emission of 'greenhouse gases' such as carbon dioxide. Society as a whole will be prepared for probable changes in our climate.
- Flood warnings and sustainable defences will continue to minimise injury, damage and distress from flooding. The role of wetlands in reducing flood risk will be recognised and the environmental benefits from natural floods will be maximised.

Source: Environment Agency (2000) *Our Vision*

Scotland

- People will have peace of mind from knowing that they live in a clean, safe, and diverse environment that they can use, appreciate, and enjoy.
- Both urban and rural areas will have an obvious and overall improvement in the extent and quality of their habitats and the wildlife that they support.
- Industry and businesses generally will be managed in a way that fully protects human health and the environment.

- Waste and wasteful behaviour will no longer be a major environmental threat because of the re-use of resources and the adoption of sustainable waste management practices.
- Neither human health nor the natural and man-made environments will be damaged by emissions to the atmosphere.
- There will be sufficient clean and healthy waters to support people's needs and those of wildlife.
- The natural resources provided by the land will be enhanced, harm to people and wildlife will be avoided, and a wide range of land uses will be supportable.
- Flood warnings and sustainable defences will continue to prevent deaths from flooding; property damage and distress will have been minimised; and all the benefits to be derived from natural floods will be exploited.
- Greenhouse gas emissions will have been greatly reduced and society will have adapted efficiently to climatic change and be prepared for further changes.

Source: Scottish Environment Protection Agency (1998) *Environmental Strategy*

BATNEEC, BPEO and BPM

In their regulation of pollution role environmental bodies have generally sought to achieve the *best practicable means* (affectionately known as BPM) of dealing with problems – 'means' that will go as far as seems reasonable towards meeting desirable standards but which do not involve too great a strain on the polluter's resources. This approach has a long history: indeed, it has been the cornerstone of industrial air pollution control since the Alkali Act 1874. Its modern version has been expanded to the *best practicable environmental option* (BPEO), which retains the element of negotiation but involves a wider consideration of environmental factors and an openness which was foreign to its predecessor (RCEP 1988: para. 1.3).[25] Central to this principle is the recognition of the need for a coordinated approach to pollution control, taking into account the danger of the transfer of pollutants from one medium to another, as well as the need for prevention. The Environment Protection Act 1990 introduced a requirement for the regulating authority to ensure that the *best available techniques not entailing excessive cost* (BATNEEC) are being used.

(1) for preventing the release of prescribed substances into an environmental medium, or, where that is not practicable, for reducing the release to a minimum; and

(2) for rendering harmless any other substance which could cause harm if released into any environmental medium.

BATNEEC is the concept favoured by and introduced in EU Directives which have been adopted in UK environmental law. It is the responsibility of the operator to demonstrate that the requirements of BATNEEC are met and also to demonstrate their competence and experience, and that effective environmental management controls are in place. Additionally, certain statutory environmental standards ('quality objectives'), specified emission limits or national quotas have to be met.

Where a process involves the release of harmful substances to more than one medium, BPEO must also be demonstrated – thus there may be trade-offs among the effects in one environmental medium against another. In order to judge the effects of different emissions in different media an integrated permit process has been adopted.

Integrated pollution prevention and control

Environmental regulation is progressively adopting *integrated pollution control* (IPC) as required by the EU.[26] This is the administrative apparatus for implementing the BPEO. It contrasts with the customary British method of operating different controls in isolation, with separate approaches to individual forms of pollution. The crucial problem with this is that pollution does not abide by the boundaries of air, land and water: pollution is mobile. In the jargon, it is a 'cross-media' problem.

A 1996 EU Directive (96/61) extended the regulatory regime and controls when implemented by the Pollution, Prevention and Control (PPC) Act and Regulations in 2000. The new regime, *integrated pollution prevention and control* (IPPC), is being implemented sector by sector and will involve regulation of 6,500 industrial processes with integrated controls over pollution, noise, waste reduction, energy efficiency and site restoration. All operators of installations covering any of the listed processes require a permit which will cover all controls where the applicant will need to demonstrate that best available techniques are being used.

Annex 1 of *PPS 23 Planning and Pollution Control* (2004) provides a review of who does what in pollution control, and is especially helpful in explaining where and how the planning system should tackle pollution issues. It should be remembered that many of the 'polluting problems' that planners have to deal with will not come under the pollution regulation regime. It points out that

> the majority of planning applications where pollution issues are potentially a consideration will not relate to [the PPC Act] processes but to smaller-scale site and industrial/commercial premises (e.g. backstreet vehicle re-spraying, existing low level flues, dry cleaning establishments etc.) . . . Smaller scale processes may be less well managed and more likely to cause problems, though of a lower intensity or more localised scale.

Thus planning will need to consider a wider range of developments whose polluting activities may be relevant to the PPC regime, or not come within the regime but constitute a *statutory nuisance*, or not come under either of these but still result in a loss of amenity which the planning system may be able to deal with.

In many cases close cooperation is required between the planning authority and the Environment Agency.

Penalties for pollution

A striking feature of the environmental legislation is the severity of the penalties for polluting (Harris 1992a). One feature in particular is noteworthy: the use of 'strict liability'. Generally, under English law, the prosecution has the burden of proving that a defendant is guilty beyond reasonable doubt. The 1990 Act provided that where it is alleged that BATNEEC has not been used in a prescribed operation, 'it shall be for the accused to prove that there was no better available technique not entailing excessive cost than was in fact used'. This makes an offence one of 'strict liability', in contrast to the traditional one of 'fault-based'.[27] Though its use is likely to be rare, it is indicative of the change in official attitudes to pollution (documented in Rowan-Robinson and Ross 1994). It will also involve highly technical matters which may present severe difficulties for the existing courts. Indeed, some have argued that there is a need for a specialised court (Carnwath 1992; Department of Land Economy, Cambridge University 1999) and the RCEP Report mentioned earlier recommended the creation of environmental tribunals. There has been no action on this so far.

Economic instruments of environmental policy

Public opinion is in favour of regulatory standards because of their apparent fairness: all are required to meet the same target. Polluters may also like them because of the certainty which they give to the market. In fact, the fairness is illusory. Fixed standards impose quite different costs on different firms depending for example on the state of their machinery and processes. More important in terms of effective environmental improvement, firms will tend not to seek anything beyond the regulatory standard even if they can achieve a higher standard at relatively low cost. They have no incentive to do so, unless they thereby obtain other benefits.

There are considerable advantages to be derived from designing pollution controls in a way that gives firms economic incentives to reduce pollution to the maximum extent. If, for example, a tax is levied for every ton of waste produced, a firm will be motivated to review its processes to reduce its waste to the minimum. Positive market incentives may also overcome the reluctance of some firms to meet regulatory standards and reduce the costs of regulation. Since administrative resources are typically inadequate, this is a significant issue. Overstretched agencies may well know that some firms are in default, but they may have some difficulty proving it, or they may have to accept a firm's assurance that it is doing the best it can. Particularly bad cases may be prosecuted, but this takes even more time and resources, and the courts can be unpredictable. In all, as the UK Round Table on Sustainable Development (2000) and others have pointed out, there is a strong case for further developing the use of economic instruments for implementing environmental policy and sustainability in the UK, especially when used as part of complementary packages including regulation, negotiated agreements, and changes to mainstream spending programmes.

Economic instruments can take many forms. The simplest economic instrument is a tax – either to deter negative actions (waste) or to promote positive ones (technological developments). For example, a tax may be levied on pollution at a rate determined in relation to the damage caused and the costs of clean-up. Such a tax could be levied on lead or carbon content. (Several European countries have such a carbon tax.) The tax provides an immediate incentive to firms to reduce their use of the pollutant – and it is a continuing incentive. The difficulty arises in setting an equitable rate – a problem which also arises with marketable pollution permits which the government are intending to introduce.

Economic incentives can be applied to some types of waste with a deposit-refund system. This is essentially the same as the charges on returnable bottles, though rather more complicated. The producer of something which would become a waste after it has been used in a manufacturing process (a solvent for instance) would be required to pay a charge for each unit produced. This would increase its price (thereby introducing an incentive for reduction in its use). A refund of the charge would be payable to anyone who returned the solvent after its use. This system has the advantage of providing a disincentive to illegal tipping. The same system can be applied to motor vehicles.

All the advisory bodies have spent considerable time in debating and recommending the use of economic instruments and they recognise progress made while pointing out that there is undoubtedly much more that can be done.[28] Some possible innovations, such as road pricing, have been debated for many years, but the technical and political difficulties constitute a major obstacle (discussed in Chapter 11). Progress has been made in the fields of landfill tax implemented in 1996 (following the EU Directive) and the creation of environmental trusts (including one that supports cars with alternative fuels). Satellite national accounts have been prepared that address economic, social and environmental costs but so far are separate from the main national accounts. The budget statement now includes a note on its environmental impacts.

Proposals have been mooted in consultation papers,[29] but there is also lots of scope to make improvements by amending existing mainstream spending especially in relation to procurement and subsidies. Environmentally damaging subsidies have been estimated at £20 billion per year (Government Panel on Sustainable Development Third Annual Report 1997). Some, including company car tax benefits and road fund licensing, are now being amended to reflect environmental costs. In the 2005 Sustainable Development Strategy, the government said it would press for EU air traffic to be included in an emissions trading scheme from 2008 or as soon as possible afterwards.

Local environment agency plans

The requirement for the production of catchment management plans previously held by the NRA was transferred on their creation to the Environment

Agencies. In England and Wales these have been supplemented by local environmental agency plans (LEAPs) which are of smaller scale covering a small or sub-catchment area and cover the full range of topics for which the Environment Agency is responsible – primarily pollution, waste, water and air quality. Local environment agency plans are non-statutory documents and progress in their preparation has been slow but they may be a material consideration in development control. Local planning authorities are encouraged to take them into account in the review of development plans. There is wide consultation with local authorities, other bodies and the public during their preparation. Note that although the catchment type of boundary is particularly useful for water management, it is less relevant for air quality and waste management which are traditionally much closer to local authority boundaries (Farmer *et al.* 1999). Nevertheless they offer possibilities for the better integration of environmental policy in a territorial plan and in this sense perhaps may develop in a similar way to the German landscape plans.

The EU Water Framework Directive will require the preparation of river basin management plans for geographical areas around catchments, coasts and estuaries. The plan will need to link water management to other environmental and economic activities in the area, for example in relation to impacts on the demand for water and the water environment. Almost certainly river basin management plans will be developed from existing catchment plans (and in England and Wales LEAPs). They will have statutory force and may be binding, with yet further implications for town and country planning.

Clean air

Concern about air pollution is not new: it was as early as 1273 that action in Britain was taken to protect the environment from polluted air. A royal proclamation of that year prohibited the use of coal in London and one man was sent to the scaffold in 1306 for burning coal instead of charcoal. Those who pollute the air are no longer sent to the gallows, but, though gentler methods are now preferred, it was not until the disastrous London smog of 1952 (resulting in 4,000 deaths) that really effective action was taken. The Clean Air Acts of 1956 and 1968 introduced regulation of emissions of dark smoke, grit and dust from furnaces, chimney heights and domestic smoke. Local authorities were empowered to establish *smoke control areas* which were very effective (coupled to the switch from coal fires to central heating).

Air quality has improved considerably since the early 1960s: smoke emissions have fallen by 85 per cent since 1960, the notorious big-city smogs are a thing of the past, and hours of winter sunshine in central London have increased by 70 per cent. In matters of the environment, however, problems are never 'solved': they are merely replaced by new ones – and there is now a long list of damaging air-borne pollutants that is the subject of new research, policies and actions.[30] Current trends show that the improvements made in respect of industrial and domestic sources of air pollution are being eroded by the damaging effect of increased traffic sources (Banister 1999; Stead and Nadin 2000). Moreover, severe problems in the shorter run can be expected in 'hotspots', particularly in congested urban centres.

Government has been very active on air quality and in this field the UK is a leader. The 1994 and 1999 UK sustainable development strategies both gave prominence to improving air quality. A UK *National Air Quality Strategy* (1997) has been agreed with national standards and targets and is already under review in the light of new research findings. A comprehensive network of air quality monitoring stations is in place and much longitudinal data available.[31] Local authorities have to undertake periodic reviews of air quality and identify the areas where national targets are not likely to be met.[32] They then produce *local air quality management plans* for the specified areas. Air quality management plans will seek to reduce emissions through addressing the sources and distribution, especially traffic (which is discussed further in Chapter 11). They may, in principle, designate areas which should be closed to traffic or be restricted to low emission vehicles – although care will be needed to avoid displacement effects and some 'local' pollution will have a non-local source. In practice they are not proving to be so radical (Miller 2000). Recent advice to planning authorities suggests that the long-held separation of the role of land use planning from pollution control is being eroded. The 1997 paper on *Air Quality and Land Use Planning* says that

> Where the impact of development is likely to be significant in air quality terms, then, provided the impact relates to the use and amenity of land, the planning application may be refused or the impact mitigated by the imposing of conditions.
>
> (DETR 1997: para. 370)

The question of significance is obvious in air quality management areas where the local authority is seeking to lead improvement in air quality but it may also be necessary to consider the 'cumulative impacts of a number of smaller developments on air quality, and the impact of development proposals in rural areas with low levels of background air pollution' (PPS 23, Appendix A). Although restricted to questions of the use and amenity of land (as opposed to health) there is a clear signal to authorities to use planning powers to generally improve air quality conditions. But air quality standards are not easy to determine: the scientific base is inadequate, and a great deal of judgement is necessary. The governmental response to this has been to work towards two measures: a long-term goal and an operational threshold which indicates when quality conditions are so low as to require an immediate response. (Confusingly, these are both termed 'standards'.) Local planning authorities are expected to have regard to the local air quality management plans and to the national standards in preparing land use development plans and in carrying out other duties such as transport planning.

The EU has played a critical role in bringing about air quality initiatives, largely supported by the UK. The Air Quality Framework Directive of 1996 sets target values for 12 air pollutants which are elaborated and revised under daughter directives. The first daughter directive was agreed under the UK Presidency in 1998 and adopted in 1999. It covers sulphur dioxide, nitrogen dioxide, particles and lead.

The second addresses carbon monoxide and benzene, and a third ozone.

The water environment

The Environment Agencies hold the main regulatory powers over the water environment, although they have no operational responsibilities (these are carried out by the water service companies or in some cases local authorities). The agencies have statutory functions in relation to water resources, and the control of pollution in inland, underground and coastal waters. Their powers are wide but there are three critical issues in relation to planning: water quality and pollution, the maintenance of water supplies and flooding.[33]

On water quality the agencies can take preventive action to stop pollution, take remedial steps where pollution has already occurred, and recover the reasonable costs of doing so from a polluter. The agencies have inherited and continued to develop a sophisticated and relatively public regulatory system which involves the setting of water quality objectives and a requirement that consent is obtained for discharges of trade and sewage effluent to controlled waters. Extensive monitoring programmes include surveys of the quality of rivers, estuaries and coastal waters. The highly detailed figures produced from these surveys are not easy to summarise or to interpret and performance is mixed. River quality is improving steadily. In 2003 only 4 per cent of rivers monitored (across 7,000 sites) were considered to be of poor quality and 1 per cent were bad, compared with 10 per cent in these categories in 1990. Although the majority of rivers still have high levels of phosphate (53 per cent) and nitrate (27 per cent), this was down on 1990 rates (64 per cent and 30 per cent). Mostly as a result of investment in sewerage works, bathing water quality is improving. In 2000 44 per cent of the 471 beaches tested in England and Wales and in 2004, 80 per cent of 491 beaches, met the standards of the EU Bathing Waters Directive,[34] and 95 per cent passed the mandatory tests.

As the UK is a country surrounded by water and with an annual rainfall of around 1,100 millimetres, one might expect that there would be no question about adequate supply of water. However, rain falls unevenly over both area and time. In the mountainous areas of the Lake District, Scotland and Wales, average annual rainfall exceeds 2,400 mm, and for most of the country there is a significant margin between effective rainfall and abstraction. But in the Thames estuary rainfall is less than 500 mm and for much of the Thames and Anglian regions licensed abstractions are more than two-thirds effective annual average rainfall. This is of great concern, even given the high level of reuse, because these are also the regions with the highest demands for new development. The drought of 1988–92 and long hot dry summer of 1995 raised awareness about the impact of demand with unacceptably low levels in some rivers and supply constraints. As a result there has been a stream of official reports and consultation papers, and development of academic studies in this area which had received remarkably little attention previously (at least in the UK).

The need for a major programme of new investment is now widely recognised, not only to replace outworn facilities, but also to meet new demands for water, for environmental protection, and for sustainability. At the same time, increased concerns about water supply have come from developers and the public. The result is a renewed awareness of the importance of the relationship between water and land use planning (Slater *et al.* 1994: 376). In addition government has made a requirement for twenty-five-year resource plans from water companies and targets for reduction of leakage of 25 per cent over three years (in 1997 about 25 per cent was lost through leakage). Demand management is certainly coming to the fore in relation to water supplies, but some areas have more water than they can cope with.

Flooding has moved up the priority list of critical issues during the 1990s and especially since the floods of Easter 1998. This prompted a review of planning guidance on flooding and development but no sooner was this completed than the worst floods since records began (400 years) devastated much of the country in 2000. The wettest autumn on record (457 mm in three months) resulted in floods in England and Wales affecting 7,406 homes at an estimated cost of £500

million and two deaths (but let's not forget that at the same time defences protected more than 400,000 homes). The severity of the floods is reflected in one report that 'RNLI lifeboats operated on the High Street to rescue residents trapped in the upper floors of buildings'.[35] Flooding is inevitable of course, it is part of a natural environmental cycle and cannot be prevented, but the risk of flooding is increasing. A *Foresight Future Flooding Report* suggested that river and coastal flood risk could increase by twenty times. The effect would be that 'the number of people at high risk could more than double from 1.6 million today, to between 2.3 and 3.5 million by the 2080s' (*Environment Agency Annual Report* 2003–4: 10).[36] It is not surprising given recent experience and forecasts for the future that flood risk has lifted up the government's agenda. Much can be done to either reduce flood risk or mitigate the consequences of flooding

The reasons for flooding are complex and very much dependent on the conditions of particular catchments and coastal cells. Global warming and associated sea level rise (and land movements) with greater and more intense periods of rainfall play a part. So do engineering works to drainage systems, rivers and coastlines (flood defences in one location can cause problems elsewhere) and agricultural practices that increase the rate of run-off.

Another principal cause is the erosion of flood plain through new development and this is a major issue for the planning system, as is location of development along the coast. Even locations distant from rivers and the coast can have an effect by increasing the amount and speed of run-off. The planning system has come in for considerable criticism during the latest round of floods. There is no doubt that building on flood plains in particular has had an impact, though the environment agencies and their predecessors will have been consulted on these developments. The Select Committee on Environment, Transport and Regional Affairs undertook a review of *Development on or Affecting the Flood Plain* in the aftermath of the 2000 floods and prior to the finalisation of new government guidance. Its conclusions were clear on the critical effect of increased run-off caused by new development, development in the flood plain and particularly development in the functional flood plain or washlands that are used for storage during floods. There are now more than 1.8 million homes in flood risk areas. The Environment Agency estimate that 8 per cent of the land area of England and Wales is at risk from flooding and that if current development patterns persist a further 342,000 homes may be added to those at risk by 2021 (*Creating an Environmental Vision: Progressing the Environmental Agency's Contribution to Sustainable Development*, 2000). The Agency objects to

> inappropriate development in the flood plain and reports that it successfully influenced 87 per cent of all flood risk applications, [but] a number of authorities are still not undertaking flood risk assessments and a number of residential developments were approved against our advice.
>
> (Environment Agency Annual Report 2003–4: 10)

(The proposed developments in the Thames Gateway under the Sustainable Communities Plan, which is a major area of flood risk, will present a considerable challenge to the Agency and local planning authorities.)

Planning guidance on flood risk has for some years emphasised that it is a material consideration and that it is appropriate to refuse permission in cases where risk is unacceptable. Recent revisions have strengthened this advice with reference to sustainable development and the precautionary principle.[37] In the words of the Scottish guidance:

> planning authorities should first, seek to avoid increasing the flood risk by refusing permission where appropriate, and secondly, seek to manage the threat of flooding only in cases where other reasons for granting permission take precedence over flood risk.
>
> (NPPG 7, 1995: para. 42)

Particular care is promoted in dealing with development proposals that lie just beyond existing flood defences where a breach may involve a high risk of loss of life. In these cases the advice is even stronger: 'development should not be permitted where the existing flood defences would not provide an acceptable level of safety'. The same applies to caravans in areas of high risk.

Responsibility for determining the extent of risk formally lies with the landowner, although all planning authorities have been issued with flood risk maps by the environment agencies. A particular problem applies where intensification may result from development that does not require planning permission, in which case planning authorities should consider Article 4 directions to remove permitted development rights. Development plans should take into account flood risk, especially where there is a history of flooding, and the environment agencies are important consultees on this matter. Policies and decisions need to be consistent with shoreline management plans and local environmental agency plans.

Although the current and proposed guidance is firm, it did not satisfy the Select Committee which made clear recommendations for stronger national guidance. The Committee called for a presumption against development in the flood plain and the adoption of a sequential approach (as for retail developments and town centres). Land already allocated for development which does not pass stringent new tests should be deallocated in plans. All this points to major costs in improving flood defences over coming years, with the Thames Barrier for one becoming redundant before the middle of the century.

Waste planning

The UK produces over 400 million tonnes of waste each year. Details of the recycling and disposal of the 116 million tonnes of industrial commercial and domestic waste are given in Table 7.4. The rest is mostly agricultural, mining and quarrying waste and sewage sludge. The legislation covering waste management is immense, with twenty-eight relevant EU directives alone. Essentially, the 1990 Act imposes a *duty of care* on all who are concerned with controlled waste. This duty, similar to that imposed on employers by the Health and Safety at Work Act 1974, is designed to ensure that waste is properly managed. It

Table 7.4 *Estimated waste production recycling and disposal 1998–9*

Waste type	Generation (m tonnes)	Recovered and recycled	Disposal	
			Land fill	Other
Inert, inhouse construction	2	39%	56%	5%
Paper and card	7	76%	22%	1%
Food	3	69%	7%	13%
Other general and biodegradable	9	42%	26%	11%
Metals and scrap equipment	6	89%	10%	1%
Contaminated and health care	5	34%	42%	22%
Mineral waste and residues	6	38%	62%	0%
Chemicals	4	21%	45%	27%
General commercial	23	18%	78%	0%
General industrial	13	11%	86%	1%
Municipal (household) waste	28	9%	83%	8%
Total	106			

Source: Waste Strategy 2000, England and Wales (Part 2: 13–14)

should be collected, transported, stored, recovered and disposed of without harm to human health or the environment. The law also ensures that the responsible authorities develop plans for managing and disposing of waste. But they first have to know what it is. The definition of waste gives rise to problems of a Byzantine character. The legal definitions in the UK now follow that in the EU Waste Framework Directive which describes waste as 'any substance or object . . . which the holder discards or intends or is required to discard'. The list includes sixteen categories of waste and these are summarised in Annexe B to the *Waste Strategy 2000* for England and Wales. An alternative definition by Mary Douglas is 'waste is matter in the wrong place' (quoted in Worpole 1999: 24). Worpole goes on to say that 'a newspaper on the café table is a highly esteemed cultural artefact; blowing around the street an hour later, it becomes a threat to our very sense of meaning and belonging. Ten newspapers scattered on the pavement and there goes our neighbourhood'.[38]

Waste regulation functions are the responsibility of the environment agencies. Waste collection remains with local government. Waste planning is the responsibility of local planning authorities (in two-tier areas it is the county) and the regional bodies also have responsibility for waste planning policy for the region. Waste planning authorities must identify suitable sites for the disposal and handling of waste in the context of BPEO, the integrated approach to environmental management and the government's national objectives for waste. Guidance for waste planning authorities in England is provided in PPS 10 *Planning for Sustainable Waste Management* and its *Companion Guide* (2005).[39] National waste policy was initially set out in the 1995 strategy for sustainable waste management *Making Waste Work*, which has subsequently been superseded by the *Waste Strategy 2000* for England and Wales. The strategy and the targets and indicators it promotes are material considerations in planning.

Given the nature of waste policy, regional bodies are especially encouraged to address waste policy in their regional spatial strategies. Local authorities will rarely be able to address waste issues independently and there is much to be gained from cooperation at the regional level, although self-sufficiency within regions is encouraged. The waste planning authority (the authority that deals with waste planning applications) will be the county in two-tier areas and the unitary council, national park or London borough elsewhere. Where there is a structure plan this will also include policies on waste, which will be superseded in due course by policies in the regional spatial strategy and local development documents. But the county will continue with a specific waste local plan. Most unitary authorities include waste policies within their unitary development plans (see Chapter 4). Minerals come under a different provision, but, since a significant proportion of waste arises from mineral workings, waste and mineral plans can be combined.

Waste policies deal with all types of waste, including scrap yards, clinical and other types of waste incinerator, landfill sites, waste storage facilities, recycling and waste reception centres, concrete crushing and blacktop reprocessing facilities, and bottle banks. National policy and targets are playing an increasingly important role. National policy is now very comprehensive (if not always very ambitious) and includes the general principles of moving away from landfill towards recycling, composting and recovering energy from waste. Nevertheless, there is still a requirement for making a realistic assessment of the need for waste facilities and 'ensuring that there is adequate scope for the provision of the right facilities in the right places'. Planning authorities have the responsibility of ensuring that waste facilities are not developed in locations where they would be harmful or otherwise unacceptable for land use reasons. In this they need to work closely with the environment agencies to ensure that planning and pollution regulation are consistent. A closely integrated 'twin-track' approach is being promoted by the agencies.

Planning authorities also have an important positive planning role in waste management through promoting 'the proximity principle' and the 'regional self-sufficiency principle'. These stem from the desirability of waste recovery or disposal being close to the place where it is produced. This 'encourages communities to take more responsibility for the waste which they – either themselves as householders or their local industry – produce. It is their problem, not someone else's'. It also limits environmental damage due to the transportation of waste.

The potential for recycling slipped down the policy agenda for some years after peaking in the 1960s, but is now being renewed presenting more challenges to the planning system. The provision of waste disposal sites was relatively problem free for the system which relied on the availability of mineral workings to provide suitable sites. The big issues were the responsibility of pollution control (Davoudi 1999). But as stricter controls, EU policy, and rising development pressures to be accommodated in a plan-led process, together with more demand for waste sorting and bulking depots and recovery facilities, the planning issues became more complex. Waste generates considerable public concern and waste plans and policies are among the most contentious. As a result planning is playing a more central role in the waste management process.

Again, the EU has played an important part in stimulating action in the UK including the preparation of the national strategies. The EU Landfill Directive (99/31) requires ambitious national targets to be set for the reduction of biodegradable municipal waste sent to landfill, banning the disposal of hazardous and non-hazardous wastes together; and banning the landfill of tyres and liquid wastes. Other objectives for waste management that the planning system needs to take into account are summarised in Box 7.5. It should be stressed that waste planning is one of the most critical issues for planning (and the review of PPG 10 to PPS 10 has been a priority). There has been resistance to locating waste facilities as major LULUs (locally unwanted land uses) and proposed further EU legislation will increase the cost and restrict the use of landfill sites. Waste facilities are always the neighbouring authority's problem – the big conurbations often dispose of waste in the territory of neighbouring shire authorities. So this is very much an appropriate topic for the new stronger arrangements for regional and sub-regional planning, though a difficult one.

PPS 10 describes waste as a 'key component of the regional spatial strategy'. The regional body has to assess the 'waste management capacity' in its region taking into account its monitoring information, national forecasts, the need for waste management in its own and neighbouring regions, and the capacity of

BOX 7.5 SOME OF THE KEY TARGETS FOR WASTE MANAGEMENT

Landfill of municipal waste

- By 2010 to reduce biodegradable municipal waste landfilled to 75 per cent of that produced in 1995
- By 2013 to reduce biodegradable municipal waste landfilled to 50 per cent of that produced in 1995
- By 2020 to reduce biodegradable municipal waste landfilled to 35 per cent of that produced in 1995

Landfill of industrial and commercial waste

- By 2005 to reduce the amount of industrial and commercial waste sent to landfill to 85 per cent of that landfilled in 1998

Recovery of municipal waste

- To recycle or compost at least 25 per cent of household waste by 2005
- To recycle or compost at least 30 per cent of household waste by 2010
- To recycle or compost at least 33 per cent of household waste by 2015

- To recover value from 40 per cent of municipal waste by 2005
- To recover value from 45 per cent of municipal waste by 2010
- To recover value from 67 per cent of municipal waste by 2015

(Recover means recycling, composting and energy recovery)

Source: Waste Strategy 2000

the area to accommodate facilities. They will then apportion the waste tonnage for three sectors – commercial, municipal and construction – to the waste management authorities. These allocations are then carried over into the local waste management plans and local development documents, which will identify particular sites and locations. Interestingly applicants for waste management facilities do not have to demonstrate a market need for their proposal. This top-down approach is partly a response to the difficulty of getting local authorities to (politically) accept the need for waste management in their areas.

Noise

'Quiet costs money . . . a machine manufacturer will try to make a quieter product only if he is forced to, either by legislation or because customers want quiet machines and will choose a rival product for a lower noise level.' So stated the Wilson Committee in 1963. This, in one sense, is the crux of the problem of noise. More, and more powerful, cars, aircraft, portable radios and the like must receive strong public opprobrium before manufacturers – and users – will be concerned with their noise level. Similarly, legislative measures and their implementation require public support before effective action can be taken.

As with other aspects of environmental quality, attitudes to noise and its control have changed in recent years, partly as a result of the advent of new sources of noise such as portable music centres, personal stereos, and electric DIY and garden equipment, as well as greatly increased traffic. (Developments in electronics have also provided easier methods of obtaining data on noise.) The increased concern about noise is reflected in a succession of inquiries and planning policy (PPG 24: *Planning and Noise*). More substantively, two Acts have been passed to provide stronger measures for dealing with the problems. The Noise and Statutory Nuisance Act, which was passed in 1993, strengthened local authority powers to deal with burglar alarms, noisy vehicles and equipment, and various other noise nuisances. Second, the Noise Act of 1996 provided a summary procedure for dealing with noise at night (11 p.m. to 7 a.m.). This includes powers for local authorities to serve a warning notice, and to seize equipment which is the source of offending noise. The 1996 Act does not require local authorities to use its provisions, but the situation is to be reviewed in the light of experience.

There are three ways in which noise is regulated: by setting limits to noise at source (as with aircraft, motorcycles and lawnmowers), separating noise from people (as with subsidised double glazing in houses affected by serious noise from aircraft or from new roads) and exercising controls over noise nuisance. Where intolerable noise cannot be reduced and reduces property values, an action can be pursued at common law or, in the case of certain public works, compensation can be obtained under the Land Compensation Act 1973.

Noise from neighbours is the most common source of noise nuisance and complaints. This is a difficult problem to deal with, and official encouragement is being given to various types of neighbourhood action, such as 'quiet neighbourhood', 'neighbourhood noise watch', noise mediation and similar schemes (Oliver and Waite 1989). There is provision under the Control of Pollution Act 1974 for the designation by local authorities of *noise abatement zones*, though the statutory procedures for these are cumbersome and, in any case, they are not well suited to dealing with neighbourhood noise in residential areas (though they are useful for regulating industrial and commercial areas).

Traffic noise takes many forms and is being tackled in various ways (conveniently summarised in Chapter 4 of the Royal Commission on the Environment 1994 report on *Transport and the Environment*). Road traffic noise is the most serious in the sense that it affects the most people. Here emphasis is being put on the development of quieter road surfaces and vehicles. Aircraft noise has long been subject to controls both nationally and (with the UK in the lead) internationally. The principal London airports are required by statute to provide sound insulation to homes seriously affected by aircraft noise, and similar non-statutory schemes apply to major airports in the provinces.

Noise is a material consideration in planning decisions and development plans may contain policies on noise particularly where there are major noise generators such as airports (although the reproduction of detailed noise contours in plans is not recommended). PPG 24 sets out four noise exposure categories (NECs) and in the worst case (category D) permission should normally be refused. The definition of boundaries between categories is difficult for non-experts, but they are clearly insufficient to prevent the building of houses adjacent to motorways which continues regardless. Such decisions aside, local authorities are taking more interest in noise and one – Birmingham City Council – with the support of central government (and building on practice in other European countries) has produced a noise map of the whole of the city, including the impact of road, rail and air traffic and ambient noise levels during both the day and night. The exercise anticipates legislation that may require such noise mapping for all urban areas. The CPRE has already produced a map of tranquil areas for the whole of England comparing the 1960s with the 1990s, which demonstrates the extensive intrusion of noise. The CPRE has now joined with the Environment Agency, the Countryside Agency and Countryside Council for Wales to designate *tranquil areas*.[40] They estimate that England has lost 21 per cent of its tranquil areas since the 1960s.

It will not come as a surprise that the EU has a noise directive in preparation which includes a requirement for noise mapping together with action plans to address identified problems and reduce the number of people exposed to excessive noise, and the provision of information on noise levels to the public.

Environmental impact assessment

As environmental issues have become more complex, ways have been sought to measure the impacts of development. Cost–benefit analysis was at one time seen as a good guide to action. By taking into account non-priced benefits such as the saving of time, and the reduction in accidents, it can 'prove' that developments such as the Victoria underground line are justified. Useful though this technique is for incorporating certain non-market issues into the decision-making process, it has serious limitations. In particular (quite apart from the problems of valuing 'time'), some things are beyond price, while others have quite different 'values' for different groups of the population. Reducing everything to a monetary price ignores factors such as these. Alternatives such as Lichfield's *planning balance sheet* and Hill's *goals achievement matrix* attempt to take a much wider range of factors into account.

Environmental impact assessment (EIA) is a procedure introduced into the British planning system as a result of an EC Directive.[41] Although it might appear that environmental assessment is nothing new on the British planning scene (hasn't this always been done with important projects?), it is in fact conceptually different in that it involves in theory a highly systematic quantitative and qualitative review of proposed projects – though practice is somewhat different (Wood and Jones 1991). Nevertheless, unlike some European countries, Britain has had, since the 1947 Act, a relatively sophisticated system which involves a case-by-case review of development proposals. Indeed, the UK government resisted the imposition of this scheme through the Directive. A summary of the procedure is given in Figure 7.1.

It is important to appreciate that EIA is a *process*. The production of an *environmental statement* (ES) is one part of this. The process involves the gathering of information on the environmental effects of a development. This information comes from a variety of sources: the developer, the local planning authority, statutory consultees (such as the Countryside Agency and environment agencies) and third parties (including environmental groups). There are now many evaluations of practice both in the UK and elsewhere.

For some types of development an EIA is mandatory. These are listed in Schedule 1 of the regulations (and are therefore inevitably known as 'Schedule 1 projects'). These include large developments such as power stations, airports, installations for the storage of radioactive waste, motorways, ports, and such like. Projects for which EIA *may* be required ('Schedule 2 projects') are those which have *significant* environmental impacts. There are three main types of development where it is considered that an EIA is needed:

- for major projects which are of more than local importance, principally in terms of physical size
- 'occasionally' for projects proposed for particularly sensitive or vulnerable locations, for example, a national park or a SSSI
- 'in a small number of cases' for projects with unusually complex or potentially adverse effects, where expert analysis is desirable, for example, with the discharge of pollutants.

There is a marked resemblance between this and the circumstances in which the Secretary of State may exercise the powers of 'call-in' – they both relate to developments of particular importance which require more than a normal scrutiny for planning and environmental purposes. In 1995, *permitted development rights* were withdrawn from projects listed in Schedule 1, and also for projects having likely significant environmental effects (DoE Circular 3/95). Good practice guidance based on evaluation of the implementation of the EIA process is extensive (see list of DETR publications) with useful comparisons among countries (CEC 1996).

Wood and Bellinger (1999) record that in the first ten years of the implementation of the Directive in

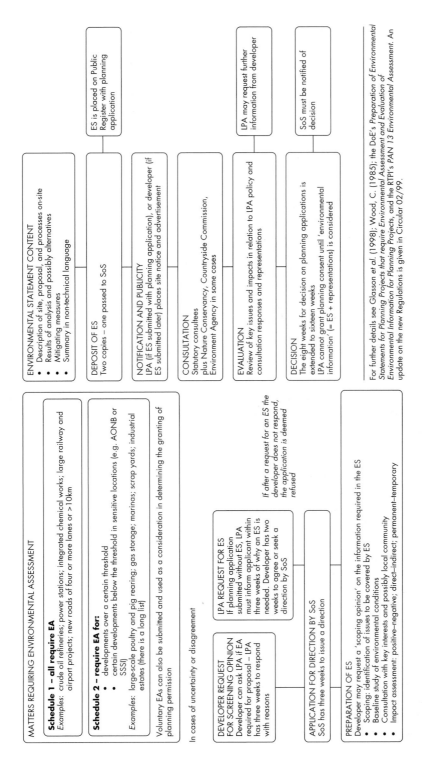

MATTERS REQUIRING ENVIRONMENTAL ASSESSMENT

Schedule 1 – all require EA
Examples: crude oil refineries; power stations; integrated chemical works; large railway and airport projects; new roads of four or more lanes or >10km

Schedule 2 – require EA for:
- developments over a certain threshold
- certain developments below the threshold in sensitive locations (e.g. AONB or SSSI)

Examples: large-scale poultry and pig rearing; gas storage; marinas; scrap yards; industrial estates (there is a long list)

Voluntary EAs can also be submitted and used as a consideration in determining the granting of planning permission

In cases of uncertainty or disagreement

DEVELOPER REQUEST FOR SCREENING OPINION
Developer can ask LPA (if EA required for proposal – LPA has three weeks to respond with reasons

LPA REQUEST FOR ES
If planning application submitted without ES, LPA must inform applicant within three weeks of why an ES is needed. Developer has two weeks to agree or seek a direction by SoS

APPLICATION FOR DIRECTION BY SoS
SoS has three weeks to issue a direction

If after a request for an ES the developer does not respond, the application is deemed refused

PREPARATION OF ES
Developer may request a 'scoping opinion' on the information required in the ES
- Scoping: identification of issues to be covered by ES
- Baseline study of environmental conditions
- Consultation with key interests and possibly local community
- Impact assessment: positive–negative; direct–indirect; permanent–temporary

ENVIRONMENTAL STATEMENT CONTENT
- Description of site, proposal, and processes on-site
- Results of analysis and possibly alternatives
- Mitigating measures
- Summary in non-technical language

DEPOSIT OF ES
Two copies – one passed to SoS

ES is placed on Public Register with planning application

NOTIFICATION AND PUBLICITY
LPA (if ES submitted with planning application), or developer (if ES submitted later) places site notice and advertisement

CONSULTATION
Statutory consultees plus Nature Conservancy, Countryside Commission, Environment Agency in some cases

LPA may request further information from developer

EVALUATION
Review of key issues and impacts in relation to LPA policy and consultation responses and representations

DECISION
The eight weeks for decision on planning applications is extended to sixteen weeks
LPA cannot grant planning consent until 'environmental information' (= ES + representations) is considered

SoS must be notified of decision

For further details see Glasson *et al.* (1998); Wood, C. (1985); the DoE's *Preparation of Environmental Statements for Planning Projects that require Environmental Assessment and Evaluation of Environmental Information for Planning Projects,* and the RTPI's PAN 13 *Environmental Assessment.* An update on the new Regulations is given in Circular 02/99.

■ **Figure 7.1 *The environmental impact assessment process***

the UK (from 1988) 3,000 environmental statements had been prepared. Drawing on other evaluations Glasson notes that

> EIA is a more structured approach to handling planning applications . . . projects and the environment benefit greatly from EIA . . . and consultants feel that EIA has brought about at least some improvements in environmental protection, in project design and the higher regard given to environmental issues.
>
> (Glasson 1999: 367)

Against this there are problems connected to the 'dual consent procedure': EIA in the planning process takes place alongside IPPC and leads to duplication of effort, the lack of attention to alternative options (although the revised Regulations address this) and the blueprint 'build it and forget approach' that tends not to consider the environmental effects of development over its full lifetime and cumulative impacts. Other evaluations have also shown that EIA may have only marginal effect on some projects (Blackmore *et al.* 1997).

Strategic environmental assessment

After many years of debate a political agreement was reached at the end of 1999 on the proposed *Directive on the Assessment of the Effects of Certain Plans and Programmes on the Environment* (the SEA Directive).[42] The Directive will require member states to establish procedures to ensure that environmental consequences of plans and programmes are identified before they are adopted, and that effective consultation is undertaken on the environmental implications. The main requirements in the amended proposed Directive are

- assessment of the environmental effects of 'all plans and programmes'[43]
- preparation of an environmental report identifying and evaluating the environmental effects of implementing the plan or programme
- consultation with relevant authorities, NGOs and

the public and with other member states if there are cross-boundary impacts
- statements summarising how the environmental considerations have been integrated into the plan or programme alongside the plan or programme itself.

The main change from the draft Directive was the link made to projects which require environmental assessment under the EIA and Habitats Directives. Plans that do not deal with this relatively significant scale of project will not be subject to SEA. In effect, similar provisions for SEA were already in place in the UK for development plans through the requirement for environmental assessment, and considerable expertise has been developed since publication of the DoE Good Practice Guide. This requirement has been carried forward into the post-2004 system of regional spatial strategies and local development documents but widened so that these plans now require a *sustainability appraisal* (SA). This was made mandatory by the 2004 Act for the regional spatial strategy, local development documents whether parts of the development plan or supplementary documents, and minerals and waste documents. Some local authorities had already made the choice to subject plans to a wider sustainability appraisal which incorporates the strategic environmental assessment. They now have the benefit of a *Practical Guide to the Strategic Environmental Assessment Directive* (2005),[44] and more specific guidance *Sustainability Appraisal of Regional Spatial Strategies and Local Development Frameworks* (2004). If local authorities follow this guidance they should also be meeting the requirements of the SEA Directive and Regulations. The Guidance defines sustainability appraisal as

> an iterative process that identifies and reports on the likely significant effects of the plan and the extent to which implementation of the plan will achieve the social, environmental and economic objectives by which sustainable development can be defined.
>
> (para. 1.2.2)

There are four stages: setting the context and baseline for assessment including constructing the framework and testing the plan's objectives against it; appraising the options and their impacts; appraising the effects of the preferred plan; and consulting on the sustainability appraisal report alongside the plan. It has been argued (prior to implementation of the SEA Directive) that the appraisal process has not contributed to policy-making; that it is highly subjective; and that it leads to inconsistent conclusions (Russell 2000). However, it is widely applied and there is a strong recognition of its importance in the delivery of sustainable development (Short *et al.* 2004). There are questions about the availability of expertise to conduct appraisals, especially in view of the demanding timescales for preparing plans. 'The bottom line is that if the government is committed to making the plan-making system more efficient, it needs to consider how to accommodate increasingly complex appraisals' (Holstein 2002: 219). It is unlikely, therefore, that the Directive in its current form will make a big impression on town and country planning process or outcomes. The next stage may be of more consequence as experience from other European countries becomes better known in the UK. A more radical and effective approach would be to require environmental or ecological compensation schemes (Wilding and Raemaekers 2000). Both are already employed in Germany.

Further reading

For a history of pollution control and much else on the origins of environmental policy, see Ashby and Anderson (1981) *The Politics of Clean Air*. Also strongly recommended is Ashby's (1978) reflective *Reconciling Man with the Environment*. A detailed legal source book is *Encyclopaedia of Environmental Law* edited by Tromans *et al.* (loose-leaf; updated regularly). Less daunting is Hughes (1996) *Environmental Law*. Miller's (2000) background paper for the RCEP on *Planning and Pollution Revisited* and Wood (1999) 'Environmental planning' both trace the history of the relationship. PPS 23 provides a summary of government policy and has a very useful annex.

Sustainability

Only the briefest indication of the mass of publications on sustainability can be given here. The Brundtland Report (1987) (Report of the World Commission on Environment and Development, *Our Common Future*) is perhaps the most quoted and misquoted source on sustainability; although its interest is increasingly historical, it is still an important original source. Major UK official references on sustainability are cited in the text, and it is certainly worthwhile to start with *Securing the Future: Delivering UK Sustainability Strategy* (DEFRA 2005), the related national strategies and *Sustainability Counts* on indicators (bearing in mind that this is a government interpretation of sustainability). For a more critical review see Jacobs (1999) *Environmental Modernisation*, Owens (1994b) 'Land, limits and sustainability', Khan (1995) 'Sustainable development', Real World Coalition's *From Here to Sustainability* edited by Christie and Warburton (2001); and Church and McHarry (1999) *One Small Step: A Guide to Action on Sustainable Development in the UK*. For an American perspective see Board on Sustainable Development Policy Division (1999) *Our Common Journey*. There is an extremely long list of Web resources on sustainability at the World Wide Web Virtual Library www.ulb.ac.be/ceese/meta/sustvl.html.

References that specifically address planning's contribution to sustainability are Layard *et al.* (2001) *Planning for a Sustainable Future*, Blowers (1993) *Planning for a Sustainable Environment*, Breheny (1992) *Sustainable Development and Urban Form*, Williams *et al.* (2000) *Achieving Sustainable Urban Form*, Buckingham-Hatfield and Evans (1996) *Environmental Planning and Sustainability*, Selman (1996) *Local Sustainability*, Selman (1999) *Environmental Planning*, IDeA (1998) *Sustainability in Development Control*, World Health Organisation (1997) *City Planning for Health and Sustainable Development*, Kenny and Meadowcroft (1999) *Planning Sustainability*, Counsell (1998) 'Sustainable development and structure plans in England and Wales' and Hales (2000) 'Land use development planning and the notion of sustainable development'. For evaluation of the application of sustainability in other countries, see Burke and Manta (1999) *Planning for Sustainable Development* in the USA and Berke *et al.* (2004) *Plan-making for Sustainability* for a New Zealand example.

On urban and regional sustainability, see Elkin *et al.* (1991) *Reviving the City: Towards Sustainable Development*, Gibbs (1994) 'Towards the sustainable city', Haughton and Hunter (1994) *Sustainable Cities*, the EU Expert Group on the Urban Environment (1996) *The European Sustainable Cities Report*, Barton (2000) *Sustainable Communities* and Ravetz (2000) *City Region 2020*.

Agenda 21

Start at the top with the UN sustainable development website at www.un.org/esa/sustdev/csdgen.htm which also has a surprisingly useful summary of action on sustainability in the UK. For government publications see LGA *et al.* (1998) *Sustainable Local Communities for the 21st Century: Why and How to Prepare an Effective Local Agenda 21 Strategy* and LGA and IDeA (1998) *Integrating Sustainable Development into Best Value*. Other main sources include Wilkes and Peter (1995) 'Think globally, act locally' and IDeA (1997) 'Local Agenda 21 in the UK: The First 5 Years'. *EG Magazine* is a monthly publication concerned with practice on LA21.

Environmental politics and the impact of the EU

The subject of environmental politics is also well covered by many textbooks, including Fischer and Black (1995) *Greening Environmental Policy: The Politics of a Sustainable Future*, Lowe and Goyder (1983) *Environmental Groups in Politics*, Worpole (1999) *Richer Futures* and Doyle and McEachern (1998) *Environment and Politics*. See also Newby (1990) 'Ecology, amenity, and society' which shows that environmental politics are not simply a modern fad. Beckerman (1995) gives an iconoclastic appraisal of 'environmental alarmism' in *Small is Stupid: Blowing the Whistle on the Greens*. On Europe see Vogel (1995) 'The making of EC environmental policy' and Shaw *et al.* (2000) *Regional Planning and Development in Europe*.

Economic instruments of environmental policy

Cairncross (1993) *Costing the Earth* (Chapter 4) is a good non-technical discussion and there is a more recent book by O'Riordan (1997) *Ecotaxation*. There is a discussion of economic instruments in Chapter 16 of Cullingworth (1997a) *Planning in the USA* on which parts of the text are based.

Air, water and waste

There are a number of general sources for information and statistics and of particular interest is McLaren *et al.* (1998) *Tomorrow's World: Britain's Share in a Sustainable Future* which sets out the theory of 'environmental space' and explains how the UK could realistically but drastically cut its use of resources. See also the DEFR eDigest of Environmental Statistics, the Environment Agency's Strategy and State of the Environment Report at www.environment-agency.gov.uk/strategy/strategy.html and Stead and Nadin (1999) 'Environmental resources and energy in the United Kingdom'.

Air pollution policy and an explanation of trends are set out in the *National Air Quality Strategy* (1999) and DETR (1999) *Economic Analysis of the National Air Quality Strategy Objectives*. In Scotland see SEPA (2000) *Air Report* and Scottish Executive (2000) *Local Air Quality Management General Guidance Series*. See also Elsom (1996) *Smog Alert: Managing Urban Air Quality* and Colls (2002) *Air Pollution: An Introduction*.

The environment agencies' websites are probably the best starting point for policy on the water environment. A number of official publications have arisen from both the drought and flooding crises although they have little to say about planning. Of interest are DETR (1998) *Water Resources and Supply: Agenda for Action*, DETR (2000) *Water Quality in England: A Guide to Water Protection in England and Wales*, DETR (2000) *Code of Practice on Conservation Access and Recreation* and in Scotland, SEPA (2000) *Improving Scotland's Water*. See also Slater *et al.* (1994) 'Land use planning and the water sector'. On flooding the relevant policy guidance notes are comprehensive and are cited in the text. They should be read in conjunction with the Select Committee Report (2000) *Development Affecting the Flood Plain*.

On waste, both volumes of the *Waste Strategy 2000* (or the Scottish Executive's *National Waste Strategy*) together with PPG 10, *Planning and Waste Management* (1999) (or the equivalents) provide a very comprehensive source. On Scottish policies see NPPG 10, *Planning and Waste Management* (1996). For a critical review of the policy of encouraging the recycling of paper products, see Collins (1996) 'Recycling and the environmental debate'. See also Samuels (2004) 'Waste' for a discussion of complying with the law on waste.

Noise

PPG 24 (1994) dealt with *Planning and Noise*. The Batho Report (1990) by the Noise Review Working Party examined a wide range of issues concerned with noise. Later reports have dealt with particular aspects such as the Mitchell Report (1991) *Railway Noise and the Insulation of Dwellings* and the Building Research Establishment report on *The Noise Climate Around Our Homes* (Sargent 1993).

Environmental assessment and appraisal

The principal texts on environmental impact assessment are Glasson *et al.* (1998) *Introduction to Environmental Impact Assessment* and Wood (1995) *Environmental Impact Assessment: A Comparative Review*. See also Elvin and Robinson (2000) 'Environmental impact assessment', Jones *et al.* (1998) 'Environmental assessment in the UK planning process', Glasson (1999) 'The first 10 years of the UK EIA system', Weston (2000) 'Reviewing environmental statements' and Wood (2000) 'Ten years on: an empirical assessment of UK environmental statement submissions'. On appraisals see Short *et al.* (2004) 'Current practice in the strategic environmental assessment of development plans in England' and Russell (2000) 'Environmental appraisal of development plans'.

Notes

1 In a public survey in Scotland in 1995 only 12 per cent of respondents could define sustainable development and only 2 per cent could explain Agenda 21. However, most respondents (64 per cent) thought that protecting the environment is more important than economic growth; that technological development is not a solution to resource depletion; and that government intervention to improve sustainability is welcome (McCaig *et al.* 1995).

2 See, for example, the monumental 1995 report of the Select Committee of the House of Lords on Sustainable Development for the range of definitions.

3 Interestingly, the British Government Panel on Sustainable Development, in its first report (1995), commented that the term was 'not so much an idea as a convoy of ideas'. It is a rallying cry, a demand that environmental issues need to be taken into account; but it provides little guide to action.

4 On the other hand, there are some formidable (if not popular) economic arguments which more prosaically point to the differences between notions of sustainability, optimality, and ethical superiority. The fact that a particular path of development is unsustainable does not necessarily mean that it is undesirable or sub-optimal. In the words of Beckerman (1995: 126), 'most definitions of sustainable development tend to incorporate some ethical injunction without apparently any recognition of the need to demonstrate why that particular ethical injunction is better than many others that one could think up'.

5 See for example Baker *et al.* (1997). The introduction to Layard *et al.* (2001) provides a fuller account of strong and weak sustainability.

6 Blowers and Leroy (1994) have argued that there is a process of 'peripheralisation' as locally unwanted land uses (LULUs) (in their case hazardous and polluting industries) are exported to areas beyond the main metropolitan centres to peripheral areas that have less power or will to resist them.

7 The framework was developed for use in an EU Fifth R&D Framework funded project on Sustainability, Development and Spatial Planning.

8 Of particular note here are LGMB (1995b), Barton *et al.* (1995), Barton (2000), the DETR good practice guide, *Planning for Sustainable Development: Towards Better Practice* (1998) and the DETR research report on *Millennium Villages and Sustainable Communities* (Llewelyn-Davies *et al.* 2000). See also Friends of the Earth (1994a), Ravetz (2000), Levett and Christie (1999) and the Town and Country Planning Association's *Tomorrow Series* of booklets on environmental planning issues including P. Hall (1999c), Hooper (1999), Marsden (1999), O'Riordan (1999) and Winter (1998). See also relevant research findings on urban intensification, particularly Breheny (1992), Breheny and Ross (1998), Rudlin (1998) and Williams, K. (1999).

9 The indicators were first published as *Indicators of Sustainable Development for the United Kingdom* but were revised for the 1999 Strategy after consultation on proposed headline indicators through the document *Sustainability Counts* (1998).

10 The broad definition of sustainable development taken in the UK is in distinct contrast to the approach in some other countries such as Sweden or New Zealand, where the ecological dimension is given much more prominence. In Sweden the objective of sustainability is defined in national legislation as being protection of the environment, to conserve the supply of environmental resources and to make most efficient use of natural resources (see Seaton and Nadin 2000).

11 DEFRA (2004) *Achieving a Better Quality of Life: Review of Progress towards Sustainable Development, Government Annual Report 2003.*

12 When faced with public protest in 2000, the government did not hesitate to withdraw from one commitment: 'to increase duty on petrol and diesel each year by 6 per cent above inflation to reduce carbon dioxide emissions from road transport, 1 per cent higher than the previous Government's commitment' (para. 5.8).

13 There is a separate Climate Change Programme (2000).

14 *Guidance on Preparing Regional Sustainable Development Frameworks* (London: DETR, 2000). See also the UK Round Table on Sustainable Development report on *Sustainable Development Opportunities for Devolved and Regional Bodies* (1999) and *Building Partnerships for Prosperity: Sustainable Growth, Competitiveness and Employment in the English Regions* (London: DETR, 2000).

15 See, for example, the LGMB reports (1993a, 1993b, 1995a, 1995b).

16 For local indicators of sustainable development see the DETR report *Local Sustainability Counts*, which is a handbook of twenty-nine indicators for LA21 and local community planning. Local authority Agenda 21 strategies are also a good source.

17 Although there was her remarkable conversion to the environmental cause in 1988 when she surprised everybody by testifying her personal 'commitment to science and the environment'. With resounding words, she rallied her followers to environmentalism, declaring that Conservatives were 'not merely friends of the earth' but also 'its guardians and trustees for generations to come'.

18 See *Guardian*, 15 September 2004 for a report on the speech; also available on the 10 Downing Street website.

19 The SDC is chaired by Jonathan Porritt, covers the whole of the UK and is sponsored by the Cabinet Office. It reports to the Prime Minister, the First Ministers in Scotland and Northern Ireland and the First Secretary in Wales.

20 Background papers were prepared for the Review of Environmental Planning addressing five main themes: the extent to which planning supports environmental sustainability; the barrier effect of administrative boundaries; the extent of integration and coordination of environmental policy and action; subsidiarity and democracy in environmental policy and assessment approaches. Background papers are available on the RCEP website: www.rcep.org.uk.

21 The EEA was established in 1993 with the objective of providing 'a seamless information system' on the environment for policy-makers. It does this by collecting and presenting in compatible format existing information through the European Environment Information and Observation Network (EIONET) which comprises 600 environmental bodies and agencies across Europe. Its membership includes EFTA

countries as well as EU. A major achievement was the preparation of the Dobris Assessment of Europe's Environment (1995) and the Second Assessment (1998) which cover forty-six countries and are the principal sources for state of the environment information in Europe. The EEA website is www.eea.eu.int.

22 CEC (2001) *The Sixth Environmental Action Programme of the EC: Environment 2010: Our Future, Our Choice*, COM (2001) 31 Final, available at http://europa.eu.int/eur-lex/en/com/pdf/2001/en_501PC0031.pdf.

23 For an explanation of the creation of the Agencies see HC Environment Committee (1992) *The Government's Proposals for an Environment Agency*, which points to the important comments made by the Advisory Committee on Business and the Environment and the Institute of Directors. The Environment Committee argued that the Agency should have more functions than was actually given by the government. The TCPA went further and argued the need for integration between environmental planning and land use planning, a point taken up in the review of environmental planning by the RCEP.

24 These quotations are taken from the draft *Environmental Vision* of the Environment Agency for England and Wales. A similar (though only five-year) vision document, *State of the Environment Report* was published by the Scottish Environment Protection Agency in 1996.

25 ALARA should also be mentioned – the principle of 'as low as reasonably achievable' which applies in the regulation of emissions from radioactive sources.

26 In fact IPC was first recommended by the Royal Commission on Environmental Pollution but was largely ignored by government until 1987 (Miller 2000). The EU has been partly responsible for the creation of a more integrated pollution control regime.

27 From 1997 to 1999 the Environment Agency successfully prosecuted 1,700 people for pollution offences including fifteen prison sentences and a $4 million fine in the case of the *Sea Empress* (reduced to £750,000 on appeal) (Environment Agency, *Creating an Environmental Vision*, 2000).

28 The Advisory Committee on Business and the Environment reports annually (to the president of the Board of Trade and the Secretary of State for the Environment) on economic instruments. Its recent reports have dealt with tradeable permits for water pollution, the landfill tax, and the promotion of alternative fuels. Proposals for alternative fuels, as well as various economic instruments, were discussed by the Royal Commission on Environmental Pollution in its report on *Transport and the Environment* and *Energy*. The Round Table on Sustainable Development has reported on the use of Economic Instruments (2000).

29 For example, in *Economic Instruments and the Business Use of Energy* (1998), *Economic Instruments for Water Pollution* (1997) and *Economic Instruments in Relation to Water Abstraction* (2000).

30 For a summary of the main air pollutants and recent trends see Stead and Nadin (2000), the UK National Air Quality Strategy and the DETR Digest of Environmental Statistics (published annually). The main air pollutants are carbon dioxide (CO_2), a 'global pollutant' thought mainly though not solely responsible for global warming; sulphur dioxide (SO_2) which contributes to acidification of soil and water; nitrogen oxides (NO_X) which also contribute to acid deposition and with other pollutants give rise to smog and poor air quality; ozone which is created in the atmosphere by chemical reactions involving sunlight, NO_X and volatile organic compounds (VOCs) which have impacts on health of the lungs; particles or particulates (PM_{10}) of many types which contribute to respiratory and cardiovascular health problems; carbon monoxide (CO); benzene and 1,3- butadiene, which are human carcinogens, the former associated particularly with leukaemia; and lead, which has many negative health effects. Proposed EU Directives will also introduce dioxins into this list.

31 There are more than 1,500 monitoring sites across the UK, most of which are automatic, and all of which make up the most sophisticated monitoring system of any EU state. Considerable information is available via the Internet (www.airquality.co.uk), and you can identify air quality from a monitoring station near you.

32 Standards are set out in the Air Quality (England) Regulations 2000.

33 The environment agencies also have certain powers to prevent flooding, as well as responsibilities for the licensing of salmon and freshwater fisheries, for navigation, and for conservancy and harbour authority functions.

34 The blue diamond is awarded to beaches where between May and September, seawater contains less than 10,000 coliforms (tiny living organisms) per 100 ml. A green circle is awarded where quality is moderate and red square where it fails the test.

35 Select Committee on Environment, Transport and Regional Affairs Second Report 2000: *Development Affecting the Flood Plain*: para. 1.

36 PPG 25 (see note 37) suggests that the 1-in-100 year high water level on the east coast may be exceeded every twenty years on average by 2050, and that rainfall will increase by 0–10 per cent by 2050. These changes add up to increases in peak flow of up to 20 per cent in the Thames and Severn catchments by 2050, although the uncertainty of forecasting is acknowledged.

37 In England planning guidance is to be found in PPG 25, *Development and Flood Risk* (2000). In Wales it is Technical Advice Note 15, *Development and Flood Risk* (1998). Scotland has taken the lead in providing a planning policy guidance note on the subject: NPPG 7, *Planning and Flooding* (1995). These statements are particularly useful with explanations of the causes of flooding and bibliographies of research and other guidance.

38 The lengthy DoE Circular 11/94 provides useful guidance on the definition of waste.

39 In Scotland there is a parallel NPPG 10, *Planning and Waste Management* (1996) and in Wales a Technical Advice Note, *Planning Pollution Control and Waste Management*. See also the *Re-inventing Waste: Towards a London Waste Strategy* (London: LPAC, 1998). *The Landfill Campaign Guide* (London: Friends of the Earth, 1997) and the Select Committee on Environment, Transport and Regional Affairs report on *Sustainable Waste Management* (1998). Park (2000) has found that the landfill tax is being implemented but that little funding is finding its way to local clean-up projects as intended.

40 Only a crude measure is possible: tranquil areas are

'4km from the largest power stations, 3km from roads with dense traffic, 2km from most other motorways and major roads, 1km from medium disturbance roads and outside the noise zone of airports and very intensive open cast mining'. See the Environment Agency's website.

41 Environmental assessment was introduced by the 1985 EC Directive on the Assessment and Effects of Certain Public and Private Projects on the Environment (85/337). The Directive was amended in 1997 by the *Amending Directive* (97/11). The amendment extended the range of projects that are subject to EIA and made other requirements in relation to the need for the planning authority to tell the developer what should be included in the EIA (scoping), the provision of information on alternative options and other procedural matters. The Directives are implemented through regulations in the UK – for England and Wales the Environmental Impact Assessment Regulations 1999 SI no. 293, and in Scotland by the Environmental Impact Assessment Regulations 1999 SSI no. 1, which are explained in Scottish Executive Development Department PAN 58 and Circular 15/1999.

42 An amended text was formally adopted in March 2000, and the Directive came into force in 2001; member states had three years to implement the Directive, and in the UK this was achieved by the *Environmental Assessment of Plans and Programmes Regulations 2004*, and there are supplementary forms of these regulations for Northern Ireland, Scotland and Wales (the main regulations cover any plan that includes England, e.g. the UK Sustainable Development Strategy). The text of the Directive does not use the term 'SEA', though it is commonly known by this name.

43 Plans coming within the terms of the Directive include agriculture, forestry, fisheries, energy, industry, transport, waste management, water management, telecommunications, tourism, town and country planning or land use *and* which set a framework for future development consent of projects, and which fall under the Environmental Assessment and Habitats Directives. Plans and programmes dealing only with finance and budgets, or serving

national defence, or which determine the use of small areas at the local level and minor modifications, will be exempt, unless the member state determines that they are likely to have significant environmental effects. Plans and programmes for the Structural Funds are also exempt, although there are other provisions in their regulations which require that environmental considerations are taken into account.

44 The ODPM had previously published *The Strategic Environmental Assessment Directive: Guidance for Planning Authorities* (2003), prepared by Levett-Therivel Consultants and the current document reproduces much of this.

8 Heritage planning

It is time to build a new future from England's past. Conservation is not backward looking. It offers sustainable solutions to the social and economic problems afflicting our towns and cities. It stands in the vanguard of social and economic policy, capable of reversing decades of decay by injecting new life into familiar areas.

Conservation-led Regeneration: The Work of English Heritage, 1998

Early actions to preserve

Britain has a remarkable wealth of historic buildings, but changing economic and social conditions often turn this legacy into a liability. The cost of maintenance, the financial attractions of redevelopment, the need for urban renewal, the roads programme, and similar factors often threaten buildings which are of architectural or historic interest.

The first state action came in 1882 with the Ancient Monuments Act, which acknowledged the interest of the state in the preservation of ancient monuments. Such preservation as was achieved under this Act (and similar Acts passed in the following thirty years) resulted from the goodwill and cooperation of private owners. A major landmark in the evolution of policy in this area was the establishment, in 1908, of the three Royal Commissions on the Historical Monuments (of England, Scotland, and Wales).[1] They had (and still have) the same purpose, exemplified by the original terms of reference of the English Commission:

> to make an inventory of the Ancient and Historical Monuments and constructions connected with or illustrative of the contemporary culture, civilisation and conditions of life of the people of England, from the earliest times to the year 1700 and to specify those that seem most worthy of preservation.

The quotation is instructive: the emphasis is on preservation and on 'ancient'. There was no concern for anything built after 1700, a prejudice which Ross (1996: 14) notes was typical of the time. Slowly changing attitudes were reflected in 1921 when the year 1714 was substituted for 1700! The date was advanced to 1850 after the end of the Second World War, and in 1963 an end-date was abolished.

The Commissions were established to record monuments, not to safeguard them. It was not until 1913 that general powers were provided to enable local authorities or the Commissioners of Works to purchase an ancient monument or (a surprising innovation in an era of sacrosanct property rights) to assume 'guardianship' of a monument, thereby preventing destruction or damage while leaving 'ownership' in private hands. Major legislative changes were made in the 1940s though, in practice, the most important innovation was the establishment of a national survey of historic buildings. This was a huge job (quite beyond the capabilities of the slow-moving Royal Commissions). It was undertaken, county by county, by so-called 'investigators' and by 1969 gave statutory protection to almost 120,000 buildings, and non-statutory recognition (but not protection) to a further 137,000 buildings.[2]

Statutory protection, however, is not sufficient by itself: the owners of historic buildings often need financial assistance if the cost of maintaining old structures is to be met.[3] Grants were introduced in 1953 for preserving houses which were inhabited or 'capable of occupation'. Further big changes were made in 1983, and later most of the provisions relating to heritage properties were consolidated in England and Wales in the Planning (Listed Buildings and Conservation Areas) Act 1990.[4]

Preservation, conservation and heritage

In considering the role of this regulatory system, it is important to appreciate what is meant by the term 'conservation'. Though often used synonymously with 'preservation', there is an important difference. Preservation implies maintaining the original in an unchanged state, but conservation embraces elements of change and even enhancement. To provide an economic base for the conservation of an old building, new uses often have to be sought. It is quite impossible to conserve all buildings in their original state irrespective of cost, and there frequently has to be a compromise between 'the value of the old and the needs of the new' (Ross 1991: 92). Thus 'new uses for old buildings' is a major factor in conservation, and it necessarily implies a degree of change, even if this is restricted to the interior.[5] Again, for conservation purposes it may be necessary to enhance a site to cater for public enjoyment. The difference is more than one of name.

'Heritage' is the fashionable term, although its use is not always welcomed.[6] Heritage takes the conservation idea further and embraces consideration of the use of what is conserved. It includes 'the process of evaluation, selection and interpretation – perhaps even exploitation – of things of the past' (Larkham 1999a: 105). For some, heritage presents historical buildings and places as commodities to be traded, packaged and marketed. And much of the UK is now neatly packaged into heritage products, carefully denoted by the brown signs marking entrances to 'Shakespeare's County', 'Lawrence Country' and many more. On the positive side, the notion of heritage draws attention to the economic potential of conservation, but it has been argued that the commodification process pays much less attention to authenticity and accuracy. Use of the term, and indeed action on conservation, was given a boost by European Architectural Heritage Year in 1974 and since then has been used widely, although government policy documents have stuck to the more appropriate 'historic environment'.[7]

Delafons (1997: 168–71) reviews conservation policy in England (as set out in PPG 15) in view of conservation doctrine built up since the nineteenth century, and in comparison with its predecessor, Circular 8/87. Despite views to the contrary, the presumption in favour of preservation remains in place but a more realistic and flexible approach is given to alternative uses for historic buildings. While the PPG was prepared during the time of deregulation and emphasis on economic development, the emphasis of PPG 15 is tipped heavily in the direction of conservation – so much so that it tends to downgrade the potential of well designed replacement buildings. It says that claims about their architectural merits cannot justify the demolition of any listed building.[8] Since the PPG (and in a different political context) English Heritage and the other agencies have put more emphasis on the regeneration potential of conservation and the concept of 'conservation for everyone'. However, it will take some time for conservation to rid itself of the well-deserved criticism of elitism, if indeed it can. The whole ethos of conservation policy in the UK has been about selection.

The latest development in terminology (if not necessarily in action) is the idea of 'sustainable conservation' or to put it more accurately, conservation for sustainability. It has been argued that the concepts are two sides of the same coin. The historic environment is a finite resource that should not be depleted. Conservation encourages the recycling of existing buildings and materials, the use of local resources and diversity in the environment. It can be argued that the historic city in many ways is a model for a sustainable city (Manley and Guise 1998: 86). But there is still much to be debated on the relationship between sustainability and conservation, not least in the widening gap between the quality of 'sheltered' historic areas and the rest of the public realm.

Heritage responsibilities

Responsibility for heritage lies with the Department of Culture, Media and Sport in England and with the devolved administrations in Scotland, Wales and Northern Ireland. Executive agencies have been created to manage the heritage – English Heritage, Historic Scotland, Cadw (Welsh Historic Monuments), and in Northern Ireland the Environment and Heritage Service.[9]

Many governmental and voluntary organisations play a role in heritage planning and the main ones are shown in Table 8.1. The executive agencies manage most government funding for the heritage (except lottery funding), maintain historic buildings and sites in government ownership and advise government on heritage matters including planning decisions. The Royal Commissions survey and compile the historic monuments records (in England the Royal Commission has been merged with English Heritage). The historic buildings councils advise government in heritage matters notably listing buildings. The advisory body, the Commission for Architecture and the Built Environment, has a wide remit 'to inject architecture into the bloodstream of the nation' (England) and in this role will often advise about the impact of new development on the heritage.

This is a field in which voluntary organisations

Table 8.1 *Government departments, agencies and advisory bodies for heritage in the UK*

	England	Northern Ireland	Scotland	Wales
Government department	Department of Culture, Media and Sport	Northern Ireland Executive	Scottish Executive	Welsh Assembly
Executive agencies	English Heritage Royal Parks Agency	Environment and Heritage Service	Historic Scotland	Cadw (Welsh Historic Monuments)
Other advisory bodies	Commission for Architecture and the Built Environment	Historic Buildings Council for Northern Ireland Historic Monuments Council for Northern Ireland	Historic Buildings Council for Scotland	Historic Buildings Council for Wales
Royal commissions	(Royal Commission on the Historical Monuments of England was merged with English Heritage in 1999)		Royal Commission on the Ancient and Historical Monuments of Scotland	Royal Commission on the Ancient and Historical Monuments of Wales
Other	Historic royal palaces	Monuments and Building Record NI		
Other funding bodies	National Heritage Fund and Heritage Lottery Fund (with separate committees in Northern Ireland, Scotland, Wales and the English regions).			

have been particularly active. The first of these dates back to 1877 when William Morris (horrified at the proposed 'restoration' of Tewkesbury Abbey) inspired the founding of the Society for the Protection of Ancient Buildings (Ross 1996). Many others have followed; the National Trust with 2.5 million members is the largest. Others with more specialist concerns include the Georgian Group, the Victorian Society and the Twentieth Century Society.[10] Other organisations have a wider remit – the Civic Trust champions improvement in all places where people work but supports heritage conservation and interpretation through, for example, heritage open days, when buildings normally closed are opened for visitors. The main organisation for planners (both officers and members) is the English Historic Towns Forum formed in 1987, and there is also a Conservation Officers' Society.

Archaeology

One indicator of the increase in the public popularity of archaeology is the number of television programmes now devoted to the subject. *Rescue archaeology* has been widely publicised through such finds as the streets from Saxon Lundenwic at the Covent Garden Opera House site and the Dover Boat – 'the Bronze Age cross-channel ferry' – unearthed during road works in Dover. Planning provisions have provided an opportunity for investigation, recording and removal of these archaeological remains prior to proposed development. But planning has also reduced such situations by recording remains and anticipating problems in local plans.

PPG 16 *Archaeology and Planning*, and similar policy statements outside England,[11] make it clear that there is a presumption in favour of the preservation of important remains, whether or not they are scheduled. There is thus a measure of protection over the large number of unscheduled sites that are on the lists maintained by county archaeological officers. (These are known as SMRs: *county sites and monuments records*.) Such sites are a 'material consideration' in dealing with planning applications.

Planning authorities make provision in their development plans for the protection of archaeological interests, often with good cooperation from large developers. What is perhaps surprising is the extent to which some developers are prepared to go to assist rescue archaeology, and even to fund it. Funding from developers for archaeological work is now four times that available from other sources. Such funding is generally welcomed but there are criticisms that this gives the developer control over the work rather than the archaeologists. Also, developers may pay for the production and/or dissemination of reports. Much archive material is being produced but is not widely available. A useful mechanism for liaison is provided by the Code of Practice of the British Archaeologists and Developers Liaison Group.[12]

The Secretary of State can also designate *areas of archaeological importance*. In these areas, developers are required to give six weeks' notice (an *operations notice*) of any works affecting the area, and the 'investigating authority' (e.g. the local authority or a university) can hold up operations for a total period of up to six months. The powers have been used very sparingly, and only five areas have been designated, comprising the historic centres of Canterbury, Chester, Exeter, Hereford and York.

A 1996 consultation paper *Protecting Our Heritage* argued that the powers are now redundant (see the discussion on the heritage review, p. 307).

Ancient monuments

The term *ancient monument* is defined very widely: it is 'any scheduled monument' and 'any other monument which in the opinion of the Secretary of State is of public interest by reason of the historic, architectural, traditional, artistic or archaeological interest attaching to it'. This is so broad a definition that it could include almost any building, structure or site of archaeological interest made or occupied by humans at any time and from 2002 also includes underwater archaeology.[13] It includes, for instance, a preserved Second World War airfield complex at East Fortune (near Haddington, in Lothian).

Legislation[14] requires the Secretary of State to prepare a schedule of monuments 'of national importance', which are then given special protection through the planning system. This 'scheduling' is a selective and continuing process. It has been underway for over a century and for many years proceeded at a very slow rate.[15] The pace of recording monuments has quickened over recent years following initiatives by the Royal Commissions, but it will still be many years before the schedule could be described as complete. At the end of 2003 there were 33,900 scheduled sites in the UK (see Table 8.2). Estimates of the total number of archaeological sites in Britain vary but it is in the region of 1 million. Since there is such a huge number of known archaeological sites and monuments, it is not

surprising that estimates differ.[16] The number is in decline, with one estimate suggesting that one site has been lost every day since 1945 (Bryant 1999). During 1998 and 1999 a Monuments at Risk Survey (MARS) revealed that at least 70,000 monuments are at risk, the main culprit being damage by ploughing.[17] In Scotland, 6,500 monuments have been scheduled though it considered that there may be over 10,000 other monuments yet to be assessed. There is more work to be done across the UK to ensure that a more complete record is created and conservation can be ensured.

PPG 16 advises that 'where nationally important archaeological remains, whether scheduled or not, and

Table 8.2 Listed building categories in the UK

England and Wales		Northern Ireland		Scotland	
Grade	Criteria	Cat	Criteria	Cat	Criteria
I	Buildings of outstanding or exceptional interest	A	Of national importance	A	Buildings of national or international importance, either architectural or historic, or fine and little altered examples of some particular period, style or building type
II*	Particularly important buildings of more than special interest but not in the outstanding class	B+	Of national importance but with minor detracting features or of national importance with some exceptional features	B	Buildings of regional or more than local importance, or major examples of some period, style or building type which may have been somewhat altered
		B1	Of national or local importance, or good		
		B2	examples of some period or style		
II	Buildings of special interest which warrant every effort being made to preserve them	C	Of positive architectural interest or historic interest but are not 'special' and including those that contribute to the value of groups of buildings	C(S)	Buildings of local importance; lesser examples of any period, style or building type, whether as originally constructed or as the result of subsequent alteration; simple well proportioned traditional buildings often forming part of a group

their settings, are affected by proposed development there should be a presumption in favour of their physical preservation'. The Acts also provide further protection from damage by users of metal detectors, although they also require monuments to be open to the public. But scheduling does not mean that a site will automatically be preserved under all circumstances. The need to preserve is a material consideration in development control (whether a monument is scheduled or not) and planning authorities may seek Article 4 directions to remove permitted development rights.

Any works have to be approved by the Secretary of State (who receives advice from the agencies, commissions and other advisory bodies). Such approval is known as *scheduled monument consent*. Where consent is refused, compensation is payable (under certain limited circumstances) if the owner thereby suffers loss.[18] In practice, the great majority of applications for consent are approved, often with conditions attached. The issue here is seen as one of balancing the need to protect the heritage with the rights and responsibilities of farmers, developers, statutory undertakers and other landowners. The legislation also empowers the Secretary of State to acquire (if necessary by compulsion) an ancient monument 'for the purpose of securing its preservation' – a power which applies to any ancient monument, not solely those which have been scheduled.

Though most heritage properties remain in private ownership, a small number are managed by the Heritage Departments – officially known as being 'in care'. These are generally of important historical, archaeological and architectural significance.[19]

Listed buildings

Under planning legislation, and quite separate from the provisions relating to monuments, the central departments maintain lists of buildings of *special architectural or historic interest*.[20]

Although the national listing survey is now substantially complete, listing is a continuing process, not only for additional buildings but also for updating information on existing listed buildings, particularly in terms of their condition. Existing listed buildings can be up- or downgraded. In addition, individual buildings can be spot-listed. This arises because of individual requests, often precipitated by the threat of alteration or demolition. The majority of these requests are made by local authorities.

At one time listing often came as a surprise to owners who were not aware that their property was under consideration for listing. Since 1995 the departments have consulted on listing although there is still no duty to consult anyone (including owners) and no right of appeal. Listing has been described as 'a fearful prospect' for owners of younger commercial and industrial properties because of the costs and delays in making changes to what are often obsolete and inefficient buildings (Derbyshire *et al.* 1999). The costs of retaining the building or any financial consequences are not considered in the listing process. An owner can apply for a certificate of immunity from listing which lasts for five years, but this may simply raise government's awareness of the need to list.

There are two objectives in listing. First, it is intended to provide guidance to local planning authorities in carrying out their planning functions. For example, in planning for redevelopment, local authorities will take into account listed buildings in the area. Second, and more directly effective, when a building is listed, no demolition or alteration that would materially alter it can be undertaken by the owner without the approval of the local authority.[21] This is *listed building consent* and is separate to planning permission but there is no fee. There have been numerous celebrated cases where people have been caught out because it was not recognised that the works require listed building consent. This arises because it is not 'development' (as defined for planning permission) that is controlled, but any works to a listed building *that affect its character as a building of special architectural or historic interest*. Thus painting a building (or even a door) may need consent if it affects architectural or historic character. Furthermore, the definition of what is listed is very wide and includes certain fixtures and fittings.[22]

The procedure for obtaining listed building consent is summarised in Figure 8.1. Applications have to be

MATTERS REQUIRING LISTED BUILDING CONSENT
- Works involving demolition, alteration or extension that would affect the character of a listed building, or object fixed to it or some structures within its curtilage
- Need for LBC extends to permitted development granted by GPDO if it affects the character of the listed building
- Separate applications are needed for LBC and planning consent

APPLICATION
is made to LPA and must include
- plans
- certificate that those with an interest have been notified

ADVERTISEMENT
If proposal involves demolition or alteration, applicant must advertise and provide a site notice

NOTIFICATION

- Grade I or II*
- involves demolition

- Grade II

LPA must notify:
English Heritage (HBMC) and national heritage organisations (see Circular 8/87)/PPG 15

No special notification unless the development involves 'substantially all' demolition of interior or has received a grant, in which case SoS notified

DECISION
LPA must have special regard to the desirability of preserving the building or its setting or any features of special interest

If LPA wish to grant consent for a grade I or II* building SoS must be notified

SoS may call-in for decision

Development must be begun within five years or as otherwise specified

REFUSED   **APPROVED**

APPEAL
Applicant may appeal to SoS if consent refused or subject to conditions

▪ **Figure 8.1 *The procedure for listed building consent in England***

advertised, and any representation must be taken into account by the local authority before it reaches its decision. Where demolition is involved, English local authorities have to notify English Heritage, the appropriate local amenity society, and a number of other bodies.[23] If, after all this, the local authority intends to grant consent for the demolition (or, in certain cases, the alteration) of a listed building, it has to refer the application to central government so that it can be considered for 'call-in' and decision by the Secretary of State. The Heritage bodies advise on these questions, and in most cases their advice is accepted.

Conditions can be imposed on a listed building consent in the same way as is done with planning permissions.[24] All these provisions apply to listed buildings, but local authorities can serve a *building preservation notice* on an unlisted building. This has the effect of protecting the building for six months, thus giving time for ODPM to consider (on the advice of English Heritage) whether or not it should be listed. For owners and developers who wish to be assured that they will not be unexpectedly made subject to listing, application can be made to the LPA for a *certificate of immunity from listing*.

With a listed building, the presumption is in favour of preservation. It is an offence to demolish or to alter a listed building unless listed building consent has been obtained. This is different from the general position in relation to planning permission where an offence arises only after the enforcement procedure has been invoked. Fines for illegal works to listed buildings are related to the financial benefit expected by the offender.

The legislation also provides a deterrent against deliberate neglect of historic buildings. This was one way in which astute owners could circumvent the earlier statutory provisions: a building could be neglected to such an extent that demolition was unavoidable, thus giving the owner the possibility of reaping the development value of the site. In such cases, the local authority can now compulsorily acquire the building at a restricted price, technically known as *minimum compensation*. If the Secretary of State approves, the compensation is assessed on the assumption that neither planning permission nor listed building consent would be given for any works to the building except those for its restoration and maintenance in a proper state of repair; in short, all development value is excluded.

The strength of these powers (and others not detailed here) reflects the concern which is felt at the loss of historic buildings. However, they are not all of this penal nature. Indeed, ministerial guidance has emphasised the need for a positive and comprehensive approach. Grants are available towards the cost of repair and maintenance. Furthermore, an owner of a building who is refused listed building consent can, in certain circumstances, serve a notice on the local authority requiring it to purchase the property. This is known as a *listed building purchase notice*. The issue to be decided here is whether the land has become 'incapable of reasonably beneficial use'. It is not sufficient to show that it is of less use to the owner in its present state than if developed. Local authorities can also purchase properties by agreement, possibly with Exchequer aid. Exceptionally, a neglected building can be compulsorily acquired.[25]

In spite of all these (and other) provisions, many listed buildings are at risk. In Scotland, (according to the 1995 National Audit Office report on *Protecting and Presenting Scotland's Heritage Properties*), the ongoing *Buildings at Risk Register* contained, in 1994, 860 listed properties which were unoccupied or derelict and which had a dubious future (over 2 per cent of the total). The position is relatively worse in England: an English Heritage report showed that 36,700 listed buildings (7 per cent of the total) are at risk from neglect; twice as many are in a vulnerable condition and need repair if they are not to fall into the 'at risk' category. Of course, most listed buildings are in private ownership, and the owners may well not feel the respect for their buildings which preservationists do; or they simply may be unable to afford to maintain them adequately. Advice, grants and default measures cannot achieve all that might be hoped and, though a precious building can be taken into public ownership, this is essentially a matter of last resort. Sharland (2000) has put the case for more careful scrutiny of how preservation can be put into effect so that we list buildings that can be preserved, and a statutory duty on owners to keep those that are listed in good repair.

Criteria for listing historic buildings

Criteria for listing historic buildings are divided into four groups according to the date of building and are shown in Box 8.1.

In choosing buildings, particular attention is paid to 'special value within certain types, either for architectural or planning reasons or as illustrating social and economic history', to technological innovation or virtuosity (for instance, cast-iron prefabrication or the early use of concrete), to any association with well-known characters or events, and to 'group value', especially as examples of town planning such as squares, terraces or model villages (DoE Circular 8/87). Buildings are graded according to their relative importance. The grading systems are set out in Table 8.3.

Scotland has for long had a rolling thirty-year rule under which any building of that age could be considered for listing. This was initially thought to be too problematic in relation to the much larger number of buildings that would be covered by such a rule in England.[26] Many buildings have been demolished which would today attract vociferous defence. On the

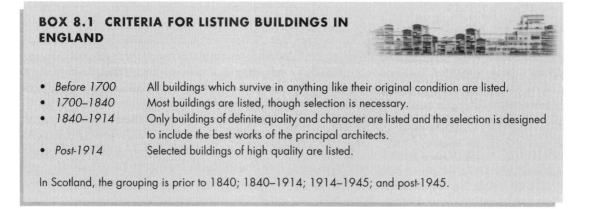

BOX 8.1   CRITERIA FOR LISTING BUILDINGS IN ENGLAND

- *Before 1700*   All buildings which survive in anything like their original condition are listed.
- *1700–1840*   Most buildings are listed, though selection is necessary.
- *1840–1914*   Only buildings of definite quality and character are listed and the selection is designed to include the best works of the principal architects.
- *Post-1914*   Selected buildings of high quality are listed.

In Scotland, the grouping is prior to 1840; 1840–1914; 1914–1945; and post-1945.

▓ ***Table 8.3*** *Numbers of listed buildings, scheduled monuments, conservation areas and world heritage sites in the UK*

	Listed buildings	Scheduled monuments (in care)	Conservation areas	World heritage sites	Historic parks and gardens
England	371,891	19,446 (400)	9,080	15	1,563
Scotland	44,462	7,035 (330)	674	3	275
Wales	22,308	3,000 (129)	400	1	n/a
Northern Ireland	8,563	1,525 (181)	40	1	150
Total UK	447,224	31,006	10,194	20	

Source: Compiled from English Heritage (2003) *Heritage Counts*

Notes: There are also three world heritage sites in overseas dependent territories and eleven additional world heritage sites are on the 'tentative' list. The numbers of ancient monuments and listed buildings in England refer to registry entries; the number of actual monuments is about double the figure here and the number of listed buildings in England is nearer 460,000.

other hand, some more recent architecture would have difficulty in finding a place in the hearts of those who support the protection of good interwar buildings. Clearly, this is an area where attitudes differ and firm guidelines are far from easy to determine – as is also the case with contemporary design and amenity guidelines.

Matters were suddenly accelerated when Sir Albert Richardson's Bracken House in the City of London was threatened with demolition. The Secretary of State decided to list this grade II*, thus copying the Scottish principle that buildings under thirty years old could be listed. At the same time, going one better than the Scots, it was decided that outstanding buildings that were only ten years old could be listed if there was an immediate threat to them (see Box 8.2).

Public participation in listing

Until 1995 most listings were proposed by the Commissions and decided by the central departments and there was no advance publicity. In 1995, the public were invited to comment on proposals from English Heritage for the listing of forty modern buildings and thirty-seven textile mills in the Manchester area.[27]

There was concern that such a highly publicised process of listing might incite owners to demolish their earmarked buildings at speed (as the Firestone building had been). Spot-listing is one answer to this (if it is done quickly enough) or the imposition of a building preservation order (though this renders the local authority liable to compensation if the building is not in fact eventually listed). A better solution would be a new power for an instant listing which carried no compensation penalties for the local authority.[28] Public opinion (when aroused) can play an important part in listing or when listed buildings are under threat, as is well exemplified by the successful campaign to save St Pancras Station (Lane and Vaughan 1992).

Some questions remain, however, and they are likely to come to the fore as local planning authorities continue to develop their competences in the area. One question is the justification for the existence of two regimes: one for the listing of historic buildings and the other for the scheduling of ancient monuments and archaeological remains. Other questions relate to the division of responsibilities between planning authorities and central government and, in particular, the degree of the integration between heritage planning and the other functions of local planning authorities (Redman 1990; Scrase 1991). Some of these

---

BOX 8.2 A SAMPLE OF THE YOUNGER LISTED BUILDINGS IN THE UK

- 1,000 red telephone boxes
- 123 cinemas
- Essex County Cricket Club Pavilion
- City Hall, Cardiff
- Jodrell Bank Radio Telescope
- Alexandra Palace, London
- 12,000 churches
- The Rotunda, Birmingham
- a petrol pump at Oxton, Nottinghamshire
- Ribblehead Viaduct, North Yorkshire
- Coventry Cathedral
- Carrickfergus Castle, Co. Antrim

issues have been taken up in 'the heritage review' described below.

Conservation, market values and regeneration

A major barrier to the conservation of some listed buildings is finding a contemporary use for them that is compatible with the character which it is desired to preserve. Research shows that listed office buildings have a 'market performance' which is generally as good as other buildings, and sometimes better. On the other hand, listing can reduce market value, particularly of small buildings in areas of high development outside conservation areas. However, the reduction is a one-time cost which is borne by the owner at the time of listing: future 'market performance' is not affected. But listing can also increase values because of the 'prestige' thereby accorded, and this can also raise neighbouring values. Like all such issues, much depends on local factors (Scanlon *et al.* 1994).

In deciding whether or not to list a building, the Secretary of State is required to have regard only to the special architectural or historic interest. No account can be taken of economic issues (such as the condition of the building and the cost of conserving it, or the possibilities of finding a viable use for the building). Nor can the personal circumstances of the owner be considered. (Such issues become relevant only when an application is made for listed building consent to demolish or alter a listed building.)

The heritage agencies have strongly promoted the regeneration potential of conservation. The 1998 English Heritage report, *Conservation-led Regeneration*, includes numerous successful and inspiring urban regeneration schemes across the country. Major projects such as the Albert Dock, Liverpool, Saltaire, West Yorkshire, and Dean Clough Mills, Halifax are well known, but there are very many smaller schemes that are equally impressive.[29]

Looking at this issue the other way around, development and regeneration can often enable restoration and conservation of the historic environment. But 'enabling development' often calls for considerable adaptation of historical assets. English Heritage define enabling development as that 'which, while it would achieve significant benefit to a heritage asset, would normally be rejected as clearly contrary to other objectives'. The argument is that the benefits of safeguarding the heritage asset – a country house for instance – offset the negative impact (and detriment to the asset itself) of say new housing development within the grounds of the house. The development makes up the 'conservation deficit' – the difference between the cost of repair and renovation to bring it into viable use and the resulting value of the property on the market. While English Heritage agrees that enabling development can be a useful planning tool, it has concluded that too often they 'destroy more than they save'. Therefore in 1999 English Heritage adopted a presumption against enabling development unless it meets strict criteria, including the development must not detract from the heritage asset or its setting, and if it is demonstrated that it is the minimum necessary to secure benefits for the asset.

In circumstances where 'enabling' is resisted, and thus private investment deterred the only answer is subsidy from public or charitable sources. English Heritage have begun to target grant assistance on more deprived areas, and where investment in existing buildings will contribute to economic and social regeneration. Other funding bodies (described later) are doing the same. Nevertheless, there are still questions about the extent to which the government's urban renaissance policy has taken conservation fully on board.

Conservation areas

Of particular importance in heritage planning is the emphasis on areas, as distinct from individual buildings, of architectural or historic interest. Statutory recognition of the area concept was introduced by the Civic Amenities Act 1967.[30] Local planning authorities have a duty 'to determine which parts of their area are areas of special architectural or historic interest, the character of which it is desirable to preserve or enhance', and to designate such areas as *conservation areas*. When a conservation area has been designated, special attention has to be paid in all planning decisions to the preservation or enhancement of its character and appearance. Demolition of all buildings (unlisted as well as listed) is controlled.[31] There are also special provisions for preserving trees.

But owners of unlisted buildings have 'permitted development rights': they are not subject to the restrictions applied to owners of listed buildings. However, local planning authorities can withdraw these permitted development rights by use of an *Article 4 direction* (discussed in Chapter 5). Indeed, this is the common use of such Directions (Tym *et al.* 1995a). They are typically intended to prevent piecemeal erosion of the character of an area through the cumulative effects of numerous small changes. Local planning authorities also have a duty to seek 'the preservation and enhancement' of conservation areas. Although some authorities take this duty seriously, it is generally poorly implemented, often on the grounds of inadequate resources.

The statutory provisions relating to the establishment of conservation areas are remarkably loose: there is no formal designation procedure, there is no requirement for a formal public inquiry (though proposals have to be put before a public meeting), and there is no specification of what qualifies for conservation area status. Circular 8/87 notes that 'these areas will naturally be of many different kinds':

> They may be large or small, from whole town areas to squares, terraces and smaller groups of buildings. They will often be centred on listed buildings, but not always. Pleasant groups of other buildings, open spaces, trees, and historic street patterns, a village green or features of historic or archaeological interest may also contribute to the special character of an area. Areas appropriate for designation as conservation areas will be found in almost every town and many villages. It is the character of areas, rather than individual buildings, that [section 60 of the Planning (Listed Buildings and Conservation Areas) Act 1990] seeks to preserve or enhance.

In practice, Larkham (1999a: 113) points out that the scope of designations has widened over recent years and there is also a trend to much larger areas.[32] The 1996 English Heritage consultation paper proposed that designations should 'include a statement identifying the specific features of the area that it is considered desirable to preserve or enhance'. In 2000 a similar call was made for better assessment of the qualities of conservation areas both existing and proposed (see p. 308).

The number of conservation areas has grown dramatically, and by the end of 1999 there were more than 9,000 in the UK (Table 8.1).[33] Over 1 million buildings are in these areas. Indeed, it was suggested some time ago that perhaps a 'saturation point' had been reached in that the resources are simply not available for 'enhancing' such a large number of areas (Morton 1991; Suddards and Morton 1991). There continues to be a widespread view that more attention should be given to managing existing conservation areas, and less to designating additional ones (Larkham and Jones 1993) but there has been little government action on this so far. Townshend and Pendlebury (1999) point to the continuing poor performance of professionals in involving residents in designation and management of conservation areas. While an expert-led approach may be required for conservation areas of national significance, it may be that a more community-led approach, facilitated by the expert, is more appropriate in many thousands of conservation areas. This leads to the tentative suggestion that there may be a need for a grading scheme for conservation areas, in a similar way to listed buildings, so as to allow for different forms of control.

English Heritage has targeted its resources in conservation on priority areas through *town schemes* and *conservation area partnerships*, where it jointly funded works with planning authorities and others. The latest in this form of initiative is the Heritage Economic Regeneration Scheme (HERS) and this succeeds the area partnerships. The schemes give more emphasis to economic and community regeneration as well as physical improvements and provided £15 million over three years from 1998. The idea is to concentrate on neighbourhood businesses, high streets and corner shops and 'where areas based assistance through building repair and enhancement will tip the balance in favour of continued local employment, new homes and inward investment'.

World heritage sites

The UNESCO World Heritage Convention established a World Heritage List of sites that UN member states are pledged to protect. The 24 sites within UK jurisdiction (of the 582 world-wide) are listed in Box 8.3. In 2000 the UK updated its 'tentative list' of sites that may be nominated for the world heritage status over the next five to ten years and four nominations were made and accepted.[34] The tentative list was a requirement of the Committee that oversees designation. The Committee had expressed the wish to consider further

---

BOX 8.3 WORLD HERITAGE SITES UNDER THE JURISDICTION OF THE UK WITH DATE OF DESIGNATION

- 1986   Giant's Causeway and Causeway Coast
          Ironbridge Gorge
          Stonehenge, Avebury and associated sites
          Durham Castle and Cathedral
          Fountains Abbey and St Mary's, Studley Royal
          The castles and town walls of Edward I in Gwynedd
          St Kilda
- 1987   Blenheim Palace
          City of Bath
          Hadrian's Wall Military Zone
          Palace of Westminster, Westminster Abbey and St Margaret's Church
- 1988   Henderson Island (Pitcairn Islands)
          Tower of London
          Canterbury Cathedral, St Augustine's Abbey and St Martin's Church
- 1995   Edinburgh Old and New Towns
          Gough Island Wildlife Reserve (St Helena Islands)
- 1997   Maritime Greenwich
- 1999   The heart of Neolithic Orkney
- 2000   Historic town of St George and related fortifications, Bermuda
          Blaenavon industrial landscape, South Wales
- 2001   Dorset and east Devon coast
          Derwent Valley Mills, Derbyshire
          New Lanark, Scotland
          Saltaire, West Yorkshire
          Royal Botanic Gardens, Kew

Current nominations
          Liverpool Commercial Centre and Waterfront

Source: DCMS website www.dcms.gov.uk

natural and industrial sites to provide a better balance with the large number of architectural sites. Thus the current UK tentative list of sites includes, for example, the Lake District, the New Forest, Shakespeare's Stratford, and the Mount Stewart Gardens, Northern Ireland. In future, national governments are able to nominate only one site each year.

The inclusion of a site on the World Heritage List carries no additional statutory controls though, of course, it underlines its outstanding importance. This is a relevant material consideration in planning control. Local planning policies should, in the words of PPG 15, 'reflect the fact that all these sites have been designated for their outstanding universal value, and they should place great weight on the need to protect them for the benefit of future generations as well as our own'. Significant development proposals affecting a World Heritage Site generally require an environmental assessment. This has not protected sites in the UK from the pressure of new development – even Stonehenge. The World Heritage Committee now requires a management plan for all listed sites, and English Nature published in 2000 *The Stonehenge World Heritage Site Management Plan*.

Historic parks and gardens

The 5,000 or more public parks in the UK have played an important role in quality of life in towns and cities, and many have great historical and landscape significance.[35] But their quality is generally deteriorating – lamentably so. The neglect of public parks is not only one manifestation of the general decline in local authority services, but also reflects changing social needs and behaviour. There is no statutory duty on local authorities to provide or maintain parks and open spaces. Indeed there seem to be no clear responsibilities in relation to parks. The Urban Parks Forum claim that neither the government has recognised the problem and the lack of even basic statistics on the amount of parkland and its quality would seem to bear that out. But Britain also boasts some of the finest historical parks in the world; most, other than the Royal Parks, are in private ownership or under the management of

the National Trust and other charitable organisations.[36] The eight Royal Parks are managed by an executive agency of the DCMS, the *Royal Parks Agency*, with a budget of £20 million.

A *Register of Parks and Gardens of Special Historic Interest in England* is compiled by English Heritage in county volumes.[37] The register uses a grading system similar to that for listed buildings with grades I, II* and II but unlike listed buildings there are no additional consents required. Given that the register is a relatively recent innovation, it may be that statutory controls will be imposed in the future (Pendlebury 1999). In the mean time, the register is a material consideration in development control and planning authorities must consult English Heritage on applications likely to affect grade I or II* parks and gardens, and consult the Garden History Society on development that may affect any site on the register. There are about 1,450 sites on the register, 120 of which are urban parks.[38] The register was first compiled in the 1980s and reviewed between 1994 and 1999. A programme of upgrading the register to include more details of historic gardens began in the early 2000s. Ironically, it is the poor condition of many parks that may be a reason that they do not appear on the register and thus miss out on the benefits that the register might give. The register includes historic cemeteries, although until 2001 they numbered only 14. The number rose to about 110 following a review of historic cemeteries for inclusion in the register by English Heritage in 2001. The following year, English Heritage published guidance for those involved in the conservation and enhancement of historic cemeteries with the apt title of *Paradise Preserved*.

Churches

The situation regarding 'ecclesiastical buildings' is exceptional and also complicated. In essence, there is what is technically termed 'the ecclesiastical exemption' from listed building and conservation area controls. The exemption may apply to many buildings – the Church of England alone has more than 16,700 churches.[39] The Church introduced measures to control demolition more than 700 years ago, and has been regularly inspecting churches for 300 years. It spends a large amount each year on the upkeep and maintenance of its buildings (mainly funded by its congregations). The result is that 'a listed Church of England church has a chance of avoiding demolition nearly three times better than a listed secular building'.

There are two parallel statutory systems of control over Church of England churches: the Church's system and the secular system. The Church's system is much stricter and more comprehensive. It involves regular inspection of every church, and embraces not only the fabric of the buildings, but also their contents and churchyards. There are two separate statutory procedures applying to parish churches (whether listed or unlisted), according to whether they are in use or redundant. Churches in use are subject to a system of inspection and reporting at the local level, and to monitoring at higher levels: by Diocesan Advisory Committees at diocesan level, and by the Council for the Care of Churches at the national level.

Redundant churches are safeguarded by the Pastoral Measure 1983, which provides procedures for deciding whether a church is still required for worship, and, if not, what the future of the building should be. The *Churches Conservation Trust* (formerly the Redundant Churches Fund) finances the management, maintenance and repair of churches judged of sufficient architectural or historic importance. The fund receives 70 per cent of its funds from the DCMS and the remainder from the Church Commissioners; in 1999 it had 300 redundant churches in its care.

Until recently, cathedrals were outside any planning procedure and, despite their huge popularity with visitors (and contribution to tourism), were not eligible for grant aid. A separate system of controls over building works was introduced by the Care of Churches Measure 1990. This is administered by the individual cathedrals jointly with a Cathedrals Fabric Commission in consultation with English Heritage, which also provides grant aid.

All church buildings are subject to normal planning control over, for example, changes of use and significant alterations. They are also listed in exactly the same way as other buildings of special historic or architectural interest. However, because of the Church's separate statutory procedure, listed building consent is not required for churches where the primary use is as a place of worship. Such consent is required, however, for alterations to redundant churches, though not if demolition is carried out pursuant to a scheme under the Pastoral Measure 1983.

A government review of the ecclesiastical exemption completed at the beginning of 1993 led to a decision to extend it to churches of all denominations where an acceptable system of control operates on principles set out in a code of practice.[40] The 1994 Ecclesiastical Exemption Order revoked the exemption for religious bodies that had not adopted their own regulation systems. For those that do, it temporarily extended exemption to other buildings within the churches' estates, but this too will be revoked unless the churches introduce their own controls over such buildings. In 1999, the Church of England adopted the Care of Places of Worship Measure which empowers the Council for the Care of Churches to compile a list of the other buildings used for purposes of the Church (such as school and college chapels), and that it wishes to fall within the protection of the Church. At some point the ecclesiastical exemption will be removed for those religious bodies that have not been included within a 'self-regulatory regime'.

Funding for conservation of the historic environment

Following the outcry over the controversial sale of the assets of the Mentmore estate in 1977, a *National Heritage Memorial Fund* (NHMF) was established by the National Heritage Act 1980. This is dedicated as 'a memorial to those who have died for the United Kingdom'.[41] The Fund gives financial assistance 'towards the cost of acquiring, maintaining or preserving land, buildings, works of art and other objects of outstanding interest which are also of importance to the national heritage'. It is doing an important job in preventing heritage assets from being exported. But given the value of some assets, it is a relatively small fund. During 1999–2000 it made grants worth

£2.7 million in respect of eleven items.[42] In addition to normal Exchequer payments into the fund, further payments can be made in relation to property accepted in satisfaction of tax debts.

The NHMF now also distributes the *Heritage Lottery Fund* (HLF). The scale of conservation work made possible by the Lottery was probably not imagined in the early 1990s. The HLF allocated £148 million in 1,872 grants in 1999–2000.[43] As the Chairman of the Funds has said, 'every age needs its patrons and the private patronage on which we have largely relied in this country is now vigorously supported (but happily not supplanted) by lottery money' (*HLF and NHMF Annual Report* 1999–2000: 2). The HLF now has four general priorities: conservation, national heritage, local heritage and access and education. Its formal aim reflects current concerns with equal access and sustainability. It is

> to improve quality of life by safeguarding and enhancing the heritage of buildings, objects and the environment, whether man-made or natural, which have been important in formation of the character and identity of the United Kingdom, in a way which will encourage more sections of society to appreciate and enjoy their heritage and enable them to hand it on in good heart to future generations.
>
> (*HLF and NHMF Corporate Plan* 2000)

It is difficult to do justice here to the range of projects that have received a contribution from the HLF, since its impact is so pervasive. The better known projects include the National Maritime Museum at Greenwich, the American Air Museum at Duxford, St George's Market in Belfast, Robert Owen's School in New Lanark (part of the World Heritage Site) and the Big Pit Mining Museum, Blaenafon (the National Mining Museum for Wales). Planning authorities have been important players in both large and small projects as initiator and sometimes co-funder. Of particular interest is the townscape heritage initiative. This provides small grants towards heritage scheme feasibility studies, and more substantial funding to improve the vitality of many towns, some of which are not traditionally seen as heritage centres.

After early criticism about elitism and unequal distribution of funding across the country, the HLF now strongly emphasises the need for funded projects to deliver wider public benefits, for example, through economic regeneration and social inclusion. There are more small scale 'community grants', providing a simpler application process. The HLF has contributed £3.9 million in 1,284 awards in 1999–2000. Regional offices have been set up in Northern Ireland, Scotland, Wales and the English Regions. A special study is underway to examine how the regeneration of coalfield communities may be supported (which have done badly in the distribution of funds so far). Another study is looking at the needs of urban parks (discussed earlier).[44]

Lottery funding aside it should not be forgotten that some local authorities and many voluntary organisations have a long-standing record of funding conservation. Also, in England there is also a relatively small (£546,000 in 1999–2000) fund now administered by English Heritage for innovatory or experimental projects that contribute to government's objectives for the heritage. Projects have been funded involving new records of the historic environment, promoting access and improving management practice. The fund supported the Civic Trust's Heritage Open Days initiative.

Preservation of trees and woodlands

Trees are a delight in themselves; they also have the remarkable quality of hiding developments which are best out of sight. Trees are clearly, so far as town and country planning is concerned, a matter of amenity. Indeed, the powers which local authorities have with regard to trees can be exercised only if it is 'expedient in the interests of amenity'. Where a local authority is satisfied that it is expedient, it can make a *tree preservation order* (TPO) applicable to trees, groups of trees, woodlands, and trees planted as a result of a planning condition. Such an order can prohibit the cutting down, topping or lopping of trees except with the consent of the local planning authority.[45] Orders are made according to a model given in the Regulations. People affected must be consulted and the planning authority must consider all objections and other comments before confirming the TPO. Subsequently any proposals to cut or lop the protected trees need consent from the planning authority. In conservation areas, trees otherwise not protected by TPOs are also subject to a special regime. The planning authority must be given six weeks' notice of any works, during which it can consider the need for a TPO.

Mere preservation, however, can lead eventually to decay and thus defeats its object. To prevent this, a local authority can make replanting obligatory when it gives permission for trees to be felled. The aim is to avoid any clash between good forestry and the claims of amenity. But the timber of woodlands or orchards always has a claim to be treated as a commercial crop, and though the making of a tree preservation order does not necessarily involve the owner in any financial loss (isolated trees or groups of trees are usually planted expressly as an amenity), there are occasions when it does.

Yet, though woodlands are primarily a timber crop from which the owner is entitled to benefit, two principles have been laid down which qualify this. First, the national interest demands that woodlands should be managed in accordance with the principles of good forestry, and second, where they are of amenity value, the owner has a public duty to act with reasonable regard for amenity aspects. It follows that a refusal to permit felling or the imposition of conditions on operations which are either contrary to the principles of good forestry or destructive of amenity ought not to carry any compensation rights. But where there is a clash between these two principles, compensation is payable. Thus, in a case where the principles of good forestry dictate that felling should take place, but this would result in too great a sacrifice of amenity, owners can claim compensation for the loss which they suffer. Normally, a compromise is reached whereby the felling is deferred or phased. The commercial felling of timber is subject to a licence by the Forestry Commission.

Planning powers go considerably further than simply enabling local authorities to preserve trees. Planning permission can be made subject to the condition that trees are planted, and local authorities themselves have power to plant trees on any land in their area. With the increasing vulnerability of trees and woodlands to urban development and the needs of modern farming, wider powers and more Exchequer aid have been provided by successive statutes. Local planning authorities are now *required* to ensure that conditions (preferably reinforced by tree preservation orders) are imposed for the protection of existing trees and for the planting of new ones.

In 1994 the DoE reviewed the TPO system and then consulted on new regulations to overcome certain anomalies. Following the change in administration the proposals were put on hold until 1998 when a further consultation paper was published. New regulations were made in 1999.[46] The changes are generally quite minor and relate to the simplification of the model order by which TPOs are made; maps; inspection; and clarification of the exemptions afforded to statutory undertakers following the wave of privatisation. Guidelines were published in 1995 on consultation between statutory undertakers and planning authorities, and a Code of Practice is proposed. Local authorities can undertake work on trees on any land, which obviously puts considerable onus on their internal consultation processes between say, the planning officers and those who undertake the work. New legislation is proposed to amend the provisions further, mainly in relation to offences committed in damaging protected trees.

A different approach has been taken to the protection of 'important hedgerows', which, because of devastating losses, have been given protection through Regulations made in 1997 (see also Chapter 9).[47] The Regulations apply only to hedgerows with a continuous length of 20 metres or more and which meet another hedgerow at either end. They need to be growing in or adjacent to common land, protected land or land used for agriculture, forestry or for keeping horses, etc. The Regulations do not apply to hedgerows around houses. If all these conditions are satisfied, before removing a hedgerow the owner must notify the planning authority, which, if it wants the hedge retained, has forty-two days to serve a 'hedgerow retention notice'. But the planning authority can do this only for an 'important hedgerow', which means it must have been in existence for thirty years or more and have some archaeological, historical, wildlife or landscape qualities.

Tourism

Heritage is an important factor in tourism. Together with 'culture' and the countryside, heritage stimulates about two-thirds of the visits made by foreign tourists. (They certainly do not come to enjoy the weather!) Heritage attractions are also the most important reason that domestic tourists give for having made their visit within the UK.[48] Since tourism plays a very significant role in economic prosperity it follows that heritage is very important in the UK's economy, and of course, in the perception of the UK that many visitors gain. But since the early 1980s, the UK's share of world tourism has generally declined (although there has been some improvement since 1994).[49]

There is an abundance of figures that demonstrate the impact of tourism.[50] There is, however, a big downside: tourism can lead to excessive wear and tear on the fabric of buildings, to congestion, to litter, and even to open hostility by residents to visitors. The generally accepted implication is that tourism has to be 'managed'. Several organisations are now devoted to this: the Historic Towns Tourism Management Group, the Heritage Cities Association (a marketing consortium) and the English Historic Towns Forum. A concern for 'vital and viable town centres' has grown, and other specialist bodies have been established, such as the Association of Town Centre Management.

That tourism, as well as heritage, is a matter of importance for local planning authorities is self-evident. However, it is not a policy area which can be isolated from related ones. It is interesting to note that the PPG on tourism (PPG 21) refers to a long list of other relevant PPGs. This list is an eloquent testimony to the interconnectedness of planning issues; and it also points to the inherent difficulty of reconciling numerous considerations – or even giving adequate consideration to all of them.

Responsibilities for tourism, like most of government, have shifted considerably over recent years with a view to addressing the 'untidy structure' of tourism organisations and linkages with heritage bodies.[51] The creation of the DNH, now DCMS, brought together UK government tourism and heritage from six departments. The Scottish Executive and the Welsh and Northern Ireland Assemblies have responsibilities for tourism and fund 'national tourist boards'. VisitBritain (*sic*) was created through a merger of the British Tourist Authority (BTA) and the English Tourism Council in 2003. It leads the marketing of Britain overseas and markets England to the British. In that first role it is answerable to the DCMS and UK Parliament and also the Scottish Parliament and Welsh Assembly for its marketing of those countries. It provides core services, intelligence and strategy development while the national tourist boards for Scotland and Wales and its own marketing England division deal with matters of implementation. (There is also an agreement with the Northern Ireland Tourist Board, although here some international marketing is undertaken in partnership with the Republic of Ireland.) Steps towards providing a more strategic perspective on British tourism were first taken in 1999 when the English Tourism Council replaced the English Tourist Board. In England much tourism activity is undertaken at the regional level. Implementation is concentrated in the regional tourist boards (RTBs) which are limited companies, funded by VisitBritain, and other commercial sources; and the strategic lead on tourism has been the responsibility since 2003 of the regional development agencies.[52] Local authorities also promote and manage tourism, and this is particularly important in the heritage towns.

The 1997 Labour government established a Tourism Forum which contributed to a major review of government policy for Tourism. The results were published in the 1999 DCMS report *Tomorrow's Tourism*. A further report *Tomorrow's Tourism Today* was published by DCMS in 2004, and recorded the reform of organisational structures for tourism summarised above. It signals more intense activity to increase the productivity of UK tourism, including more emphasis on the potential of sport, exemplified by the London 2012 Olympics bid, and reform of the licensing laws to support the tourism and hospitality industries. Similar reviews are underway elsewhere in the UK.[53] The reports have the same emphasis on promoting tourism for its economic and regeneration potential; in particular there is a need to ensure that the tourism industry in the UK performs at a similar rate to its competitors, since it is losing ground. An action programme was produced for England, and later discussed at a national tourism summit. Of particular relevance for planning are action points to provide a sustainable blueprint for tourism to safeguard the countryside and heritage; to encourage more integrated promotion of the heritage; to develop niche markets in such areas as 'film tourism' so as to 'unlock the potential of Britain's unique cultural and natural heritage'; and to encourage regeneration of traditional tourist resorts.

As with all matters concerning sustainability there is more than a hint of contradiction in the action points. On the one hand, business and economic priorities mean that the heritage has to earn a return on the public funds invested. On the other hand there is a need to preserve the integrity of the properties and places visited. The relevant documents all note this problem, and much needs to be addressed in tourism management at the local level. It is well recognised that the presentation and management of historic properties cannot be a purely commercial operation. In any case, too great a success in attracting custom could place unsustainable pressures on the very experience which the customers are seeking. For example, marketing efforts have helped to increase visitor numbers at historic houses from 3 million in 1970 to nearer 11 million at the end of the 1990s. But 'it has been estimated that the wear and tear due to visitors in one year exceeds the previous domestic wear and tear of two decades and in some cases up to a whole century' (Lloyd 1999: 1).[54]

Perhaps the most controversial subject arising from the aggressive development of tourism, and one that is of major concern to planning authorities, is the 'regional casino'. A fact sheet available from DCMS on the Gaming Bill 2005 usefully summarises the current position on casinos and the proposed changes. It notes how the planning and licensing of casinos operates separately – a planning permission does not

guarantee a license and vice versa. In 2004 Great Britain had 134 relatively small casinos (by US standards) operating in 53 'permitted areas'. The proposals will allow casinos in all areas and large casinos on the US model. Initially eight regional casinos were proposed but in 2005 the proposal was amended to one 'super' regional casino, eight large and eight small casinos. DCMS established a Casino Advisory Panel chaired by Stephen Crow, the former Chief Planning Inspector, to make recommendations on the location of these casinos in 2006. Much (inevitably abortive) work has gone into to many bids, most linked to wider regeneration proposals such as the East Birmingham 'sport village' led by Birmingham City Football Club.

The ETC has taken a lead role on the idea of 'sustainable tourism' and has created a Sustainable Task Force (*sic*). A report on *Sustainable Tourism at the Local Level* was published in 2002. Unfortunately, the English Tourist Board 'vision for sustainable tourism' hardly touched on the concept of sustainability. It reads 'England will promote and develop tourism that exceeds visitor expectations, ensures the long term viability of the industry, benefits local communities and helps to protect and enhance the places in which it takes place'.[55]

Some critics have been very outspoken about so-called 'sustainable tourism'. As far back as 1955, Croall presented strong arguments for greater protection of the environment against the effects of tourism. An even stronger case is made by Minhinnick (1993), who argues that 'the idea of making tourism an environmentally sustainable activity is at best an exciting pipe-dream and at worst a deceit'. This elegantly written essay is merciless in its criticisms: 'the trouble with tourism is that moderation is not part of its language' and 'local distinctiveness is erased and replaced by mediocre uniformity'. While these fundamental conflicts barely figure in official reports, there are lots of practical suggestions, but these are predominantly concerned with minimising environmental impacts.

The 1998 consultation on the national sustainability strategy included a paper on *Tourism: Towards Sustainability*, which considers some of the issues. These include the potential of tourism to benefit local communities, the need to manage visitor flows, the transport impacts, and planning. Indeed, planning figures very prominently in the responses to this consultation paper, particularly the need to amend planning guidance to ensure that tourism development meets 'sustainability criteria' and to strengthen plans and development control powers to ensure that tourism investment is concentrated where it can do most benefit for regeneration and least damage (see Box 8.4). But much of the agenda for tourism is no more than the agenda for everyone else – better strategic coordination of policy through planning, closer integration of transport modes, quality public transport, opportunities for cycling and walking, effective reuse and renewal of the heritage, and the need to find a way to spread tourism and its benefits beyond the mainstream 'honeypots' and out to the regions.

The heritage review

In November 1999, the government announced its intention to carry out a systematic review of policy for the historic environment in England, led by the DETR and DCMS. Early in 2000, English Heritage was instructed to undertake stage one – a review of current policies in consultation with all interests. The terms of reference for the review were restrictive: the principles of PPG 15 were to stay in place, resources would remain much the same, and there would be no major structural reorganisation of responsibilities. The review was to consider in particular the relationship between heritage and tourism and the roles of the numerous bodies involved in conservation. In addition, English Heritage was asked to review the condition of the historic environment, the need for dealing with heritage at risk, possible simplification of the procedures, and connections with urban regeneration and emerging policy on sustainable development.

The report, *Power of Place*, was published at the end of 2000, as a step towards a strategy for the historic environment in England. The report's many recommendations are mostly precise and send a strong message about how to strengthen conservation in all areas of public policy. This is to be done by, for example, ensuring that conservation is reflected in all government sectoral policy, ensuring that tax and funding regimes support conservation at least as well as new build,[56] strengthening regulation and powers for designated areas and improving information and skills. Despite the limitations imposed by the terms

---

BOX 8.4 SIGNIFICANCE OF TOURISM 2004

- 25.5 million overseas visitors visited the UK
- UK residents took 122 million trips of one night or more in the UK
- Overseas tourists spent about £13 billion in the UK
- Expenditure on travel to British operators was £3.3 billion
- Total expenditure on tourism was £61 billion
- Employment in tourism was about 18 million
- There were 125,000 tourism businesses
- The share of GDP attributable to tourism was 6 per cent
- Over fifty historic towns attracted over 20,000 overseas overnight visitors
- Six historic towns received over 150,000 overseas overnight visitors
- The government provided almost £100 million grant to the tourist boards
- UK visitors abroad spent £8.2 billion more than overseas visitors spent in the UK

of reference, the report makes a strong case for organisational rationalisation saying that government

> lacks an effective historic environment dimension to wider policy objectives and DCMS has never given the issue the attention or priority it deserves. These problems are in addition to the split responsibility for planning and listed building and other consent procedures.
>
> (p. 43)

The theme of getting a more consistent approach to conservation across all arms of government appears also in relation to ensuring that conservation figures in the policies of RDAs and local strategic partnerships.[57] Another strong theme is the promotion of the economic value of the historic environment in terms of investment returns (at least when it can be used as offices), job creation and tourism ('pound for pound, repair and maintenance create more employment than new-build').

On the role of the planning system, the report notes that designations have been more successful in relation to buildings and monuments than with areas of land: conservation areas, parks and gardens and battlefields. One area that is lagging behind badly is marine heritage. Of the 34,000 known marine archaeological sites in English territorial waters, only 38 are afforded statutory protection – thus the report argues that marine heritage should be brought within the remit of English Heritage.

The report argues for more systematic evaluation of all buildings and sites that are identified for conservation, and that character appraisal or assessment of historical assets should be normal practice and could form the basis of 'spatial masterplans' for their future development.[58] Similarly, conservation plans should be prepared for historic sites and can provide the basis for management agreements with owners. Capacity studies are also mentioned, especially in relation to tourism impacts. But for all this work there is a considerable shortfall of relevant skills and qualified staff. Many planning authorities (22 per cent in England) have no staff in the conservation field (which says a lot about some local authority attitudes to the

historic environment). Therefore, English Heritage recommend that appropriate performance criteria on heritage management should be included in the Best Value regime (see Chapter 3). The need to improve the information about historic environments, complete records and provide easier access to them is also given attention.

Some recommendations are more challenging, notably that permitted development rights should be withdrawn in all conservation areas. This recommendation and the effective withdrawal of rights of owners is unlikely to be acceptable without significant tightening of the criteria and procedure for designation.[59] It would also have the effect of increasing the difference between quality of the built environment in conservation areas and elsewhere, when it might be argued that the deteriorating quality of all environments is the more significant issue.

One significant message for government is that it needs to put its own house in order, including a number of public bodies such as the Ministry of Defence, whose actions in some cases suggest that it sees the heritage as an obstacle to realising a quick return rather than an asset. It is stated unequivocally that 'examples of best practice in estate management are found in the large private estates and not the public sector'.[60]

Further reading

History and general context

Informative accounts of the history of historic preservation are given by Hobson (2004) *Conservation and Planning* (with case studies of conservation in practice today), Ross (1996) *Planning and the Heritage* and Delafons (1997) *Politics and Preservation*. (Ross was former head of the Listing Branch of the DoE and Delafons was for twelve years Deputy Secretary responsible for land use planning at the Department of the Environment, and so they give an insider's account, and inevitably concentrate on the government view.) Pickard (1996) *Conservation in the Built Environment* is a standard text. Larkham (1996) *Conservation and the City* sets conservation in the context of urban morphology and international comparisons, and his 1999 paper, 'Preservation, conservation and heritage', is a most useful summary and contains advice on further reading. There is an extensive list of sources at the Historic Environment Information Resources Network: http://www.britarch.ac.uk/HEIRNET/. The DCMS and Royal Commissions' annual reports are useful sources on current activities. The *Heritage Review* (noted above) will be an important source.

Conservation law

General texts covering many points in this chapter are Mynors (1998) *Listed Buildings, Conservation Areas and Monuments* and Suddards and Hargreaves (1996) *Listed Buildings*. An excellent summary of the legal provisions covering heritage is given in Moore (2002) *A Practical Approach to Planning Law*.

Policy for the historic environment

In England, policy guidance is given in PPG 15 (*Planning and the Historic Environment*), PPG 16 (*Archaeology and Planning*) and DoE Circular 8/87, *Historic Buildings and Conservation Areas – Policy and Procedures*. See also the DETR paper on *Contemporary Issues in Heritage and Environment Interpretation*. English Heritage publish a *Conservation Bulletin* three times a year.

For Northern Ireland see PPS 6, *Planning, Archaeology and Built Heritage* (1999). Two useful references are Hendry (1993) 'Conservation in Northern Ireland', in RTPI, *The Character of Conservation Areas*, vol. 2: *Supporting Information*.

For Scotland see NPPG 18, *Planning and the Historic Environment* (1999); PAN 42, *Archaeology*; Scottish Office (1998) *Planning and the Historic Environment*; Historic Scotland Circular 1/1998, *The Memorandum of Guidance on Listed Buildings and Conservation Areas*; National Audit Office (1995) *Protecting and Presenting Scotland's Heritage Properties*; and the Scottish Executive's *Report of the Conservation Control Working Group*.

For Wales see Planning Guidance Wales (First Revision 1999) Section 5, WO Circular 60/96 *Planning and the Historic Environment: Archaeology*, WO Circular 61/96 *Planning and the Historic Environment: Historic Buildings and Conservation Areas*, and the National Audit Office, Wales, *Protecting and Conserving the Built Heritage in Wales*

Archaeology and ancient monuments

In addition to PPG 16 (and variants outside England), see Tym *et al.* (1995b) *Review of the Implementation of PPG 16: Archaeology and Planning*; English Heritage (1992) *Development Plan Policies for Archaeology*. A legal text is Pugh-Smith and Samuels (1996a) *Archaeology in Law*. Many examples are given on the CDRom of the Royal Commission on Historical Monuments in England (1998) *Monuments on Record: Celebrating 90 Years of the Royal Commissions on Historical Monuments, CD Rom* (Swindon: RCHME, now available from English Heritage in England and the Royal Commissions elsewhere in the UK).

Conservation areas

The Character of Conservation Areas (RTPI 1993) is the most recent comprehensive survey and still provides a very useful overview and (in the second volume) some useful supplementary material including a bibliography.

Historic parks and gardens, and trees

Pendlebury (1999) traces the history of controls over historic parks and gardens; see also Pendlebury (1997) 'The statutory protection of historic parks and gardens'. There is a growing literature on parks: see Comedia (1999) *Park Life: Urban Parks and Social Renewal*, DETR (1996) *People, Parks and Cities: A Guide to Current Good Practice in Urban Parks* and the extensive evidence in the House of Commons Environment, Transport and Regional Affairs Committee, *Twentieth Report: Town and Country Parks*. On trees see DETR, *Protected Trees: A Guide to Tree Preservation Procedures*.

Economic aspects and regeneration through conservation

There have been a number of studies of the consequences of listing and conservation for returns on property; the most recent is the Investment Property Databank report for English Heritage (1999) *Investment Performance of Listed Buildings*. See also Scanlon *et al.* (1994) *The Economics of Listed Buildings* and Drury (1995) 'The value of conservation'. On regeneration see the English Heritage reports (1999) *Heritage Dividend: Measuring the Results of English Heritage Regeneration* and (1998) *Conservation-led Regeneration: The Work of English Heritage*.

Tourism

The government reviews of tourism policy consider the link with heritage. See DCMS (2000) *Tomorrow's Tourism: A Growth Industry for the New Millennium*, (1998) *Tourism: Towards Sustainability*, Scottish Executive (2000) *A New Strategy for Scottish Tourism* and the Welsh Assembly (2000) *Achieving Our Potential*. Croall (1995) *Preserve or Destroy: Tourism and the Environment* presents a well-argued case for greater protection of the environment against the effects of tourism. Some of these issues are taken up in a special edition of *Built Environment* 26(1) (2000) edited by MacDonald. Also invaluable is the annual *English Heritage Monitor*, published by the English Tourist Council.

Notes

1   The Royal Commission for Historical Monuments of England was merged with English Heritage in April 1999.

2   Ross gives an interesting account of how this mammoth job was done, often on a voluntary or near-voluntary basis. The survey took twenty-two years and, even then, it was incomplete. Given the attitudes of the time, Victorian architecture was almost totally neglected.

3   The issue was highlighted by the Gowers Report (1950) on *Houses of Outstanding Historic and Architectural Interest*. The Historic Buildings and Ancient Monuments Act 1953 followed, which established the Historic Building Councils for England, Scotland and Wales and the first system of grants.

4   Curiously those relating to ancient monuments are still separate in the Ancient Monuments and Archaeological Areas Act 1979; both are separate from the planning legislation and the need for consolidation is widely recognised. The separation of the conservation and planning legislation is reflected in that the primacy of development plans does apply in the case of listed building consent.

5   It was no doubt with this necessity for change in mind that the CPRE altered the second word in its name from 'preservation' to 'protection'.

6   The phrase 'national heritage' was in fact used in arguing for a statutory duty to list buildings of merit in the 1947 Act (Delafons 1997: 60, quoted in Larkham 1999).

7   Policy guidance for the different countries of the UK is listed at the end under further reading.

8   Delafons gives a review of the case of No. 1 Poultry, an application for demolition of grade II buildings that found its way to the House of Lords before consent was granted. While the Lords' decision supported the discretion of the Secretary of State to allow demolition in certain cases, paradoxically the outcome was a further limitation of discretion by reinforcing the preservation doctrine.

9   During the 1990s there was much reorganisation of responsibilities with devolution and the transfer of heritage from the Department of Environment (responsible for planning) to first the Department for National Heritage and then in 1997 the DCMS. English Heritage was created in 1983. Its formal name is the Historic Buildings and Monuments Commission.

10 It was Octavia Hill who founded the National Trust in 1895. The Georgian Group was founded in 1937, after the Commissioners for Crown Lands demolished Nash's Regent Street and threatened to do the same with Carlton House Terrace. The widespread destruction of Victorian and Edwardian buildings, particularly the Euston Arch, led to the creation of the Victorian Society in 1957. The Twentieth Century Society, which was set up in 1979 (originally

as the Thirties Society), to safeguard interwar architecture, nearly saved the Firestone factory on the Great West Road, but was thwarted by the developers (Trafalgar House) who moved the bulldozers in over the August 1980 holiday weekend before the procedure for 'spot-listing' had been completed. SAVE Britain's Heritage was formed in 1975, the European Architectural Heritage Year, with a special emphasis on finding alternative uses for historic buildings. There are now many heritage organisations, several of which have statutory consultee status on proposals to demolish listed buildings.

11 Policy guidance for archaeology is given under further reading.

12 This is an organisation promoted by the British Property Federation and the Standing Conference of Archaeological Unit Managers (itself a representative body of some seventy-five professional archaeological units).

13 The National Heritage Act 2002 extended the definition of ancient monument to include those under the seabed within the seaward limits of United Kingdom territorial waters. This effectively extends the competence of English Heritage.

14 The Ancient Monuments and Archaeological Areas Act 1979.

15 The term 'schedule' originates from the Ancient Monument Protection Act 1882, which provided for the protection of twenty-nine monuments which were set out in a *schedule* to the Act; the term has persisted. The Royal Commissions and English Heritage undertake many surveys on specific types of buildings such as churches, hospitals and even monuments of the Cold War, such as the 368 metre wide 'listening post' antenna at Cruicksands in Bedfordshire, which was demolished in 1996. A large selection of examples of monuments for the whole of Britain is available on the CDRom *Monuments on Record: Celebrating 90 Years of the Royal Commissions on Historical Monuments* (1998). Databases and catalogues are also available from the relevant agencies' websites.

16 The problem of estimating is further complicated by the fact that some figures refer to register entries (which cover more than one site), and others to individual sites.

17 MARS was established by English Heritage and was based on a sample of 1,300 1 km by 5 km transects of land and examination of photographic records since the late 1940s (Bryant 1999).

18 A proposed development at Abbey Mead near Swindon revealed a Roman water temple and gardens. The developers stopped work and have subsequently been paid £1 million in compensation by English Heritage.

19 A very high proportion are of great antiquity, including prehistoric field monuments such as Maiden Castle, prehistoric structures such as Stonehenge, Roman monuments such as Wroxeter and parts of Hadrian's Wall, and a large number of medieval buildings. Properties in care of Historic Scotland include Edinburgh Castle, Stirling Castle, Fort George and Urquhart Castle (near Loch Ness). Welsh Historic Monuments manage Chepstow Castle, the Blaenavon Ironworks, the Welsh Slate Museum (Llanberis), Neath Abbey and Tintern Abbey. In Northern Ireland, 166 monuments are in the care of the DoENI, including Londonderry's City Walls, Newtownards Priory, Enniskillen Castle, Tully Castle, Carrickfergus Castle, and perhaps surprisingly, the Carrickfergus Gas Works.

20 In England this is the Department for Culture, Media and Sport. Consideration was given recently to devolving control over listing to English Heritage but it was thought that this would lead to extra costs and was not taken up.

21 In 1997 the House of Lords issued 'the Shimizu judgment' which affected the meanings of demolition and alteration (*Shimizu (UK) Limited v Westminster City Council* ([1997] 1 All ER 481). DETR Circular 14/97 (DCMS Circular 1/97) summarised the implications, the primary effect being to remove the need for conservation area consent for partial demolition of a building. See also DETR's June 2000 consultation paper, *The Impact of the Shimizu Judgement*.

22 In March 2000, Ian Hislop, editor of *Private Eye*, found himself in the news because he had to retrospectively apply for listed building consent after planning officers (visiting in respect of a major proposal to extend the building) noticed that recent repairs had been made to the wooden frame of his

sixteenth-century cottage. Scrase (2000) refers to various cases where the removal of fixtures and fittings has been determined to be alterations to the listed building, including in one case the removal of paintings.

23 Consultees may include the Ancient Monuments Society, the Council for British Archaeology, the Georgian Group, the Society for the Protection of Ancient Buildings, the Victorian Society and the Royal Commission on Historical Monuments. Scottish and Welsh authorities are required to notify their respective Royal Commissions on Ancient and Historical Monuments. Planning authorities must consult the Secretary of State and English Heritage on all applications affecting grades I and II* buildings, and all buildings if demolition is involved. In Greater London the authorisation of English Heritage is required for all listed buildings before consent can be granted.

24 The type of conditions that can be imposed are set out in DoE Circular 8/87, and include the preservation of particular features, the making good of damage caused by works of alteration, and the granting of access (before work commences) to a named body to enable a photographic record or measured drawings to be made.

25 There is only one case of this: the St Ann's Hotel building in Buxton, which is part of a late-eighteenth-century crescent. It had had a long history of neglect which continued through various ownerships. After all alternatives had been exhausted, the Secretary of State served a compulsory purchase order in 1993.

26 There have been marked changes in attitudes to different architectural styles. Perhaps the most famous comes from Paris, where the Eiffel Tower was once described in terms of 'the grotesque mercantile imaginings of a constructor of machines', but is now 'the beloved signature of the Parisian skyline and an officially designated monument to boot' (Costonis 1989: 64). Perhaps the Millennium Dome might follow the same path?

27 Among the modern buildings proposed for listing were the Centre Point office block in central London, Millbank Tower, the John Lewis warehouse at Stevenage, and the signal box at Birmingham New Street station.

28 The 1996 consultation paper discusses a proposal for a new power for the Secretary of State to provisionally list an endangered building to allow consultation to take place. Such a procedure (as with the current listing regime) would not involve any compensation.

29 The Albert Dock, Liverpool comprises five grade I dock buildings now converted into 93,000 square metres of television studios, galleries, offices and shops. The Salt Mills and Saltaire Village projects have created 1,800 jobs from mixed public–private investment of £50 million. Dean Clough Mills, Halifax was a private sector scheme where under the direction of Sir Ernest Hall 3,500 jobs have been housed in redundant carpet mills. The SAVE Britain's Heritage paper on *Catalytic Conversion* (1998) makes a similar argument with a set of examples of good practice.

30 The Act was promoted as a private member's bill by Duncan Sandys, President of the Civic Trust, and passed with government backing.

31 The Shimizu judgment (referred to earlier) has had a particular impact on conservation areas. Previously, the understanding was that the Act provided for all forms of demolition to all buildings in conservation areas to be subject to control, including the demolition of any part of a building (which, for example, could include boundary walls). The Shimizu judgment confirmed that demolition 'refers to pulling down a building so that it is destroyed completely or at least to a very significant extent' (DETR/DCMS *Consultation Paper on the Impact of the Shimuzu Judgement*, 2000). In practice therefore, partial demolition no longer required conservation area consent. Thus substantial control is lost over piecemeal changes to conservation areas, especially to dwelling houses, where householders' permitted development rights allow them to undertake partial demolitions (which would allow, for example, demolition of boundary walls to create a parking space). A number of solutions are being considered in England, including withdrawal of permitted development rights in conservation areas. In Scotland, the matter has been dealt with through Circular 1/200.

32 Larkham (1999a: 113) suggests that perhaps the largest conservation area is the Yorkshire Barns and

Walls area which covers tens of square kilometres, but the most complex must be Bath where the single conservation area now covers 1,914 hectares (66 per cent of the city's area). In another article Larkham (1999b) considers the recent trend for the conservation of residential suburbs.

33 The lower figure for Northern Ireland is probably a reflection that only central government can make designations. Five new designations in the Edwardian suburbs of Belfast were made by the minister in 2000.

34 See DCMS (1999) *World Heritage Sites: The Tentative List of the United Kingdom of Great Britain and Northern Ireland*.

35 We are grateful to the Garden and Landscape team at English Heritage for pointing out the omission of historic parks and gardens in the twelfth edition.

36 The National Trust boast that its 161 gardens comprise the largest and most diverse group of private gardens in the world.

37 Similar lists are being compiled in the other countries of the UK. In Scotland it is known as the Register of Historic Gardens and Designed Landscapes and in Wales the Register of Landscapes, Parks and Gardens of Special Historic Interest in Wales and the Register of Landscapes of Outstanding Historic Interest in Wales. The register can be consulted at www.english-heritage.org.uk/.

38 The figures are taken from English Heritage's evidence to the Select Committee on Environment, Transport and the Regional Affairs Enquiry into *Town and Country Parks* (1999) www.publications. parliament.uk/pa/cm/cmselect.htm. The numerous memoranda of evidence provide interesting reading on the demise of urban parks generally, as well as the ones of historic value. The comments from the Urban Parks Forum are taken from their submission to the inquiry.

39 The Environment Committee found during a 1986–7 inquiry that of the Church of England's 16,700 churches, 8,500 were pre-Reformation and 12,000 were statutorily listed (2,675 in the highest grade).

40 The code includes the requirement that all works to a listed church building which would affect its character are submitted for the approval of an independent body, and that there is consultation with the LPA,

English Heritage, and national amenity societies. The religious bodies to which the exemption now applies because they have an agreed self-regulatory system are the Church of England, the Church in Wales, the Roman Catholic Church, the Methodist Church, the Baptist Union of Great Britain, the Baptist Union of Wales and the United Reformed Church.

41 A National Land Fund had been established with similar intent in 1946 with a fund of £50 million, part of which was used for the purchase by the Secretary of State of buildings of outstanding architectural or historic interest, together with their contents. The fund was raided by the Exchequer in 1957 and became moribund.

42 The largest grant was £1.1 million to the British Library (the second of two instalments) towards the acquisition of the Sherborne Missal (one of the finest illuminated Gothic manuscripts in the world, with 694 pages of information on medieval life). Grants relating to the built environment have previously included £4.9 million for the purchase of Croome Park, Worcester, and support to the Barlands Farm Romano-Celtic Boat (found during development of a supermarket).

43 In total 1,200 grant applications were made requesting £620 million and 450 projects were completed. There are 44 projects worth more than £5 million; about one-quarter of the funding goes to 'conservation and restoration' and one-third on 'historic building repair or refurbishment'. In 1999–2000 in Northern Ireland 87 grants were awarded worth £4.7 million, in Scotland 167 grants worth £19.8 million and in Wales 129 grants worth £20.2 million.

44 £16.3 million is directed to the *Urban Parks Programme* and another £4.9 million to the *Places of Worship Scheme*.

45 This discussion is based on the system in England. The arrangements are similar elsewhere, but there are significant differences for Northern Ireland. Note that the cutting down of a tree is not 'development' and then not subject to the normal controls – thus the need for special provisions. There is an 'informal' tree register (a registered charity) which is compiling a list of notable trees – www.tree-register.org.

46 *Tree Preservation Orders Draft Regulations: A Consultation Paper* (1998).

47 The Hedgerows Regulations 1997 (SI no. 1160).

48 In a 1993 survey of overseas visitors to London, undertaken by the British Tourist Authority and the London Tourist Boards, 69 per cent of holidaymakers gave heritage as the main reason for visiting London (the figure for visitors from North America was 80 per cent). With domestic tourists, the most frequently mentioned main purpose of a holiday (after walking holidays) was to visit heritage attractions and sites.

49 In 1997, Britain was the fifth most visited country in the world, after France, the United States, Spain and Italy.

50 The statistics in Box 8.4 are taken from a House of Commons research paper (Bardgett 2000) (which gives a very useful account of recent trends and policies) and the ONS datasets on tourism.

51 The quote comes from the House of Commons National Heritage Committee (1994: para. 54), which complained that 'there is a serious lack of coherence about policy for the preservation of our heritage and its very important links with the tourist industry', which involved thirty-four quangos.

52 Each RTB takes the lead on a specific issue – sustainability and rural tourism is led by the Cumbria RTB, resorts by the North West RTB, and walking by the South West RTB.

53 See the Scottish Executive Report (2000) *A New Strategy for Scottish Tourism* and National Assembly for Wales (2000) *Achieving Our Potential*. In Northern Ireland an Action for Tourism Task Force has been created.

54 Environmental capacity studies have been undertaken at numerous visitor attractions; the best known is the Chester Study undertaken by the local authority and English Heritage. The studies can go into great detail. At Stonehenge it has involved mathematical model-ling of visitor movements and studies on the best type of grass.

55 The quote is taken from the ETC's website, www.wisegrowth.org.uk, which has been set up to provide a national source of expertise, advice and information on sustainable tourism. The English Historic Towns Forum are investigating best practice advice for historic towns on sustainable tourism.

56 The issue raised most by consultees was value added tax (VAT). Conservation and repair are subject to VAT, but replacement is not. Thus, for example, it may pay to replace windows rather than repair them. The main recommendation in relation to tax is to equalise VAT at 5 per cent for all building work.

57 Perhaps because of the restrictive brief, another very long-standing issue does not appear to be addressed: the separate systems and lists for buildings, monuments, gardens and battlefields, first mentioned in 1950 (Larkham 1999a: 108).

58 Parish or village appraisals are well established (see Moseley 1997). The Heritage Review draws attention to the work of the Urban Design Alliance, which is carrying out ten pilot 'placecheck' projects which have similar objectives.

59 One in ten planning authorities have used Article 4 Directions to withdraw permitted development rights in conservation areas. This recommendation is justified with reference to the many changes that can take place in conservation areas with no control at all. For example, 'the local authority can resurface streets and pavements in new materials and patterns, install traffic signs, and together with over 30 private utility companies install cabinets, kiosks and other street furniture without any control or coordination' (p. 35).

60 In relation to the management of historic public buildings see the English Heritage (2002) report *Better Public Buildings*.

9   Planning and the countryside

> The challenge for rural communities is clear. Basic services in rural areas are overstretched. Farming has been hit hard by change. Development pressures are considerable. The environment has suffered.
>
> White Paper, *Our Countryside: The Future – A Fair Deal for Rural England*, 2000

The changing countryside

Many changes have taken place in the British countryside since the early postwar policies were forged. Suburban commuter residential development, roads and transport, people seeking recreation, the changing economy, forestry, conservation, and a host of other pressures have grown beyond any expectation. The changes show no sign of abating: they never have. The British countryside has been subject to continual change: the 'natural' scenery which is now the concern of conservationists is the human-made result of earlier economic change. The changes continue: the most recent are those which come with the crisis in the agricultural industry, and with the growth of both population and economic activity.

A major plank of postwar policy was that a prosperous agriculture not only would be of strategic economic value but also would provide the best means of preserving the countryside. Aided by European policies, and by technological advances, the promotion of agricultural production has been a huge success. Unfortunately, as so often happens with policy successes, the solution of one problem gives rise to another. In place of the need for increased agricultural production is the problem of dealing with surpluses and finding ways of reducing output. Matters are further complicated (throughout Europe) by more productivity increases resulting from a number of factors

including continuing technological advances (notably biotechnological developments) and agricultural development in eastern Europe (historically a major food producing region). The pressures for change in agricultural policy have been increased by mounting concern over the rural landscapes which have changed in response to newer production methods, and by growing demands for conservation and recreation.

A reversal of long-established, popular policies does not come easily, and the difficulties are increased when so many interests benefit from the subsidised regime. Above all is the overwhelming impact of the EU Common Agricultural Policy. Reform is a long and difficult process, though support for commodity production has at last changed to direct payments to farmers, particularly for environmental benefits. The issues are complex both in economic terms (reform of the CAP may have a major impact on agricultural land values) and politically (the problems and political muscle of farmers vary across the EU). UK policies are severely constrained and the scale and speed of change is highly uncertain.

As a result, change is taking place more slowly than is required. Nevertheless, changes in policy have been made. A milestone was the Agriculture Act 1986 which required agricultural ministers to maintain a balance between the interests of agriculture and wider rural and environmental interests, and particularly to give more attention to 'the conservation and enhancement of the natural beauty and amenity of the countryside', and to 'the promotion of the enjoyment of the countryside by the public'. (More tangibly, the legislation provided for the establishment of *environmentally sensitive areas*, which are discussed later.) A subsequent DoE Circular (16/87) noted that 'the need now was to foster the diversification of the rural economy so as to open up wider and more varied employment opportunities'. The Circular continued:

> The agricultural quality of the land and the need to control the rate at which land is taken for development are among the factors to be considered [in assessing planning applications affecting agricultural land], together with the need to facilitate development and economic activity that provides jobs, and the continuing need to protect the countryside for its own sake rather than primarily for the productive value of the land.

The circular was a mere twelve paragraphs long, but it represented a dramatic change in policy and began a debate which still rages today. Since the late 1940s the object of rural policy has shifted dramatically from the maintenance of the agricultural economy and farming land to managing the 'balance' between the natural qualities of the countryside and quality of life in rural communities.

A significant number of the 20 per cent of people who live in the countryside face problems of a particular nature. Public transport is limited (75 per cent of rural parishes have no daily bus service), there is a shortage of shops (40 per cent have no shop or post office, and 70 per cent have no general store), most (80 per cent) have no general practitioner based in the parish, half have no school, and there is a shortage of affordable housing for local people. On the other hand, unemployment in rural districts is lower than the national average, employment growth is higher (with a fall in land-based industries, and a growth in service sectors) and the rural economy is becoming increasingly diversified, housing conditions in rural areas are generally better than in urban areas and car ownership is higher in rural areas. (Some general trends in rural change are given in Box 9.1.)

Of course, one can play around with such figures to 'prove' different points, but there is clearly a range of conditions in rural areas, and some of the problems require specifically rural policies. Housing and transport in particular require distinctly rural programmes. Examples of these are the rural exceptions policy which permits rural sites to be released for affordable housing as an exception to normal development plan policies, and from 2005 the allocation of small sites for social housing development and rural traffic strategies along the lines suggested by the Countryside Commission's *Rural Traffic: Getting it Right* (1998).[1] So while agriculture has declined there is still a very distinctive stream of rural policy in government.

The shifting rural policy agenda raised questions about the division of central government countryside responsibilities between departments, including those concerned with agriculture, countryside conservation and recreation, the environment and planning. There have been, as a consequence, several organisational changes. The Ministry of Agriculture Fisheries and Food was once a very powerful part of government, quite separate from environmental and planning responsibilities. Its influence slowly dwindled in parallel with the decline in agriculture as an economic sector and in competition with other environmental and rural development concerns. In its 1997 General Election manifesto, the Labour Party announced its commitment to create a new department to 'lead renewal in rural areas': a Department of Rural Affairs. Thus a major shake-up saw the environmental protection function of government move from its place in the Department of Environment, Transport and the Regions (DETR) (which included planning) to join with MAFF in the Department for Environment, Food and Rural Affairs (DEFRA). There was a general support for the idea from rural interest groups, but also anxieties about the huge area which fell within its purview.

> The new Department's remit is extremely broad. It stretches from the administration of subsidy payments for farmers to overall responsibility for the implementation of the Kyoto Protocol in the United Kingdom, taking in rural development, fisheries, waste disposal, water and flooding,

BOX 9.1 CHANGING COUNRYSIDE IN ENGLAND

Rural areas are in constant change. The main features over recent years are as follows:

- *population growth:* net migration of 60,000 people per year into wholly or predominantly rural districts between 1991 and 2002
- *an ageing population:* the number of people aged 65 or over in wholly or predominantly rural districts increased by 161,000 (12 per cent) between 1991 and 2002, while the number aged 16–29 decreased by 237,000 (18 per cent)
- *relative prosperity especially in more accessible areas:* higher income per head than the national average – but with a disadvantaged minority amidst prevailing affluence
- *economic weaknesses, with associated social deprivation, in a minority of 'lagging' rural areas:* characteristically in areas adjusting to a decline in mining, agriculture and fishing, and tending to be in more peripheral areas
- *convergence between the urban and rural economies:* though agriculture is still at the core of the rural economy and society, employment in agriculture has decreased by 30 per cent (151,000) since the early 1980s, employees in rural businesses are now more likely to be in manufacturing (25 per cent), tourism (9 per cent) or retailing (7 per cent) than in agriculture (6 per cent)
- *increased mobility through the car:* bringing benefits for many but reducing the customer base for public transport and thus creating difficulties for those without access to a car; half a million (14 per cent) rural households do not have a car and many people in households which do have a car do not have access to it when they need to travel
- *pressures on the countryside – especially through demand for housing and transport:* rural areas remain a rich resource, valued by both residents and visitors for fine landscapes, biodiversity and open space; these contribute to enjoyment and general well-being as well as enhancement for the benefit of all.

Source: DEFRA (2004) *Rural Strategy*

conservation, animal health, pollution and some elements of food safety. Much of its remit is covered by European legislation, and DEFRA therefore has a significant role to play in negotiations within Europe.

The criticisms did not stop there. Many complaints focused on the implied interpretation of 'sustainable development' which appeared to focus on environmental sustainability, rather than on an overarching approach to integrate social, economic and environmental thinking into all decision-making across

government. The RSPB went further and argued that the creation of DEFRA raised that danger of sidelining it into a policy ghetto for green issues. The CPRE observed that 'the environment overall is being divorced from other Government policy decisions'. More widely, DEFRA was accused of being jack of all trades and master of none. From another perspective, there was concern that DEFRA has neither the name of 'agriculture' nor 'farming' in its title: a matter of deep concern to the National Farmers' Union. DEFRA emphasised that it was not a ministry for farmers:

we are the ministry for rural affairs and the environment; . . . it is important that DEFRA makes clear the central role played by agriculture in delivering a host of its objectives, and in particular those relating to rural communities, the countryside and sustainable development.

This has been insufficient for some interests, who refer to DEFRA as the Department for the Elimination of Farming and Rural Areas.[2] All this goes to show that it is very difficult to generalise about rural or countryside interests; there are many different interests or factions.

The creation of DEFRA and the department's objectives are explained in Chapter 3. Here, mention should be made of the public service agreement target 4 which is 'to reduce the gap in productivity between the least well performing quartile of rural areas and the English median by 2006, and improve the accessibility of services for rural people'. In order to meet this DEFRA must engage in 'rural proofing' to ensure that all government policy is examined for its impacts on the countryside. Though this is not new it has been formalised as 'a process by which the potential impacts of policy and decision-making on rural areas are evaluated, taking the needs of those who live and work in the countryside fully into account'. The purpose is to make sure that the needs of rural areas are not sidelined, and indeed that they are reflected at the heart of all policy-making. An abbreviated 'rural proofing checklist' is set out in Box 9.2. The success of rural proofing is assessed annually in a report by the Countryside Agency.

The Countryside Agency came into being through a merger of the Countryside Commission and most of the functions of the Rural Development Commission (RDC) (other RDC functions went to the regional development agencies). The rationale for this was that it would allow the development of a more strategic and integrated approach to rural policy. It is one of four major non-departmental bodies sponsored by DEFRA. The others are English Nature, British Waterways and the Environment Agency. The latter is discussed more fully in Chapter 7. DEFRA is taking forward proposals from the Haskins review of Rural Delivery Mechanisms to merge the Countryside Agency with English Nature and also the Rural Development Service (part of DEFRA).

Planning policy and the countryside

The new dimensions of countryside policy introduced from the 1980s involve issues of land management for which the planning system is not fully adequate. The planning system, even given the major changes in 2004, is designed to deal with land use, not its management, and its powers are concentrated on urban land uses. It has little control over agriculture and forestry, and until the later part of the twentieth century paid much less attention to the countryside than towns. It was assumed when the system was established in the postwar years that a prosperous agriculture would by itself deal with any problems of the rural economy.

As interest in land use and spatial planning for rural areas has grown, the planning system has become more active in rural policy but has been criticised for being too defensive about what sort of development is suitable in the countryside and for taking a one-dimension environmental view of the countryside. The (former) Rural Development Commission is one example, claiming that the planning system took sustainable development

> to mean environmental protection and reducing travel needs by concentrating development into larger settlements. Those strands of sustainable development relating to economic development and social equity tend to be overlooked . . . Emphasis should be given to the social and economic implications of not providing for development.
>
> (*Rural Development and Land Use Policies* 1998)

Despite its limitations, the planning system is seen as a critical component in the management of the countryside and rural economy. The Countryside Agency set out a positive and challenging statement of its conception of future planning policy in its *Planning for Quality of Life in Rural England* (1999). This stressed

BOX 9.2 RURAL PROOFING CHECKLIST

Rurality

- Few service outlets – access to proposed rural beneficiaries?
- Higher service delivery needs – costs for service providers in reaching rural clients; lost economies of scale.
- Greater travel needs – further distances to travel.
- Few information points – fewer libraries, rural businesses, etc.

Rural economies

- Small (economic) markets – markets small and/or scattered.
- Weak infrastructure – transport and telecommunications less attractive; less competition among providers.
- Small firm economy – more businesses are micro-business.
- Land-based industries – will a policy tackle both rural and urban concerns?

Rural communities

- Needs not concentrated – will policies be targeted at the deprived?
- Different types of need – poor access to services; low wages; limited jobs; lack of affordable housing.
- Low institutional capacity – private, public and voluntary bodies are smaller and struggling.

Rural environment

- Few sites for development – few brownfield sites in acceptable locations.
- Landscape quality and character – highly valued: likely policy impact.
- Countryside amenity and access – impact of people wishing access.

'the essential interdependence between a thriving rural economy, sustainable communities and the proper care and enjoyment of the countryside'. It argued that, in judging development proposals, the new philosophy should be 'is it good enough to approve?' not 'is it bad enough to refuse?' This calls for a proactive approach by local planning authorities: their policies need to be 'criteria based' and clearly state 'the qualities the plan wishes to pursue'. The 2004 Act changes to the system may go some way to help meet these aspirations. The Countryside Agency was quick off the mark in providing guidance to local authorities on how they might use the new system to best effect in rural areas through, for example, the use of *concept statements*. The Agency has experimented with this approach in partnership with South Hams Council and produced simple expressions of 'the kind of place that new development should create' (2003: 4). The same message is at the core of this and other guidance on rural economic development offered by the Agency: 'local authorities should use the planning system in positive ways to encourage the types of sustainable development that bring lasting economic, social and environmental benefits to their particular rural areas' (2003).

Government has for some time been seeking to 'free up' planning restraint in rural areas, especially for economic development (a reminder may be needed here that unemployment is generally lower in rural areas). The Cabinet Office Performance and Innovation Unit's *Rural Economies* (1999) report addressed the role of the planning system. This was offered as a contribution to the debate on 'modernising the policy framework for rural economies'. (The plural is used because the diversity across the country makes it inappropriate to think of a single 'rural economy', separate from and different to economic activity in urban areas.) Among the ideas put forward for discussion were the extension of planning controls to agricultural development, the relaxation of development control in rural areas to allow 'sympathetic and appropriate economic activity, and a change in the presumption against development of the best and most versatile agricultural land' (but with appropriate protection for areas of high environmental value).

The government in England through the then DETR and MAFF issued a White Paper in 2000, *Our Countryside: The Future – A Fair Deal for Rural England*, which expressed the same views about planning, saying that its one priority had been to protect the countryside, important landscapes and the environment. It called for a more flexible and positive attitude to development in the countryside, especially where this supports the provision of services and affordable homes and diversification of the rural economy. These views are based on the (largely erroneous) assumption that planning is the problem. Studies have shown that the planning system has not generally blocked rural diversification (Shorten and Daniels 2001) although there is room for improvement in positive action. The problems of providing for affordable housing have long been recognised in the planning system (as discussed

in Chapter 6). Nevertheless, serious questions are raised about the implications for planning policy for the countryside and how it can contribute to other commitments to retain services in rural areas (which has been pursued with a 50 per cent rate relief for services that offer community benefit) rural transport and accessibility and market town regeneration. National policy is being amended to facilitate more positive action by local planning authorities, for example, to secure more affordable homes through planning obligations; and making access to the countryside easier.

The 2000 Rural White Paper was 'refreshed' (to use DEFRA's term) in 2004 with the publication of the *Rural Strategy*. It sets out three main priorities for rural policy:

- supporting enterprise and targeting resources at areas in greatest need (lagging areas), by fostering business development, raising skill levels and building local institutional capacity
- tackling rural social exclusion and providing fair access to services and opportunities so that no one is disadvantaged by living in a rural area (although expectations about services will obviously be different in rural and urban areas)
- protecting the natural environment through integrated management and ensuring that more people from a wider range of backgrounds are able to visit and enjoy the countryside.

Farming figures in the objectives and there is a separate strategy for *Sustainable Farming and Food* (2002), but the absence of agriculture from the headline objectives demonstrates just how far rural policy has come since the 1940s. The themes of rural policy are similar to those for urban (though government continues to keep them apart) particularly on the common themes for policy and programme integration, more local choice and more flexibility in providing for economic and social development. The proliferation of initiatives in the countryside underlines the need for some integration of policies. An indication of what this might involve was first provided by Scottish Natural Heritage, which combines responsibilities for conservation, amenity, and recreation. In *An Agenda for Investment in Scotland's Natural Heritage* (1992), SNH pointed out that each economic activity related to the countryside is dealt with independently: agriculture, fisheries, forestry, mineral extraction, recreation and tourism, country sports, rural industries. Yet 'all of these activities are based on use, in one way or another, of the natural heritage: *the natural heritage is the common resource*'. It is also a declining resource, since many of these uses contribute to 'a draw-down of Scotland's natural capital'. The deterioration is substantial and 'calls into question the capacity of the natural heritage to sustain the range of uses to which it is subjected'. All this (and more) clearly indicates the need for an integrated approach to the rural environment.

The Scottish Office followed the SNH lead with a *Rural Framework* strategy, which unequivocally states that 'tackling rural issues in a sectoral manner does not work'. The keynote of the preferred approach is partnership. An early example is to be found in the Cairngorms and the Trossachs, where joint machinery is being established to deal in a comprehensive way with the complex problems of these famous areas, now established as national parks. This theme was taken up in White Papers in the mid 1990s,[3] which stressed the need for a new and integrated approach to meet varied objectives for the countryside. Local participation too, has assumed great significance in Scotland following devolution:

> The process of developing a new approach to policy-making will involve engaging with the policy-making community across Scotland and reaching out beyond those already well represented in established channels of power, to include smaller organisation at national, regional and local level.
>
> (Hassan 1999: 14)

This appears to be widely accepted, as the discussion of Scottish local government in Chapter 3 shows. The policy statement *Towards a Development Strategy for Rural Scotland* (1998) reiterated the philosophy: to 'not set rural Scotland apart; reflect the diversity of rural Scotland; work through an integrated approach; and facilitate community involvement'. It was soon

followed by *Rural Scotland: A New Approach* in 2000. The title of one chapter sums up the thinking about policy for rural Scotland: 'Integral dynamic and with a thirst for change'. The 'new approach' follows similar lines to the Rural Strategy in England (and those in Wales) seeking to change the view of the countryside so as to recognise its contribution to the economy. While continuing to value the natural and cultural heritage there is emphasis on promoting a more diverse and dynamic rural economy, to investing in improving skills of the rural labour force and ensuring that young people have opportunities for employment without moving away and addressing social cohesion – access to services and opportunity. A Scottish National Rural Partnership was established to take forward the proposals of Rural Scotland and a *Taking Stock* exercise was conducted in 2003 which documents, or rather promotes, many case studies illustrating successful intervention.

The National Assembly for Wales has followed a similar path with a series of national statements on rural Wales. The main document is the *Rural Development Plan for Wales 2000–06* (2000) which gives a broad lead to the spending of EU Objective 1 funding for West Wales (see Chapter 4) as well as a general framework for Assembly rural policy. Inevitably other national policy statements take up the rural theme, including the *Plan for Wales* (2001), *The National Economic Development Strategy: A Winning Wales* (2001), *The Strategic Agenda for the Welsh Assembly Government: A Better Country* (2003) and the *National Spatial Plan: People, Places and Futures* (2003). This list could be repeated for the other nations of the UK. It is used here to illustrate that rural policy is not confined to one department or strategy. Integration needs to start at the national level. In Wales the underlying theme of all rural policy is the relatively poor economic performance of the rural region (thus its designation as an Objective 1 area) and the desire to bring it up to levels of similar regions in the UK. There are four general themes for rural policy that are common across the strategies:

- the need to address disparities in economic performance and social cohesion . . . and to create a better balanced distribution of economic growth;

- a concern to diversify the economy and employment which is linked to assisting agriculture with structural change but the decline of particular industrial sectors;
- capacity building within local communities, the development of partnership approaches, and to improve economic performance while cultivating Welsh culture; and
- improving the quality of labour supply through developing skills, closely linked to the desire to improve the quality of job opportunities and the need to retain and regain skilled workers and young people.

(Brown, *et al*. 2004: 3)

The national policy documents across the UK set out a vision. The implementation is left to the local level, and though the planning systems do not figure prominently in these statements, they are centre stage when it comes to resolving the 'tensions' or contradictions in these policy statements. New housing, economic activity and in-migrants are not always welcomed, even where there are problems. That is partly because they have not in the past solved the problems. New jobs in the countryside are not always taken by local people, especially if they are better paid, although they might indirectly create more jobs in basic services. While unemployment in rural areas is relatively low, jobs may be located many miles from home. New housing may be taken by in-migrants, some of whom will be long distance commuters. Local people on low incomes tend to be channelled into social housing, which is concentrated in particular locations. These are generalisations of course, but serve here to illustrate the complexity of seeking to intervene in the rural economy.

The principal national planning policy for the countryside in England is given in PPS 7 *Sustainable Development in Rural Areas*; there are similar statements for the devolved administrations (listed at the end of the book). This is one of the commendably shorter policy statements. It is unequivocal that new development in the countryside should be strictly controlled and that it should be 'located in or next to towns or other service centres that are accessible by public transport, walking and cycling' (para. 1). This should not prevent the authority from supporting 'a wide range of economic activity in rural areas', the identification of new economic locations and the expansion of existing business. This approach is already practised in much of rural Britain, but not without struggles. Over much of the past half-century, planning policies for the countryside have been protectionist (especially in the south) and part of this has been to see town and country as separate entities engaged in a perennial series of battles, both large and small. Those who have moved from the town to the country quickly join the defenders of the countryside against further 'urban invasion'. Successive governments have been very one-sided in this perpetual struggle, backed by a highly popular green belt policy, and coalitions of interests committed to saving the countryside. In the mean time, the roles of town and country have changed dramatically; economic development in the countryside is now little different from that of towns, and people's lifestyles are more or less the same whether living in the suburbs or much of the countryside. A 'positive' planning policy for the countryside will have to address entrenched attitudes, which will be much more resistant to change. A research report for the National Assembly for Wales found that employment growth in rural areas is growing rapidly while housing growth continues to be concentrated in small towns, illustrating the relative influence that the planning system has on the location of employment compared with housing. It also points to the interdependency of town and country especially in determining their economic performance. Towns in prosperous rural areas tend to do well, and once they are doing well they tend to continue to do so. The reverse is also true. It will be very difficult to tackle existing disparities because of the 'path dependency' of economic success (Brown *et al*. 2004).

As is often the case, the substantive policy content of rural policy is accompanied by reform of the administration, devolution to the regions, merger of agencies, integration of functions and a tidying up of disparate funding streams (see Box 9.3). The Strategy devolves responsibility for DEFRA's economic and social regeneration activities (and the requirement to deliver on ambitious objectives) to the regional development agencies. RDAs already have the main responsibility for economic development in both urban and rural areas, but this will mean a transfer and thus 'slimming down' of the economic regeneration activities in DEFRA and in the Countryside Agency. To assist in the process DEFRA has brought its disparate funding streams for regeneration into a single *Rural Regeneration Funding Programme*, which will be devolved to the RDAs. This will not make a great difference in spending since it is worth £21 million but it will clear up overlapping functions. The Strategy also introduces yet another body at the regional level – the *rural priority board*. This, like the greater use of RDAs, was a recommendation of the Haskins Review, but DEFRA have given the regions the choice about if and how these boards will be established to the regions. At the top level the integration of the landscape, access and recreation parts of the Countryside Agency with English Nature and the

environmental functions of the Rural Development Service will be implemented with new primary legislation, probably for establishment at the beginning of 2007. In the mean time these organisations must form a strong partnership in a 'confederated integrated agency' and a common vision. The purpose of integration is to bring the objectives 'of conserving and enhancing the resource of nature together with realising the social and economic benefits for people of so doing' (p. 35). The agency will be the champion for integrated resource management, nature conservation, biodiversity, landscape and access and recreation.

The national parks

Perhaps the most notable long-term policy relating to the countryside has been the establishment of the national parks, areas of outstanding natural beauty, and

---

BOX 9.3 COUNTRYSIDE POLICY FRAMEWORK

Our Vision is of:

- a living countryside, with thriving rural communities and access to high quality public services
- a working countryside, with a diverse economy giving high and stable levels of employment
- a protected countryside in which the environment is sustained and enhanced, and which all can enjoy
- a vibrant countryside which can shape its own future and with its voice heard by Government at all levels.

Our aim is to sustain and enhance the distinctive environment, economy and social fabric of the English countryside for the benefit of all.

Source: DETR and MAFF (2000) *Our Countryside: The Future – A Fair Deal for Rural England*

The government's three priorities for rural policy are:

1 Economic and social regeneration – supporting enterprise across rural England, but targeting greater resources at areas of greatest need.
2 Social justice for all – tackling rural social exclusion wherever it occurs and providing fair access to services and opportunities for all rural people.
3 Enhancing the value of our countryside – protecting the natural environment for this and future generations.

Source: DEFRA (2004) *Rural Strategy*

other areas that were designated for protection. The national parks were a response to a very long-term public demand. This stretched from the early-nineteenth-century fight against enclosures, James Bryce's abortive 1884 Access to Mountains Bill and the attenuated Access to Mountains Act 1939, to the promise of the National Parks and Access to the Countryside Act 1949, an Act which, among other things, poetically provided powers for 'preserving and enhancing natural beauty'. Many battles have been fought by voluntary bodies such as the Commons, Open Spaces and Footpaths Preservation Society and the Campaign to Protect Rural England, but they worked in a largely legislative vacuum until the Second World War. The mood engendered by the war augured a better reception for the Scott Committee's emphatic statement that 'the establishment of national parks is long overdue' (1942: para. 178). The Committee had very wide terms of reference, and for the first time an overall view was taken of questions of public rights of access to the open country and the establishment of national parks and nature reserves within the context of a national policy for the preservation and planning of the countryside.

Government acceptance of the necessity for establishing national parks was announced in the series of debates on postwar reconstruction which took place during 1941 and 1943, and the White Paper on *The Control of Land Use* referred to the establishment of national parks as part of a comprehensive programme of postwar reconstruction and land use planning. Not only was the principle accepted but, probably of equal importance, there was now a central government department with clear responsibility for such matters as national parks. There followed the Dower (1945) and Hobhouse (1947) reports on national parks, nature conservation, footpaths and access to the countryside, and in 1949, the National Parks and Access to the Countryside Act which established the National Parks Commission and gave the main responsibility for the parks to local planning authorities.

The administration of the national parks has been a matter of controversy throughout their history. Dower (1945) had envisaged that there would be ad-hoc committees with members appointed in equal numbers by the Commission and the relevant local authorities. Local representation was necessary since the well-being of the local people was to be the first consideration, but the parks were also to be *national*, and thus wider representation was essential. The lengthy arguments on this issue were eventually resolved by the 1949 Act in favour of a local authority majority, with only one-third of the members being appointed by the Secretary of State. (In line with his conception of truly national parks, Dower had proposed that the whole cost of administering them should be met by the Exchequer – an idea which was never accepted.) Increasing pressures on the countryside have led to a succession of policy and legislative changes. In 1968, the Countryside Act replaced the National Parks Commission with a more powerful Countryside Commission. A Countryside Commission for Scotland was established under the Countryside (Scotland) Act 1967. The Wildlife and Countryside Act 1981 strengthened the provisions for management agreements and introduced compensation for farmers whose rights were restricted (a major change in principle). Later there were major structural changes in the organisation of agencies responsible for countryside matters, including the establishment of a separate Countryside Council for Wales and the merging of the countryside Commission for Scotland with the Nature Conservancy for Scotland as the Scottish Natural Heritage. The Environment Act 1995 established independent national park authorities, which took over the responsibilities previously exercised by local government. These are now the sole local planning authority for a national park area. In addition to the normal plans, a national park authority is required to prepare a *national park management plan* (Countryside Commission 1997). This goes further than the scope of development plans: in addition to establishing policies, it is intended to spell out how the park is to be managed. In 2004 DEFRA conducted *An Evaluation of Planning Policies* in national parks in England which noted, among many other things, that 90 per cent of planning applications are approved in national parks, which may be surprising, but is explained with reference to careful pre-application discussions and applicants' prior knowledge of the designation. The main purpose of the review was to consider the implications of the 2004 planning reforms, which have a special impact on national parks since, for example, they may be covered by more than one regional strategy and a number of community strategies.

From their inception, the national parks have had two purposes: 'the preservation and enhancement of natural beauty' and 'encouraging the provision or improvement, for persons resorting to national parks, of facilities for the enjoyment thereof and for the enjoyment of the opportunities for open air recreation and the study of nature afforded thereby'. There is inevitably some conflict between these twin purposes, and the National Parks Review Panel (Edwards 1991), set up by the Countryside Commission, recommended that they be reformulated to give added weight to conservation – as did the earlier Sandford Report (1974). This argument, which continues, was a major issue in the debates on the sections of the Environment Act 1995, which established independent national park authorities. Controversy centred on the need 'to promote the *quiet* enjoyment and understanding in national parks'. The final outcome is set out in Box 9.4. Most of the land in the national parks is in private ownership (74 per cent), and only 2 per cent is owned

by the park authorities. The National Trust and the Forestry Commission each own 7 per cent. Three-quarters of the funding for the parks comes from the Exchequer.

There are now eleven national parks in England and Wales with the designation of the New Forest National Park in March 2005 and a twelfth for the South Downs proposed (discussed on p. 326). The other national parks (shown in Figure 9.1) are the Brecon Beacons, Dartmoor, Exmoor, Lake District, Northumberland, North York Moors, Peak District, Pembrokeshire Coast, Snowdonia and the Yorkshire Dales. There are also two national parks in Scotland, which are discussed separately later.

The Broads were also proposed as a national park by the Dower Report (1945), but at that time were rejected because of their deteriorated state and the anticipated cost of management (Cherry 1975: 54). Since 1968 they have been managed by the Norfolk and Suffolk Broads Authority, which was initially a voluntary consortium formed by the relevant public authorities (with powers and financial resources under the provisions of the Local Government Act 1972, and with 75 per cent Exchequer funding). Discussions continued over several years among the large number

---

BOX 9.4 NATIONAL PARKS: PURPOSES

The Environment Act 1995 provides that the purposes of national parks shall be 'conserving and enhancing the natural beauty, wildlife and cultural heritage of the areas', and 'promoting opportunities for the understanding and enjoyment of the special qualities of those areas by the public'. If there is a conflict between these purposes, any relevant authority 'shall attach greater weight to the purpose of conserving and enhancing the natural beauty, wildlife and cultural heritage of the area'.

A national park authority, in pursuing these purposes

shall seek to foster the economic and social well-being of local communities within the national park, but without incurring significant expenditure in doing so, and shall for that purpose cooperate with local authorities and public bodies whose functions include the promotion of economic and social development within the area of the national park.

Source: Environment Act 1995

of interested bodies and, in 1984, the Countryside Commission reviewed the problems of the area and the progress that had been made by the Broads Authority. Its conclusion was that, despite some achievements, the authority had not made significant improvements in water quality. Moreover, an effective framework for the integrated management of water-based and land-based recreation had not been established and the loss of traditional grazing marsh was continuing. The outcome was the designation of the area as a body of equivalent status to a national park, but with a constitution, powers and funding designed to be appropriate to the local circumstances. A new Broads Authority (with the same name as its predecessor) was established by the Norfolk and Suffolk Broads Act 1988. The duties of the authority are extensive with very similar duties of the national parks and the same level of protection. It is the local planning authority and the principal unit of local government for the area. It has strong environmental responsibilities, and is required by the Act to produce a plan which has a wider remit than those required under the planning Acts: it is more akin to a national park management plan. The separate legislation, it is argued, has advantages in dealing with the special circumstances of the Broads – the very extensive navigation of the waterways.

The New Forest has, in addition to the public authorities in the area, a corporate body of 'Verderers' which is responsible for managing the grazing and commoning within the forest. As well as the protection provided in these ways, many parts of the Crown land in the Forest are designated as *sites of special scientific interest*, and its southern fringe is within the South Hampshire Green Belt. It might thus appear that the New Forest was adequately (even if somewhat confusingly) protected prior to designation as a national park. However, quite apart from questions of coordinating all these protectors of the forest, there was a further need to safeguard the surrounding grazing lands, which are under pressure for development, and also to ensure that adequate provision is made for recreation in a manner which is in harmony with conservation considerations. It is presumably because of the conflicting interests in the area that its

designation has proved to be long drawn out. The New Forest has for long been a candidate for designation as a national park, and in 1998 the Countryside Commission (as it then was) urged that special legislation be passed to deal with the particular needs of the area.

Even more lengthy has been the controversy over the South Downs, which though proposed for designation by the Hobhouse Committee was rejected as a national park in 1950 because of 'the perceived lack of recreational potential'. Pressure continued and in the light of the stronger emphasis on the recreational role of parks and the need for special measures to manage the area, a designation order was made in 2003 and an inquiry held into the 5,000 representations in 2005. The final decision rests with the Secretary of State for Environment, Food and Rural Affairs.

Landscape designations

Both the Dower and Hobhouse Reports proposed that, in addition to national parks, certain areas of high landscape quality, scientific interest and recreational value should be subject to special protection (see Table 9.1). These areas were not considered, at that time, to require the positive management which it was assumed would characterise national parks, but 'their contribution to the wider enjoyment of the countryside is so important that special measures should be taken to preserve their natural beauty and interest'. The Hobhouse Committee proposed that such areas should be the responsibility of local planning authorities, but would receive expert assistance and financial aid from the National Parks Commission. A total of fifty-two areas, covering some 26,000 sq. km, was recommended including, for example, the Breckland and much of central Wales, long stretches of the coast, the Cotswolds, most of the Downland, the Chilterns and Bodmin Moor (Cherry 1975: 55).

The 1949 Act did not contain any special provisions for the care of such areas, the powers under the planning Acts being considered adequate for the purpose. It did, however, give the Commission power to designate *areas of outstanding natural beauty* and provided for Exchequer

Key

▨	National parks
▨	Areas of outstanding natural beauty
—	Heritage coasts
☐	Other special protected areas
▨	National scenic area (Scotland)
----	National trails

Figure 9.1 *Selected protected areas in the UK*

Table 9.1 *National parks, areas of outstanding natural beauty and national scenic areas*

	Number	Area (sq. km)	Percentage of total land area
England			
National parks*	8	10,507	8
Areas of outstanding natural beauty	37	20,510	16
Northern Ireland			
Areas of outstanding natural beauty	9	2,849	20
Scotland			
National parks	2	5,680	7
National scenic areas	40	10,018	13
Wales			
National parks	3	4,129	20
Areas of outstanding natural beauty	5	727	4

Note: * Includes New Forest National Park but not South Downs designated area.

grants on the same basis as for national parks. Forty-one areas have been designated in England and Wales, covering some 21,000 sq. km (over 15 per cent of the area of England, and 4 per cent of Wales). In Northern Ireland, there are nine AONBs, covering 2,800 sq. km. The Scottish equivalent of AONBs number forty and cover 10,000 sq. km.

Areas of outstanding natural beauty are, with some notable exceptions, generally smaller than national parks. They are the responsibility of local planning authorities, which have powers for the 'preservation and enhancement of natural beauty' similar to those of park planning authorities. There has been continuing debate on the question as to whether the designation of areas of outstanding natural beauty serves any useful function. Presumably, attitudes have changed, as indicated by the new legislative provisions of the Countryside and Rights of Way Act 2000. These require management plans for all AONBs and also the creation of conservation boards that would take over the management function from the local authorities, particularly for some of the larger AONBs which cross a number of local authority boundaries 'where unified management of the AONB would bring benefits' (the quotations are from DETR Circular 04/2001).

In addition to AONBs, there are many local authority designations designed to assist in safeguarding areas of the countryside from inappropriate development; some of these have been given additional status through inclusion in structure and local plans. Although these are like AONBs in that they involve the application of special criteria for control in sensitive areas, they do not imply any special procedures for development control. The DoE Consultation Paper on *The Future of Development Plans* (1986) made reference to *areas of landscape quality, areas of great landscape value, landscape conservation areas, coastal preservation areas* and *areas of semi-national importance*. The same paper proposed a new statutory designation, the *rural conservation area* which, it was suggested, would provide a more coherent framework. This idea, however, found little favour during the consultation process, and it was therefore abandoned. It was thought that the desired objectives could be achieved through statements of policy in development plans. The advice in PPS 7 takes a strong line against local designation which may unduly restrict development and economic activity without identifying the particular features of the local countryside which need to be respected or enhanced. It says that local designations (which are now widespread) need careful justification with reference to special circumstances. This is more than a hint that some of these designations should disappear.

Hedgerows

A significant feature of the countryside landscape is the hedgerows. Mainly because of agricultural economics, there has been a dramatic loss in hedgerows: since 1947 over half of these have disappeared. Although hedges are protected (under section 97 of the Environment Act 1995 and the Hedgerow Regulations 1997), this protection is limited. A Hedgerow Review Group recommended in 1998 that, in the long term, the statutory provisions should be amended. The practical effect of the current legislation is to centralise the decision about what are important hedgerows. Moreover, the criteria are complex and difficult to operate. As a result, local authorities are severely constrained. Unfortunately, to deal with this adequately would require primary legislation and, given the current legislative programme, this is unlikely in the short term. In the mean time, however, revisions could be made to the regulations. (This requires less parliamentary time.) The review therefore concentrated on how the regulations and the associated guidance might be changed to deliver stronger hedgerow protection. The detailed recommendations provide an alternative set of criteria. These are currently under consideration.

Scottish designations

Scotland contains large areas of beautiful unspoiled countryside and wild landscape. It has the majority of Britain's highest mountains, with nearly 300 peaks of over 900 metres; it has the great majority of the UK islands and its coast is over 10,000 km in length. Despite expectations to the contrary, there were, until recently, no national parks in Scotland. Though a Scottish committee (the Ramsay Committee)

recommended, in 1945, the establishment of five Scottish national parks, no action followed. The reasons for this inaction were partly political and partly pragmatic (Cherry 1975: Chapter 8). A major factor was that (with the exception of the area around Clydeside and, in particular, Loch Lomond) the pressures which were so apparent south of the border were absent.

Nevertheless, the Secretary of State used the powers of the Planning Act 1947 to issue *National Parks Direction Orders*. These required the relevant local planning authorities to submit to the Secretary of State all planning applications in the designated areas (which included Loch Lomond/Trossachs, the Cairngorms and Ben Nevis/Glen Coe). In effect therefore, in an almost Gilbertian manner, while Scotland at this time did not have any national parks, it had an administrative system which enabled controls to be operated as if it did! But, of course, this approach was inherently negative, and it was not until the Countryside (Scotland) Act 1967 that positive measures could be taken on a significant scale. This Act provided for the establishment of the Countryside Commission for Scotland – later joined with the Nature Conservancy Council for Scotland to form Scottish Natural Heritage. It also enabled the establishment of regional parks and country parks. A policy framework for these was set out in the Commission's 1974 report *A Park System for Scotland*. The report also recommended the designation of national parks in Scotland, though the term *special park* was used. Until recently, this has not been accepted, though objectives similar to those of national parks have been achieved under other designations. Despite the reluctance to establish national parks in Scotland, increased pressure for them has mounted (Rice 1998) and in 1997 the Secretary of State for Scotland announced that

> National parks would be the correct way forward for Loch Lomond and the Trossachs, quite probably in the Cairngorms, and possibly in a few other areas as well. I see national parks in Scotland as integrating economic development with proper protection of the natural heritage. Scottish Natural Heritage was asked to provide advice for action by the Scottish Parliament.

This it did by undertaking wide consultations, inviting views, and commissioning. In 1998, Scottish Natural Heritage produced a consultation paper *National Parks for Scotland* outlining its initial ideas. These emphasised the importance of an integration of social and economic purposes along with the protection and enhancement of the natural and cultural heritage, and the enjoyment, understanding and sustainable use of natural resources. Also stressed were local community involvement in national parks, and a strong park plan 'prepared through consensus with a zoning system to help reconcile differing needs'. These and other proposals make this document an outstanding statement on participatory planning.[4]

In 2000, the National Parks (Scotland) Act was passed, and two national parks were designated: Loch Lomond and the Trossachs (2002) and the Cairngorms (2003). Areas proposed for national parks are large in area and small in number; the Cairngorms Park is 1,467 square miles and the largest national park in the UK. Scottish national parks have wider powers than those south of the Border, including statutory responsibilities for the economy and rural communities. They are central government bodies and wholly funded by the Scottish Executive: 20 per cent of the membership of the two parks is directly elected and the other 80 per cent are chosen by the Secretary of State, half of whom are nominated by the constituent local authorities.

There are numerous other designations in Scotland, including forty national scenic areas, four regional parks and thirty-six country parks (see Table 9.2). The *national scenic areas* are of similar status to areas of outstanding natural beauty in England. They extend over an area of more than 1 million hectares, and include such marvellous sites as Ben Nevis and Glen Coe, Loch Lomond and the World Heritage Site of the islands of St Kilda. Development control in these areas is the responsibility of the local planning authorities, which are required to consult with Scottish Natural Heritage for certain categories of development. As in England and Wales, there is an increasing concern for 'positive action to improve planning and land use management' in these areas, and for dealing with the erosion of footpaths. There is also a similar complaint about the lack of resources. A *regional park* is statutorily

defined simply as 'an extensive area of land, part of which is devoted to the recreational needs of the public'. The four parks are Clyde-Muirshiel, Loch Lomond, the Pentland Hills and Fife. The regional parks, which cover 86,000 hectares, are primarily recreational areas, and each has a local plan which sets out management policies. Emphasis is laid on *integrated land management* schemes to ensure that public access is in harmony with other land uses. In this, they give effect to Abercrombie's green belt philosophy, articulated in the Clyde Valley Regional Plan. He conceived these *outer scenic areas* not only as recreational areas but also as a means of protecting the rural setting of the conurbations (Smith and Wannop 1985). Since the passing of the 1967 Act, Scottish local authorities have provided thirty-six *country parks* spread across the central belt and north east. The parks are 'registered' with SNH, which makes grants for capital development expenditure and also towards the cost of a ranger service. Country parks not only are of direct benefit to their 11 million annual visitors, but also have a conservation objective of 'drawing off areas that are sensitive due to productive land uses and fragile wildlife habitats.'

Northern Ireland designations

Northern Ireland boasts some of the finest countryside in the UK, often with special value as wildlife habitat. One reason for this is that farm and field sizes are smaller than on the mainland, and almost all the farms are owner occupied (Glass 1994). Much of the countryside remains unspoiled, but development pressures are increasing, and there has been extensive building of isolated houses in the countryside.[5] Progress with planning for landscape and nature conservation has been slower than in the rest of the UK. Legislation is far less developed, and there has been much criticism about the delays in designating areas needing protection and management (Dodd and Pritchard 1993). Criticisms of the backwardness of countryside and nature conservation led to a review on behalf of the Secretary of State by Dr Jean Balfour, whose 1983 report, *A New Look at the Northern Ireland*

Table 9.2 *Scottish designations*

	Number	Area (ha)	% land area of Scotland
Statutory sites designated under international conventions and directives			
Candidate special areas of conservation	238	962,667	9.6
Special protection areas (Birds Directive)	137	634,780	7.8
Ramsar wetlands sites	51	313,208	3.9
Natural World Heritage sites	1	853	0.01
Biogenic reserves	2	2,388	0.03
Non-statutory site designations of international importance			
Biosphere reserves	5	11,199	12.9
European diploma areas	2	5,848	0.07
Statutory sites designated under national statute			
Sites of special scientific interest	1,451	1,005,152	12.5
Areas of special protection	8	1,518	0.02
National scenic areas	40	1,001,800	12.5
National parks	2	567,994	7.1
Regional parks	4	86,160	1.1
Country parks	36	6,481	0.08
Long distance routes	5	(731 km)	n/a
Local nature reserves	36	9,410	0.12
National nature reserves	66	117,228	1.5
Other non-statutory site designations			
Historic gardens and designated landscapes	328	66,765	0.83
Marine conservation areas	29	111,895	n/a

Source: Scottish Natural Heritage (2005) *Facts and Figures 2003–04*

Countryside, confirmed the low priority given to conservation and its lack of status in the work of rebuilding as part of the 'Rebuilding of Northern Ireland' (Northern Ireland Department of the Environment). The result was the setting up of a new unit within the department and an advisory Council for Nature Conservation and the Countryside. Legislation extending nature conservation powers soon followed. Nevertheless, designations remained 'pitifully slow'. A new initiative was taken in the early 1990s, culminating in the publication in 1993 of a comprehensive *Planning Strategy for Northern Ireland*.

Statutory designations are much the same as in England and Wales, and the DoENI has all powers in respect of designation and management of special areas (including those that in England are exercised by English Nature and the Countryside Agency). It can designate national parks (though it has not done so), areas of outstanding natural beauty (of which there are nine, covering 285,000 hectares), nature reserves (of which there are forty-five, covering 4,300 hectares), and *areas of special scientific interest* (the equivalent of the SSSI), of which there are 196, covering 92,000 hectares. An additional non-statutory designation is the *countryside policy area*, which is employed to restrict building in the countryside.

The coast

A few figures underline the particular significance of the coast, and therefore of coastal planning: nowhere in the UK is more than 135 km from the sea; the coastline is 18,600 km in length and the territorial waters extend over about a third of a million sq. km. About a third of the coast of England and Wales is included in national parks and AONBs and large areas of the coast are owned or protected by the National Trust. Following the *Enterprise Neptune* fund-raising appeal, the Trust protects 920 km of the coastline in England, Wales and Northern Ireland (mainly by ownership, the remainder by covenant). The National Trust for Northern Ireland and the National Trust for Scotland also own large stretches of the coastline and protect further parts by way of conservation agreements. In addition, there are *marine nature reserves* (discussed later).

In spite of all this protection, the pressures on the coastline are proving increasingly difficult to cope with. Indeed the complexity of responsibilities, statute and policy covering the coast is part of the problem. Between a quarter and a third of the coastline of England and Wales is developed and this has been increasing.[6] Growing numbers of people are attracted to the coast for holidays, for recreation and for retirement. There are also economic pressures for major industrial development in certain parts, particularly on some estuaries (which have international importance for nature conservation). The problem is a difficult one which cannot be satisfactorily met simply by restrictive measures: it requires a positive policy of planning for leisure provision. This has long been accepted, and the *heritage coast* designation, introduced in 1972, implies recreational provision as well as conservation. The Countryside Commission (now Agency) has urged that every heritage coast should have a management plan. It has also established the *Heritage Coast Forum* as 'a national body to promote the heritage coast concept and to act as a focus and liaison point for all heritage

coast organisations'. This is seen as a needed addition to the activities of the Agency whose capacity to promote all the initiatives that are necessary is limited.

There are forty-five heritage coasts in England and Wales, protecting some 1,500 km: about a third of the total length of the coastline (*Britain 2000*: 323). In Scotland, twenty-six *preferred coastal conservation zones* have been defined with a total length of 7,546 km, covering three-quarters of Scotland's mainland and islands coastlines. The Environment Committee, in its 1992 report, *Coastal Zone Protection and Planning*, complained of the lack of coordination among the host of bodies concerned with coastal protection, planning and management. In England, there are over eighty Acts which deal with the regulations of activities in the coastal zone, and as many as 240 government departments and public agencies involved in some way. In Scotland, 'in 1995 there were 79 Acts of Parliament relating to the Scottish coastal zone and marine environment (Cleator & Irvine, 1995); and this has increased in the intervening years especially following devolution' (Firn and McGlashan 2001). Not surprisingly, there have been suggestions that action is required to simplify, rationalise, coordinate or consolidate matters. Although an apparently obvious and sensible idea, it is remarkably difficult to see how the situation can be significantly changed, and the Environment Committee contented itself by asking for a review of legislation and responsibilities. The government response was negative. It was pointed out that, though there were many Acts relating to the coast, the same could be said about the land! Indeed, it was neither possible nor desirable to treat the coast separately from the adjoining land or from the territorial and international waters. Moreover, the suggestion that the town and country planning system might be extended seaward was not persuasive, though it was agreed that 'it is now time to take this debate further'.

The EU has also been active in funding coastal management initiatives through the *Demonstration Programme on Integrated Coastal Zone Management* (ICZM). The programme involved thirty-five projects throughout the EU, including several in the United Kingdom. It is questionable if the Community has competence in this area, but there is no doubt that this is an issue that requires coordination at the transnational as well as national and regional levels. The Demonstration Programme reported in 1999 with proposals for a European-wide strategy, and support for rethinking the future arrangements for coastal zone management in the member states.[7]

Two discussion papers were issued by the DoE in 1993 (*Managing the Coast* and *Development below Low Water Mark*). These drew upon the Countryside Agency's experience with Heritage Coasts, and underlined the usefulness of management plans drawn up by local authorities in liaison with the relevant bodies concerned. The Commission responded that there was a need for guidance on the form and content of such plans, and for integration with the shoreline management plans for flood and coastal defence which are being promoted by DEFRA. In 1994, an organisation of maritime local authorities, the National Coasts and Estuaries Advisory Group, published a *Directory of Coastal Planning and Management Initiatives in England* (1994). This was followed, in 1995, by a DoE guide to government policies affecting the coastal zones (*Policy Guidelines for the Coast*). The latest proposal is for the designation of *Marine Environmental High Risk Areas* covering no more than 10 per cent of the coastline. They are intended to address the potential pollution from marine shipping. In 2000 government commissioned *Research into Integrated Coastal Planning in the North West Region*. Elsewhere, many local authorities have built considerable experience in coastal zone management, not least Dorset, where a World Heritage Site has been designated.[8] Planning policy for the coast was last revised in 1992.

Waterways

The advent of the Labour government proved to be a turning point for the waterways. After years of inadequate funding, a 1999 DETR paper *Unlocking the Potential: A New Future for British Waterways* announced precisely what its title promised. This was one sector of public policy which was in dire need of funding, and which benefited from the government's Comprehensive Spending Review. Additional funding was provided both for current expenditure and to deal with outstanding debt in relation to uneconomic expenditure on uneconomic activities which British Waterways is no longer carrying out (such as freight carrying). Its status as a nationalised industry was considered to be no longer appropriate since its trading activities had become a small and declining part of its operations. But it was still responsible for 2,000 miles of the 3,700 miles of waterways in Britain. This involves more than one might expect. In the words of a British Waterways annual report,

> the waterways offer an outstanding historical, environmental and ecological resource; a rare example of eighteenth century technology still working today to perform its original purpose, and a focus for communities to build their shared understanding of the past, whilst at the same time working to secure a future.

An important source of funding for improvements to the waterways is the Board's property portfolio. In conjunction with other organisations, and with extensive use of partnerships, the Board has undertaken several major urban renewal developments, such as in the development of Sheffield Basin, Paddington Basin and the comprehensive development of Gas Street Basin in Birmingham (which gained the city the Excellence on the Waterfront award – jointly with Boston and New York!). Since about a quarter of the total length of waterways falls within the boundaries of the (former) metropolitan counties, there is considerable potential for waterside development (though it should be noted that most waterside property is not owned by British Waterways). Some development has attracted criticism from the pleasure craft operators because it has been seen as destroying the ambience of the canals and thus making them less attractive for cruising. As with many leisure pursuits, there is a problem of satisfying conflicting interests. Those who love the often closed and secretive world of much of the urban canal system may not welcome the new focal point developments, but these have proved to be extremely popular with others. Another conflict arises between the use of the waterways for leisure and their function as an aquatic habitat. Such conflicts cannot be prevented, but it is an explicit policy of British Waterways to achieve an appropriate balance among differing interests and uses.

The waterway network has been described as a county park which is two thousand miles long by ten yards wide, but it is much more than that: British Waterways owns over two thousand listed structures and ancient monuments and sixty-four sites of scientific interest. It has a major programme of management and conservation in relation to these and other heritage features. Over 8 million people use the waterways in the course of a year, mainly for informal recreation rather than boating. In its 1993 report on *The Waterway Environment and Development Plans*, British Waterways comments that this informal recreation 'offers the greatest prospect for increased use of the waterways', and this is recognised in its initiative in seeking joint study and action with local authorities and a host of relevant agencies. Although British Waterways cannot impose direct charges, related leisure facilities are income-generating, for example shops, public houses, hotels, restaurants and museums. A good example is the National Waterways Museum at Gloucester, which has over 100,000 visitors a year. In 1999, an initiative was taken to affiliate three waterway museums (Ellesmere Port, Stoke Bruerne and Gloucester) to a new charitable Waterways Trust. This has gained national designation from the Museums and Galleries Commission, and also the Heritage Lottery Fund.

The range of activities of British Waterways is interesting. Historically, of course, the canals were essentially an effective means of freight transport, until the advent of the competitive railways after which the canals seemed likely to expire. Certainly, as a means of moving freight, the canals had little prospect against rail and more so road traffic, particularly since, given a political resistance to charging for road use (then as now!) there was an economic incentive to choose a cheaper form of transport. The waterways, however, assumed a new lease of life as their use for freight largely gave way to recreation and leisure. Indeed, this is a now a significant and growing role. The waterways also provide a focus for much urban regeneration and

brownfield housing development where a 'canal view' demands a premium.

Public rights of way

The origin of a large number of public rights of way is obscure. As a result, innumerable disputes have arisen over them. Before the 1949 Act, these could be settled only on a case-by-case basis, often with the evidence of 'eldest inhabitants' playing a leading role. The unsatisfactory nature of the situation was underlined by the Scott (1942), Dower (1945) and Hobhouse (1947) reports, as well as by the 1948 report of the Special Committee on Footpaths and Access to the Countryside. All were agreed that a complete survey of rights of way was essential, together with the introduction of a simple procedure for resolving the legal status of rights of way which were in dispute. The 1949 Act attempted to provide for both. This Act has been amended several times. Under the current provisions, county councils have the responsibility for surveying rights of way (footpaths, bridleways and 'byways open to all traffic') and preparing and keeping up to date what is misleadingly called a *definitive map*. The maps are supposedly conclusive evidence of the existence of rights of way but, in fact, they are not necessarily either complete or conclusive. They are incomplete because inadequate resources have been devoted to undertaking the necessary surveys, and they can be inconclusive because a map may wrongly identify a right of way. The latter is a legal matter which is not discussed here (see Chesman 1991), but the former is a continuing problem of planning policy and administration. The Countryside Agency has pointed out that

> the showing of a way as a footpath does not prove that there are not, for example, additional unrecorded rights for horse riders to use the way. Nor is the fact that a way is omitted from the definitive map proof that the public has no rights over it.
>
> (Countryside Agency 2003a: 11)

The definitive maps show some 225,000 km of rights of way in England and Wales. There are four categories:

footpath, bridleway, road used as a public path and byway open to all traffic. Most ways shown on maps are 'existing', that is they are not specifically designated but their dedication as a right of way is presumed from evidence of use by the public and the actions of the landowner. Few new recreational paths have been designated, though they are certainly needed in some parts of the country. To get dedication as a right of way, it is necessary to show that there has been uninterrupted use of the way by the public for twenty years. There has been a loss of access by both neglect and deliberate obstruction. Each year, some 1,500 formal proposals, affecting 500 km of the network, are made to change rights of way (by creation, diversion or extinguishment). Of these, about three-quarters are unopposed. The net change is negligible. It is difficult to establish what the overall effect is, though the Ramblers' Association has maintained that over a half of the public rights of way 'are unavailable to all but the most determined and agile person' (Blunden and Curry 1989).

Another footpath problem arises from their popularity: this is the wear and tear caused by a great intensity of use. The Pennine Way in particular has suffered from this, and 'damage limitation' experiments are under way. Recent measures have included laying flagstones which are delivered by helicopter. The Pennine Way is one of the long-distance routes which now stretch over some 2,700 km. The designation of these hikers' highways has been laborious, but they have had the attention and backing of the Countryside Agency, which has official responsibility for their establishment. Rights of way provide a structured framework for public access to the countryside, and they are of particular value to the energetic walker. However, they meet only part of the need: the number of people who enjoy a wander in the countryside is far greater than the number who hike long distances. The Hobhouse Report (1947) had argued for a public right of access to all open country: among the many benefits it foresaw was the freedom to ramble across the wilder parts of the country. Hobhouse's idyllic view was very much in line with the long-standing arguments for a 'right to roam' over all open country. The 1949 Act was much more circumspect, and provided for a right

of access only where an access agreement was made with the owner. The essential arguments on this issue have not changed, but circumstances have. There is now a much wider demand for access to the countryside, fostered both by increasing leisure pursuits and a huge increase in the ease of travel. But this very increase has strengthened the arguments of landowners about inappropriate use of the countryside. The 1997 Labour Party manifesto stated that 'our policies include greater freedom for people to explore our open countryside. We will not, however, permit any abuse of a right to greater access'. This was hardly a full-blooded commitment to legislate for a 'right to roam', and a succession of proposals and consultation papers led to more modest ideas of improving rights of way.

The government acted on this commitment with the Countryside and Rights of Way Act 2000. The rights of way provisions started to come into effect from April 2002, although full implementation was not achieved until 2005. The Act gives walkers the right to walk across mountain, moor, heath and down, and registered common land. The main exemptions are access by cycle, horse or vehicle, and to gardens, parks and cultivated land. The Countryside Agency and Countryside Council for Wales have begun preparation of maps showing where the new right of access will apply, and these will become definitive, with no additions allowed after a certain date yet to be determined. New powers for highway authorities and magistrates courts will allow for the removal of obstructions, although landowners will be able to restrict access for twenty-eight days each year. Local authorities will have to publish plans for improving rights of way, including provision for those with impaired mobility.

In Scotland, the position in relation to rights of way and access to the countryside is different from that south of the Border. The legal system is distinct, and the pressures on the countryside, until recently, have generally been lower. There is relatively freer access to the Scottish countryside: there is 'a well-established system of mutual respect between walker and landowner' (Blunden and Curry 1989: 152). Nevertheless, this is not so in areas close to the towns where access is severely restricted and, as pressures have mounted, the inadequacy of the legal situation has become apparent. In 1997, the government invited Scottish Natural Heritage to review the situation and to make recommendations to the Scottish Parliament. Given the wider concerns with land reform (discussed in Chapter 6), SNH decided that the scope of the review should be extended beyond the manifesto commitment regarding access to open country to embrace access over all land and inland water. The review was initially undertaken by an existing Access Forum, which includes representatives of the major interests such as landowners, local authorities, conservationists and recreation organisations, as well as SNH. Their proposals achieved a remarkable degree of consensus, and thus they have a legitimacy which helped to bring about legislative change. The main proposal was that there should be a non-motorised right of access to land and water for informal recreation and passage, subject to the responsible exercise of that right, to protection of the privacy of individuals, to safeguards for the operational needs of land managers, and any necessary constraints for conservation needs. In supporting this, SNH believed that there was a compelling case for modernising the current arrangements. The Land Reform (Scotland) Act 2003 provides the statutory right of responsible access for informal recreation and passage backed, and provides for a countryside access code to be drawn up. It gives powers to local authorities to uphold access rights 'over any route, waterway or other means by which access rights may be exercised', to plan for a system of core paths and to establish one or more local access forums for their area. Rangers may also be appointed to advise and assist both walkers and owners about their rights.

Greater access is consistent with a number of public policies, such as those for greater social inclusion and equity, the improvement of health and education, and the achievement of more sustainable development. The absence of a clear duty on local authorities to promote access (as distinct from discretionary powers), together with the current pressures on their resources, have led to a low priority for access needs. A significant increase in resources is required for local authorities and other public bodies involved in open-air recreation and tourism, as well as for additional incentives to landowners to improve the provision and management

of access over their land. A new countryside code of practice should be accompanied by a concerted effort to promote good behaviour and to improve visitor management. Government support for the proposals was announced in early 1999.[9]

Provision for recreation and country parks

In the early postwar years, national recreation policy was largely concerned with national parks (and their Scottish shadow equivalents), areas of outstanding natural beauty, and the coast. Increasingly, however, there has developed a concern for positive policy in relation to metropolitan, regional and country parks. The Countryside Act 1968 (following a White Paper, *Leisure in the Countryside*) gave additional powers to the then Countryside Commission for 'the provision and improvement of facilities for the enjoyment of the countryside', including experimental schemes to promote countryside enjoyment. At the same time, local authorities were empowered to provide *country parks*, including facilities for sailing, boating, bathing and fishing. These country parks are not for those who are seeking the solitude and grandeur of the mountains, but for the large urban populations who are 'looking for a change of environment within easy reach'. There is now a wide range of country parks, picnic sites, visitor-interpretative sites, recreation paths, interpretative trails, cycleways and similar facilities provided by local authorities and the Countryside Agency. There are over 250 country parks ranging in size from 11 to 1,875 hectares, and more than half of them attract at least 100,000 visits a year, totalling 57 million visits. Indeed, it has been suggested that the growth in this provision has led to or at least affected the decline of urban parks (which receive no similar funding).[10] Certainly, the deterioration of the quality and attractiveness of urban parks has been rapid and disastrous. A select committee has proposed the establishment of an Urban Parks and Greenspaces Agency as an urban equivalent of the Countryside Agency, although there are wider issues here which the committee did not address.

English Heritage maintains a Register of Parks and Gardens of Special Historic Interest in England (discussed in Chapter 8). The main purpose of this is to draw attention to those which constitute an important part of the cultural heritage, and also to encourage and advise local authorities to provide adequate protection for these sites through the development control system.[11]

More than 1,300 million day visits were made to the English countryside in 1996 but this fell to 1,126 million by 2003.[12] The impact of visitors is substantial; recreation and tourism make a significant contribution to the rural economy and total spending by all visitors was £9 billion in 1994 and £9.7 billion in 2003. Total employment supported by visitor activity has been estimated at a third of a million jobs (RDC, *The Economic Impact of Recreation and Tourism in the English Countryside*, 1997). For the most part, the impact on the environment is of manageable proportions though there are conflicts in specific areas, and traffic problems certainly can considerably reduce the 'quality of the recreational experience'; people with access to a car are twice as likely to visit the countryside. The impact of the car can be extreme; in the Peak District, on occasion, a condition of gridlock is created by the sheer volume of traffic. Although there are some areas where recreational pressures have undesirable impacts, generally leisure and tourism are less of a threat than industrial and agricultural activities. However, as the Environment Committee pointed out in its 1995 report on *The Environmental Impact of Leisure Activities*, there are difficult problems of crowding, overuse and conflict of activities in certain areas. The favourite answer is 'good management', and there is no doubt that this can help in preventing visitors 'loving to death' the beauty spots they wish to visit (to use the apt phrase adapted as the title of a report on sustainable tourism in Europe by the Federation of Nature and National Parks). So can 'countryside codes', 'visitor awareness' campaigns, and such like. But more drastic measures are inevitable in the most popular locations: Dovedale attracts 2 million visitors a year, of whom 750,000 use the main footpath. On busy Sundays no fewer than 2,000 people *an hour* can be crossing the river by the stepping stones. (The photograph on the cover of Jonathan Croall's 1995 book, *Preserve*

or Destroy: Tourism and the Environment, is the most eloquent statement of the problem of which this is only one illustration.)

In Northern Ireland, there were indications that the 'peace dividend' was having an effect with tourist pressures on attractive areas such as County Fermanagh. Various schemes are being tried: from strict control of cars (though driving cars on beaches is still common) and the provision of public transport to the expansion of facilities in new areas. The Countryside Agency is placing increased importance on funding recreational facilities close to where people live, which has the additional advantage of helping 'to reduce the number of countryside trips made by car and provides opportunities for countryside recreation for people without cars, for the young, and for those with special needs'. One particularly interesting initiative which started in the early 1980s, and is now well established, is the work of the *Groundwork Trusts*. Conceived as an *additional* resource for converting waste land to productive uses, particularly in urban fringe areas, it facilitates cooperative efforts by voluntary organisations and business, as well as public authorities. There is now a large network of trusts in the UK, involving 120 organisations in partnership. Their enterprise is wideranging and ranges from land reclamation, landscaping and environmental appreciation to provision for recreation, and many other activities seen as desirable and worth while in local communities. Examples include the development of the Taff Trail, which is a long-distance footpath and cycleway linking Cardiff and Brecon; recreating wildlife sanctuaries and access around mining villages in east Durham; the development of the Middleton Riverside Park on a totally derelict site a few miles from Manchester; and a programme which encourages owners of industrial and commercial premises 'to stand back and take a look at the external image of their premises and then to make practical landscape improvements. The scheme is clearly highly adaptable to local conditions and aspirations' (Jones, P. 1998, 1999). Some activities are simply impossible to accommodate in popular locations: high-powered recreational vehicles and boats not only are environmentally damaging, but also destroy the pleasure that others are seeking.

The Environment Committee in its 1995 report on *The Environmental Impact of Leisure Activities* argued that 'the principle of sustainability in leisure and recreation involves the provision of facilities for all activities, not only for the aesthetically pleasing and non-intrusive ones'. This is an interpretation of the concept of 'sustainability' which would not be universally accepted, but there would be more support for the Committee's proposal that sites should be selected which are 'suitable for noisy and obtrusive activities'. Whether such sites could be readily found is more problematic, even on 'derelict land or land of low amenity'. An economist would point out that this is a classic case for a charging mechanism which would enable compensation to be paid to those adversely affected.

Much of the greater willingness to provide additional opportunities and facilities for recreation in the countryside has emanated from the changed economics of agriculture. Less agricultural production and more diversified activities are desirable. Indeed, without changes in the pattern of economic activity, many rural areas will be adversely affected by the changes in agriculture. The provision of recreational facilities is a potentially lucrative business, as has been dramatically illustrated in those areas where it has been realised. A study of Center Parcs holiday villages demonstrates the substantial benefits which these had brought to the local areas: £4.5 million around Sherwood (Nottinghamshire) and £5.3 million around Elveden (Norfolk). The employment created might eventually amount to well over 1,000 jobs. Tourism also brings more indirect benefits to rural areas such as the survival of bus services and village shops. There is, of course, a cost to be borne for these advantages – in terms of changed character, and conflicts between the interests of visitors and residents (particularly in areas which have attracted new residents). Such conflicts can be reduced by 'good management', but they are inherent in the dynamics of social and economic change.

Countryside grant programmes

Changes in policies relating to farming have had a more tangible effect than the heated arguments of contenders

for and against freer countryside access. The lower priority for food production has led to attempts to broaden the role of landowners as 'managers' of the countryside. Within this changing framework, 'access becomes a means of diversifying the agricultural economy'. Successive measures have reflected the changing priorities, and there has been increased emphasis on the role of farmers as 'stewards of the countryside' which has led to a greater concentration of funds on environmental schemes.[13]

The Countryside Agency has been in the lead in promoting conservation and recreation as explicit objectives of agricultural policy. Its 1989 policy statement *Incentives for a New Direction in Farming* argued that the diminishing need for agricultural production provided an opportunity for 'environmentally friendly' farming. It presented a menu of incentives for farmers and landowners to provide environmental and recreational benefits. These ideas were translated into the *countryside premium*: an experimental scheme which gave incentives for land to be set aside for recreation. It was followed by the *countryside stewardship* which provides incentives for the protection and enhancement of valued and threatened landscapes. This scheme proved to be a successful one: in its first four years, some 5,000 agreements were concluded covering 91,000 hectares. The Countryside Commission for Wales has a parallel scheme (*Tir Cymen*). Other 'countryside' bodies also have schemes: the Forestry Commission operates a *woodland grant scheme* which includes payments for the management of woods to which the public have access, and under its *countryside access scheme* DEFRA makes payments for land which is 'set-aside' for public access.

The Agriculture Act 1986 made provision for *environmentally sensitive areas* where annual grants were given by MAFF to enable farmers to follow farming practices which will achieve conservation objectives. The introduction of ESAs marked a fundamental policy change (Bishop and Phillips 1993: 325). They provide financial support for practices which result in environmental benefits, in contrast to earlier schemes which gave compensation for forgone profits. ESAs have developed into the main plank of the ministry's countryside protection policy. There were nineteen ESAs in the original scheme introduced in 1987. By 1999, the number (in the UK) had increased to forty-three. Under an EU Directive to prevent nitrate pollution in sensitive areas, the government administers two schemes (nitrate sensitive areas and nitrate vulnerable zones) which involve payments to farmers to compensate them for making changes to farming practices which reduce nitrate leaching, thus protecting public drinking water sources.

The government has also operated grant schemes for farm diversification and for farm woodlands. The former (now partly amalgamated with the EC-funded *farm and conservation grant scheme*) is mainly aimed at encouraging environmentally beneficial investments. The latter (the *farm woodland scheme*) began as an experiment in 1991, but the results were disappointing and a comprehensive review led to the introduction of a new *farm woodland premium scheme* in 1992. The objectives are to enhance the farmed landscape and environment and to encourage a productive land use alternative to agriculture.

During the 1990s new schemes were introduced at a bewildering rate. In addition to those already mentioned, there is the *wildlife enhancement scheme*, *parish paths partnership*, and the *countryside employment programme* and *redundant building grant*. Indeed, it is sometimes difficult to be clear where one scheme ends and another begins. Box 9.5 lists most of the 'agri-environment schemes designed to encourage environmentally friendly farming and public enjoyment of the countryside'. Nevertheless, it is apparent that a determined effort is being made to offset some of the effects of the stimulation to excess production and degradation of the countryside. The effectiveness of these various schemes is a matter of controversy (Winter 1996; Adams, W. M. 1996). The 2004 *Rural Strategy* promises to cut down the one hundred or so rural funding schemes to three major programmes: rural regeneration, agriculture and food industry regeneration, and natural resource protection. The agri-environmental element of funding is addressed by the 2002 *Strategy for Sustainable Food and Farming: Facing the Future*. The Strategy not only noted the progress made on slowing the deterioration in the quality of the rural environment, but also recognised that despite the introduction of legislation protecting habitats,

stronger planning policies and funding to encourage more environmentally friendly practices:

> significant problems remain, including continued attrition of the historic environment, serious overgrazing in some upland areas, declines in the population of widespread species and the loss of biodiversity within some surviving habitats.
>
> (p. 27)

A thorough reform of the funding system is now underway, which involves an entry level and broadly based agri-environment scheme that will reward general agricultural management practices that go beyond good farming practice but 'which are less arduous than the prescriptions of the main schemes', together with proposed integration of 'higher-level schemes such as countryside stewardship and environmentally sensitive areas'. The entry level scheme has been piloted in Lincolnshire.

Nature conservation

The concept of wildlife sanctuaries or nature reserves is one of long standing and, indeed, it antedates the modern idea of national parks. In other countries, some national parks are in fact primarily sanctuaries for the preservation of big game and other wildlife, as well as for the protection of outstanding physiological features and areas of outstanding geological interest. British national parks were somewhat different in origin, with an emphasis on the preservation of amenity and providing facilities for public access and enjoyment (though, as noted earlier, the trend has been to give increasing priority to conservation). The concept of nature conservation is primarily a scientific one concerned particularly with the management of natural sites and of vegetation and animal populations. The Huxley Committee argued in 1947 that there was no fundamental conflict between these two areas of interest:

> their special requirements may differ, and the case for each may be presented with too limited a vision:

BOX 9.5 SOME AGRI-ENVIRONMENT SCHEMES

Countryside access scheme

Aims to increase the benefits from land which is set-aside by offering incentives to farmers to increase public access on the best located sites.

Countryside stewardship scheme

Aims to protect and restore targeted landscapes, their wildlife habitat and historical features, and to improve opportunities for public access; currently being developed as the government's main incentive scheme for the wider countryside.

Environmentally sensitive areas

Incentives to farmers to safeguard the area and to improve public access in areas of particularly high value because of landscape, wildlife or history which are threatened by changes in farming practices.

Farm woodland premium scheme

Grants towards cost of planting trees on agricultural land.

Habitat scheme

Incentives to create or improve wildlife habitats.

Moorland scheme

Aims to protect or improve moorland by reduction of grazing.

Nitrate sensitive areas

For reduction of nitrate leaching.

but since both have the same fundamental idea of conserving the rich variety of our countryside and sea-coasts and of increasing the general enjoyment and understanding of nature, their ultimate objectives are not divergent, still less antagonistic.

However, to ensure that recreational, economic and scientific interests are all fairly met presents some difficulties. Several reports dealing with the various problems were published shortly after the war (Dower 1945; Huxley 1947; Hobhouse 1947). The outcome was the establishment of the Nature Conservancy, which was later replaced by English Nature, Scottish Natural Heritage and the Countryside Council for Wales. In Northern Ireland, Scotland and Wales, central responsibility for nature conservation and access to the countryside rests with a single body. In England they are separate, but with proposals for their amalgamation, as described above.

Legislation often emerges as a response to new perceptions of problems; but sometimes legislation itself fosters such perceptions. So it was with the Wildlife and Countryside Act 1981. Introduced as a mild alternative to the Labour government's aborted Countryside Bill (and stimulated by the need to take action on several international conservation agreements), the Conservative government expected no serious trouble over the Bill. It was very mistaken: the Bill acted as a lightning rod for a host of countryside concerns that had been building up over the previous decade or so – moorland reclamation, afforestation and 'new agricultural landscapes', loss of hedgerows, damage to SSSIs, and such like. The Bill had a stormy passage through Parliament, with an incredible 2,300 proposed amendments. Though most of these failed, the Bill was considerably amended during the process. The major focus of argument (with the strong National Farmers Union and the Country Landowners Association holding the line against a large but diffuse environmental lobby) was the extent to which voluntary management agreements could be sufficient to resolve conflicts of interest in the countryside. The government steadfastly maintained that neither positive inducements nor negative controls were necessary. Indeed, it was held that controls would be counterproductive in that

they would arouse intense opposition from country landowners.

Of particular concern was the rate at which SSSIs were being seriously damaged, the speed at which moorland in national parks was being converted to agricultural use or afforestation, and the adverse impact of agricultural capital grants schemes both on landscape and on the social and economic well-being of upland communities. On the first issue, the government finally made a concession and provided for a system of 'reciprocal notification'. This required the then Nature Conservancy Council (NCC) to notify all landowners, the local planning authority and the Secretary of State of any land which, in their opinion, 'is of special interest by reason for any of its flora, fauna, or geological or physiological features', and 'any operations appearing to the [NCC] to be likely to damage the flora or fauna or those features'. Landowners were required to give three months' notice of intentions to carry out any operation listed in the SSSI notification. This was intended to provide the NCC with an opportunity 'to discuss modifications or the possibility of entering into a management agreement'. This much vaunted voluntary principle did not work: sites were damaged while consultations were under way. Amending legislation, passed in 1985, was designed to prevent this.

Another issue of contention arose when, during the debates, attempts to extend grants from 'agricultural business' to countryside conservation were defeated, and a host of amendments divided the Opposition and confused the issues. As passed, the amendments did little more than exhort the Minister of Agriculture, when considering grants in areas of special scientific interest, to provide advice on 'the conservation and enhancement of natural beauty and amenities of the countryside' and suchlike, 'free of charge'. There is, however, power to refuse an application for an agricultural grant on various 'countryside' grounds, but such a refusal renders the objecting authority (the county planning authority in national parks and the NCC in SSSIs) liable to pay compensation. This is a return to the pre-1947 planning system (even though it applies to only a small part of the country) and it has, not surprisingly, given rise to a considerable amount of debate. William Waldegrave (when Minister for the Environment, Countryside and Local Government) argued:

> I think that the moral and logical position of the farmer who finds that his particular bit of flora or fauna is now rare is such that there should be no hesitation in saying that he deserves public money if he is asked to do better than those who have been allowed to extinguish their bits, and if it is expensive for him to do so. I believe such flows of money, from taxpayer to land-user, for conservation expenses, are thoroughly justified and should become a useful and permanent adjunct to farm incomes for quite a considerable number of farmers, often in the rather more marginal farming areas where the inherent difficulty of farming has prevented our predecessors from extirpating species which may have gone for good elsewhere.
>
> (Waldegrave *et al.* 1986)

The issue here goes much further than appears at first sight, since it raises questions about the ownership of development rights. While the postwar planning legislation nationalised rights of development in land, it effectively excluded agriculture and forestry. The owner of a listed building receives no compensation for the restrictions which are imposed, and may even be charged for repairs deemed necessary by the local authority in default. The farmer, on the other hand, expects – and obtains – payment for 'profits forgone' in 'desisting from socially undesirable activity, or merely for departing from what is conventionally regarded as good agricultural practice'. Thus, to quote Hodge (1999), there was 'the irony of one UK government agency being obliged to buy out the subsidies being offered by another'. Moreover, there was concern that some landowners threatened changes merely as a means of extracting compensation: a few very large payments had much publicity. Not surprisingly, the Labour government objected in principle to this system and proposed that payments should be made to landowners only where this was in furtherance of conservation management.

Public support for the proper management of SSSIs is essential and appropriate, but it should be given where positive management prescriptions are required. The Government is not prepared, in future, to pay out public money simply to dissuade operations which could destroy or damage these national assets.[14]

The main element in the new policy approach is more effective protection and better management. The Countryside and Rights of Way Act 2000 imposes on public bodies a statutory duty 'to secure the positive management' of SSSIs; the powers of conservation agencies have been strengthened; and the courts have been given powers in relation to penalties for damaging an SSSI. However, experience shows that good management is more generally achieved by negotiation and agreement than by legislative fiat (though this can be helpful as a reserve power). A good example is the working partnership that has been agreed with farmers in Bowland where there has been damage to SSSIs through overgrazing. Overgrazing is encouraged by the CAP, which pays farmers according to the numbers of livestock on the land. Although this policy maximises production, it does so at the expense of wildlife. The problem can be met by reducing grazing and channelling agricultural support in the uplands to make hill farming both economically viable and environmentally efficient. The Bowland initiative involves a partnership between government departments and agencies, local authorities, voluntary conservation organisations and farmers. This partnership includes a complex funding package which enables farmers to address a number of conservation issues.[15]

Biodiversity

Nature conservation has risen on the political agenda with world-wide concern for biodiversity. Though the terms are different, they mean very much the same thing: the variety of life on the earth. Official biodiversity policy was set out in *Biodiversity: The UK Action Plan* and *Sustainable Development: The UK Strategy*, both published in 1994.

Policies take two main forms: the protection of particular species of flora and fauna, and the designation and protection of conservation sites. It is the latter which is of particular relevance to this book. Statutory designations have multiplied in recent years, particularly with the impact of European directives. The indigenous designations include nature reserves, SSSIs, marine nature reserves and sites of importance for nature conservation. Protection is provided by various obligations, agreements with owners and occupiers, and through acquisition by local authorities, English Nature, and other bodies.

In March 2004, there were 395 *national nature reserves* in the UK, and 1,046 *local nature reserves* (see Table 9.3). The former are, by definition, sites of national importance. Nature reserves are not sanctuaries: they are preserves where the conditions provide 'special opportunities for the study of, and research into, matters relating to the fauna and flora'. Most of these are in private ownership subject to a management agreement, but voluntary organisations such as the Royal Society for the Protection of Birds and county wildlife trusts own and manage their own non-statutory reserves.

Sites of special scientific interest number 6,569 and cover about 8 per cent of the land area of Britain. In these protected areas, occupiers must obtain permission before certain listed activities (known as potentially

damaging operations or PDOs) can be carried out; there is also stricter planning control. Draft new guidelines on protecting, managing and conserving SSSIs in England were published in 2000 to minimise widespread damage (see discussion below). The designation of *marine nature reserves* was introduced by the Wildlife and Countryside Act 1981. There are three statutory reserves: Lundy Island (off the coast of Devon), Skomer (Dyfed) and Strangford Lough (Northern Ireland). There are also a number of non-statutory marine reserves which have been established by voluntary conservation groups. As required by EC Directives on Conservation, special measures are to be taken to conserve certain habitats. These include a number of classifications: *special protection areas* (SPAs) under the Birds Directive and *special areas of conservation* (SACs) under the Habitats Directive. The European network of these SPAs and SACs is known as the Natura 2000 Network. SACs automatically become SSSIs subject to consultation. *Ramsar sites* are protected wetlands (so called after the town in which the convention was signed). Provisions relating to *limestone pavement areas* (which is really a geological designation) were introduced by the Wildlife and Countryside Act 1981. These cover no more than 5,000 acres in England and Wales, but are 'of great natural beauty and scientific interest'. They are popular with gardeners

Table 9.3 Protected areas in the UK 2004

	Number	Area (sq. km)
National nature reserves	395	234
Local nature reserves (GB)	1,046	45
Sites of special scientific interest (GB)	6,569	2,341
Areas of special scientific interest (NI)	211	93
Marine nature reserves	3	19
Special areas of conservation (SACs)	340	17,659
Special protection areas (SPAs)	242	1,470
Ramsar wetland sites	144	759
Environmentally sensitive areas	43	3,190
Biosphere reserves	9	43
Biogenic reserves	18	8

Source: DEFRA (2004) *eDigest of Environmental Statistics* (Table 16)

looking for stone for rockeries: hence the need for protection.[16] This bewildering (though incomplete recital of conservation instruments) suggests that the time might be near when some rationalisation could be considered appropriate. However, they are testimony to a heightened regard for 'this common inheritance'. This has been evolving for some time. It was in 1968 that the Countryside Act provided that 'in the exercise of their functions relating to land under any enactment, every minister, government department and public body shall have regard to the desirability of conserving the natural beauty and amenity of the countryside'. Since 1986, there has been a statutory duty to balance the interests of agriculture with rural and environmental interests, and 'agri-environmental regulation' has entered the planning lexicon. A wide range of countryside initiatives are being introduced. Indeed, as this discussion illustrates, it is difficult to keep pace with the changing policies and programmes.

There are several aspects of conservation policy which are worth noting. The first is that it is not restricted to designated sites (despite their large number): in addition to these nationally (and inter-nationally) important sites, there are many more which are of local importance. The now superseded PPG 7 drew attention to the importance of these sites for local communities, often affording people the only oppor-tunity of direct contact with nature, especially in urban areas. A second, related, point is that (since wildlife does not respect human-made boundaries) it is impor-tant to safeguard 'wildlife corridors, links or stepping stones from one habitat to another'. The Habitats Directive specifically requires member states of the EU

> to encourage the management of features of the landscape which . . . by virtue of their linear and continuous structure (such as rivers with their banks or the traditional systems for marking field bound-aries) or their function as stepping stones (such as ponds or small woods), are essential for the migra-tion, dispersal and genetic exchange of wild species.

What is particularly significant about this approach (which is only briefly illustrated here) is that it is mandatory on local authorities, and that, in addition to land use designations, actual *management* is involved. (The mandate comes from a combination of two legal instruments: the requirement for plans to include policies in respect of 'the conservation of natural beauty and amenity of the land' and the provision in the Conservation (Natural Habitats etc.) Regulations 1994 that these policies 'shall be taken to include policies encouraging the management of features of the land-scape which are of major importance for wild flora and fauna'. UK work on biodiversity is coordinated by the UK Biodiversity Group. There are also country groups for each of the four constituent countries.

Forestry

In 2004, forest and woodland covered some 2.8 million hectares in the UK or 12 per cent of the land area: about 8 per cent of land area of England, 16 per cent of Scotland, 12 per cent of Wales, and 6 per cent of Northern Ireland. The woodland is made up of about 59 per cent conifer and 41 per cent broadleaved, though this ratio is changing as the majority of new planting is broadleaved. There has been a steady increase in the forest area as shown in Table 9.4; during the 1980s the increase was of some 300,000 hectares and from 1990 to 2004 another 400,000 hectares has been added; that is a 29 per cent increase since 1980. It should be remembered here that forestry is not gen-erally within planning control, though, as PPS 7 points out, it can prevent the conversion of woodland to urban and other uses.

About two-fifths of productive forestry is managed by Forest Enterprise, the development and manage-ment arm of the Forestry Commission. A substantial reorientation of forestry policy has emerged since the late 1970s. In particular, there has been a major move away from a preoccupation with production, to a more balanced approach which places importance on amenity and environmental factors. This has come about after a lengthy period of debate and scrutiny. There was, for example, much argument both before and after the Countryside and Wildlife Act 1981, which centred on the effects of hill farming and forest policies. A

Table 9.4 *Area of woodland in the UK 1924–2003*

	1924	1947	1965	1980	1999	2003
England	660	755	886	948	1,097	1,110
Scotland	435	513	656	920	1,282	1,327
Wales	103	128	201	241	287	285
N Ireland	13	23	42	67	81	85
	1,211	1,419	1,785	2,176	2,747	2,807

Source: DEFRA (2004) *eDigest of Environmental Statistics*

succession of reports concluded that these policies, far from sustaining the economies and landscapes of the uplands of England and Wales, were major factors in their decline (MacEwen and MacEwen 1982; Sinclair 1992).

The predominant concern for production was badly affecting the vitality of rural communities and the conservation of the countryside. It was argued that much more employment could be created by coordinated policies sensitively directed to the problems of the uplands as a whole, rather than to separate aspects of them. This was essentially a call for 'integrated' policies which began to emerge later. The immediate result, however, was a provision in the Wildlife and Countryside (Amendment) Act 1985 requiring the Forestry Commission to attempt a reasonable balance between the interests of forestry and of conservation and enhancement of the countryside. Forestry policy in the UK has been increasingly influenced by international commitments such as the 1992 Rio Earth Summit which led to a statement on forestry principles, the Rio Declaration and Agenda 21 which provided an agenda for sustainable development. A review of forestry policy in 1994 (*Our Forests: The Way Ahead*) proclaimed that 'UK forestry policy is based on the fundamental tenet that forest resources and forest lands should be sustainably managed to meet the social, economic, ecological, cultural and spiritual human needs of present and future generations'. In making changes to existing land use policies, however, there are constraints imposed by the Common Agricultural Policy, which is (as always?) under negotiation.

In addition to the broader economic implications of forestry policy for rural areas, there are several issues

which are of particular relevance to countryside policy: amenity, wildlife, access and recreation. Problems arise because, though forest production is essentially a very long-term enterprise, there is need, in the words of the National Audit Office, *Review of Forestry Commission Objectives and Achievements* (1986), 'to have regard to a number of broadly drawn secondary objectives [which] can produce conflicts with and constraints upon the Commission's primary aim of increasing the supply of timber'. One of these secondary objectives is the preservation (and enhancement) of the landscape and of the wildlife it sustains.

Concern for such wider issues has increased as environmental awareness has grown. Current policy is one of 'multiple-purpose forestry'. This embraces a wide range of approaches. Planting of broadleaves is being expanded. Access to forests is being extended, with improved arrangements for access agreements. New *national forests* are being established in the Midlands and in central Scotland.[17] A new national *forest park* (in addition to the sixteen previously established by the Forestry Commission) has been opened at Gwydir. A *community forests* programme (operated jointly by the Forestry Commission and the Countryside Agency, with local authority support) is under way which aims to create attractive green settings (rich in wildlife and easily accessible) for the enjoyment and health of residents, and to encourage economic regeneration. Additionally there is a *community woodlands* scheme which promotes new woodlands near centres of population (with assistance from the *woodlands grant scheme*).

In 1999, a wide-ranging review was published under the title of *A New Focus for England's Woodlands*. This describes the two main aims of forestry policy in

England as being the sustainable management of existing woods and forests, and a continued steady expansion of the woodland area to provide more benefits for society and for the environment. The strategy has four components: rural development, economic regeneration, recreation, access and tourism, and the environment and conservation. The rural development policy area lays stress on the creation of a higher proportion of well-designed larger woodland planting. Larger woodlands are usually capable of providing more public benefits than is possible with smaller woods, though the latter 'can make a significant contribution of local biodiversity, amenity, environmental health and sustainable development'. Economic regeneration involves promoting forestry in the restoration of former industrial land (the amount of which in England is some 175,000 hectares). The Forestry Commission's Land Regeneration Unit, which was established in 1997, has demonstrated the potential, as has the National Urban Forestry Unit, which works in partnership with local authorities, the private sector and non-governmental organisations. There are also the Community Forests and woodlands schemes mentioned above as well as the Woodlands by the Motorways project (partly sponsored by the Highways Agency and Esso). The recreational aspects of forestry are well documented: some 3 million visits a year are made to the woods and forests of England. The current *Forestry Strategy for England* (1998) includes better information about the opportunities, improving facilities and developing the Forestry Commission's Woodland Park network. There are many positive environmental benefits of forestry including the improvement of air quality, and the promotion of biodiversity. Among the actions on this front are better management of woods and forests, more research on the environmental benefits of forestry, and the promotion of greater appreciation of the value of trees, woodlands and forests. Much of this, of course, is not new, but there is now a very positive approach to the expansion and management of forestry in the UK. Not surprisingly this has called for another regional strategy: *regional forestry frameworks*.

Further reading

General

The Countryside Agency and its equivalents elsewhere in the UK offer a wealth of information covering all aspects of the countryside including planning. The address for publications is Countryside Agency Sales, PO Box 125, Wetherby, West Yorkshire LS23 7EP. A free catalogue of publications is available from this address and at www.countryside.gov.uk/. Reference is made to some of the key Agency publications in the text. A new textbook is now available: Bishop and Phillips (2004) *Countryside Planning: New Approaches to Management and Conservation*. See also Hodge (1999) 'Countryside planning', Murdoch (2003) *The Differentiated Countryside*, Ilbery (1998) *The Geography of Rural Change*, Gilg (1999) *Perspectives on British Rural Planning Policy* and (1997) *Rural Planning in Practice*, and Owen (2002b) 'Locality and community'.

There is a long history of writing on countryside and its planning: important books include Champion and Watkins (1991) *People in the Countryside: Studies of Social Change in Rural Britain*, Cherry (1994b) *Rural Change and Planning: England and Wales in the Twentieth Century*, Cloke *et al.* (1994a) *Lifestyles in Rural England*, and Newby (1985) *Green and Pleasant Land: Social Change in Rural England*. Detailed accounts of change can be found in the series *Progress in Rural Policy and Planning* (5 vols, 1991–5), Gilg (1996) *Countryside Planning: The First Half Century* and other titles by Gilg. A short account of countryside issues in Northern Ireland is given by Lipman (1999). There are many popular books bemoaning the fate of the countryside: a highly readable and informative one by an agricultural journalist is Harvey (1997) *The Killing of the Countryside*. The damage done to the countryside by modern methods of farming is analysed in a famous text by Shoard (1980) *The Theft of the Countryside*.

A good statement of the problems of the rural economy and of government policies in relation to this is set out in DETR (1998) *Guidance to the Regional Development Agencies on Rural Policy*. There are numerous studies by the former Rural Development Commission (1998a) *Rural*

Disadvantage: Understanding the Processes and (1998b) *Rural Development and Land Use Planning Policies.*

National parks, access to the countryside and rights of way

For accounts of the background to and the implementation of the 1949 Act, see Cherry (1975) *National Parks and Recreation in the Countryside* (Volume 2 of *Environmental Planning 1939–1969*) and Blunden and Curry (1989) *A People's Charter? Forty Years of the National Parks and Access to the Countryside Act 1949.* Two major reviews of national parks policies are the Sandford Report (1974) and the Edwards Report (1991). A passionate critique of the restrictions on access (and much else) is given by Shoard (1987) *This Land is our Land* and (1999) *A Right to Roam.*

The Countryside Agency published in 2003 a most useful *Guide to Definitive Maps and Changes*, which is a most complicated topic. Official publications on access to the countryside proliferated in 1998 and 1999: *Access to the Open Countryside: Consultation Paper* (1998), *Options on Access* (1999), *Improving Rights of Way in England and Wales* (1999) and *Access to Open Countryside of England and Wales: The Government's Framework for Action.* Sandwiched between all these DETR publications was a report commissioned from the Countryside Commission on *Rights of Way in the 21st Century* (1998). For a foreign comparison see Peter Scott Planning Services (1998) *Access to the Countryside in Selected European Countries.* For Scotland, the most important document is *Access to the Countryside for Open-air Recreation* (SNH, 1998).

Coastal issues and waterways

On coastal issues, see DoE (1995) *Policy Guidelines for the Coast*, PPG 20 (1992) *Coastal Planning*, House of Commons Environment Committee (1992) *Coastal Zone Protection and Planning*, SODD *Coastal Planning* (NPPG 13) and Cleator (1995) *Review of Legislation Relating to the Coastal and Marine Environment of Scotland.* Two EC reports were issued in 1999: *Lessons from the European Commission's Demonstration Programme on Integrated Coastal Zone Management* and *Towards a European Integrated Coastal Management Strategy: General Principles and Policy Options.*

The relationship between planning and waterways is dealt with in BWB (1993) *The Waterway Environment and Development Plans.* More recent is British Waterways (1999) *Our Plan for the Future 2000–2004* and DETR (1999) *Unlocking the Potential: A New Future for British Waterways.*

Recreation

Recreation in the countryside is discussed at length in HC Environment Committee, *The Environmental Impact of Leisure Activities* (1995). For a review of the impact of different recreation activities on the environment, see Sidaway (1994) *Recreation and the Natural Heritage.* A 1995 report by the Countryside Commission and others is devoted to exploring the concept of *Sustainable Rural Tourism* and the ways in which it can be translated into practice. See also PPG 21 *Tourism* (1992), Segal Quince Wicksteed (1996) *The Impact of Tourism on Rural Settlements*, Curry (1997) 'Enhancing countryside recreation benefits through the rights of way system in England and Wales' and Bell (1997) *Design for Outdoor Recreation.* A statement on *Countryside Recreation: Enjoying the Living Countryside* (1999) is briefly presented in a Countryside Commission publication with this title. Further references on tourism are given in Chapter 8.

Nature conservation and biodiversty

Current official guidance on nature conservation is given in PPG 9 (1994) while there is a planned PPS on *Biodiversity and Geological Conservation.* Additionally see NPPG14 on *Natural Heritage (1999)*, DoENI PPS *Planning and Nature Conservation* and TAN5 *Nature Conservation* (1996).

On the implementation of biodiversity policies see RTPI (1999a) *Planning for Biodiversity*, UK Biodiversity Group and Local Government Management Board (1997) *Guidance for Local Biodiversity Action Plans*, Scottish Biodiversity Group (1998) *Biodiversity in Scotland: The Way Forward* and (1998) *Local Biodiversity Action Plans: A Manual.*

Two publications of the Countryside Council for Wales are *The Welsh Landscape: A Policy Document* (1996) and

Protecting Our Natural Heritage: A Guide to the Designated Sites and Landscapes of Wales (1997).

Despite the special controls, many SSSIs have been damaged: see, for example, reports of the National Audit Office: (1994) *Protecting and Managing Sites of Special Scientific Interest in England*, (1997) *Protecting Environmentally Sensitive Areas* and (1999) *SSSIs: Better Protection and Management: The Government's Framework for Action*. On local nature reserves see Barker and Box (1998) 'Statutory local nature reserves in the United Kingdom'.

Forestry and woodlands

There was a major policy statement issued after the Rio Summit: *Sustainable Forestry: The UK Programme* (1994), following the review reported in *Our Forests: The Way Ahead* (1994). This has been followed by a *New Focus for England's Woodlands* (1998), the *Forest Strategy for England* (1999) and the National Forest Company (2004) *The National Forest: The Strategy*. For Scotland, see SODD (1999) *Indicative Forest Strategies* (Circular 9/1999).

Notes

1 PPG 3 Housing and its Annex A deal with affordable housing and the exceptions policy. The update to PPG 3 that deals with the allocation of small sites is considered in Chapter 6 on Land Policies. At the time of writing PPG 3 is about to be replaced with a PPS.

2 This name is often used by 'Muckspreader' in his or her 'Down on the Farm' column in *Private Eye*.

3 *Rural England: A Nation Committed to a Living Countryside* (1995), *Rural Scotland: People, Prosperity and Partnership* (1995) and *A Working Countryside for Wales* (1996).

4 The Natural Heritage (Scotland) Act 1991, which provided for the establishment of Scottish Natural Heritage also introduced *natural heritage areas*. It was intended that they would be designated for a wide range of situations both in upland and lowland Scotland were there is both a landscape and a nature conservation interest, and where there is therefore

a need for integrated management. However, since the government now supports the introduction of national parks, it is unlikely that any NHAs will be designated (NPPG 14, *Natural Heritage*, 1998). However, the SNH is developing a Natural Heritage Zones Programme which will identify twenty-one zones reflecting the diversity of Scotland's unique natural heritage. For each zone, a statement is being prepared reviewing trends, opportunities and pressures. Finally, in conjunction with partner organisations, future action will be determined (SNH 1997/98). See also the 1998 consultation paper *National Parks for Scotland* which emphasised the importance of social and economic purposes along with the protection and enhancement of natural and cultural heritage, and the enjoyment, understanding and sustainable use of natural resources. Also stressed was local community involvement in national parks.

5 In 2004 the Department for Regional Development, Northern Ireland consulted on a proposed Planning Policy Statement 14 on Sustainable Development in the Countryside, which addressed mostly the issue of dispersed rural housing.

6 As part of the Neptune Coastline Campaign, an assessment was undertaken by Reading University of changes to developed and undeveloped coastline. This showed that the length of coastline of high landscape quality in each region had decreased from 5 to 9 per cent between 1965 and 1995. However, a few counties had increased lengths of high landscape coastline, probably arising from the demise of industrial activities (Burgon 2000).

7 *Lessons from the European Commission's Demonstration Programme on Integrated Coastal Zone Management* (Luxembourg: OOPEC) and *Towards a European Integrated Coastal Zone Management (ICZM) Strategy: General Principles and Policy Options* (Luxembourg: OOPEC).

8 Dorset County Council pioneered a planning and management coastal strategy which was commended by the EU, and helped to lead to successful designation of World Heritage Site status. See *Planning* 5 November 1999: 6. Information can be obtained from the Dorset Coast Forum at the Dorset CC website www.dorset-cc.gov.uk. For general infor-

mation about the coastal zone, see the website www.theukcoastalzone.com.

9 Scottish Office Press Release, 2 February 1999.

10 See the discussion in the HC Environment Subcommittee on Town and Country Parks (1999), p. xi. There is no proof for this, but it is pointed out that the deterioration of urban parks coincided with the establishment of funded countryside parks. Of course, there was also an accompanying growth in car ownership and thus increased opportunities for countryside visits. The Heritage Lottery Fund has an Urban Parks Programme, which has been very popular: by April 1999, it had received 462 applications and awarded £117 million in grants (op. cit., vol. II, p. 137).

11 Further details are given in the HC Environment Subcommittee report *Town and Country Parks* (1999) vol. II, pp. 50–3. The University of York has compiled an online searchable database of historic parks and gardens http://www.humbul.ac.uk/.

12 Countryside Commission (1999) *Countryside Recreation: Enjoying the Living Countryside* (the various figures quoted relate to 1996) and Countryside Agency (2004) *The State of the Countryside 2004*.

13 These schemes have a positive economic benefit, even though it is not easily measured. See Hanley *et al.*

(1999) 'Assessing the success of agri-environmental policy in the UK'.

14 DETR (1999) *Sites of Special Scientific Interest: Better Protection and Management*, p. 4. See also the consultation document with the same title (DETR 1998).

15 The issues are the loss of herb-rich hay meadows and rushy grazing pastures, which are important to wading birds. The project also promotes the natural regeneration of woodlands by fencing to prevent sheep grazing as well as improving heather regeneration (English Nature press release, Bowland Farmers Take Action to Work for Wildlife, 2 July 1999).

16 Limestone Pavement Orders are designated for the most valuable parts of limestone pavements. There is, however, still an illegal trade in these pavements. The Limestone Pavement Working Group is examining the problems. See HC Select Committee of the Environment, Transport and Regional Affairs Committee, Fifteenth Report (1998/99), p. xii.

17 The English 'national forest' is being planted by the National Forest Company over some twenty years in about 520 sq. km of Derbyshire, Leicestershire and Staffordshire. It is funded by government and the aim is to get woodland cover of about a third of the area. By 1999, nearly 3 million trees had been planted and this had increased to 9 million by the end of 2004.

10 Urban policies and regeneration

The problem is that for too long urban policy has acted as 'a filler in of gaps', mopping up the worst cases of fallout produced by wider economic and policy changes. It has functioned as both a form of symbolism and crisis management.

Atkinson 1999b: 84

Introduction

The breadth of public intervention embraced by the term 'urban policies' denies any simple summary: indeed, the enduring feature of urban policy has been the endless experimentation with new and often disconnected initiatives. What is consistent is the fragmentation of effort, lack of a strategy, weak involvement of local communities, the marginal impacts (in contrast to mainstream public spending and private investment) and bias to property development and economic development. The criticisms have not been ignored. Since the early 1990s, there has been a welcome increase in the influence of local government and local communities, though spending is still relatively low and the baffling range of ad-hoc initiatives remains. The evaluation of urban policies that underpins these critical comments is taken up at the end of the chapter. Here, two preliminary points need to be made.

First, the title may be somewhat misleading in that a number of the policies discussed extend to rural areas (housing for example) or may have non-spatial dimensions (e.g. economic development). Nevertheless, the focus of the discussion is on urban areas in general and inner cities in particular. Second, the title is in the plural since there is no such thing as a single urban policy or a set of policies, rather urban policy has

been described as 'incoherent' and not deserving of the appellation 'policy' (Atkinson 1999b: 84). Moreover, it is not a simple matter to define what are, and what are not, urban policies. National economic, welfare and housing policies may play a more significant role than urban aid or urban regeneration policies. Nevertheless, there is a group of policies which are officially labelled urban (or increasingly regeneration) that has sufficient urban identity to justify discussing them together.

The chapter opens with a discussion of inadequate housing: the starting point of urban policy (and planning) in the nineteenth century, and still a major concern for policy-makers. From the 1960s attention shifted from physical conditions to the social aspects of housing and, later, to areas of social need. A further shift took place in the late 1970s when economic issues were seen as being the key to urban regeneration. By the mid 1980s this had become the conventional wisdom, with an accent on large-scale property development undertaken in partnership with the private sector. By the early 1990s, the value of these large projects was increasingly questioned with recognition of the need to invest in people as well as places. From 1997, the Labour government has experimented with many initiatives, though attention has now moved to the difficult but much needed task of creating a more coordinated and consistent set of programmes.

Inadequate housing: from clearance to renewal

Britain has a very large legacy of old housing which is inadequate by modern standards. This results from the relatively early start of the industrial revolution in this country and the rapid, unplanned and speculative urban development which took place in the nineteenth century. (The contrast with, for example, the Scandinavian countries, whose industrial revolutions came later when wealth was greater and standards higher, is marked.) As a result, British policies in relation to clearance and redevelopment are of long standing, though it was the Greenwood Housing Act 1930 which heralded the start of the modern slum clearance programme. Over a third of a million houses were demolished before the Second World War brought the programme to an abrupt halt.

By 1938, demolitions had reached the rate of 90,000 a year: had it not been for the war, over 1 million older houses would (at this rate) have been demolished by 1951. The war, however, not only delayed clearance programmes, but also resulted in enforced neglect and deterioration. War damage, shortage of building resources and (of increasing importance in the period of postwar inflation) crude rent restriction policies increased the problem of old and inadequate housing. It was not until the mid 1950s that clearance could generally be resumed, and well over 2 million slum houses have been demolished since then. But the problem is still one of large dimensions, as illustrated in Box 10.1.

BOX 10.1 STATE OF HOUSING IN THE UK

England

The housing stock

- There were 21.1 million homes in England in 2003 (and 20.5 million households); the number increases by about 160,000 per year, which is half the rate of growth in the mid 1960s; 80 per cent houses and 20 per cent flats; 80 per cent urban and 20 per cent rural locations.
- The average size for a home is 88 square metres (pre-1980) and 83 square metres (post-1980), but average living space is increasing because of smaller households; retired households tend to have the largest living space (58 square metres per person); ethnic minority households tend to have the smallest (22 square metres per person).
- 70 per cent of homes were owner occupied, 10 per cent private rented and 20 per cent social housing (13 per cent rented from local authorities and 7 per cent from registered social landlords); 3 per cent of homes were vacant (700,000 homes).

The standard of housing

- There were 5.3 million private and 1.4 million social sector houses non-decent homes (30 per cent and 35 per cent of their stock respectively); 900,000 were statutorily unfit for human habitation (4.2 per cent) and 69 per cent were in need of some repair.
- The number of non-decent homes fell from 9.4 million (46 per cent) in 1996 to 7.1 million (33 per cent) in 2001 to 6.7 million (31 per cent) in 2003, and in the eighty-eight most deprived areas from 1.4 million to 0.8 million between 1996 and 2003 (about two-thirds of the overall improvement).

- The cost of achieving a decent standard for all homes would be about £50 billion, on average £7,200 per dwelling; 40 per cent need less than £1,000 and 10 per cent need £20,000 or more spending.
- Between 1996 and 2001 there were 99,000 demolitions, with 10,000 in the year 1996–7 compared with 54,000 in 1974–5, and 90,000 in 1969.

Liveability

- 3.3 million (16 per cent) of all households occupy homes with liveability problems; those with poor quality environments are most concentrated in city and other urban centres.
- Poor quality environments are concentrated in the most deprived areas; 21 per cent of households in the eighty-eight most deprived districts have poor quality environments compared to 12 per cent elsewhere. These households are also more likely to live in non-decent homes (40 per cent) compared with 28 per cent in other areas.

Sources: ODPM (2003) *English House Condition Survey 2001*; ODPM (2005) *English House Condition Survey 2002–03: Key Findings*; ODPM (2005) *Survey of English Housing 2002–03*.

Note: The English House Condition Survey has been published every five years since 1967, but from 2001 became a 'continuous survey' with annual updates. The Survey of English Housing is published annually. They are both published by ODPM.

Northern Ireland (2001)

- There were 646,000 homes in Northern Ireland in 2001 (and 616,000 households); the number is increasing by about 9,000 per year; 92 per cent houses (24 per cent of total were 'bungalows') and 8 per cent flats; 67 per cent urban and 33 per cent rural locations (the proportion of stock in isolated rural areas is growing).
- 67 per cent of homes were owner occupied, 8 per cent private rented and 21 per cent social housing (18 per cent from the Housing Executive and 3 per cent housing association); 5 per cent of homes were vacant (and are excluded from tenure classes).
- 4.9 per cent of homes (31,600) do not meet the fitness standard (down from 7.3 per cent in 1996); 44 per cent of unfit dwellings were vacant.
- 33 per cent of homes needed urgent repairs and 59 per cent needed general repairs (down from 76 per cent in 1996); the total repair bill is £728 million for urgent repairs and £934 million for general repairs.

Source: Northern Ireland Housing Executive (2003) *Northern Ireland House Condition Survey 2001*

Scotland (2002)

- There were 2.2 million homes in Scotland in 2002 (and 2.2 million households); 62 per cent houses and 38 per cent flats, of which 60 per cent were in tenements (23 per cent of all stock); 84 per cent urban and 16 per cent rural locations.

- 62 per cent of homes were owner occupied (5 per cent more than in 1996), 8 per cent private rented and 30 per cent social housing (24 per cent from local authorities and the public sector, 6 per cent from housing associations or cooperatives); 4 per cent of homes were vacant (87,000).
- There were about 20,000 homes (just less than 1 per cent) that did not meet the 'tolerable standard'; 40,000 homes had inadequate kitchen provision; 7,000 lacked an adequate bathroom; 111,000 households did not have sufficient bedrooms.
- 21 per cent of the housing stock was in need of 'critical repair' (wind and weather proofing) and the total cost of bringing dwellings up to the tolerable standard, making improvements and fixing repairs is about £1.8 billion.

Source: Communities Scotland (2003) *Scottish House Condition Survey 2002*

Wales (2003 and 1998)

- There were 1.29 million occupied homes in Wales in 2003 (the surveys do not cover vacant dwellings or holiday homes); 91 per cent houses (35 per cent terraced houses) and 9 per cent flats; 81 per cent urban and 9 per cent rural locations.
- 74 per cent of homes were owner occupied (up from 71 per cent in 1996), 18 per cent social rented (13.6 from local authorities, 4.1 per cent from housing associations) and 9 per cent private rented.
- There were 98,000 homes (8.5 per cent) that did not meet the fitness standard in 1998 (down from 13.4 per cent in 1993); 38,300 did not have sufficient bedrooms for the household. The total repair cost was estimated at £1.1 billion in 1998.

Source: National Assembly for Wales (2004) *Welsh Housing Statistics* and (2001) *Welsh House Condition Survey 1998*

Note: Percentages may not add to 100 because of rounding.

Since the 1950s, wholesale clearance of housing has given way to 'renewal'. The emphasis gradually shifted from individual house improvements, first to the improvement of streets or areas of sub-standard housing, and later to the improvement of the total environment. Initially, it was assumed that houses could be neatly divided into two groups: according to the 1953 White Paper *Houses: The Next Step*, there were those which were unfit for human habitation and those which were essentially sound. As experience was gained, the improvement philosophy broadened, and it came to be realised that there was a very wide range of housing situations related not only to the presence or otherwise of plumbing facilities and the state of repair of individual houses, but also to location, the varying socio-economic character of different neighbourhoods and the nature of the local housing market. A house lacking amenities in Chelsea was, in important ways, different from an identical house in Rochdale: the appropriate action was similarly different. Later, it was better understood that appropriate action defined in housing market terms was not necessarily equally appropriate in social terms. A middle-class invasion might restore the physical fabric and raise the quality and character of a neighbourhood, but the social costs of this were borne largely by displaced low income families (or gentrification). The problem thus became redefined.

Growing concern for the environment also led to an increased awareness of the importance of the factors causing deterioration. It became clear that these are more numerous and complex than housing legislation had recognised. Through traffic and inadequate parking provision were quickly recognised as being of physical importance. The answer, in appropriately physical terms, was the rerouting of traffic, the closure of streets and the provision of parking spaces (together with floorscape treatments and the planting of trees). Most difficult of all is to assess the social function of an area, the needs it meets, and the ways in which conditions can be improved for (and in accordance with the wishes of) the inhabitants.

For a considerable time, this issue of the social function of areas was largely ignored, although a strong shift towards improvement rather than clearance was heralded by the Housing Act 1969. This increased grants for improvement, and introduced *general improvement areas* (GIAs), which were envisaged as being areas of between 300 and 800 'fundamentally sound houses capable of providing good living conditions for many years to come and unlikely to be affected by known redevelopment or major planning proposals'. The enhanced grants and the GIAs made a significant contribution to the reduction in the number of unfit properties, though in some areas gentrification unexpectedly took place, reducing the amount of privately rented housing, and affecting the existing communities (Wood 1991: 52).

The Housing Act 1974 made a major reorientation of policy and brought social considerations to the fore. There was a new emphasis on comprehensive area-based strategies implementing a policy of 'gradual renewal'. The powers (and duties) conferred by the Act focused upon areas of particular housing stress. Local housing authorities were required to consider the need for dealing with these as *housing action areas* (HAAs). Though these were conceived in terms of housing conditions, particular importance was attached to 'the concentration in the area of households likely to have special housing problems – for instance, old-age pensioners, large families, single-parent families, of families whose head is unemployed or in a low income group' (DoE Circular 13/75).

The intention was that intense activity in HAAs would significantly improve housing conditions and the well-being of the communities within a period of about five years. In the event, HAA designation lasted much longer in many cities. Various additional powers were made available to local authorities within housing action areas, for compulsory purchase, renewal, and environmental improvement; grant aid for renewal was targeted to these areas.

Housing renewal areas

The area-based approach to private sector housing renewal was retained, although substantially altered, by the Local Government and Housing Act 1989. In addition to individual income-related house renovation grants, the Act introduced *renewal areas* (RAs), which replaced GIAs and HAAs. There are also powers for local authority support for *group repair schemes* to renovate the exteriors of blocks of houses.

The thrust of the changes reflects a concern for a broader strategic approach including economic and social regeneration as well as housing renewal over a longer ten-year period of designation. It involved the resumption of clearance; the use of partnerships to bring together the initiatives of local authorities, housing associations, property owners, and residents; and a system of grants which were mainly both mandatory and income related.

A *neighbourhood renewal assessment* is a central part of the renewal area concept, and has to precede designation. This is, in effect, a plan-making and implementation programme combined, including an assessment of conditions, an estimate of the resources available, and the selection of the preferred options. The procedure goes much further than the typical land use planning process to incorporate a cost–benefit analysis of different alternative policies over a thirty-year period, including the qualitative social and environmental implications. Consultation is a requirement both during and after the declaration process, with twenty-eight days given for responses to an explanatory summary of the proposals which must be delivered to every address in the area. An interesting requirement is that all who make representations which are not accepted must be provided with a written explanation.

Renewal areas are larger than the former HAAs and GIA, with at least 300 properties as a minimum. More than 75 per cent of the dwellings must be privately owned, and at least 75 per cent must be considered unfit or qualify for grants. At least 30 per cent of the households must be in receipt of specified state benefits, thus ensuring that 'a significant proportion of residents in an RA should not be able to afford the cost of the works to their properties'.

By 1999 more than a hundred neighbourhood renewal assessments had been undertaken and renewal areas designated. C. Wood (1996) provides an overview of the programme, and a further evaluation was published by the DoE in 1997 (Austin Mayhead 1997).[1] The average size of RA was 1,526, or about twice the size of GIAs and HAAs. Wood notes that this is slower progress than might have been expected and points to the variation in enthusiasm across the country. He provides three case studies of RA designation in Birmingham, a city that at that time had 47,000 unfit privately owned dwellings and another 80,000 considered borderline. The city council has estimated that the total bill for the repair and improvement of the private housing stock in the city would be £750 million. It is perhaps not surprising therefore that Wood concludes that 6,400 properties designated in RAs in Birmingham will remain in the same condition as they started (or have further deteriorated) at the end of the ten-year plan period.

The difference between the scale of the problem and funding available was one reason why authorities were not generally enthusiastic about the RA concept, as the 1997 evaluation points out. This was more than just a negative perception. The evaluation considered ten case study RAs in detail and only two had made a significant impact on the full range of objectives, especially in gaining private sector investment. Nevertheless, the conclusion was that the neighbourhood renewal assessment process is sound, and the partnership elements in particular had an impact on generating successful solutions. The government provides a *Neighbourhood Renewal Assessment Guidance Manual* (2004).[2] More

emphasis has been put on the partnership dimension and links to the local strategic partnership; the need for sustainability assessment; forward strategies to ensure that improvement is sustained; and understanding of the wider context and forces shaping the housing market in the region and sub-region. This latter point reflects concerns that in some places the number of renewed properties has not been matched by demand (considered in Chapter 6).

Action in renewal areas is supported by an improvement grants regime that has changed substantially over the fifteen years of HRA operation to date. From 2002 local authorities are given much more discretion in the awarding of improvement grants subject to having an explicit policy in place which has been agreed with key stakeholders. Mandatory grants were available for homes that failed the statutory fitness standard until 1996, with discretionary grants on such matters as adaptations to enable elderly people or people with disabilities to stay in their homes. Mandatory status led to a big backlog of applications in some authorities. Changes were made in 1996 to focus grants more on the poorest households and to give more discretion to the local authority to pay above the minimum rates for alterations for people with disabilities and those moving from clearance areas.[3]

Public sector homes: estate action

In 1979, the DoE set up a *priority estates project* to explore ways in which problem council estates could be improved. The problems of these estates varied, but all had become neglected and run down; some had been vandalised. A 1981 report *Priority Estates Project 1981: Improving Problem Council Estates* considered three experiments in the improvement of such estates, and concluded that the task of improvement involved a great deal more than mere physical renovation: social and economic problems needed to be addressed at the same time. In 1985, the DoE established an *Urban Renewal Unit* later called *Estate Action* to encourage and assist local authorities to develop a range of measures to revitalise run-down estates. Measures included transfers of ownership and/or management to tenants' cooperatives or *management trusts* involving tenants; sales of tenanted estates to private trusts or developers; and sales of empty property to developers for refurbishment for sale or rent. Estate Action funding was allocated on a competitive basis and required participation of tenants and consideration of the use of right to buy and involvement of the private sector. It began the trend to a broader urban regeneration approach to estate renewal rather than simply housing improvement.

A research report evaluating six early Estate Action Schemes (Capita Management Consultancy 1996) evaluated the achievement of Estate Action objectives: improving the quality of life of residents, bringing empty properties quickly back into use, reducing levels of crime and 'incivility', diversifying tenure and attracting private investment. By then, £2 billion had been spent over eight years but with disappointing results. Indeed, the report makes very depressing reading, although physical change had been made, this did not lead to the expected (or hoped for) social and economic improvements and there were difficulties in involving residents in the development and implementation of the schemes. More generally, the report demonstrates the difficulty of using physical plans for achieving social and economic goals. Nevertheless Estate Action marked a change in the way government funded housing renewal by introducing competition for funding around government priorities, establishing mixed tenure estates as a shared policy goal, and widening the focus from housing improvement to wider social and economic change (Kintrea and Morgan 2005). The last schemes were approved in 1995 and in total Estate Action cost nearly £3 billion on 317 schemes involving the improvement of 490,000 dwellings and the transfer of 93,000 dwellings.[4]

Housing action trusts

The 1987 White Paper *Housing: The Government's Proposals* announced the creation of *housing action trusts* (HATs) to tackle the management and renewal of badly run-down housing estates. HATs were to be the housing equivalent of the urban development corporations, but were introduced only with the consent of a majority of the tenants. Thus, like the urban development corporations, HATs were non-departmental public bodies responsible directly to the Secretary of State, thus usurping the powers of the local authority, but with substantial additional investment for improving the physical, social and economic conditions of estates. Funding was allocated for HATs as early as 1988, but there was considerable delay in getting the first ones started because of fierce opposition by affected local authorities and tenants, because of the loss of control to the private sector, and fears of reduced availability of social housing when some was sold to private owners after improvement (Rao 1990). Only six HATs were designated, but almost £1 billion was spent together with PFI funding up to 2004. Two HATs, Birmingham and Liverpool, completed in 2005, Brent (Stonebridge) will complete in 2007. The Birmingham HAT at the Castle Vale Estate will have run for twelve years and has invested £270 million in the construction or improvement of 3,400 homes (including demolition of a number of tower blocks). It levered in £93 million private investment (with the ubiquitous Sainsbury store) and unemployment on the estate has fallen from 26 per cent to less than 5 per cent.[5]

An evaluation (Capita Management Consultancy 1997) reported that the HAT designation was transforming the physical structure of estates, with substantial replacement or reworking of postwar system built, high-rise blocks and their complex walkways. They also changed the social make-up in some cases with very substantial transfers and development of housing for owner occupation, with some impact on the negative stereotyping. Low demand for housing on such estates continued to be a problem and has become the current focus for research and policy for problem estates (Niner 1999). Low demand resulting in high vacancies and rapid turnover affects all parts of the country and all tenures but is particularly acute in council estates in the North, where local authorities are demolishing homes that are impossible to rent or sell. Low demand arising from unpopularity is more important than need in determining the use of housing (Power and Mumford 1999; Burrows and Rhodes

1998). How an estate gets into this position is explained more by the severe poverty and unemployment than the quality of housing, although the original break-up of stable communities through the slum clearance programme may be the origin of the problem.

Decent homes for all

The *English House Condition Survey 2002–03* found that 900,000 dwellings (4.2 per cent of the dwelling stock) were still unfit for human habitation according to the statutory fitness standard (see Box 10.2 for definitions). This was down from 1.5 million (7.4 per cent of the stock) in 1996. The sharp improvement from 1996 followed little progress during the 1980s and 1990s in improving the overall quality of the housing stock (despite the building of some 650,000 new homes between 1991 and 1996). Relatively stable economic conditions, low interest rates and the attractiveness of housing as an investment in the wake of the dot.com crash of 2000 have played a part in the turn around of housing conditions. The government would also point to housing policy and improved financing of local government and spending on housing. Nevertheless, in addition to the 900,000 unfit dwellings (13 per cent

of the stock), 7 million do not meet the broader standard of a 'decent home', 2 million (27 per cent of all non-decent dwellings) are in disrepair, and 500 thousand (7 per cent) require modernisation (*English House Condition Survey* p. 5). The most common problem is providing a 'reasonable degree of thermal comfort' with 5.6 million homes (80 per cent) failing on this criterion, which requires relatively little expenditure. It is also the most vulnerable that suffer poor housing conditions – poverty, poor housing and poor neighbourhoods go together – and the worst housing is associated with higher proportions of ethnic minority, young and unemployed people and lone parents.

The Survey also considers the extent to which housing is located in 'decent places' and concludes that 11 per cent of the housing stock (2.4 million homes) is in poor neighbourhoods, most of which are either private sector terraced housing near city centres or council housing estates in the suburbs. Just over half the housing in these neighbourhoods is rated non-decent. Heavy traffic and parking affects 2.5 million homes, and half a million homes are affected by 'vandalism, graffiti and other forms of anti-social behaviour' and/or concentrations of vacant and boarded up buildings.

BOX 10.2 DECENT HOMES STANDARD

The government aims to bring all social housing up to the decent homes standard by 2010 and aims to get 70 per cent of vulnerable households (those in receipt of specified means-tested benefits or tax credits) in private sector housing, whether rented or owner occupied in decent homes by 2010.

A decent home is one that

- meets the current statutory minimum standard for housing (the current *fitness standard*, this is the statutory measure)*
- is in a *reasonable state of repair*
- has reasonably *modern facilities and services*
- provides a *reasonable degree of thermal comfort*.

The current statutory minimum fitness standard was introduced by the Housing Act 2004 – the *Housing, Health and Safety Rating System*. This is not a standard but an evaluation framework of twenty-nine categories of 'housing hazard' which are to be assessed for their impact on *actual* or *potential* residents. The hazards include, for example, excess heat and cold, crowding and space, water supply and ergonomics. The extent of a hazard will depend on the type of occupant – for example, stairs will present more of a hazard for elderly people.

The previous fitness standard defined fitness according to the condition of the housing only – in terms of serious disrepair, structural stability, dampness prejudicial to health, and the availability of basic services and facilities. These requirements were a minimum standard and if one was not met the house was not considered to be fit for human habitation. In Scotland the Housing (Scotland) Act 1969 introduced the concept of a tolerable standard, which differs in detail from the fitness standard in England and Wales.

A reasonable state of repair means that the dwelling does not have one or more key building components that are old and, because of their condition, need replacing or major repair; or two or more other building components are old and need replacing or major repair. Key components include walls, roofs, windows, doors, chimneys, electrics and heating systems.

- a kitchen which is twenty years old or less, and with adequate space and layout
- a bathroom and WC which is thirty years old or less and which is appropriately located
- adequate external noise insulation
- adequate size and layout of common entrance areas for blocks of flats.

A reasonable degree of thermal comfort means efficient heating (programmable central heating or electric storage heaters) or similar, and effective insulation, which varies according to type of heating (e.g. from 50 mm to 200 mm loft insulation).

Note: * The fitness standard was made by the Local Government and Housing Act 1989 amendments to the Housing Act 1985.

'Decent homes' is a simple and captivating idea that has become very complex in execution. The Housing Green Paper 2000 began a process of updating and broadening the standards applied to test the fitness of housing by introducing the *decent homes standard*. A policy statement later the same year committed the government to bringing all social housing up to this standard by 2010.[6] Box 10.2 gives a summary of the decent homes standard and its four components – the statutory fitness standard, state of repair, facilities and 'thermal comfort'. The statutory fitness standard was earmarked for radical change in the same Green Paper and the Housing Act has duly brought changes into play. The new system known as the housing, health and safety ratings system (HHSRS) is not a standard, but an assessment of the hazards that homes present for the health and safety of residents. The Act also provides new powers and discretion for local authorities to intervene on the basis of the hazard assessment in the most appropriate way.

Evidence given to the ODPM Select Committee Inquiry into the decent homes initiative confirmed that the HHSRS is likely to be more complex, difficult to record, open to considerable interpretation and, in the case of required facilities, too prescriptive.[7] It concluded that the standard will inevitably be more expensive to operate and require new skills of housing officers. The Committee also pointed out that aspects of the system provide a surprisingly good example of lack of joined-up government. The new 'thermal comfort' criterion does not complement the Fuel Poverty Strategy (FPS) and its methods of assessment, or the Energy Efficiency Commitment (EEC). Introducing the thermal efficiency of homes into the standard caused many homes to be 'non-decent', though they could be brought up to standard quite cheaply through the installation of minimum levels of insulation. The government rebuttal was that it is this standard that triggers action and improvements would be at a higher level. Nevertheless, improved homes may not meet the FPS and may also require more work to contribute to the EEC. The Committee conclude that 'there is little cross-departmental joined-up thinking, leaving disparate policies in need of alignment and integration in order to maximise the benefit of resources spent' (p. 16), and that generally the decent homes standard is set too low. There is also a suspicion that the government is using the standard to push more social housing out of local government control, since in order to meet targets, local authorities will have to consider arm's length management, transferring to other providers and PFI schemes.[8]

Scottish housing

Scottish housing is different from that south of the Border in significant ways. There is a high proportion of tenement properties, dwellings tend to be smaller, rents are lower and a higher proportion of the housing stock is owned by public authorities. These and other differences reflect history, economic growth and decline, local building materials, and climate. Above all, Scotland has for long faced a major problem of poor quality tenement housing. Despite the large amount of clearance in the postwar years, there still remains much poor quality housing in both the private and the public sectors.

The 2002 *Scottish House Condition Survey* suggests that, because of the dominance and relatively young age of much of the public housing stock, a high proportion of housing has the basic amenities, but the condition is often poor, and requires high levels of expenditure. Historically low council rents contributed to the relatively low demand for private housing which, coupled with massive public house building programmes, gave Scotland the highest proportion of public sector housing in Western Europe – and made Glasgow City Council the largest public sector landlord (McCrone 1991). (It should be noted that the Scots use the term 'house' in the English sense of 'dwelling', i.e. it embraces a flat or tenement.) The Right to Buy has shifted the balance somewhat between the owner occupied and public housing sectors: owner occupation rose from 35 per cent in 1979 to 57 per cent in 1995 and by 2002 reached 62 per cent; while the social rented sector fell from 54 per cent to 30 per cent. Nevertheless, in spite of the sale of nearly a quarter of the public housing stock, the level of public renting in Scotland is still much higher than in England and Wales.

The Conservative government believed that the large-scale public ownership of housing was at the root of much of the Scottish housing problem. Its strong desire to reduce the public housing sector (and particularly to break up the public ownership of the large peripheral estates) was an important background issue in Scottish housing policy for many years. Equally distinctive has been the institutional context of Scottish housing policy. The relationship between the Scottish Office and Scottish local authorities has always been much closer than in England, and there has been much easier coordination. This has resulted in what Carley (1990a: 51) describes as 'a much more clearly defined and integrated housing-neighbourhood renewal policy, which covers housing and planning issues together'. The Scottish Office has tended to work with local authorities (which have a single local authority association), rather than exerting central control through such mechanisms as UDCs and HATs. Instead, there have been centrally sponsored bodies which have worked in cooperation with local government.

The integrated approach to housing and community or neighbourhood development is evident in the form and goals of the housing body, *Communities Scotland*, which replaced Scottish Homes (which incorporated the Housing Corporation in Scotland) in 2001. Communities Scotland brings together responsibilities for leading urban regeneration and housing improvement with tackling poverty and disadvantage, increasing the supply of housing and affordable housing, Fuel Poverty Strategy, the regulation of social landlords and research on housing such as *Scotland's House Condition Survey*. It works with local authorities and other providers especially in the production of *local housing strategies* and providing funding for their implementation.

As in England and Wales policy in relation to the older private housing stock placed more emphasis on rehabilitation than on clearance. But, with much of the older tenement properties, the scope for improvement is severely restricted by the decayed fabric of the buildings, their internal layout and the high cost of alteration, as well as the practical problems of multiple ownership. Some of these difficulties have been met by the use of powers of compulsory improvement of a whole tenement structure, and by the establishment of ad-hoc housing associations. However, the Scottish legislation has long provided for more flexibility than the English. *Housing action areas* were not superseded in the same way as in England, since they were more flexible tools which required involvement of residents in housing renewal.

In 2001 the Housing (Scotland) Act introduced a requirement for all local authorities to prepare a local housing strategy (LHS) including an action plan showing how it will be implemented. The strategies effectively replace the old housing plans and provide much broader assessment of housing markets and the condition of housing. At the time, Scotland's only benchmark was the 'tolerable standard' below which houses should be condemned. Relatively few houses failed this standard, so the Scottish Executive in 2004 adopted the broader Scottish Housing Quality Standard (SHQS) with a target that all social housing should meet the standard by 2015. All authorities have now prepared a strategy and additional 'standard delivery plans' (SDPs) were required to show how the SHQS target would be met.

Emphasis on area policies

Several streams of thinking are apparent in policy on deprived areas in Britain: inadequate physical conditions, the perception of 'large' numbers of immigrants (many of whom were born in Britain), educational disadvantage, and a multiplicity of less easily measurable social problems concentrated in particular areas. For a very long time, there was a preoccupation in policy with inadequate physical conditions particularly in relation to plumbing). Indeed, British housing policy developed from sanitary policy (Bowley 1945), and it still remains a significant feature. Area policy in relation to housing was almost entirely restricted to slum clearance until the late 1960s, when concepts of housing improvement widened, first to the improvement of areas of housing and then to environmental improvement. Despite a number of social surveys, the policy was unashamedly physical: so much so that increasing powers were provided to *compel* reluctant owners and tenants to have improvements carried out. Not until the 1970s was attention focused on the social character and function of areas of old housing. A series of reports made recommendations relating to the recognition of specific areas of housing stress and social and economic disadvantage.

The Milner Holland Committee (1965) looked favourably on the idea of designating the worst areas as *areas of special control* in which there would be wide powers to control sales and lettings, to acquire, demolish and rebuild property, and to make grants. The National Committee for Commonwealth Immigrants (NCCI 1967) argued for the designation of *areas of special housing need* to control overcrowding, insanitary conditions and the risk of fire. These proposals were not accepted by the government, though increased powers to control multi-occupation and abuses were provided. The Plowden Committee (1967) reported on primary education in very broad terms, but underlined the complex web of factors which produced seriously disadvantaged areas. It recommended positive discrimination for designated schools in the most deprived areas. The Seebohm Committee (1968) reported on personal and family social services and recommended designation of *areas of special need* to be given priority in the allocation of resources.

More helpful was its reference to citizen participation, which underlined a point hardly recognised by the Skeffington Committee (1969) even though it was specifically concerned with it. It was Seebohm, not Skeffington, who clearly saw that, if area action was to be based on the wishes of the inhabitants and carried out with their participation, 'the participants may wish to pursue policies directly at variance with the ideas of the local authorities . . . Participation provides a means by which further consumer control may be exercised over professional and bureaucratic power'.

This is an issue which will be discussed in a broader context in Chapter 12. Here, we briefly survey some of the ways in which the development of thinking on deprived areas has been translated into policy from the early 1970s to the Blair government.

Urban programme

Area policies in relation to housing improvement, however inadequate they may have been, were based on long experience of dealing with slum clearance and redevelopment. With other area policies there was no such base upon which to build, and both legislation and practice were hesitant and experimental. The approach, however, has remained consistently a spatial one, focusing on particular cities and areas within cities.

The *educational priority areas programme* was established in 1966 (Halsey 1972), and the *urban aid programme* in 1972. The *urban aid programme* (later recast as the *urban programme*) funded mainly social schemes, but it was progressively widened in scope to embrace voluntary organisations, and to cover industrial, environmental and recreational provision. The 1977 White Paper *Policy for the Inner Cities* and the Inner Urban Areas Act 1978 brought about significant changes. The new policy was 'to give additional powers to local authorities with severe inner area problems so that they may participate more effectively in the economic development of their areas'.

It is difficult to give a coherent account of this programme since its objectives were never clearly spelled out, and its extreme flexibility gave rise to a great deal of confusion (McBride 1973). With the development of inner city partnerships and programme authorities, the position became even more confused. By 1990, the number of individual programmes had swollen to thirty-four (National Audit Office, *Regenerating the Inner Cities*, 1990). The urban programme was the major plank of the 'deprived area policy' stage. Additionally there were the community development projects. These produced a veritable spate of publications ranging from carefully researched analyses to neo-Marxist denunciations of the basic structural weaknesses of capitalist society, though the original aim was 'to overcome the sense of disintegration and depersonalisation felt by residents of deprived areas'.[9]

The urban programme continued with an increasing emphasis on economic development. It became a valuable source of funding for the many thousands of projects and organisations that have been supported. At its height, about 10,000 projects were funded each year in the fifty-seven programme areas, costing £236 million in 1992–3. In its later years, almost half the expenditure was devoted to economic objectives, and the rest was shared roughly equally between social and environmental objectives. Urban programme funding was largely taken over by the Single Regeneration Budget (SRB) operated from the Government Offices for the Regions from 1995.

Policy for the inner cities and Action for Cities

The increasing emphasis on economic regeneration objectives came to dominate urban policy during the 1980s and 1990s. The return of the Conservative government led to a review of inner city policy, which concluded that a much greater emphasis needed to be placed on the potential contribution of the private sector. Ten years later the rhetoric was much the same. After the 1987 election, Margaret Thatcher, then Prime Minister, announced her intention to 'do something about those inner cities'. The immediate result was the publication of a glossy brochure entitled

Action for Cities (Cabinet Office 1988). This maintained that, though the UK had benefited during the 1980s by embracing the ethic of enterprise, this change in attitude had not reached into the inner city. The aim of urban policy therefore was to establish 'a permanent climate of enterprise in the inner cities, led by industry and commerce'. *Action for Cities* programmes were claimed to have involved central government expenditure of £3 billion in 1988–9 and £4 billion in 1990–1 but, in fact, little additional money was involved. Critics argued that the package merely gave the appearance that a determined effort was being made to get to grips with the problems. However debatable this might be (Lawless 1989: 155), it did indicate some reorientation of thinking on urban policy.

Following Michael Heseltine's 1981 visit to Merseyside, in the wake of the Toxteth riots, the Merseyside Task Force Initiative was created. Initially, this was a task force of officials from the DoE and the then Department of Industry and Employment, established to work with local government and the private sector to find ways of strengthening the economy and improving the environment in Merseyside. This proved exceptionally difficult partly because of the multiplicity of agencies involved.

As a result, an attempt was made in later initiatives to obtain a greater degree of coordination through City Action Teams and Inner City Task Forces. First set up in 1985, City Action Teams (CATs) were to take a broader, even regional, view of the coordination of government programmes. Each team was chaired by the regional director of one of the main departments involved: the DoE, the DTI and the Training Agency.[10] Their funding was limited, reflecting their role as coordinators rather than direct providers. Lawless (1989: 61) argues that they were 'unable to devise anything that might be termed a corporate central-government strategy towards inner-city areas'. They were disbanded at the time of the setting up of the Government Offices for the Regions, which took on the coordinating role.

A total of sixteen Inner City Task Forces were established in 1986–7. Their role was essentially one of trying to bend existing programmes and private sector investment and priorities to the inner city. They

were initially expected to have a life of two years, but in the event the lifespan has been variable. Their general objective was to increase the effectiveness of central government programmes in meeting the needs of the local communities.

One of Michael Heseltine's early initiatives was the establishment of the *Financial Institutions Group* under his leadership and staffed by twenty-five secondees from the private sector. Their most important proposal was for an *urban development grant* (UDG) on the lines of the American *urban development action grant*. This was introduced in 1982 'to promote the economic and physical regeneration of inner urban areas by levering private sector investment into such areas'. It was flexible in terms of the area covered, local authorities contributed 25 per cent of grant aid and the private sector contribution to a project had to be significant. It was replaced by the *urban regeneration grant* in 1986. In its lifetime, it supported 296 projects at a cost of £136 million with a corresponding private sector investment of £555 million. This represented a leverage ratio of about 1:4.

The urban regeneration grant supported ten schemes at a cost of £46.5 million, with private sector investment of £208 million. Together, they provided and estimated 31,966 jobs, 6,750 new homes and 1,456 acres of land brought back into use (Brunivels and Rodrigues 1989: 66). *City grant* was launched by the *Action for Cities* initiative in 1988, and replaced several existing grants including the urban development grant. Its aims were similar but it was paid directly to the private sector developers to support the provision of new or converted property for industrial, commercial or housing development. The final evaluation of these grant regimes noted that they were successful in assisting private sector investment but the focus on job creation in evaluation did not 'address the basic causes of poor regional or local growth' (Price Waterhouse 1993: 63). (See Table 10.1.)

Urban development corporations

The 1977 White Paper on inner cities considered the idea of using new town style development corporations

Table 10.1 Selected regeneration and inner city expenditure and plans 1987–8 to 2001–2 (£m)

	1987–8	1988–9	1989–90	1990–1	1991–2	1992–3	1993–4	1994–5	1995–6	1996–7	1997–8	1998–9	1999–2000	2000–1	2001–2
UDCs and DLR	160.2	255	476.7	607.2	601.8	515	341.2	287.1	217.4	196.1	168.8	0	0.2		
Estates Action	75	140	190	180	267.5	348	357.4	372.6	315.9	251.6	173.5	95.7	66.9	63.9	39.4
Housing Action Trusts					10.1	26.5	78.1	92	92.5	89.7	88.3	90.2	86.4	88.4	88.4
City Challenge						72.6	240	209.0	204.9	207.2	142	9.8	1.7		
Challenge Fund									125	265.1	483.1				
New Deal for Communities												0.2	48.5	120.7	450.0
City Grant/Derelict Land Grant[a]	103.5	95.7	73.5	100.6	113.8	145.0	128.7								
English Partnerships (URA)[b]				16.8	–16.4	6.5	24.2	191.7	211.1	224.0	258.8	294.2	225.5	212.2	78.2
Urban Programme[c]	245.7	224.3	222.7	225.8	237.5	236.2	166.5	67.8							
City Action Teams			4	7.7	8.4	4.6	3.4	0.2							
Inner City Task Forces	5.2	22.9	19.9	20.9	20.5	23.6	18	15.4	11.9	8.7	6.0	1.6	1.2		
Single Regeneration Budget[d]									136.4	277.5	458.8	560.9	181.4		
Regional development agencies[e]												12.2	593.0	628.5	820.0
London Development Agency														241.8	228.5
Manchester Regeneration Fund/Olympic bid/city centre				0.8	12.2	26.8	30.2	2	0.3	1.1	5.4	1.7			
Coalfield Areas Fund							2.3	2	0.4	0	0	0	10.0	15.0	10.0

Sources: DoE *Annual Report 1993* (Figure 45), DoE *Annual Report 1995* (Figure 43), DoE *Annual Report 1996* (Figure 31, p. 50) and DETR *Annual Report 2000* (Table 10a)

Notes:

a City Grant and Derelict Land Grant became part of English Partnerships funding from November 1993 and April 1994 respectively

b Regional funding for English Partnerships transferred to the regional development agencies from April 2000

c Urban Programme funding became Single Regeneration Budget from 1995–6

d Single Regeneration Budget transferred to regional development agencies from April 1999, and to the London Development Agency from April 2000

e Regional development agencies' funding also includes rural development programmes funding from April 1999

to tackle inner areas, but the then Labour government concluded that it was inappropriate for the inner cities, and that local government should be the prime agency of regeneration. The Conservative government from 1979 thought differently, mainly because it had little faith in the capabilities of local government. The manifest argument, however, was that the regeneration of major areas of our cities was 'in the national interest, effectively defining a broader community who would benefit from the regeneration' (Oc and Tiesdell 1991: 313). The effect of imposing centrally directed agencies into the hearts of major cities created considerable conflict (although not all the local authorities were against the designations), and long lasting bad feeling about the role of central government in urban regeneration.

The Local Government, Planning and Land Act 1980 made the necessary legislative provisions, and defined the role of an urban development corporation as being:

> to secure the regeneration of its area . . . by bringing land and buildings into effective use, encouraging the development of existing and new industry and commerce, creating an attractive environment, and ensuring that housing and social facilities are available to encourage people to live and work in the area.

Though their structure and powers were based on the experience of the new town development corporations, the UDCs were different in several important respects. Their task was a limited one and they had relatively short lives, of ten years or so. The first were designated in 1981 and three subsequent 'generations' followed, the last in 1993. All four generations of UDCS had wound up operations by 1998, but a 'fifth generation' is now planned to help drive forward the implementation of the *Sustainable Communities Plan* and its growth areas in the South East (see Table 10.2). (Chapter 6 discusses the *Sustainable Communities Plan*.) The designation procedure is rapid, being made by statutory instrument. Fourteen UDCs were created initially, twelve in England, one in Northern Ireland (Laganside) and one in Wales (Cardiff Bay).[11] Except

for the London Docklands, their areas were not large, though they suffered especially severe derelict land or plant closure problems (Bovaird 1992). The UDCs had extraordinary powers of land acquisition and vesting of public sector land. They also (unlike the new town development corporations) usurped the local authority's development control functions (except in Wales), for determining planning applications (including their own proposals), enforcement and other matters. In short, they had very wide planning responsibilities and freedom from local authority controls. This was not accidental or incidental: it was an essential feature of their conception. Furthermore, they were run by Boards of Directors drawn primarily from business, and they were accountable only to the central government.

Expenditure on UDCs rose to a peak of over £600 million in 1990–1, though in most years stood around £200 million. At one time, this was by far the largest share of spending on regeneration: in 1992–3 it amounted to half of all inner city spending, reflecting the significant shift away from local authority directed expenditure. The major share of public funding, and similarly high levels of private investment, went to London Docklands, and within that area to major projects. Just one project, the Limehouse Link road, required the demolition of over 450 dwellings and cost £150,000 per metre, the most expensive stretch of road in the UK (Brownill 1990: 139).

Final evaluation of the first urban development corporations confirms that property development, land reclamation and physical improvement dominated their objectives and outputs. They made a significant impact on the geography of our cities through the transformation of largely derelict and degraded land created largely through massive industrial restructuring. Powers of land acquisition, site assembly, reclamation, and financial incentives were important tools. But they have not met any wider objectives. Furthermore, their attention rarely strayed from the designated areas to considerations about the wider economies in which they were situated, a problem exacerbated by narrow performance indicators. Most UDCs could quote impressive figures (Tym *et al.* 1998) but it has been argued that most of the new activity that was generated would otherwise have occurred

Table 10.2 *Urban development corporations in England: designation, expenditure and outputs*

			Annual gross expenditure (£m)		Cumulative outputs		
			1992–3	1995–6	Jobs created	Land reclaimed (ha)	Private investment (£m)
First generation	London Docklands	1981	293.9	129.9	63,025	709	6,084
	Merseyside	1981	42.1	34	14,458	342	394
Second generation	Trafford Park	1987	61.3	29.8	16,197	142	915
	Black Country	1987	68	43.5	13,357	256	690
	Teesside	1987	34.5	52.8	7,682	356	837
	Tyne and Wear	1987	50.2	44.9	19,649	456	758
Third generation	Central Manchester	1988	20.5	15.5	4,909	33	345
	Leeds	1988	9.6	0	9,066	68	357
	Sheffield	1988	15.9	16.4	11,342	235	553
	Bristol	1989	20.4	8.7	4,250	56	200
Fourth generation	Birmingham Heartlands	1992	5	12.2	1,773	54	107
	Plymouth	1993	0	10.5	8	6	0
Total			621.4	398.2	165,716	2,713	11,240

The new urban development corporations

			Proposed end date	Comments
Fifth generation: sustainable communities	Thurrock	2003	2010	Same boundary of LB of Thurrock
	London Thames Gateway	2004	2014	Two separate areas in six London Boroughs
	West Northamptonshire	*2004	2016	Three separate areas around the towns of Northampton, Daventry and Towcester
	Milton Keynes UDA	*2004	2016	Designated UDA but managed by a Partnership Board of the local authority, English Partnerships and others

Sources: DoE *Annual Reports 1993, 1995* and *1996*

Notes: Leeds UDC was wound up on 31 March 1995. Bristol UDC was wound up on 31 December 1995. The figures for outputs are cumulative over the lifetime of the UDCs and include all activity within the urban development area, not just those of the UDC. Designation orders proposed at time of writing.

elsewhere. Overall, the most important legacy of the 'first round' UDCs is perhaps the lesson that sustained regeneration can be achieved only by the combined efforts of all agencies and stakeholders working to the same objectives.

The partnership approach figures more prominently in the latest round of UDCs – a fifth generation. In 2003, the ODPM began consultation on three new urban development areas and UDCs to support the urban growth areas around London. Thurrock UDC was designated in 2003 and was followed by London Thames Gateway and West Northamptonshire in 2004. The contentious issue of relations with local authorities and other stakeholders has been handled more sensitively than some of the earlier designations. Local authorities will be represented on the boards and the UDCs will be the local planning authority only for 'applications directly relevant to its purposes (which are defined as strategic and significant)'. The consultation generally returns support for the proposals, including the local authorities accepting the loss of DC powers for strategic developments, while householder and minor applications will remain with the councils. Development planning powers stay with the councils although the UDCs will prepare their own development strategies that, the minister has said, will have to take account of the development plan. An agency agreement will be agreed between the UDC and local authorities, whereby the local authority may help deliver the UDC's planning functions (as happened in some of the earlier UDCs). The lifetime of the UDCs varies in a seven to ten year band, with planned dates for reviews to consider progress.

In the Milton Keynes area a unique solution has been found in the designation of an urban development area under powers of the Urban Regeneration Agency (English Partnerships, see p. 367) but instead of a corresponding UDC, planning in the area will be controlled by a partnership committee made up of ten members from the local authority, English Partnerships and other partners. In this case the 'strategic developments' that will be business of the partnership have been defined as 'those involving 10 or more new dwellings, office, industrial and retail developments involving 1,000 square metres or more of floor space and any development involving a site area of 1 hectare or more'.[12]

This new generation of UDCs will also add yet another institution to a more complex web of regional and sub-regional governance, and in particular will need to work alongside the regional development agencies. Inter-regional boards are proposed to bring together the many agencies involved in regeneration.

City Challenge

A major switch in regeneration funding mechanisms was announced in May 1991, in the form of City Challenge. This marked a significant change in policy, and one that later evaluations regarded as 'the most promising regeneration scheme so far attempted' (Russell et al. 1996). Thus it has provided the model for its successors, in particular the Single Regeneration Budget Challenge Fund.

Though the emphasis on land and property development remained, City Challenge recognised that this should be more closely linked to the needs of local communities and the provision of opportunities for disadvantaged residents. It was also intended to encourage a long-term perspective on change, and to integrate the work of different programmes and agencies.

Only fifteen authorities were invited to bid for funds in the first round, eleven of which were selected to start on their five-year programmes in 1992.[13] Subsequently, a second round was opened to all urban programme authorities and a further sixteen five-year programmes were approved in 1993. Most City Challenge partnership programmes were in inner city locations, but a few were on the urban fringe. The Dearne Valley was unique in being 52 sq. km in area, covering a number of smaller settlements in the South Yorkshire conurbation. City Challenge encouraged an integrated approach, with a focus on property development but cutting across a range of topic areas, including economic development, housing, training, environmental improvements, and social programmes relating to such matters as crime, and equal opportunities.

The competition prize was substantial: £7.5 million for each area for each of five years' funding which together with levered-in funding meant that partnerships have spent on average £240 million. The total central government expenditure amounted to £1.15 billion over eight years. This was not new money. City Challenge was a different approach to spending rather than an allocation of new funds: it was to be large scale, holistic, strategic and based on partnership.

Successful authorities prepared an action programme setting out projects which were funded through existing programmes, with similar, if simplified, procedures for making and considering applications. The private sector was expected to play a significant role, and its involvement had to be demonstrated before projects were agreed.

Local communities were seen to be important partners, but their place in the management structures was variable, sometimes involved in management, sometimes only as consultees (Bell 1993). Nevertheless, City Challenge gave impetus to bring together different sectors and the creation of new and positive relationships between them, shifting attitudes and mainstream investment (de Groot 1992). City Challenge was an undoubted success and provided a model for subsequent regeneration efforts. Evaluations (Russell *et al.* 1996; Oatley and Lambert 1995; KPMG Consulting 1999a) have emphasised the value of partnership involving local communities and business, and conclude that it was relatively good value with leverage ratios of challenge funding to private investment and to other public funding of 1:3.78 and 1:1.45. Needless to say, some areas had particular problems of sustaining improvement after the end of the programme. The problems of deprivation, educational opportunity, crime and unemployment are deep seated and, as the KPMG report argues, warrant sustained long-term action. The challenge was taken up by the Single Regeneration Budget.

English Partnerships

English Partnerships (EP) is the operating name of the Urban Regeneration Agency and Commission for New Towns and is a non-departmental public body. It was launched in 1993 with the objective of promoting the regeneration of areas of need through the reclamation or redevelopment of land and buildings; the reuse of vacant, derelict and contaminated land (see Chapter 6) and the provision of floorspace for industry, commercial and leisure activities, and housing. Its main job is to fund the gap between the costs of undertaking development and the end value. In this it takes on the role of the previous land grants (city grant and derelict land grant described above) and responsibilities of English Estates. The changes were made to place

disparate individual projects within a strategic framework. It operated through six regional offices until April 1999 when its regional operations were separated out and transferred to the RDAs. From May 1999 the national operations combined with the Commission for New Towns but retained the name English Partnerships.

English Partnerships had a turnover of £444 million in 2003–4 rising to £540 million in 2004–5, and about 190 staff. The Commission for New Towns part of EP had an operating surplus of £73 million in 2003–4 because of disposal of land and property. The creation of English Partnerships and its flexible investment fund has provided more flexibility than previous grant regimes, and can be linked to provision of advice, joint ventures with the private sector and others, provision of loans or loan guarantees, and direct funding support of development.

English Partnerships contributes to wider strategies, which has meant working within the planning framework established by regional guidance and development plans. Thus, collaboration, especially with local authorities, is an essential feature of its work. The partnership element of English Partnerships has been taken seriously, not least because the agency can realise its regeneration objectives only by coordinating contributions from a variety of sources. Comprehensive information has been provided for potential partners on the qualities that English Partnerships want to see in project proposals, notably on design and the creation of mixed use developments.

English Partnerships has concentrated on action in four areas: developing its land assets and portfolio of strategic sites (which originate mostly from the new towns programme), creating development partnerships (mostly with RDAs and local authorities), improving the environment through land renewal and development, and finding new sources of funding to match public resources. The Agency reviewed its purpose in 2003 in the light of the government's Sustainable Communities Plan and Spending Reviews that have committed more funding to delivering development objectives, and now aligns itself closely with ODPM's public service agreement to achieve 'a better balance of housing availability and demand'. Its objectives remain broadly the same but from 2004 in line with the government's agenda it proposes to concentrate its core business on priority areas: the 20 per cent most deprived wards, the housing market renewal areas and the four major growth areas around the South East.[14] (For an explanation of the last two see Chapter 6.) It also provides extensive assistance to urban regeneration companies (explained later) and the Coalfields, and provides advice to the government on brownfield land, notably through leadership of the National Land Use Database, which provides a detailed schedule of all land previously in urban use in England.[15]

Many EP-supported projects have promoted housing and commercial development in the new towns. Others include some very high profile projects, notably the regeneration of the Greenwich Peninsula (£180 million) with the Millennium Dome, and the Millennium Villages' experimental development incorporating many energy saving and lifestyle innovations (see Chapter 7). Less well known is the English Coalfields regeneration programme, which includes eighty-two sites and 'boasts' one of the largest portfolios of contaminated land in Europe – 3,400 hectares or the equivalent space taken by some 95,000 houses, three times the annual total land reclamation of English Partnerships.[16]

An interim evaluation of English Partnerships (PA Consulting Group 1999) reported that it has made an effective contribution to land-based regeneration. Over the years it has become more positive in involving local communities in its work as well as the big players, and has created innovative ways of maximising the potential to claw back surpluses into the public purse where developments are successful. Additional functions have been taken on, especially in relation to the coordination of foreign inward investment at the national level. It scored less well on the environmental and social dimensions and other aspects of market failure outside the land question. Recommendations were made about the need to consider more fully the environmental impacts of development which the RDAs have had to take on board. The strategic dimension to land reclamation has improved, but continued difficulties of coordination are recognised. This is one area where the transfer of some function to the RDAs

should help to improve because of their wider remit to create a strategic focus for regeneration within the regions.

Millennium Communities

English Partnerships is leading the Millennium Communities Programme, which is 'to set the standard' for mainstream design and construction.[17] The programme was launched in 1997 and will in total involve about 6,000 homes in seven locations across England (shown in Figure 10.1). They are demonstration projects aimed to influence practice of the house building industry and local authorities by showing how the design and construction of new communities can incorporate sustainability principles including good public transport links, energy efficiency, a mix of housing and employment opportunities and community involvement. The aim is to reduce the consumption of natural resources and ensure that homes use at least 50 per cent less energy, 20 per cent less water and 20 per cent less waste disposal than comparable homes elsewhere. All the communities are on difficult brownfield sites where English Partnerships has led remediation. An 'action research project' on sustainable communities (Llewellyn-Davies et al. 2000) compared two of the millennium communities (Greenwich and Allerton Bywater) with three other places where sustainability had been a key criterion of regeneration and development. The conclusions pointed to the importance of initial site selection in determining many of the sustainability characteristics of built development, especially in relation to the local provision of services and car dependency. Both compared well on resource consumption with, for example, Poundbury. But the programme overall was criticised for 'seeking trend-breaking results through a fairly conventional large scale top-down commercial development process', such that non-commercial outcomes could be achieved only through the imposition of conditions and use of subsidies. The programme, however, is about influencing mainstream volume builders' approach to more difficult urban development sites (see Table 10.3).

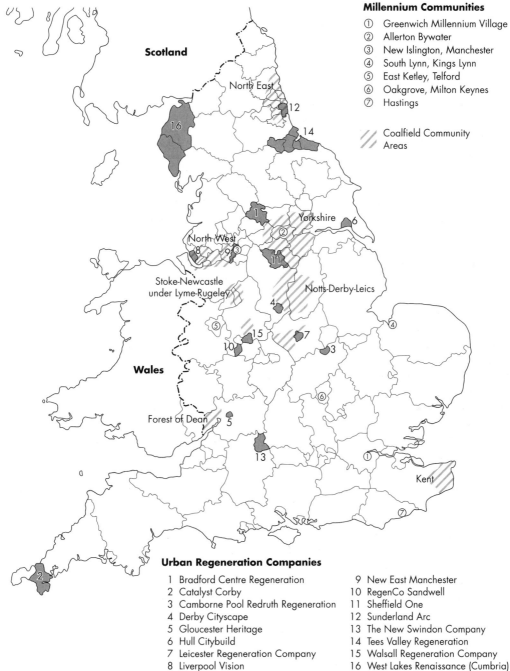

Millennium Communities

① Greenwich Millennium Village
② Allerton Bywater
③ New Islington, Manchester
④ South Lynn, Kings Lynn
⑤ East Ketley, Telford
⑥ Oakgrove, Milton Keynes
⑦ Hastings

Coalfield Community Areas

Scotland

North East

Yorkshire

North West

Stoke-Newcastle under Lyme-Rugeley

Notts-Derby-Leics

Wales

Forest of Dean

Kent

Urban Regeneration Companies

1 Bradford Centre Regeneration
2 Catalyst Corby
3 Camborne Pool Redruth Regeneration
4 Derby Cityscape
5 Gloucester Heritage
6 Hull Citybuild
7 Leicester Regeneration Company
8 Liverpool Vision
9 New East Manchester
10 RegenCo Sandwell
11 Sheffield One
12 Sunderland Arc
13 The New Swindon Company
14 Tees Valley Regeneration
15 Walsall Regeneration Company
16 West Lakes Renaissance (Cumbria)

Figure 10.1 *Urban initiatives in England*

Table 10.3 English Partnerships outputs and targets

		2002–3 Outturn	2003–4 Target	2004–5 Estimate	2005–6 Estimate
Brownfield land reclaimed (ha)		125	200	300	400
Housing units facilitated (completions)	Affordable	196	300	500	900
	Other tenure	1,260	1,500	1,800	2,400
Housing units facilitated (starts on site, all tenures)		n/a	2,500	3,000	3,800
Employment floorspace created (000 sq. m)		235	280	320	350
Private sector investment attracted (£m)		304	330	380	450

Source: English Partnerships Corporate Plan Summary 2003–06

Single regeneration programme

In response to criticisms of the fragmented nature of funding programmes for regeneration, and building on the competitive and partnership approach of City Challenge came the Single Regeneration Budget. This was introduced in 1994 with the intention of promoting integrated economic, social and physical regeneration through a more flexible funding mechanism. It was administered for the first of six rounds by the government offices and for the last two (one in London) by the regional development agencies, although central government maintained considerable influence in both writing the rules and making final decisions on funding allocations. The programme enabled local regeneration partnerships to bid for funding against a list of priorities with considerable flexibility in the specific objectives and measures.[18] From 2002, responsibility for allocating this regeneration funding has been devolved to RDAs within their single programme funding. So SRB has developed first as an integrated approach to funding regeneration administered by the centre, through regional administration and central decision, to complete devolution to the regions.

Four principles underpinned SRB's design and execution: the need for a strategic approach, partnership among the public, private, community and voluntary sectors, competitive bidding for available funds, and payment by results. The SRB addressed criticisms of short-termism and narrow compartmentalised approaches, and brought together twenty previously separate funding programmes. Schemes could include projects needing funding for up to seven years (or much less), but they had to be 'strategic' and establish links to other investment plans, such as those required for European Structural Funds and economic development strategies. In the early rounds there was no mention of the relationship between the SRB and development plans or regional guidance, but now there is a strong direction that single programme funding must contribute to achievement of the regional strategies. Ambitious projects were promoted, requiring in bids a long list of anticipated outputs in terms of jobs created or safeguarded, number of people trained, new business start ups, and hectares of land to be improved (Foley *et al.* 1998).

From Round 5 (the first under the Labour administration) there was an attempt to refocus and target the funding more directly to places in most need. A two-tier funding approach was adopted which required 80 per cent of funding to be channelled to large comprehensive schemes in the fifty most deprived local authority areas defined by four measures on the *Index of Local Deprivation*. The remaining 20 per cent was targeted at pockets of deprivation in the coalfields, rural areas and coastal towns. A further innovation was the provision for financial support for capacity building, indeed from Round 6 it was envisaged that this will be a component of most bids and that much of the first year of operation should be devoted to capacity building so that local communities can play an active and effective role in the creation and management of schemes. Even so, the SRB investments are still relatively minor compared with the main programmes of participating departments (Hill and Barlow 1995). Only 22 per cent of the resources required for each project come from the SRB, which puts considerable pressure on other public and private sources which they may not be able to meet (Hall and Nevin 1999). On the positive side, there is no doubt that the SRB has promoted more strategic thinking and an increase in operational partnerships. Indeed, the new ways of working across public, private and community sectors may be its most important outcome (Fordham *et al.* 1998).

The value of the funds incorporated into the SRB varied considerably from year to year: £6 billion in 1993–4 but falling to £1.36 billion in 1995–6. Funding increased under the Labour administration, but nevertheless, the expenditure in 2001–2 was £1.7 billion, or in real terms a 30 per cent fall from the 1993–4 figure. Total funding over six rounds was £26 billion, £9 billion of which came from the private sector. Rounds 1 to 6 generated 1,027 approved bids (one-quarter in London) worth over £5.7 billion and estimated £8.6 billion of private sector investment (and not including EU co-financing). The government's own evaluation of the SRB is upbeat, with claims that the partnerships forecast they will create or safeguard some 790,000 jobs, complete or improve 296,000 homes and support 103 community organisations and 94,000 businesses.[19]

ODPM commissioned monitoring of SRB Challenge Fund through examination of twenty case study partnerships over an eight-year period.[20] The interim findings (Rhodes *et al.* 2002: 13) indicate that most partnerships covered very small areas 'consisting of a number of wards', although a further 20 per cent covered the whole local authority area and in just over half of cases the local authority was lead partner. Earlier

evaluations of the SRB noted the domination of employment and economic development related output measures (though this is largely in the form of human resource development); the centralised and opaque decision-making on bids; the limited resources in comparison with the scale of the problem which leads to a thin distribution of funding; and the need for more effort to involve local communities (Hall, S. 1995; Brennan *et al.* 1998; Hall and Nevin 1999). Nevertheless, they agree that the Challenge Fund and SRB have had a positive impact on promoting more strategic thinking and partnership working in regeneration. Rhodes *et al.* (2002) make a direct comparison of City Challenge and SRB value for money (noting the difficulties in making such comparisons) and conclude that

> each net additional job created in the City Challenge basket was costing approximately £28,000 alongside a cost per qualification provided to a trainee of £3,450. The broad SRB equivalents are £25,000 per net additional job and £4,200 respectively per qualification provided and, on this broad basis VFM looks very similar between the two schemes.
>
> (Rhodes *et al.* 2002: 22)

New Deal for Communities and neighbourhood renewal

The incoming Labour administration of 1997 maintained the approach to urban policy that it inherited, including the competitive elements in the SRB and the emphasis on partnership with the private and community sectors. But it has also brought forward a raft of new initiatives with the intention of making a concerted attack on social exclusion in the worst parts of cities and to redirect funding through local authorities. In 1998, the Secretary of State announced extra funding for housing renewal and urban regeneration of £5 billion over three years, most of which was to be at the disposal of local authorities.[21] An extra £3.6 billion over three years was allocated to local authorities through their Housing Investment Programmes to start tackling the estimated £10 billion backlog in housing renovation, with the priority to improve the quality of the public housing stock. The enhanced urban regeneration programme has two main elements – the refocused SRB (explained previously) and the *New Deal for Communities* (NDC).

The initiative for the latter came from the Cabinet Office's Social Exclusion Unit report *Bringing Britain Together: A National Strategy for Neighbourhood Renewal* (1998).[22] The report sets out the intention to concentrate regeneration efforts on the most deprived neighbourhoods. At the same time it introduced 'floor targets' meaning that the performance of government departments and agencies would be evaluated on the basis of the worst cases in their areas as well as the averages. As well as the NDC, the National Strategy brought forward the Community Empowerment Fund to help local people take part in the new local strategic partnerships (explained in Chapter 3), the Community Chest Scheme to provide small grants for local projects and numerous other schemes to support tenant participation, involvement of faith communities and more (all of which were generally targeted at the eighty-eight most deprived areas). The overall goal was to deepen involvement of communities in urban policy (Chanan 2003).

The NDC provides £1.9 billion for seventeen first round (pathfinder) partnerships and twenty-two second round partnerships to be spent over ten years.[23] With this rather limited funding, it is perhaps obvious that the major objective is as much to bend and improve coordination of existing spending programmes as to provide new money. Eligible local authority areas were invited to establish a partnership and prepare a delivery plan for neighbourhoods of between 1,000 and 1,400 households. The principal objectives of the scheme are to tackle 'worklessness' (poor job prospects), high levels of crime, educational under-achievement and poor health, but the pathfinder partnerships have in fact addressed a much wider range of issues. The revised guidance includes supplementary objectives including a better physical environment, improved sports and leisure opportunities, and better facilities for access to the arts. The programme gives complete flexibility to the local partnership to define its objectives (within the priorities listed), its ways of working, and its

actions, though its plan requires approval by central government.

The first evaluations reveal in part just what little progress has been made on establishing local capacity to address regeneration in a holistic way.[24] The pathfinder partnerships have reported delays in community involvement because no structures were in place, a lack of trust among stakeholders and agencies, and difficulties in even understanding what is already spent in the neighbourhood under other mainstream programmes. Nevertheless, the New Deal for Communities has generated considerable interest and enthusiasm in communities, not least for the freedom it gives to the neighbourhoods to define their own approach. But it needs to be remembered that the opportunities and resources it provides are limited to areas with a population of less than 60,000. One way of looking at this is to say that it is a targeting of effort on the most deserving cases. Another is to say that it is no more than an experiment.

After publication of the 1998 report *Bringing Britain Together*, eighteen policy action teams (PATs) were created bringing together practitioners, academics and residents from deprived neighbourhoods. The policy action teams made almost 600 recommendations, and were used in the preparation of a consultation document on the National Strategy for Neighbourhood Renewal published in April 2000 (see Box 10.3). Considering the 'completely new approach to policy-making' (as the government described the policy action team approach) there is little new thinking in the proposals. Indeed Oatley (2000) points out the strong resemblance to the 1977 White Paper on *Policy for the Inner Cities*. Oatley makes a damning critique of the strategy's reliance on area-based policy (discussed below) and concludes that there is little hope for a significant impact on urban poverty, though they may 'soften the damaging consequences' of economic and social change.

Delivering an Urban Renaissance

The Labour government delivered its long awaited Urban White Paper at the end of 2000 – *Our Towns and Cities: The Future – Delivering an Urban Renaissance*

BOX 10.3 OBJECTIVES OF URBAN AND NEIGHBOURHOOD RENEWAL POLICY

The Single Regeneration Challenge Funding Round 6 (1999)

The overall priority is to improve the quality of life of local people in areas of need by reducing the gap between deprived and other areas, and between different groups. The objectives are

- improving the employment prospects, education and skills of local people
- addressing social exclusion and improving opportunities for the disadvantaged
- promoting sustainable regeneration, improving and protecting the environment and infrastructure, including housing
- supporting and promoting growth in local economies and businesses
- reducing crime and drug abuse and improving community safety.

Our Towns and Cities: The Future – Delivering the Urban Renaissance (2000)

Our vision is of towns, cities and suburbs which offer a high quality of life and opportunity for all, not just the few.

We want to see:

- people shaping the future of their community, supported by strong and truly representative local leaders
- people living in attractive, well-kept towns and cities which use space and buildings well
- good design and planning which makes it practical to live in a more environmentally sustainable way, with less noise, pollution and traffic congestion
- towns and cities able to create and share prosperity, investing to help all their citizens reach their full potential
- good quality services – health, education, housing, transport, finance, shopping, leisure and protection from crime – that meet the needs of people and businesses wherever they are.

(2000). The paper rambles across most areas of government policy and their impacts on urban areas, though a significant part is related to the planning system and explaining the need for a 'complete physical transformation' of our towns and cities. The main challenges are:

- to accommodate the new homes we will need by 2021 making best use of brownfield land

- to encourage people to remain and move back into urban areas
- to tackle the poor quality of life and lack of opportunity in certain urban areas
- to strengthen the factors in all urban areas which will enhance their economic success
- to make sustainable urban living practical, affordable and attractive.

A White Paper *Implementation Plan* was published in 2001 with a long list of short-term actions, many of which illustrate the difficulty of turning vague but ambitious objectives of the White Paper into practical action. On the physical transformation, much of the content and proposed White Paper actions reflect the findings of the Urban Task Force: *Delivering an Urban Renaissance: The Report of the Urban Task Force Chaired by Lord Rogers of Riverside* (1999).[25] This report presented 105 recommendations to government, many of which, as the White Paper explains, have found a positive response. Recommendations of particular significance for the planning system are shown in Box 10.4.

BOX 10.4 SELECTED RECOMMENDATIONS OF THE URBAN TASK FORCE 1999

- Require local authorities to prepare a single strategy for the public realm and open space, dealing with provision, design, management, funding and maintenance.
- Revise planning and funding guidance to discourage local authorities from using 'density' and 'over-development' as reasons for refusing planning permission, and to create a planning presumption against excessively low density urban development.
- Make public funding and planning permissions for area regeneration schemes conditional upon the production of integrated spatial masterplan.
- Develop and implement a national urban design framework, disseminating key design principles through land use planning.
- Place local transport plans on a statutory footing and include explicit targets for reducing car journeys.
- Introduce home zones using tested street designs, reduced speed limits and traffic calming.
- Set a maximum standard of one car parking space per dwelling for all new urban residential development.
- Develop a network of regional resource centres for urban development, coordinating training and encouraging community involvement.
- Produce detailed planning policy guidance to support the drive for an urban renaissance.
- Strengthen regional planning – provide an integrated spatial framework for planning, economic development, housing and transport policies; steer development to locations accessible by public transport; and encourage the use of sub-regional plans.
- Simplify local development plans and avoid detailed site-level policies.
- Devolve detailed planning policies for neighbourhood regeneration into more flexible and targeted area plans, based upon the production of a spatial masterplan.
- Review employment land designations and avoid over-provision.
- Reduce the negotiation of planning gain for smaller developments with a standardised system of impact fees.
- Review planning gain to ensure developers have less scope to buy their way out of providing mixed tenured neighbourhoods.
- Oblige local authorities to carry out regular urban capacity studies as part of the development plan-making process.
- Adopt a sequential approach to the release of land and buildings for housing.

- Require local authorities to remove allocations of greenfield land for housing from development plans, where they are no longer consistent with planning objectives.
- Retain the presumption against development in the green belt and review the need for designated urban green space in a similar way.
- Prepare a scheme for taxing vacant land.
- Modify the General Development Order so that advertising, car parking and other low-grade uses no longer have deemed planning consent.
- Streamline the compulsory purchase order legislation and allow an additional 10 per cent payment above market value to encourage early settlement.
- Launch a national campaign to 'clean up our land' with targets for the reduction of derelict land over five, ten and fifteen years.
- Introduce new measures to encourage the use of historic buildings left vacant.
- Facilitate the conversion of empty space over shops into flats.

Source: Towards an Urban Renaissance: The Report of the Urban Task Force Chaired by Lord Rogers of Riverside (1999). The Urban Task Force Report makes 105 recommendations in total, and many not listed here are of interest to planning. The full list of recommendations together with explanations of the government's response is given in an annexe to DETR (2000) *Our Towns and Cities: The Future – Delivering an Urban Renaissance.*

The most significant planning related action on the Task Force report is the revision in 2000 of PPG 3, *Housing* (discussed in Chapter 6). Many other promised actions are underway or delivered, such as the changes to the planning obligations system to widen the range of community benefits that can be supported; a revision of PPG 1, now PPS 1; provisions for the use of master-plans in regeneration; the creation of home zones; initiatives on the training of planners, designers and developers; the creation of regional centres of excellence to address skills improvement in each region; and the creation of urban regeneration companies (as described below).

The Urban Task Force published a review of progress under the title *Towards a Strong Urban Renaissance* in 2005 noting that government initiatives and 'sustained economic growth and stability' had led to some success and that 'For the first time in 50 years there has been a measurable change in culture in favour of towns and cities, reflecting a nationwide commitment to the Urban Renaissance' (p. 2). The review makes more than fifty recommendations to strengthen progress ranging widely from those that are very likely to be followed soon such as the provision of more design guidance to highway engineers; enforcing reviews of parking standards; and measures to promote the development of small infill sites especially for affordable housing; and others that are more ambitious such as an energy efficiency obligation on developers. On some matters the Task Force are not unanimous. In a footnote, Professor Sir Peter Hall expresses regret that he cannot support increasing minimum densities (the case for which, he argues, is unproven) and requirements to develop brownfield land first in the growth areas, pointing to the way current policies are 'causing an unprecedented increase in apartment construction, unsuitable for families with children and undesired by potential residents' (p. 19).

Delivering an Urban Renaissance is less of a presentation of government policy for discussion prior to bringing forward legislation, and more of a stock-take. The paper itself notes that it 'completes the first phase of the Government's long term programme'. Proposals for substantial new legislation or action are thin on the ground. It explains the performance of the UK economy since the late 1940s, and numerous vignettes illustrating examples of good practice across the UK and further afield. It is therefore, useful reading for students, although they should beware the obvious gloss.

The State of English Cities

The importance of strategy and cross-cutting or 'joined-up' action is stressed throughout the Urban White Paper. So too is the variation in experience across England and thus the importance of regional bodies. These are issues raised in the accompanying report on *The State of English Cities*.[26] This report notes that the principles of policy integration, partnership and local authority leadership that have been developed within urban policy since the early 1980s are also right for the future. However, it also stressed the need for 'rethinking scales of intervention' so as to 'tie city policies to the broader frameworks of regional and sub-regional strategy' and for consideration of urban–rural interdependencies (p. 46). Urban policy should be based on relatively large areas – perhaps with specific planning and fiscal regimes as also promoted by the Urban Task Force. The city-region concept looks set for a come back, having been promoted, and followed up to some extent in the 1960s. Parkinson and Boddy (2004) describe how the idea of cities as 'dynamos of the UK national economy . . . rather than the economic basket cases as they were sometimes portrayed, [has] seized the imagination of politicians, pundits, academic researchers and business leaders alike' (p. 1). In the conclusion to the same volume they explain how 'the effectiveness of strategic planning at a sub-regional scale 'is critical to realising the long run success of cities' (p. 407).[27]

In practical terms the government has provided some (limited) extra funding for the regional development agencies and more flexibility in their operation. The Implementation White Paper also recognised the role of regional planning strategies, and the 2004 reforms have done much to strengthen this level of planning, but very little was said about the relationship

between planning and economic strategies. *The State of English Cities* report points to the weakness of regional economic development strategies in identifying the distinctively regional aspects of the strategy. The same could be said for regional planning guidance; and both still tend to treat urban and rural issues separately, though guidance on regional spatial strategies in PPS 11 has addressed this to some extent.

Despite the liberal use of the word 'strategy' and the obvious intention to link many disparate themes of public policy and private action to bring about improvement, the long lists of actions taken (and to be taken) in many different sectors, reveal just how difficult this job is. This point is emphasised by the recognition that many of the 'urban problems and policies' discussed here are not peculiar to urban areas at all. Nevertheless, in 2005, the number and variety of urban policy initiatives continued to grow.

The monitoring and analysis of *The State of English Cities* continues with the creation of a Town and City Indicators Database (University of Liverpool *et al.* 2002) and a first monitoring report to the 2005 Urban Summit (Parkinson *et al.* 2005). The findings in this latter report are set out with some caution though they mirror similar comments about urban trends and policy discussed at the end of this chapter, notably the main problems of urban policy in England: its fragmentation across departments (with questions about the commitment of some departments such as Transport to the urban renaissance and sustainable communities agendas), the number of special initiatives and targets, the short-termism of policy and resource constraints. On the other hand the research team identify welcome changes in the wider recognition of the role of cities in economic competitiveness: a greater awareness of the spatial impacts of sectoral policies and signs of improvement in national–local government relations. Despite the caveats about these being tentative findings, most points are well known and understood. A less circumspect position is taken in the related report on *Competitive European Cities*, which compares the competitiveness of the eight English core cities with their European neighbours.[28] While the findings show that the English cities are recovering from a period of decline they 'lag behind their European

counterparts . . . [in] innovation, workforce qualifications, connectivity, employment rates, social
composition and attractiveness to the private sector'
(Parkinson *et al.* 2004: 50). This leads the research team
to conclude that

> There will have to be a wider debate about the
> consequences of a very powerful implicit territorial
> policy, whereby substantial public resources flow
> into different areas of the nation through a range of
> disconnected policies and institutions in higher
> education, research and development, transporta-
> tion, housing, health – even the Sustainable
> Communities plan. Often these policies and
> programmes support already prosperous regions as
> much as – if not more than – the less prosperous
> places, which need critical support and intervention.
> (Parkinson *et al.* 2004: 69)

Urban regeneration companies

This proposal from the Task Force Report has been
taken forward energetically and the initial White Paper
target of fifteen urban regeneration companies (URCs)
was easily met. A first phase of three pilot companies
in England: Liverpool Vision, New East Manchester
and Sheffield One was followed by thirteen more
accepted applications by mid 2004. Twenty-one URCs
were approved in England and Wales by early 2005
and further applications are being considered (see Table
10.4).[29] This scale of activity is new but the type of
approach is well established though the arrangements
may be more informal. A regeneration company was
established in Nottingham in 1998, and some years
before a number of councils established strong inter-
agency working arrangements, as in the case of the
Heartlands initiative in Birmingham (Wood, C. 1994).
In Scotland the approach is particularly well established
through the eight local development companies (LDCs)
in Glasgow designated from 1986 and involving a
partnership of the City Council, Scottish Enterprise,
the social inclusion partnerships and other stakeholders. Following the report on the three pilot
companies in England, in 2003 the Scottish Executive

consulted on a Scottish variant of the urban regeneration company. Three pathfinder URCs were designated
in Scotland in 2004 with a share of £20 million additional funding from the Scottish Executive. Similar
arrangements are also in place in Londonderry,
Northern Ireland.

The companies are established by a partnership of
local authorities, the regional development agency and
other business and community stakeholders. Local
strategic partnerships and local inclusion partnerships
(Scotland) are also considered important partners. In
England, English Partnerships is also a partner in many
URCs and provides guidance and support to them all.
URCs are independent companies whose operation
costs (revenue funding) come from the partners. Private
sector funding of running costs has been encouraged
through a tax relief incentive. They do not have capital
funding separate from their sponsoring bodies and
do not undertake development activity themselves.
The public sector development investment comes
from English Partnerships who make acquisitions for
development schemes that support the URC strategy,
and remediate land in need of improvement. The
regional development agencies also commit funding to
co-finance development. In both cases the URCs
provide a focus for existing funding rather than
generating new money

The rationale for URCs flows from the analysis of
market failure and barriers to implementing physical
urban regeneration noted in the Urban Task Force
Report and numerous other sources. Their main task
is to create a favourable climate for private sector investment in places where there are 'latent development
opportunities'. The ODPM's view is that

> URCs provide an added impetus and focus for the
> delivery of a core series of physical development
> projects, which – allied with other regeneration and
> community activities – set out to attract inward
> investment, address deprivation, create economic
> activity and reverse the process of decline. Although
> not exclusively, there should be a primary focus on
> physical development projects and on the re-use of
> brownfield land where opportunities exist.
> (*URCs: Guidance and Qualification Criteria* p. 6)

Table 10.4 Urban regeneration companies in the UK

Company name	Established
Bradford Centre Regeneration	2002
Catalyst Corby	2001
Central Salford	2004
Clydebank	2004
CPR Regeneration (Camborne, Poole and Redruth)	2002
Craigmillar, Edinburgh	2004
Derby Cityscape	2003
Gloucester Heritage	2004
Hull Citybuild	2002
Leicester Regeneration Company	2001
Liverpool Vision*	1999
New East Manchester*	1999
Newport Unlimited	2002
Peterborough City	2005
Raploch, Stirling	2004
ReBlackpool	2005
Regenco Sandwell	2003
Renaissance Southend	2005
Sheffield One*	2000
Sunderland Arc	2004
Tees Valley Regeneration	2002
The New Swindon Company	2004
Walsall Regeneration Company	2003
West Lakes Renaissance (Furness and West Cumbria)	2003
Other local initiatives have the same characteristics as URCs	
ILEX (Londonderry and the Derry City Council area)	2003
Nottingham Regeneration Limited	1998
Eight local development companies in Glasgow City Council	1986 on

Notes:
List updated February 2005
* indicates the first three pilot companies in England

Evaluations of urban policy and planning implementation over many years point to the importance of leadership, skills and capacity, coordination of activity, and building market confidence, as well as the need for funding. URCs address barriers by engaging with the private sector in regeneration within a wider strategic framework (independent from the local authority), champion the potential of areas, and coordinate plans and actions.

The planning system plays an important role in underpinning the URC agenda (Amion Consulting 2001). One of the essential requirements of URC operation that the ODPM is looking for in its *Guidance and Qualification Criteria* (2004) is a set of

customised planning procedures and coordinated internal administrative arrangements in and among member local authorities in support of URC activities and projects. For example, planning committees focused on the URC area . . . identified personnel in local authorities to act as primary points of contact on URC matters and to act as URC 'champions' internally . . . [and] memoranda of understanding to cement arrangements.

(p. 9)

Again, such special arrangements for implementation are not confined to the URCs (see for example Carmona *et al.* 2003).

URCs present an interesting example of national, regional and local relations in urban regeneration (see Box 10.5). The initiatives are local and make use of regional funding, but it is a 'nationally led' programme with copious guidance from ODPM and a formal requirement for approval from ODPM and DTI. Indeed, the ODPM's own *Policy Stocktake* in 2004 (undertaken by ODPM) concluded that more central guidance was required. Nevertheless, URCs are an illustration of the steady 'localisation' of urban regeneration policy and action but in the traditional UK context where central government is unable to relinquish central supervision and control. Raco *et al.* (2003) explain how a more local focus for urban regeneration policy is argued to provide more flexibility and responsiveness. It may help to deliver more a effective management role in local service delivery (the one-stop shop), a local economic development role in acting as development agency and a political role in acting as champions for the area. Their evaluation of the Govan Local Development Company in Glasgow, perhaps the largest such example with 170 employees, suggests that URCs may be able to deliver on all three roles, but only if they are able to work through the institutional complexity and, critically, influence spending; and then we should not expect too much:

On the one hand, they can act as local coordinating agencies, thereby expanding local institutional capacities and enhancing the effectiveness of service delivery (Healey 1997). On the other hand, their existence (re)produces institutional fragmentation and complexity which often hinders the develop-

ment and implementation of coherent strategic agendas. LDCs are not a panacea for the failures and limitations of existing urban policy, nor are they a substitute for well-resourced, strategic, and coherent urban policy initiatives. Yet they do have a role to play within urban-regeneration policy.

(p. 301)

Coordination of urban initiatives

The most consistent feature of urban policy is the never-ending stream of policies and initiatives, and the current phase probably offers more than any other. Since coming to power in 1997, the Labour administration have taken many new initiatives, often with an area or spatial focus, including health action zones, employment zones, education action zones and crime reduction programmes (zones). By 2001, the ODPM was 'signposting' eighty-seven other urban policy initiatives not to mention other commitments in the White Paper. To this needs to be added the work of other organisations such as the Local Government Association's twenty-two partnerships for regeneration. Many of these initiatives have a bearing on town and country planning – although few mention in their supporting documentation the statutory planning framework for guiding the spatial distribution of investment and activities.

It is hard to reconcile the claim that Labour's new deal promotes more strategy in the face of such a complex web of initiatives and competences. Even the regeneration minister, Lord Rooker has described urban policy as like 'a bowl of spaghetti'.[30] In 1999, the Environment, Transport and Regional Affairs Committee voiced concern that intervention in urban areas is 'confused and badly coordinated' and 'that there should be fewer initiatives and they should be better coordinated locally'.[31] The reasoning for complexity is that differing conditions of each area require variety and flexibility in the policy response. Much guidance is on offer (and some is recommended in the further reading at the end of this chapter) but getting some sensible coordination of actions that ensures complementarity of actions is a considerable task. This hasn't

BOX 10.5 EXAMPLE OF URBAN REGENERATION COMPANY: GLOUCESTER HERITAGE

Established in 2004, Gloucester Heritage URC has a focus on 343 hectares of the central area of the city with a vision to improve the gateway areas, tackle brownfield regeneration and improve public transport. Key projects include linking the Docks areas with the central retail core and Cathedral Quarter.

Objectives

To facilitate

- reclamation and development of 40.5 hectares of brownfield land
- repair and reuse eighty-two historic buildings
- development of 41,800 sq. metres of retail floor space
- building of 3,000–3,500 new homes
- improvement of infrastructure with a new mainline railway station and inner relief road
- levering in £1 billion of private sector investment over ten years.

Funding

Initial revenue funding will be £750,000 per annum, provided by the South West Regional Development Agency and English Partnerships: £250,000 each, Gloucester City Council and Gloucestershire County Council.

Implementation

English Partnerships purchased the 2.45 hectare Gloucestershire College of Art and Technology campus. The proceeds are being used to finance a new campus in the Docks area and the old campus is being used for major mixed use development in a strategically important location near the city centre.

Sources: English Partnerships (2004) *Urban Regeneration Companies: Coming of Age*; ODPM fact sheet at www.odpm.gov.uk (December 2004)

gone unrecognised in government. The main response is the local strategic partnership to provide 'a single overarching local coordination framework within which other, more specific local partnerships can operate'. The background to LSPs and their evolution so far is examined in Chapter 3. It is a development of existing good practice, and may well lead to the reduction of other partnerships which will be subsumed under the new arrangements. But the early indications are that LSPs and their main instrument, the community strategy, are not connecting other initiatives in the way intended. All partnerships should demonstrate how they relate to other area based initiatives, and work within the regional economic development strategy led by the RDAs, other regional strategies, and the community strategy.

Employment, training and enterprise agencies

Employment is, of course, one of the principal economic considerations in town and country planning, yet it cannot be said that employment policies have ever been successfully integrated with physical planning policies. At least in part, this is due to organisational separatism and the fact that local authorities have little responsibility for employment and training.

Initiatives in these areas have been a function largely of central government devolved to regional offices. These grew in number and administrative complexity until a major reorganisation was made following the 1990 White Paper *Employment in the 1990s*. This produced a network of seventy-two *training and enterprise councils* (TECs) in England and Wales, and twenty-two *local enterprise companies* (LECs) in Scotland. This was a 'privatised' organisation with objectives to localise and decentralise training policy (Bennett 1990). Copied from the USA (where they were known as *private industry councils*), the underlying rationale was that if local businesses take a central place in guiding the support programmes for employment and training, the result will be a better response to local employment needs, a more business-like mode of operation, and increased leverage of private sector support for training (Lewis, N. 1992).

In 2001 the TECs were merged with the Further Education Funding Council in the Learning and Skills Council (LSC) following recommendations in the 1999 White Paper *Learning to Succeed*. The LSC has an annual budget of £8 million (2003–4) and is responsible for all education and training post-16 other than higher education (universities), including youth and employment training. It works through a network of forty-seven 'regional' offices with the intention of

tailoring policy to local conditions. Their activities include maintaining a knowledge base of local labour markets and training needs and provision, providing information to employers, employees and the unemployed, encouraging employers to meet training needs, and directly providing or commissioning training for employees or the self-employed.

European dimension to urban policy

The Community does not have a mandate from the Treaties to develop an urban policy (Nadin and Seaton 2000), but it has been argued that since more than 80 per cent of the EU population live in urban areas, and that cities and towns are the motors of economic growth, it is sensible for the Community to take a view on the impact and potential of its actions in them.

In 1990 the Commission, led by Directorate General for the Environment, published a *Green Paper on the Urban Environment* as 'a first step towards debate and reflection, and attempts to identify possible lines of action' (CEC 1990). One of the first actions was to set up an Expert Group on the Urban Environment made up of national representatives and independent experts to advise the Commission. The Expert Group has taken forward many initiatives and played a central part in setting up the Sustainable Cities and Towns Campaign.

The Directorate General for Regional Policy has also had a considerable involvement in urban policy because of its impact on urban areas through structural funding. Community co-financing has played an important role in many urban regeneration initiatives in the UK, primarily in areas covered by Objectives 1 and 2 (see Chapter 4). In addition the *URBAN* initiative provided funding for coordinated action for 'neighbourhoods in crisis' and *urban pilot actions* promoted innovation in tackling urban regeneration problems. *URBAN* is one of only four thematic initiatives under the 2000–6 round of Structural Funding. The Community has also funded networks of cities and towns which concentrate on exchanging experience – notably *Eurocities* (a network of more than sixty non-capital cities) (Griffiths, R. 1995) and *Metrex* – the network of metropolitan planning authorities.

Urban areas have been recipients of considerable Community funding that has helped to fill the gap as national funding has declined (Chapman 1995) but doubts have been raised about the coordination of that investment. In 1997 the Commission published *Towards an Urban Agenda in the European Union*. This received very positive support and encouragement from the Committee of the Regions and English Partnerships. It was followed in 1998 by *Sustainable Urban Development in the European Union: A Framework for Action*. The main objective of this paper is to stimulate better coordination of existing Community actions that affect urban areas. Its impact was considerable. Many important but 'unanticipated' Community initiatives in the environmental field were announced in the Fifth Framework on the Environment document, which provides a template for the urban paper. The Framework sets an unambiguous agenda for Community action on urban matters, so the principal actions are summarised in Box 10.6. Of particular note is the emphasis on area-based initiatives which will need to be 'essential constituents' of the plans and programming documents which guide the spending of Community funds. The UK approach has been to promote further intergovernmental cooperation rather than formal Community action.[32]

Future action on urban matters by DG Environment will be guided by the proposed *Thematic Strategy on the Urban Environment*, which is a commitment of the Sixth Environmental Action Programme. As explained in Chapter 7, four themes for the Strategy have been identified:

- *sustainable urban management*: concerning, for example, the adoption of explicit environmental targets, actions and monitoring by local authorities in an integrated urban management system
- *sustainable construction*: concerning, for example, the minimisation of resource inputs to construction, recycling of construction materials, and maximising energy efficiency in new construction
- *sustainable urban design*: concerning, for example, the appropriate physical form of urban areas for more sustainability, redesigning and retro-fitting existing urban areas and building on brownfield land
- *sustainable transport*: including, for example, the types of measures to promote more sustainable mobility and tools for evaluating the impacts of transport measures.

The most controversial proposals are for each city across the EU with populations of more than 100,000 (about

---

BOX 10.6 *SUSTAINABLE URBAN DEVELOPMENT IN THE EUROPEAN UNION: A FRAMEWORK FOR ACTION* (1999) – SELECTED RECOMMENDATIONS

- Explicit urban programming for Structural Fund support to provide support to area-based action for urban regeneration protecting and improving the urban environment
- Promotion of inter-urban cooperation to exchange experience on the urban environment
- Development of know-how and international exchange of experience on discrimination, exclusion and urban regeneration
- Better implementation of existing environmental legislation at the urban level
- Further EU legislation concerning waste, air quality, water and noise
- Strengthening pollution control and clean-up in towns and cities
- Measures to reduce the environmental impact of urban transport
- Sustainable urban energy management
- Climate protection (moving towards a Directive on taxation of energy products)
- Improving comparative information on urban conditions

500 cities in the EU) to prepare an environmental management plan and a sustainable transport plan, with strong support from some member states, including Germany and France. At the time of writing a Directive is proposed though it may be that the Open Method of Coordination is used whereby each member state signs up to common systems and targets. At the beginning of 2005, more evaluation of the proposals was underway with the UK government leading a review of the elements that new urban management plans might contain, and to identify possible difficulties with a legal requirement to prepare such plans (Atkinson and Mills 2005).

Scottish urban policies

The evolution of urban policy in Scotland is different in important ways to experience in England, though the challenges are the same. There is a longer history of central–local partnerships in urban regeneration (Boyle 1993) and less involvement of free-standing agencies because of Scotland's distinctive governmental organisation, and the close relationship that exists between local and central government. The turn to social and community issues in the 1970s and 1980s persisted in Scotland while the property development approach took a firm hold in England. Nevertheless, Scotland has struggled to break from 'a persistent and damaging failure to link regeneration schemes to wider economic, infrastructure and land use planning policies operating at larger spatial scales' (Turok 2004: 112). Urban policy has also tended to emphasise the housing problem at the expense of broader economic and social regeneration, although in one area, partnership working, it has led developments elsewhere in the UK. Of particular note are the *Glasgow Eastern Area Renewal* (GEAR) project and the establishment of the Scottish Development Agency (now Scottish Enterprise) which has played an important part in partnership working.

The rationale for the SDA was that economic regeneration required strong measures to promote local growth that local authorities were not equipped to deliver. But importance was attached to retaining the role of local authorities as well as the many other agencies involved (McCrone 1991: 926). No urban development corporations were established in Scotland, since the SDA and now Scottish Enterprise fulfil a similar function and can operate anywhere in the country.

The GEAR project had economic, environmental and social objectives and pioneered multi-agency partnership working. Despite initial teething problems, the physical impact on the area was massive. Economic revitalisation, however, proved much more elusive; although up to 2,000 jobs were retained or created, the benefits to the resident disadvantaged population were limited. The experience of the GEAR project was used subsequently in tackling disadvantaged areas facing the additional problem of severe local employment crises – in Glengarnock in North Ayrshire, Clydebank and Bathgate. The accomplishments were impressive, in terms of job creation, the provision of new infrastructure, and widespread environmental improvement. This was largely attributed to the establishment of a clear strategy bringing together environmental, employment objectives and the participation of local authorities in the scheme (McCrone 1991: 929). Further partnerships followed with the SDA assuming a coordination role in Leith, Dundee, Motherwell, Monklands and Inverclyde, all areas with long-standing deprivation. A more ambitious scheme was undertaken jointly by SDA and the City of Glasgow in the old Merchant City area, with the aid of housing improvement grants and the SDA's *local enterprise grants for urban projects* (LEG-UP). The outcome has been judged a success, which together with other initiatives in the city has contributed to bringing an about-turn in both the city's morale and outside perceptions of Glasgow as a place for investment.

The Scottish statements *New Life in Urban Scotland* (1988) and *Urban Scotland into the 1990s: New Life Two Years On* (1990) reviewed the achievements and the priorities for further urban action. Much attention has been devoted to the large housing estates on the periphery of Scottish cities. Those of Paisley (Ferguslie Park), Glasgow (Castlemilk), Edinburgh (Wester Hailes) and Dundee (Whitfield) have undergone regeneration through *Urban Partnerships* established by the Scottish Office. These involved the local authorities, the local community organisations, the Scottish Office, Scottish Enterprise, and other agencies operating in the areas, but were criticised for being inward looking with little strategic context (Hall 1997a). The final evaluation of the urban partnerships (Tarling *et al.* 1999) reveals that the £485 million invested over ten years has been cost-effective: 3,726 new homes have been built and 9,253 improved, employment has improved in two of the partnerships (but fallen in one), and crucially, the partnerships have improved the image of the estates. But the fundamental problems of poverty and disadvantage remain, and a key reason is the 'churning' of the population as newly employed residents move on (to more attractive locations) to be replaced by unemployed households. In 1993, small urban regeneration initiatives (SURIs) were set up taking forward the partnership approach in eleven council estates across Scotland. The depth of the 'partnership' thinking is reflected in key publications at that time, including *Progress in Partnership* (1993), *Programme for Partnership* (1995) and a report on *Partnership in the Regeneration of Urban Scotland* (1996). As in England and Wales, a competitive system of allocating funds to partnership projects was established. The main share of urban regeneration resources was ring fenced for twelve *priority partnership areas*.[33]

Questions have been raised subsequently about insufficient funding given the tasks; the difficulty of demonstrating social and economic disadvantage in the relatively sparsely populated areas; and the negative effects of the pressure of competitive bidding (McCarthy 1999). For Turok (2004),

> the rhetorical formula of a comprehensive, coordinated approach shifted responsibility onto local partnerships, without any guidance about the relative importance of, and links between, people, place and economic policies. The message was that if agencies worked together more closely in consultation with the community, neighbourhood deprivation could be removed.
>
> (Turok 2004: 116)

Post-devolution policy has continued in the same way. The PPA approach was expanded to other areas and

renamed social inclusion partnerships (SIPs). The Urban Programme in Scotland was replaced by the *Social Inclusion Partnership Fund* in 1999 with £1.3 million extra funding and extra money for the priority partnership areas to develop much needed dedicated support units. But these initiatives downplayed the geographical dimension and physical infrastructure in preference of a concentration on tackling social disadvantage. Devolution has given greater discretion to Scotland, but this has not been to the advantage of urban policy. Policy review groups considered and made recommendations for developing distinctive Scottish policies; the outcome for urban policy was the *Community Regeneration Statement*, *Better Communities in Scotland: Closing the Gap*. As in England this signalled a shift towards more strategic thinking and implementation allied to the community planning process and a 'migration' of the social inclusion partnerships with community planning partnership funding.

But those close to Scottish urban policy are not optimistic about the impacts of the statement. Turok (2004) suggests that:

> It would be premature to say that the Executive has an urban policy, or has even fully recognised the distinctive challenges and opportunities of cities. Different national strategies and policy approaches co-exist, some of which have an explicit urban aspect, while others are neutral or indifferent to cities.
>
> (Turok 2004: 126)

Northern Ireland

Urban policy in Northern Ireland has followed a similar pattern to that in the rest of the UK with early emphasis on social problems giving way to a concentration on property-led regeneration and, in recent years, a shift of attention to place marketing, public–private partnerships and community development (Berry and McGreal 1995b). However, the special circumstances, notably the violence and subsequent central government control of policy and implementation, have been important in shaping the problem and responses.

Urban conditions and unfairness in employment and housing allocations were important factors in the start of 'the troubles' from 1969, and terrorism has subsequently accentuated the difficulties of tackling them.

Urban problems are concentrated in Belfast and Londonderry, although there is a separate *Community Regeneration and Improvement Programme* for smaller towns. The urban policy budget in Northern Ireland amounts to more than £48 million (1990–2000) but to this must be added the considerable funding (£63 million p.a.) that came by virtue of the designation of the whole of the Province as an ERDF Objective 1 region until 1999 (with transitional funding to 2006) and the significant inward cash flows for employment and housing investment through mainstream funding programmes. There is also a Northern Ireland *Urban Initiative for Peace and Reconciliation* which provides over £10 million for urban regeneration to improve the quality of life, enhance the environs of sectarian interface areas and support a wide range of regeneration projects.

Urban regeneration is the responsibility of the Department for Social Development (DSD) which was set up in 1999. The Northern Ireland Housing Executive contributes on housing renewal. The DSD's *Neighbourhood Renewal Strategy for Northern Ireland: People and Place* launched in June 2003 provides an urban regeneration policy framework for both departments. The Strategy aims for physical, social, economic and community renewal in the most deprived urban communities: the worst 10 per cent of wards in terms of deprivation, which are designated as *urban renewal areas*. Partnerships have been created in each area and will prepare vision statements and *neighbourhood action plans*. As in England, the emphasis is on coordinating sectoral programmes and spending, though with special mention of the need to avoid segregation.

Urban policy in the most deprived areas of Belfast is managed by the Belfast Regeneration Office (BRO) and implemented through four regeneration teams which coordinate the activities in thirteen urban renewal areas. The BRO manages allocation of the Neighbourhood Renewal Fund, urban development grants and EU funding. Urban renewal in Belfast has

a long history: the BRO previously managed the programme *Making Belfast Work*, which covered the thirty-two most deprived wards in Belfast and committed £275 million on 350 projects between 1988 and 2004. A separate Londonderry Initiative was established in 1988 and ran until 2004, spending £42 million in the most deprived parts of the city. Regeneration here and in neighbouring towns is now managed by the North West Development Office (NWDO). A separate 'People and Place' renewal strategy is being developed for Londonderry. Both Belfast and Londonderry also have comprehensive renewal schemes which allow for the compulsory acquisition of land and property after consultation. Elsewhere in Northern Ireland the Regional Development Office (RDO) coordinates regeneration activity.

There is one urban development corporation, Laganside, which was established in 1989 following closure of shipbuilding yards. After development investment of £800 million and considerable change to the areas alongside the River Lagan, the Corporation is due to close in March 2007.

Evaluation of urban and area based policy

Evaluation of the effectiveness of government initiatives has been a feature of urban policy since the early 1970s, but has gained particular importance since the mid 1990s as part of the search for 'evidence-based policy'. Modern management techniques have also come to the fore with the formulation of more specific objectives, targets and indicators which can be monitored and evaluated. This has made the easy option of vagueness unacceptable in principle, but practice is a different matter, as specific targets are mostly poor surrogates for the rhetorical and vague goals of 'regeneration', 'renaissance', 'attractive places', 'sharing prosperity' and the like (see Box 10.7). Measures or outputs such as training places, jobs, visitors, roads and reclaimed land can be counted but are often disputed. Worse still, they do not give a real assessment of outcomes. To what extent have the fundamental objectives of regeneration been realised? And do such measures

BOX 10.7 ASSESSING THE IMPACT OF URBAN POLICY: CONCLUSIONS FOR FUTURE URBAN POLICY

1 There are clear indications of the importance of creating effective coalitions of 'actors' within localities and that these are most likely to result from . . . mechanisms which encourage or require long-term collaborative partnerships.
2 Local authorities – in their newly emergent roles as enablers and facilitators – need to be given greater opportunities to play a significant part in such coalitions.
3 Local communities equally need to be given opportunities to play roles in such coalitions. The evidence of increasing polarisation suggests the need for specific resources to address the scope for community capacity-building within deprived areas.
4 There remains a need to improve the coherence of programmes both across and within government departments. This requires a greater emphasis on the identification of strategic objectives which can guide departmental priorities. Area targeting has played an important part in those cases where separate programmes have been successfully linked so as to create additionality, thereby suggesting the value of giving even greater emphasis to area based approaches . . .
5 An important part of such coherence must derive from less ambiguity in the targeting of resources. There is a strong argument for the development of an urban budget which might be administered at regional level so as to reflect the varying constraints and opportunities across different regions, and to improve coordination across programmes and departments.

Source: Robson *et al.* (1994)

1 Lack of linkage between physical development and economic regeneration which benefits the socially excluded.
2 Lack of regional frameworks to support city and local regeneration and link key policies (such as transport, and education and vocational training), to derive maximum benefit to regeneration.
3 Lack of integration of short-term initiatives within a long-term vision on the future role of cities and their hinterlands, and of an investment framework to support that vision.

Source: Carley and Kirk (1998)

give a real account of the improvement of places, and social and economic life?

In answering these questions, there is an extraordinary difficulty at the outset: how to define clearly what the objectives of policy are. The problem is very familiar to policy analysts (Rittel and Webber 1973), but it has to be constantly tackled anew by policy-makers. A good example (and this discussion has to be illustrative rather than comprehensive) is the apparently simple matter of increasing employment. The

problem is one of finding a satisfactory definition (or even concept) of the term 'new employment'. Other complications arise when account is taken of the 'life' of new jobs created: many of the jobs created in the course of regional development programmes later disappeared (Hughes 1991). Moreover, there may be a lag in the growth of jobs which could be difficult to take into account. Again, policy may be directed not to the objective of short-term job creation, but towards increasing the long-term competitiveness of the area in a changing national and world economy. This poses obvious difficulties of evaluation. Going further, if the aim of policy is wealth creation (a term that was popular for a short time in the mid 1980s), or sharing prosperity (as used in *Delivering an Urban Renaissance*), any thought of evaluation becomes mind-boggling. How is wealth or prosperity to be defined (particularly in these environmentally conscious days)? Is the object to raise the average level of wealth, or the level of those who are the poorest? Such questions quickly undermine attempts at evaluation. Similar questions can be asked of current concerns with equality of opportunity and social cohesion.

Despite all the conceptual and practical difficulties, researchers have been able to draw some important conclusions from their evaluations of urban policy. For example, many of the jobs that have been 'created' would have arisen without any intervention. Indeed, if the projects that have been evaluated are typical, 'such programmes are unlikely to make more than a modest contribution to the economic regeneration of the inner cities' (Martin, S. 1989: 638). On reflection, this is perhaps unsurprising. In a complex interdependent society (and, increasingly, an interdependent world) 'local' issues are elusive. In Kirby's words (1985: 216), 'we cannot attempt to understand the complexities of local economic affairs *in situ*'. Much research corroborates this view. The experience of the SDA in Glasgow showed that, though the provision of premises attracted some firms to the area, this was 'at the expense of other parts of the city, and most jobs were filled by inward commuters anyway' (Turok 1992: 372).

Another issue in which difficulties of assessment abound (and in which myths live on) is that of the impact of property-led development. Much urban policy has been based on the assumption that property development will somehow or other stimulate economic growth and social improvements. How this is to happen has not been articulated, and there has been little detailed research on the subject. Such research as has been undertaken offers no clear conclusions, though studies 'suggest that access to markets, management abilities, and the availability of finance are more important than buildings' and 'levels of investment in product development and production technology,

together with differences in the way human resources are managed, are most significant' (Turok 1992).

Many other factors other than property will play a part in successful regeneration, such as the availability of a skilled labour force, ready finance, and an attractive environment. Moreover, property development can present its own problems: this became clear in 1990, when rental values fell as the economy dipped. The slowing down of property investment in 1990 turned into a spectacular collapse, with catastrophic effects on the construction industry and local economies.

Urban development corporations above all have failed to consider the relationship between the local economy and property development in adopting a policy of privatism: 'the attracting into the inner city of private developers whose activities can in turn demonstrate that regeneration is taking place' (Edwards and Deakin 1992: 362). We have learned that this type of urban policy can have a detrimental effect on local economic activity as, for example, when the precarious position of small local firms is challenged with competition from outside the locality. Urban policy has been characterised by short-term thinking, centred on getting the best return from particular sites (Centre for Local Economic Strategies (CLES) 1992b). There is little to support the view that property-led urban regeneration produces a trickle-down of benefits for the local disadvantaged community as the local economy improves. Conclusions from a comprehensive evaluation of urban policy suggest other policy priorities would have greater impact (Robson *et al.* 1994).

These observations on property-led urban regeneration are not to suggest that physical improvements are not needed, and to some extent there has been a return to a concern with the physical quality of places in government policy. It will need to remain a central part of urban regeneration. The point is to understand the ways in which physical regeneration opportunities link to social and economic development. This entails ensuring that 'non-physical policy interventions . . . keep the momentum of regeneration rolling forward once physical rebuilding is complete' (Carley and Kirk 1998).

Much policy takes the form of targeting resources on the most deprived areas. In practice this has had only marginal impact, because the cuts in mainstream public spending tended to fall more heavily on some of the most deprived areas, leaving them even worse off per capita (Robson *et al.* 1994). Not surprisingly therefore, there was increasing concentration of the most disadvantaged in the worst off areas. Also, as some people in these areas benefit from intervention and gain employment they are more likely to move out and be replaced with others in greater need. Housing allocation policies tend to reinforce these effects leading to increasing concentration of unemployment in the worst areas. Thus there is growing polarisation within the conurbations, with the benefits of targeting being felt most by the surrounding areas that are better placed to take advantage.

Providing an overview of urban policy impact is daunting. Researchers have underlined the interlocking nature of urban policy, continual change in programmes and the difficulty of identifying a single unambiguous set of objectives against which to measure progress, as the main difficulties. On the positive side Robson *et al.* (1994) found that regeneration funding has had a positive impact on residents' perceptions of their area. There has also been some general 'limited success for government policy', particularly in smaller cities and the outer districts of conurbations. Where well-coordinated multi-agency approaches have been taken, some policy instruments have worked well. But 'the amount of money going into urban policy is minuscule compared to the size of the problems which are being tackled . . . many of the poorer areas are not improving or at least not nearly as much as the better-off areas within the districts'. The five main conclusions of Robson's research are listed in Box 10.7, together with three principal conclusions of a review by Carley and Kirk (1998).

Does government take heed of the messages from evaluation? The speed at which new initiatives are introduced would suggest not. Ho (2003: 2) explains how 'new regeneration initiatives have often been introduced before the evaluation studies of an ongoing research initiative were completed'. City Challenge replaced the Urban Programme before evaluation was completed; the New Deal was announced while SRB evaluation was still underway. Moreover, there is little

evidence to support some aspects of policy and action. A review of the evidence base for regeneration policy found that there was much evidence on physical and economic outcomes and little on health and education outcomes. Moreover, they found that much of the evidence base was not robust and relied on a small number of case studies of 'best practice' (Tyler *et al.* 2001, quoted in Ho 2003: 3). Moreover, the influence of learning 'what works where' is probably greater now than ever and the large number of initiatives and speed of change arise in part from a desire to experiment. The major initiatives are routinely accompanied by complementary (action) research projects and other steps have been taken to improve feedback on performance such as the *Urban Sounding Board*, development of town and city indicators (Wong 2002), research seeking transferable lessons from 1990s initiatives such as the enterprise zones, and the *Towns and Cities: Partners in Urban Renaissance Project* (URBED 2002). This last project was intended 'to take the pulse of urban renaissance delivery' and involved twenty-four cities and towns in action research to identify and tackle barriers to progress. All very well, but after publishing five main reports and yet more case studies for the 2002 Summit, no more was heard. As Ho reminds us: 'the purpose of evaluation is not necessarily to learn lessons' (2003: 4).

Some lessons have been taken on board by the post-1997 government. In a review of the prospects for change, Lawless (1996) points to the need for a context which encourages local strategy building rather than centrally directed ad-hoc responses. Extraordinary as it may seem, and admittedly with the important exception of jobs and employment, problems which most affect the urban disadvantaged have received minimal attention. But this in turn calls for a new form of urban governance which would be stronger at both the regional and local community levels, and a more politically mature programme. Bailey *et al.* (1995) conclude more radically that

> cities need to be seen as an important element of the national economy and that the growth, redevelopment and improvement of these assets can and should be linked with redistributive welfare policies

as part of a strategic and comprehensive national economic policy driven by the public sector.

(Bailey *et al.* 1995: 229)

This consensus of opinion on the need for change prompted more attention to the quality of cities and their regions seeking a more integrated view of their physical, economic, social and governance qualities). The Single Regeneration Budget and integrated government offices for the regions started a trend to more coordination and consistency in action among departments. City Challenge too took a longer-term view and gave a leading role to local authorities, and sought to incorporate (although not without some difficulty) local communities. The experience of promoting partnerships has been taken forward such that the partnership working, first introduced in urban policy in 1995, and now managed competition have become requirements for almost all urban initiatives in England (Oatley 1995; Atkinson 1999a). In Scotland, the competitive element is not so fierce, but the process of negotiation between partnerships and central government is still competitive and has suffered from a lack of transparency. The Labour administration has not only continued with competition but also brought a much more concerted effort on the worst areas of deprivation, an equal emphasis on investment in people as well as the physical environment, moves towards more strategic thinking and joined-up policy on regeneration, and to some extent increased resources.

Underlying the transition from a physical policy concerned with deteriorated housing to a social policy in aid of deprived areas and then to an economic policy for strengthening the base for local growth, and most recently to policy addressing social exclusion, is the dramatic change which has taken place in the character of cities. There has been a relentless decline of population and employment in the urban areas generally, and particularly in the inner areas. (The other side of the coin is the growth in smaller towns and rural areas.) The economy of the older industrial British cities has been transformed: their manufacturing base has been eroded, and there has been little of the expanding tertiary industries to take its place. Cameron (1990: 486) suggests 'the aggregate decline of many

major British cities is inevitable and indeed desirable. Probably a growing percentage of British consumers and producers will seek locations for living and producing outside such cities.'

The notion of area-based solutions has never been stronger, partly because it is thought to be the best way to solve the problems, but also for financial reasons (there is not enough money to go round). Britain is a European leader in the design of area-based programmes (Parkinson 1998). However, a continuing deficiency of urban policies is that they assume the issue to be tackled lies in particular parts of particular cities: but 'inner city' problems is a misleading abstraction. Adapting a passage from Marc Fried (1969) (who was commenting upon the concept of poverty), 'the inner city [is] an empirical category, not a conceptual entity, and it represents congeries of unrelated problems'.

Progress will not be made by 'comprehensive action' but by identifying priority fields in which effort should be concentrated. (This, of course, is precisely what the Conservative government did, even if its choice of focus was debatable.) Most of the problems identified in inner cities or in poor suburban estates are matters of national policy relating to all areas. Thus, though poverty is undoubtedly a problem that is spatially concentrated in particular neighbourhoods, most of the residents in them are not in poverty; and most poverty is not in those neighbourhoods. The arguments *against* any area-based policy are strong (Townsend 1976). Oatley (2000) makes the point most strongly in his critique of the *Neighbourhood Renewal Strategy*.

Area-based policies are notoriously unsuccessful in addressing 'people poverty'. Concentrating resources on a small number of neighbourhoods is both administratively and politically convenient, masking the widespread nature of deprivation within society and allowing us to feel that the problem is being dealt with. These responses may at best concentrate resources in areas with high need for the wrong reasons, and at worst, seriously mislead us into thinking that we are tackling the problems when in fact we are only producing palliatives to alleviate the worst symptoms.

(Oatley 2000: 89)

To the extent that the problems relate to the deprived, it makes more sense to channel assistance to them directly, irrespective of where they live. Only to the extent that the problems are locationally concentrated, should remedies focus on specific locations – as in the case of renewal areas. Oatley goes on to call for radical alternative solutions that might 'break out of the dismal cycle of unfulfilled promises' (2000: 94). None of this is to deny the importance of directly tackling those problems of decay and disadvantage which are all too apparent in many deprived neighbourhoods. Nor is there any argument against the desirability of attempting better organisation of services at local levels, or improved coordination both within and between agencies. But such approaches are not going to solve the problem. Kintrea and Morgan (2005) remind us that

> the character of problem neighbourhoods has not changed much over 25 years, in spite of many different types of intervention. At best, neighbourhood renewal policy has perhaps stopped the worst areas becoming even worse than they would otherwise have been, and helped to sustain the quality of life for residents at a basic level.
>
> (Kintrea and Morgan 2005: 6)

What is crucial is to identify the forces which have created the problems and to establish means of stemming or redirecting them.

While the current rhetoric of urban policy is about partnership and strategy, the reality is an agglomeration of initiatives and agencies which even the professional is hard pressed to comprehend. Initiatives continue to give the impression that they are no more than short life laboratory experiments. The institutional arrangements being put into place in England are of particular concern with regional development agencies, the government offices, sectoral government departments, the regional chambers and local authorities all having a role in developing strategy and implementation, while local communities are being given the flexibility to invent their own responses. Whether these will be moulded into a responsive, integrated and efficient governmental machine remains to be seen.

Further reading

Urban policy and regeneration

There is a wide literature on urban policies. An edited review by Johnston and Whitehead (2004) *New Horizons in British Urban Policy* is one of the few sources on the post-1997 Labour government's performance; see in particular Jones and Ward 'Neo-Liberalism, crisis and the city' and Healey (2004) 'Towards a social democratic policy agenda for cities'. Other general texts include Atkinson and Moon (1994) *Urban Policy in Britain*, which provides a historical review and evaluation, and Blackman (1995) *Urban Policy in Practice*. Earlier texts (and still relevant since the core problems are not so different) are Lawless (1989) *Britain's Inner Cities*, Robson (1988) *Those Inner Cities* and MacGregor and Pimlott (1990) *Tackling the Inner Cities*. See also Turok and Shutt (1994), 'Urban policy into the 21st century' (special issue of *Local Economy*), Lawless (1996) 'The inner cities: towards a new agenda'; Oatley (1998) *Cities, Economic Competition and Urban Policy*; Atkinson (1999) 'Urban crisis', Hambleton and Thomas (1995) *Urban Policy Evaluation: Challenge and Change*. Healey *et al.* (1995a) *Managing Cities* explores the links between local problems and global economic forces. A special edition of *Planning Practice and Research* (10 (3–4)) was devoted to urban policy. For a practical view see Burwood and Roberts (2002) *Learning from Experience: The BURA Guide to Achieving Effective and Lasting Regeneration*. The nature of 'community' and how it can be 'empowered' in urban regeneration is considered at length in Chanan (2003) *Searching for Solid Foundations: Community Involvement in Urban Policy*.

Housing renewal

The ODPM commissioned a series of reports, *The Evaluation of English Housing Policy 1975–2000* from a consortium of universities between 2003 and 2004; of particular value here are Kintrea and Morgan (2005) *Housing Quality and Neighbourhood Quality* and the general overview of findings by Stevens and Whitehead (2005) *Lessons from the Past, Challenges for the Future for Housing Policy*. All the reports are available on the ODPM website under housing research. There are few recent texts on housing renewal. Bramley *et al.* (2004a) consider the relationships among housing, regeneration and planning issues; Balchin (1995) *Housing Policy* places renewal within its wider context. See also Balchin and Rhoden (1998) *Housing: the Essential Foundations* and Wood (1991) 'Urban renewal', in Alterman and Cars (eds) *Neighbourhood Regeneration: An International Evaluation*.

Urban development corporations and regeneration companies

English Partnerships is producing material on URCs though so far it has been mostly promotional, for example (2004) *Urban Regeneration Companies: Coming of Age* and it also hosts a website for the companies. See also Amion Consulting (2001) *Urban Regeneration Companies: Learning the Lessons* and Parkinson and Robson (2000) *Urban Regeneration Companies: A Process Guide*. There is a large number of writings on urban development corporations, predominantly of a highly critical nature. They include Brownill (1990) *Developing London's Docklands*; N. Lewis (1992) *Inner City Regeneration*; National Audit Office (1988) *Urban Development Corporations* and (1993) *Regenerating the Inner Cities*; and Thornley (1993) *Urban Planning under Thatcherism* (Chapter 8). Imrie and Thomas (1999) *British Urban Policy: An Evaluation of the Urban Development Corporations* contains case studies. The Final Evaluation was undertaken in three projects reporting in 1998: Roger Tym and Partners (1998) *Urban Development Corporations: Performance and Good Practice*, Centre for Urban Policy Studies (1998) *The Impact of Urban Development Corporations in Leeds, Bristol and Central Manchester* and Cambridge Policy Consultants (1998) *Regenerating London Docklands*.

City Challenge and competitive bidding

Among the many useful studies are Bailey and Barker (1992) *City Challenge and Local Regeneration Partnerships*, de Groot (1992) 'City Challenge', Davoudi and Healey (1994) *Perceptions of City Challenge Processes*, Oatley (1995) 'Competitive urban policy and the regeneration game' and Oatley and Lambert (1995) 'Evaluating competitive urban policy'. Numerous case studies of particular city challenge

partnerships have been written including Davoudi (1995a) 'City Challenge: a sustainable mechanism or temporary gesture', Oc *et al.* (1997) 'The death and life of City Challenge' and Taussik and Smalley (1998) 'Partnerships in the 1990s'.

Single Regeneration Budget

The bidding guidance documents are still available on the ODPM website and provide a full explanation of how the SRB operated together with information on progress of schemes agreed during previous rounds. A recent evaluation is by Rhodes *et al.* (2002) *Lessons and Evaluation Evidence from Ten Single Regeneration Budget Case Studies*. Early evaluation of SRB is provided by the HC Environment Committee report (1995) on the *Single Regeneration Budget* and Mawson (1995) *The Single Regeneration Budget*. Other evaluations include Foley *et al.* (1998) 'Managing the challenge', Hall and Nevin (1999) 'Continuity and change' and Brennan *et al.* (1998) *Evaluation of the Single Regeneration Challenge Fund Budget: A Partnership for Regeneration*.

Partnerships

The concept of partnership and its progress in urban policy is recorded in Bailey *et al.* (1995) *Partnership Agencies in British Urban Policy*, which also contains a number of case studies. See also Lawless (1994) 'Partnership in urban regeneration in the UK', Hastings and McArthur (1995) 'A comparative assessment of government approaches to partnership with the local community', Nevin and Shiner (1995) 'Community regeneration and empowerment', the DETR *Local Strategic Partnerships: Consultation Document* (2000) and Carley *et al.* (2000b) *Urban Regeneration through Partnership: A Study of Nine Regions*. For a theoretical perspective on the value of partnerships see Mackintosh (1992) 'Partnership: issues of policy and negotiation' and Hastings (1996) 'Unravelling the process of "partnership" in urban regeneration policy'. A 1999 report of the National Audit Office is *English Partnerships: Assisting Local Regeneration*.

Scotland, Wales and Northern Ireland

Many of the references cited above cover practice across different parts of the UK. A critical account of Scottish urban policy is given by Turok (2004) 'Scottish urban policy'; earlier reviews include McCarthy (1999) 'Urban regeneration in Scotland', McCrone (1991) 'Urban renewal' and Boyle (1993) 'Changing partners'. For a critical review of the setting up of Scottish Enterprise and Highlands and Islands Enterprise, see Danson *et al.* (1989) 'Rural Scotland and the rise of Scottish Enterprise' and Hayton (1993) 'Scottish Enterprise' (the annual reports of these organisations give a different perspective). A research report by McAllister (1996) reviews *Partnership in the Regeneration of Urban Scotland* and Tarling *et al.* (1999) provide *An Evaluation of the New Life for Urban Scotland Initiative*. See also Scottish Office, *Programme for Partnership: Urban Regeneration Policy* (1995) and McCarthy and Newlands (1998) *Governing Scotland: Problems and Prospects – The Economic Impact of the Scottish Parliament.*

For reviews of urban policy in Wales see Alden and Romaya (1994) 'The challenge of urban regeneration in Wales'. See also Welsh Office Paper, *Programme for the Valleys: Building on Success* (1993), and a more critical account by Morgan (1995) 'Reviving the valleys?'. Berry and McGreal (1995a) 'Community and inter-agency structures in the regeneration of inner-city Belfast' give a perspective of urban policy in Northern Ireland.

Evaluation

Many of the references mentioned so far include some evaluation of particular programmes or projects but there is a body of literature which takes evaluation of urban policy as its central theme, in particular Ho (2003) *Evaluating British Urban Policy: Ideology, Conflict and Compromise.* Other principal sources are Robson *et al.* (1994) *Assessing the Impact of Urban Policy*, Tyler *et al.* (2001) *A Review of the Evidence Base for Regeneration Policy and Practice*, and the DETR report by Robson *et al.* (2000) *The State of English Cities.* A broader perspective, including consideration of the methodological problems of evaluation, is given in the edited text by Hambleton and Thomas (1995) *Urban Policy Evaluation.* See also Pacione

(1997) 'The urban challenge', Pantazis and Gordon (2000) *Tackling Inequalities: Where Are We Now and What Can Be Done?*, Edwards (1997) 'Urban policy' and Shaw and Robinson (1999) 'Learning from experience?'.

On the contribution of property development to urban regeneration, see Healey *et al.* (1992b) *Rebuilding the City: Property-led Urban Regeneration.* Evaluations of recent initiatives are made by Oatley (2000) 'New Labour's approach to age-old problems' (and other articles in the same volume: *Local Economy* 15(2)) and Lawless and Robinson (2000) 'Inclusive regeneration?'.

Notes

1 The evaluations of renewal areas make use of surveys undertaken by Couch and Gill (1993) and Austin (1995). The 2000 Green Paper on Housing *Quality and Choice: A Decent Home for All*, notes that 132 renewal areas have been designated with proposed investment of £1.75 billion.

2 This updates 1992 guidance and complements Circular 17/96 *Private Sector Renewal: A Strategic Approach*, Circular 5/03 *Housing Renewal Guidance* and *Running and Sustaining Renewal Areas: Good Practice Guide* (1999). See also ODPM (2002) *What Works? Reviewing the Evidence Base for Neighbourhood Renewal.*

3 C. Wood (1994) notes that Birmingham City Council had a backlog of 8,000 applications in 1993 which would have used up the entire budget for many years to come. The 1996 changes were implemented by the Housing, Grants, Construction and Regeneration Act 1996, and most recent changes by the 2002 Regulatory Reform Order (Housing Assistance).

4 The figures are from Kintrea and Morgan (2005), who draw on Wilcox (2002). DETR published *Sustainable Estate Regeneration: A Good Practice Guide*, which compares success through Estate Action, the SRB and mainstream funding.

5 ODPM Update Newsletter 5 (2004): 8.

6 In 2002 the target was amended to include the goal that 70 per cent of vulnerable households in private housing would be in decent homes by 2010. A series of factsheets on the decent homes standard and the

Housing Act 2004 is available on the ODPM website. The 2000 policy statement was *Quality and Choice: A Decent Home for All – The Way Forward for Housing*.

7 House of Commons ODPM: Housing, Planning, Local Government and the Regions Committee Fifth Report, *Decent Homes* (2004). See also the *Government's Response to the Report*. For an explanation of the HHSRS see Housing Health and Safety Rating System Guidance (Version 2) (2004), ODPM.

8 Local authorities were required to complete an appraisal of options during 2005, to determine the method of delivery of 'decent homes'.

9 There were twelve community development projects in all: in Birmingham, Coventry, Cumbria, Glamorgan, Liverpool, Newcastle, Newham, Oldham, Paisley, Southwark, Tynemouth and West Yorkshire. Additionally, 1974 saw the introduction of a small number of *comprehensive community programmes* in areas of 'intense urban deprivation'. In the wake of these, large numbers of studies were undertaken.

10 There were eight CATs: Birmingham, Cleveland, Leeds/Bradford, Liverpool, London, Manchester/Salford, Nottingham/Leicester/Derby, and Tyne and Wear.

11 The twelve English UDCs were (with their date of designation) London Docklands (1981); Merseyside (1981); Trafford Park (1987); Black Country (1987); Teesside (1987); Tyne and Wear (1987); Central Manchester (1988); Leeds (1988); Sheffield (1988); Bristol (1989); Birmingham Heartlands (1992); Plymouth (1993). Details of their expenditure and outputs are given in the twelfth edition of this book.

12 The partnerships committee is likely to comprise ten members: two independent private sector appointments (one will chair), three Milton Keynes Council members, two English Partnerships Board members and three Local Strategic Partnership members (one each from the Chamber of Commerce, the Health and the Community Sectors); ODPM, *Sustainable Communities: An Urban Development Area for Milton Keynes: Consultation Summary and Decision Document* (London: ODPM, 2004).

13 The first round authorities completed their five-year programmes in 1997 and were Bradford, the Dearne Valley Partnership (Barnsley, Doncaster and Rotherham), Lewisham, Liverpool, Manchester, Middlesbrough, Newcastle, Nottingham, Tower Hamlets, Wirral and Wolverhampton. The second round five-year programme (which was open to all fifty-seven urban programme areas) ended in 1998. The authorities were Barnsley (the only authority to win in both rounds), Birmingham, Blackburn, Bolton, Brent, Derby, Hackney, Hartlepool, Kensington & Chelsea, Kirklees, Lambeth, Leicester, Newham, North Tyneside (ended in 1999), Sandwell, Sefton, Stockton-on-Tees, Sunderland, Walsall and Wigan.

14 EP had previously favoured deprived areas, for example those in receipt of City Challenge funding. See English Partnerships, *Corporate Plan 2003–06* (2003).

15 Information on NLUD is available at www.nlud. org.uk.

16 This comparison is calculated at twenty-eight houses to the hectare, which is the average rate for development on previously developed land according to the DETR's report on *Land Use Change in England* (1999).

17 The Millennium Communities are at Allerton Bywater (near Leeds), Greenwich Millennium Village, Hastings, New Islington (Manchester), Oakgrove (Milton Keynes), South Lynn (King's Lynn) and Telford. The English Partnerships website has more details of each community, www.englishpartnerships. co.uk, and some have their own website.

18 This quotation is taken from the SRB Bidding Guidance Round 6 (para. 1.3.1). The guidance for the earlier rounds one to six was extensive and can be found on the ODPM website.

19 DETR, SRB Bidding Guidance Round 6 (1999). See DETR/DTLR/ODPM Annual Reports and Rhodes *et al.* (2002) for a more detailed breakdown of outputs.

20 The first bidding round was evaluated by the HC Select Committee on the Environment in its 1995 report Single Regeneration Budget, where the concept was generally supported subject to a number of recommendations to reduce bureaucracy, improve consistency, and increase the involvement of voluntary and community groups.

21 Most government claims about new funding are contested and it gets increasingly difficult to untangle

22 The Social Exclusion Unit has established a programme to produce a *National Strategy for Neighbourhood Renewal* that will be based on the findings of eighteen Policy Action Teams which are undertaking an intensive programme of policy development.

government funding streams as reports on spending tend to be linked to cross-cutting objectives rather than programmes.

23 The seventeen first-round pathfinder partnerships are in the local authority areas of Birmingham, Bradford, Brighton and Hove, Bristol, Hackney, Hull, Leicester, Liverpool, Manchester, Middlesbrough, Newcastle-upon-Tyne, Newham, Norwich, Nottingham, Sandwell, Southward and Tower Hamlets. The twenty-two second round partnerships are in Birmingham, Brent, Coventry, Derby, Doncaster, Hammersmith and Fulham, Haringey, Hartlepool, Islington, Knowsley, Lambeth, Lewisham, Luton, Oldham, Plymouth, Rochdale, Salford, Sheffield, Southampton, Sunderland, Walsall and Wolverhampton.

24 Findings from the pathfinder authorities are given in the DETR report, *New Deal for Communities: Learning Lessons: Pathfinders' Experiences of NDC Phase 1* (1999).

25 Another important source for the White Paper is the New Commitment to Regeneration approach developed by the Local Government Association and the Cabinet Office. Details are at http://www.cabinet-office.gov.uk/seu/index/national_strategy.htm.

26 Robson *et al.* (2000). See also Parkinson *et al.* (2005) and the ODPM's glossy *A Tale of Eight Cities* (2004).

27 See also the government's statement on the core cities *Making it Happen: Urban Renaissance and Prosperity in our Core Cities* (ODPM, 2004) and Parkinson *et al.* (2004) *Competitive European Cities: Where do the Core Cities Stand?*, which has a very useful bibliography.

28 The core cities are Birmingham, Bristol, Leeds, Liverpool, Manchester, Newcastle, Nottingham and Sheffield.

29 The performance of the three pilot companies is being evaluated and a research report is due. An earlier 'stocktake' of progress of all the companies was published in 2004.

30 Reported in the *Guardian*, quoted in Johnson and Whitehead (2004: 5).

31 Select Committee on Environment, Transport and Regional Affairs Eleventh Report on the Proposed Urban White Paper (1999). The Select Committee report explains why other bodies have made such criticisms.

32 An Urban Exchange Initiative was started under the UK Presidency in 1998 with the aim of preparing a non-binding informal framework for national approaches to urban regeneration. Papers and case studies were prepared but the overall outcome is unknown.

33 The Partnership Priority Areas were designated by the Scottish Office in response to proposals from local authority led partnerships, covering the whole of their area. The twelve areas are Great Northern (Aberdeen), Ardler (Dundee), Craigmillar and North (Edinburgh), East End, North and Easterhouse (Glasgow), Inverclyde, Motherwell North (North Lanarkshire), Paisley (Renfrewshire) and North Ayr (South Ayrshire).

11 Transport planning

What nobler agent has culture or civilization than the great open road made beautiful and safe for continually flowing traffic, a harmonious part of a great whole life?

Frank Lloyd Wright 1963

Perhaps the most important contribution to reduction in congestion is to acknowledge the close links between land use and transport, and to make a greater use of the planning system to control transport growth. The principal objective must be to maximize accessibility and minimize trip lengths. These twin policy objectives would guarantee the greatest levels of demand for public transport, cycling and walk modes.

Banister 2002: 255

Mobility and accessibility

Transport is many things: it is a means of getting from one place to another. It includes a range of very different forms of travel – walking, cycling, and travelling by car, bus, train or aircraft. A journey to work on the London Underground is very different from a country holiday tour. Except perhaps for the tourism type of journey, transport is unlike other goods in that it is a means to an end: it is not an end in itself. Indeed, much transport is an impediment to the enjoyment of something else. It is a means of providing access. Mobility is not important of itself: its importance is in providing access. Thus transport should not be considered in isolation from the social and economic activities that drive the demand for it and the spatial distribution of activities which determines the pattern of journeys. Yet the debate on transport often forgets this elementary point, and focuses on mobility: faster roads, faster trains, and more frequent buses.

The advantage of focusing on accessibility rather than mobility is that it opens up the possibility of alternative means: changing land use relationships for

example. As an advertisement on a condominium tower above a Toronto metro station neatly pointed out, 'If you lived here, you would be home now'. The greater is the accessibility, the lesser the need for 'transport'. Thus, transport planning is much more than the building of roads, even though this has not always appeared to be the case. It should involve a consideration of the relationship between different land uses, and between land uses and transport feasibilities, as well as the relationships between different transport modes and their relative effectiveness in meeting economic, financial, social, and environmental goals. There is nothing profound in these observations but, until recently, transport policy appeared to deny their validity. Roads have formed the major focus of policy. It is therefore fitting that we start by considering road traffic.

The growth of traffic

Between 1950 and 1960, the number of vehicles on the roads of Britain more than doubled, from 4.0 million to 8.5 million. The number more than doubled again by the end of 1980, to 19 million. By 1997 the number had risen to 27 million. The most dramatic increase was in cars, from around 2 million in 1950 to 25 million in 2003 (Table 11.1). The proportion of households owning a car rose from a mere 14 per cent in 1951 to 73 per cent in the year 2003 (Table 11.2). In terms of total road traffic (measured in billion vehicle kilometres) the increase has been from 53 in 1950 to 490 in 2003 (Table 11.3). Despite a large road-building programme,

including some 3,476 km of motorway, the increase in the length of the road network has been far less than the increase in traffic. There was 392,321 km of road in the UK in 2003, an increase of only about a third over the 1951 total of 297,466 km (which was only fractionally greater than the 1909 figure of 282,380). The consequence, of course, has been that roads have become far more crowded, though as we will see, this should not be seen as an argument that more roads would have reduced congestion.

Table 11.1 *Number of vehicles, Great Britain 1950–2003 (thousands)*

	1950	1960	1970	1980	1990	2000	2003
Private cars	1,979	4,900	9,971	14,660	19,742	23,196	24,985
Goods vehicles	439	493	545	507	482	418	426
Motor cycles	643	1,583	923	1,372	833	825	1,005
Public transport vehicles	123	84	93	110	115	86	96
All vehicles	3,970	8,521	13,548	19,199	24,673	28,898	31,207

Source: Transport Statistics Great Britain 2004 (Table 9.1) (other special categories of vehicles are not included in this table)

Table 11.2 *Proportion of households with cars, Great Britain 1951–2003 (%)*

	1951	1960	1970	1980	1990	2000	2003
No car	86	71	48	42	33	27	26
1 car	13	27	45	44	44	45	44
2 cars	1	2	6	13	19	23	24
3+ cars	—	—	1	2	4	5	5

Source: Transport Statistics Great Britain 2004 (Table 9.14)

Table 11.3 *Road traffic, Great Britain 1950–2003 (billion vehicle kilometres)*

	1950	1960	1970	1980	1990	2000	2003
Cars and taxis	25.6	68.0	155.0	215.0	335.9	376.8	393.0
Light vans	7.8	14.7	18.9	23.1	35.7	52.3	57.9
Goods vehicles	11.2	15.7	19.0	22.6	24.9	28.2	28.5
All motor vehicles	53.1	112.3	200.5	271.9	396.5	467.1	490.3
Pedal cycles	19.9	12.0	4.4	5.1	5.3	4.2	4.5

Source: Transport Statistics Great Britain 2004 (Table 7.1)

Traffic forecasts are, of course, only estimates, and no more reliable than weather forecasts – less so in fact. (Forecasts based on other forecasts are particularly suspect: the traffic forecast is based mainly on assumed economic growth and, of course, on an absence of serious impediments to car ownership and use.) The *1997 Road Traffic Forecast* exercise suggested an increase in total traffic of between 3 per cent and 15 per cent by 2001, and between 36 per cent and 84 per cent by 2031. The increase for car traffic is forecast at between 3 per cent and 14 per cent by 2001, and between 30 per cent and 75 per cent by 2031 (Table 11.4). In fact the increase in total traffic between 1997 and 2001 was 5.4 per cent and the increase in car traffic was 4.4 per cent. Current forecasts for the period between 2000 and 2010 are for an increase of between 22 and 29 per cent in car and taxi traffic and between 23 and 29 per cent increase in all traffic. Bus and coach traffic is forecast to fall slightly.[1]

With the emphasis which is so often placed on increases in cars and traffic, it is easy to forget that over a quarter of households do not have a car. The proportion without a car is higher in the North and in Scotland. It is also higher for the economically inactive and for unskilled manual workers. Generally, car ownership increases in inverse proportion to size of town, and is highest in rural areas where 85 per cent of households have a car.

Car forecasts have two components: car ownership and car use. Car ownership is still increasing. Indeed, the 'saturation' level has not yet been reached in any country (not even in the USA). It is theref[...] to guess what this level may be, and, of course, it may well differ among countries. The saturation level is assumed to be the level observed in the highest income households of each type in recent years. (It can therefore change.) Car use is also difficult to predict. It fell during the period 1973–6 when GDP fell and real fuel prices rose, but there was no fall when similar conditions applied during 1979–82. Use of second and third cars is not lower than the use of first cars: in fact it is higher. Moreover, of the vehicles on the road, four-fifths are cars. Most of the remainder are goods vehicles. The forecasts for these are calculated separately for light vans and heavy goods vehicles. Light goods traffic is forecast to increase by between 85 per cent and 251 per cent by 2031. Heavy goods traffic is more of a problem. Previous forecasts had proved to be far too low (the 'high' forecast for the period 1982 to 1987 was 4 per cent; the actual was 22 per cent). A recent forecast gave an increase for rigid heavy goods vehicles of between 15 per cent and 56 per cent by 2031. For articulated heavy goods vehicles the range was from 96 per cent to 165 per cent. The point does need further elaboration: making traffic forecasts is a hazardous venture!

The forecasts did not deal with rail traffic. Total passenger mileage has fluctuated but recent years have seen an increase to a level slightly above the rate in the 1950s (Table 11.5). However, in terms of journeys by the constituent systems a different picture emerges. The national rail network experienced a large though

Table 11.4 *National road traffic forecasts by vehicle type 1996–2031 (1996=100)*

	Cars			Total traffic		
	Low	Central	High	Low	Central	High
2001	103	109	114	103	109	115
2011	116	127	137	117	128	139
2021	126	143	159	129	146	163
2031	130	153	175	136	160	184

Source: DETR (1997) *National Road Traffic Forecasts (Great Britain)*.

Note: The 10-Year Transport Plan (2000) predicts greater reductions in car and lorry traffic.

Table 11.5 Passenger travel by mode, Great Britain 1952–2003 (in billion passenger km (bpk) and as percentage of total)

	1952		1960		1970		1980		1990		2000		2003	
	bpk	%	bpk	%	bpk	%	bpk	%	bpk	%	bpk	%	bpk	%
Bus	92	42	79	28	60	15	52	11	46	7	47	6	47	6
Car	58	26	139	49	297	74	388	79	588	85	639	85	678	85
Motor cycle	7	3	11	4	4	1	8	2	6	1	5	1	6	1
Pedal cycle	23	11	12	4	4	1	5	1	5	1	4	1	5	1
All road*	180	82	241	86	365	91	453	92	645	94	695	93	736	93
Rail	39	18	40	14	36	9	35	7	39	6	47	8	93	49
Air	0.2	0.1	0.8	0.3	2.0	0.5	3.0	0.6	5.2	0.8	7.6	1.0	9.1	1.2
All modes	219	100	282	100	403	100	491	100	689	100	749	100	838	100

Source: Transport Statistics Great Britain 2004 (Table 1.1)

Note: * Includes vans and taxis.

erratic fall from 1950, but this has been redressed since 1995 with now just over 1 billion journeys in 2003–4, an increase of more than 25 per cent over the 1990–1 performance. Thus the number of journeys by rail was higher in 2003–4 (1.014 billion) than in 1950 (1.01 billion), and so too is the distance travelled (40.9 billion kilometres against 32.5 billion), though the journeys are taken over a network which is about half the size of the 1950 equivalent: it is now 16,652 km and was 31,336 km in 1950. There have been increases in journeys on London Underground, and on new systems such as London Docklands Light Railway, Manchester Metrolink (Altram) and Sheffield Supertram (Stagecoach). Glasgow Underground has seen no change, while Tyne and Wear (Nexus) has returned to early levels of use experienced during the 1990s after a considerable dip to 2001. What stands out from these passenger journey figures shown in Table 11.6 is the significance of the London Underground, which is almost on a par with the whole of the national rail network in terms of numbers of journeys, although passenger kilometres travelled are much less at 7.3 billion in 2003–4 compared with 40.9 billion for the rail network. It is rather disappointing to see the very slow increase in light rail traffic which reflects the difficulty of implementing these projects; this is discussed later in this chapter.

Freight traffic by rail has declined steadily, until very recently when there has been a small increase. By contrast, well over half of freight (measured in billion tonne kilometres) went by road (Table 11.7). It has been held for some time that since road and rail serve mainly different markets, there is little scope for transferring road freight to the railways. As is discussed later in the context of integrated transport policies, this view has now changed. One dramatic example was the decision in 2004 of the Post Office to abandon rail completely in favour of road haulage of post.

The huge increase in car traffic and the relative decline in bus and rail travel do not signify a mass transfer from public transport. On the contrary, the figures show that most of the increase in car usage is newly generated traffic. Total passenger travel has increased enormously: people are travelling more than they used to. Though this has been subject to little research, most of the increase must have arisen from the dispersed pattern of activities and the increased separation of home and work. This has been facilitated by improvements to transport infrastructure, illustrating the impact of 'transport supply' on demand. Moreover, since this new traffic is based on dispersal, it may be very difficult to change it to a public transport mode.

Transport policies

Public policy on transport has a long history (Barker and Savage 1974), but postwar policy began with a plan for a network of new trunk roads (which was not implemented) and a plan for the nationalisation of road haulage and the railways (which was). Much energy was dissipated in the nationalisation and denationalisation processes, and more attention was paid to ownership and control than to transport policy. Experience with the centralised and, later, the decentralised British Railways left a legacy of unease about railway spending in the Transport Department which persisted (Truelove 1992; Kay and Evans 1992). The postwar history of the railways is complex and, at the time of writing, still uncertain. Nevertheless, there are some signs for hope with both the national railways and the slowly increasing local urban networks. With road haulage, the role of government since denationalisation has been largely restricted to safety controls, though there has been acrimonious argument over axle weights (Juggernauts). Bus services have been particularly affected by conflicting political philosophies. Indeed, fights over fares policy were a significant factor in the Conservative government's decision to abolish the Greater London Council and the Metropolitan County Councils.

Until recently, cycling and walking have been given relatively little attention: in fact neither had a place of any significance in the directory of travel. The major focus of transport policy has been on roads, and only in recent years has it become generally accepted that either cycling or walking have to be considered within a wider context of integrated transport planning. The starting point for any discussion of this must be the Buchanan Report (1963) on *Traffic in Towns*.

Table 11.6 Rail travel in Great Britain 1990–2004 (million passenger journeys)

	1990–1	1995–6	1996–7	1997–8	1998–9	1999–2000	2000–1	2001–2	2002–3	2003–4
National rail network	809	761	801	846	892	931	957	960	976	1,014
London Underground	775	784	772	832	866	927	970	953	942	948
Docklands Light Railway	8	14	17	21	28	31	38	41	46	49
Glasgow Underground	14	14	14	14	15	15	14	14	13	13
Tyne and Wear Metro	44	36	35	35	34	33	33	33	37	38
Altram Manchester Metrolink		13	13	14	13	14	17	18	19	19
Sheffield Supertram		5	8	9	10	11	11	11	12	12
West Midlands Metro						5	5	5	5	5
Croydon Tramlink							15	18	19	20
All rail	1,627	1,660	1,771	1,858	1,967	2,061	2,054	2,068	2,117	
All light rail	68	73	79	85	94	120	127	136	142	

Sources: Focus on Public Transport 1999 (Table 5) and Transport Statistics Great Britain 2004 (Table 6.2)

Table 11.7 Domestic freight transport by mode, Great Britain 1953–2003

| | Goods moved in billion tonne kilometres | | | | | | |
	1953	1960	1970	1980	1990	2000	2003
Road	32	49	85	93	136	158	159
Rail	37	30	25	18	16	18	19
Water	0	20	23	54	56	67	*67
Pipeline	0.2	0	4	10	11	11	10
All modes	89	100	136	175	219	254	256

Sources: *Transport Trends 2000* (Table 1.14) and *Transport Statistics Great Britain 2004* (Table 4.1)

Note: * 2002 figure; water transport figures not comparable with later ones, but are included in the total.

Buchanan Report 1963

It is traffic in towns which strongly shows that the car is a 'mixed blessing' (to use the title of an earlier book by Buchanan). As a highly convenient means of transport, it cannot, other things being equal, be bettered. But its mass use restricts its benefits to car users, imposes penalties (in congestion, pollution and reduction of public transport) on non-motorists, involves huge expenditure on roads, and at worst plays havoc with the urban environment. A major landmark in the development of thought in this field was the 1963 Buchanan Report. This eloquent survey surmounted the administrative separatism which had prevented the comprehensive coordination of the planning and location of buildings on the one hand, and the planning and management of traffic on the other. With due acknowledgement to the necessarily crude nature of the methods and assumptions used, the report proposed, as a basic principle, the canalisation of larger traffic movements on to properly designed networks that would service areas within which environments suitable for a civilised urban life could be developed. The two main ideas here were for primary road networks and environmental areas.

There must be areas of good environment – urban rooms – where people can live, work, shop, look about and move around on foot in reasonable freedom from the hazards of motor traffic, and there must be a complementary network of roads – urban corridors – for effecting the primary distribution of traffic in the environmental areas.

The simplicity of this concept is in stark contrast to the complexity and huge cost of its application. But what were the alternatives? Buchanan stressed that the general lesson was unavoidable: 'if the scale of road works and reconstruction seems frightening, then a lesser scale will suffice provided there is less traffic'. The great danger, in Buchanan's view, lay in the temptation to seek a middle course between a massive investment planning and a curtailing of the use of vehicles 'by trying to cope with a steadily increasing volume of traffic by means of minor alterations, resulting in the end in the worst of both worlds: poor traffic access and a grievously eroded environment'. (This, of course, is precisely what has happened.)

An improvement of public transport is no answer to these problems, though it must be an essential part of an overall answer. The implication is that there must be a planned coordination between transport systems, particularly with regard to journeys to work in concentrated centres. On this, Buchanan recommended that transport plans should be included as part of the statutory development plans. This was accepted and passed into legislation by the Town and Country Planning Act 1968.

Road policies in the 1980s

Despite the fact that the Thatcher government (1979–90) had no doubts on the economic and social value of roads, its concern for reducing public expenditure took priority during most of the 1980s. White Papers on the trunk road system stressed the importance of roads for economic growth, but 'national economic recovery' demanded a close rein on public expenditure. The top priority within a restricted road-building programme was for 'roads which aid economic recovery and development' (foremost among which was the M25). Other priorities were for environmental improvement (by the building of bypasses), road maintenance ('preserving the investment already made') and improved road safety. It was expected that the balance of the programme was likely to change as the major inter-urban routes were completed. Increasingly (so it was thought) the emphasis would shift to schemes which were required to deal with specific local problems. This perception was dramatically altered by the 1989 traffic forecasts, and a White Paper of that year, *Roads for Prosperity*, which announced a massive increase in road-building.

Between 1980 and 1990, the road network increased by 18,400 km. By 1989, investment in trunk roads was nearly 60 per cent higher in real terms than ten years earlier. The policies continued to follow earlier ones in emphasising the importance of roads to economic growth (despite little evidence on the matter). Traffic on new trunk roads and motorways often exceeded forecasts, and this 'success', as it was eccentrically called by the Transport Department, demanded more investment in further lanes and traffic systems as for example in the case of the M52. This 'massive' investment (as it was advertised by the government) was of course a response to the 1989 traffic forecasts, although it was stressed that the forecasts were not really forecasts at all.

> They are in no sense a target or an option; they are an estimate of the increase in demand as increased prosperity brings more commercial activity and gives more people the opportunity to travel, and to travel more frequently and for longer distances.
> (White Paper, *Trunk Roads, England: Into the 1990s*, 1990)

Yet, if the official forecasts were not to be used as a basis for policy, what is the alternative? The forecasts were based on the assumption that the demand for roads would be met with an appropriate supply, that there would be no significant policy of traffic restraint, and that attitudes towards motoring (and its cost) would not change. Though the official stance on such issues was a coy one, they raise important questions which rose to the forefront of debate (Goodwin *et al.* 1991). Increasing concern about the impact of specific road construction schemes, coupled with a more general concern about traffic congestion, resulted in a near-paralysis of policy. This had the tangible advantage of assisting in the restraint of public expenditure, which was further helped by the 1994 roads review. Before discussing this, however, it is useful to examine how the need for roads was approached by the Department of Transport. Successive governments have discovered that 'need' (whether for roads, health services, or houses) is an elusive concept, and many have tried to find ways of giving it an objective basis. Not only is this appealingly rational, but also it changes the nature of the debate. Argument can be settled by recourse to the 'facts'. In a democracy, such nonsense encounters stiff resistance. The history of assessing the need for roads is a good illustration of this.

Trunk Road Assessment 1977

An independent assessment by the newly established Advisory Committee on Trunk Road Assessment of the methods used for assessing the need for roads was highly critical. The conventional methodologies were judged to be essentially 'extrapolatory', 'insensitive to policy changes' and partly self-fulfilling. Public concern about road planning was shown to be well founded.[2] There were, however, no easy solutions: indeed the issues were inherently complex. The way forward lay in a more balanced appraisal, 'ongoing monitoring arrangements', and more openness – with no attempt 'to disguise the uncertainties inherent in the whole process.

The Labour government's response was positive, particularly since the Buchanan Report was followed by an expanded road-building programme, reports from the newly established advisory Committee on Trunk Road Assessment in 1977, and from the Standing Advisory Committee on Trunk Road Assessment (SACTRA) report on Urban Road Appraisal in 1986. More were to come, but as the list of official publications in the appendix clearly shows, reviews of this nature seemed to become part of the transport scene! Indeed, it seems that this is a field of constant concern, an alternative to 'not in my back yard'. It would be interesting (if perhaps somewhat boring) to analyse the various reports as an example of government by perpetual inquiry.

The Labour government's 1978 White Paper, *Report on the Review of Highway Inquiry Procedures*, represented a marked change in approach. National policies were to be set out for parliamentary debate in White Papers: these would 'also serve as an authoritative background against which local issues can be examined at public inquiries into particular road schemes'. It was hoped that this would avoid the confusion at local inquiries between national policies and their application in specific areas. It was pointed out, however, that this would work only if the methods of assessing national needs (what the Committee termed 'a highly esoteric evaluation process') were acceptable. Since these methods could not be properly examined at local inquiries (or, indeed, by Parliament), they were to be subject to 'rigorous examination' by the Committee (now elevated to the Standing Advisory Committee, SACTRA). The Committee's report was published in 1979, under the title *Trunk Road Proposals: A Comprehensive Framework for Appraisal*.

This report examined the techniques used to evaluate the economic value of proposed road schemes. The department's system of cost–benefit analysis (the COBA programme) was criticised for the narrowness of its approach, and certain changes followed. For example, instead of using only one traffic forecast, high and low levels were introduced.

Urban Road Appraisal 1986

The 1979 SACTRA report dealt with inter-urban roads; in 1984 it was given the task of assessing the traffic, environmental, economic and other effects of road improvements within urban areas. Some urgency for a review was added by the abolition of the GLC and the metropolitan county councils: this gave the Secretary of State, the London boroughs and the metropolitan district councils new responsibilities for tackling the transport problems of major urban areas. The Committee's report, *Urban Road Appraisal*, together with the government response, was published in 1986. Much of the report sets out recommendations as principles rather than as detailed prescriptions, and it thus seems to have more than a fair share of platitudes. As a result, though the government accepted many of these, it was inevitable that they 'can only be applied once detailed guidance on their application has been prepared'. Moreover, 'where further development or research is needed, the detailed guidance required to implement them will take some time to prepare, and progress must be subject to the availability of resources'. The Committee could hardly have found this a very encouraging response!

There were four broad areas which the Committee saw as contributing to the nature and extent of change needed; the Committee concluded that one reason why many road schemes took so long to bring to completion was that 'the opportunity to debate their justification comes too late in the procedural chain'. This was accepted, as was a recommendation that national and local objectives should be treated separately, 'so that conflicts and common denominators can be readily seen'. However, a proposal that there should be a two-stage inquiry process for the larger and more complex schemes 'presented difficulties'. Though it was regarded as 'a constructive proposal', the government was 'not convinced of the practicality of separating the examination of policy options from consideration of detailed design and local issues'. The *Urban Appraisal Report* also echoed the widespread unease about the methods used to assess the economic value of road schemes, but its recommendation was couched in such broad terms that the Government had no difficulty in side-stepping it by maintaining that it reflected existing practice.

Environmental impact and NATA

The issues, which successive governments might have hoped would be settled by the various inquiries, refused to disappear: in fact, they became more problematic as public attention widened to encompass more and more matters which had not traditionally been regarded as pertaining to roads. Above all, concern had grown enormously on the environmental effects of roads and, indeed, of all forms of traffic. Inevitably, further inquiries were commissioned including another SACTRA Report (*Assessing the Environmental Impact of Road Schemes*), which was published in 1992.

By this date, the arguments about the limitations of COBA had intensified. Not only were environmental considerations now at the forefront of the debate (particularly after the shock of the 1989 traffic forecasts), but also it was being argued that any sharp distinction between economic and environmental impacts was false (Pearce *et al.* 1989). Earlier reports had advised that environmental benefits and costs should not be evaluated in money terms but should be subject to 'professional judgement'. The rationale for this is a simple one: there is no acceptable way to estimate the 'value' of a cathedral, a marvellous view, or other such 'non-economic goods', though this has not stopped economists from trying (Schofield 1987). But this leads to a host of mind-boggling questions. What is the value of land which is safeguarded from development? Is it the 'economic' value for development, or the lower 'social' value which is determined by planning controls? Which value should be used in evaluating alternative routes for a road? Other questions are equally baffling: if environmental factors may be important, which should be taken into account and which should be ignored? (The SACTRA Report has a long list of local, regional, national and global factors.) How are the cumulative effects of a multiplicity of apparently unimportant decisions to be dealt with? Will future increases in traffic increase environ-

mental damage, or will technological innovations more than offset these? The range and number of questions seem endless. No wonder that cost–benefit analysis is having a hard time![3]

By 1998 the DETR had formulated a *New Approach to Appraisal* (NATA) which was used in the annual Roads Review. The new approach assesses transport investments against the government's five objectives for transport: environmental impact, safety, economy, accessibility and integration. The emphasis is on transparency in the presentation of the evaluation, and requires a one-page summary table, drawing together economic, environmental and social factors. COBA is still present in the analysis, and monetary values are assigned where they can be, but this is presented alongside without giving prominence to any one type of effect or to benefits expressed in money terms. It is accepted that different effects cannot be aggregated or compared directly, and so the analysis does not make judgements; it only provides information for decision-makers. Thus it is now clearer that it is the politicians and their advisers and not the appraisal technique that makes the decision. A NATA is now required of all transport authorities, and extensive guidance on how it should be applied is available.[4]

Do new roads generate traffic? (1994)

A major issue in the debate on forecasting traffic needs is the extent to which new roads generate traffic. Certainly, their immediate effect on pre-existing roads can be dramatic, but this may be short-lived. Traffic seems to increase faster than new roads can be built. American studies have argued that, typically, the creation of new road space is eventually (and it may be sooner rather than later) taken up by increased traffic. Where does this traffic come from? Downs (1992) has put forward an elegant explanation in his theory of 'triple convergence'. This is based on the simple fact that since every driver seeks the easiest route, the cumulative result is a convergence on that route. If it then becomes overcrowded, some drivers will switch to an alternative route which has become relatively less crowded. These switches continue until there is an equilibrium situation (which like any human equilibrium is not stable – conditions constantly change).

On this theory, building a new road, or expanding an existing one, will have a 'triple convergence'. First, motorists will switch from other routes to the new one ('spatial convergence'); second, some motorists who avoided the peak hours will travel at the more convenient peak hour ('time convergence'); third, travellers who had used public transit will switch to driving since the new road now makes the journey faster ('modal convergence'). The outcome depends upon the total amount of traffic (actual and potential) in relation to the available roads. If the increase in traffic stimulated by the new road is modest, there will be an observable benefit for all. Though peak-hour traffic may be congested, this is simply because so many drivers are travelling at the time which is most convenient to them. (There may, however, be a loss to transit passengers if the 'modal convergence' leads to a reduction in service.) On this argument, new roads can generate traffic by diversion from public transport. Are there other ways in which extra traffic can be generated?

Intuitively, it would seem obvious that there are other ways: the more congested and difficult a road journey is, the more likely a potential traveller will seek an alternative. Conversely, ease of road journeys must generate increased trips. Of course, if this were self-evident, it would not have taken SACTRA over two hundred pages to discuss it; nor would the Department of Transport have been so resistant to it. Indeed, the matter is a complex one, mainly because it is difficult to establish cause and effect over time on matters where there are many variables. However, the department accepted the thrust of the SACTRA report, even though it held that 'clear evidence was lacking'. A *Guidance Note on Induced Traffic* was issued (DoT 1/95), and the department's research programme was augmented. At the risk of over-extending the issue, a note on the effects of the Newbury Bypass indicates that the argument is still very much alive. This bypass aroused an extremely bitter controversy and violent opposition, resulting in a huge expenditure on security and ejection of protesters. (The cost of building was

£74 million; the cost of policing was an extra £26 million). A report by the local authority (West Berkshire) less than a year after its opening stated that reduction of peak traffic had been only 25 per cent (the prediction was 40 per cent). Three factors are suggested for the limited impact of the bypass:

> People have retimed their journeys through the town; local people who had suppressed car use altogether may have taken to their cars again in the belief that congestion had been reduced; and more traffic has been pushed on to the ring road by the pedestrianisation scheme [in the town centre].
>
> (*Guardian* 12 July 1999)

This is a telling example of the acute problems raised by road-building, and there is a wider point: new roads not only induce traffic, but also encourage car ownership and use. As car use increases, other methods of transport are used less and, as a result, standards of service fall (thereby further increasing the attraction of car use). Moreover, road-building and ease of car use have major impacts on the location of new developments (of all kinds – housing, employment, shopping and leisure). Many new locations are car dependent, and therefore may increase the demand for road travel, and hence the need for more road construction. In this cumulative way roads certainly generate more traffic. However, it is extremely difficult to forecast patterns of land use, travel, and the interactions between land use and transport. Nevertheless, by the early 1990s, it was clear that some fundamental changes in transport policy were needed. Against this background it is not surprising that there was a temptation to fashion a number of new policy instruments. Among these was a review of trunk road-building, traffic management in London, traffic calming, a cycling strategy, and a step towards the integration of transport and land use planning with the publication of national government policy for planning in transport in Planning Policy Guidance Note 13.

Trunk roads review 1994

The early 1990s witnessed increasing recognition of the impossibility of catering for a continuation in the growth of road traffic, though acceptable alternative policies seemed elusive. A bumper crop of reports in 1994 (including the SACTRA report discussed in the previous section) provided conflicting advice and a massive excuse for further delays in taking positive decisions. However, the *Trunk Roads in England: 1994 Review* (and its 1995 successor *Managing the Trunk Road Programme*) did signal a significant shift in road-building policy. It detailed a reduced road programme (announced in the previous year); a total of forty-nine road schemes were withdrawn completely, and many others were postponed.[5]

This was one of the first steps in a major re-orientation of transport policy. Although

> it is no part of this Government's policies to tell people when and how to travel . . . we must be aware of the consequences if people continue to exercise their choices as they are at present. There is no realistic possibility of simply halting traffic growth . . . The Government's policy for sustainable development is to strike the right balance between securing economic development, protecting the environment, and sustaining future quality of life.

Resources were now to be devoted to the improvement of sections of existing key routes which were likely to experience congestion in the near future (primarily by adding lanes to existing motorways) and to providing urgently needed bypasses. Existing proposals for new trunk routes were to be reduced still further, and the programme of major urban road improvements would be a very limited one. Clearly, this represented a sea change in transport policy.

Roads policy since 1997

The 1998 White Paper *A New Deal for Trunk Roads* (discussed on p. 408) apparently put the seal on the shift in emphasis away from road-building, stating

that 'the priority would be maintaining existing roads rather than building new ones . . . Simply building more roads was not the answer to traffic growth'. The essence of the new roads policy was essentially to build as few as possible. Since new roads can lead to more traffic, adding to the problem (rather than reducing it) all plausible options needed to be considered before a new road was built. The Highways Agency was given new strategic aims of giving priority to better maintenance, making better use of existing roads, and putting greater emphasis on environmental and safety objectives.[6] Responsibility for some 40 per cent of existing trunk roads was to be transferred to local authorities ('de-trunked'). The other 60 per cent were identified as the nationally most important routes (the 'core network') and remain the responsibility of the Highways Authority. Important improvements are now planned through the regional spatial strategy system (formerly regional guidance) 'to ensure integration across all forms of transport with land use planning'.

During the 1990s, transport policy widened to deal with such issues as better safety, better driver information, tackling noise and environmental protection. The most radical proposal was the tentative introduction of tolling on trunk roads, which was later realised for the first time in December 2003 with the opening of the Birmingham Northern Relief Road or 'M6 Toll' in the West Midlands.[7] As with the road user and workplace charges which were proposed for local authorities (outlined later), it was intended that a start would be made with small-scale pilot charging schemes. Technical trials of electronic systems and their impacts were put in hand. However, intentions are one thing, actions are another. The Environment Transport and Regional Affairs Committee questioned the government commitment to the shift in transport policy away from roads, pointing to the plan to invest £21 billion in new roads over ten years.[8] While there have been shifts in government attitudes on road-building, the road lobby has always found a willing ear at the ministry (and continues to do so), primarily, it is suggested, because of the continuing influence of the outdated idea that new roads stimulate economic growth (Woolmar 1997).

Nevertheless, the 1990s did witness a significant change in direction in transport policy and in the linkage with land use planning. In 1994 the Departments of the Environment (then responsible for planning) and Transport jointly published PPG 13 *Transport* providing guidance on the integration of transport and land use planning. There had been many calls for such integration, and these were given weighty support in the 1994 report of the Royal Commission on Environmental Pollution (*Transport and the Environment*). The key aim of PPG 13 was to ensure that local authorities carried out their land use policies and transport programmes in ways which help to 'reduce growth in the length and number of motorised journeys; encourage alternative means of travel which have less environmental impact; and hence reduce reliance on the private car.'

Thus transport planning explicitly became a major component not only of land use planning, but also of environmental policy and of the *UK Sustainable Development Strategy*. Indeed, the PPG underlined the government commitment to providing a policy framework which will help to ensure that people's transport decisions are compatible with environmental goals. This was to be facilitated by the policy of increasing the real level of fuel duty by at least 5 per cent each year (a pledge that was dropped in 2000 in the face of protests about the increasing costs of fuel). Electronic tolling on motorways was also envisaged 'when the appropriate technology became available'.

It was stressed throughout in the original PPG 13 that the relationships between transport and land use planning had to be carefully examined at all levels, and that integration and coordination had to be promoted by regional planning guidance (through the regional conferences of local authorities) and in development plans. Strategies were required that would reduce the need to travel and maximise the opportunities for travel by public transport. Car parking was also a strategic matter (with policies to be set out in regional guidance and development plans) 'to avoid the destructive potential for competitive provision of parking by neighbouring authorities'. Other matters dealt with included plans for safe and attractive areas for pedestrians; provision for cyclists; traffic management;

provision of park and ride schemes; and 'accessibility profiles for public transport in order to determine locational policies designed to reduce the need for travel by car'.

A 1999 revision of PPG 13 made by the Blair government continued in the same vein, though with stronger emphasis on guaranteeing access by public transport to new developments, ensuring forms of development that encouraged non-motorised transport, with implementation through green transport plans, transport assessments and national car parking standards. The guidance also included preferred locations for particular types of development (housing, shopping, leisure and services) but, as the Civic Trust has noted, the guidance is very general, mostly in the form of situating new development where it is accessible, and adds little to what is said in other PPGs.[9]

National transport debate and the 'New Deal'

The PPG was full of ideas, but lacked guidance on how they might be implemented. Further advice is given in PPG 13, *Guide to Better Practice*, which quickly found its way into development plans, and influenced the outcome of appeals. The PPG had an immediate strong influence on development plans, and subsequently on office and other major developments. Though the first PPG 13 and similar policy statements for other parts of the UK attempted to settle some major policy issues, they merely provided a guide to the issues which needed to be debated. The Conservative government was baffled by the complexities, by the need to reduce pollution and the difficulties of persuading a largely car-owning electorate to use less convenient forms of transport, and by its preoccupation with privatising British Rail. In a valiant attempt to reclaim the initiative the Secretary of State launched a series of speeches on salient transport issues (Glaister *et al.* 1998).

The threads were drawn together in a Green Paper *Transport: The Way Forward*. This was a well-documented paper, but quite indecisive. Many of the ideas in the paper appear in some form or other in the Blair government's consultation paper *Integrated Transport Policy*, and subsequently found their way to the 1998 White Paper, *A New Deal for Transport: Better for Everyone*. The White Paper is an introduction and summary of a series of 'daughter documents' on such issues as roads, buses and parking charges. The range of issues covered was unprecedentedly wide, from wheel clamping to railways, from safety to public transport links to airports, from vehicle emissions to measures for a more inclusive society. The *New Deal for Transport* promised to improve the urban environment by creating the conditions for people to move around more easily. More road space and priority is to be given to pedestrians, cyclists and public transport. The meaning of 'integrated transport policy' is described in Box 11.1. Clearly, this is a very broad conception, and the details are spelled out over many pages. Some of the major features are summarised below.

Separate papers were published for Northern Ireland, Wales and Scotland, where transport problems are rather different.[10] For instance, the pattern of car ownership is significantly different in Scotland: in 1997 there were thirty motor vehicles for every hundred people, compared to forty-eight in England and Wales. There is therefore 'a stronger need in general to cater for people who do not have access to a car' (though, of course, there is a 'greater potential for further expansion in car ownership'). There are also marked differences in car ownership between different parts of Scotland, reflecting geography, wealth and economic activity. Rural Scotland, however, has a relatively high rate of car ownership – more a matter of necessity than wealth.[11] In Northern Ireland, 'current levels of congestion, even in Belfast, are not at levels that justify dramatic action to actively restrain the use of the private car'. However, 'this provides greater opportunity to . . . ensure that the growth of car dependence is limited', though restraining future traffic growth 'will take a long time and it will raise many complex issues'. For the immediate future, the priority will be for measures designed to encourage a change in travel behaviour away from car dependence, rather than restrictions on choice'.[12]

The White Paper provides an indication of the ways in which integrative policies can be implemented. These include the establishment of a new Independent Commission for Integrated Transport, the introduction of local transport plans and revised regional policy, extensive partnerships and cooperation between transport providers and, 'where necessary', strengthened local authority powers to secure integration. A £180

BOX 11.1 INTEGRATION OF TRANSPORT POLICY

An integrated transport policy means

- integration within and between different types of transport – so that each contributes its full potential and people can move easily between them
- integration with the environment – so that our transport choices support a better environment
- integration with land use planning – at national, regional and local level, so that transport and planning work together to support more sustainable travel choices and reduce the need to travel
- integration with our policies for education, health and wealth creation – so that transport helps to make a fairer, more inclusive society

Source: DfT (1998) *A New Deal for Transport*

billion transport investment programme was set out in *Transport 2010: The 10 Year Plan* (DETR 2000) and many proposals were realised through the *Transport Act 2000* and a revision of PPG 13 in 1999.[13] The 10 Year Plan promised £60 billion each for rail, roads and local transport. The objectives in the plan are very ambitious: a 50 per cent increase in passenger use on the railway, twenty-five new light rail projects, and 10 per cent increase in bus use, among others.

The Commission for Integrated Transport (CIT) was established 'to provide independent advice to government on the implementation of integrated transport policy, to monitor developments across transport, environment, health and other sectors, and to review progress towards meeting our objectives'. Media coverage of the 'integration problem' has undoubtedly increased as a result of the CIT's activities. Among the specific issues being addressed by the CIT are the setting of national road traffic and public transport targets, revisions to the National Road Traffic Forecast, lorry weights, the development of rail freight, review of transport safety arrangements, progress with green transport plans, the new rural bus partnership fund in England, and research needs. The CIT is the first significant measure on integrating transport following the abolition of the British Transport Commission in the early 1950s. Since then the government failed to produce any coherent transport policy and instead 'transport has been treated as a collection of separate and independent modes' (Bagwell and Lyth 2002: 209).

The 10 Year Transport Plan was updated through to 2014–15 in a 2004 White Paper *The Future for Transport*, which responded to continued criticism on lack of progress on national transport problems and, particularly, the demise of Railtrack. The main message is that instant solutions are not possible. It draws attention to various government actions (mostly in place) and the commitment to sustained investment over a long period to make headway in consequence of lack of investment the 1980s and 1990s. The White Paper repeatedly explains the difficulty of tackling transport problems in the short term and pushes further on the need to 'manage the demand for transport', referring to the 2003 DfT report on *Managing Our Roads*. Nevertheless, it is still timid on demand management measures, suggesting for example, that national road charging may be 'technically feasible' by 2015 or 2020.[14] In the mean time it promises that government will 'lead the debate' on road pricing which, the cynics might say, will not speed its implementation. There are other mixed messages. It makes reference to the *UK Sustainable Development Strategy*, but goes no further on the environmental costs of roads, airports and other infrastructure than mitigation and compensation measures. A Transport Innovation Fund is to be set up to support innovative measures such as road pricing and promoting modal shift. Planning figures in the White Paper both as part of the problem ('past planning policies have added to the challenges presented by increasing mobility') and the solution: 'it is essential that planning and transport policies are closely coordinated to produce more sustainable patterns of development' (para. 1.6). But this is about as far as it goes. A 2002 Select Committee Report on the *10 Year Plan for Transport* had called for more to be done with regional strategies, noting that 'only one half page of the 107 pages of the 10 Year Plan is set aside to describe them' (para. 67). The Committee went on to recommend 'greater clarity . . . about the linkages between the different levels of regional planning and transport decisions, local transport plans and multi-modal studies' (para. 70). The inquiry also revealed that little thinking had gone into the role of land use change and regeneration in meeting the government's transport objectives. Some of these criticisms may have been addressed by new policy on regional planning, though the update of the Plan does not go much further on the relationship between land use and transport.

Regional planning for transport

Planning for transport at the regional level was strengthened in parallel with the changes made to regional planning guidance and the introduction of regional spatial strategies and the elaboration of national strategies by the devolved administrations (which are discussed in Chapter 4). In England, a regional transport strategy was created as an integral, but clearly identifiable, part of regional planning guidance. It set out the regional priorities for all forms of transport; guidance on the integration of different services; accessibility criteria for regionally significant forms of development; and the strategic context for demand management such as road pricing. The regions were also encouraged to undertake multi-modal studies (MMSs) to examine the role of different transport modes within an area or corridor. The regional spatial strategies introduced by the 2004 reforms contain, like the regional guidance before them, a regional transport strategy. *Planning Policy Statement 11: Regional Spatial Strategies* stresses that 'better integration between transport and spatial planning is critical to the development of an effective regional strategy' (para. 1). On the one hand, transport policies must take account of the spatial strategy and on the other land use planning must take account of the existing transport strategy. It is an indictment of our regional planning system that this truism needs to be written down in the policy

statement. One of the reasons for this is that transport investment has been strongly controlled at the centre – and thus beyond the scope of regional planners, but while promoting more strategy for transport at the regional level, the government are promising to make a clearer statement of what funding is available for transport investment and that more decisions will be made at the regional level. The devolved administrations have already taken on much of the competence for transport policy. The aims of the regional transport strategy are set out in Box 11.2.

Doubts remain. At the national level, there is still insufficient attention to the role and planning of transport infrastructure in relation to the economic development of regions outside the South East, and helping to redress the gap in growth rates among the regions. The regional development agencies jointly commissioned a study in 2004 (ECOTEC and Faber Maunsell 2004) to identify criteria to assist in decision-making on 'surface infrastructure of national economic

---

BOX 11.2 AIMS OF THE REGIONAL TRANSPORT STRATEGY

The RTS should provide:

- regional objectives and priorities for transport investment and management across all modes to support the spatial strategy and delivery of sustainable national transport policies
- a strategic steer on the future development of airports and ports in the region consistent with national policy and the development of inland waterways
- guidance on priorities for managing and improving the trunk road network, and local roads of regional or sub-regional importance
- advice on the promotion of sustainable freight distribution where there is an appropriate regional or sub-regional dimension
- a strategic framework for public transport that identifies measures to improve accessibility to jobs and key services at the regional and sub-regional level, expands travel choice, improves access for those without a car, and guides the location of new development
- advice on parking policies appropriate to different parts of the region; and
- guidance on the strategic context for local demand management measures within the region.

Source: ODPM (2004) PPS 11 *Regional Spatial Strategies* (p. 58)

importance' and to develop a framework for the identification of new infrastructure that is needed. The study concluded that while the UK is effective at assessing transport schemes once identified,

> we lack a means of systematically identifying surface transport infrastructure and interventions that are of national economic importance in the first place. Unlike a number of other European countries we do not have an overall national view that integrates the regional and national picture.
>
> (ECOTEC and Faber Maunsell 2004: 8)

The researchers made a tentative application of criteria to identify a network that would support economic growth in the regions and came up with an 'illustrative' proposal that 'the existing largely London-centric, network should be expanded to form a national grid with stronger east–west links, high speed rail links from the regional capitals to London and rail links into Heathrow'. The minister made a polite response but the reality is that most infrastructure spending is needed to support the Communities Strategy in and around the South East (see Chapter 6).

At the regional level, and given the current inadequacy of means for central government to ensure that local authorities comply with regional policy,[15] there must be major concerns about implementation, particularly in view of the great unpopularity of some of the measures proposed, and fears of competitive 'inaction' by neighbouring authorities. It remains to be seen whether the new regional planning system will be sufficiently effective to overcome such problems.

Local transport plans

Local transport plans (LTPs) (local transport strategies (LTSs) in Scotland) replaced transport policies and programmes (TPPs). Initial guidance on local transport plans was published by the DETR in 1998 and provisional one-year plans were published in 1999. Revised guidance and a good practice guide were published in 2000, and the Transport Act 2000 made five-year local transport plans a mandatory requirement.[16]

The emphasis of the plan should be on integration through involvement of all relevant interests. The guidance is lengthy and detailed, but a selective summary is given in Box 11.3. Local transport plans should be consistent with appropriate development plans, and eventually the two should be integrated with each other. They should also be consistent with regional transport strategies.[17]

Buses and light rail

Buses are seen as 'the workhorses of the public transport system', but as was shown in Table 11.5, they are not very popular. Only 6 per cent of total journey distance travelled in Great Britain in 2003 was by bus and coach, compared with 11 per cent in 1980 and 42 per cent in 1952. For this to change in a positive direction, major improvements are needed to bus design and comfort, cleanliness, regularity, reliability and interconnection (with both other buses and other forms of public transport), greater priority on the road, and improved information for passengers. Experience to date (given the background of privatisation, deregulation and competition) points to the limitations of improving the attractiveness of bus transport. Banister (2002: 90) explains: 'the net effect of deregulation in the metropolitan areas has been significant increases in fares (+31.8 per cent in real terms between 1985 and 1991) and a decline of 26.2 per cent in passenger journeys'.

The *New Deal* White Paper stated that 'quality partnerships' (already in operation in cities such as Aberdeen, Brighton, Leeds and Swansea) will be developed by local authorities against the background of their local transport plans. The details of these partnerships are spelled out in a separate document (*From Workhorse to Thoroughbred: A Better Role for Bus Travel*, 1999). Quality partnerships are given a statutory (but inadequate) basis 'so that all concerned can have the confidence to invest'. Local authorities will have a range of powers to ensure service stability, good timetable information, and systems of flexible joint ticketing. In one of their persuasive passages, the Transport Committee pointed to the hub of the public policy

BOX 11.3 CONTENT OF LOCAL TRANSPORT PLANS

LTPs must demonstrate consistency with the government's transport objectives, and cover all travel modes, including

- voluntary or community transport (or its potential) particularly in rural areas
- local strategy for cycling and walking targets
- traffic management and demand restraint
- enforcement of emission standards
- proposals for pilot schemes for road user charging and taxation of workplace parking
- cooperation with major retailers and leisure operators on car access and alternative means of access
- integrated strategy on parking, planning policies and transport powers
- local road casualty reduction target
- interchange improvements
- bus-based park and ride schemes and related reduction in town centre parking or pedestrianisation
- proposals for capital expenditure on public transport information schemes
- promotion of green transport plans by employers
- integrated strategy for travel to school
- planning and management of the highway network
- strategy for rural transport, and for countryside traffic management schemes
- issues connected with freight distribution
- promotion of social inclusion, including disability issues and extension of bus access for welfare to work
- action on climate change, air quality and noise.

Source: DETR (1998) *Guidance on Local Transport Plans*

---

issue with transport deprived areas: not surprisingly it is basically a financial issue (though there are additional matters such as those of labour availability). These 'traditional' subsidy schemes are workable within a normal financial context: matters are very much more difficult where there is no normal bus service in operation upon which additional features can be harnessed. In the Transport Committee's words:

> The traditional bus service is not a cost-effective public transport solution for many areas, particularly rural areas, where demand for public transport is low. Public transport today includes taxis, share-taxis, and demand-responsive buses operating on flexible routes, car share clubs and community

transport minibuses among others as well as the conventional bus, tram and rail systems. Many innovative examples of ways of tackling public transport shortages have already been developed in the UK and elsewhere in Europe.

For this purpose, several schemes have been introduced by the Department of Transport for funding needy areas. These include two 'challenge fund' initiatives to provide support for new transport services. The *Rural Bus Challenge* was introduced in 1998 to stimulate the provision and promotion of rural public transport. In 2001 an *Urban Bus Challenge* was aimed at improving public transport for deprived areas. Another scheme is the *Rural Bus Subsidy*, which is aimed at encouraging the development of new innovative public transport services, particularly for areas with little or no public transport provision. In such areas a conventional bus service is unlikely to be a cost-effective solution. The new scheme enables parish councils and local groups to work in partnership with local authorities with the objective 'to support schemes which reduce rural isolation and social exclusion through enhanced access to jobs and services'.

The popularity of light rail systems or 'rapid transit' is spreading. Unfortunately, they tend to be very expensive (particularly if they use a fixed rail) and they were out of favour for many years. Schemes had been established earlier for several cities, including Glasgow, Tyne and Wear, Merseyside, and London (the Jubilee Line), but many more were shelved. Increased road congestion (and prospects of much more in the future), the model of the London Docklands Light Railway (promoted as part of the Docklands renewal strategy) and increasing experience of foreign systems re-awakened political interest in light rail. In the early 1990s there were forty urban areas with proposals for rapid transit. Despite government predictions few have been realised. Britain does not compare well with other European countries on rapid transit, and this may in part be due to the fact that public transport generally is expected to cover a large proportion of its operating costs. This makes the outlook for rapid transit in Britain less certain, though there is some comfort in the fact that most foreign public transport networks have improved their 'revenue-operating cost ratio'. There is, however, the added difficulty in Britain that any rapid transit system would find itself in competition with deregulated bus services.

Rapid transit attracts considerable vocal support, but little in the way of the necessary funding. Government policy statements suggest that the potential for rapid transit will improve for some selected cities, but the overall position is that less expensive and risky investment in bus services and priority measures are favoured. (This of course assumes that light rail and buses address the same tasks and travellers. Will they tempt drivers to give up their cars?) The 10 Year Plan provided for up to twenty-five new rapid transit lines in major cities and conurbations. The extended plan

in the 2004 *Future of Transport* White Paper confirmed that the government had no intention to fund this many schemes. Although it gives no details it makes plain its dissatisfaction with relatively poor performance of some light rail lines, pointing the finger particularly at Sheffield Supertram, Croydon Tramlink and Midland Metro, where passenger numbers have not met targets. The local plans see these as the beginning of networks and long-term strategies integrated with development plans. The Midland Metro, for example, is part of a much wider programme of work opening up areas of economic decline to new investment and changing the character and role of Birmingham's city centre. The government continue to measure short-term benefits and seems to regard them as 'one-off' infrastructure projects.

The bane of overcrowding

The major problem with much of the public transport system (particularly on the buses) is gross overcrowding (typically due to the inadequate amount of funding). The issue is neatly posed by the title of a House of Commons Transport Committee on *Overcrowding on Public Transport* (2003). For this Committee, a study was carried out by Oxford Economic Forecasting (2003) on *The Economic Effects of Transport Delays on the City of London*. This concluded that public transport will be attractive only if it meets people's needs. 'That means that there must be adequate capacity to cope with peak flows into most, if not all, major urban areas'.

A similar conclusion emerged regarding the railways and, of course, the London Underground. To quote again from the Transport Committee's report 'some crowding can be inevitable at peak times, but our inquiry has convinced us that the level of overcrowding is so great that many travellers face daily trauma on their journeys. Passengers are unable to board vehicles, or if they can, are forced into intolerable conditions'. Though there are well-known perceptions of these problems being of particular severity in London, they are not restricted to that clearly problematic area. Several of the reports for or by the Transport Committee make this very clear, such as *Standing Room*

Only: Overcrowding in Railways: A Report by the Rail Passenger Users Committee North Western England (2000) and *Railways in the North of England* (Transport Committee, 2002). The problems are not esoteric; they are quite simple; the major thing that is needed is adequate investment. Would that it was so simple politically!

Road user and workplace parking charges

Governments are very reluctant to levy charges on motoring (as distinct from taxes on vehicles and on petrol). One may wonder why charges on car parking, and tolls on roads which go over or under a river are acceptable, but congestion charges and road charges are not. The issue is, of course, a political one, and there is no simple rationale for these distinctions. The political nature of the issue is apparent in the extreme caution with which it is being broached. The 1998 White Paper laid the path for the Transport Act 2000 which provides for local authorities to charge road users so as to reduce congestion, as part of a package of measures in a local transport plan that would include improving public transport. This neatly delegates responsibility and political risk to local authorities. Central government will be responsible only for introducing pilot schemes on motorways and trunk roads to assess what lessons they provide. The first two road charging schemes were implemented by the Greater London Authority (discussed below) and by the City of Durham which has a tortuous central one-way high street; they make a curious contrast! It would be tempting fate to summarise the lessons to be learned: it would be a complex challenge to establish these and, in any case, it is still early days. Nevertheless the Commission for Integrated Transport (CfIT) has an explicit responsibility here, and its 2002 Annual Report has a brief summary. Certainly, it attracted widespread media interest, though the Commission sadly noted that 'the view still exists within the media that congestion charging is an anti-motorist policy.' CfIT is pushing forward its contrary view which implies that the only true anti-motorist policy would be not to tackle congestion and allow congestion to continue to grow. It is pointed out that the British seem to have a special relationship with their cars. British people make more use of cars than any other European country, despite having below average car ownership.

> Almost nine out of ten motorised journeys (car, bus, motorbike) in the UK are by car, compared with the EU average of just over eight out of ten. We are also travelling further in our cars. The annual distance travelled by car increased by 45 per cent between 1985/6 and 1997/9 (DTLR 2000). Indeed, the average British household now spends almost 15 per cent of expenditure on motoring.
>
> (CfIT, *Paying for Road Use* 2003)

A DfT consultation document, *Breaking the Logjam*, details the proposals for local congestion and workplace charges. It is stressed that the charges will be optional ('it will be up to those local councils which think it would help in their area to put up well reasoned proposals') and they will initially be of a pilot nature (to encourage fresh thinking and to learn from practical experience). Contrary to the usual dictates of public finance, the income from these charges will be retained by local authorities to be spent on local transport improvements. This is a good way of making charges more acceptable to both local authorities and motorists.[18] Road user charges will have to be in keeping with the local transport plan as drawn up to reflect the regional transport strategy and the *National Air Quality Strategy*. Levies on private non-residential parking at the workplace face the danger of simply displacing parking on to adjacent streets. Consequently, the introduction of the levy will require the enforcement of on-street parking controls.

Walking and cycling

Interestingly, the first issue discussed in the *New Deal* White Paper is 'making it easier to walk'. This nicely emphasises the priority for pedestrians. Measures include more pedestrian crossings, more direct and convenient routes for walking, and increased pedestrianisation (illustrated by a striking photograph of a pedestrianised Trafalgar Square).[19] Speed limits and 20 mph zones are already being introduced, 'with markedly beneficial effects'.[20] The Transport Act 2000 gives local authorities the opportunity to make orders to create 'home zones' or quiet lanes to govern traffic and reduce speeds. It spells out that a strategic view of how walking and cycling can be encouraged should be set out in the regional spatial strategies and that implementation of practical measures should be included in local transport plans and development plans. Many local planning authorities have some supplementary documents on cycling and walking.

Almost one in five car trips on the urban network at 8.50 in the morning are taking children to school. Walking and cycling to school are to be encouraged by planning safer routes. A £50 million *Safe Routes to School* scheme was established in 2003 funded jointly by the DfT and Department for Education.[21]

School travel plans are to be produced by local authorities and schools. A new School Travel Advisory Group has produced a best practice guide and a volume of thirty case studies.[22] The 2004 *Future of Transport* White Paper sets a tangible, if not ambitious, target to roll out school travel plans to every school in England by 2010. In Scotland, a Scottish Walking Strategy Forum has been established to consider how walking could be made more popular. There is also a Scottish Cycle Challenge Initiative which promotes projects which encourage cycling to work and to school.

Britain lags behind many other countries in the use of cycles.[23] The world has twice as many bicycles (around 800 million) as cars, and bicycle production outnumbers cars by three to one. There are over 13 million cycles, and over a third of all British households have at least one. About 11 million people use their cycles at least once a year; in an average week about 3.6 million are used. Over 1 million people use a bicycle as their main means of transport to work. But only 2 per cent of total trips are made by cycle. The total distance travelled by cycles is between 5 million and 6 million kilometres a year, compared with around 350 million kilometres for cars. On the other hand, a greater distance is travelled by cycle than by bus and train combined. Despite this apparent abundance of statistics, it is difficult to obtain an accurate picture of cycle use (and still less of any potential increase).[24]

A tangible boost to cycling provision came with the Millennium Commission's grant of £43 million towards the 6,000-mile National Cycle Network organised by the charity Sustrans. The objective is 'to create at least one quality cycle route through every town in the UK' and to link 'towns and cities from Dover to Inverness to Belfast' (Sustrans 1996). The benefits of the scheme have been prominently advertised: 6,000 miles of high quality cycle and pathways within two miles of 21 million people by 1996, and by September 2005 more than 10,000 miles of the network will be in place, much of which has drawn additional funding from a wide range of sponsors. Half of the network is entirely free of vehicular traffic. This is one of the great success stories. Sustrans estimated that there would be more than 100 million journeys a year on the network and this now looks like a conservative estimate in the long term since, by 2002, 97 million trips were being made, the equivalent of 540 million miles of cycling and 260 million miles by other users.[25] Cycling and walking initiatives are now being supported by the Department of Health as part of its programme for a more healthy population, but the greater long-term benefit will be an improvement in the image of the cycle as a means of transport.

It was, however, the 1996 *National Cycling Strategy* which marked a genuine change in governmental attitudes to cycling. This claims to represent 'a major breakthrough in transport thinking in the UK'. The target was to double the number of trips by cycle by 2002, and quadruple the number by 2012. It is not clear where these figures come from, but they exhibit an eagerness which to date has been restricted to cycling enthusiasts. The Strategy covers a wide range of relevant issues, including appropriate planning measures, safety, provision of parking for cycles (and accommodation for them on public transport), integrating cycling with traffic management, cycle security, and the 'communication programme' needed to change attitudes to cycling. Practical guidance is given in the 2004 *Walking and Cycling Action Plan*. Targets for increased cycling were included in the *New*

Deal White Paper and supersede those in the 1996 *National Cycling Strategy*. The *10 Year Transport Plan* changed the target to trebling cycling trips from 2000 to 2010. They are ambitious and this may be the reason that neither target has been included in the formal list of government objectives.

Three towns, Worcester, Darlington and Peterborough, have also been selected as demonstration projects and will seek radical changes in travel behaviour, and in particular, increases in walking and cycling, with the backing of a £10 million government fund. The targets are to reduce traffic by between 7.5 and 10 per cent over five years. Money is perhaps not the main problem in bringing forward planning initiatives for walking and cycling. McClintock (2001) points out that

> one obstacle to effective take-up [of walking and cycling] has been the reluctance of some, especially senior and more experienced professionals, to accept that these modes are really now to be taken seriously, rather than as before, tending to give priority in transport planning to motor traffic and its demands.
>
> (McClintock 2001: 200)

Green transport plans

Bringing about major changes in travel behaviour is 'a shared responsibility', requiring cooperation on the part of travellers, employers, hospitals and educational establishments and any organisation or company that can have an impact on travel patterns. Major employers are urged to consider preparing green transport plans which will integrate 'the various ways in which an organisation uses transport to ensure that they complement each other and benefit the strategic business objectives' (DETR, *The Benefits of Green Transport Plans*, 1999). These can include measures that encourage travel to work by public transport, cycling or walking, a flexible benefits package to provide attractive alternatives to a company car, a review of standard working hours, a car-sharing scheme, using video-conferencing and other IT equipment to reduce

business travel, and enhancing the fuel efficiency of the vehicle fleet.[26]

The Advisory Committee on Business and the Environment has recommended that companies seek to reduce by 10 per cent the total number of people commuting to work alone by car. Whether there are adequate incentives for the development of green transport plans is, however, in some doubt (Potter 1999). Nevertheless, all government departments are expected to have green transport plans in operation by March 2000. Hospitals are singled out for special mention in the White Paper because of their place in travel generation. PPG 13 notes that some or all of a green transport plan may be made binding through either conditions attached to a planning permission or through a related planning obligation. Planning applications must be accompanied by a green travel plan for all major developments over certain thresholds for smaller developments that will generate significant amounts of travel in areas where traffic reduction and alternative modes are priorities, and where it will help to address a particular local problem. The objective in all cases is to deliver more sustainable transport. School travel plans promoting safe non-car routes to school are also required where they are to be expanded.

Traffic calming

Traffic calming is an expressive term which, though used in different ways, essentially refers to measures for reducing the harmful effects of motor traffic. In its limited sense, it refers to speed reductions, parking restrictions, pedestrianisation schemes, and such like. In a wider sense, it is synonymous with overall traffic policy, including car taxation and land use measures designed to reduce the need for car journeys. Advocates of traffic calming can make some telling points in its support. For instance, a 50 kph speed limit (about 30 mph) in residential areas is 'acknowledged in many European countries' to be 'far too high'; at speeds of 30 kph or below additional road space is created since cars need less space. If traffic calming is restricted to a few streets, its benefits are reduced: traffic simply redistributes itself to neighbouring streets. Complete exclusion of traffic can have a dramatic impact on town centres, and has been widely adopted on grounds of safety, amenity and increased turnover for shops in pedestrianised streets – though the economic benefits are far from certain.

The term 'traffic calming' was introduced by Dr Carmen Hass-Klau as the translation of the German term *Verkehrsberuhigung*. Her book, *Civilised Streets* (Hass-Klau *et al.* 1992), contains detailed technical descriptions of well-established methods such as speed bumps, chicanes (kinks in a road to slow down traffic) and pinch points, as well as some less well known techniques. It also presents an assessment of traffic calming experience in Germany, the Netherlands, Denmark and Sweden. It describes and comments on some forty British traffic calming schemes. The traditional approach has been to segregate traffic and pedestrians:

with reductions in traffic speed, they can both be accommodated, but with the pedestrian instead of the car being master. (See Box 11.4.)

A major shortcoming of many traffic calming schemes is that they are essentially local in concept and operation. Rather than being parts of a comprehensive transport policy, they are typically reactions to vocal residents. As a result, the effect of calming in some areas is to move the problem elsewhere. Indeed, Banister (1994b: 212) has suggested that 'positive responses from those living in the traffic-calmed area are more than outweighed by anger from those living in adjacent areas where traffic levels (and accidents) have increased'. It is unfortunate that calming does not simply reduce the total amount of traffic; perhaps with proper planning, it can? The same point arises in relation to parking policy, which is arguably the simplest and

BOX 11.4 BYPASS DEMONSTRATION PROJECT

Traffic calming is particularly appropriate after the completion of a bypass. Although the town's inhabitants may feel that the bypass has solved their local problems, in fact it can bring new problems in its wake. The old route will have all the features of a heavily trafficked route: it will bear all the marks of a road which has been adapted (and perhaps mutilated) to accommodate high levels of traffic. Quite apart from the poor appearance of the place, the reduced traffic will facilitate higher speeds, and paradoxically, new traffic hazards may appear. And (the final irony) though traffic will initially decrease significantly, it can soon build up again. To explore how these problems can be dealt with, the DoT in conjunction with local authorities mounted a Bypass Demonstration Project (announced in the White Paper *This Common Inheritance*, and completed in 1995). A major object of this was to demonstrate how the benefits of a bypass can be enhanced by an overall improvement scheme. Six towns were selected and, with some financial support from the DoT, major traffic calming and other 'town enhancement' works were undertaken.

The report on the study revealed the problems and opportunities. The removal of through traffic allows a radical change in the street space. Inevitably the benefits are not equally shared, but the six project towns have shown how pedestrians, visitors, cyclists, disabled people, and civic uses in general can benefit substantially, while still maintaining vehicular access in a traffic-calmed environment. Such schemes demand a great deal of professional input, coordination of effort, public involvement, and cost: the expenditure in the six towns ranged between £1m and £2m. The benefits are striking, and can amount to a transformation of the area.

Source: DoT (1995) *Better Places through Bypasses: Report of the Bypass Demonstration Project* (the six towns were Berkhamsted, Dalton in Furness, Market Harborough, Petersfield, Wadebridge and Whitchurch)

most effective method of reducing private car use. A further problem is the damage that some traffic calming measures have on the appearance of attractive towns and villages, with a clutter of intrusive signs and roadworks. Many of the measures discussed under the heading of traffic calming have more traditionally been known as traffic management, though the concern is now with wide environmental and amenity issues as well as with traffic flow. This is becoming an increasingly sophisticated area of policy. Additional legislation is a testament to the importance now attached to it: the Traffic Calming Act 1992 extended the statutory provisions for 'the carrying out on highways of works affecting the movement of vehicular and other traffic for the purposes of promoting safety and of preserving or improving the environment'.

Parking restrictions and standards

Parking restrictions are the simplest and the most acceptable of traffic controls, which certainly was not true when parking meters were first introduced (Plowden 1971). Until recently, parking restrictions were largely confined to cars entering congested areas, and charges have been raised as demand exceeded capacity. A favoured measure is to escalate the charging rate for long-stayers.[27]

National policy on parking in England is set out in the 1999 PPG13, and is elaborated in other regional strategies (especially so for London) and local transport plans. However, studies have shown that compliance with the guidance is poor, particularly in the outer London Boroughs. Moreover, though it is advised that strategies for parking should be developed in conjunction with neighbouring authorities, there is 'little evidence' of this. Indeed, cooperation 'may even be restricted in order to preserve the use of parking as an independent counter in order to attract development in competition with other authorities'. Planning authorities have allowed parking provision 'well in excess even of peak time demand.' The quotations are from a report commissioned by DETR, which concluded that in the South East, 'Government policy

guidance relating to the use of parking standards as a demand management tool is not reflected in the majority of standards adopted by local authorities' (Llewelyn-Davies and JMP Consultants 1998). Other studies have reached the same conclusion.[28] This is a neglected area of planning policy, but this report forcibly shows that the neglect can jeopardise policies or even render them ineffective. As policy has become increasingly concerned to restrict cars (rather than their use in particular areas) there will need to be a dramatic change in the implementation of parking standards. Of course, these can only apply in the future: Thus, 'considerable reduction compared with present norms of provision will therefore be needed to prompt modal shift away from the car' (Llewelyn-Davies and JMP Consultants 1998).

It is perhaps not surprising, therefore, that the 1999 version of PPG 13 for England introduced national maximum standards for car parking. It notes that levels of parking can be more significant than levels of public transport provision in determining means of travel, even for locations very well served by public transport. It requires development plans to set maximum levels of parking for broad classes of development.

Traffic management in London

The Road Traffic Act 1991 provided a new legislative framework for traffic in London to combat congestion through special parking and other traffic management measures. The Act empowered the Secretary of State to designate a network of priority routes (commonly known as red routes because of their distinctive markings and controls). They are aimed at reducing traffic congestion and improving traffic conditions on main routes, particularly for buses, without encouraging additional car commuting into central London. A pilot scheme in 1991 proved successful: overall journey times improved by 25 per cent, bus journey times were reduced by more than 10 per cent, and reliability increased by 33 per cent. More people used buses and road casualties fell significantly. As a result, a permanent scheme was introduced in 1992. This network covers all trunk roads in London as well as local roads which are of strategic importance. The planning, coordination, implementation, maintenance and monitoring of traffic management on the network are the responsibility of Transport for London and a Traffic Director for London.[29]

The network plan forms the framework for detailed local plans, which are the responsibility of the London local authorities. Central to this new system was a package of traffic management schemes and a reform of on-street parking in Greater London. Traffic management measures include increased priority for buses, improved pedestrian crossings, enforcement of parking regulations, and encouragement to cyclists to use alternative roads to red routes except where separate cycle tracks can be provided. A range of traffic calming measures was implemented on side roads which might be affected by the red route traffic. These regulate speed and deter motorists from using side roads as 'rat runs'. In 20 mph zones, there is the additional advantage that the road hump regulations are far more relaxed: for example, warning signs are not required. Doubts about the legality of traffic calming measures were settled by the Traffic Calming Act 1992, which provided for the making of regulations governing them.

London has also led the way on road pricing, better known in the city as the congestion charge. Ken Livingstone made this a central issue in his campaign to be elected as Mayor, and it has to be stressed that this is one of those (many) problems that could not be tackled effectively without the right institutional arrangements, in this case provided by the creation of the Greater London Authority. It enabled a long-term, strategic, London-wide analysis of the problem and potential remedies on which the congestion charge was justified. The election of Livingstone in the knowledge that he was determined to implement the charge suggests that road pricing can be an acceptable if not widely popular policy, as it has been in other places such as Singapore and Oslo. There is a point at which the disadvantages of chronic congestion and the need for action become obvious, and this is patently the case in parts of London. Road pricing in the city takes the form of a cordon charge; drivers pay a fee, set initially at £5 but proposed to rise to £8, when crossing entrance points to the charging zone, which includes only the central area. Cameras and computer systems monitor visitors to the zone and police the system. Many concessions were made to reduce opposition for residents, people with disabilities, firms operating large fleets of vehicles and others. This short explanation does not give a real sense of the complexity of implementing and managing such a system, which is a major factor deterring other potential cities from taking up charging.

The London scheme has been watched very closely (and not only by other cities in the UK). Grey and Begg (2001: 2) argue that 'the *perceived* success of congestion charging in London is likely to determine whether it will be delivered successfully in areas like Bristol, Edinburgh and Leeds'. The emphasis on 'perceived' is theirs, and understandably so. They illustrate the problem of 'winning hearts and minds' with national newspaper headlines and articles: 'Motorists to face charges for road use' and 'Majority opposed to road tolls'. Understandably, motorists tend to see the charge as another tax and many others take some winning over, thus consultation and public relations are as important as the technical problems. Perceptions in London are now more positive in the light of experience.

By early 2005, after two years of operation, the charge had reduced traffic by 15 per cent and congestion by 30 per cent, and as a result disruption to bus services had fallen by 60 per cent. This is one of those few situations where forecasts have been met; indeed, the figures are very close to predictions. The charge is expensive to operate, but in 2003–4 it raised a net £80 million for investment in the transport system. Monitoring also suggests that the charge has had no discernible effect on the location of business or property values, and an initial decline in footfall in retail areas has been reversed. London now proposes to implement a low emissions zone by 2007.[30] The 10 Year Transport Plan expected to see at least eight schemes in place by 2010, but despite positive signals from the London experience, Durham is the only other city with road pricing. Edinburgh chose not to go forward with a scheme in 2005. (See Box 11.5.)

For many years, the government have been wrestling with the problem of modernising the London Underground. Proposals for splitting up the operation among

BOX 11.5 IMPACT OF TRANSPORT POLICIES IN FIVE CITIES

There is a wealth of experience on the impact of different types of transport policy on road traffic, but there is a bewildering range of possibilities. An indication is given by a 1994 report from the Transport Research Laboratory (Dasgupta *et al.*). This investigated the effects of various policies on urban congestion in five cities (Leeds, Sheffield, Derby, Bristol and Reading). It was found that halving public transport fares increased bus use by between 7 per cent and 20 per cent, but the proportionate effect on car use was slight: only 1 to 2 per cent. Other options examined included raising fuel costs by 50 per cent, doubling parking charges, halving the number of parking places, and applying a cordon charge of £2 in the peak and £1 in the off-peak period. The latter two measures had the greatest effect: they reduced car use in the central areas by about a fifth (and increased it in the outer area by between 3 and 5 per cent). Different types of policies have different effects, but they also vary among cities, and between peak and off-peak periods. The study concludes that, when interpreting the results, it is important to take into account the complicated interrelationships among modal transfer, redistribution, changes in vehicle-kilometres, and changes in trends.

Source: Dasgupta *et al.* (1994) *Transport Research Laboratory Impact of Transport Policies in Five Cities*

---

various public–private partnerships (PPP) were strongly contested, but the government put the PPP in place before transferring competence for the Underground to the Mayor for London in 2003. The PPP will provide £16 billion over sixteen years for investment in the trains, signalling and track. London Underground is still a public sector organisation under Transport for London which has responsibility for running the trains and organising the service. The infrastructure is leased to three companies for thirty years. This investment will have far-reaching effects on the demand for land and property and thus its value. A study by Jones Land LaSalle in 2004 estimated that the extension to the Jubilee line stations had increased land value by about £2.8 billion. It was by this method of raising land values that much of the suburban railway network around London was constructed in the nineteenth and early twentieth centuries.

Railways

The government's view, as expressed in the *New Deal White Paper*, that there is the potential for a 'railway renaissance', predated the unprecedented chaos that enveloped the railway system in the winter of 2000–1. The parlous state of British railways was not, of course, created overnight. It is the product of many years of neglect, weak investment and poor management, and was well known to many rail commuters, especially in the South East of England. But it became a national scandal almost overnight following the Hatfield rail crash.[31] In the aftermath of the crash Railtrack, which was then responsible for the track infrastructure, was forced to bring forward its programme of repair and maintenance and imposed hundreds of speed restrictions. The effects on timetables was exacerbated by severe flooding. The ensuing chaos revealed the depth of the demise of the railway system.

The Conservative government had privatised and broken up British Rail into many separate companies so as to stimulate competition and investment. The current government claims that investment actually fell until 1997. Such claims are contested, but it is a fact that standards have fallen. After Hatfield, the collapse of the rail performance was so great as to render 'any meaningful comparison impossible'.[32] There is no question of renationalising the railways, but a national *Strategic Rail Authority* for Great Britain was established to provide 'a clear, coherent and strategic programme' for development. Investments are needed in infrastructure works and rolling stock improvements. Some lines can be substantially improved at relatively short notice and at moderate cost, using more and longer trains, extended platforms and improved signalling. Others face 'pinch points' that restrict capacity, such as the Glasgow Central approaches and the East Coast main line between Finsbury Park and Peterborough. In all fifteen key bottlenecks were identified and a programme for solving these could be completed by 2006. The 10 Year Plan proposes £49 billion investment in the railways, £26 billion of which will come from the public sector. These figures look impressive but are not so different from what has gone before. Much reliance is placed on levering huge sums of private sector investment. Little is said in all these plans about the relationship to regional spatial planning or local planning.

It is difficult to be enthusiastic about *The Future of the Railway* (to take over the sarcastically named report of the House of Commons Transport Committee, published in April 2004). There has been an incredible series of reorganisation: these have been more concerned with 'strategic leadership to the rail industry', 'regulating the administration and controlling financing of the railways', 'customer protection' matters, and a consequent labyrinth of administration. Currently, however, the outlook is brightened by the 10 Year Plan, which promises completion of modernisation of the West Coast route and (rather optimistically) a 50 per cent increase in passenger-kilometres while also reducing overcrowding.

Freight traffic

Freight traffic is considered at length in *Sustainable Distribution: A Strategy* published a year after the 1998 White Paper. As the title suggests, great importance is attached to environmental aspects of freight movement, and many of its proposals are concerned with such matters as pollution, pressures on the landscape, noise and disturbance, and accidents. Stress is laid on the importance of the entire supply chain and its management ('logistics'). The analysis is interesting (not a common notable feature of such documents) and well worth studying. The package of measures, however, contains little that is new. The annual increases in fuel duty were intended to continue, though this has now been significantly amended through measures in response to protests over fuel costs. Vehicle excise duty rates for lorries are under review ('to reflect the environmental damage they cause'). In preparing development plans, local authorities are to consider and, where appropriate, protect sites that provide opportunities for the transfer of freight from road to rail and to consider opportunities for new developments which are served by waterways.

The Strategic Rail Authority's Freight Strategy seeks explicitly to influence the planning system at national, regional and local levels, recognising that its 80 per cent growth target will require extensive development of rail infrastructure in locations such as 'quarries, steel stockholding points, open cast sites, major manufacturing and production plants'. It is particularly concerned with protecting sites of strategic importance, though because of the inevitable objections it will be no guarantee that planning permission will be forthcoming.

Scottish guidance on transport and planning

The Scottish guidance on transport and planning was the first to be published. The Planning Advice Note *Transport and Planning* set out 'good practice advice' on measures which local planning authorities may consider in fulfilling their integrated land use and transport planning responsibilities in a sustainable manner. Developers are required to produce a transport assessment for significant travel-generating developments (which is also now required in England). Though this is distinguished from a formal environmental assessment, it may form part of it. The coverage of the transport assessment is dependent on the scale, travel intensity and travel characteristics of the proposal. Essentially, it provides information to enable the local planning authority to determine the suitability of the location for the proposed use. This is assessed 'in terms of both the potential and likely accessibility for people and freight by all modes'. It enables the local planning authority 'to determine whether the location has the potential to minimise travel, particularly by private car'. Where a transport assessment is required, the developer has to demonstrate that:

- The site, as existing or as a result of the development works, is physically accessible by a network of footpaths and cycle routes, and public transport will deposit passengers within a short and easy walk of the development.
- For non-residential developments, the network of public transport, walking and cycle routes serving the site links with the majority of the forecast catchment population, with public transport being regular and frequent throughout the opening hours of the development.
- For residential developments, a high degree of accessibility to local day-to-day services such as convenience shops, schools, clinics, libraries and community centres, particularly by walking and cycling, and accessibility to significant urban centres providing a range of services and employment, by walking, cycling and public transport.

Transport assessments should set out the likely effect of the developer's proposals, particularly on reducing the level of car use, and should indicate how these measures relate to any specific targets in the development plan, or in the local transport strategy. The development plan will outline the transport priorities for particular parts of the local authority's area and the likely nature and scope of contributions that would be expected as part of development on key sites in the plan. Development proposals, related to levels of travel demand or to thresholds stated in the plan, will be expected to help deliver the transport objectives of the plan.

Major changes in thinking about transport policy in Scotland have emerged through publication of revised consultation drafts of Scottish planning policy papers in January 2004. At about the same time (September 2003) a consultation paper was published on *Proposals for a New Approach to Transport in Scotland*. With a price tag increasing to over £1 billion a year over the current spending period in 2005–6, it was envisaged that a significant improvement in the transport system would be created. The consultation paper set out 'proposals which could change the way in which large parts of the transport systems of Scotland are managed and improved'. The envisaged investment required long-term planning 'stepping outside our normal financial planning horizons'; in turn, this would need an organisation which can focus on the long term and be able to keep a goal in sight despite short-term glitches. The investments will focus in particular on railways: these offer great scope for improving links and alleviating congestion. Presumably with the experience south of the Border, it is proposed that the main goals are 'reliability, being on time, connectivity information and simple access arrangements, like ticketing, are more important than incremental improvements in speed'. The new body will be charged to deliver integration. It is against this type of thinking that a new executive agency is proposed, tentatively called *Transport Scotland*, which would:

- be a centre of excellence in delivering transport
- provide a foundation for the development of Scotland's largest transport projects
- work for an integrated, multi-modal approach to services
- take social justice and sustainable transport as central goals
- aim to achieve reliable and improving services across Scotland.

EU transport policy

A common transport policy, furthering the free movement of people and goods, has been an objective of the EU since the Treaty of Rome. However, little progress was made on this until the mid 1980s when the European Parliament challenged the Council of Ministers in the European Court for failing to meet its transport obligations. The Commission's 1992

BOX 11.6 TRANS-EUROPEAN TRANSPORT NETWORKS

The objective is to increase the integration of existing networks for transport, telecommunication and energy as a means of improving the competitiveness of the European economy. Priority projects were identified, including the high-speed rail links Paris–Brussels–Cologne–Amsterdam–London, Cork–Dublin–Belfast–Larne–Stranraer, and the Channel Tunnel Rail Link. Other priority routes extend the TGV from France into Italy and Germany, and across southern Europe, and also the Øresund fixed link between Denmark and Sweden. Some 30,000 km of new and upgraded high-speed rail track and 12,000 km of motorways are planned by 2020. The UK was awarded £28 million in 2000 to support the TEN-Ts, £18.4 million of which has gone to the Channel Tunnel Rail Link. The theory that investment of EU transport funding can have a positive impact in the core of Europe where congestion is highest and investment already greatest raises some questions.

White Paper *The Future Development of the Common Transport Policy* emphasised the positive role that a coordinated transport policy could play in promoting economic growth through the creation of the Single Market and what is described as 'sustainable mobility'.

Until the 1990s, policy has been directed primarily at measures to deregulate cross-frontier movements and to increase competition in the transport sectors. A more explicit spatial dimension has since been added with the identification of the Trans-European Networks, explained in Box 11.6. There has also been a shift in emphasis on the contribution that policy can make to reducing the impact of pollution on the environment. Implementation of policy relating to new infrastructure is the responsibility of member states. (However, the EU has made contributions to major projects, for example, in the improvement of links between the UK and Ireland and the West Coast main rail line.) The EU is also concerned with *Developing the Citizens Network*, which is promoting alternative transport modes in areas that are dominated by car use, and clean urban transport, which it supports through the CIVITAS initiative, which seeks to support radical integrated policies and modal shift.

Some American ideas

At first sight it may seem perverse to call in aid from the country which invented and developed the motor car and its relations to such an extent that it has devised a wide range of measures to keep it under some form of control. It may be that the EU is already following in the footsteps of the USA with the emphasis on the TENs providing continental scale routes, provided by federal action in the USA many years ago. Today and in a curious way, the USA has succeeded in introducing a planning system to do battle with the car. This is not to say, of course, that it has won the battle: far from it! But it has introduced some interesting ideas which are of interest to the British planning system.

The first to be mentioned goes by the very un-British name of 'smart-growth': the antidote to suburban sprawl. Much of this is well known to us such as concentrating new growth in selected areas, but it also uses tax benefits and a 'live near your work' (LNYW) programme that can drastically reduce commuting time. In Pittsburgh, 'the city with the lid off', an aggressive clear-up of brownfield sites has transformed a major area of the city. There are many variations on this theme, which are recounted in Motavalli's (2001) *Breaking Gridlock: Moving toward Transportation that Works*. A major thrust from the federal government came with the Intermodal Surface Transportation Efficiency Act 1991 (rapidly shortened to ISTEA (or ICD TEA). It was followed by the Transportation Equity Act for the 21st Century 1998. This is known as TEA-21 and has a remarkable range of provisions including 'providing state and local government's flexibility to pay for bike trails and pedestrian walkways instead of more and more roads'.

Another dimension of this wide approach is 'congestion pricing', telecommuting (which numbered 2 million in 1992, and between 7.5 million and 15 million in 2000), traffic calming (e.g. changes in street alignment which seeks to alter the liveability of a community) and a 'flexiplace' option (as in the US Environmental Protection Agency, which allows workers to spend up to two days a week at an alternative work location). Probably better known are the yellow school buses which constitute a veritable military convoy on the interstates. It goes perhaps without saying that the USA has a long way to go on transport policy; for example, 90 per cent of commuters do not pay for parking (Banister 2002: 200).

Public attitudes and the future

Measures such as those outlined above would have been inconceivable without public support for stronger controls, and it is by no means certain that all of them will prove to be so now. This question of public acceptability is a crucial factor in transport policy. It is also one that changes over time, particularly as the impacts of increased traffic or stricter restrictions are experienced. There is a very real problem in reconciling private and public interests. Each car owner regards congestion problems as being created by other motorists; the individual's contribution to the total is negligible. This zero marginal cost for the individual imposes high costs on the collectivity of users, but car users have no incentive to economise in their use of road space: to them it is a free good. The car can be more than a means of transport. It can be an extension of a driver's personality, a symbol of affluence or power, an object to be loved as well as used. The 'love affair' with the car is, however, under strain: mass ownership

and use have made it less pleasant (though not necessarily less appealing) than it was (Goodwin *et al.* 1991: 144). Whether the disenchantment has gone far enough to warrant more penal methods of controlling its use is the basic political question. Recent surveys are helpful in showing the nature of public opinion and the scope that might exist for radical changes in policy.

Some drivers may support better public transport, for example, because they believe that other drivers will use it, and so clear the roads for them. In a study commissioned by the Oxford Transport Studies Unit, Cullinane (1992) noted the extent of car-dependence: about a half of households in the survey perceived a car to be essential to their lifestyle and a further 13 per cent would not want to be without one. However, a quarter of households did not have a car and had no intention of getting one. The overall conclusion was that car dependence was increasing, and that, if things are allowed to continue as they are, it would become increasingly difficult to persuade owners to reduce their car usage. Reflecting on the survey as a whole, Cullinane (1992) concludes that congestion seems likely to increase,

> and that there will be some voluntary reduction in traffic as the problems intensify. However, the level of attachment of most people to their car is such that it will take some positive action from outside to force any real reduction in traffic, and this positive action will have the most impact if it hits people's purses.

A number of studies have pointed in the same direction. Cars are highly valued by those who can afford them, but the problems of congestion are becoming increasingly burdensome. There is support for better public transport and, though no massive changeover by car users is to be expected, 'all changes take place at the margin'. Although it is very apparent that motorists do not like the idea of road pricing, there is good evidence that they would accept it (reluctantly) if it were part of a package which provided them with some offsetting benefits particularly in the form of good public transport. This is perhaps the most important finding of research both in Britain and abroad. Cervero (1990) has reviewed North American studies of transit pricing and concludes:

> For the most part, riders are insensitive to changes in fare levels, structures, or forms of payments, though this varies considerably among user groups and operating environments. Since riders are approximately twice as sensitive to changes in travel time as they are to changes in fares, a compelling argument can be made for operating more premium quality transit services at higher prices. Such programs could be supplemented by vouchers and concessionary programs to reduce the burden on low-income users.

Even the most ardent supporter of road pricing admits that there are many unknowns in the matter. Theoretical studies may be suggestive, but many of the issues are empirical – or at least need empirical testing. Unfortunately, it is one of the dilemmas of research on traffic restraint that though empirical evidence is needed, this is difficult to obtain since governmental authorities are unwilling to experiment without adequate predictions (May 1986: 120). This is a case where both academic and political considerations call for more research. The bibliography of research reports seems set to expand. A major missing element in the analysis of the latest White Papers is that of the car as much more than a means of mobility. In Susan Owens' words:

> The most casual review of car advertising reveals a parallel universe to that of the White Paper: it is a world in which cars confer identity and status, speed thrills, drivers enjoy increasing levels of protection and creature comforts, and, in a parody of the re-allocation of road space, the boundaries of where vehicles may go are seen as more and more permeable. None of the relevant measures (differential taxation, better enforcement of traffic laws, or education to improve driver attitudes) seem any more than Canute-like in the face of this tide of material and the very considerable interests behind it.
>
> (Owens 1998: 331)

One final point is that adjustments in one direction can be difficult to reverse because of the many consequential changes which have taken place. Life has adapted to the incredible flexibility and freedom afforded by the motor car. Jobs, shops, schools, leisure facilities (indeed a wide range of activities) have dispersed: the car has made this possible and even necessary. It is significant that a quarter of journeys and travelling time is for leisure activities (Reid and Margatroyd 1999: 9). Much development is still in the pipeline (and has been given planning permission), and forces for dispersal will continue to operate. Two topical illustrations can be cited. First, in the health service there is a trend towards large multipurpose hospitals, and the closure of small ones such as accident and emergency units: 'although these may mean longer journey times, care will be at the highest level on arrival, with no need for a transfer' (*The Times* 30 September 1999). Second, Boots the Chemists is developing two hundred edge-of-town stores, of which forty have already been built (*Planning* 1 October 1999).

Other social forces are strong: the growth of two-income households can lead to compromise decisions on housing, necessitating two cars and two journeys in different directions. Affluent teenagers demand a car as soon as they are legally allowed to drive. To many the car has greatly widened opportunities. But the forces that have widened choices for many have reduced them for those without a car. The government has an almost missionary-like desire to enable the car-less to have at least some of the freedom enjoyed by car owners. By reducing car traffic, travel is made easier, or so the philosophy maintains: 'better for everyone' as the title of the 1998 White Paper has it. Yet there are doubts: congestion is democratic; rationing by price benefits the richer and penalises the poorer. Car pollution could be attacked by promoting the development of emission-free electric cars (see the note on California in Box 11.7). And, of course, there is the defeatist view that congestion provides its own solution!

Rationing space by congestion controls is, on one argument, no different from rationing education or health care: in Graham Searjent's words, 'it is about providing for the few or not providing for all'

BOX 11.7 CALIFORNIA AIR QUALITY EXPERIENCE

California, which has acute air quality problems, requires car manufacturers to produce specified numbers of 'clean', that is, electric, cars. These have to meet severe standards: how they are met will depend on the results of a research and development effort (which so far has been disappointing). A number of other states have adopted similar policies, which are sometimes described as 'technology-forcing': the Standards involved are stricter than can be met with existing technology. Evidence that such an approach can be effective (even if not as quickly as its protagonists would wish) is suggested by the successful development of catalytic convertors. But it may seem dangerous to rely on a technological quick-fix to environmental problems. Only slowly is opinion coming around to recognising the need to do something about fragmented patterns of development, extensive urban sprawl and using planning to achieve 'smart growth'.

(*The Times* 30 September 1999). This line of argument is unappealing to some, but it does give pause for thought, particularly if a major wider policy objective is social inclusion.[33]

Further reading

General and transport statistics

Two useful overall annual summaries of transport policy are to be found in the *Department of Transport Annual Report* and, more interestingly, in National Statistics, *UK 2004: The Official Yearbook of the United Kingdom and Northern Ireland* (TSO 2003, Chapter 21). Publications of statistics and statistical commentaries on transport have mushroomed since the late 1990s. In addition to the annual *Transport Statistics for Great Britain* and (much broader in scope) *Social Trends*, several series have been launched, including *Focus on Personal Travel* (1998), *Focus on Public Transport* (1999) and *Transport Trends* (annual).

Transport policy and planning

The Buchanan Report (*Traffic in Towns*, 1963) is historically a landmark, but see also Buchanan (1958)

Mixed Blessing: The Motor Car in Britain. Truelove (1992) *Decision Making in Transport Planning* provides an excellent overview of transport policies and politics. A principal textbook is Banister (2002) *Transport Planning* and see also his (1994b) *Transport Planning in the UK, USA, and Europe*, which has a broader canvas, with some comparative analysis and a useful bibliography, and *Transport, the Environment and Sustainable Development* jointly edited by Banister and Button (1993). Another general work on transport is Glaister *et al.* (1998) *Transport Policy in Britain*.

An influential report by Goodwin *et al.* was published in 1991: *Transport: The New Realism*. See also Goodwin's *The End of Hierarchy? A New Perspective on Managing the Road Network* (1995) and the introduction to the recent literature on the car is provided by David Banister's (1999) review essay 'The car is the solution, not the problem?'

The report of the Royal Commission on Environmental Pollution, *Transport and the Environment* (1990), and the various reports of the Standing Advisory Committee on Trunk Road Assessment (SACTRA) and the Select Committee on the Environment, Transport and Regional Affairs are referred to in the chapter. A broader economic assessment of *The True Costs of Road Transport* is given by Maddison *et al.* (1996). SACTRA's latest report is a highly technical economic analysis of *Transport and the Economy*

(1999). Quite different is Whitelegg's wide discussion of transport issues in *Critical Mass: Transport, Environment and Society in the Twenty-first Century* (1997). See also the Institution of Highways and Transportation's *Transport in the Urban Environment* and *Guidelines for Developing Urban Transport Strategies*; the DETR's *A Good Practice Guide for the Development of Local Transport Plans* (2000) and Transport 2010, *Meeting the Local Transport Challenge* (2001).

There is a large library of American books, of which two are particularly recommended as offering an approach which is refreshing to British readers: Downs (1992) *Stuck in Traffic: Coping with Peak-Hour Traffic Congestion* and Dunn (1998) *Driving Forces: The Automobile, its Enemies, and the Politics of Mobility* (1998).

Only slight reference is made in the text to railways. This was not unintentional. The postwar history of the railways is confused (and confusing), with little long-term certainties: it is certainly far too complex to summarise here, particularly since future changes may well change the scene significantly. An informative, comprehensive account is to be found in Gourvish (2002) *British Rail 1974–97*.

Freight transport

Dealing specifically with freight transport are the Strategic Railway Authority's (2001) *Freight Strategy*, DETR (1999) *Sustainable Distribution: A Strategy*, National Audit Office (1997) *Regulation of Heavy Lorries*, Plowden and Buchan (1995) *A New Framework for Freight Transport*, Royal Commission on Environmental Pollution (1994) *Transport and the Environment* (Chapter 10) and DoT (1996) *Transport: The Way Forward* (Chapter 15).

Traffic calming and management

See DoE (1995) PPG 13, *A Guide to Better Practice: Reducing the Need to Travel through Land Use and Transport Planning*, DoT (1995) *Better Places through Bypasses: The Report of the Bypass Demonstration Project*, Hass-Klau (1990) *The Pedestrian and City Traffic* and Hass-Klau *et al.* (1992) *Civilised Streets: A Guide to Traffic Calming*. On the

extension of special parking areas to local authorities outside London, see DoT (1995) *Guidance on Decriminalised Parking Enforcement Outside London* (DoT Local Authority Circular 1/95). For a study of the *Community Impact of Traffic Calming Schemes* in Scotland, see Ross Silcock Ltd and Social Research Associates (1999).

A rare comparative study of different approaches to transport planning in large cities is London Research Centre (1992) *Paris, London: Comparisons of Transport Systems*. A special edition of *Built Environment* 25(2) (1999), edited by Stephen Marshall, deals with travel reduction.

Congestion charging

The amount of writing on charging road users is in striking contrast to the amount of action. Action was promised in *Breaking the Logjam: The Government's Consultation Paper on Fighting Traffic Congestion and Pollution through Road User and Workplace Parking Charges* (DETR 1998). The *Smeed Report* (1964) is an early milestone; in the 1980s attention was focused on attracting private sector involvement in the provision of roads (*New Roads by New Means: Bringing in Private Finance*, 1989), then in the 1990s it switched to congestion with the 1993 Green Paper *Paying for Better Motorways*, which was followed by a HC Transport Committee report *Charging for the Use of Motorways* (1994), and the governmental response, *Government Observations* (1994). The Transport Committee launched a wider inquiry later that year: *Urban Road Pricing* (1995). The government has commissioned a number of studies in this and related fields: see, for instance, MVA Consultancy (1995) *The London Congestion Charging Research Programme: Principal Findings*, Lewis (1994); Grieco and Jones (1994) 'A change in the policy climate? Current European perspectives on road pricing'; Jones (1991b) 'UK public attitudes to urban traffic problems and possible countermeasures: a poll of polls'; London Boroughs Association (1990) *Road Pricing for London*, 1990; and Nevin and Abbie (1993) 'What price roads?' Transport for London now publishes annual reviews of the London Congestion Charge.

Walking and cycling

A two-volume report of the Environment, Transport and Regional Affairs Committee deals at length with *Walking in Towns and Cities* (HC 167–1 and 167–11, TSO, 2001). The second volume contains sixty-three memoranda. The DETR Traffic Advisory Leaflet 3/00 *Walking Bibliography* is useful, though it is mostly limited to official publications. See also the background papers to *Developing a Strategy for Walking* (DETR 1997). The Scottish Executive Central Research Unit has published a report by System Three on *Research on Walking* (1999). The Environment, Transport and Regional Affairs Committee held an inquiry into *Walking in Towns and Cities* in 2000–1.

The principal source on cycling is McClintock (2002) *Planning for Cycling*. The *National Cycling Strategy: A Factual Summary* (DoT 1996) and later annual reports are slender but important official documents in this field. See also McClintock (1992) *The Bicycle and City Traffic*, Tolley (1990b) *The Greening of Urban Transport: Planning for Walking and Cycling in Western Cities* and the earlier HC Transport Committees report on *Cycling* (1991). Several publications by Mayer Hillman deal with these areas, such as *Children, Transport and the Quality of Life* (Hillman 1993b) and also issues such as safety. Sustrans produces much useful information including Network News which reports on the National Cycle Network; see www.sustrans.org.uk. The Cyclists' Public Affairs Group (C-PAG) publishes widely on cycling: www.ctc.org.uk.

Notes

1 Table 7.5 of the 2004 Transport Statistics Great Britain summarises the current forecasts, which are made by the Department for Transport.

2 Although it was the opposition to specific roads that received most publicity (as at Westway, Airedale, Twyford Down, Archway), there was a broadly based lack of confidence in the system by which the need for roads was addressed. Although some extremists took an extreme stance, an eloquent justification for this was provided by Tyme (1978) *Motorways versus Democracy*.

3 The report emphasises the importance of clear policy objectives and 'strategic' and long-term effects. These, of course, are precisely the policy issues with which the political process has difficulty. The problem falls into that class of which Rittel and Webber (1973) have neatly termed 'wicked problems': problems that are unique to a specific place and time, which defy definitive formulation, and which can be 'resolved' only by political judgement.

4 DETR, *Guidance on the New Approach to Appraisal* (2000) and DETR, *Understanding the New Approach to Appraisal* (2000).

5 Four programmed new routes were abandoned: M12–M25 (Chelmsford), A5–M11 (Stansted), a new motorway to the south and west of Preston, and the M55–A585 (near Blackpool). These were in addition to the motorway links between the M56 and M62 (Manchester) and the M1 and M62 (Yorkshire) which had been abandoned earlier because of the difficulty in finding an environmentally acceptable route.

6 Ove Arup and University of Reading (1999) *Planning Policy Guidance on Transport (PG 13): Implementation 1994–96*.

7 The M6 Toll was built after extensive public examination of the proposal and much intensive opposition. The need arose from congestion on the M6 through Birmingham and Staffordshire where the design capacity of 72,000 vehicles a day had to cope with actual traffic of 180,000 vehicles a day. Midland Expressway Limited has a concession to run the M6 Toll for fifty-three years to 2054.

8 *Transport: The Way Forward* (Cm 3234, 1996).

9 Tewdwr-Jones (1998: 523) suggests that 'the enthusiasm with which the Labour government has retained in place a set of policy documents released by Conservative governments undoubtedly indicates the non-political nature of their contents.'

10 *A Transport Statement for Northern Ireland* (1998), *Travel Choices for Scotland* (Cm 4010, 1998) and *Transporting Wales into the Future* (Welsh Office 1998).

11 *Travel Choices for Scotland*, para. 2.1.18. See also Farrington *et al.* (1998).

12 *A Transport Strategy for Northern Ireland*, pp. 6, 13 and

39. See also *Moving Forward: Northern Ireland Transport Policy Statement* (1998).

13 The main elements of the investment programme were £700 million more for local transport, £300 million for local bus services, more than £300 million for the rail industry, and over £400 million for the trunk road and motorway network. See the supplementary memorandum by the Minister of Transport (Dr John Reid) to the HC Environment, Transport and Regional Affairs Committee on *Integrated Transport White Paper* (20 January 1999, HC 32, p. 246).

14 The DfT commissioned research in 2003 on *Feasibility of Road Pricing in the UK*, available on the DfT website.

15 See the discussion above on parking standards. Some proposals for 'ensuring compliance' with government policy are given in Llewelyn Davies and JMP Consultants (1998: 30). These include more active involvement of government offices in the development plan process, and clearer direction to the Planning Inspectorate to ensure compliance of plans with government guidance. The position has now changed with the new machinery for regional planning guidance.

16 *Guidance on Full Local Transport Plans* (DETR 2000) and *A Good Practice Guide for the Development of Local Transport Plans* (DETR 2000).

17 Two Acts preceded this: the Road Traffic Act 1997 and the Road Traffic Reduction (National Targets) Act 1998. These required the publication of reports on current and future levels of road traffic. At the local level, these will now become part of local transport plans. The national reports will be published by the central government.

18 The proposals envisage that local authorities will be able to retain net income only if there are 'worthwhile transport-related projects to be funded'. Moreover, central government will have the power to require a portion of the revenue to be paid to the Treasury.

19 Policies in the White Paper draw on earlier discussions on *Developing a Strategy for Walking* (DETR 1997). A national walking strategy has been promised for some time but seems to have been overtaken by the other initiatives recorded here. The

DETR has published *Encouraging Walking: Advice to Local Authorities* (2000).

20 See Transport Research Laboratory reports on *Review of Traffic Calming Schemes in 20mph Zones* and *Urban Speed Management Methods* (both of these are summarised in DETR Traffic Advisory Leaflet 9/99 *20mph Speed Limits and Zones*, available from the DETR Local Transport Division).

21 *Travelling to School: An Action Plan* and *Travelling to School: A Good Practice Guide* (2003).

22 DETR, *School Travel: Strategies and Plans – A Best Practice Guide for Local Authorities* (1999). See also Cairns (1999).

23 While cycling in the UK accounts for less than 2 per cent of trips (and is declining), the proportions are 10 per cent in Sweden, 11 per cent in Germany, 15 per cent in Switzerland and 18 per cent in Denmark (DoT 1996: 7).

24 Cycling first hit the higher political agenda in 1994 when the DoT published its June 1994 Cycling Statement.

25 *Sustrans Network News* (summer 2004, p. 8).

26 See Solesbury (1999), Worpole and Greenhalgh (1999) and also the working paper *Good Connections: Helping People to Communicate in Cities*.

27 Thus the central Cambridge car park has a rate of £1.50 for the first hour, £2.90 for the second hour, £4.30 for three hours and £7.00 for four hours. It then jumps to £13.00 for five hours and £20.00 for over five hours. There are many variations on this theme.

28 See also Ove Arup and University of Reading (1997). This showed that nationally the parking policies of local authorities did not reflect PPG 13 guidance. Other studies demonstrating the inadequacy of parking controls include London Transport Planning (1997). This showed that neither provision of parking at typical office developments, nor the standards included in some boroughs exceeded the RPG maximum level by a factor of ten or more.

29 The Traffic Director for London was established by the 1991 Act, and responsibility was later transferred to the Mayor and Transport for London. The Mayor decided to reinstate the post of Traffic Director in 2003, especially to sort out problems

with roadworks. When announcing this he said: 'The reality is that the main cause of disruption in London is the 190 utility companies who are responsible for 40,000 days per month of roadworks' (TfL Press Release, 13 January 2003).

30 GLA Press Release, 12 October 2004. This is part of a £10 billion investment programme for transport in London, which will go to improvements to the Underground and extension of the Docklands Light Railway and many other initiatives.

31 The Hatfield crash was caused by one rail breaking into 300 pieces with the train travelling at 115 mph; four passengers were killed and seventy injured.

32 *On Track 2* (2000) (the newsletter of the Strategic Rail Authority).

33 The Transport Foresight Panel for the White Paper on integrated transport gave a 'vision of what lies ahead'. This includes a zero-emission car:

> By 2020 many urban buses and minibuses will be powered by 'clean diesels' using redesigned fuels. . . . The ECO-CAR will have arrived. Many inner-city and city-centre businesses and residents will hire, for periods as short as an hour, specialised zero-emission hybrid-powered cars.
>
> (*The Role of Technology in Implementing an Integrated Transport Policy*, Office of Science and Technology, DTI, 1998)

Planning, the profession and the public

Planning proposals are generally presented to the public as a *fait accompli*, and only rarely are they given a thorough *public* discussion

Cullingworth 1964: 273

The year 2003 will prove to have been a defining moment in the history of British planning. A convergence of wills of planners and politicians throughout the British Isles has created a momentum for change which recognizes planning to be at the heart of the future well being of our society.

Goodstadt 2003: 2

Introduction

The right of the public to have a direct say in planning decisions and the inherently political nature of planning are now taken for granted. The formal machinery for objections and appeals, initially devised only for specified uses by a restricted range of interests, is now employed much more widely. Many informal mechanisms have been created by planning authorities and others to improve the capacity of and opportunity for local communities and interest groups to play a part in formulating and implementing planning policy. Even so, many questions remain about the effectiveness of public participation, whose interests are served by planning, and the relationship between professional and political decisions. The first part of this chapter explains the history of public participation in planning, the nature of 'interests' and the mechanisms that enable them to influence the planning process. The second part then considers the professionalisation of planning and the relationship between the profession, politicians and the public.

Skeffington and participation in planning

The lack of concern for public participation indicated in the first quotation (which is taken from the first edition of this book) was a result in part of the political consensus of the postwar period, and in part of the trust that was accorded to 'experts' – which, by definition, included professionals. The time was perceived to be one of rapidly expanding scientific achievement, and the methods that had made such progress in the physical sciences were thought to be transferable to the problems of social and political organisation (Hague 1984). This, together with the advent of new social security, health and other public services, led to a rapid growth in professions and the bureaucracies in which they worked. Town planners, though having identity problems which took many years to settle, had a good public image: they were to be the builders of the Better Britain which was to be won now that the military battle was over. In the same spirit as established professions, they sought to establish a strong, scientific and objective knowledge base. Armed with the right techniques in manipulating the environment, they were to address the physical spatial development problems of the nation and, at least by implication, the underlying social and economic forces which drive physical development.

In retrospect, the approach implied a depoliticising of issues which were later appreciated to be of intense public concern. This was further obscured by professional techniques and language which the public could not be expected to understand (Glass 1959). Of course, planners were not alone in this: on the contrary, they simply took the same stance as other 'disabling professions' – to use Illich's (1977) term. At the time, however, the lack of political debate and participation was not widely recognised as a problem. Professionals were perceived as acting in everyone's interest – the general public interest.

It was in the 1960s that these ideas were effectively challenged in the UK, closely following experience in the USA. (See for example Broady (1968) on the UK and Gans (1968, 1991) on the USA.) By this time, the political consensus had broken down, and there was widespread dissatisfaction both with the lack of access to decision-making within government, and with the way in which benefits were being distributed. Although it claimed to serve the public interest, the planning system began to be seen as an important agent in the distribution of resources – frequently with regressive effects (Pickvance 1982). The idea of an objective, neutral planning system was increasingly challenged. The physical bias of the planning system had failed to address social and economic problems: perhaps it sometimes even made them worse. There was growing concern for a new type of 'social planning' which would seek to redress the imbalance in access to goods, services, opportunities, and power. To achieve this, some saw the need for 'advocacy planning' (Goodman 1972) which would provide experts to work directly with disadvantaged groups. This critique has had consequences for planning practice of greater permanence than that achieved by the intellectual arguments themselves. Changes were made in the statutory planning procedures, and consultation and participation gradually became an important feature of the planning process.

The Skeffington Report (1969) is sometimes celebrated as the turning point in attitudes to public participation in planning, though its recommendations are mundane and rather obvious, for example, on keeping people informed throughout the preparation of plans, and asking them to make comments. This is testimony to the distance which British local government had to go in making citizen participation a reality. The Committee was aware but did not report on more fundamental issues, noting for example, the danger of participation becoming an 'abstraction if it becomes identified solely with planning procedures rather than with the broadest interests of people'. It was many years before the public were offered engagement in the wider local policy-making through Agenda 21 and community strategies. Skeffington's proposals for the appointment of 'community development officers . . . to secure the involvement of those people who do not join organisations' and for 'community forums' had little impact at the time. What was conspicuously lacking in the debate on public participation was an awareness of its implications for the transfer of some power from elected members to groups of electors. But it was not Skeffington but the Seebohm Committee (1968) which highlighted the tension between participation and traditional representative democracy.

Participation cannot be effective unless it is organised, but this, of course, is one of the fundamental difficulties. Though a large number of people may feel vaguely disturbed in general about the operation of the planning machine, and particularly upset when they are individually affected, it is only a minority who are prepared to do anything other than grumble. The minority may be growing, but as far as can be seen, public participation will always be restricted: 'the activity of responsible social criticism is not congenial to more than a minority' (Broady 1968). Despite the increasing numbers actively participating in plan-making in the 1980s and 1990s, the general point still holds (Edmundson 1993) and furthermore, it is the same situation in other countries (Barlow 1995). The 'public' are much more likely to be engaged with the system on site specific issues which affect them.

Despite its failure to address more fundamental

questions, the reforms introduced as a result of Skeffington were generally acclaimed by the profession. It was the academic commentators who first questioned the underlying assumptions. Neo-Marxists drew attention to the more fundamental divisions of power in the political and economic structure of capitalist society, and how these continued to be evident in the outcomes from the planning system. Practitioners too began to see that extensive participation exercises produced only limited gains, and some advocate planners were among the first to reject the approach for its weaknesses. The critics argued that, like all 'agents of the state', planners operate within a 'structural straitjacket' and, irrespective of their own values, will inevitably serve the very interests which they are supposed to control (Ambrose 1986). This was supported by research findings demonstrating that planning had operated systematically in the interests of property owners. There was also substantial theoretical work concerned with the role played by the planning system in the interests of capital (Paris 1982).

The critiques were powerful but, by their very nature, they could offer little guidance to planners working in a professional, politically controlled system. Indeed, how could they respond to allegations that a fundamental purpose of planning in society is the legitimation of the existing order? If participation merely supports a charade of power sharing, leaving entrenched interests secure, what alternatives do planners have? Planning and planners became the primary explanation for the failures of urban and rural development: the postwar housing estates that were built as quickly as possible (and with few resources left over for 'amenities'); the motorways that were belatedly built to cope with the great increase in traffic congestion, but which destroyed the social and physical fabric of towns; the demise of village amenities in areas of development restraint; and the participation processes which raised hopes which were dashed by the outcomes. Certainly, planning played a part in all these, but was it the determining factor or was it, in Ambrose's (1986) words, the scapegoat?

In response to these failures, some planners and community activists tried more radical approaches. Popular planning aims

to democratise decision-making away from the state bureaucrats or company managers to include the workforce as a whole or people who live in a particular area . . . empowering groups and individuals to take control over decisions which affect their lives, and therefore to become active agents of change.

(Montgomery and Thornley 1988: 5)

Not surprisingly, the few examples of such practice are to be found in the left-wing strongholds of London and some metropolitan district councils, and with only limited success.[1] Despite the rhetoric of empowerment these initiatives have tended to involve the usual mixture of meetings, publicity and leaflets. There has been little lasting impression on communities' capacity to get involved, though they have improved the resource base of communities – an admirable, but different objective (Thomas 1995).

The reality of participation in most planning practice has been somewhat different. The Planning Act 1968 (and its Scottish equivalent of 1969) made public participation a statutory requirement in the preparation of development plans. The main stimulus for this came, not from the grass-roots, but from central government which was keen to divest itself of the responsibility to consider all development plans and appeals and what the then permanent secretary called 'the crushing burden of casework' for the government department. From 1968 many powers were devolved and with this came requirements for opportunities for other interests (predominantly development interests) to participate and object during the plan-making and adoption process. Although ministers have consistently expressed their desire to see local communities making their own decisions on planning matters, the procedures also ensure that local choice, however democratically arrived at, does not transgress 'the general public interest' or other private interests which central government deems to be important.

The new consultation provisions had limited effect over much of the country because during the 1970s and 1980s, many local authorities avoided preparing statutory development plans, in part because they believed the costs of taking a plan through the formal procedures of consultation and objection outweighed any benefits (Bruton and Nicholson 1983). As a result, the legitimacy which plans provided to decision-making was limited. The failure of local authorities to keep plans up to date exacerbated this. When planning authorities did seek public 'involvement' they tended to adopt a 'prepare reveal and defend' strategy or even 'attack and response' (Rydin 1999: 188 and 193).

The perverse outcome of all this was an over-reliance on informal policy (or in some cases no policy at all), ad-hoc decision-making, and consequently considerably less accountability. Much of the debate during the 1980s and 1990s over 'the future of development plans' has turned on this issue. Planning practitioners have argued that in order to facilitate the production of plans, the procedures should be streamlined, for example by removing the requirement for a public inquiry. In fact, this was not the problem. Many local authorities demonstrated little commitment to plan production until the adoption of statutory plans for the whole of their areas was made mandatory and their significance in development control increased.

Other planning authorities demonstrated commitment to enabling community participation in planning from the 1970s on. But during the 1980s the dominant influence was 'business'. Housing, industrial, commercial and minerals interests all effectively enjoyed special treatment through the planning system. The then Secretary of State Nicholas Ridley argued that in the interests of the country as a whole, local concerns needed to be set aside in favour of a presumption in favour of new development. Many decisions were taken out of the hands of local planning authorities (never mind communities) so as to allow for major development of all forms. There was 'a consistent diminution of the significance accorded to general public participation in policy formulation, as part of an effort to "streamline" the system and reduce delays' (Thomas, H. 1996: 177).

Disregard to local opinion gave rise to fierce criticism of the Conservative government from its own party members in the shires, and it was forced into an about-turn on community involvement under the slogan 'local choice'. This was not a signal to local authorities that they could respond as they wished to local demands.[2] Local autonomy was to be exercised only where it was within parameters laid down by the centre in policy statements. In 1989 the government began what was to become a frustrated campaign to increase certainty in planning by ensuring that both national and local planning policy was in place. Developers were warned that they would have to bear costs if they pursued applications contrary to up-to-date statutory plans while local planning authorities were told to be 'realistic about the overall level of provision'. The 1992 version of PPG 1 confirmed this change by noting the 'presumption in favour of proposals which are in accordance with the development plan'; the 1997 version eschewed any presumption and instead noted simply that 'an application for planning permission or an appeal shall be determined in accordance with the plan, unless material considerations indicate otherwise'.

During the 1980s and 1990s numerous amendments were made to plan-making and development control procedures. Their general effect was to reduce the emphasis on early public participation and consultation in the statutory procedure, while increasing opportunities for formal objection. In the early stages of the process, a statutory requirement for consultation before the planning authority has adopted its preferred view was replaced with discretion to decide on the appropriate publicity for individual plans. In contrast, the rights of formal objection after deposit of the plan were extended. Government introduced a provision for objections where the local authority had not accepted the inspector's or panel's recommendations (introduced by the 1991 Act) and more recently, a move to a two-stage deposit (introduced by the 1999 Regulations) and an end to the requirement for consultation with statutory consultees. The rationale was that two opportunities for formal objection would draw out significant problems earlier in the process and allow more time for their resolution. But it might also have encouraged local authorities to firm up their proposals very early in the process. PPG 12 (1999) talked of consultation 'based on a key issues approach . . . which focuses on those local communities, businesses, organisations and individuals relevant to the proposals being put forward' (para. 2.8).

In Scotland, experience in both the principle and practice of participation followed a similar trajectory, although it is often recorded anecdotally, that the small size of the country and proximity of central government, local authorities and communities has led to closer informal linkages. Scotland has for some time given greater emphasis to the benefits to be gained from early consultation with a wide range of interests.

During the 1990s consultation and participation practice resumed generally along the Skeffington model though with considerable variation among planning authorities and innovation in participation methods (De Montfort University and University of Strathclyde 1998). Planning, like other local public services, was influenced by the shift to the community-oriented enabling role of local government (Higgins and Allmendinger 1999). Attention turned to service quality review and the great raft of initiatives originating in John Major's administration, notably the Citizen's Charter. Building on pioneering work in some local authorities, the Citizen's Charter applied six principles: monitoring performance standards; the provision of full and accurate information about how services are run; the creation of greater choice between service providers; courtesy and helpfulness in service provision; making sure things are put right where they go wrong; and value for money.[3] A central aim was to furnish more information to the public about the performance of service providers.

The concept was taken up with enthusiasm by planners and specific charters were created for aspects of planning such as development control and individual planning authorities, all alongside the *Planning Charter* for the entire service.[4] The charters focused on easily measurable facets of service delivery, such as responding to inquiries. The impact of charters depended on how challenging they were. The charters did nothing to extend the rights of the citizen in planning matters, but rather spell out existing rights and, in so far as this helps to increase understanding, they were welcomed. But the relationship between citizen and government is not fundamentally changed: indeed, in its original form it was defined in a one-dimensional way – 'the citizen as consumer' (LGMB 1992). Some 'customer oriented' improvements to specific aspects

of the planning service, such as the one-stop shop initiative, are discussed in Chapters 4 and 5. The Service First initiative under which the charters were created has been subsumed within the Blair government's Office of Public Services Reform (OPSR) which seeks to deliver public services designed around the needs of their customers. Four principles guide reform: setting national standards that are meaningful to the public, devolving and delegating responsibility and accountability for delivery of services to the local level, allowing for flexibility to match the range of customer aspirations, and expanding choice for the consumer.[5]

A more important stimulus to participation in local government during the 1990s was the promotion of sustainable development through the Local Agenda 21 initiative (as described in Chapter 7). Local authorities are the principal actors in translating and implementing global and national objectives for sustainable development at the local level, and participation is central to the formulation of strategies. LA21 embraces a far wider range of concerns than planning including for example, green accounting and purchasing, energy conservation and recycling, but always at its heart is the objective of raising awareness and engaging citizens in promoting good environmental practice (Scott 1999). LA21 has become an important stimulus for promoting more creative ways of engaging local people in policy-making, some of which are finding their way into the separate arrangements for consultation on planning matters.

Community parish and neighbourhood planning

By the mid 1990s government emphasis on the 'consumer' had given way to a 'rediscovery of community'. As part of its general commitment to 'democratic renewal' and decentralisation, government at all levels sought to establish more direct linkages with citizens and thus strengthen accountability, and in some circumstances, to seek to empower communities by devolving policy and decision-making to local levels. Giddens sees this as a much needed response to 'the new circumstances of the global age; it is a deepening or democratising of democracy' (1998: 72). Given the centralisation of government in the UK it could also be described as a welcome application of the principle of subsidiarity – that is placing competence (or perhaps leaving it) at the most appropriate level (Nadin and Shaw 1999). Inspiration for planners' response to this challenge has come from Healey (1997), Forester (1999) and others. Strong government incentives have been established by linking democratic renewal to urban and rural regeneration funding and the practical test of Best Value (explained in Chapters 3 and 5). A raft of practical initiatives and advice on how to engage communities has come forward, sometimes initiated by those who have been champions for community planning for many years (Wates 2000).

At the centre of the government's drive for modernising government is the community strategy – a new instrument to provide for more local community involvement in policy-making and implementation. The Local Government Act 2000 placed a duty on all local authorities in England and Wales to prepare a community strategy. Similar arrangements apply in Scotland and Northern Ireland. The community strategy should 'allow local communities to articulate their aspirations, needs and priorities; coordinate [and focus] the actions of the council, and of the public, private, voluntary and community; locally . . . and contribute to the achievement of sustainable development both locally and more widely'.[6] A community strategy should include a long-term vision across all sectors of activity together with projected outcomes and an action plan with priorities. The process should deliver a shared commitment to implementation and implementation of the strategy must be monitored. The government have recognised that it may be difficult to engage particular interests and build up mutual trust in the process, so the community strategy is seen in a developmental way with early publication of an initial document, then reviews of performance and continued efforts to engage a wider range of interests and to embed the strategy into partners' plans and programmes. How will the community strategies relate to development documents? The formal answer supplied by the legislation is that they 'will have regard' to each other. The Entec (2002) study on relationships

between community strategies and local development frameworks for the ODPM was diplomatic in pointing out that the linkages 'are only beginning to emerge', though it was able to pull out some 'good practice' – sharing the same 'branding' or more critically, sharing the same vision.

A local strategic partnership is the favoured means of formulating and applying the strategy. Chapter 3 explained how 'partnership working' has become the norm for policy development at the local level, indeed many authorities were already suffering from 'partnership overload' well before community strategies came forward. LSPs were intended to be umbrella partnership and local authorities were encouraged to use this arrangement to replace or streamline existing arrangements. The LSP should comprise different elements of government, the private, voluntary and community sectors and its core task is to prepare the community strategy and make special efforts to involve those who are normally under-represented in policy-making – faith, black, minority ethnic communities and young people. A five year evaluation of LSPs is underway. Interim findings are inconclusive but point to the problems of finding appropriate representatives and managing such a diverse body both internally and in relations with the other principal actors and agencies. One issue must be the local authority base and boundary which is too small to deal with some cross-boundary issues and too large to really engage with local communities. One aspect of community plans for some partnerships are sub-authority visions for regeneration areas which are much closer to the Town and Country Planning Association's (1999c) suggestion of neighbourhood plans. Anecdotal evidence suggests more streamlined partnerships and more attention to the neighbourhood level in future exercises.

Community involvement is a long-standing feature of urban regeneration efforts and the 1997 administration's commitment to 'community' was first evident here. The government's central aim in the *New Deal for Communities* (described more fully in Chapter 10) is to engage local community development and local action. It has prompted ambitious experiments in community or neighbourhood action, which addresses the need to establish capacity for broader and long-term participation, including the use of visioning processes to engage a wider representation (Carley and Kirk 1998; Carley 1999). Evaluation of community involvement in urban policy by the Community Development Foundation points to the difficulties of participation especially among disadvantaged neighbourhoods but points to some success in the government's drive for community involvement. The report uses the loose concept of 'social capital' to explain the principal barriers to involvement, that is, the weakness of networks and linkages among individuals which reduce the willingness to mobilise and engage.[7]

Initiatives to give communities a more effective voice in government are evident in rural areas too. The 2000 Rural White Paper advocated parish plans and other mechanisms to give local communities a more effective voice, and drew attention to successful experience with village appraisals, action plans and design statements.[8] The practice of preparing village design statements was by the end of the 1990s well established and understood. Owen (2002a) explains how experience with design statements might inform the preparation of parish or neighbourhood plans, not least in the uncertain relationships with local government, statutory development plans and community plans. The proposed thinning out of statutory local planning policy set in train by the 2004 Act may well leave room for more influence from the community level, with some of the current proposals for one thousand parish plans becoming supplementary planning guidance (there are more than 8,000 parish and community councils). Alternatively, where substantial change is envisaged the parish plan might become an area action plan (Owen and Moseley 2003). To be considered for this status the plans would need to demonstrate that they have effectively engaged a wide spectrum of local opinion. If so, the preparation of such plans provides a much more local and meaningful way for people to be involved in policy and decision-making, than local plans, development frameworks and community strategies, given the large territories covered by UK local authorities.

It should not be assumed that more locally oriented plans necessarily mean wider or deeper involvement.

Neighbourhood actions plans, parish plans and other very local instruments are just as inclined to manipulation by powerful interests as any other plan. The mechanisms (and political will) to engage people in the process are critical. Community visioning exercises are gaining popularity in providing an effective channel between the plan-makers and community. Visioning can help to bring together disparate policy concerns that exist in compartmentalised departmental bunkers and tackle problems in the way that communities see them, and as Murtagh (2001) shows in his study of visioning in Derry/Londonderry, in dealing with deeply entrenched alternative community views. Visioning has its roots in North America where experience suggests that it can be a valuable method of producing shared but abstract goals while community participants are more interested in concrete outcomes and are not easily convinced that they will have a bearing on eventual decisions (Shipley *et al.* 2004; Shipley and Newkirk 1999; Solop 2001).

Community involvement and the 2004 reform of the planning system

A primary component of the planning reform agenda for government and its programme for 'sustainable communities' has been improving community involvement. The reforms include a number of changes with this in mind, including the requirement for local authorities to publish its relevant policy in a statement of community involvement. This and other changes in the formal opportunities for involvement in the plan-making and development control processes are set out in Chapters 4 and 5. But it is useful here to complete the discussion on participation in planning by reviewing the general change of approach and likely impacts of the current reforms.

The very large package of ODPM publications that herald reform includes a statement on *Community Involvement in Planning: The Government's Objectives* (2004). It enlarges on principles that are set out in Planning Policy Statement 1 at para. 1.39. To paraphrase (and in part interpret these principles), community involvement

- should be appropriate to the level of planning – while there is always a need to base proposals on an understanding of the needs of the community different types of plan will suggest different levels of involvement
- should be 'front loaded' – involving participation at the earliest stages of a proposal when it might make a difference
- should employ methods that are appropriate to the experience of anticipated participants, and offer assistance such as Planning Aid
- should provide opportunity for continued involvement and incorporate effective feedback
- should be transparent in terms of the opportunities and rules for participation
- should be planned in from the start of the process.

The principles also make reference to the need to involve 'hard-to-reach' groups. They do not make reference to the need for participation efforts to be proportionate to the potential outcome, but elsewhere PPS 1 does emphasise that involvement in no way replaces the decision-making responsibilities of local planning authorities (or presumably, the Secretary of State). The Statement on Community Involvement recognises that many local authorities already apply these principles and have adopted innovative and effective techniques for participation, sometimes involving effective liaison with developers on major schemes (Carmona *et al.* 2003). The very laudable intentions come through strongly – an open planning system with active involvement of all interests. It underplays the important distinction between consultation phases of plan-making when the planning authority has not determined its final position and should be working cooperatively and creatively with other interests on shaping the plan, and the representation stage when the authority has determined its position. Representations will challenge the position and the adversarial process of inquiry or examination places objectors in direct opposition to the local authority which is defending its position developed in full participation with a much wider range of interests. At this stage the authority is defending its preferred plan, and its role is as *the* major interest. Too often

the formal objection and representation stages are confusingly described as opportunities to participate or be consulted.

The inevitable and often extensive conflict which often comes more fully to the surface in the objection or representation stage is noted. No doubt there is scope to find solutions and build consensus as the paper suggests, but much opposition will remain. Indeed while the paper calls for a simplification of the planning process, the reality from the community perspective is a (more) complex arrangement of plans, programmes, strategies and initiatives. Planning does not necessarily address problems as communities see them (for example, crime, unemployment and congestion) which gives rise to more uncertainty and in turn, opposition or apathy.

We should not expect a government publication to draw out the fundamental and inevitable imbalances in power around particular development proposals. There are few examples of the impact of participation or community involvement in planning but they tend to illustrate just how difficult it has been for local communities to make substantial gains in the process. Certainly real gains are often made, but these are often at the margins. Coulson (2003) illustrates this with a case study of a major development at Selly Oak in Birmingham. The principal actors were Sainsburys, Birmingham University and a hospital trust. These major players made concessions and the community won important gains in environmental improvements and the detail of the development which were 'by no means inevitable'. Consultation with very well organised Neighbourhood Forum gave the planners a comprehensive understanding of their concerns. The planners 'had been mediating while negotiating as an interested party' and got what they thought were the best proposals from the developers (Coulson 2003: 193). The development interests/landowners got what they wanted and the residents were left with a feeling of failure. The reason for this is that the planning system could not deliver on their problems – 'much of what is needed to make a community a better place may lie outside the process of land-use planning' (p. 194). Coulson (2003) concludes that more explicit recognition is needed of the limitations of planning to deliver what communities want and the inherent biases in favour of property owners and developers. Coulson's case study contrasts with the government's principles and statement on community involvement, and perhaps make it seem rather naive. There is undoubtedly a lack of awareness of the real operation of power among participants (never mind all interests) in the planning and development processes.

Public participation in development control

About 700,000 applications for planning permission are made each year to local planning authorities in Britain (up from 500,000 in a few years) of which over four-fifths are granted. This enormous spate of applications involves great strains on the local planning machinery which, as the government have recognised, has not been adequately staffed to deal with them and at the same time has to undertake the necessary work involved in preparing and reviewing development plans. Yet full consideration by local planning staffs is needed if planning committees are to have the requisite information on which to base their decisions.

The 2004 Act requires the local planning authority to prepare a statement of community involvement (discussed above) which must include its policies for consultation on planning applications. At the least, when a planning application is lodged, statutory consultees (listed in Chapter 5), other organisations and neighbouring occupiers will be invited to comment. Others may also make representations if they are aware of the proposal. In England and Wales, the system of notification in the case of 'bad neighbour' developments (such as sewage works, dance halls and zoos) was replaced in 1992 with a requirement for notification of owners and other interests in land which is the subject of a planning application, but this does not necessarily extend to neighbours. Provision for notification of neighbours is dealt with under the provisions for publicity.[9]

Local authorities have the responsibility of deciding, on a case by case basis, what type of publicity to require. Major developments, as defined in the GDPO, require *either* site notices *or* neighbour notification, *and* a newspaper advertisement (see Chapter 5). Developments involving listed buildings or conservation areas require newspaper advertisements in lieu of neighbour notification. A survey of London authorities found that direct contact by letter was most effective, although the letters used by about a third of authorities were written in ways which meant they were unlikely to be understood by two out of three people (Edmundson 1993: 13). The results have been confirmed by more recent research. The requirements of the GDPO have been described as 'overkill' and unnecessarily expensive, especially in the need for newspaper advertisements, which have questionable effect (Harrison 1994). Moreover, in its Circular on the matter,[10] the government stresses that obligations to publicise applications should not jeopardise the target of deciding 80 per cent of applications within a period of eight weeks. In this case, speed is thus apparently to have higher priority than public participation. In the case of some telecommunications development which was allowed as permitted development under the prior notification scheme, the government explicitly said that it is reluctant to bring it within normal development control because of the possible delaying effect for the developers.

Scotland and Northern Ireland have different arrangements for neighbour notification and publicity about planning applications. In the Scottish system, notification is the responsibility of the applicant, who certifies to the local authority that neighbours, as well as owners and lessees, have been notified. This can be problematic for the applicant, and can lead to false certification (whether inadvertent or deliberate) and has been a constant source of complaint to the Ombudsman. A study of the Scottish system (Edinburgh College of Art *et al.* 1995) found that policing the notification efforts of applicants would merely increase the workload for councils; the effectiveness could be improved only by transferring responsibility for notification to them. Overall, it advocated the Northern Ireland system of notification where there is a two-tier approach. A non-statutory system requires the Planning Service to notify neighbours, but the identification of neighbours (through presentation of a list of 'notifiable interests') is undertaken by the applicant. The merits of the different approaches across the UK have been compared, with the Northern Ireland system coming out on top.

Once comments are gathered, it is up to the planning authority to consider them in making the decision. Should planning permission be granted and it later emerges that a notifiable neighbour has not been notified, the local authority cannot revoke the planning permission. The only recourse open to the third party is a private action against the applicant, but it has to be shown that failure to notify was carried out 'knowingly and with deceitful intention' (Berry *et al.* 1988: 806). This brief and by no means complete review gives a flavour of the complexity of the arrangements which the ODPM's 2004 Review of the Publicity Requirements for Planning Applications (in England) found to create confusion for non-statutory consultees. The study also found that newspaper advertisements were ineffective and not value for money so recommends that these be replaced with more use of planning authority website advertisements. Indeed many of the recommendations relate to the improved use of ICT, together with standardisation of the periods of time for notification in different forms.

Over recent years there has been a considerable increase in opportunities for presentations to be made to planning committees, both by the applicant and objectors, with evidence of very positive results in terms of the 'customer's' perception of the service (Shaw 1998; Darke 1999; Manns and Wood 2001). However, planning committees often have remarkably little time during a meeting in which to come to a decision. Agendas for meetings tend to be long: an average of five to six minutes for consideration of each application is nothing unusual, and in some cases the time spent on an application may be much less. It cannot, therefore, be surprising that in a large proportion of cases (in the bigger authorities at least) the recommendations of the planning officer are approved *pro forma*. Also, many applications are now dealt with through delegated powers and will not be discussed by committees (see Chapter 5). Limited open discussion of applications may be a reflection of harmonious relationships between councillors and officers, although there

have been a number of well-publicised cases where elected members have consistently acted against the recommendations of officers, as discussed below.

Several important implications follow from this. First, and most obvious, is the danger that decisions will be given which are 'wrong' – that is, they do not accord with planning objectives. Second, good relationships with the public in general and unsuccessful applicants in particular are difficult to attain: there is simply not sufficient time. Third, this lack of time corroborates the view of many (unsuccessful) applicants that their case has never had adequate consideration: a view which is further supported by the manner in which refusals are commonly worded. Phrases such as 'detrimental to amenity' or 'not in accordance with the development plan', and so on, mean little or nothing to individual applicants. They suspect that their cases have been considered in general terms rather than in the particular detail which they naturally think is important. Therefore, many applicants and objectors are taking advantage where it is available to present their case to committee. Townsend's (2002) study of public speaking rights in one authority supported earlier conclusions by Darke (1999). Speaking to an application led to a more cautious approach being taken by councillors, leading to deferrals of decision-making to allow for site visits to take place. Speakers' cases also were more likely to lead to rejection of planning officers' recommendations, although these tended to be the most controversial cases, where this was more likely. Overall, the effect is limited and most decisions were still made in line with officers' recommendations.

The value of public participation in development control needs to be considered in the light of the very different attitudes that people will take about a planning principle that they generally agree with (for example preventing unnecessary development in the countryside, when it is applied to their own land). This natural human failing is encouraged by the curious compromise situation which currently exists in relation to the control of land. On the one hand, it is accepted in principle (and law) that there is no right to develop land, unless the development is publicly acceptable (as determined by a political instead of a financial decision). On the other hand, though the allocation of land to particular uses is determined by a public decision, the motives for private development are financial, and the financial profits which result from the development constitute private gain. This unhappy circumstance (which is discussed at length in Chapter 6) involves a clash of principles which the unsuccessful applicant for planning permission experiences in a particularly sharp manner. It follows that local planning officials may have a peculiarly difficult task in explaining to a landowner why, for example, a particular field needs to be 'protected from development'.

Nevertheless, the success which attends this unenviable task does differ markedly among different local authorities. The question is not simply one of the great variations in potential land values in different parts of the country or in the relative adequacy of planning staffs. Though these are important factors, there remains the less easily documented question of attitudes towards the public. There is considerable variation in the efforts that planning authorities make to assist and explain matters to the public, though the recent reforms and Planning Delivery Grant are expected to contribute to a more consistently positive experience for the public and applicants.

The more extensive introduction of mediation in the planning process has also been on the agenda for some time though with little concrete progress. A 2000 DETR study on *Mediation in the Planning System* confirmed the potential for mediation in dealing with appeals especially for householders and where both parties entered the process voluntarily (Welbank *et al.* 2000). It recommended a 'mediation service' and a best practice guide. A report on a second project on mediation by the same team led by Michael Welbank reported in 2002. More specific recommendations were made for a national mediation service as part of the Planning Inspectorate and a team of mediators to be available for a 'stakeholder dialogue service' (Welbank *et al.* 2002). The potential of mediation was again confirmed in the 2004 Review of the Planning Inspectorate, which recommended further study. Mediation surfaces again in the ODPM's 2002 cross-national comparative study of *Participatory Planning for Sustainable Communities* undertaken by Heriot-Watt University. The main recommendation in the report

is yet again to experiment with the use of mediation to resolve or reduce objections to planning proposals, though this report sees mediation as not a bolt-on to the planning process but part of a 'participatory planning' approach putting negotiation and engagement at the heart of the process. Such approaches are often seen to be most relevant at the local level dealing with particular proposals in development control, but this report gives mediation a wider remit – starting with regional spatial strategies. This is long overdue. The formulation of regional planning guidance has tended to involve only an elite network of regional stakeholders in an exclusionary process (Pattison 2001). Applying these recommendations will require changes throughout the system, to procedure, to institutions and, as we have heard so often over recent years, to the culture of planning. This in turn requires a shift in the ideology from one of public interest to one of participation. But is this what the government means when talking of community involvement? We return to this question below.

Rights of appeal

An unsuccessful applicant for planning permission can, of course, appeal to the Secretary of State and, as already noted, a large number do so. Each case is considered by the department on its merits. This allows a great deal of flexibility, and permits cases of individual hardship to be sympathetically treated. At the same time, however, it can make the planning system seem arbitrary, at least to the unsuccessful appellant. Although broad policies are set out in such publications as the planning policy guidance notes and planning policy statements, the general view in the central departments is that a reliance on precedent could easily give rise to undesirable rigidities.

Other issues relevant to this view are the flexibility of the development plan, the wide area of discretion legally allowed to the planners in the operation of planning controls, and the very restricted jurisdiction of the courts. All these necessitate a judicial function for the department. However, this function is only quasi-judicial: decisions are taken not on the basis of legal rules as in a court of law or in accordance with case-law, but on a judgement as to what course of action is, in the particular circumstances and in the context of ministerial policy, desirable, reasonable and equitable. By its very nature this must be elusive, and the unsuccessful appellant may well feel justified in believing that the dice are loaded. Many are also unaware of the rules of the game and how they have changed. The sudden increase in the rate of appeals (explained in Chapter 5) is largely related to the increase in the rate of refusals, which in turn is in part related to the significant shifts in national and local policy since 1991 (Arup 2004). Applicants have not kept pace with the changes and are likely to feel aggrieved by more restrictive policies. Furthermore once appealed they find that the hearing or public inquiry is heard by a 'ministerial inspector' (and probably in the town hall of the authority whose decision is being appealed) which may not make for confidence in a fair and objective hearing.

Of course, part of the expressed dissatisfaction comes from those who are compelled to forgo private gain for the sake of communal benefit: the criticisms are not really of procedures, and they are not likely to be assuaged by administrative reforms or good public relations. Fundamentally, they are criticisms of the public control of land use – in particular, if not in principle. Alternatives are regularly debated, perhaps the most significant example is the environmental court which is discussed below.

Third party interests

The government has now confirmed in reforming the planning system that it has no intention of introducing third party rights of appeal. Nevertheless, it remains an issue for the system and especially those third parties who are disadvantaged. Third parties are those affected by planning decisions, but who have no legal interest, not being the applicant or the authority. Over the years there have been calls to extend to them the right to appeal should planning permission be granted as is already the case in Ireland. The rights of third parties were highlighted in the so-called Chalk Pit case (*Public*

Law (summer 1961): 121–8; Griffith and Street 1964). This, in brief, concerned an application to 'develop' certain land in Essex by digging chalk. On being refused planning permission, the applicants appealed to the minister, and a local inquiry was held. The inspector recommended dismissal partly because of the impact on the neighbouring property of a Major Buxton. The minister disagreed and allowed the appeal.

Major Buxton then appealed to the High Court, partly on the ground that in rejecting his inspector's findings of fact, the minister had relied on further information supplied by the Minister of Agriculture without giving the objectors any opportunity of correcting or commenting upon it. But Major Buxton now found that he had no legal right of appeal to the courts: indeed he apparently had no legal right to appear at the inquiry. (He had only what the judge thought to be a 'very sensible' administrative privilege.) In short, Major Buxton was a 'third party': he was in no legal sense a 'person aggrieved'. Yet clearly in the wider sense of the phrase Major Buxton was very much aggrieved, and at first sight he had a moral right to object and to have his objection carefully weighed. But should the machinery of town and country planning be used for this purpose by an individual? Before the town and country planning legislation, landowners could develop their land as they liked, provided they did not infringe the common law, which was designed more to protect the right to develop rather than to restrain it. However, as the judge stressed, the planning legislation was designed 'to restrict development for the benefit of the public at large and not to confer new rights on any individual member of the public'.

This, of course, is the essential point. It is the job of the local planning authority to assess the public advantage or disadvantage of a proposed development, subject to a review by the Secretary of State if those having a legal interest in the land in question object. Third parties cannot usurp these government functions. Nevertheless, it is generally agreed that third parties should have a right to let their views be known to the planning committee as discussed previously.

Use of public inquiries and examinations

Public inquiries into major planning appeals and called-in planning applications have had a stormy passage for many years, particularly those held in connection with highways and major developments such as Stansted, Windscale and Sizewell. Since the early 1990s similar difficulties are routinely experienced with local plan inquiries which are keenly contested affairs.

The reforms of the planning system have brought some changes to the inquiry procedure (explained in Chapter 4). Henceforth inquiries into plans will be known as examinations, the inspector's report will be binding and the examination will address not only objections but also the overall soundness of the plan. But these changes do not challenge the fundamental rights of objectors to make their views known, to appear at the examination and to question the planning authority. The inquiry/examination will remain a cornerstone of the system. The planning inquiry/examination is a microcosm of the land use planning system, and it reflects many of its competing positions and underlying conflicts of interest. It is perhaps here where the clash of planning ideologies is most easily seen. McAuslan (1980) used the example of road inquiries:

> The disenchantment with public inquiries into road proposals is only the most public and publicised manifestation of a general disenchantment with the system of land use planning, to which the conflict of ideologies within and over the use of the law is an important contributor . . . What this use of the inquiry has shown is that the reforms introduced as a result of the Franks Report twenty years ago based on the principle of openness, fairness and impartiality, and concentrating on procedures did not change (perhaps were not designed to change) the overriding purpose of the public local inquiry which was and is to advance the administration's version of the public interest.
>
> (McAuslan 1980: 72)

Using McAuslan's terminology, this is a triumph of 'the public interest ideology' over 'the ideology of public participation'. The important point here is that the inquiry/examination is not an extension of public participation, but 'a limited and carefully controlled and confined discussion of specific proposals . . . inimical to the kind of wide-ranging discussion that participators are demanding'.

The precise role of the inquiry/examination has long been a subject of debate. The procedure is a long-standing feature of British government administration, with its origin in the Parliamentary Private Bill Procedure that provided an opportunity for objections to government proposals to be heard by a Parliamentary Committee (Wraith and Lamb 1971). The procedure has grown as much by accident as design since it has been successively amended to take into account changes elsewhere in the system. Its twofold purpose has continued: to gather information for government and to provide a route for individual redress. Like appeal inquiries, local plan inquiries involve the same balancing of private and public interests through a procedure which, although essentially administrative, has many of the hallmarks of judicial courtroom practice. However, the essential nature of the planning procedure is administrative. Final decisions are taken by government, at either the central or the local level. The 2004 reforms change the balance here in that the inspector's report will be binding in all cases, whereas previously it was only so for appeals. The inspector (and Secretary of State) has wide discretion. As the Franks Report (1957: para. 272) noted, the process 'must allow for the exercise of a wide discretion in the balancing of public and private interests'. The legitimacy of the decisions rests with the political accountability of the decision-maker (Parliament or the local council) rather than on the weighing and testing of evidence as in a court of law. But many aspects are akin to a judicial process. Objectors have a statutory right to appear, and although round table and more informal hearings are commonly used, the evidence may be tested through a process of adversarial questioning before the inspector. There is inherent ambiguity in a system which has as its main objective the gathering of evidence to assist in the making of a governmental decision, while at the same time operating in the manner of a judicial hearing (Wraith and Lamb 1971). The essential dilemma of this quasi-judicial process was described in the Franks Report:

> If the administrative view is dominant the public inquiry cannot play its full part in the total process, and there is a danger that the rights and interests of the individual citizen affected will not be sufficiently protected . . . If the judicial view is dominant there is a danger that people will regard the person before whom they state their case as a kind of judge provisionally deciding the matter, subject to an appeal to the Minister.
>
> (Franks 1957: paras 273–5)

The difficulties were increased by the 1991 reforms. In confirming the role of districts as the responsible authority for adopting local plans, the administrative role of the inquiry is reinforced. In introducing the provision which allows for objections where the local authority does not accept the recommendations of an inspector or panel, the changes lend weight to the judicial role. Thus, the central questions have changed little over the years following the Franks Committee, not least because while identifying the ambiguity in objectives, the report found in favour of neither, and fell back instead on the need for balance, and consistent application in the inquiry procedure of the principles of 'openness, fairness and impartiality' (Bruton et al. 1980: 377).

The three principles have guided inspectors with some success, and the courts have played a relatively small part in the planning process (although this is growing). Nevertheless, each of the three principles requires some qualification. The Franks Report (1957) itself recognised that impartiality needed to be qualified since in some circumstances central government was both a party to the debate, perhaps putting forward a proposal, and at the same time the decision-maker. How, in this situation, can the procedure be impartial? Here, one of the parties to the dispute will make the final decision, giving at least the appearance of being the judge and jury in its own court.[11]

The openness and fairness of the inquiry also need to be qualified. First, there is widespread misunderstanding of the procedure, especially the respective roles of inspector and local authority. The adversarial nature of the inquiry, with the inspector playing a passive role while objectors and the local authority exchange evidence and questions has important implications for the way in which the agenda is structured, and it limits potential outcomes. In his case study of the Belfast Urban Areas Plan inquiry, Blackman (1991b) points out how an adversarial hearing focuses attention on the evidence brought forward to support the position of particular interests. The inquiry becomes moulded into a battle about which interest should prevail, and this precludes debate about alternative and potentially shared solutions, which may be in a 'common or generalisable social interest'.

All this has now to be considered in the context of the 'plan-led system'. More emphasis on statutory plans means that more development interests, neighbouring authorities and service providers will be concerned to influence the content of plans and thus the outcome of inquiries. The number of objections to plans has increased dramatically over recent years.[12] Many have questioned the system's capacity to cope with this burden, particularly since the biggest test for contentious plans has been later in the modifications stages. Government's response has been to emphasis the need for involving stakeholders prior to final proposals or plans coming forward. A two-stage representations procedure has also been introduced. After the first round planning authorities are encouraged to negotiate with objectors and agree changes (as explained in Chapter 4). In future, those who want to contribute to shaping development plans will have to put considerable effort into the early stages of participation. But authorities will still be under pressure to get plans in place. This could be a recipe for tension, conflict and frustration and it will be interesting to see how this works out.

Inquiries and examinations continue to be a source of dissatisfaction. Some criticism is to be expected given that in a predominantly adversarial debate there will almost always be a losing side. But despite increasing flexibility in the way they are run by the Inspectorate, there are continued criticisms of their expense and complexity. The problem mainly lies with local authorities who are insufficiently prepared and resourced for the scale of the plan inquiry task and furthermore pay insufficient attention to the advice offered. Research has shown that many are unable or unwilling to explore the possibility of compromise (or provide information and clarify uncertainties) with objectors (Steel *et al.* 1995) which does not bode well for the government's new participatory approach. Few authorities have offered the support to objectors that would have reduced the effort needed from inspectors to clarify objections at the inquiry stage.[13] Procedural problems are often compounded by the nature of the plans being examined, which are sometimes overly complex and detailed, and which are not conducive to effective communication in public consultation.

Research in Scotland has concluded similarly that management of the process has been poor in too many authorities. The *Review of Development Planning in Scotland* (Hillier Parker *et al.* 1998) makes similar points about a lack of focus to consultation, and the need for a greater sense of urgency and commitment to the process by local authorities. However, its recommendations concentrate on changing the procedures, in particular making the report of the inquiry binding on the authority.

The role of planning inspectors (and in Scotland reporters) is critical in the consideration of formal objections. Research on inquiries and the Inspectorate's own commissioned studies confirm that inspectors provide a high quality service, although there are inconsistencies in practice arising from the wide discretion available to inspectors in managing proceedings. However, when problems arose in the procedure, inspectors had few powers and resources to put them right.

A difficulty with many inquiries or examinations is determining where the boundaries of discussion are to be drawn: there is always the danger that argument will spill over into a broader policy framework. It is common at inquiries into particular matters for the most general questions of policy to arise. This is hardly surprising since typically the development being debated is, in fact, the application of one or more policies to a particular situation: this readily offers the opportunity for questioning whether the policy is intended to apply to the case at issue – or whether it should. Even wider issues arise, such as the desirability of supporting a particular way of generating nuclear power, or the need for more roads, or the role of the planning system in providing affordable housing. Pressure groups which, for example, may be opposed to the building of new roads or out-of-town shopping centres anywhere, irrespective of the merits (or otherwise) of particular projects, will want to use the inquiry as a platform on which to make their wider case. That they are able to do this is sometimes a reflection on the lack of national policy on certain issues. The Heathrow Terminal 5 inquiry was the latest in a long line of inquiries which have spent much time and public money debating what the national policy should be.[14]

This raises the question as to whether the provisions for national policy debate are adequate. It makes sense, of course, to argue that Parliament should be the arena for the national policy debate, and the local authority for debate on local policies. It also seems reasonable to maintain that it is quite inappropriate for major issues of principle to be argued when they are simply being applied locally. But issues are not so easily packaged, in reality there is a sharing of competences among different jurisdictional levels. And in the UK, the lack of a regional tier has often meant that national and local levels have had to work cooperatively on strategic issues. Some site-specific proposals raise acute issues of national policy which have not been settled or adequately discussed, and sometimes government may avert proper discussion at the national or regional levels because of the complexity and sensitivity of the issues involved. This may be the reason that one type of public inquiry, for which legislative provision was made in 1968, has never been used: the Planning Inquiry Commission (PIC).[15]

There have been several proposals for the funding of third parties at major public inquiries, though they differ on the form that this should take – and the difficulties to which it could give rise. The government has taken the narrow approach that 'most objectors participate in public inquiries to defend their own interests'. This is a perfectly proper activity, but there is no reason why it should be financed out of public funds; however; the government is supporting Planning Aid's efforts to advise poorly resourced objectors. In Canada, the Berger Commission on the Mackenzie Valley Pipeline Inquiry arranged for a funding programme which cost nearly $2 million, for 'those groups that had an interest that ought to be represented, but whose means would not allow it'. The federal government has an 'intervener funding programme' which was used in the Beaufort Sea environmental assessment review (Cullingworth 1987).[16]

Human Rights Act 1998

The Human Rights Act 1998 came into force on 2 October 2000. The Act enables citizens to take action under the European Convention on Human Rights (dating from 1950), through the domestic courts. The Convention was used in court cases involving planning well before the Act came into force. Thus the Act does not confer new rights but its enactment raises awareness about the rights set out in the Convention and should ensure that they are taken into account by government and the courts. All public bodies are required to act in accordance with the Convention rights. Local planning authorities and other planning agencies, including the Planning Inspectorate are thus bound by the Act.

The main issue for planning is the Convention right arising from Article 6 which guarantees that 'In the determination of his civil rights and obligations or of any criminal charge against him, everyone is entitled to a fair and public hearing within a reasonable time by an independent and impartial tribunal established by law'. This has raised questions for the UK since ministers make policy, apply it and decide on appeals. While this duty is mostly undertaken through the Planning Inspectorate or Recorder's Office, their independence from ministers is in question (inspectors represent and 'stand in the shoes of the minister'). In any event the minister has the right to 'recover' appeals for his or her decision.

The Scottish Minister has already accepted that neither ministers nor reporters are independent and impartial (in the meaning of the Convention) but has argued that the provision to further right to challenge decisions in the courts does meet the Convention. Although the right of challenge in the courts is severely restricted (being limited, generally, to procedural issues rather than the policy merits of the case) this argument has been accepted in at least one case so far. Nevertheless, the future of the Planning Inspectorate has been questioned for a while, although that matter now seems to be resolved. Indeed the flood of court cases that some warned about (Baker 2000) has not materialised.

A related issue is the idea of an environmental court. The merits or otherwise of an environmental court system, as currently practised, for example, in Australia and New Zealand, have been discussed in the DETR's *Environmental Court Project* led by Professor Malcolm Grant. The report proposes a two-tier approach with a first level court comprising a revised Planning Inspectorate as an independent tribunal and a second level environmental court that would deal with the same sort of appeals now considered by ministers. At the time of writing, the government has confirmed its intention to continue with the current arrangements. The second stage of the review of the Inspectorate (discussed in Chapter 3) and more cases brought under the Human Rights Act will influence developments in the longer term.

Interests in planning

'Interests' in planning are usually thought of in terms of the organisations, groups and individuals who are actively engaged in the planning arena: they are identified by their participation in the land development and planning processes. These include land and property owners, developers, special interest groups, national government and its agencies, and local authorities themselves (in both their landowning and regulatory capacities). Some organisations have become particularly skilled in presenting their views at both national and local levels. The Home Builders' Federation, for example, is an important national organisation that regularly presents evidence at public inquiries in support of the interests of the house building industry.

Voluntary organisations operate in a variety of ways, from lobbying to education, and from active participation in planning processes to the ownership and management of protected land and property. Membership of such organisations has increased dramatically since the early 1970s.[17] National voluntary organisations obviously do not command the resources available to commercial interests, but they do employ experts and can be very effective in promoting their causes. Some of them have become increasingly sophisticated since the late 1970s and can be appropriately described as 'major elites' (Goldsmith 1980). There are innumerable 'minor elites' of small groups who become involved in an ad-hoc way with particular issues. The evolution of participation and consultation in planning has favoured these self-defined interest groups, and given them a relatively privileged position in the planning process.

It has long been appreciated that such groups are not necessarily representative of anything wider than the interests of their active supporters. Many people are not able or willing to take the time to engage in 'participation', and some groups who have a clear stake in planning outcomes (such as home buyers or job seekers) are too diffuse to have become effective participators, and 'rarely if ever emerge as definable actors in the development process' (Healey *et al.* 1988: Chapter 7). Three 'interests' that permeate almost all planning activity – race, gender and disability – yet which are rarely represented in disputes over particular development projects are discussed briefly in the following sections. Other specific interests are beginning to receive special attention, notably older people (Gilroy 1999), while local authorities are strongly encouraged in planning reform agendas to consider the needs of specific interests and those that are hard to reach in the creation and implementation of planning policy.

The revised PPS 12 (2004) says little about these interests, but states that when preparing plans, planning authorities

> must comply with the general duty in the Race Relations (Amendment) Act 2000 to promote race

equality. This duty means that authorities must have due regard to the need to eliminate unlawful racial discrimination and promote equality of opportunity and good relations between persons of different racial groups . . . Local planning authorities should also comply with the Disability Discrimination Act 1995 which places a duty on all those responsible for providing a service to the public not to discriminate against disabled people by providing a lower standard of service. Service providers now have to consider making reasonable adjustments to the way they deliver their services so that disabled people can use them.

(para. 3.1.10)

This is perhaps a step back from the previous PPG 12, which made a more general statement encouraging planning authorities 'to consider the relationship of planning policies and proposals to social needs and problems' and to 'address social exclusion'. It also made specific reference to the impacts of planning on 'ethnic minorities, religious groups, elderly and disabled people, women, single parent families, students, and disadvantaged people living in deprived areas' (para. 4.13). PPS 12 is, of course, a slimmed down version of the previous guidance and much emphasis is now placed on the statement of community involvement (a statutory document subject to independent examination) in which the local authority must identify particular interests affected by its proposals.

Advice to those who wish to address issues of exclusion through the new system comes from Hill and Salter (2004):

> The planning process is fundamentally adversarial in nature, and as a result is inevitably more concerned with 'exclusion' than 'inclusion' . . . In promoting more inclusion in planning, the profession would need to move from its current position as the arbiter of rights in law to one in which it administers and responds to a collective management process of multiple interests and views.

(p. 1)

The difficulty of achieving such change is illustrated by the quotations highlighted on the cover of this report calling for inclusion in planning which date from 1972.

Race and planning

Questions of equal racial opportunities have figured increasingly on the town planning agenda since the early 1970s, though with questionable impact on practice. The Race Relations Act 1976 places a duty on all local authorities to eliminate unlawful racial discrimination in their activities and to promote 'good relations between persons of different racial groups'. It was in 1978 that the RTPI established a joint working party with the Commission for Racial Equality (CRE) to investigate the multiracial dimension of planning, and to make recommendations for any necessary changes in practice. Their 1983 report, *Planning for a Multi-Racial Britain*, was a frank assessment of the inadequacies of the then current thinking on race and planning (RTPI and CRE 1983). It is perhaps only a little less relevant today, and its recommendations apply to all planners, wherever they work.

The working party's deliberations were spurred by the deteriorating race relations in Britain's cities, and by numerous reports calling for more action from central government and other bodies (Scarman 1981). These and other studies demonstrated that:

> Black people in Britain share all the problems of social malaise and multiple deprivation with their white neighbours. In addition, however, they experience the additional difficulties of racial discrimination and disadvantage, because of the colour of their skin.
>
> (CRE 1982: 9)

The RTPI and CRE report (1983) argued that, despite its record of innovation in participation, the profession had failed to address the issue of race. Indeed, the RTPI's own 1983 report on participation failed to explicitly consider the racial dimension. Ethnic minorities were mentioned only once, with reference to the need to provide interpretation at public meetings. *Planning for a Multi-Racial Britain* sought to improve sensitivity to racial issues 'and show how planning practice can be modified to avoid racial discrimination, promote equality of opportunity, and improve race relations for the benefit of the whole community'. It identified three elements in the racial dimension of planning:

(1) reviewing the impact of current policies, practices and procedures upon different racial groups, with a view to ascertaining actual or potential racial discrimination;
(2) building racial distinctions into surveys, analyses and monitoring with a view to identifying the special needs of different racial groups; allowing the impact of policies to be assessed, and providing a basis for any appropriate positive action;
(3) positive action in planning policies, procedures standards and decision-making, partly by directing positive non-racial policies and actions towards groups containing high proportions of black people, and partly by taking special steps to ensure that black people have equal access to the benefits offered by town planning.

The report also considered how the profession could be made more representative in terms of its membership, and the implications of this for the education of planners. It became an important guide for committed practitioners and teachers, and formed the basis for other more detailed guidance, foremost of which was the GLC's *Race and Planning Guidelines*. Publication of this was squeezed in just before abolition (when perhaps the most vigorous supporter of equal opportunities was lost).

The GLC report, however, has subsequently provided a rich source of material for guiding good practice in the London boroughs and elsewhere. Its recommendations illustrate the inadequacy of the 'colour-blind' approach identified as commonplace in the RTPI/CRE report. Such an approach, though holding that all people should be treated equally, is insensitive to the cultural traditions and needs of particular ethnic groups. Unintentional discrimination can therefore result. It is perhaps surprising therefore that Thomas and Krishnarayan (1993) are able to note the almost complete absence of reference to racial matters in major planning reports such as the Audit Commission's *Building in Quality* and the DoE's *Development Plans: A Good Practice Guide*. Despite evidence of innovative practice in a few places (Best and Bowser 1986) planning has made only a weak contribution to challenging racial discrimination and disadvantage. Thomas and Krishnarayan (1994a: 1891) explain this with reference to the 'socially conservative' nature of planning: the tendency to focus on technical problems of land use management rather than radical social goals which has been strongly reinforced by professionalisation, the narrow interpretation of the objectives of planning by government and the courts, and the reproduction of existing social and spatial divisions in planning policy.

During the 1990s various surveys showed that there continued to be little progress. A survey for the Local Government Association revealed that, in general, equal opportunity issues had a lower priority than in the 1980s, that there was little policy guidance on the issue and the colour blind approach still dominated practice. The effect has been to institutionalise indirect discrimination within the planning system (Loftman and Beazley 1998a, 1998b) and to stereotype different groups by oversimplifying their internal diversity (Ratcliffe 1998). However, the same survey also revealed examples of good practice, though these were very much the exception.

New duties on local authorities under the Race Relations (Amendment) Act 2000 were hoped to provide added impetus for effective action. Public bodies now have a 'general duty' to promote race equality. The driver for this requirement was the Macpherson Report (1999), which followed the tragic murder of Stephen Lawrence. Public bodies including local authorities must eliminate unlawful discrimination, promote equality of opportunity and promote good race relations. This is a shift from reaction to instances of discrimination to a more active role in promoting change for the better relations. As part of the duty each body must prepare a race equality scheme (RES) which sets out its corporate approach to promoting race equality, and assesses its performance.

Thomas (2004) has surveyed the Welsh local authority race equality schemes and found that the planning service remains peripheral to race relations issues and in one authority was not even recognised as relevant to the new statutory duty. He is pessimistic about the impact of the 2000 Act:

> the assessment presents a sobering picture which is consistent with other early assessments (CRE and Schneider-Ross, 2003). It finds schemes which are, at best, formally competent, but generally lacking in explanation or justification . . . there is a widespread commitment to data gathering and monitoring, which may provide a firmer and more sophisticated basis for future race equality schemes. But this source of optimism is balanced by the sizeable minority of authorities whose schemes evince a lack of interest in and commitment to promoting race equality as a central component of their activities – i.e. as a mainstream activity. For them, this appears to be a further statutory duty with the letter of which they may comply, but the spirit of which seems to have left them unmoved. Perhaps most worrying of all is the lack of transparency in the schemes, and in particular the almost universal lack of reasons for choices.
>
> Thomas 2004: 44

Many more authorities need to look towards the examples of good practice and past recommendations on how to address race and ethnic mix in planning.

Women and planning

The absence of policy explicitly related to concerns of women has attracted little attention until recently. One of the many reasons for this is the inadequacy of the classic texts on social theory and urban studies. In reviewing these, Greed (1993) writes:

> Women appear in studies of working class communities as a variety of oversimplified stereotypes, based on observing them as mono-dimensional residents tied to the area rather than as people with jobs, interests and aspirations beyond its boundaries. Young and Willmott give emphasis to women

in their study, but their fondness for seeing them in the role of 'Mum', as virtually tea machines, and almost as wall paper to the main action of life, is open to question.

(Greed 1993: 233)

Greed (1994) has described the male domination of the profession as 'only a temporary intermission'. Women were primary contributors to the social movement which promoted town planning early in the twentieth century. She argues that the professionalisation of planning, its institutionalisation within the government structure, and the limited access to qualifying courses has 'kept most women out'. The gender bias reflects, and in part perpetuates, the patriarchal structure of British society, and continues to influence the recruitment and education of planners. Women generally tend 'to under-achieve and under-aspire as regards a career', and they 'hesitate to embark upon a professional career', although once on a planning course women tend to outperform men (Fitzsimmons 1990).

Awareness of women's issues in planning grew strongly during the 1980s and was reflected in important and influential reports from the GLC (1986) and the RTPI (1987, 1988d). The negative impacts of increasing mobility and planning policies that encouraged this have come in for particular criticism (Hamilton 1999).[18] While issues have been identified, their impact has been marginal. One difficulty (in addition to the power of traditional attitudes and ways of thinking and perceiving) is the general lack of explicit social policies in plans. Thus, the provision of sporting facilities and the open space standards applied to them are routinely regarded as legitimate land use matters. These predominantly male activities are contrasted by Greed (1993: 237) with crèches, which are commonly regarded as social issues, even though they 'may have major implications for central area office development' – a fact which is explicitly taken into account in some US cities (Cullingworth 1993: Chapter 7). Another factor is the limited extent to which these issues have been addressed in the curricula of planning schools (Loevinger Rahder and O'Neill 1998).

Calder *et al.* (1993) have identified the scope of 'women's specific needs policies' and their research and a study by L. Davies (1996) show some specific policies are being incorporated in development plans, even weathering DETR scrutiny. However, this has not been taken up widely and where relevant policies have been included they have often been deleted following requests from the DETR after plan scrutiny, or in response to inspectors' reports after the inquiry. L. Davies (1996) notes the removal of a policy from the Hammersmith and Fulham UDP which required sheltered lockable spaces for buggies in new large scale shopping developments.

Concerns for the needs of women are generally much less developed than those of race or disability, especially outside London where some authorities simply 'didn't consider this an issue' (L. Davies 1996). Even where they are, 'child care' is the predominant 'women's issue', reflecting the assumption of women's primary role as carers, and demonstrating

the limited nature of the majority of planning initiatives which it may be argued are designed to ameliorate current constraints on women rather than to challenge the status quo in the drive for greater equality for women.

(Little 1994b: 266)

Planning and people with disabilities

One group that has received more attention from the planning system is that of people with disabilities. The 2004 Act (and its predecessor) requires local authorities to draw to the attention of planning applicants the need to consider the requirements of the Chronically Sick and Disabled Persons Act 1970, the Disability Discrimination Act 1995 and associated Regulations.

Disability has a broad definition including (as the RTPI's Practice Advice Note 3 points out) those who suffer breathlessness or pain, who need to walk with a stick, are partially sighted, have difficulty in gripping because of arthritis, or are pregnant. The 1995 Act defines a disabled person as someone with 'a physical or mental impairment which has a substantial and long-term adverse effect on his ability to carry out normal day-to-day activities'. Access policies can also generally make life easier for parents with children and elderly people. As Gilroy with Marvin (1993: 24) point out, most people are or will be physically impaired at some time during their life. That impairment or disability can be turned into a handicap by the environment.

The needs of people with disabilities have been considered largely in terms of the design of the built environment, and planning has been at the centre of this. In 1999 service providers and employers had to make 'reasonable adjustments for disabled people', such as changing the way the service is provided. From 2004 (2002 for education providers) regulations made under the 1995 Act require employers and service providers to alter their premises or services when 'a physical feature makes it impossible or unreasonably difficult for disabled persons to access its service'. This is resulting in many relatively small changes to buildings which are sometimes also very controversial, especially where the heritage is concerned.[19]

It is primarily the Building Regulations and not the planning system which is the means by which access requirements are enforced (L. Davies 1996). Important though this is, it leads to an overly simplistic stereotyping of the problems faced by individuals with disabilities. H. Thomas (1992) has made a strong critique of typical attitudes:

> The 'regs' can become a checklist which defines the needs of disabled people, ignoring, indeed disallowing, the possibility that individual professionals dealing with particular cases need to learn from the experience of disabled people themselves. The British legislation which relates specifically to planning with its references to practicality and reasonableness, reinforces a strand in planners' professional ideologies which emphasises the role of the planner in reaching optimum solutions in situations involving competing needs or interests. Thus might a fundamental right to an independent and dignified life be reduced to an 'interest' to be balanced against the 'requirements' of conservation or aesthetics.
>
> (Thomas 1992: 25)

L. Davies (1996) is critical of central government advice, which although often only very recently updated, fails to give the necessary impetus. The DoE *Good Practice Guide*, for example, is described as 'woefully lacking'. She also observes that, whereas negotiations between interest groups and planners have often strengthened plans on these matters, further objection by other interests and scrutiny by the DETR later in the process have resulted in a 'watering down of policies'. Above all, policies apply only to new building whereas much of our environment, including the headquarters of the RTPI and universities (as Greed (1996c: 233) ruefully notes), present major access problems for many people.

Access to information

Information is power; so it not surprising that undemocratic societies guard it jealously. It is surprising, however, that secrecy is so prevalent in Britain. The fiasco over the Crossman Diaries revealed the absurdities in striking detail. British secrecy is a legacy of old styles of government (Cullingworth 1993: 191). These persist in many ways, and the elitist origins are still apparent. Government is carried on in an elitist atmosphere of determining what actions are in the public interest, as viewed through the eyes of the government. Traditionally, participation has been limited, and information restricted, though there are clear signs of a significant change in attitudes.

Ironically, though traditional attitudes are more entrenched at the centre, it is central government that has forced local government to provide greater 'freedom of information'. A major milestone in this is the Local Government (Access to Information) Act 1985, which imposes a duty on local authorities, the National Parks and other bodies such as passenger transport authorities (PTAs), 'to publish information . . . about the discharge of their functions and other matters'. The underlying principle of this Act is that all meetings of local authorities, their committees and subcommittees should be open to the press and the public. There is a power to exclude only in narrowly defined circumstances. The Act also provides for public access to agendas, reports and minutes of all meetings, and opens for public inspection certain background papers which relate to the subject matter of reports to council, committee or subcommittee meetings.

A review of the Act after ten years of operation (Steele 1995) was generally positive, finding that the Act had established 'minimum standards of openness and accountability', that the provisions were exceeded by four out of five councils, and that this had been done at little extra cost. Planning figures prominently where rights are exercised with 60 per cent of planning committees attended by more than ten people (against 10 per cent average for council committees). While rights are being exercised, take up is limited to making information *available*, and most authorities do not actively promote information and thus make it truly *accessible*.

Similar conclusions were reached following a survey of public access to planning information in Scotland undertaken for the Scottish Office (School of Town and Regional Planning, University of Dundee 1997). The best practice recommendations indicate the less than positive approach taken to access in some authorities – including having literature explaining how the system operates and ensuring that professional staff are available at suitable times. The Cabinet Office Service First initiative is going to require a much more thorough approach, including opening of offices beyond the usual working day.

A major influence on improving accessibility is the EU, particularly with its 1990 *Directive on Freedom of Access to Information on the Environment* (Birtles 1991). This was a product of the commitment in the EC's Fourth Action Programme to enable groups and individuals to take a more effective part in protecting and promoting their interests. The objective of the Directive was 'to ensure freedom of access to, and dissemination of, information on the environment held by public authorities and to set out the basic terms and conditions on which such information should be made available'.

The regulations define environmental information as 'the state of any water or air, the state of any flora or fauna, the state of any soil, or the state of any natural site or other land', together with activities or measures adversely affecting, or designed to protect these states. Revised *Access to Environmental Information Regulations* came into force in January 2005 following a new EU Directive 2003/4/EC of the same name. A Code of Practice has been published by DEFRA for public authorities explaining how they can meet the requirements of the regulations.[20] The new regulations widen the scope of organisations that come under the provisions, especially to address public–private partnerships and private companies that provide public services such as water and transport. The new regime introduces a 'public interest test' in that access to information can be refused only if the 'public interest in refusing the request outweighs the public interest in disclosing information'. For planning authorities such a situation might arise where potential damage to the environment might result where a designation is planned but not in force, for example, where information provided to a landowner might result in damage to a proposed site of special scientific interest.

Organisations affected have twenty working days to provide the information or request further clarification from those who are requesting it. Previously organisations were to produce a list of all information sources, and to make it publicly available, but the accent has now switched much more strongly to positive actions to provide information routinely and widely, in particular making use of websites. This shift is intended in part to address the critical issues of 'knowing that there is someone to ask, and knowing that there is something to ask for' (Clabon and Chance 1992: 25). Once an issue is identified, the problem is mostly one of *navigation* around complex information systems (Moxen *et al.* 1995). Under the new regulations, bodies are expected to give advice about what sort of information is available and indicating further information that might help with their inquiry.

In order to encourage an active approach by governments the Treaty on European Union also enables any citizen to make a complaint to the Commission that an EU provision is not being applied. Krämer (1991) argues that this casts the EC in the role of Ombudsman for the environment. Additional rights to 'environmental information' arise from the Directives on environmental assessment, though the use of such measures to date is limited. Important factors are the limited knowledge of existence of information gaining access, the costs of retrieval, and understanding what is found.

In 1994 central government adopted a *Code of Practice on Access to Government Information* which, alongside the charter initiatives and requirements to justify expenditure more fully, has led to the publication of more detailed reports from departments and agencies. In 1999 the DETR also published a *Code of Practice on the Dissemination of Information during Major Infrastructure Developments*. The opening up of government has been increased following the Freedom of Information Act 2000.

The Act was preceded by the White Paper *Your Right to Know* (1997), explaining that the aim was to make all government information available 'unless it would clearly cause harm to national security, personal privacy, safety, business activities and law enforcement'. The Freedom of Information Act 2000 supersedes the Code and provides for a general right of access to information held by all public authorities except the secret services. As well as primary legislation requiring disclosure, it is also intended that much more government information is routinely published, although the public authorities will be allowed to charge for it.

The cost issue is worth further note in respect of planning documents. Central government publications are now highly priced (presumably at market levels), as a result of which their cost is prohibitive to ordinary citizens. There is a striking contrast with many US government and European Union publications, which the public can obtain free of charge. A good example has been set by the Scottish Office, which announced at the end of 1992 that it had decided to issue its planning guidance (NPPGs, Circulars and PANs) generally free of charge. although retaining the discretion to make a charge for PANs where appropriate.

With local government, a bigger problem is the paucity of publications, but the cost of planning documents tends to be very high. No doubt, these expensive reports are intended for a small high-price market, but high quality (or quantity) of plan production is not important in relation to the needs for public information and participation. A number of authorities have printed plans in newspaper or poster format. These are models well worth copying, as is the policy of North Kesteven District Council, whose plan is free; the London Boroughs of Tower Hamlets and Barnet gave plans away free to residents.

e-Planning

Another means of providing information is via the Internet, and this is growing rapidly. Under the modernising government initiative, the government aimed to ensure that 25 per cent of all interactions with the citizen should be made available by electronic means by 2002, and 100 per cent by 2005.[21] Good examples of local authority websites now contain details of planning applications and the local plan. Examples are easy to find through the Planning Portal.

Like many other public services (and life generally) the delivery of the planning service is fundamentally affected by the widespread access to information and communications technologies. The government has established targets for public services to go online and the modernising planning agenda includes a strong commitment to making more use of ICT and to address the generally outdated approach in many local authorities (Land Use Consultants 2002). Online access to key documents has been available to academics for some years but only recently is becoming more common in practice. The ODPM sponsored Planning Portal is taking shape as the central online reference for planning practice with access to information about the planning system and local authority plans and more.[22] Local planning authority websites will be the prime target for inquiries and on this another ODPM project led by Wandsworth Borough Council will provide considerable support. PARSOL or Planning and Regulatory Services Online is preparing tools and guidelines and schemes to assist local authorities provide more effective online access for planning, building regulations, environmental health and trading standards. For government this may mean a speedier and more efficient planning system, but online access also broadens access to a wider range of interests.

Maladministration, the Ombudsman and probity

Most legislation is based on the assumption that the organs of government will operate efficiently and fairly. This is not always the case but, even if it were, provision has to be made for investigating complaints by citizens who feel aggrieved by some action (or inaction). As modern post-industrial society becomes more complex, and as the rights of electors and consumers are viewed as important, pressures for additional means of protest, appeal and restitution grow.

At the parliamentary level, the case for an ombudsman was reluctantly conceded by the government, and a Parliamentary Commissioner for Administration was appointed in 1967. The Commissioner is an independent statutory official whose function is to investigate complaints of maladministration against central government departments and other government bodies acting on their behalf, and referred to him through Members of Parliament (Gregory and Pearson 1992). Powers of investigation extend over all central government departments, and there is an important right of access to all departmental papers. In 1994, the powers were extended to cover the *Code of Practice on Access to Government Information*.

Only a small fraction of the Parliamentary Commissioner's cases relate to planning matters and, of course, the concern is with administrative procedures, not with the merits of planning decisions. The Commissioner's reports give full but anonymised texts of reports of selected cases which have been investigated. Illustrative cases include a complaint that the Secretary of State for the Environment failed to understand the grounds on which a request had been made for intervention; a complaint that following a motorway inquiry, the inspector called for further evidence from the DoE on which objectors were not given the opportunity to cross-examine; and a complaint by a group

of local residents that an appeal decision to allow a gypsy caravan site paid little heed to local residents' objections, ignored important relevant facts and was taken on the basis of inconsistent attitudes. In all these cases, the Parliamentary Commissioner concluded that the complaint could not be upheld. This is not always the case, however, and the Commissioner's subsequent criticisms have led to changes in internal administrative procedures in the DETR. The cases in which the Commissioner does find 'maladministration' are often of extraordinary complexity, if not real confusion. Indeed, complexity and confusion can be major factors in the failures in communication and the misunderstandings which result in 'maladministration'.

The popularity of the Parliamentary Commissioner led to pressures for the establishment of a similar institution for local government. In the mid 1970s, *Commissioners for Local Administration* were set up for England, Scotland and Wales. With good sense, they recently decided that they should be known as the *Local Government Ombudsmen* (using the Swedish word; in Sweden the office dates back to 1809: Renton 1992). This is not only their popular name, but also makes explicit that their responsibilities are confined almost entirely to local government (though it is hardly gender sensitive!). The Ombudsmen also deal with police authorities.

As with the Parliamentary Commissioner, complaints have to be referred via an elected member (although in exceptional circumstances the Ombudsman can accept a complaint direct), a requirement on which there is considerable controversy. There is also concern about the situation which arises when a local authority refuses to 'remedy' a case in which maladministration or injustice is found by a Commissioner. To date, however, only limited legislative changes have been made. These include a power for local authorities to incur expenditure to remedy injustice without specific authorisation by the Secretary of State; a requirement that local authorities must notify the Ombudsman of action taken in response to an adverse report; a power for the Ombudsman to publish in a local newspaper a statement concerning cases in which a local authority has refused to comply with the Ombudsman's recommendations; and a new responsibility for the Ombudsmen to provide local authorities with advice on good practice, based on the experience of their investigations.

Not all have taken kindly to the 'interference' of the local ombudsman, and their annual reports (while noting with satisfaction a general improvement in the handling of complaints by local authorities) often name authorities which have refused to remedy cases of maladministration and personal injustice. The irony is that the Ombudsman is wholly funded by contributions from local authorities.

A high proportion of complaints concern planning matters: of the more than 18,658 complaints received in England in 2003–4 4,096 (21 per cent) were on planning matters, which is second only to housing. More than half of these (2,529) were about neighbour amenity issues raised by third parties in relation to planning applications, with enforcement the next most important planning issue (672).[23] The Ombudsmen have constantly noted that aggrieved objectors to planning permissions (third parties) have little or no redress – unlike the aggrieved applicant, who can appeal to the Secretary of State. The Ombudsmen have no power to deal with the merits of planning decisions, but they have difficulty in explaining the difference between a planning decision which constitutes maladministration and one which is simply disputed (Hamersley 1987). Some typical cases include the case where the local authority failed to notify a telecommunications company within the fifty-two-day prior notification period which meant that a mast was deemed planning approval. The Ombudsman found maladministration and recommended that the council should either pay for the mast to be moved 10 metres to the west (where it was acceptable) or pay compensation equivalent to any loss of value to the properties of the complainants. The council agreed to this suggestion and the Ombudsman commended the council for its positive response. Other examples include complaints where the local authority approved applications under delegated powers even though the parish council had objected, and thus required discussion at committee; failure to make a condition on an approval to meet the concerns of objectors even though they were assured it would be made; and failure to identify from

site visits or to bring to the attention of the planning committee impacts on neighbours' amenity which may have led to refusal of applications. Many other examples are available in the annual digest of cases. Where maladministration is found (about half the cases) the authority is usually asked to value the cost of the failure (for example, in loss of amenity) and to pay a sum for the trouble taken in bringing the case. The Ombudsmen have certainly had an influence in encouraging local authorities to improve their planning procedures, and to go beyond minimum statutory requirements (see Box 12.1).

Despite improvements to procedure in many local authorities, there have been recent well-publicised cases of extreme maladministration leading to fraud and corruption. Although small in number, such cases raise more general concerns about the integrity of officers and probity of councillors, especially in the national context where the Scott Inquiry (and the dubious activities which gave rise to it) have brought these general questions to the attention of a very wide audience. The best known case is that of North Cornwall District Council, though a number of other planning authorities including Doncaster have been the subject of investigations. (*Private Eye* has a constant supply of alleged misdemeanours in local government.) In North Cornwall complaints were first taken up by the Ombudsman, then the district auditor, the police and Channel 4 television. Finally, the DoE set up an official inquiry (Lees 1993), which unequivocally condemned the local councillors for granting permissions to local people for development in the open countryside.[24]

The Local Government Act 1992 and the Code of Conduct for Councillors established the principle that councillors should not take part in proceedings if they have a direct or indirect pecuniary interest in the issue under discussion. But such a simple distinction does not cover the many ways that councillors and officers can be influenced or themselves influence decisions. At the heart of the issue is lobbying. Applicants and objectors lobby councillors (and sometimes officers). Councillors may lobby colleagues (although not taking part in the decision themselves). Committee chairs can put pressure on officers, and so on. Whether or not such

BOX 12.1 CODE OF BEST PRACTICE IN PLANNING PROCEDURES

- Members and officers should avoid indicating the likely decision on an application or otherwise committing the authority during contact with applicants and objectors.
- There should be opportunities for applicants and objectors, and other interested parties such as parish councils, to make presentations to planning committee.
- All applications considered by planning committee should be subject to full, written reports from officers incorporating firm recommendations.
- The reasons given by planning committee for refusing or granting should be fully minuted, especially where these are contrary to officer advice or local plan.
- Councillors and planning officers should make oral declarations at planning committee of significant contact with applicants and objectors, in addition to the usual disclosure of pecuniary and non-pecuniary interests.
- No member should be appointed to planning committee without having agreed to undertake a period of training in planning procedures as specified by the authority.

Source: Standards of Conduct in Local Government (1997: 75)

practices constitute improper activity is not always easy to discern.

Because of the number of cases of alleged impropriety or 'sleaze' in government generally during the Conservative administration, the Nolan Committee was established to consider Standards in Public Life. The Nolan Report (1997) on local government included a chapter on planning. The report criticised the national code for being inadequate, complicated and in parts, inconsistent and even impenetrable. Building on the report, the government has published a 'new ethical framework' to govern the conduct of elected members and also local government employees (who were not covered by the code). In 2001 a Standards Board for England was established to promote high ethical standards and to investigate allegations where elected members' behaviour may have fallen short of the required standards. The Standards Board has published model codes of conduct and their provisions must be included in local codes for councillors. In 2003–4 3,566 allegations of breaches of the code were referred to the Board, which alone seems to justify its existence. Unfortunately, the reports don't tell us how many are related to planning.[25] However, we do know

that planning figures significantly in problem cases. So much so that planning was seen to require extra measures including the need for councillors to undertake training in planning because of the difficulties in dealing fairly with planning law and its implementation.[26] There should also be a greater degree of openness in the planning process; this would among other things, assist in dealing with the problems facing local authorities in granting permission for their own proposed developments and 'the potential for planning permission being bought and sold'.

In coming to these conclusions, the Nolan Report noted that in 1947 'the need for postwar reconstruction was clear. Development enjoyed broad public support'. Things have now changed.

> Development is now a term which has a pejorative ring, and the planning system is seen by many people as a way of preventing major changes to cherished townscapes and landscapes. If the system does not achieve this (and it is a role which it was not originally designed to perform), then the result can be public disillusionment.
>
> (para. 277)

In Scotland, a 1998 consultation paper on the Nolan Report, *A New Ethical Framework for Local Government in Scotland*, broadly accepted its recommendations, but took issue with a number of them. The outcome was one of the first pieces of legislation enacted by the Scottish Parliament, the Ethical Standards in Public Life etc. (Scotland) Act 2000. It established a national Standards Commission, created with objectives similar to the Standards Board in England. A single code has been published for all local government (instead of a model code) together with a Guidance Note for Local Authorities.

The professionalisation of planning

As the quotation at the head of this chapter suggests, the professional body has made some fundamental changes over recent years, redefining the core of its expertise, attempting to broadening its membership and reforming the way that planners are educated. Other professional bodies have found the same need for major reform in response to major challenges to the professions. Some background is needed on how the planning profession has evolved in the UK before coming back to these recent changes.

The emergence of the planning profession in the early part of the twentieth century has been well documented (for example, Cherry 1974; Healey 1985). The profession emerged in part as a response to the fragmentation of policy, expertise and professions that were contributing to urban and regional development (Sutcliffe 1981b). The emphasis of the movement for a planning profession was on the need for coordination of actions among different sectors and disciplines, largely through physical plan-making and design skills. The expansion of planning activity created demands for planners who initially were provided by other professions, although planning education and the profession expanded rapidly. Impetus for establishing planning practice slowly moved from social reform movements to the profession. However, Healey (1995: 496), among others, argues that there was little intellectual underpinning to the conception at that time of town planning within the context of social and economic development.

The strength of the professional body in the UK owes much to its place in government. Planning is a state activity and the state has given legitimacy to the profession through formal recognition the designation of the Royal Charter. Wilding (1982) describes this as a 'profession–state alliance' where the state uses professions to assist with fulfilling its responsibilities and to legitimate state intervention, while state sponsorship has enabled the occupation of planning to 'gain control of the substance of its own work' (quoted in Low 1991: 26). Thus the Royal Town Planning Institute is incorporated by Royal Charter 'to advance the science and art of town planning for the benefit of the public'. It exhibits many of the recognised 'traits' of professions: a body of knowledge which it seeks to consolidate and reproduce, control over the recruitment, education and training of its members, a measure of autonomy and self-regulation and maintenance of a common code of ethics. Working as a planner is not restricted to members of the RTPI, but many employers require membership.[27]

The professionalisation of planning has been challenged by numerous authors who draw on wider debates about the role of professions in society. Reade (1987), Evans, B. (1995), and Evans and Rydin (1997) have been most critical from outside the profession in the UK. Hague (1984) and Healey (1985) have made more constructive proposals from within. Similar arguments have been made in other countries (Hoch 1994). There are two main parts to their arguments: justification for planning action in the public interest is flawed, and the lack of distinctive intellectual base for the profession. The argument that planners work in the general public interest is more important to planning than most other professions, and it is increasingly difficult to make. This was certainly a stronger argument in the immediate postwar years under conditions of a strong political consensus (Hague 1984), but there is much less justification today. As Giddens points out, 'fifty years ago everyone was a planner, now no one seems to be' (1998: 15). Reade (1987) has made a comprehensive case against the professional mystique that is created around planning by claims that it provides objective technical expertise in the general interest on matters which he says should be resolved through the political process. Low (1991: 26) concludes that 'in practice urban planning is a disguised form of political decision-making'. Nevertheless, Campbell (1999: 302), drawing on Thompson (1985), suggests that 'the ethic of neutrality . . . is still deeply rooted in conceptions of the planner's professional role'. Reade (1987) has also challenged claims to an identifiable distinctive intellectual base or competence for planning. Certainly there is little agreement about what it is: claims are made for a design base, the social science base, environmental management and (probably nearest to our view) policy coordination.

The Schuster Report (1950) addressed this issue with recommendations to replace the design approach to planning education with a blend of skills in economy, society and governance. This was initially opposed by the Institute and was not applied until the 1970s. In the mean time education built on ideas from the USA and the scientific rational approach became influential in planning education and, to some extent, practice. These scientific rational approaches demanded more social science skills and brought geography and sociology into planning education

Despite very significant uncertainties about the political objectives of planning and the intellectual basis of planning expertise, planning as a profession in some ways strengthened in the later years of the twentieth century. Planning services have continued to be in demand and planning has retained its professional status. But societal trends are tending to question claims for and the value of professionalism: the role of the state as a provider of goods and services has been eroded, and so too the professions which provide public services. Planning action is now much more overtly concerned with public and private sector cooperation, which is a further challenge to the public interest goals of planning. Traditional departments of local government dominated by single professional groups have been broken down in favour of cross-authority working. The relations and tensions between planning officers and elected members are changing, with the Nolan Report (1997) firmly placing the

politicians' interests before professional judgement (Campbell 1999).

The impacts of the rise of neo-liberalism during the 1980s, its antagonism toward the welfare state and its adherence to individualism has perhaps now receded somewhat though the primacy of individual freedom and personal choice has also been embraced by new social democratic parties. Hague (1997: 142) argues that this means the 'standardised perception of needs' is no longer relevant; the public interest ethos 'must be revamped to protect the interest of minorities and to deliver equal opportunities'. Planning practice has a major challenge in dealing with changed public attitudes that reject collectivist solutions assert individual rights promote non-professional voluntary action. The 'state–professions alliance' is undermined along with the traditional public interest and service arguments which have underpinned professional legitimacy. Evans (1995) describes the effect as planning changing from a 'welfare profession' serving the public interest to a skills-based profession, selling a service. Practice has responded of course with great changes to its 'soft infrastructure' – the values, norms and standards which have guided practice (Hull 2000). The question is whether the intellectual base for planning can respond to these challenges. Campbell (1999) argues that 'planning has not really advanced to meet these intellectual changes'. Evans and Rydin (1997), Tewdwr-Jones (1996) and others argue that the profession has largely acquiesced to the management orientation of local government planning into little more than 'bureaucratic proceduralism'.

There is considerable role confusion in the planning profession about the difference of planning skills from those of other professions, and in differentiating the roles of professionals, politicians and communities. The evidence suggests that citizens also have the same difficulty in identifying the contribution of planning and planners. But planning skills are in demand, because of their ability to play the system which they primarily gain through experience in working in it. They possess the means but until very recently the profession said very little about the ends, the substantive objectives and outcomes of planning. The similarity with the situation in the USA is striking.

Hoch (1994) observed that the dominance of the social science approach in America has tended to make planners analysts rather than doers, and thus less effective. Planning tends to ignore the bigger moral, political and aesthetic questions.

After the critique, it is perhaps surprising to note the remarkable extent and success of professionalisation of planning in the UK. Planning practice and education in the UK is more professionalised than any other European country (Healey 1985: 493). The RTPI has 18,185 members (October 2003) of whom more than 1,000 were based overseas in more than ninety countries, the largest contingents being Hong Kong with 315 members and Australia with 108. Previous surveys suggest that about half the professional staff working on planning in the public and private sectors are members of the RTPI (LGMB 1993c).[28] The overwhelming majority of the membership is young, white and male. Women and ethnic minorities are under-represented, although there has been growth in the proportion of women members from 15 per cent in 1990 to 27 per cent in 2003. The RTPI also supports the Planning Aid service which provides free independent advice on planning (through the voluntary effort of members) to people, communities and voluntary groups.

Planning education

The profession maintains its status primarily through controlling entry to the profession and the qualifications that provide entry. The RTPI held its own examinations for membership until 1992 but, from the earliest days, specific courses were set up to train planners (the first being at Liverpool University in 1909, followed by University College London in 1914: Batey 1993; Collins 1989). By 1945, there were nine courses in town and country planning, all of which were postgraduate. Student numbers in town and country planning increased sharply to fifty-four separate courses in 1978 (Thomas 1990a, 1990b) and fifty-seven (twenty undergraduate) in 1981.

A government-inspired review of *Manpower Requirements for Physical Planning* forced some courses to cease recruitment during the 1980s (Amos *et al.* 1982), although the impact on the number of planning students was reduced because of increased intakes in the remaining schools. In 1988, a total of 766 students was recruited into 31 courses. Student numbers have increased substantially since then, spurred on by the promotion of participation in higher education generally, although recruitment has fluctuated and in the late 1990s was particularly difficult for a number of schools. The number of providers has fallen now to twenty accredited schools, including Hong Kong and the Joint Distance Learning Course, which is managed from the University of the West of England. The Institute reviews the accreditation of planning schools at least once every five years and publishes guidelines setting out policy on the education of planners together with a core curriculum. Students completing accredited courses can (after two years of practical experience) apply for corporate membership of the Institute. There are more than thirty other institutions that offer courses with a substantial planning component but which are not recognised by the RTPI.

In the year 1999–2000 there were 3,127 students enrolled on planning courses in recognised schools. This compares with 3,715 in 1991–2. This is far less than the demand for graduates. A 2003 survey showed that '87 per cent of local planning authorities nationally, and 94 per cent in London, reported experiencing recruitment and retention problems that were affecting their ability to deliver an effective planning service' (Edmundson 2004: 7). The same survey confirmed that the short-term response to such problems is to employ agency staff many of whom originate in other countries, and non-planning qualified graduates. This, in turn, created a five-fold increase in demand for part-time postgraduate planning courses in London between 2000 and 2004. The ODPM has recognised the shortage of qualified planners as an important issue for planning reform and is making efforts with partners such as the RTPI to improve the attractiveness of planning as a career.

The vision and manifesto for planning

Over recent years a programme of fundamental reform of the RTPI has been fashioned and implemented complementing the government's reform of the planning system. The objective of reform is described in the Institute's *A New Vision for Planning* as a 'radical evolution which will lead to a body so different that it will be seen as a New Institute' (RTPI 2000: 1). The *Vision* document goes on to say that it 'is built around the core ideas of a planning that is spatial, sustainable, integrative, inclusive, value-driven, and action-oriented'. This is a tall order. The *New Vision* recognises the increasing complexity of the professional environment, and the demands for planners to take on board 'a wider set of issues'. Its practical effects so far have been a reform of the education requirements for membership (discussed below) and to broaden opportunities for membership and involvement of the Institute. Related professionals in transport, environment, community planning and regeneration are being encouraged to get involved. The use of terms such as 'spatial action' and 'mediating place' may not make the Institute any more accessible but they try to summarise a view of planning that stresses its role in addressing the interrelationships of all government policy with spatial impacts by bringing together a mix of knowledge and skills to do this. The *New Vision* sees the professional planner as a facilitator. The planner will recognise and work with conflicting sets of values, and competing objectives. Therefore most attention is given to the process of planning as an inclusive procedure working towards negotiated outcomes.

The objectives of planning are given much less attention in the *New Vision* and the Institute has filled this gap with a bold and campaigning *Manifesto for Planning* (RTPI 2004a). The *Manifesto* presents ten main campaign priorities for including a national spatial development framework and 'an end to simplistic targets based on the speed of decision-making'. The issues raised reflect thinking in the wider membership. In a 2001 survey of members, the five top priorities for action were to support research assessing good outcomes from planning, to promote lifelong learning, to sponsor research to show how spatial planning supports sustainable development, to challenge government openly where planning policy fails tests of social equity, and to establish think tanks to promote research and creative thinking in planning. Underlying these priorities is the wish of the membership for a more positive portrayal of the benefits of planning and the profession.

Perhaps the most significant practical impact of new thinking in the Institute has been the changes to planning education. In 2001 the RTPI instituted the most thoroughgoing review of town planning education, qualifications and training in fifty years. It was undertaken by a Planning Education Commission chaired by Professor Peter Fidler and reported in 2003 (Brown *et al.* 2003). The Commission's report reflects the changes in planning practice and the profession noted above. Its most radical recommendation was to reduce the minimum duration of study for postgraduate courses leading to membership from two academic years full time to bring it into line with the standard masters requirement in the UK of one calendar year. This would be complemented by more emphasis on learning in the workplace and continuing professional development. The Institute welcomed the report and is working with the planning schools on implementing the changes on the basis of a renewed policy statement on planning education (RTPI 2004b).

The Town and Country Planning Association should also be mentioned here. It was established as the Garden City Association by Ebenezer Howard in 1899 to realise the first garden city. It took on a wider remit to campaign for effective town planning in the early part of the century and took the TCPA name in 1941. It is not a professional association and membership is open to anyone. The TCPA has been a consistent advocate for a rational and humane form of spatial development.[29] It is also a mine of critical yet constructive commentary on planning, notably through its journal *Town and Country Planning*. (See Box 12.2.)

BOX 12.2 WORLD TOWN PLANNING DAY

World Town Planning Day was founded in 1949 by the late Professor Carlos della Paolera of the University of Buenos Aires. To promote the day Paolera founded the International Organisation for the World Town Planning Day with the objective of advancing public and professional interest in town planning throughout the world. Arrangements should be made on or near 8 November each year for meetings, lectures, exhibitions, broadcasts or celebrations relating to town planning.

Celebrations of the day were held in a different country every year (designated as president for that year) from 1950 until the late 1970s. Britain was the focus in 1965. Paolera designed a symbol for the day in the shape of a flag whose top half is blue, and the bottom half green, respectively standing for air and land, and with a bright gold sun in the middle.

Source: Town and Country Planning Association (the TCPA occasionally flies the flag from its offices overlooking the Mall)

In conclusion

The planning scene has been dominated for many years by a veritable orgy of institutional change. The pace of change has accelerated under the reforming Blair administration. Although all this was intended as a means of facilitating better planning, it is possible that it has had the opposite effect of restraining the development of policies appropriate to changing conditions and perceptions. If the filing cabinets are being constantly moved, it is difficult to bring their contents up to date. Furthermore, some of the institutional changes (even if promising in the longer run) may have added to the confusion over the role of 'town and country planning' in relation to regional and national economic planning, to the management of the economy, to the increasingly strident demands for environmental protection, to the place of public participation in the planning process, and to even more intractable issues such as 'the energy question', the distribution of incomes and 'access to opportunity'. It is, however, a nice question as to whether a more stable institutional structure would have facilitated the formulation of more appropriate and effective policies in the context of the baffling economic and social problems of the time.

What does seem clear is that the faith in the efficacy of institutional change was misplaced. Consecutive attempts at the reorganisation of local government seem to have created as many problems as they solved. In the words of Matthew Arnold, 'faith in machinery is our besetting danger'.

The basic problems lie deeper: they relate to the functions, scope and practicability of 'town and country planning'. The crucial issues with which 'planning' is concerned do not fall within the responsibility or competence of the planning authority, or even within that of local government – jobs and poverty being the two most obvious ones. Hence central government wrestles with the political pressures to which problems in such areas give rise, though typically with disappointing results.

From a cynical viewpoint, much effort is wasted at both local and central levels in attempting to control the uncontrollable. The proclamations of politicians are given a credibility which is unwarranted. It also has unfortunate consequences, since the illusion that problems can be 'solved' turns easily into a delusion, and constant failure debases the political process and breeds cynicism.

More positively, there has been a continued discussion of the limits, role and purpose of planning.

There has been a steady succession of reviews and studies: the RTPI (1976) discussion paper on *Planning and the Future*, the Nuffield Foundation's (1986) report *Town and Country Planning*, the TCPA's (2000) review of the planning system in the late 1990s, *Reinventing Planning*, and the 2002 *Review of Environmental Planning* by the Royal Commission on Environmental Pollution. It is interesting to speculate why this untypically deep questioning began when it did. Perhaps it was a sign of the coming of age of planning. Two factors were of particular importance: an awakening of concern for making government more responsive (what was inadequately termed 'public participation') and a sea change in the economy.

The first started with protests against unwanted developments, big and small, particular and generalised. Some of these protests led to gargantuan 'inquiries', which continue from Roskill and Windscale to Terminal 5 at Heathrow. Others were more modest and localised, but also much more numerous. With hindsight, the most important were those which in reality were protests not simply against a particular development (though that was the manifest objective), but against the policies which these represented.

Typically, these were not the responsibility of planners, but of other professions and, above all, of politicians who forged the policies. Politicians, at both central and local levels, perceived problems (understandably) in the terms in which they were presented. Problems labelled as housing shortages, road congestion, slum clearance and redevelopment portrayed the obvious solutions: build houses quickly, build more roads, clear the slums and redevelop the worn-out parts of the inner city. The political responses were to 'solve' these clearly articulated problems. But policies involve choices and, again with hindsight, some of the choices had undesirable results: more houses involved high densities and few amenities, new roads increased the attraction of private transport and the decline of public alternatives, slum clearance destroyed communities, and so on.

The perceived 'failures' of planning – high rise development, difficult-to-let council housing schemes, urban motorways, inner city decline, and the like – added to the mounting concern about the role and character of planning. Whether, or to what extent, these were 'failures' and, if so, the degree to which 'planning' was to blame are questions which were seldom raised, let alone answered in their historical context. But they were seen to symbolise the inadequacy of planning. An alternative interpretation would lay emphasis on the growth of real public participation. Public participation is not a subsidiary process which can be held in check: once it begins to work effectively it transforms the nature of the planning process. On occasion, it can get 'completely out of control', as it did in some well-publicised highway inquiries. Though disruptive, these led to a major reappraisal of both highway inquiry procedures and highway planning. Here the point is that the lesson was learned: it had become apparent that participation could work.

The professional acceptance of public participation (though by no means unanimous) was a remarkable feature of the 1970s. That it came first in planning, but not in other fields such as education or health, may be related to the transformed nature of planning education and the changed character of the 'intake' to the profession (Cherry 1974; Centre for Environmental Studies 1973). Indeed, it may be that it is this which above all explains the new humility, the introspective questioning and the new intellectualism which was so marked a feature of the time. In this respect, planners departed from the norms of professionalism, though not without internal strife.

The profession's commitment to public participation continued into the 1980s, despite an increasingly hostile political framework. The growth of a participatory ethic, however, may have been of lesser importance than the impact of economic change. A new humility grew in response to a gradual realisation that changes in the economy were structural rather than cyclical. Policies based on the assumption that the task in hand was to channel the forces of economic growth were increasingly perceived to be misplaced. Planning was no longer to be preoccupied with controls over the location of growth: it was to be remoulded to assist in the actual promotion of growth.

Certainly, the 1980s and 1990s saw a remarkable change in the political scene. A new and clear political philosophy emerged: a major objective of planning was now to facilitate enterprise with the minimum of constraints. Planning controls were reduced, most of the new town development corporations disbanded, the GLC and the metropolitan county councils abolished, and statutory requirements for public participation in the preparation of plans were reduced. Local government was increasingly bypassed in favour of ad-hoc bodies designed to promote private sector involvement.

Much of this dramatic change stemmed from the explicit political stance of the Thatcher government, and the belief that, somehow or other, planning itself was part of the problem – an attitude encapsulated by the remark that jobs were being locked up in the filing cabinets of planners. But there were also some deeper undercurrents. Above all, economic conditions highlighted the importance of the *promotion* of development in contrast to its *control*. Successive public expenditure crises also took their toll.

The advent of the Blair government in 1997 (following eighteen years of Conservative control) promised great changes, within a framework of what the election manifesto termed 'a new centre and centre-left politics'. Proposals for planning were largely subsumed under radical proposals for devolution, and 'good local government', but commitments were made in relation to a number of policy areas such as an integrated transport policy, 'life in our countryside' and a right to greater access. There followed an avalanche of publications, initiatives and experiments, and now the new framework of planning instruments brought forward under the 2004 Act.

More generally, there is increased confusion about the role of planning. The promises held out by the 'new' structure plan system failed to materialise. It appeared to be no more effective, speedy and flexible or satisfying than the system it was designed to replace. Whether the latest changes to strategic planning will prove to be more effective remains to be seen. When the needs for economic growth and for planning clash, the former is likely to win.

A major unresolved issue is that of the allocation of land for housing, and the government are now taking concerted action to deliver more housing through the Sustainable Communities 'plan'. Public opinion in the past has been very hostile, at least in the southern and

more prosperous parts of the UK, and the apparent lack of any determined, or at least effective, resistance to the new proposals is a surprise. This time, there are promises to deal with the great infrastructure inadequacies that have in the past virtually precluded much needed housing development. Unfortunately, the much heralded integrated transport system is conspicuous by its absence.

It is always difficult to see current events in perspective, and there is abundant scope for debating whether the changes that have taken place are fundamental or not. More likely they will be overtaken by new problems, or by the redefinition of old problems which cannot readily be foreseen.

Further reading

Planning and politics

This is an enormous topic and many of the central texts on planning address it. Some recommendations are Albrechts (2003) 'Reconstructing decision-making'; for recent relations between political ideologies and planning see Tewdwr-Jones (2002) *The Planning Polity* and Allmendinger and Tewdwr-Jones (2000) 'New Labour, new planning?' Earlier work includes Ambrose (1986) *Whatever Happened to Planning?*, Reade (1987) *British Town and Country Planning*, Blowers (1980) *The Limits of Power*, Low (1991) *Planning, Politics and the State*, Thornley (1993) *Urban Planning under Thatcherism*, Healey (1997) *Collaborative Planning* and Taylor (1998) *Urban Planning Theory since 1945* (Chapter 5). See also the further reading for Chapter 1.

Participation in planning

Overviews of the historical development of participation in planning are given by Rydin (1999) 'Public participation in planning', Thomas, H. (1996) 'Public participation in planning'. Early titles include Dennis (1970) *People and Planning*, Davies (1972) *The Evangelistic Bureaucrat: A Study of a Planning Exercise in Newcastle-upon-Tyne*, Broady (1968) *Planning for People*, Levin and Donnison (1969) 'People and planning' and, from the

USA, Gans (1968) *People and Plans*, and his more recent collection of essays (1991) *People, Plans and Policies*. Some titles give a flavour of their content: Wates (1977) *The Battle for Tolmers Square*, Forman (1989) *Spitalfields: A Battle for Land*, Anson (1981) *I'll Fight You for It: Behind the Struggle for Covent Garden* and Christensen (1979) *Neighbourhood Survival*. Other studies include Colenutt and Cutten (1994) 'Community empowerment in vogue or vain?', the DoE study *Community Involvement in Planning and Development Processes* (1995) and the DETR study on *Sustainable Local Communities for the 21st Century: Why and How to Prepare an Effective LA21 Strategy* (1998).

Good practice guidance has been issued as part of the 2004 reforms as *Participatory Planning for Sustainable Communities* (ODPM 2002) and see also *Community Involvement in Planning: The Government's Objectives* (2004). A broader review of citizen engagement with public services was undertaken by Aspden and Birch (2005) *New Localism* and the Home Office published *Citizen Engagement and Public Services: Why Neighbourhoods Matter* (2005). The Scottish Executive published a White Paper on *Your Place, Your Plan: Public Involvement in Planning* (2003).

The DETR *Guidance on Enhancing Public Participation in Local Government* (1998) includes as an annexe a list of guides on public participation; see also the Planning Officers' Society (1998) *Public Involvement in the Development Control Process: A Good Practice Guide*.

Reviews of practice in participation are given by Coulson (2003) 'Land-use planning and community influence', Loftman and Pratt (1994) 'Public participation in the UDP process' (on Birmingham), Reeves (1995) 'Developing effective public consultation' (on Sheffield), Hall (1999) 'Town expansion: constructive participation' (on Stevenage); and Shaw (1998) 'Who's afraid of the double whammy?'. For international comparisons see Barlow (1995) and the report of the Council of Europe's conference hosted by the Planning Inspectorate on *Public Participation in Regional/Spatial Planning in Different European Countries*. For an inside look at the role of Planning Aid see Reeves and Burley (2002) 'Public inquiries and development plans in England'.

Community, neighbourhood and parish planning

Two useful starting points explaining the meaning of 'community involvement' are the Community Development Foundation's (2003) *Searching for Solid Foundations: Community Involvement in Urban Policy* and papers by Abram and Cowell on their ESRC project website at www.shef.ac.uk/communityplanning/. The Oregon Chapter of the American Planning Association (1993) have provided *A Guide to Community Visioning*. On rural community planning see Owen (2002a) 'From village design statements to parish plans' and Moseley (2002) 'Bottom-up "village action plans"'. For a Scottish perspective see also Stevenson (2002) *Getting 'Under the Skin' of Community Planning*. An international perspective is given by Heriot-Watt University *et al.* (2003) *Participatory Planning for Sustainable Communities*.

Inquiries and examinations

The DoE study on *The Efficiency and Effectiveness of Local Plan Inquiries* (1997) contains an extensive bibliography. See also the Planning Inspectorate (1996) *Development Plan Inquiries: Guidance for Local Authorities*. On EIPs see Phelps (1995) 'Structure plans' and Baker and Roberts (1999) *Examination of the Operation and Effectiveness of the Structure Planning Process*. On participation in development control see Edinburgh College of Art (1995) *Review of Neighbour Notification* and Planning Aid for London (1995) *Publicity for Planning Applications*. The implications of the Human Rights Act are considered by Hart (2000).

Race and planning

Reeves (2005) *Planning for Diversity* covers much of the material for this and other sections. Thomas (2000) provides a thorough examination of *Race and Planning: The UK Experience*. In addition to the more general ODPM literature review on *Equality and Diversity in Local Government in England* (2003) and the RTPI's (1993) report on *Ethnic Minorities and the Planning System*. See also the DETR's *New Deal for Communities: Race Equality Guidance* (1999) and the earlier RTPI and CRE (1983) *Planning for a Multi-Racial Britain* is still relevant. See also

Thomas and Krishnarayan (1994b) *Race, Equality and Planning: Policies and Procedures* but also see the reflections on this research by one of the researchers, Thomas, H. (1997) 'Ethnic minorities and the planning system: a study revisited' and the LGA survey by Loftman and Beazley (1998a) *Race, Equality and Planning*.

Women and planning

Two textbooks provide a comprehensive analysis of theory and practice. Greed (1994) *Women and Planning: Creating Gendered Realities* is a mine of interesting examples, and provides a guide to reading. Little (1994a) *Gender, Planning and the Policy Process* links the issue to wider debates. Other key sources are Gilroy with Marvin (1993) *Good Practices in Equal Opportunities* and Greed (1991) *Surveying Sisters: Women in a Traditional Male Profession*. See also London Women and Planning Group (1991) *Shaping our Borough: Women and Unitary Development Plans*, RTPI (1988d) *Planning for Choice and Opportunity* and Practice Advice Note No. 12 *Planning for Women* (1995), Little (1994b) 'Women's initiatives in town planning in England' and L. Davies (1996) 'Equality and planning: gender and disability'.

People with disabilities

The British Standards Institution (1979, 1978) publishes the *Code of Practice for Access for the Disabled to Buildings (BS 5810)* and *Code of Practice for Design for the Convenience of Disabled People (BS 5619)*. See also Development Control Policy Note 16 (1985), RTPI (1988a, 1988b) Planning Advice Notes, *Access for Disabled People* and *Access Policies in Local Plans*, and London Boroughs' Disability Resource Team (1991) *Towards Integration: The Participation of Disabled People in Planning*. The government's website on disability and the 1995 Act is a useful resource at www.disability.gov.uk

Access to information

On the general topic a starting point is the Policy Studies Institute report *Public Access to Information* (Steele 1995), updated with the White Paper *Your Right to Know: Explanatory Notes on the Freedom of Information Bill*, and University of Dundee (1997) *Public Access to Planning Information*. On access to environmental information see European Environmental Bureau (1994) *Your Rights under European Union Environment Legislation* and Moxen *et al.* (1995) *Accessing Environmental Information in Scotland*. The ODPM published a paper on the framework for *Access to Information in Local Authorities* in 2002.

Maladministration, the Ombudsman and probity

The annual reports of the separate Commissioners for England, Scotland and Wales provide all the facts plus a flavour through thumb nail sketches of the cases being heard. See also CPRE (1999e) *The Local Government Ombudsman*.

The planning profession

The intellectual development of the planning profession is explained by Healey (1985); see also Taylor, N. (1992) 'Professional ethics in town planning', Blau *et al.* (1983) *Professionals and Urban Form*, Evans (1993) 'Why we no longer need a town planning profession' and Grant (1999b) 'Planning as a learned profession'. For an explanation of the development of the planning curriculum during the twentieth century, see Rodriguez-Bachiller *et al.* (1992) 'The English planning lottery'. Evidence on the problems of staffing planning departments and how this is being addressed is provided in Edmundson (2004) *Recruitment and Retention of Planners*.

Notes

1   One successful example that is often quoted is Coin Street on the south bank of the Thames in central London, where the local community prepared their own plan for a highly valued commercial site including affordable housing and other amenities (Brindley *et al.* 1996). Other less successful stories of community participation are told in Wates (1977) and Forman (1989).

2   The incoming Secretary of State, John Patten (1989), confirmed the commitment to 'local choice', noting

that in planning 'many of the important choices are decisions which can and should be made locally, to reflect the values which local communities place on their surroundings'.

3 Three principles were later added to make nine in total: a constant search for innovation and improvement, working with other providers to increase effectiveness and coordination, and consulting and involving present and potential users.

4 There are also *Local Environment Charters* that address the right of access to environmental information held by public authorities, the right to participate in decision-making on environmental issues, and the right to seek remedies in the event of shortcomings in environmental services.

5 See ODPM, *Reforming Public Services: Principles into Practice* (2002).

6 ODPM, *Preparing Community Strategies: Government Advice to Local Authorities* (2001). Other advice and good practice case studies were also published by the Local Government Association, and are available at www.lga.gov.uk/index.htm.

7 The Office for National Statistics project on social capital includes a useful literature review of social capital and explains how it is being measured in the UK.

8 The Countryside Agency has published a Guidance Note on the Preparation of Parish and Town Plans.

9 Neighbours are defined as those having conterminous boundaries (either at the side or above and below) and within 4 metres of the boundary. Most local authorities go beyond the statutory requirements and take a broader definition of neighbouring properties, and organisations who should be consulted (Spawforth 1995, cited in Thomas, K. 1997).

10 Circular 15/92, Publicity for Planning Applications. See also the GDPO (1995: Article 8).

11 The findings of research (outlined in Chapter 3) show that local authorities make modest use of their power to reject inspectors' recommendations (only in one out of ten cases). However, it is the much smaller number of rejected recommendations that receive the most publicity.

12 The average number of objections in 1997–8 was 1,250, down from 1,400 in 1994–5. The Planning Inspectorate's work on local plan inquiries grew threefold between 1988 and 1991. This was largely due to an increase in the average length of inquiries, from just over two weeks in 1988 to eight weeks in 1993 (sitting for three and half days per week). Since then the average length of inquiries has remained about the same, although there is great variation from one plan to another. The full duration from the opening of the inquiry to receipt of the inspector's report was forty-nine weeks in 1994–5. (This is still quite a bit less than the time the authority spends in preparing plans prior to deposit.)

13 The findings on the objectors' understanding of the procedure were very similar to those of Bruton *et al.* (1982) who had examined the issue ten years before. The Planning Inspectorate publish *Guidance for Local Planning Authorities* (1996), which will be updated for the 2004 arrangements.

14 The inquiry into Terminal 5 at Heathrow (under construction) was the longest planning inquiry; it sat for 525 days, heard 700 witnesses and received 6,000 documents (but it will be Europe's third largest airport, after Frankfurt and the rest of Heathrow). Long inquiries are not a new phenomenon; the inquiry into the Greater London Development Plan considered 28,000 objections over 22 months in the years 1970 to 1972.

15 This was heralded in the 1967 White Paper which preceded the 1968 reforms of the planning system. It was argued that for planning cases which raised wide or novel issues of more than local significance, a Planning Inquiry Commission should be set up consisting of three to five members appointed by the Secretary of State to make recommendations.

16 On the basis of Canadian experience, Purdue and Kemp (1985: 685) have advocated limited state funding because some objectors 'genuinely contribute to the wider understanding of the issues involved'.

17 Between 1971 and 1998 membership of the Royal Society for the Protection of Birds (established 1889) increased from 98,000 to over 1 million, the National Trust (established 1895) from 278,000 to 2.6 million, and the relative newcomer, Friends of the Earth (established 1971) from a mere 1,000

to more than 114,000 in the UK and 1 million worldwide. However, recent trends show a levelling off.

18   The impact of changes in transport and mobility has been identified as particularly important for women and a Women's Transport Network has been established within the DETR's Mobility Unit to secure wider understanding of women's specific transport needs.

19   The Disability Rights Commission publishes a Code of Practice on Rights of Access to Goods, Facilities, Services and Premises (2004) which explains the effect of the regulations. See www.drc-gb.org.

20   A very helpful explanation of the Access to Environmental Information with sources, including the code of practice, is given on the DEFRA website: www.defra.gov.uk/corporate/opengov/.

21   The ways in which the government intend to achieve these targets are set out in the Cabinet Office Channels Implementation Policy, available at www.citu.gov.uk/moderngov/cppolicy.htm

22   www.planningportal.gov.uk/wps/portal

23   The Commission for Local Administration in England Annual Report 2003–4.

24   In fact the rate of decisions going against officers' recommendations was no higher and even less than that for some other authorities across the country. The publicity brought to North Cornwall has brought major changes and for the new committee, extensive awareness raising through training on both issues of good conduct in local government and the operation of the planning system.

25   *Modern Local Government: In Touch with the People* (Cm 4014, 1998, Chapter 6), *Modernising Local Government: A New Ethical Framework* (1998) and *Local Leadership, Local Choice* (Cm 4298 1999, Chapter 4). See also the Standards Board for England Annual Reports.

26   The DETR has subsequently published a suggested syllabus for *Training in Planning for Councillors* (1998) in cooperation with the RTPI, LGA and IDEA. See also the LGA publication *Probity in Planning: The Role of Councillors and Officers* (1997) and Cowan (1999).

27   The [1998] Members Survey confirmed that 63.8 per cent of members said an RTPI qualification was unnecessary to gain employment in planning when they first entered the profession; and 54.4 per cent reported that it was not necessary in order for them to retain their present job.

(Grant 1999b: 6)

28   Non-members working in planning include professionals who are not eligible for membership, such as architects who are working principally on planning matters, or students studying for qualifications that lead to membership, as well as those eligible who choose not to join. The total 'professional body' in its widest sense is thus more than 35,000 strong. In addition, there are support staff that, in 1991, totalled over 8,000 in local government and an equal number elsewhere. In total, therefore, there is a planning workforce of some 50,000. Three-quarters work in the public sector (with about two-thirds in local government) and about one-fifth were in private consultancies and the development industry. The dominant activities undertaken by planners are in development control and development planning, though they are engaged in a very wide range of other jobs.

29   For a review of the hundred years' work of the TCPA see the anniversary edition of its journal *Town and Country Planning* 68(6) (1999) and Hardy (1991a, 1991b).

Bibliography

Abel-Smith, B. and Townsend, P. (1965) *The Poor and the Poorest*, London: Bell

Abercrombie, P. (1945) *Greater London Plan*, London: HMSO

Adair, A. S., Berry, J. N. and McGreal, W. S. (1991) 'Land availability, housing demand and the property market', *Journal of Property Research* 8: 59–69

Adams, B., Okely, J., Mogan, D. and Smith, D. (1976) *Gypsies and Government Policy in Britain*, London: Heinemann

Adams, D. (1992) 'The role of landowners in the preparation of statutory local plans', *Town Planning Review* 63: 297–323

Adams, D. (1994) *Urban Planning and the Development Process*, London: UCL Press

Adams, D. (1996) 'The use of compulsory purchase under planning legislation', *Journal of Planning and Environment Law* 1996: 275–85

Adams, D. and Pawson, G. P. (1991) *Representation and Influence in Local Planning*, Manchester: Department of Planning and Landscape, University of Manchester

Adams, D. and Watkins, C. (2002) *Greenfields, Brownfields and Housing Development*, Oxford: Blackwell Science

Adams, D., Russell, L. and Taylor-Russell, C. (1994) *Land for Industrial Development*, London: Spon

Adams, D., Disberry, A., Hutchison, N. and Munjoma, T. (2001) 'Managing urban land: the case for urban partnership zones', *Regional Studies* 35(2): 153–62

Adams, D., Disberry, A., Hutchison, N. and Munjoma, T. (2002) 'Land policy and urban renaissance: the impact of ownership constraints in four British cities', *Planning Theory and Practice* 3(2): 195–217

Adams, J. G. U. (1990) 'Car ownership forecasting: pull the ladder up, or climb back down?' *Traffic Engineering and Control* 31: 136–41

Adams, J. G. U. (1995) *Risk*, London: UCL Press

Adams, J. G. U. (1998) 'Carmageddon', in Barnett, A. and Scruton, R. (eds) *Town and Country*, London: Cape

Adams, W. M. (1986) *Nature's Place: Conservation Sites and Countryside Change*, London: Allen and Unwin

Adams, W. M. (1996) *Future Nature: A Vision for Conservation*, London: Earthscan

Addison and Associates (2004) *Evaluation of the Planning Delivery Grant 2003–04*, London: ODPM

Advisory Committee on Business and the Environment (1993) *Report of the Financial Sector Working Group*, London: Department of Trade and Industry

Advisory Committee on Business and the Environment (annual) *Progress Report*, London: Department of Trade and Industry

Albrechts, L. (1991) 'Changing roles and positions of planners', *Urban Studies* 28: 123–37

Albrechts, L. (2003) 'Reconstructing decision making: planning versus politics', *Planning Theory* 3(3): 249–68

Albrechts, L., Moulaert, F., Roberts, P. and Swyngedouw, E. (eds) (1989) *Regional Policy at the Crossroads: European Perspectives*, London: Jessica Kingsley

Alden, J. (1992) 'Strategic planning guidance in Wales', in Minay, C. L. W. (ed.) 'Developing regional planning guidance in England and Wales: a review symposium', *Town Planning Review* 63: 415–34

Alden, J. and Boland, P. (eds) (1996) *Regional Development Strategies: A European Perspective*, London: Jessica Kingsley

Alden, J. and Offord, C. (1996) 'Regional planning guidance', in Tewdwr-Jones, M. (ed.) *British Planning Policy in Transition: Planning in the 1990s*, London: UCL Press

Alden, J. and Romaya, S. (1994) 'The challenge of urban regeneration in Wales: principles, policies and practice', *Town Planning Review* 65: 435–61

Aldous, T. (1989) *Inner City Urban Regeneration and Good Design*, London: HMSO

Aldridge, H. R. (1915) *The Case for Town Planning*, London: National Housing and Town Planning Council

Aldridge, M. (1979) *The British New Towns*, London: Routledge and Kegan Paul

Aldridge, M. (1996) 'Only demi-paradise? Women in garden cities and new towns', *Planning Perspectives* 11: 23–39

Aldridge, M. and Brotherton, C. J. (1988) 'Being a programme authority: is it worthwhile?', *Journal of Social Policy* 16: 349–69

Alexander, A. (1982) *Local Government in Britain since Reorganisation*, London: Allen and Unwin

Alexander, C. (1965) 'A city is not a tree', *Architectural Forum* 122: 1 (reprinted in Bell, G. and Thywitt, T. (eds) (1972) *Human Identity in the Urban Environment*, Harmondsworth: Penguin)

Alexander, E. R. (1981) 'If planning isn't everything, maybe it's something', *Town Planning Review* 52: 131–42

Alexander, E. R. (1992) *Approaches to Planning: Introducing Current Planning Theories, Concepts and Issues* (second edition), Philadelphia, PA: Gordon and Breach

Alker, S., Joy, V., Roberts, P. and Smith, N. (2000) 'The definition of brownfield', *Journal of Environmental Planning and Management* 43(1): 49–69

Allaby, M. (1994) *Macmillan Dictionary of the Environment*, London: Macmillan

Allanson, P. and Whitby, M. (eds) (1997) *The Rural Economy and the British Countryside*, London: Earthscan

Allen, H. J. B. (1990) *Cultivating the Grass Roots: Why Local Government Matters*, The Hague: International Union of Local Authorities

Allen, J. and Hamnett, C. (eds) (1991) *Housing and Labour Markets: Building the Connections*, London: Unwin Hyman

Allen, K. (1986) *Regional Incentives and the Investment Decision of the Firm* (DTI), London: HMSO

Allen, R. (1995) 'Policy and grassroots action: a vital mix', in *First Steps: Local Agenda in Practice*, London: HMSO

Allinson, J. (1999) 'The 4.4 million households: do we really need them anyway?' *Planning Practice and Research* 14: 107–13

Allmendinger, P. (1996a) *Thatcherism and Simplified Planning Zones: An Implementation Perspective*, Oxford Planning Monographs vol. 2, no. 1, Oxford: Oxford Brookes University

Allmendinger, P. (1996b) 'Twilight zones', *Planning Week* 4(29): 14–15

Allmendinger, P. (1997) *Thatcherism and Planning: The Case of Simplified Planning Zones*, Aldershot: Avebury

Allmendinger, P. (2001) *Planning Theory*, Basingstoke: Palgrave Macmillan

Allmendinger, P. (2003) 'Integrating planning in a devolved Scotland', *Planning Practice and Research* 18(1): 19–36

Allmendinger, P. and Chapman, M. (eds) (1999) *Planning Beyond 2000*, Chichester: Wiley

Allmendinger, P. and Prior, A. (2000) *Introduction to Planning Practice*, London: Wiley

Allmendinger, P. and Tewdwr-Jones, M. (2000) 'New Labour, new planning? The trajectory of planning in Blair's Britain', *Urban Studies* 37(8): 1379–402

Allmendinger, P. and Thomas, H. (eds) (1998) *Urban Planning and the British New Right*, London: Routledge

Alonso, W. (1971) 'Beyond the inter-disciplinary approach to planning', *Journal of the American Institute of Planners* 37: 169–73 (reprinted in Cullingworth, J. B. (ed.) 1973 *Planning for Change* (vol. 3 of *Problems of an Urban Society*), London: Allen and Unwin)

Alterman, R. (2001) *National-level Planning in Democratic Countries: an International Comparison of City and Regional Policy-making*, Liverpool: Liverpool University Press

Alterman, R. and Cars, G. (eds) (1991) *Neighbourhood Regeneration: An International Evaluation*, London: Mansell

Alterman, R., Harris, D. and Hill, M. (1984) 'The impact of public participation on planning: the case of the Derbyshire structure plan', *Town Planning Review* 55: 177–96

Ambrose, P. (1974) *The Quiet Revolution: Social Change in a Sussex Village 1871–1971*, London: Chatto and Windus

Ambrose, P. (1986) *Whatever Happened to Planning?*, London: Methuen

Ambrose, P. (1994) *Urban Process and Power*, London: Routledge

Ambrose, P. and Colenutt, B. (1975) *The Property Machine*, Harmondsworth: Penguin

Amery, C. and Cruickshank, D. (1975) *The Rape of Britain*, London: Paul Elek

Amin, A. and Tomaney, J. (1991) 'Creating an enterprise culture in the North East? The impact of urban and regional policies of the 1980s', *Regional Studies* 25: 479–87

Amion Consulting (2001) *Urban Regeneration Companies: Learning the Lessons*, London: DTLR

Amos, F. J. C., Davies, D., Groves, R. and Niner, P. (1982) *Manpower Requirements for Physical Planning*, Birmingham: Institute of Local Government Studies

Amundson, C. (1993) 'Sustainable aims and objectives: a planning framework', *Town and Country Planning* 62: 20–2

Anderson, M. A. (1981) 'Planning policies and development control in the Sussex Downs AONB', *Town Planning Review* 52: 5–25

Anderson, M. A. (1990) 'Areas of outstanding natural beauty and the 1949 National Parks Act', *Town Planning Review* 61: 311–39

Anderson, S. S. and Eliassen, K. A. (2001) *Making Policy in Europe*, London: Sage

Anderson, W. P., Kanaroglou, P. S. and Miller, E. J. (1996) 'Urban form, energy and the environment', *Urban Studies* 33: 7–35

Anon (1956) 'Ye olde English green belt', *Journal of the Town Planning Institute* 42: 68–9

Anson, B. (1981) *I'll Fight You for It: Behind the Struggle for Covent Garden*, London: Cape

Archbishop of Canterbury's Commission on Urban Priority Areas (1985) *Faith in the City: A Call for Action by Church and Nation*, London: Church House

Archbishop of Canterbury's Commission on Urban Priority Areas (1990) *Living Faith in the City: A Progress Report*, London: General Synod of the Church of England

Archbishops' Commission on Rural Areas (1990) *Faith in the Countryside*, Stoneleigh Park, Warwickshire: Acora

Armitage, A. (chair) (1980) *Lorries, People and the Environment*, Armitage Report, London: HMSO

Armstrong, J. (1985) 'The Sizewell inquiry', *Journal of Planning and Environment Law* 1985: 686–9

Arnell, N. W., Jenkins, A. and George, D. G. (1994) *Implications of Climate Change for the NRA* (National Rivers Authority), London: HMSO

Arnold, C. (1997) 'Planning gain: how are off-site liabilities passed back to landowners', PhD dissertation, Department of Land Economy Library, University of Cambridge

Arnold, C. (1999) 'Planning gain: how are off-site liabilities passed back to land owners', *Journal of Planning and Environment Law* October: 869–77

Arnstein, S. R. (1969) 'A ladder of citizen participation', *Journal of the American Institute of Planners* 35: 216–24

Arup (2004) *Review of the Publicity Requirements for Planning Applications*, London: ODPM

Arup Economic Consultants (1990) *Mineral Policies in Development Plans* (DoE), London: HMSO

Arup Economic Consultants (1991) *Simplified Planning Zones: Progress and Procedures* (DoE), London: HMSO

Arup Economics and Planning (1994) *Assessment of the Effectiveness of Derelict Land Grant in Reclaiming Land for Development* (DoE), London: HMSO

Arup Economics and Planning (1995a) *Coastal Superquarries: Options for Wharf Facilities on the Lower Thames* (DoE), London: HMSO

Arup Economics and Planning (1995b) *Derelict Land Prevention and the Planning System* (DoE), London: HMSO

Arup Economics and Planning (1999) *Control of Outdoor Advertising: Fly-posting*, London: DETR

Arup Economics and Planning (2003) *The Planning Service: Costs and Fees*, London: ODPM

Arup with Nick Davies Associates (2004) *Standard Application Forms: Final Report*, London: ODPM

Ashby, E. (1978) *Reconciling Man with the Environment*, Stanford, CA: Stanford University Press

Ashby, E. and Anderson, M. (1981) *The Politics of Clean Air*, Oxford: Clarendon Press

Ashford, D. E. (ed.) (1979) *The Politics of Urban Resources*, Chicago, IL: Maaroufa

Ashworth, W. (1954) *The Genesis of Modern British Town Planning*, London: Routledge and Kegan Paul

Askew, J. (2004a) 'Mast location debate rages', *Planning* 30 July: 15

Askew, J. (2004b) *Mobile Phone Masts: Report of an Inquiry by the All Party Mobile Group*, London: All Party Parliamentary Mobile Group

Aspden, J. and Birch, D. (2005) *New Localism: Citizen Engagement, Neighbourhoods and Public Services – Evidence from Local Government*, London: ODPM

Assembly of Welsh Counties (1992) *Strategic Planning Guidance in Wales*, Mold: Clwyd County Council

Association of Conservation Officers (ACO) (1992) *Listed Buildings Repair Notices*, Brighton: ACO

Association of County Archaeological Officers (1993) *Archaeological Heritage*, London: Association of County Councils

Association of County Councils (ACC) (1992) *National Parks: The New Partnership*, London: ACC

Association of County Councils (1993) *The Enabling Authority and County Government*, London: ACC

Association of London Government (ALG) (1996a) *Paying the Wrong Fare: Affordable Public Transport*, London: ALG

Association of London Government (1996b) *Red Routes: Do They Work?*, London: ALG

Association of National Park Officers (ANPO) (1988) *National Parks: Environmentally Favoured Areas? A Proposal for a New Agricultural Policy to Achieve the Aims of National Park Designation*, Bovey Tracey: ANPO

Association of Town Centre Management (ATCM) (1992) *Working for the Future of our Towns and Cities*, London: ATCM

Association of Town Centre Management (1994) *The Effectiveness of Town Centre Management*, London: ATCM

Association of Town Centre Management (1997) *Managing Urban Spaces in Town Centres*, (DoE), London: The Stationery Office

Astrop, A. (1993) *The Trend in Rural Bus Services since Deregulation*, Crowthorne: Transport Research Laboratory

Atkinson, R. (1999a) Discourses of partnership and empowerment in contemporary British urban regeneration', *Urban Studies* 36(1): 59–72

Atkinson, R. (1999b) 'Urban crisis: new policies for the next century', in Allmendinger, P. and Chapman, M. (1999) *Planning Beyond 2000*, Chichester: Wiley

Atkinson, R. and Mills, L. (2005) 'The thematic strategy on the urban environment: will it make a difference?', *Town and Country Planning* March: 106–8

Atkinson, R. and Moon, G. (1994) *Urban Policy in Britain: The City, the State and the Market*, London: Macmillan

Audit Commission: *see* appendix on *Official Publications*

Austin, J. (1995) 'Renewal areas: the research findings', London: Austin Mayhead

Austin Mayhead (1997) *Neighbourhood Renewal Assessment and Renewal Areas*, London: DETR

Automobile Association (AA) (1995a) *Transport and the Environment: The AA's Response to the Royal Commission on Environmental Pollution*, London: AA

Automobile Association (1995b) *Shopmobility: Good for People and Towns*, London: AA

Bagwell, P. and Lyth, P. (2002) *Transport in Britain: From Canal Lock to Gridlock*, London: Hambledon and London

Bailey, N. (2003) 'Local strategic partnerships in England: the continuing search for collaborative advantage, leadership and strategy in urban governance', *Planning Theory and Practice* 4(4): 443–57

Bailey, N. and Barker, A. (eds) (1992) *City Challenge and Local Regeneration Partnerships: Conference Proceedings*, London: Polytechnic of Central London

Bailey, N., Barker, A. and MacDonald, K. (1995) *Partnership Agencies in British Urban Policy*, London: UCL Press

Bailey, S. J. (1990) 'Charges for local infrastructure', *Town Planning Review* 61: 427–53

Baily, M. N. and Kirkegaard, J. F. (2004) *Transforming the European Economy*, London: Eurospan

Bain, C., Dodd, A. and Pritchard, D. (1990) *RSPB Planscan: A Study of Development Plans in England and Wales*, Sandy, Beds: Royal Society for the Protection of Birds

Baker, L. (2000) 'Lawyers forecast a log jam of appeals', *Planning* 10 November: 1

Baker, M. (1998) 'Planning for the English regions: a review of the Secretary of State's regional planning guidance', *Planning Practice and Research* 13(2): 153–69

Baker, M. and Roberts, P. (1999) *Examination of the Operation and Effectiveness of the Structure Planning Process: Summary Report*, London: DETR

Baker, S., Kousis, M., Richardson, D. and Young, S. (1997) *The Politics of Sustainable Development*, London: Routledge

Baker Associates (1999) *Proposals for a Good Practice Guide on Sustainability Appraisal of Regional Planning Guidance*, London: DETR

Baker and Associates (2001) *Review of the Use Classes Order and Part 4 of the GDPO (Temporary Uses)*, London: DTLR

Balchin, P. (1995) *Housing Policy: An Introduction*, London: Routledge

Balchin, P. (1996) *Housing Policy in Europe*, London: Routledge

Balchin, P. (1999) 'Housing', in Cullingworth, J. B. (ed.) *British Planning: 50 Years of Urban and Regional Policy*, London: Athlone

Balchin, P. and Rhodes, M. (eds) (1997) *Housing: The Essential Foundations*, London: Routledge

Balchin, P., Sýkora, L., with Bull, G. (1999) *Regional Policy and Planning in Europe*, London: Routledge

Baldock, D., Cox, G., Lowe, P. and Winter, M. (1990) 'Environmentally sensitive areas: incrementalism or reform?', *Journal of Rural Studies* 6: 143–62

Baldock, D., Bishop, K., Mitchell, K. and Phillips, A. (1996) *Growing Greener: Sustainable Agriculture in the UK*, London: Council for the Protection of Rural England

Balfour, J. (1983) *A New Look at the Northern Ireland Countryside*, Belfast: Department of the Environment, Northern Ireland

Ball, M. (1983) *Housing Policy and Economic Power*, London: Methuen

Ball, R. M. (1989) 'Vacant industrial premises and local development: a survey, analysis, and policy assessment of the problems in Stoke-on-Trent', *Land Development Studies* 6: 105–28

Ball, R. M. (1995) *Local Authorities and Regional Policy in the UK: Attitudes, Representations and the Local Economy*, London: Paul Chapman

Ball, S. and Bell, S. (1994) *Environmental Law* (second edition), London: Blackstone

Banham, R., Barker, P., Hall, P. and Price, C. (1969) 'Non-plan: an experiment in freedom', *New Society* 26: 435–43

Banister, D. (1994a) 'Reducing the need to travel through planning', *Town Planning Review* 65: 349–54

Banister, D. (1994b) *Transport Planning in the UK, USA and Europe*, London: Spon

Banister, D. (ed.) (1995) *Transport and Urban Development*, London: Spon

Banister, D. (ed.) (1998) *Transport Policy and the Environment*, London: Spon

Banister, D. (1999) 'Review essay: the car is the solution, not the problem?', *Urban Studies* 36: 2415–19

Banister, D. (2002) *Transport Planning*, London: Spon

Banister, D. and Button, K. (eds) (1993) *Transport, the Environment and Sustainable Development*, London: Spon

Banister, D. and Watson, S. (1994) *Energy Use in Transport and City Structure*, London: Planning and Development Research Centre, University College London

Banister, D., Capello, R. and Nijkamp, P. (eds) (1995) *European Transport and Communications Networks*, Chichester: Wiley

Bannon, M. J., Nowlan, K. I., Hendry, J. and Mawhinney, K. (1989) *Planning: The Irish Experience 1920–1988*, Dublin: Wolfhound Press

Bardgett, L. (2000) *The Tourism Industry*, House of Commons Research Paper 00/66, London: House of Commons Library

Barker, G. M. A. and Box, J. D. (1998) 'Statutory local nature reserves in the United Kingdom', *Journal of Environmental Planning and Management* 41(5): 629–42

Barker, T. C. and Savage, C. I. (1974) *An Economic History of Transport*, London: Hutchinson

Barkham, J. P., MacGuire, F. A. S. and Jones, S. J. (1992) *Sea-Level Rise and the UK*, London: Friends of the Earth

Barkham, R., Gudgin, G., Hart, M. and Harvey, E. (1996) *The Determinants of Small Firm Growth* (Regional Studies Association), London: Jessica Kingsley

Barlow, J. (1982) 'Planning practice, housing supply and migration', in Champion, A. G. and Fielding, A. J. (eds) *Migration Processes and Patterns, vol. 1: Research Progress and Prospects*, London: Belhaven

Barlow, J. (1986) 'Landowners, property ownership, and the rural locality', *International Journal of Urban and Regional Research* 10: 309–29

Barlow, J. (1988) 'The politics of land into the 1990s: landowners, developers, and farmers in lowland Britain', *Policy and Politics* 16: 111–21

Barlow, J. (1995) *Public Participation in Urban Development: The European Experience*, London: Policy Studies Institute

Barlow, J. and Chambers, D. (1992) *Planning Agreements and Affordable Housing Provision*, Brighton: Centre for Urban and Regional Research, University of Sussex

Barlow, J. and Duncan, S. (1992) 'Markets, states and housing provision: four European growth regions compared', *Progress in Planning* 38(2): 93–177

Barlow, J. and Duncan, S. (1994) *Success and Failure in House Provision: European Systems Compared*, Oxford: Elsevier Science

Barlow, J. and Gann, D. (1993) *Offices into Flats*, York: Joseph Rowntree Foundation

Barlow, J. and Gann, D. (1995) 'Flexible planning and flexible buildings: reusing redundant office space', *Journal of Urban Affairs* 17: 263–76

Barlow, J. and King, A. (1992) 'The state, the market, and competitive strategy: the house building industry in the United Kingdom, France, and Sweden', *Environment and Planning A* 24: 381–400

Barlow, J., Cocks, R. and Parker, M. (1994a) 'Delivering affordable housing: law, economics and planning policy', *Land Use Policy* 11: 181–94

Barlow, J., Cocks, R. and Parker, M. (1994b) *Planning for Affordable Housing* (DoE), London: HMSO

Barlow, M. (chair) (1940) *Report of the Royal Commission on the Distribution of the Industrial Population*, Cmd 6153, Barlow Report, London: HMSO

Barnekov, T., Boyle, R. and Rich, D. (1989) *Privatism and Urban Policy in Britain and the United States*, Oxford: Oxford University Press

Barnekov, T., Hart, D. and Benfer, W. (1990) *US Experience in Evaluating Urban Regeneration* (DoE), London: HMSO

Barnett, A. and Scruton, R. (eds) (1998) *Town and Country*, London: Cape

Barrett, S. and Fudge, C. (eds) (1981) *Policy and Action*, London: Methuen

Barrett, S. and Healey, P. (eds) (1985) *Land Policy: Problems and Alternatives*, Aldershot: Avebury

Barrett, S. and Whitting, G. (1983) *Local Authorities and Land Supply*, Bristol: School for Advanced Urban Studies, University of Bristol

Barrett, S., Stewart, M. and Underwood, J. (1978) *The Land Market and Development Process*, Bristol: School for Advanced Urban Studies, University of Bristol

Barton, H. (ed.) (2000) *Sustainable Communities: The Potential for Eco-neighbourhoods*, London: Earthscan

Barton, H. and Bruder, N. (1995) *A Guide to Local Environmental Auditing*, London: Earthscan

Barton, H., Davis, G. and Guise, R. (1995) *Sustainable Settlements: A Guide for Planners, Designers and Developers*, Bristol: University of the West of England; and Luton: Local Government Management Board

Barton, H., Guise, R. and Grant, M. (2002) *Shaping Neighbourhoods*, London: Spon.

Bassett, K. (1993) 'Urban cultural strategies and urban regeneration: a case study and critique', *Environment and Planning A* 25: 1773–88

Bassett, K. (1996) 'Partnerships, business elites and urban politics: new forms of governance in an English city?', *Urban Studies* 33: 539–55

Bastrup-Birk, H. and Doucet, P. (1997) 'European spatial planning from the heart', *Built Environment* 23(4): 307–14

Bateman, M. (1985) *Office Development: A Geographical Analysis*, New York: St Martin's Press

Bateman, M. and Riley, R. (eds) (1987) *The Geography of Defence*, London: Croom Helm

Batey, P. (1985) 'Postgraduate planning education in Britain', *Town Planning Review* 56: 407–20

Batey, P. (1993) 'Planning education as it was', *The Planner* 79(4): 25–6

Batho, W. J. S. (chair) (1990) *Report of the Noise Review Working Party*, Batho Report, London: HMSO

Batley, R. (1989) 'London docklands: an analysis of power relations between UDCs and local government', *Public Administration* 67: 167–87

Batley, R. and Stoker, G. (eds) (1991) *Local Government in Europe: Trends and Development*, London: Macmillan

Batty, M. (1984) 'Urban policies in the 1980s: a review of the OECD proposals for managing urban change', *Town Planning Review* 55: 489–98

Batty, M. (1990) 'How can we best respond to changing fashions in urban and regional planning?', *Environment and Planning B: Planning and Design* 17: 1–7

Baxter, J. D. (1990) *State Security, Privacy and Information*, Hemel Hempstead: Harvester Wheatsheaf

BDP Planning and Leighton Berwin (1998) *The Use of Permitted Development Rights by Statutory Undertakers*, London: The Stationery Office

BDP Planning and Oxford Institute of Retail Management (1994) *The Effects of Major Out-of-Town Retail Developments*, London: HMSO

Bean, D. (ed.) (1996) *Law Reform for All*, London: Blackstone

Beckerman, W. (1974) *In Defence of Economic Growth*, London: Cape

Beckerman, W. (1990) *Pricing for Pollution: Market Pricing,*

Government Regulation, Environmental Policy (second edition), London: Institute of Economic Affairs

Beckerman, W. (1995) *Small is Stupid: Blowing the Whistle on the Greens*, London: Duckworth

Bedford, T., Clark, J. and Harrison, C. (2002) 'Limits to new public participation practices in local land use planning', *Town Planning Review* 73(3): 311–32

Bedfordshire County Council (1996) *A Step by Step Guide to Environmental Appraisal*, Sandy, Beds: Royal Society for the Protection of Birds

Beesley, M. E. and Kain, J. F. (1964) 'Urban form, car ownership and public policy: an appraisal of *Traffic in Towns*', *Urban Studies* 1: 174–203

Begg, H. M. and Pollock, S. H. A. (1991) 'Development plans in Scotland since 1975', *Scottish Geographical Magazine* 107: 4–11

Begg, I. (1991) 'High technology location and the urban areas of Great Britain: development in the 1980s', *Urban Studies* 28: 961–81

Begg, T. (1996) *Housing Policy in Scotland*, Edinburgh: John Donald

Bell, G. and Thywitt, T. (1972) *Human Identity in the Urban Environment*, Harmondsworth: Penguin

Bell, J. L. (1993) *Key Trends in Communities and Community Development*, London: Community Development Foundation

Bell, M. and Evans, D. (1998) 'The National Forest and Local Agenda 21: an experiment in integrated landscape planning', *Journal of Environmental Planning and Management* 41: 237–51

Bell, S. (1992) *Out of Order: The 1987 Use Classes Order – Problems and Proposals*, London: London Boroughs Association

Bell, S. (1997) *Design for Outdoor Recreation*, London: Spon

Bell, S. (1998) *Landscape: Patterns, Perception, Process*, London: Spon

Bendixson, T. (1989) *Transport in the Nineties: The Shaping of Europe*, London: Royal Institution of Chartered Surveyors

Bengs, C. (2002) *Facing ESPON*, Stockholm: Nordregio

Bengs, C. and Böhme, K. (1998) *The Progress of European Spatial Planning*, Stockholm: Nordregio

Benington, J. (1994) *Local Democracy and the European Union: The Impact of Europeanisation on Local Governance*, London: Commission for Local Democracy

Bennett, B. (2002) Ups and downs for development control, *Town and Country Planning* 71(2): 49

Bennett, R. and Errington, A. (1995) 'Training and the small rural business', *Planning Practice and Research* 10: 45–54

Bennett, R. J. (1990) 'Training and enterprise councils (TECs) and vocational education and training', *Regional Studies* 24: 65–82

Benson, J. F. and Willis, K. G. (1992) *Valuing Informal Recreation on the Forestry Commission Estate* (Forestry Commission Bulletin 104), London: HMSO

Bentham, C. G. (1985) 'Which areas have the worst urban problems?', *Urban Studies* 22: 119–31

Benyon, J. (ed.) (1984) *Scarman and After: Essays Reflecting on Lord Scarman's Report, the Riots, and their Aftermath*, Oxford: Pergamon

Berke, P., Crawford, J., Dixon, J. and Ericksen, N. (2004)

Plan-making for Sustainability: The New Zealand Experience, Aldershot: Ashgate

Berry, J. and McGreal, S. (1995a) 'Community and inter-agency structures in the regeneration of inner-city Belfast', *Town Planning Review* 66: 129–42

Berry, J. and McGreal, S. (eds) (1995b) *European Cities, Planning Systems and Property Markets*, London: Spon

Berry, J. N., Fitzsimmons, D. F. and McGreal, W. S. (1988) 'Neighbour notification: the Scottish and Northern Ireland models', *Journal of Planning and Environment Law* 1988: 804–8

Berry, J., McGreal, W. S. and Deddis, W. G. (1993) *Urban Regeneration: Property Investment and Development*, London: Spon

Bertuglia, C. S., Clarke, G. P. and Wilson, A. G. (eds) (1994) *Modelling the City: Performance, Policy and Planning*, London: Routledge

Best, J. and Bowser, L. (1986) 'A people's plan for central Newham', *The Planner* 27(11): 21–5

Best, R. (1981) *Land Use and Living Space*, London: Methuen

Beveridge, W. (chair) (1942) *Social Insurance and Allied Services*, Cmd 6404, Beveridge Report, London: HMSO

Bibby, P. R. and Shepherd, J. W. (1990) *Rates of Urbanisation in England 1981–2001* (DoE), London: HMSO

Bibby, P. R. and Shepherd, J. W. (1993) *Housing Land Availability: The Analysis of PS3 Statistics on Land with Outstanding Planning Permission* (DoE), London: HMSO

Bibby, P. R. and Shepherd, J. W. (1997) 'Projecting rates of urbanization for England 1991–2016: method, policy applications and results', *Town Planning Review* 68: 93–124

Bibby, P. R. and Shepherd, J. W. (1999) 'Refocusing national brownfield housing targets', *Town and Country Planning* 70: 302–5

Biddulph, M. (2003) *Home Zones: A Planning and Design Handbook*, Bristol: Policy Press

Bingham, M. (1998) 'A plan-led system: the potential and actual role of development plans in development control', PhD dissertation, Department of Land Economy Library, University of Cambridge

Bingham, M. (2001) 'Policy utilisation in planning control: planning appeals in England's "plan-led" system', *Town Planning Review* 72(3): 321–40

Biodiversity Group and Local Government Management Board (1997) *Guidance for Local Biodiversity Action Plans*, Luton: Local Government Management Board

Birch, A. H. (1998) *The British System of Government* (tenth edition), London: Routledge

Birkenshaw, P. (1990) *Government and Information*, London: Butterworths

Birtles, W. (1991) 'The European directive on freedom of access to information on the environment', *Journal of Planning and Environment Law* 1991: 607–10

Bishop, J. (1994) 'Planning for better rural design', *Planning Practice and Research* 9: 259–70

Bishop, K. (1992) 'Assessing the benefits of community forests: an evaluation of the recreational use benefits of two urban fringe woodlands', *Journal of Environmental Planning and Management* 35: 63–76

Bishop, K. and Hooper, A. (1991) *Planning for Social Housing*, London: National Housing Forum (Association of District Councils)

Bishop, K. D. and Phillips, A. C. (1993) 'Seven steps to market: the development of the market-led approach to countryside conservation and recreation', *Journal of Rural Studies* 9: 315–38

Bishop, K. and Phillips, A. (2004) *Countryside Planning: New Approaches to Management and Conservation*, London: Earthscan

Bishop, K. D., Phillips, A. C. and Warren, L. (1995a) 'Protected for ever? Factors shaping the future of protected areas policy', *Land Use Policy* 12: 291–305

Bishop, K. D., Phillips, A. C. and Warren, L. (1995b) 'Protected areas in the United Kingdom: Time for new thinking', *Regional Studies* 29: 192–201

Bishop, K., Tewdwr-Jones, M. and Wilkinson, D. (2000) 'From spatial to local: the impact of the European Union on local authority planning in the UK', *Journal of Planning and Environmental Management* 43(3): 309–34

Blackhall, J. C. (1993) *The Performance of Simplified Planning Zones*, Newcastle upon Tyne: Department of Town and Country Planning, University of Newcastle

Blackhall, J. C. (1994) 'Simplified planning zones (SPZs) or simply political zeal?', *Journal of Planning and Environment Law* 1994: 117–23

Blackman, T. (1985) 'Disasters that link Ulster and the North East', *Town and Country Planning* 54: 18–20

Blackman, T. (1991a) 'People-sensitive planning: communication, property and social action', *Planning Practice and Research* 6(3): 11–15

Blackman, T. (1991b) *Planning Belfast: A Case Study of Public Policy and Community Action*, Aldershot: Avebury

Blackman, T. (1995) *Urban Policy in Practice*, London: Routledge

Blackmore, R., Wood, C. and Jones, C. E. (1997) 'The effect of environmental assessment on UK infrastructure project planning decisions', *Planning Practice and Research* 12(3): 223–38

Blair, T. (1998) *Leading the Way: A New Vision for Local Government*, London: Institute for Public Policy Research

Blaker, G. (1995) 'Gypsy law', *Journal of Planning and Environment Law* 1995: 191–6

Blau, J. R., La Gory, M. and Pipkin, J. S. (1983) *Professionals and Urban Form*, New York: State University of New York

Blowers, A. (1980) *The Limits of Power: The Politics of Local Planning Policy*, Oxford: Pergamon

Blowers, A. (1983) 'Master of fate or victim of circumstance: the exercise of corporate power in environmental policy making', *Policy and Politics* 11: 375–91

Blowers, A. (1984) *Something in the Air: Corporate Power and the Environment*, London: Harper and Row

Blowers, A. (1986) 'Environmental politics and policy in the 1980s: a changing challenge', *Policy and Politics* 14: 11–18

Blowers, A. (1987) 'Transition or transformation?: environmental policy under Thatcher', *Public Administration* 65: 277–94

Blowers, A. (ed.) (1993) *Planning for a Sustainable Environment: A Report by the Town and Country Planning Association*, London: Earthscan

Blowers, A. and Evans, B. (eds) (1997) *Town Planning into the 21st Century*, London: Routledge

Blowers, A. and Leroy, P. (1994) 'Power, politics and environmental inequality: a theoretical and empirical analysis of the process of "peripheralisation"', *Environmental Politics*, vol. 3, no. 2, Summer, 197–228

Blunden, J. and Curry, N. (1988) *A Future for our Countryside?*, Oxford: Blackwell

Blunden, J. and Curry, N. (1989) *A People's Charter? Forty Years of the National Parks and Access to the Countryside Act 1949*, London: HMSO

Board on Sustainable Development Policy Division (1999) *Our Common Journey: A Transition Toward Sustainability*, Washington, DC: National Academy Press

Boddy, M. and Parkinson, M. (eds) (2004) *City Matters: Competitiveness, Cohesion and Urban Governance*, Bristol: Policy Press

Boddy, M., Lambert, C. and Shape, D. (1997) *City for the 21st Century: Globalisation, Planning and Urban Change in Contemporary Britain*, Bristol: Policy Press

Body, R. (1982) *Agriculture: The Triumph and the Shame*, London: Maurice Temple Smith

Bogdanor, V. (1993) *The Blackwell Encyclopaedia of Political Institutions*, Oxford: Blackwell

Bogdanor, V. (1999) *Devolution in the United Kingdom*, Oxford: Oxford University Press (Opus Books)

Böhme, K. and Bengs, C. (eds) (1999) *From Trends to Visions: The European Spatial Development Perspective*, Stockholm: Nordregio

Bond, M. (1992) *Nuclear Juggernaut: The Transport of Radioactive Materials*, London: Earthscan

Bongers, P. (1990) *Local Government and 1992*, Harlow: Longman

Bonnel, P. (1995) 'Urban car policy in Europe', *Transport Policy* 2: 83–95

Bonyhandy, T. (1987) *Law and the Countryside: The Rights of the Public*, Oxford: Professional Books

Booker, C. and Green, C. L. (1973) *Goodbye London: An Illustrated Guide to Threatened Buildings*, London: Fontana

Booth, C., Darke, J. and Yeandle, S. (eds) (1996) *Changing Places: Women's Lives in the City*, London: Paul Chapman

Booth, P. (1996) *Controlling Development: Certainty and Discretion in Europe, the USA, and Hong Kong*, London: UCL Press

Booth, P. (1999a) 'Discretion in planning versus zoning', in Cullingworth, J. B. (ed.) *British Planning: 50 Years of Urban and Regional Policy*, London: Athlone

Booth, P. (1999b) 'From regulation to discretion: the evolution of development control in the British planning system', *Planning Perspectives* 14(3): 277–89

Booth, P. (2002a) 'A desperately slow system? The origins and nature of the current discourse on development control', *Planning Perspectives* 17(4): 309–24

Booth, P. (2002b) 'From property rights to public control: the quest for public interest in the control of urban development', *Town Planning Review* 73(2): 153–70

Booth, P. (2003) *Planning by Consent: The Origins and Nature of British Development Control*, London: Routledge

Booth, P. and Beer, A. R. (1983) 'Development control and design quality', *Town Planning Review* 54: 265–84 and 383–404

Borchardt, K. (1995) *European Integration: The Origins and Growth of the European Community* (fourth edition), Luxembourg: Office for the Official Publications of the European Communities

Borins, S. F. (1988) 'Electronic road pricing: an idea whose time may never come', *Transportation Research A* 22A: 37–44

Borraz, O., Bullmann, U. anmd Hambleton, R. (1994) *Local Leadership and Decision Making: A Study of France, Germany, the United States and Britain*, London: Local Government Chronicle and Joseph Rowntree Foundation

Bosworth, J. and Shellens, T. (1999) 'How the Welsh Assembly will affect planning', *Journal of Planning and Environment Law* March: 219–24

Boucher, S. and Whatmore, S. (1990) *Planning Gain and Conservation: A Literature Review*, Reading: Department of Geography, University of Reading

Boucher, S. and Whatmore, S. (1993) 'Green gains? Planning by agreement and nature conservation', *Journal of Environmental Planning and Management* 36: 33–50

Bourne, F. (1992) *Enforcement of Planning Control* (second edition), London: Sweet and Maxwell

Bovaird, T. (1992) 'Local economic development and the city', *Urban Studies* 29: 343–68

Bovaird, T., Tricker, M., Martin, S. J., Gregory, D. G. and Pearce, G. R. (1990a) *An Evaluation of the Rural Development Programme Process*, London: HMSO

Bovaird, T., Gregory, D., Martin, S., Pearce, G. and Tricker, M. (1990b) *Evaluation of the Rural Development Programme Process*, London: HMSO

Bovaird, T., Tricker, M., Hems, L. and Martin, S. (1991a) *Constraints on the Growth of Small Firms: A Report on a Survey of Small Firms* (Aston Business School), London: HMSO

Bovaird, T., Gregory, D. and Martin, S. (1991b) 'Improved performance in local economic development: a warm embrace or an artful sidestep?', *Public Administration* 69: 103–19

Bowers, J. (ed.) (1990a) *Agriculture and Rural Land Use*, Swindon: Economic and Social Research Council

Bowers, J. (1990b) *Economics of the Environment: The Conservationists' Response to the Pearce Report*, Newbury: British Association of Nature Conservationists

Bowers, J. (1992) 'The economics of planning gain: a reappraisal', *Urban Studies* 29: 1329–39

Bowers, J. (1995) 'Sustainability, agriculture, and agricultural policy', *Environment and Planning A* 27: 1231–43

Bowers, J. K. and Cheshire, P. C. (1983) *Agriculture, the Countryside and Land Use: An Economic Critique*, London: Methuen

Bowley, M. (1945) *Housing and the State 1919–1944*, London: Allen and Unwin

Bowman, J. C. (1992) 'Improving the quality of our water: the role of regulation by the National Rivers Authority', *Public Administration* 70: 565–75

Boyack, S. (1999) Speech of the Minister of Transport and the Environment to the Royal Town Planning Institute National Conference, 25 November 1999: available on the Scottish Executive website http://www.scotland.gov.uk

Boydell, P. and Lewis, M. (1989) 'Applications to the High Court for the review of planning decisions', *Journal of Planning and Environment Law* 1989: 146–56

Boyle, R. (ed.) (1985) 'Leveraging urban development: a comparison of urban policy directions and programme impact in the United States and Britain', *Policy and Politics* 13: 175–210

Boyle, R. (1993) 'Changing partners: the experience of urban economic policy in West Central Scotland, 1980–90', *Urban Studies* 30: 309–24

Boyne, G., Jordan, G. and McVicar, M. (1995) *Local Government Reform: A Review of the Process in Scotland and Wales*, London: Local Government Chronicle Communications in association with the Joseph Rowntree Foundation

Boynton, J. (1986) 'Judicial review of administrative decisions: a background paper', *Public Administration* 64: 147–61

Bradbury, J. and Mawson, J. (eds) (1997) *British Regionalism and Devolution: The Challenges of State Reform and European Integration*, London: Jessica Kingsley

Bradford, M. and Robson, B. (1995) 'An evaluation of urban policy', in Hambleton, R. and Thomas, H. (eds) *Urban Policy Evaluation: Challenge and Change*, London: Paul Chapman

Bramley, G. (1989) *Land Supply, Planning, and Private House-building*, Bristol: School for Advanced Urban Studies, University of Bristol

Bramley, G. (1993a) 'The impact of land use planning and tax subsidies on the supply and price of housing in Britain', *Urban Studies* 30: 5–30

Bramley, G. (1993b) 'Land use planning and the housing market in Britain: the impact on house building and house prices', *Environment and Planning A* 25: 1021–51

Bramley, G. (1996a) *Housing with Hindsight: Household Growth, Housing Need and Housing Development in the 1980s*, London: Council for the Protection of Rural England

Bramley, G. (1996b) 'Impact of land use planning on the supply and price of housing in Britain: reply to comment by Alan W. Evans', *Urban Studies* 33: 1733–7

Bramley, G. and Pawson, H. (2002) 'Low demand for housing: incidence, causes and UK national policy implications', *Urban Studies* 9(3): 393–422

Bramley, G. and Smart, G. (1995) *Rural Incomes and Housing Affordability*, Salisbury: Rural Development Commission

Bramley, G. and Watkins, C. (1995) *Circular Projections: Household Growth, Housing Development, and the Household Projections*, London: Council for the Protection of Rural England

Bramley, G. and Watkins, C. (1996) *Steering the Housing Market: New Building and the Changing Planning System*, Bristol: Policy Press

Bramley, G., Bartlett, W. and Lambert, C. (1995) *Planning, the Market and Private House-building*, London: UCL Press

Bramley, G., Munro, M. and Lancaster, S. (1997) *The Economic Determinants of Household Formation: A Literature Review*, London: DETR

Bramley, G., Munro, M. and Pawson, H. (2004a) *Key Issues in Housing: Policies and Markets in 21st Century Britain*, London: Palgrave Macmillan

Bramley, G., Hague, C., Kirk, K., Prior, A., Raemaekers, J. and Smith, H. (2004b) *Review of Green Belt Policy in Scotland*, Edinburgh: Scottish Executive

Bray, J. (1992) *The Rush for Roads: A Road Programme for Economic Recovery: A Report by Movement Transport Consultancy for ALARM UK and Transport 2000*, London: Transport 2000

Brearley, C. (1999) 'Integrated transport policy: the implications for planning', *Journal of Planning and Environment Law* May: 408–15

Breheny, M. J. (1983) 'A practical view of planning theory', *Environment and Planning B: Planning and Design* 10: 101–15

Breheny, M. J. (1989) 'Chalkface to coalface: a review of the academic–practice interface', *Environment and Planning B: Planning and Design* 16: 451–68

Breheny, M. J. (1991) 'The renaissance of strategic planning', *Environment and Planning B: Planning and Design* 18: 233–49

Breheny, M. J. (ed.) (1992) *Sustainable Development and Urban Form*, London: Pion

Breheny, M. J. (1995) 'The compact city and transport energy consumption', *Transactions of the Institute of British Geographers* (NS) 20: 81–101

Breheny, M. J. (1997) 'Urban compaction: feasible and acceptable?', *Cities* 14: 209–17

Breheny, M. J. (1999a) 'The people: where will they work?' *Town and Country Planning* 68: 174–5

Breheny, M. J. (ed.) (1999b) *The People: Where Will They Work? Report of Town and Country Planning Association, Research into the Changing Geography of Employment*, London: Town and Country Planning Association

Breheny, M. J. and Congdon, P. (eds) (1989) *Growth and Change in a Core Region: The Case of South East England*, London: Pion

Breheny, M. J. and Hall, P. (eds) (1996) *The People: Where Will They Go? National Report of the Town and Country Planning Association; Regional Inquiry into Housing Need and Provision in England*, London: Town and Country Planning Association

Breheny, M. J. and Hooper, A. (1985) *Rationality in Planning: Critical Essays on the Role of Rationality in Urban and Regional Planning*, London: Pion

Breheny, M. and Ross, A. (1998) *Urban Housing Capacity: What can be done*, London: Town and Country Planning Association

Breheny, M., Gent, T. and Lock, D. (1993) *Alternative Development Patterns: New Settlements* (DoE), London: HMSO

Brenan, J. (1994) 'PPG 16 and the restructuring of archaeological practice in Britain', *Planning Practice and Research* 9: 395–405

Brennan, A., Rhodes, J. and Tyler, P. (1998) *Evaluation of the Single Regeneration Challenge Fund Budget: A Partnership for Regeneration – An Interim Evaluation*, London: DETR

Bridge, G. (1993) *Gentrification, Class and Residence*, Bristol: School for Advanced Urban Studies

Bridges, L. (1979) 'The structure plan examination-in-public as an instrument of intergovernmental decision making', *Urban Law and Practice* 2: 241–64

Briggs, A. (1952) *History of Birmingham* (2 vols), Oxford: Oxford University Press

Brindley, T., Rydin, Y. and Stoker, G. (1996) *Remaking Planning* (second edition), London: Routledge

British Standards Institution (BSI) (1978) *Code of Practice for Design for the Convenience of Disabled People (BS 5619)*, London: BSI

British Standards Institution (1979) *Code of Practice for Access for the Disabled to Buildings (BS 5810)*, London: BSI

British Waterways Board (BWB) (1993) *The Waterway Environment and Development Plans*, Watford: BWB

British Waterways Board (1999) *Our Plan for the Future 2000–2004*, London: BWB

Britton, D. (ed.) (1990) *Agriculture in Britain: Changing Pressures and Policies*, Wallingford: CAB International

Broady, M. (1968) *Planning for People*, London: Bedford Square Press

Bromley, M. P. (1990) *Countryside Management*, London: Spon

Bromley, R. and Thomas, C. (eds) (1993) *Retail Change: Contemporary Issues*, London: UCL Press

Brooke, C. (1996) *Natural Conditions: A Review of Planning Conditions and Nature Conservation*, Sandy, Beds: Royal Society for the Protection of Birds

Brooke, R. (1989) *Managing the Enabling Authority*, Harlow: Longman

Brooks, J. (1999) '(Can) modern local government (be) in touch with the people?', *Public Policy and Administration* 14(1): 42–59

Broome, J. (1992) *Counting the Cost of Global Warming*, Cambridge: White Horse Press

Brotherton, D. I. (1989) 'The evolution and implications of mineral planning policy in the national parks of England and Wales', *Environment and Planning A* 21: 1229–40

Brotherton, D. I. (1992a) 'On the control of development by planning authorities', *Environment and Planning B: Planning and Design* 19: 465–78

Brotherton, D. I. (1992b) 'On the quantity and quality of planning applications', *Environment and Planning B: Planning and Design* 19: 337–57

Brotherton, I. (1993) 'The interpretation of planning appeals', *Journal of Environmental Planning and Management* 36: 179–86

Brown, C., Claydon, J. and Nadin, V. (2003) 'The RTPI's Education Commission: context and challenges', *Town Planning Review*, 74(3): 333–45

Brown, C., Farthing, S., Konstadakopulos, D., Nadin, V. and Smith, I. (2004) *Dynamic Smaller Towns: Identification of Critical Success Factors*, Cardiff: National Assembly for Wales

Brownill, S. (1990) *Developing London's Docklands: Another Great Planning Disaster?*, London: Paul Chapman

Bruff, G. E. and Wood, A. P. (2000) 'Local sustainable development: land-use planning's contribution to modern local government', *Journal of Environmental Management and Planning* 43(4): 519–39

Brundtland, G. H. (chair) (1987) *Our Common Future* (World Commission on Environment and Development), Brundtland Report, Oxford: Oxford University Press

Brunet, R. (1989) *Les Villes Européennes*, Rapport pour la Délégation à l'Aménagement du Territoire et à l'Action Régionale (DATAR), Paris, la Documentation Français

Brunivells, P. and Rodrigues, D. (1989) *Investing in Enterprise: A Comprehensive Guide to Inner City Regeneration and Urban Renewal*, Oxford: Blackwell

Brunskill, I. (1989) *The Regeneration Game: A Regional Approach to Regional Policy*, London: Institute for Public Policy Research

Bruton, M. J. (ed.) (1974) *The Spirit and Purpose of Planning* (second edition), London: Hutchinson

Bruton, M. J. (1980) 'PAG revisited', *Town Planning Review* 48: 134–44

Bruton, M. J. (1985) *Introduction to Transport Planning*, London: UCL Press

Bruton, M., Crispin, G. and Fidler, P. (1980) 'Local plans: public local inquiries', *Journal of Planning and Environment Law*: 374–85

Bruton, M., Crispin, G. and Fidler, P. (1982) 'Local plans: the role and status of the public local inquiry', *Journal of Planning and Environment Law*: 276–86

Bruton, M. J. and Nicholson, D. J. (1983) 'Non-statutory plans and supplementary planning guidance', *Journal of Planning and Environment Law* 1983: 432–43

Bruton, M. J. and Nicholson, D. J. (1985) 'Supplementary planning guidance and local plans', *Journal of Planning and Environment Law* 1985: 837–44

Bruton, M. J. and Nicholson, D. J. (1987) *Local Planning in Practice*, London: Hutchinson

Bryant, B. (1995) *Twyford Down: Roads, Campaigning and Environmental Law*, London: Spon

Bryant, S. (1999) '50 years of loss', *Town and Country Planning* 68(1): 16

Buchanan, C. D. (1958) *Mixed Blessing: The Motor Car in Britain*, London: Leonard Hill

Buchanan, C. D. (chair) (1963) *Traffic in Towns*, Buchanan Report, London: HMSO

Buckingham-Hatfield, S. and Evans, B. (eds) (1996) *Environmental Planning and Sustainability*, London: Wiley

Budd, L. and Whimster, S. (eds) (1992) *Global Finance and Urban Living: A Study of Metropolitan Change*, London: Routledge

Building Design Partnership (BDP) (1996) *London's Urban Environment: Planning for Quality* (Government Office for London), London: HMSO

Bunce, M. (1994) *The Countryside Ideal: Anglo-American Images of Landscape*, London: Routledge

Bunnell, G. (1995) 'Planning gain in theory and practice: negotiation of agreements in Cambridgeshire', *Progress in Planning* 44(1): 1–113

Burbridge, V. (1990) *Review of Information on Rural Issues*, Edinburgh: Central Research Unit, Scottish Office

Burchell, R. W. and Listokin, D. (eds) (1982) *Energy and Land Use*, New Brunswick, NJ: Center for Urban Policy Research, Rutgers University

Burchell, R. W. and Sternlieb, G. (eds) (1978) *Planning Theory in the 1980s*, New Brunswick, NJ: Center for Urban Policy Research, Rutgers University

Burgon, J. (2000) *An Assessment of Changes in the Developed and Undeveloped Coast of England and Wales between 1965 and 1995*, Reading: Department of Geography, University of Reading

Burke, P. and Manta, M. (1999) *Planning for Sustainable Development: Measuring Progress in Plans*, Working Paper, Cambridge, MA: Lincoln Institute of Land Policy

Burke, T. and Shackleton, J. R. (1996) *Trouble in Store: UK Retailing in the 1990s*, London: Institute of Economic Affairs

Burns, D., Hambleton, R. and Hoggett, P. (1994) *The Politics of Decentralisation: Revitalising Local Democracy*, London: Macmillan

Burrows, R. (1997) *Contemporary Patterns of Residential Mobility in England*, York: Centre for Housing Policy, University of York

Burrows, R. and Rhodes, O. (1998) *Unpopular Places? Area Disadvantage and the Geography of Misery in England*, York: Joseph Rowntree Foundation

Burton, T. P. (1989) 'Access to environmental information: the UK experience of water registers', *Journal of Environmental Law* 1: 192–208

Burton, T. P. (1992) 'The protection of rural England', *Planning Practice and Research* 7(1): 37–40

Burwood, S. and Roberts, P. (2002) *Learning from Experience: The BURA Guide to Achieving Effective and Lasting Regeneration*, London: BURA

Business in the Community (1990) *Leadership in the Community: A Blueprint for Business Involvement in the 1990s*, London: Business in the Community

Business in the Community (1992) *A Measure of Commitment: Guidelines for Measuring Environmental Performance*, London: Business in the Community

Butler, D. and Kavanagh, D. (1997) *The British General Election of 1997*, London: Macmillan

Butler, D., Adonis, A. and Travers, T. (1994) *Failure in British Government: The Politics of the Poll Tax*, Oxford: Oxford University Press

Butt, R. (1999) 'Beyond motherhood and apple pie', *Town and Country Planning* 68: 106

Byrne, D. (1989) *Beyond the Inner City*, Milton Keynes: Open University Press

Byrne, S. (1989) *Planning Gain: An Overview – A Discussion Paper*, London: Royal Town Planning Institute

Byrne, T. (1994) *Local Government in Britain* (sixth edition), London: Penguin

CAG Consultants and Oxford Brookes University (2002) *Research into Regional Sustainable Development Frameworks: Final Report to the English Regions Network*, English Regions Network

Cairncross, F. (1993) *Costing the Earth* (second edition), London: Business Books and Economist Books

Cairncross, F. (1995) *Green Inc: Guide to Business and the Environment*, London: Earthscan

Cairns, S. (1999) 'Redirecting the school run', *Town and Country Planning* 68: 300–1

Calder, N., Cavanagh, S., Eckstein, C., Palmer, J. and Stell, A.

(1993) *Women and Development Plans*, Newcastle upon Tyne: Department of Town and Country Planning, University of Newcastle

Callander, R. F. (1987) *A Pattern of Landownership in Scotland*, Finzean: Haughend

Callander, R. F. (1998) *How Scotland is Owned*, Edinburgh: Canongate

Callies, D. (1999) 'An American perspective on UK planning', in Cullingworth, J. B. (ed.) *British Planning: 50 Years of Urban and Regional Policy*, London: Athlone

Callies, D. L. and Grant, M. (1991) 'Paying for growth and planning gain: an Anglo-American comparison of development conditions, impact fees and development agreements', *Urban Lawyer* 23: 221–48

Cambridge Policy Consultants (1998) *Regenerating London Docklands*, London: DETR

Cameron, G. C. (1990) 'First steps in urban policy evaluation in the United Kingdom', *Urban Studies* 27: 475–95

Cameron, G. C., Monk, S. and Pearce, B. J. (1988) *Vacant Urban Land: A Literature Review*, London: DoE

Cameron, S., Baker, M., Bevan, M., Hull, A. and Williams, R. (1998) *Regionalisation, Devolution and Social Housing*, London: National Housing Forum

Campbell, M. (ed.) (1990) *Local Economic Policy*, London: Cassell

Campbell, S. and Fainstein, S. S. (eds) (1996) *Readings in Planning Theory*, Oxford: Blackwell

Capita Management Consultancy (1996) *An Evaluation of Six Early Estate Action Schemes* (DoE), London: HMSO

Capita Management Consultancy (1997–8) *Housing Action Trusts: Evaluation of Baseline Conditions* (11 vols), London: DETR

Capner, G. (1994) 'Green belt and rural economy', in Wood, M. (ed.) *Planning Icons: Myth and Practice* (Planning Law Conference, *Journal of Planning and Environment Law*), London: Sweet and Maxwell

Card, R. and Ward, R. (1996) 'Access to the countryside: the impact of the Criminal Justice and Public Order Act 1994', *Journal of Planning and Environment Law* 1996: 447–62

Cardiff University and Buchanan Partnership (1997) *Slimmer and Swifter: A Critical Examination of District Wide Local Plans and UDPs*, London: Royal Town Planning Institute

Carley, M. (1990a) 'Neighbourhood renewal in Glasgow: policy and practice', *Housing Review* 39: 49–51

Carley, M. (1990b) *Housing and Neighbourhood Renewal: Britain's New Urban Challenge*, London: Policy Studies Institute

Carley, M. (1991) 'Business in urban regeneration partnerships: a case study in Birmingham', *Local Economy* 6: 100–15

Carley, M. (1997) *Sustainable Transport and Retail Vitality: State of the Art for Towns and Cities*, Stirling: Historic Burghs Association of Scotland with Transport 2000

Carley, M. (1999) 'Neighbourhoods: building blocks of national sustainability', *Town and Country Planning* 68(2): 61–4

Carley, M. and Kirk, K. (1998) *City-wide Urban Regeneration* (Central Research Unit, Scottish Executive), Edinburgh: The Stationery Office

Carley, M., Campbell, M., Kearns, A., Wood, M. and Young, R. (2000a) *Regeneration in the 21st Century: Policies into Practice*, Bristol: Policy Press

Carley, M., Chapman, M., Hastings. A., Kirk, K. and Young, R. (2000b) *Urban Regeneration through Partnership: A Study in Nine Urban Regions in England, Scotland and Wales*, Bristol: Policy Press

Carmichael, P. (1992) 'Is Scotland different? Local government policy under Mrs Thatcher', *Local Government Policy Making* 18(5): 25–32

Carmona, M. (1996) 'Controlling urban design: Part 1: A possible renaissance?' *Journal of Urban Design* 1(1): 47–73

Carmona, M. (1998) 'Residential design policy and guidance: prevalence, hierarchy and currency', *Planning Practice and Research* 13(4): 407–19

Carmona, M. (1999) 'Residential design policy and guidance: content, analytical basis, prescription and regional emphasis', *Planning Practice and Research* 14(1): 17–38

Carmona, M. and Gallent, N. (2004) 'Planning and house building: a step change from the bottom up?' *Town Planning Review* 75(1): 95–122

Carmona, M. and Sieh, L. (2004) *Measuring Quality in Planning: Managing the Performance Process*, London: Spon

Carmona, M., Carmona, S. and Gallent, N. (2003) *Delivering New Homes: Processes, Planners and Providers*, London: Routledge

Carnie, J. K. (1996) *The Safer Cities Programme in Scotland* (Central Research Unit, Scottish Office), Edinburgh: HMSO

Carnwath, R. (chair) (1989) *Enforcing Planning Control: Report by Robert Carnwath QC*, Carnwath Report, London: HMSO

Carnwath, R. (1991) 'The planning lawyer and the environment', *Journal of Environmental Law* 3: 57–67

Carnwath, R. (1992) 'Environmental enforcement: the need for a specialist court', *Journal of Planning and Environment Law* 1992: 799–808

Carole Millar Research (1999) *Perceptions of Local Government: A Report of Focus Group Research*, Edinburgh: Central Research Unit, Scottish Office

Carson, R. (1951) *The Sea Around Us*, New York: Oxford University Press

Carson, R. (1962, 1965) *Silent Spring*, Harmondsworth: Penguin

Carter, C. (1997) *Local–Central Government Relations: A Compendium of Joseph Rowntree Foundation Research Findings*, York: Joseph Rowntree Foundation

Carter, C. and John, P. (1992) *A New Accord: Promoting Constructive Relations Between Central and Local Government*, York: Joseph Rowntree Foundation

Carter, N., Brown, T. and Abbott, T. (1991) *The Relationship between Expenditure-Based Plans and Development Plans*, Leicester: School of the Built Environment, Leicester Polytechnic

Carter, N., Oxley, M. and Golland, A. (1996) 'Towards the market: Dutch physical planning in a UK perspective', *Planning Practice and Research* 11: 49–60

Cartwright, L. (1997) 'The implementation of sustainable development by local authorities in the south east of England', *Planning Practice and Research* 12(4): 337–47

Cartwright, L. (2000) 'Selecting local sustainable development indicators: does consensus exist in their choice and purpose?', *Planning Practice and Research* 15(1–2): 65–78

Casellas, A. and Galley, C. (1999) 'Regional definitions in the European Union: a question of disparities', *Regional Studies* 33: 551–8

Castells, M. (1991) *The Informational City: Economic Restructuring and Urban Development*, Oxford: Blackwell

Catalano, A. (1983) *A Review of Enterprise Zones*, London: CES

CB Hillier Parker and Saxell Bird Axon (1998) *The Impact of Large Foodstores on Market Towns and District Centres* (DETR), London: The Stationery Office

CEC (2001) *Sixth Environment Action Programme: Environment 2010 – Our Future, Our Choice*, COM (2001) 31 final, Luxembourg: OOEPC

CEC (Eurostat) (2003a) *Economic Portrait of the European Union* (Office of the Official Publications of the European Communities), Luxembourg: OOEPC

CEC (Eurostat) (2003b) *Measuring Progress Towards a More Sustainable Europe* (Office of the Official Publications of the European Communities), Luxembourg: OOEPC

CEC (Eurostat) (2003c) *Regions: Statistical Yearbook* (Office of the Official Publications of the European Communities), Luxembourg: OOEPC

CEC (Eurostat) (2003d) *Statistical Yearbook on Candidate Countries Data 1997–2001*, Luxembourg: OOEPC

Central Housing Advisory Committee (1967) *The Needs of New Communities*, London: HMSO

Central Office of Information: *see* appendix on *Official Publications*

Centre for Environmental Studies (1970) *Observations on the Greater London Development Plan*, London: CES (reprinted in Cullingworth, J. B. (ed.) (1973) *Planning for Change* (vol. 3 of *Problems of an Urban Society*), London: Allen and Unwin)

Centre for Environmental Studies (1973) 'Education for planning', *Progress in Planning*, 1: 1–100

Centre for Local Economic Strategies (CLES) (1990) *Inner City Regeneration: A Local Authority Perspective*, Manchester: CLES

Centre for Local Economic Strategies (1991) *City Centres, City Cultures: The Role of the Arts in the Revitalisation of Towns and Cities*, Manchester: CLES

Centre for Local Economic Strategies (1992a) *Reforming the TECs: Towards a Strategy – Final Report of the CLES TEC/LEC Monitoring Project*, Manchester: CLES

Centre for Local Economic Strategies (1992b) *Social Regeneration: Directions for Urban Policy in the 1990s*, Manchester: CLES

Centre for Local Economic Strategies (1993) *The Green Local Economy: Integrating Economic and Environmental Development at the Local Level*, Manchester: CLES

Centre for Local Economic Strategies (1994) *Rethinking Urban Policy: City Strategies for the Global Economy*, Manchester: CLES

Centre for Local Economic Strategies (1996) *Regeneration through Work: Creating Jobs in the Social Economy*, Manchester: CLES

Centre for Local Economic Strategies (1998) *Joining Up? The New Deal, the Public Sector and the Employer Option*, Manchester: CLES

Centre for Local Economic Strategies (1999) *Strategies for Success: A First Assessment of Regional Development Agencies' Draft Regional Economic Strategies*, Manchester: CLES

Centre for Scottish Public Policy (1999) *Parliamentary Practices in Devolved Parliaments*, Edinburgh: The Stationery Office

Centre for Urban Policy Studies (1998) *The Impact of Urban*

Development Corporations in Leeds, Bristol and Central Manchester, London: DETR

Cervero, R. (1990) 'Transit pricing research: a review and synthesis', *Transportation* 17: 117–39

Cervero, R. (1995) 'Planned communities, self-containment and commuting: a cross-national perspective', *Urban Studies* 32: 1135–61

Chadwick, G. F. (1978) *A Systems View of Planning: Towards a Theory of the Urban and Regional Planning Process*, Oxford: Pergamon

Champion, A. G. (1989) *Counterurbanisation: The Changing Pace and Nature of Population Deconcentration*, London: Edward Arnold

Champion, A. G. (1996) *Urban Exodus: Migration between Metropolitan and Non-metropolitan Areas in Britain*, Swindon: Economic and Social Science Research Council

Champion, A. G. and Atkins, D. (1996) *The Counterurbanisation Cascade: An Analysis of the Census Special Migration Statistics for Great Britain*, Newcastle upon Tyne: Department of Geography, University of Newcastle upon Tyne

Champion, A. G. and Fielding, A. J. (eds) (1982) *Migration Processes and Patterns, vol. 1: Research Progress and Prospects*, London: Belhaven

Champion, A. G. and Green, A. E. (1992) 'Local economic performance in Britain during the late 1980s: the results of the third booming towns study', *Environment and Planning A* 24: 243–72

Champion, A. G., Atkins, A., Coombes, M. and Fotheringham, S. (1998) *Urban Exodus*, London: Council for the Protection of Rural England

Champion, T. and Hugo, G. (2003) *New Forms of Urbanization: Beyond the Urban–Rural Dichotomy*, Aldershot: Ashgate

Champion, T. and Watkins, C. (1991) *People in the Countryside: Studies of Social Change in Rural Britain*, London: Paul Chapman

Chanan, G. (2003) *Searching for Solid Foundations: Community Involvement in Urban Policy*, London: ODPM

Chandler, J. A. (1996) *Local Government Today* (second edition), Manchester: Manchester University Press

Chapman, D. and Larkham, P. (1992) *Discovering the Art of Relationship*, Birmingham: Faculty of the Built Environment, Birmingham Polytechnic

Chapman, D. and Larkham, P. (1999) 'Urban design, urban quality and the quality of life: reviewing the Department of the Environment's Urban Design Campaign', *Journal of Urban Design* 4(2): 211–32

Chapman, M. (1995) 'Urban policy and urban evaluation: the impact of the European Union', in Hambleton, R. and Thomas, H. (eds) *Urban Policy Evaluation: Challenge and Change*, London: Paul Chapman

Chapman, P. (1998) *Poverty and Exclusion in Rural Britain: The Dynamics of Low Income and Employment*, York: Joseph Rowntree Foundation

Charles, HRH The Prince of Wales (1989) *A Vision of Britain: A Personal View of Architecture*, London: Doubleday

Chartered Institute of Environmental Health (CIEH) (1995) *Travellers and Gypsies: An Alternative Strategy*, London: CIEH

Chartered Institute of Housing, National Housing Federation, Room and Shelter (1999) *National Blueprint: The Delivery of Affordable Homes through the Planning System*, published jointly by the four organisations

Chartered Institute of Transport (CIT) (1991) *London's Transport: The Way Ahead*, London: CIT

Chartered Institute of Transport (1992) *Paying for Progress: A Report on Congestion and Road Use Charges: Supplementary Report*, London: CIT

Checkland, S. G. (1981) *The Upas Tree: Glasgow 1875-1975 and after 1975–1980*, Glasgow: University of Glasgow Press

Cherry, G. E. (1974) *The Evolution of British Town Planning*, London: Leonard Hill

Cherry, G. E. (1975) *National Parks and Access to the Countryside: Environmental Planning 1939–1969, Volume 2*, London: HMSO

Cherry, G. E. (1981) *Pioneers in British Planning*, London: Architectural Press

Cherry, G. E. (1982) *The Politics of Town Planning*, London: Longman

Cherry, G. E. (1988) *People and Plans: The Shaping of Urban Britain in the Nineteenth and Twentieth Centuries*, London: Edward Arnold

Cherry, G. E. (1992) 'Green belt and the emergent city', *Property Review* December: 97–101

Cherry, G. E. (1994a) *Birmingham: A Study in Geography, History and Planning*, Chichester: Wiley

Cherry, G. E. (1994b) *Rural Change and Planning: England and Wales in the Twentieth Century*, London: Spon

Cherry, G. E. (1996) *Town Planning in Britain since 1900: The Rise and Fall of the Planning Ideal*, Oxford: Blackwell

Cherry, G. E. and Penny, L. (1990) *Holford: A Study in Architecture, Planning and Civic Design*, London: Spon

Cherry, G. E. and Rogers, A. (1996) *Rural Change and Planning: England and Wales in the Twentieth Century*, London: Spon

Cheshire County Council (1995) *Cheshire County Structure Plan: New Thoughts for the Next Century*, Chester: Cheshire County Environmental Planning Service

Cheshire, P. and Sheppard, S. (1989) 'British planning policy and access to housing: some empirical estimates', *Urban Studies* 26: 469–85

Cheshire, P., D'Arcy, E., and Giussani, B. (1991) *Local, Regional and National Government in Britain: A Dreadful Warning*, Reading: Department of Economics, University of Reading

Chesman, G. (2004) 'Conditions in planning permissions requiring the subsequent completion of planning obligations', *Journal of Planning and Environment Law* December: 1649–53

Chesman, G. R. (1991) 'Local authorities and the review of the definitive map under the Countryside and Wildlife Act 1981', *Journal of Planning and Environment Law* 1991: 611–14

Chesterton and Urban Initiatives (1995) *Dwelling Provision through Planning Regeneration*, Chesterton Report, Hertford: Hertfordshire County Council

Chisholm, M. (1995) *Britain on the Edge of Europe*, London: Routledge

Chisholm, M. (2003) 'Trust in question', *Journal of Local Government Law* 2: 32–7

Christensen, T. (1979) *Neighbourhood Survival*, Dorchester: Prism Press

Christie, B. (2000) 'Finding a way forward for LA21 in Scotland', *EG* 6(5): 5–8

Christie, I. and Warburton, D. (eds for Real World Coalition) (2001) *From Here to Sustainability*, London: Earthscan

Chubb, R. N. (1988) *Urban Land Markets in the United Kingdom* (DoE), London: HMSO

Church, A. (1988) 'Urban regeneration in London docklands: a five-year review', *Environment and Planning C: Government and Policy* 6: 187–208

Church, C. and McHarry, C. (eds) (1999) *One Small Step: A Guide to Action on Sustainable Development in the UK*, London: Community Development Forum

Churchill, R., Warren, L. M. and Gibson, J. (eds) (1991) *Law, Policy and the Environment*, Oxford: Blackwell

Civic Trust (1991) *Audit of the Environment*, London: Civic Trust

Civic Trust (1999) *Brownfield Housing: 12 Years On*, London: Civic Trust

Clabon, S. and Chance, C. (1992) 'Legal profile: freedom of access to environmental information', *European Environment* 2(3): 24–5

Clapson, M. (1998) *Invincible Green Suburbs, Brave New Towns*, Manchester: Manchester University Press

Clark, A., Lee, M. and Moore, R. (1996) *Landslide Investigation and Management in Great Britain* (a support document for PPG 14) (DoE), London: HMSO

Clark, D. M. (1992) *Rural Social Housing: Supply and Trends – A 1992 Survey of Affordable New Homes*, Cirencester: Association of Community Councils in Rural England

Clark, G., Darrall, J., Grove-White, R., Macnaghten, P. and Urry, J. (1994) *Leisure Landscapes: Leisure, Culture and the English Countryside – Challenges and Conflicts*, London: Council for the Protection of Rural England

Clark, M., Smith, D. and Blowers, A. (1992) *Waste Location: Spatial Aspects of Waste Management, Hazards, and Disposal*, London: Routledge

Clawson, M. and Hall, P. (1973) *Land Planning and Urban Growth*, Baltimore, MD: Johns Hopkins University Press

Claydon, J. (1998) 'Discretion in development control: a study of how discretion is exercised in the conduct of development control in England and Wales', *Planning Practice and Research* 13(1): 63–80

Claydon, J. and Smith, B. (1997) 'Negotiating planning gains through the British development control system', *Urban Studies* 34(12): 2003–22

Clayton, A. M. H. and Radcliffe, N. J. (1996) *Sustainability: A Systems Approach*, London: Earthscan

Cleator, B. (1995) *Review of Legislation Relating to the Coastal and Marine Environment of Scotland*, Perth: Scottish Natural Heritage

Cloke, P. (ed.) (1987) *Rural Planning: Policy into Action?*, London: Paul Chapman

Cloke, P. (ed.) (1988) *Policies and Plans for Rural People*, London: Allen and Unwin

Cloke, P. (ed.) (1992) *Policy and Change in Thatcher's Britain*, Oxford: Pergamon

Cloke, P. (1993) 'On problems and solutions: the reproduction of problems for rural communities in Britain during the 1980s', *Journal of Rural Studies* 9: 113–21

Cloke, P. (1996) 'Housing in the open countryside: windows on "irresponsible" planning in rural Wales', *Town Planning Review* 67: 291–308

Cloke, P. and Little, J. (1990) *The Rural State: Limits to Planning in Rural Society*, Oxford: Clarendon Press

Cloke, P., Milbourne, P. and Thomas, C. (1994a) *Lifestyles in Rural England* London: Rural Development Commission

Cloke, P., Doel, M., Matless, D., Phillips, M. and Thrift, N. (1994b) *Writing the Rural: Five Cultural Geographies*, London: Paul Chapman

Clotworthy, J. and Harris, N. (1996) 'Planning policy implications of local government reorganisation', in Tewdwr-Jones, M. (ed.) *British Planning Policy in Transition: Planning in the 1990s*, London: UCL Press

Coates, D. (1994) 'Wanted? More land, enquire within', *House Builder* April: 29–30

Cobb, D., Dolman, P. and O'Riordan, T. (1999) 'Interpretations of sustainable agriculture in the UK', *Progress in Human Geography* 23: 209–35

Coccossis, H. and Nijkamp, P. (eds) (1995) *Sustainable Tourism Development*, Aldershot: Avebury

Cochrane, A. (1991) 'The changing state of local government: restructuring for the 1990s', *Public Administration* 69: 281–302

Cochrane, A. (1993) *Whatever Happened to Local Government?*, Buckingham: Open University Press

Cochrane, A. and Clarke, A. (1990) 'Local enterprise boards: the short history of a radical initiative', *Public Administration* 68: 315–36

Cocks, R. (1991) 'First responses to the new "breach of condition" notice', *Journal of Planning and Environment Law* 1991: 409–18

Cole, I. and Smith, Y. (1996) *From Estate Action to Estate Agreement: Regeneration and Change on the Bell Farm Estate, York*, Bristol: Policy Press

Cole, J. and Cole, F. (1993) *The Geography of the European Community*, London: Routledge

Coleman, A. (1990) *Utopia on Trial* (second edition), London: Hilary Shipman

Colenutt, B. and Cutten, A. (1994) 'Community empowerment in vogue or vain?' *Local Economy* 9: 236–50

Collar, N. A. (1999) *Green's Concise Scots Law: Planning* (second edition), Edinburgh: W. Green

Collins, L. (1996) 'Recycling and the environmental debate: a question of social conscience of scientific reason?', *Journal of Environmental Planning and Management* 39: 333–55

Collins, M. P. (1989) 'A review of 75 years of planning education at UCL', *Planner* 75(6): 18–22

Collins, M. P. and McConnell, S. (1988) *The Use of Local Plans for Effective Town and Country Planning: Report of a Research Project Funded by the Nuffield Foundation*, London: Bartlett

School of Architecture and Planning, University College London

Colls, J. (2002) *Air Pollution: An Introduction* (second edition), London: Spon

Colquhoun, I. (2003) *Design Out Crime: Creating Safe Communities*, London: Architecture Press

Comedia (1999) *Park Life: Urban Parks and Social Renewal*, Gloucestershire: Comedia

Commission for Architecture and the Built Environment (CABE) (2003) *Protecting Design Quality in Planning*, London: CABE

Commission for Architecture and the Built Environment (2004) *Design Reviewed*, London: CABE

Commission for Racial Equality (CRE) (1982) *Local Government and Racial Equality*, London: CRE

Commission for Racial Equality and Schneider-Ross (2003) *Towards Race Equality: An Evaluation of the Public Duty to Promote Race Equality and Good Race Relations in England and Wales*, London: CRE

Commission on Social Justice (1994) *Social Justice: Strategies for National Renewal*, London: Vintage

Committee on Spatial Development (1999) *European Spatial Development Perspective: Towards Balanced and Sustainable Development of the EU*, presented to the Meeting of Ministers of Regional/Spatial Planning, Potsdam

Community Development Foundation (2003) *Searching for Solid Foundations: Community Involvement and Urban Policy*, London: ODPM

Confederation of British Industry (CBI) (1988) *Initiatives beyond Charity: Report of the CBI Task Force on Business and Urban Regeneration*, London: CBI

Confederation of British Industry (1995) *Missing Links: Setting National Transport Priorities*, London: CBI

Confederation of British Industry (1996) *Winning Ways: Developing the UK Transport Network*, London: CBI

Connal, R. C. and Scott, J. N. (1999) 'The new Scottish Parliament: what will its impact be?' *Journal of Planning and Environment Law* June: 491–7

Connelly, S., Abram, S. and Richardson, T. (eds) (2001) *Perspectives on Participation*, Research Paper, Sheffield: Department of Town and Regional Planning, University of Sheffield

Connolly, M. E. H. and Loughlin, S. (1990) *Public Policy in Northern Ireland: Adoption or Adaptation?*, Belfast: Policy Research Institute

Consortium Developments (1985) *New Country Towns*, London: Consortium Developments

Cook, A. J. and Davis, A. L. (1993) *Package Approach Funding: A Survey of English Highway Authorities*, London: Friends of the Earth

Cooke, P. and Morgan, K. (1993) 'The network paradigm: new departures in corporate and regional development', *Environment and Planning D* 11: 543–64

Coombes, M., Raybould, S. and Wong, C. (1992) *Developing Indicators to Assess the Potential for Urban Regeneration* (DoE), London: HMSO

Coombes, T., Fidler, P. and Hathaway, A. (1992) 'South West regional planning review: towards a regional strategy', in Minay, C. L. W. (ed.) 'Developing regional guidance in England and Wales: a review symposium', *Town Planning Review* 63: 415–34

Coon, A. (1988) 'Local plan provision: the record to date and prospects for the future', *The Planner* 74(5): 17–20

Coon, A. (1989) 'An assessment of Scottish development planning', *Planning Outlook* 32(2): 77–85

Cooper, A. J. (1998) 'The origins of the Cambridge Green Belt', *Planning History* 20(2): 6–19

Cooper, D. E. and Palmer, J. A. (eds) (1992) *The Environment in Question*, London: Routledge

Coopers & Lybrand (1985) *Land Use Planning and the Housing Market: An Assessment*, London: Coopers & Lybrand

Coopers & Lybrand (1987) *Land Use Planning and Indicators of Housing Demand*, London: Coopers & Lybrand

Coopers & Lybrand (1994) *Employment Department Baseline Follow-up Studies: Final Overall Report to the Employment Department*, London: Coopers & Lybrand

Corkindale, J. (1998) *Reforming Land Use Planning*, London: Institute of Economic Affairs

Corner, T. (1999) 'Planning, environment, and the European Convention on Human Rights', *Journal of Planning and Environment Law* April: 301–4

Cornford, J. and Gillespie, A. (1992) 'The coming of the wired city? The recent development of cable in Britain', *Town Planning Review* 63: 243–64

Cornford, T. (1998) 'The control of planning gain', *Journal of Planning and Environment Law* August: 731–49

Costonis, J. J. (1989) *Icons and Aliens: Law, Aesthetics, and Environmental Change*, Urbana-Champaign, IL: University of Illinois Press

Couch, C. (1990) *Urban Renewal: Theory and Practice*, London: Macmillan

Couch, C. (2003) *City of Change and Challenge: Urban Planning and Regeneration in Liverpool*, Aldershot: Ashgate

Couch, C. and Gill, N. (1993) *Renewal Areas: A Review of Progress*, Working Paper 119, Bristol: School for Advanced Urban Studies, Bristol University

Couch, C., Eva, D. and Lipscombe, A. (2000) 'Renewal areas in north-west England', *Planning Practice and Research* 15(3): 257–68

Coulson, A. (2003) 'Land-use planning and community influence: a study of Selly Oak, Birmingham', *Planning Practice and Research* 18(2): 179–95

Council for the Protection of Rural England (CPRE) (1988) *Welcome Homes: Housing Supply from Unallocated Land*, London: CPRE

Council for the Protection of Rural England (1990a) *From White Paper to Green Future*, London: CPRE

Council for the Protection of Rural England (1990b) *Our Finest Landscapes: Council for the Protection of Rural England Submission to the National Parks Review Panel*, London: CPRE

Council for the Protection of Rural England (1990c) *Future Harvest: The Economics of Farming and the Environment – Proposals for Action* (Jenkins, T. N.), London: CPRE and WWF

Council for the Protection of Rural England (1990d) *Planning Control over Farmland*, London: CPRE

Council for the Protection of Rural England (1991) *Energy Conscious Planning* (Owens, S.), London: CPRE

Council for the Protection of Rural England (1992a) *Campaigners Guide to Using EC Environmental Law* (Macrory, R.), London: CPRE

Council for the Protection of Rural England (1992b) *Campaigners' Guide to Local Plans*, (Green Balance), London: CPRE

Council for the Protection of Rural England (1992c) *Our Common Home: Housing Development and the South East's Environment*, London: CPRE

Council for the Protection of Rural England (1992d) *Transport and the Environment: Council for the Protection of Rural England's Submission to the Royal Commission on Environmental Pollution*, London: CPRE

Council for the Protection of Rural England (1992e) *Where Motor Car is Master: How the Department of Transport Became Bewitched by Roads* (Kay, P. and Evans, P.), London: CPRE

Council for the Protection of Rural England (1992f) *The Lost Land: Land Use Change in England 1945–1990* (Sinclair, G.), London: CPRE

Council for the Protection of Rural England (1993a) *Index of National Planning Policies*, London: CPRE

Council for the Protection of Rural England (1993b) *Preparing for the Future: A Response by the Council for the Protection of Rural England to the Consultation Paper on the UK Strategy for Sustainable Development*, London: CPRE

Council for the Protection of Rural England (1993c) *Water for Life: Strategies for Sustainable Water Resource Management*, London: CPRE

Council for the Protection of Rural England (1994a) *Leisure Landscapes: Leisure, Culture and the English Countryside: Challenges and Conflicts* (Clark, G., Darrall, J., Grove-White, R., Macnaghten, P. and Urry, J.), London: CPRE

Council for the Protection of Rural England (1994b) *Environmental Policy Omissions in Development Plans*, London: CPRE

Council for the Protection of Rural England (1994c) *The Housing Numbers Game*, London: CPRE

Council for the Protection of Rural England (1994d) *Public Access to Planning Documents*, London: CPRE

Council for the Protection of Rural England (1995a) *Renewable Energy in the UK: Financing Options for the Future* (Mitchell, C.), London: CPRE

Council for the Protection of Rural England (1995b) *The End of Hierarchy: A New Perspective on Managing the Road Network* (Goodwin, P.), London: CPRE

Council for the Protection of Rural England (1995c) *Circular Projections: Household Growth, Housing Development, and the Household Projections* (Bramley, G. and Watkins, C.), London: CPRE

Council for the Protection of Rural England (1996a) *The Campaigners' Guide to Minerals*, London: CPRE

Council for the Protection of Rural England (1996b) *Growing Greener: Sustainable Agriculture in the UK* (Baldock, D., Bishop, K., Mitchell, K. and Phillips, A.), London: CPRE

Council for the Protection of Rural England (1997) *Planning More to Travel Less*, London: CPRE

Council for the Protection of Rural England (1998) *Rural Transport Policy and Equity*, London: CPRE

Council for the Protection of Rural England (1999a) *Fair Examination? Testing the Legitimacy of Strategic Planning for Housing*, London: CPRE

Council for the Protection of Rural England (1999b) *Hedging your Bets: Is Hedgerow Legislation Gambling with our Heritage?*, London: CPRE

Council for the Protection of Rural England (1999c) *Transport: The New Regional Agenda*, London: CPRE

Council for the Protection of Rural England (1999d) *Plan, Monitor and Manage: Making it Work*, London: CPRE

Council for the Protection of Rural England (1999e) *The Local Government Ombudsman*, London: CPRE

Council for the Protection of Rural England (1999f) *Park and Ride: Its Role in Local Transport*, London: CPRE

Council of Europe (1989) *European Campaign for the Countryside: Conclusion and Declarations*, Strasbourg: The Council

Council of Europe (1991) *The Bern Convention of Nature Conservation*, Strasbourg: The Council

Council of Europe (1992) *The European Urban Charter*, Strasbourg: Standing Conference of Local and Regional Authorities of Europe, Council of Europe

Counsell, D. (1998) 'Sustainable development and structure plans in England and Wales: a review of current practice', *Journal of Environmental Planning and Management* 41(2): 177–94

Countryside Agency (2003a) *A Guide to Definitive Maps and Changes to Public Rights of Way*, Cheltenham: Countryside Agency

Countryside Agency (2003b) *Concept Statements and Local Development Documents: Practical Guidance for Local Planning Authorities*, Cheltenham: Countryside Agency

Countryside Agency (2004) *Planning for Sustainable Rural Economic Development: Good Practice Advice for Local Authorities*, Cheltenham: Countryside Agency

County Planning Officers' Society (CPOS) (1991) *Regional Guidance and Regional Planning Conferences Progress Report*, CPOS

County Planning Officers' Society with Metropolitan Planning Officers' Society and District Planning Officers' Society (1992) *Planning in the Urban Fringe: Final Report of the Joint Special Advisory Group* (CPOS), Middlesbrough: Department of Environment, Development and Transportation, Cleveland County Council

County Planning Officers' Society (1993) *Planning for Sustainability* (CPOS), Winchester: County Planning Department, Hampshire County Council

County Planning Officers' Society (annual) *Opencast Coalmining Statistics* (CPOS) (published annually by Durham County Council)

Cowan, R. (1995) *Planning Aid Handbook*, London: Royal Town Planning Institute

Cowan, R. (1999) *The Role of Planning in Local Government*, London: Royal Town Planning Institute

Cowell, R. and Owens, S. (1998) 'Suitable locations: equity and sustainability in the minerals planning process', *Regional Studies* 32(9): 797–811

Cox, A. (1984) *Adversary Politics and Land*, Cambridge: Cambridge University Press

Cox, G., Lowe, P. and Winter, M. (1986) *Agriculture, People and Policies*, London: Allen and Unwin

Cox, G., Lowe, P. and Winter, M. (1990) *The Voluntary Principle in Conservation*, Chichester: Packard

Crabtree, J. R. and Chalmers, N. A. (1994) 'Economic evaluation of policy instruments for conservation: standard payments and capital grants', *Land Use Policy* 11: 94–106

Craig, P. and De Búrca, G. (1999) *EU Law*, Oxford: Oxford University Press

Crawford, C. (1989) 'Profitability and its role in planning decisions', *Journal of Environmental Law* 1: 221–44

Cripps, J. (1977) *Accommodation for Gypsies: A Report on the Working of the Caravan Sites Act 1968*, London: HMSO

Croall, J. (1995) *Preserve or Destroy: Tourism and the Environment*, London: Calouste Gulbenkian Foundation

Croall, J. (1997) *LETS Act Locally: The Growth of Local Exchange Trading Systems*, London: Gulbenkian Foundation

Cross, D. (1992) 'Regional planning guidance for East Anglia', in Minay, C. L. W. (ed.) 'Developing regional guidance in England and Wales: a review symposium', *Town Planning Review* 63: 415–34

Cross, D. and Bristow, R. (eds) (1983) *English Structure Planning*, London: Pion

Crouch, C. and Marquand, D. (eds) (1989) *The New Centrism: Britain out of Step with Europe?*, Oxford: Blackwell

Crow, S. (1996a) 'Lessons from Bryau', *Journal of Planning and Environment Law* 1996: 359–69

Crow, S. (1996b) 'Development control: the child that grew up in the cold', *Planning Perspectives* 11: 399–411

Crow, S. (1998a) 'Planning gain: there must be a better way', *Planning Perspectives* 13: 357–72

Crow, S. (1998b) 'Challenging appeal decisions, or the use and abuse of the "toothcomb"', *Journal of Planning and Environment Law* May: 419–31

Crow, S. (2000) 'The public examination of draft regional planning guidance: some reflections on the process', *Journal of Planning and Environment Law* October: 990–1002

Cuddy, M. and Hollingsworth, M. (1985) 'The review process in land availability studies: bargaining positions for builders and planners', in Barrett, S. and Healey, P. (eds) *Land Policy: Problems and Alternatives*, Aldershot: Avebury

Cullinane, S. (1992) 'Attitudes towards the car in the UK: some implications for policies on congestion and the environment', *Transportation Research* 26A: 291–301

Cullingworth, J. B. (1960a) 'Household formation in England and Wales', *Town Planning Review* 31: 5–26

Cullingworth, J. B. (1960b) *Housing Needs and Planning Policy: A Restatement of the Problems of Housing Need and 'Overspill' in England and Wales*, London: Routledge and Kegan Paul

Cullingworth, J. B. (1964) *Town and Country Planning in England and Wales*, London: George Allen and Unwin

Cullingworth, J. B. (chair) (1967) *Scotland's Older Houses* (Report of the Scottish Housing Advisory Committee, Subcommittee on Unfit Housing in Scotland), Cullingworth Report, Edinburgh: HMSO

Cullingworth, J. B. (ed.) (1973) *Problems of an Urban Society*, vol. 3: *Planning for Change*, London: Allen and Unwin

Cullingworth, J. B. (1975) *Reconstruction and Land Use Planning 1939–1947: Environmental Planning 1939–1969, Volume 1*, London: HMSO

Cullingworth, J. B. (1979) *New Towns Policy: Environmental Planning 1939–1969, Volume 3*, London: HMSO

Cullingworth, J. B. (1980) *Land Values, Compensation and Betterment: Environmental Planning 1939–1969, Volume 4*, London: HMSO

Cullingworth, J. B. (1987) *Urban and Regional Planning in Canada*, New Brunswick, NJ: Transaction

Cullingworth, J. B. (ed.) (1990) *Energy Policy Studies*, vol. 5: *Energy, Land, and Public Policy*, New Brunswick, NJ: Transaction

Cullingworth, J. B. (1993) *The Political Culture of Planning: American Land Use Planning in Comparative Perspective*, New York and London: Routledge

Cullingworth, J. B. (1997a) *Planning in the USA*, New York: Routledge

Cullingworth, J. B. (1997b) 'British land use planning: a failure to cope with change?', *Urban Studies* 34: 945–60

Cullingworth, J. B. (ed.) (1999) *British Planning: 50 Years of Urban and Regional Policy*, London: Athlone Press

Cullingworth, J. B. and Karn, V. A. (1968) *The Ownership and Management of Housing in the New Towns*, London: HMSO

Curry, N. (1992a) 'Nature conservation, countryside strategies and strategic planning', *Journal of Environmental Management* 35: 79–91

Curry, N. (1992b) 'Controlling development in the national parks of England and Wales', *Town Planning Review* 63: 107–21

Curry, N. (1993) 'Negotiating gains for nature conservation in planning practice', *Planning Practice and Research* 8(2): 10–15

Curry, N. (1994) *Countryside Recreation, Access and Land Use Planning*, London: Spon

Curry, N. (1997) 'Enhancing countryside recreation benefits through the rights of way system in England and Wales', *Town Planning Review* 68: 449–63

Curry, N. and Peck, C. (1993) 'Planning on presumption: strategic planning for countryside recreation in England and Wales', *Land Use Policy* 10: 140–50

Cyclists Public Affairs Group (1995) *Investing in the Cycling Revolution*, Godalming (Surrey): Cyclists Touring Club

Cyclists Touring Club (1993) *Cycle Policies in Britain: The 1993 CTC Survey*, Godalming: CTC

Dabinett, G. and Lawless, P. (1994) 'Urban transport investment and regeneration: researching the impact of South Yorkshire Supertram', *Planning Practice and Research* 9: 407–14

Dales, J. H. (1968) *Pollution, Property and Prices*, Toronto: University of Toronto Press

Dalziel, M. and Rowan-Robinson, J. (1986) 'Resurrecting the two price system for land', *Journal of Planning and Environment Law* 1986: 409–15

Damer, S. and Hague, C. (1971) 'Public participation in planning: evolution and problems', *Town Planning Review* 42: 217–24

Damesick, P. J. (1986) 'The M25: a new geography of development?', *Geographical Journal* 152: 155–60

Daniels, R. (2001) 'Housing, growth and the quality of urban life', paper presented to the Planning Summer School, Exeter, September

Daniels, S. (1993) *Fields of Vision: Landscape Imagery and National Identity in England and the United States*, Oxford: Polity Press

Danson, M. (1999) 'Economic development: the Scottish Parliament and the development agencies', in McCarthy, J. and Newlands, D. (eds) *Governing Scotland: Problems and Prospects: The Economic Impact of the Scottish Parliament*, Aldershot: Ashgate

Danson, M. W., Lloyd, M. G. and Newlands, D. (1989) 'Rural Scotland and the rise of Scottish Enterprise', *Planning Practice and Research* 4: 13–17

Danson, M. W., Lloyd, M. G. and Newlands, D. (1992) *The Role of Regional Development Agencies in Economic Regeneration*, London: Jessica Kingsley

Darke, R. (1999) 'Public speaking rights in local authority planning committees', *Planning Practice and Research* 14: 171–83

Darley, G., Hall, P. and Lock, D. (1991) *Tomorrow's New Communities*, York: Joseph Rowntree Foundation

Dasgupta, M., Oldfield, R., Sharman, K. and Webster, V. (1994) *Impact of Transport Policies in Five Cities*, Crowthorne: Transport Research Laboratory

Davidoff, P. (1965) 'Advocacy and pluralism in planning', *Journal of the American Institute of Planners* 31 (reprinted in Faludi, A. (ed.) (1973) *A Reader in Planning Theory*, Oxford: Pergamon)

Davidoff, P. and Reiner, T. A. (1962) 'A choice theory of planning', *Journal of the American Institute of Planners* 28: 103–15 (reprinted in Faludi, A. (ed.) (1973) *A Reader in Planning Theory*, Oxford: Pergamon)

Davies, A. R. (2001) 'Hidden or hiding? Public perceptions of participation in the planning system', *Town Planning Review* 72(2): 193–216

Davies, C., Pritchard, D. and Austin, L. (1992) *Planscan Scotland: A Study of Development Plans in Scotland*, Sandy, Beds: Royal Society for the Protection of Birds

Davies, H. W. E. (1992) 'Britain 2000: the impact of Europe for planning and practice', *The Planner, Town and Country Planning Summer School Proceedings* 78(21): 21–2

Davies, H. W. E. (1994) 'Towards a European planning system?', *Planning Practice and Research* 9: 63–9

Davies, H. W. E. (1996) 'Planning and the European question', in Tewdwr-Jones, M. (ed.) *British Planning Policy in Transition: Planning in the 1990s*, London: UCL Press

Davies, H. W. E. (1998) 'Continuity and change: the evolution of the British Planning System 1947–97', *Town Planning Review* 69(2): 135–52

Davies, H. W. E. (1999) 'The planning system and the development plan', in Cullingworth, J. B. (ed.) *British Planning: 50 Years of Urban and Regional Policy*, London: Athlone

Davies, H. W. E. and Gosling, J. A. (1994) *The Impact of the European Community on Land Use Planning in the United Kingdom*, London: Royal Town Planning Institute

Davies, H. W. E., Edwards, D. and Rowley, A. R. (1984) 'The relevance of development control', *Town Planning Review* 51: 5–24

Davies, H. W. E., Edwards, D., Roberts, C., Rosborough, L. and Sales, R. (1986a) *The Relationship between Development Plans and Appeals*, Reading: Department of Land Management and Development, University of Reading

Davies, H. W. E., Edwards, D. and Rowley, A. R. (1986b) *The Relationship between Development Plans, Development Control, and Appeals*, Reading: Department of Land Management and Development, University of Reading

Davies, H. W. E., Rowley, A. R., Edwards, D., Blom-Cooper, A., Roberts, C., Rosborough, L. and Tilley, R. (1986c) *The Relationship between Development Plans and Development Control*, Reading: Department of Land Management and Development, University of Reading

Davies, H. W. E., Edwards, D. and Rowley, A. R. (1989a) *The Approval of Reserved Matters following Outline Planning Permission* (DoE), London: HMSO

Davies, H. W. E., Hooper, A. J. and Edwards, D. (1989b) *Planning Control in Western Europe*, London: HMSO

Davies, J. G. (1972) *The Evangelistic Bureaucrat: A Study of a Planning Exercise in Newcastle-upon-Tyne*, London: Tavistock

Davies, L. (1996) 'Equality and planning: race' and 'Equality and planning: gender and disability', in Greed, C. (ed.) *Implementing Town Planning: The Role of Town Planning in the Development Process*, Harlow: Longman

Davies, P. and Keate, D. (1995) *In the Public Interest: London's Civic Architecture at Risk*, London: English Heritage

Davies, R. (1995) *Retail Planning Policies in Western Europe*, London: Routledge

Davies, R. L. and Campion, A. G. (1983) *The Future of the City Centre* (Institute of British Geographers), London: Academic Press

Davis, K. C. (1969) *Discretionary Justice: A Preliminary Inquiry*, Baton Rouge, LA: Louisiana State University Press

Davis, M. (1990) *City of Quartz: Excavating the Future in Los Angeles*, London: Verso

Davison, I. (1990) *Good Design in Housing: A Discussion Paper*, London: House Builders Federation and Royal Institute of British Architects

Davison, R. C. (1938) *British Unemployment Policy: The Modern Phase*, London: Longman

Davoudi, S. (1995a) 'City Challenge: a sustainable mechanism or temporary gesture', in Hambleton, R. and Thomas, H. (eds) *Urban Policy Evaluation: Challenge and Change*, London: Paul Chapman

Davoudi, S. (1995b) 'City Challenge: the three-way partnership', *Planning Practice and Research* 10: 333–44

Davoudi, S. (1999) 'A quantum leap for planners: the role of the planning system within changing approaches to waste management', *Town and Country Planning* 68(1): 20–3

Davoudi, S. (2001) 'Planning and the twin discourses of sustainability', in Layard, A., Batty, S. and Davoudi, S., *Planning for a Sustainable Future*, London: Spon

Davoudi, S. (2005) 'The ESPON – past, present and future', *Town and Country Planning* 74(3): 100–2

Davoudi, S. and Atkinson, R. (1999) 'Social exclusion and the British planning system', *Planning Practice and Research* 14: 225–36

Davoudi, S. and Healey, P. (1994) *Perceptions of City Challenge Policy Processes: The Newcastle Case*, Newcastle upon Tyne: Department of Town and Country Planning, University of Newcastle upon Tyne

Davoudi, S. and Healey, P. (1995) 'City challenge: sustainable process or temporary gesture?', *Environment and Planning C: Government and Policy* 13: 79–95

Davoudi, S., Healey, P. and Hull, A. (1997) 'Rhetoric and reality in British structure planning in Lancashire: 1993–95', in Healey, P., Khakee, A., Motte, A. and Needham, B. (eds) *Making Strategic Spatial Plans: Innovation in Europe*, London: UCL Press

Dawson, D. (1985) 'Economic change and the changing role of local government', in Loughlin, M., Gelfand, M. D. and Young, K. (eds) *Half a Century of Municipal Decline, 1935–1985* London: Allen and Unwin

Dawson, J. and Walker, C. (1990) 'Mitigating the social costs of private development: the experience of linkage programmes in the United States', *Town Planning Review* 61: 157–70

Dawson, J. A. (1994) *Review of Retailing Trends, with Particular Reference to Scotland*, Edinburgh: Central Research Unit, Scottish Office

Deakin, N. and Edwards, J. (1993) *The Enterprise Culture and the Inner Cities*, London: Routledge

Dear, M. and Scott, A. J. (eds) (1981) *Urbanization and Urban Planning in a Capitalist Society*, London: Methuen

Deas, I. and Ward, K. (2000) 'The song has ended but the melody lingers: regional development agencies and the lessons of the urban development corporation experiment', *Local Economy* 14(2): 114–32

Debenham, Tewson and Chinnocks (1988) *Planning Gain: Community Benefit or Commercial Bribe?*, London: Debenham, Tewson and Chinnocks

de Groot, L. (1992) 'City Challenge: competing in the urban regeneration game', *Local Economy* 7: 196–209

Deimann, S. (1994) *Your Rights under European Union Environment Legislation*, Brussels: European Environmental Bureau

Delafons, J. (1990a) *Development Impact Fees and Other Devices*, Berkeley, CA: Institute of Urban and Regional Development, University of California at Berkeley

Delafons, J. (1990b) *Aesthetic Control: A Report on Methods Used in the USA to Control the Design of Buildings*, Berkeley, CA: Institute of Urban and Regional Development, University of California at Berkeley

Delafons, J. (1997) *Politics and Preservation: A Policy History of the Built Heritage 1882–1996*, London: Spon

Delafons, J. (1998) 'Reforming the British planning system 1964–5: the Planning Advisory Group and the genesis of the Planning Act of 1968', *Planning Perspectives* 13: 373–87

De Montfort University and University of Strathclyde (1998) *Enhancing Public Participation in Local Government*, London: DETR

Denington, E. (chair) (1966) *Our Older Homes: A Call for Action* (Central Housing Advisory Committee), Denington Report, London: HMSO

Dennington, V. N. and Chadwick, M. J. (1983) 'Derelict land and waste land: Britain's neglected land resource', *Journal of Environmental Management* 16: 229–39

Dennis, N. (1970) *People and Planning*, London: Faber

Dennis, N. (1972) *Public Participation and Planners' Blight*, London: Faber

Department of Land Economy, Cambridge University (1999) *Environmental Court Project Final Report*, London: DETR

Department of National Heritage (1996) *Protecting our Heritage: A Consultation Paper on the Built Heritage of England and Wales*, London: Department of National Heritage

Department of Trade and Industry (DTI) (1999) *Biotechnology Clusters*, Sainsbury Report, London: DTI

Derby City Council (1989) *Sir Francis Ley Industrial Park – Simplified Planning Zoning: The First Twelve Months*, Derby: The Council

Derby City Council (1991) *Sir Francis Ley Industrial Park – Simplified Planning Zoning: The First Three Years*, Derby: The Council

Derbyshire, A., McNaught, S. and Blatchford, C. (1999) 'Listing needs much bigger alterations', *Estates Gazette* May: 145–7

Derthick, M. (1972) *New Towns In-Town: Why a Federal Program Failed*, Washington, DC: Urban Institute

De Soissons, M. (1988) *Welwyn Garden City*, Cambridge: Publications for Companies

Devereux, M. and Guillemoteau, D. (2001) 'A spatial vision for north-west Europe', *Town and Country Planning* February: 48–9

Devon County Council (1991) *Traffic Calming Guidelines*, Exeter: The Council

Dewar, D. (1999) 'Doubts over designation plans [for designation of the New Forest and the South Downs as national parks]', *Planning* 8 October: 12

Diamond, D. (1979) 'The uses of strategic planning: the example of the National Planning Guidelines in Scotland', *Town Planning Review* 50: 18–25

Diamond, D. and Spence, N. (1989) *Infrastructure and Industrial Costs in British Industry*, London: HMSO

Dicken, P. (1992) *Global Shift: The Internationalization of Economic Activity*, London: Paul Chapman

Diefendorf, J. (1990) *Rebuilding Europe's Blitzed Cities*, London: Macmillan

Dieleman, F. M. and Hamnett, C. (1994) 'Globalisation, regulation and the urban system', *Urban Studies*, 31(3): 357–64

Dinan, D. (ed.) (1998) *Encyclopedia of the European Union*, London: Macmillan

Dinan, D. (1999) *Ever Closer Union: An Introduction to European Integration* (second edition), London: Macmillan

Distributive Trades Economic Development Committee (1988) *The Future of the High Street*, London: National Economic Development Council

District Planning Officers' Society (DPOS) (1992) *Affordable Housing*, Romsey: DPOS

Doak, J. (1999) 'Planning for the reuse of redundant defence estate: disposal processes, policy frameworks and development impacts', *Planning Practice and Research* 14: 211–24

Dobry, G. (chair) (1974a) *Control of Demolition*, London: HMSO

Dobry, G. (chair) (1974b) *Review of the Development Control System: Interim Report*, London: HMSO

Dobry, G. (chair) (1975) *Review of the Development Control System: Final Report*, Dobry Report, London: HMSO

Dobry, G., Hart, G., Robinson, P. and Williams, A. (1996) *Blundell and Dobry: Planning Applications, Appeals and Proceedings*, London: Sweet and Maxwell

Dobson, A. (1990) *Green Political Thought: An Introduction*, London: HarperCollins

Dobson, A. (1991) *A Green Reader*, London: Andre Deutsch

Dobson, A. (1995) 'No environmentalism without democratisation', *Town and Country Planning* 64: 322–3

Docklands Consultative Committee (DCC) (1991) *Ten Years of Docklands: How the Cake Was Cut*, London: DCC

Docklands Consultative Committee (1992) *All That Glitters is Not Gold: A Critical Assessment of Canary Wharf*, London: DCC

Dodd, A. M. and Pritchard, D. E. (1993) *RSPB Planscan Northern Ireland: A Study of Development Plans in Northern Ireland*, Sandy, Beds: Royal Society for the Protection of Birds

Doig, A. (1984) *Corruption and Misconduct in Contemporary British Politics*, Harmondsworth: Penguin

Doig, A. and Ridley, F. F. (1995) *Sleaze, Politicians, Private Interests and Public Reaction*, Oxford: Oxford University Press

Doling, J. (1997) *Comparative Housing Policy: Government and Housing in Advanced Industrialized Countries*, London: Macmillan

Donnison, D. and Middleton, A. (eds) (1987) *Regenerating the Inner City: Glasgow's Experience*, London: Routledge

Doogan, K. (1996) 'Labour mobility and the changing housing market', *Urban Studies* 33: 199–221

Dorling, D. and Atkins, D. (1995) *Population Density, Change and Concentration in Great Britain 1971, 1981, and 1991*, Studies on Medical and Population Subjects 58, Office of Population Censuses and Surveys, London: HMSO

Doucet, P. (2002) 'Transnational planning in the wake of the ESDP', in Faludi, A. (ed.) *European Spatial Planning*, Cambridge, MA: Lincoln Institute of Land Policy

Douglas, A. (1997) *Local Authority Sites for Travellers* (Central Research Unit, Scottish Office) Edinburgh: The Stationery Office

Douglass, M. and Friedmann, J. (eds) (1998) *Cities for Citizens: Planning and the Rise of Civil Society in a Global Age*, Chichester: Wiley

Dower, J. (chair) (1945) *National Parks in England and Wales*, Cmd 6628, Dower Report, London: HMSO

Dowling, J. A. (1995) *Northern Ireland Planning Law*, Dublin: Gill and Macmillan

Downey, J. and McGuigan, J. (1999) *Technocities*, London: Sage

Downs, A. (1992) *Stuck in Traffic: Coping with Peak-Hour Traffic Congestion*, Washington, DC: Brookings Institution and Lincoln Institute of Land Policy

Doxford, D. and Hill, T. (1998) 'Land use for military training in the UK: the current situation, likely developments and possible alternatives', *Journal of Environmental Planning and Management* 41: 279–97

Doyle, T. and McEachern (1998) *Environment and Politics*, London: Routledge

Draper, P. (1977) *Creation of the DoE: A Study of the Merger of Three Departments to Form the DoE*, Civil Service Studies 4, London: HMSO

Drevet, J-F. (2002) 'The European Union and its frontiers: towards new cooperation areas for spatial planning', in Faludi, A. (ed.) *European Spatial Planning*, Cambridge, MA: Lincoln Institute of Land Policy

Drewry, G. and Butcher, T. (1991) *The Civil Service Today* (second edition), Oxford: Blackwell

Drivers Jonas (1992) *Retail Impact Assessment Methodologies*, Edinburgh: Central Research Unit, Scottish Office

Drury, P. (1995) 'The value of conservation', *Conservation Bulletin* (English Heritage), July: 20

Dubben, N. and Sayce, S. (1991) *Property Portfolio Management*, London: Routledge

Duffy, K. and Hutchinson, J. (1997) 'Urban policy and the turn to community', *Town Planning Review* 68: 347–62

Dühr, S. (2002) *The Influence of EU Policy on Land Use Planning in Britain*, Faculty of the Built Environment Working Paper, Bristol: University of the West of England

Dühr, S., Farthing, S. and Nadin, V. (2005) 'Taking forward the spatial visions', *Town and Country Planning* March: 97–9

Dundee School of Town and Regional Planning, University of Dundee (1997) *Public Access to Information*, Edinburgh: Scottish Office

Dungey, J. and Newman, I. (eds) (1999) *The New Regional Agenda*, London: Local Government Information Unit

Dunleavy, P. (1995) 'Policy disasters: explaining the UK's record', *Public Policy and Administration* 10(2): 52–70

Dunlop, J. (1976) 'The examination-in-public of structure plans: an emerging procedure', *Journal of Planning and Environment Law* 1976: 8–17

Dunmore, K. (1992) *Planning for Affordable Housing*, London: Institute of Housing

Dunn, J. A. (1998) *Driving Forces: The Automobile, its Enemies and the Politics of Mobility*, Washington, DC: Brookings Institution Press

Durman, M. and Harrison, M. (1996) *Bournville 1895–1914*, Birmingham: Article Press, Department of Art, University of Central England

Durrant, K. (2000) 'Making design decisions', *Planning Inspectorate Journal* 18: 7–10

Duxbury, R. (1999) *Telling and Duxbury's Planning Law and Procedure*, London: Butterworths

Duxbury, R. (2002) *Telling and Duxbury's Planning Law and Procedure* (twelfth edition), London: LexisNexis

Dwyer, J. and Hodge, I. (1996) *Countryside in Trust: Land*

Management by Conservation, Recreation and Organisation, Chichester: Wiley

Dynes, M. and Walker, D. (1995) *The Times Guide to the New British State: The Government Machine in the 1990s*, London: Times Books

Earp, J. H., Headicar, P., Banister, D. and Curtis, C. (1995) *Reducing the Need to Travel: Some Thoughts of PPG 13*, Oxford: School of Planning, Oxford Brookes University

Easteal, M. (1995) 'A thoroughly modern review', *Local Government Chronicle*, 22 September: 14–15

ECA and Scott Wilson (2002) *Land Use: Exploring the Scope for Action at the EU Level: Final Report* (European Commission Markets Team), Brussels: CEC DG Environment

Economist (1999) 'English devolution: regional awakening?', *The Economist*, 30 January: 34

ECOTEC and Faber Maunsell (2004) *Surface Infrastructure of National Economic Importance: A Study for England's Regional Development Agencies*, Birmingham: Faber Maunsell

ECOTEC and Roger Tym and Partners (2004) *Planning for Economic Development*, London: ODPM

ECOTEC Research and Consulting (1990) *Dynamics of the Rural Economy*, London: DoE

ECOTEC Research and Consulting (1993a) *Review of UK Environmental Expenditure: A Final Report to the DoE*, London: HMSO

ECOTEC Research and Consulting (1993b) *Reducing Transport Emissions through Planning*, London: HMSO

ECOTEC (1999) *Scoping Study of RPG Targets and Indicators*, London: ODPM

Edinburgh College of Art (School of Planning and Housing) and Peter P. C. Allan (Chartered Town Planning Consultants) (1995) *Review of Neighbour Notification*, Edinburgh: Central Research Unit, Scottish Office

Edinburgh College of Art and Brodies WS and Halliday Fraser Munro Planning (1997) *Research on the General Permitted Development Order and Related Mechanisms*, Edinburgh: The Stationery Office

Edmundson, T. (1993) 'Public participation in development control', *Town and Country Planning Summer School 1993 Proceedings*, London: Royal Town Planning Institute

Edmundson, T. (2004) *Recruitment and Retention of Planners: Towards Addressing the Need for Planners in London*, London: ALG, RTPI and ALBPO

Edwards, G. and Spence, D. (1997) *The European Commission* (second edition), London: Cartermill International

Edwards, J. (1987) *Positive Discrimination, Social Justice and Social Policy: Moral Scrutiny of a Policy Practice*, London: Tavistock

Edwards, J. (1990) 'What is needed from public policy?', in Healey, P. and Nabarro, R. (eds) *Land and Property Development in a Changing Society*, Aldershot: Gower

Edwards, J. (1997) 'Urban policy: the victory of form over substance', *Urban Studies* 34: 825–43

Edwards, J. and Batley, R. (1978) *The Politics of Positive Discrimination*, London: Tavistock

Edwards, J. and Deakin, N. (1992) 'Privatism and partnership in urban regeneration', *Public Administration* 70: 359–68

Edwards, R. (chair) (1991) *Fit for the Future: Report of the National Parks Review Panel*, Edwards Report, Cheltenham: Countryside Commission

Ekins, P. (1986) *The Living Economy: A New Economics in the Making*, London: Routledge

Elcock, H. (1994) *Local Government*, London: Routledge

Elkin, S. H. (1974) *Politics and Land Use Planning: The London Experience*, Cambridge: Cambridge University Press

Elkin, T., McLaren, D. and Hillman, M. (1991) *Reviving the City: Towards Sustainable Development*, London: Policy Studies Institute

Ellis, G. (2002) 'Third party rights of appeal in planning: reflecting on the experience of the Republic of Ireland', *Town Planning Review* 73(4): 437–66

Elsom, D. (1996) *Smog Alert: Managing Urban Air Quality*, London: Earthscan

Elson, M. J. (1986) *Green Belts: Conflict Mediation in the Urban Fringe*, London: Heinemann

Elson, M. J. (1990) *Negotiating the Future: Planning Gain in the 1990s*, Gloucester: ARC

Elson, M. J. (1991) *Green Belts for Wales: A Positive Role for Sport and Recreation*, Cardiff: Sports Council for Wales

Elson, M. J. and Ford, A. (1994) 'Green belts and very special circumstances', *Journal of Planning and Environment Law* 1994: 594–601

Elson, M. and MacDonald, R. (1997) 'Urban growth management: distinctive solutions in the Celtic countries?', in MacDonald, R. and Thomas, H. (eds) *Nationality and Planning in Scotland and Wales*, Cardiff: University of Wales Press

Elson, M. J., Walker, S. and MacDonald, R. (1993) *The Effectiveness of Green Belts* (DoE), London: HMSO

Elson, M. J., MacDonald, R., Steenberg, C. and Broom, G. (1995a) *Planning for Rural Diversification* (DoE), London: HMSO

Elson, M. J., Steenberg, C. and Wilkinson, J. (1995b) *Planning for Rural Diversification: A Good Practice Guide* (DoE), London: HMSO

Elson, M. J., Steenberg, C. and Mendham, N. (1996) *Green Belts and Affordable Housing: Can We Have Both?*, Bristol: Policy Press

Elvin, D. and Robinson, J. (2000) 'Environmental impact assessment', *Journal of Planning and Environment Law* September: 876–93

Emerson, M. (1998) *Redrawing the Map of Europe*, London: Macmillan

Emms, J. E. (1980) 'The Community Land Act: a requiem', *Journal of Planning and Environment Law* 1980: 78–86

Empty Houses Agency (1998) *Joined Up Thinking: A Directory of Good Practice for Local Authority Empty Property Strategies*, London: Empty Houses Agency

English Heritage (2002) *Paradise Preserved*, London: English Heritage

English Historic Towns Forum (1992) *Townscape in Trouble: Conservation Areas – The Case for Change*, Bath: The Forum

English Nature: *see* appendix on *Official Publications*

English Partnerships (2000) *Urban Design Compendium*, London: English Partnerships

English Partnerships (2003) *Towards a National Brownfield Strategy*, London: English Partnerships

English Partnerships (2004) *Urban Regeneration Companies: Coming of Age*, Milton Keynes: English Partnerships

English Regional Associations (ERA) (1998) *Regional Governance: Statement of General Principles*, Taunton: ERA

English Tourist Board (ETB) (1991) *Tourism and the Environment: Maintaining the Balance* (prepared in conjunction with the Employment Department Group), London: ETB

English Tourist Board (1995) *English Heritage Monitor 1995* (published annually), London: ETB

English Tourist Board and Civic Trust (1993) *Turning the Tide: A Heritage and Environment Strategy for a Seaside Resort*, London: ETB

Ennis, F. (1994) 'Planning obligations in development plans', *Land Use Policy* 11: 195–207

Ennis, F. (1997) 'Infrastructure provision, the negotiating process and the planner's role', *Urban Studies* 34: 1935–54

Entec (2002) *Delivering Affordable Housing through Planning Policy*, London: The Stationery Office

Entec (2003) *The Relationships between Community Strategies and Local Development Frameworks*, London: ODPM

Entec (2004) *Study into the Environmental Impacts of Increasing the Supply of Housing in the UK: Final Report*, London: DEFRA

Environ (1994) *Parking Provision in City Centres: A Lot to be Desired?*, Leicester: Environ

Environ (1995) *Indicators of Sustainable Development in Leicester*, Leicester: Environ

Environment Agency: *see* appendix on *Official Publications* (under National Rivers Authority)

Environmental Resources (1992) *Economic Instruments and Recovery of Resources from Waste* (DoE), London: HMSO

Environmental Resources Management (2003) *Improving the Delivery of Affordable Housing in London and the South East*, London: ODPM

Erikson, R. A. and Syms, P. M. (1986) 'The effects of enterprise zones on local property markets', *Regional Studies* 20: 1–4

Esher, L. (1981) *A Broken Wave: The Rebuilding of England 1940–1980*, London: Allen Lane (Penguin edition 1983)

Essex, S. (1996) 'Members and officers in the planning policy process', in Tewdwr-Jones, M. (ed.) *British Planning Policy in Transition: Planning in the 1990s*, London: UCL Press

Essex County Council (1973) *Design Guide for Residential Areas*, Chelmsford: The Council

Etzioni, A. (1967) 'Mixed-scanning: a "third" approach to decision-making', *Public Administration Review* 27(5): 385–92 (reprinted in Faludi, A. (ed.) (1973) *A Reader in Planning Theory*, Oxford: Pergamon)

European Conference of Ministers of Transport (1995) *Urban Travel and Sustainable Development*, Paris: OECD

European Environment Agency (EEA) (2002) *Towards an Urban Atlas: Assessment of Spatial Data on 25 European Cities and Urban Areas*, Environmental Issue Report no. 30, Copenhagen: EEA

European Environment Agency (2003) *Europe's Environment: The Third Assessment*, Environmental Assessment Report no. 10, Copenhagen: EEA

European Environmental Bureau (EEB) (1994) *Your Rights under European Union Environment Legislation*, Brussels: EEB

European Ombudsman (1999) *The European Ombudsman Annual Report for 1998*, Luxembourg: OOPEC

European Spatial Planning and Information Network (ESPRIN) (2000) *Urban and Rural Relations: Final Report to the DETR*, Newcastle upon Tyne: School of Architecture and Planning, University of Newcastle upon Tyne

European Spatial Planning Observation Network (2004a) *ESPON in Progress*, Luxembourg, ESPON

European Spatial Planning Observation Network (2004b) *ESPON Briefing 1: Diversity within the European Territory; A Selection of New European Maps*, Luxembourg, ESPON

Evans, A. W. (1973) *The Economics of Residential Location*, London: Macmillan

Evans, A. W. (1983) 'The determination of the price of land', *Urban Studies* 20: 119–29

Evans, A. W. (1988) *No Room! No Room! The Costs of the British Town and Country Planning System*, London: Institute of Economic Affairs

Evans, A. W. (1991) 'Rabbit hutches on postage stamps: planning, development and political economy', *Urban Studies* 28: 853–70

Evans, A. W. (1995) 'The property market: ninety per cent efficient?', *Urban Studies* 32: 5–29

Evans, A. W. (1996) 'The impact of land use planning and tax subsidies on the supply and price of housing in Britain: a comment', *Urban Studies* 33: 581–5

Evans, A. W. and Eversley, D. (eds) (1980) *The Inner City: Employment and Industry*, London: Heinemann

Evans, B. (1993) 'Why we no longer need a town planning profession', *Planning Practice and Research* 8: 9–15

Evans, B. (1995) *Experts and Environmental Planning*, Aldershot: Avebury

Evans, B., Joas, M., Sundback, S. and Theobald, K. (2004) *Governing Sustainable Cities*, London: Earthscan

Evans, B. and Rydin, Y. (1997) 'Planning, professionalism and sustainability', in Blowers, A. and Evans, B. (eds) (1997)

Evans, C. (1998) *Derelict Land and Brownfield Regeneration*, Oxford: Chandos

Evans, D. (1992) *A History of Nature Conservation*, London: Routledge

Evans, R. (1997) *Regenerating Town Centres*, Manchester: Manchester University Press

Eve, G. and Department of Land Economy, University of Cambridge (1992) *The Relationship between House Prices and Land Supply* (DoE), London: HMSO

Eve, G.: Gerald Eve Research (1995) *Whither the High Street?*, London: Gerald Eve Research

Eve, Grimley J. R. (1992) *Use of Planning Agreements*, London: HMSO

Everest, D. A. (1990) 'The provision of expert advice to government on environmental matters: the role of advisory committees', *Science and Public Affairs* 4: 17–40

Eversley, D. E. C. (1973) *The Planner in Society*, London: Faber

Eversley, Lord (1910) *Commons, Forests and Footpaths: The Story of the Battle during the Past Forty-five Years for Public Rights over the Commons, Forests and Footpaths of England and Wales*, London: Cassell

Fainstein, S. S. (1994) *The City Builders: Property, Politics, and Planning in London and New York*, Oxford: Blackwell

Fainstein, S. S. and Campbell, S. (eds) (1996) *Readings in Urban Theory*, Oxford: Blackwell

Fainstein, S. S., Gordon, I. and Harloe, M. (eds) (1992) *Divided Cities: New York and London in the Contemporary World*, Oxford: Blackwell

Faludi, A. (ed.) (1973) *A Reader in Planning Theory*, Oxford: Pergamon

Faludi, A. (1987) *A Decision-Centred View of Environmental Planning*, Oxford: Pergamon

Faludi, A. (2000) 'The European Spatial Development Perspective: what next?', *European Planning Studies* 8(2): 237–50

Faludi, A. (ed.) (2002) *European Spatial Planning*, Cambridge, MA: Lincoln Institute of Land Policy

Faludi, A. (2003) 'Unfinished business: European spatial planning in the 2000s', *Town Planning Review* 74(1): 121–40

Faludi, A. and Waterhout, B. (2002) *The Making of the European Spatial Development Perspective: No Masterplan*, London: Routledge

Farmer, A., Skinner, I., Wilkinson, D. and Bishop, K. (1999) *Environmental Planning in The United Kingdom: A Background Paper For The Royal Commission on Environmental Pollution*, London: RCEP

Farnham, D. and Horton, S. (1992) 'Human resource management in the new public sector: leading or following private employer practice', *Public Policy and Administration* 7: 42–55

Farrington, J. (1995) 'Military land in Britain after the cold war', *Geography* 80: 272–89

Farrington, J., Gray, D., Martin, S. and Roberts, D. (1998) *Car Dependence in Rural Scotland* (Central Research Unit, Scottish Office), Edinburgh: The Stationery Office

Farthing, S. (1996) *Evaluating Local Environmental Policy*, London: Avebury

Farthing, S. and Ashley, K. (2002) 'Negotiations and the delivery of affordable housing through the English planning system', *Planning Practice and Research* 17(1): 45–58

Feldman, P. (1999) *Closing Doors: Examining the Lack of Housing Choices in London*, London: National Housing Federation

Fergusson, A. (1973) *The Sack of Bath: A Record and an Indictment*, Salisbury: Compton Russell

Ferris, J. (1972) *Participation in Urban Planning: The Barnsbury Case – A Study of Environmental Improvement in London*, London: Bell

Field, B. and MacGregor, B. (1987) *Forecasting Techniques for Urban and Regional Planning*, London: Hutchinson

Fielden, G. B. R., Wickens, A. H. and Yates, I. R. (1994) *Passenger Transport after 2000 AD*, London: Spon (for the Royal Society)

Fielder, S. (1986) *Monitoring the Operation of the Planning System: The National Dimension*, Reading: Department of Land Management and Development, University of Reading

Fielding, T. and Halford, S. (1990) *Patterns and Processes of Urban Change in the United Kingdom* (DoE Reviews of Urban Research), London: HMSO

Firn, J. and McGlashan, D. (2001) *A Coastal Management Trust for Scotland: A Concept Development and Feasibility Study*, Edinburgh: Scottish Executive

Fischer, F. and Black, M. (1995) *Greening Environmental Policy: The Politics of a Sustainable Future*, London: Paul Chapman

Fischer, F. and Forester, J. (eds) (1993) *The Argumentative Turn in Policy Analysis and Planning*, London: UCL Press

Fit, J. and Kragt, R. (1994) 'The long road to European spatial planning: a matter of patience and mission', *Tijdschrift voor Economische en Sociale Geografie* 85: 461–5

Fitzsimmons, D. (1990) 'Training for planning', *The Planner*, 76(14): 29–30

Fitzsimmons, D. S. M. (1995) 'Removal of agricultural occupancy conditions: the Northern Ireland controversy', *Journal of Planning and Environment Law* 1995: 670–8

Flowers, Lord (chair) (1981) *Coal and the Environment*, Flowers Report, London: HMSO

Flynn, A. and Marsden, T. K. (1995) 'Rural change, regulation, and sustainability', *Environment and Planning A* 27: 1180–92

Flynn, A., Leach, S. and Vielba, C. (1985) *Abolition or Reform? The GLC and the Metropolitan County Councils*, London: Allen and Unwin

Foley, P. (1992) 'Local economic policy and job creation: a review of evaluation studies', *Urban Studies* 29: 557–98

Foley, P., Hutchinson, J. and Fordham, G. (1998) 'Managing the challenge: winning and implementing the Single Regeneration Budget Fund', *Planning Practice and Research* 13: 63–80

Fontaine, P. (1995) *Europe in Ten Lessons* (second edition), Luxembourg: Office for the Official Publications of the European Communities

Fordham, G. (1995) *Made to Last: Creating Sustainable Neighbourhood and Estate Regeneration*, York: Joseph Rowntree Foundation

Fordham, G., Hutchinson, J. and Foley, P. (1998) 'Strategic approaches to local regeneration: the Single Regeneration Budget Challenge Fund', *Regional Studies* 33(2): 131–41

Fordham, R. (1989) 'Planning gain: towards its codification', *Journal of Planning and Environment Law* 1989: 577–84

Fordham, R. (1990) 'Planning consultancy: can it serve the public interest?', *Public Administration* 68: 243–8

Forester, J. (1980) 'Critical theory and planning practice', *Journal of the American Planning Association* 46: 275–86

Forester, J. (1982) 'Planning in the face of power', *Journal of the American Planning Association* 48: 67–80

Forester, J. (1989) *Planning in the Face of Power*, Berkeley, CA: University of California Press

Forester, J. (1999) *The Deliberative Practitioner: Encouraging Participatory Planning Processes*, Cambridge, MA: MIT Press

Forman, C. (1989) *Spitalfields: A Battle for Land*, London: Hilary Shipman

Forsyth, J. (1992) 'Tower Hamlets: setting up a regeneration corporation in Bethnal Green', in Bailey, N. and Barker, A. (eds) *City Challenge and Local Regeneration Partnerships: Conference Proceedings*, London: Polytechnic of Central London

Fothergill, S. and Gudgin, G. (1982) *Unequal Growth: Urban and Regional Change in the UK*, London: Heinemann

Fothergill, S. and Guy, N. (1990) *Retreat from the Regions: Corporate Change and the Closure of Factories*, London: Jessica Kingsley and Regional Studies Association

Fothergill, S., Kitson, M. and Monk, S. (1985) *Urban Industrial Change: The Causes of the Urban–Rural Contrast in Manufacturing Employment Change* (DoE), London: HMSO

Fothergill, S., Kitson, M. and Perry, M. (1987) *Property and Industrial Development*, London: Hutchinson

Fox, M. and Turner, S. (1994) 'Northern Ireland', in Freshfields Environment Group (eds) *Tolley's Environmental Handbook*, Croydon: Tolley

FPD Savills Research (2001) *The Value of Housing Design and Layout: A Research Report*, London: ODPM

Franks, O. S. (chair) (1957) *Report of the Committee on Administrative Tribunals and Enquiries*, Cmnd 218, Franks Report, London: HMSO

Freestone, D. (1991) 'European Community environmental policy and law', in Churchill, R., Warren, L. M. and Gibson, J. (eds) *Law, Policy and the Environment*, Oxford: Blackwell

Freestone, R. (2000) *Urban Planning in a Changing World: The Twentieth Century Experience*, London: Spon

Freilich, R. H. (1999) 'Final report for compulsory purchase: an appropriate power for the 21st century', *Journal of Planning and Environment Law* December: 1076–86

Freilich, R. H. and Bushek, D. W. (eds) (1995) *Exactions, Impact Fees and Dedications*, Chicago, IL: American Bar Association

Frey, H. (1999) *Designing the City: Towards a More Sustainable Urban Form*, London: Routledge

Fried, M. (1969) 'Social differences in mental health', in Kosa, J., Antonovsky, A. and Zola, I. K. (eds) *Poverty and Health: A Sociological Analysis*, Cambridge, MA: Harvard University Press

Friends of the Earth (FoE) (1989) *Action for People: A Critical Appraisal of Government Inner City Policy*, London: FoE

Friends of the Earth (1991) *Local Responses to 1989 Traffic Forecasts*, London: FoE

Friends of the Earth (1992) *Less Traffic, Better Towns*, London: FoE

Friends of the Earth (1994a) *Planning for the Planet: Sustainable Development Strategies for Local and Strategic Plans*, London: FoE

Friends of the Earth (1994b) *Working Future: Jobs and the Environment*, London: FoE

Friends of the Earth (1997) *Stopping the Sprawl: The Housing Campaign Guide*, London: FoE

Frost, M. and Spence, N. (1993) 'Global city characteristics and Central London's employment', *Urban Studies* 30: 547–58

Froud, J. (1994) 'The impact of ESAs on lowland farming', *Land Use Policy* 11: 107–18

Fry, G. K. (1984) 'The attack on the civil service and the response of the insiders', *Parliamentary Affairs* 37: 353–63

Fudge, C., Lambert, C., and Underwood, J. (1982) 'Local plans: approaches, preparation and adoption', *The Planner* 68 (March–April): 52–3

Fudge, C., Lambert, C., Underwood, J. and Healey, P. (1983)

Speed, Economy and Effectiveness in Local Plan Preparation and Adoption, Bristol: School of Advanced Urban Studies, University of Bristol

Fulford, C. (1998) *The Costs of Reclaiming Derelict Sites*, London: Town and Country Planning Association

Fuller Peiser and Reading University (1999) *Development of the Redundant Defence Estate*, London: Thomas Telford

Fulton, Lord (chair) (1968) *Report of the Committee on the Civil Service*, Cmnd 3638, Fulton Report, London: HMSO

Funck, B. and Pizzati, L. (eds) (2003) *European Integration, Regional Policy and Growth*, London: The World Bank

Fyson, A. (1999a) 'Iron out defence land policy to get the full benefits', *Planning* 30 July: 13

Fyson, A. (1999b) 'Inner cities will not hold growing housing numbers', *Planning* 28 May: 15.

Gahagan, M. (1992) 'City challenge: a solution to regeneration through partnership?', in Bailey, N. and Barker, A. (eds) *City Challenge and Local Regeneration Partnerships: Conference Proceedings*, London: Polytechnic of Central London

Gallent, N. (2000) 'Planning and affordable housing: from old values to New Labour', *Town Planning Review* 71(2): 123–47

Gallent, N. and Bell, P. (2000) 'Planning exceptions in rural England: past, present and future', *Planning Practice and Research* 15(4): 375–84

Gallent, N., Mace, A. and Tewdwr-Jones, M. (2002) 'Delivering affordable housing through planning: explaining variable policy usage across rural England and Wales', *Planning Practice and Research* 17(1): 465–83

Gallent, N., Mace, A. and Tewdwr-Jones, M. (2004) 'Second homes: a new framework for policy', *Town Planning Review* 75(3): 287–308

Gans, H. J. (1968) *People and Plans: Essays on Urban Problems and Solutions*, New York: Basic Books

Gans, H. J. (1991) *People, Plans and Policies: Essays on Poverty, Racism and Other National Urban Problems*, New York: Columbia University Press

Gardner, B. (1996) *Farming for the Future: Policies, Production and Trade*, London: Routledge

Garmise, S. (1997) 'The impact of European regional policy on the development of the regional tier in the UK', *Regional and Federal Studies* 7: 1–24

Garside, P. L. and Hebbert, M. (eds) (1989) *British Regionalism 1900–2000*, London: Mansell

Gatenby, I. and Williams, C. (1992) 'Section 54A: the legal and practical implications', *Journal of Planning and Environment Law* 1992: 110–20

Gatenby, I. and Williams, C. (1996) 'Interpreting planning law', in Tewdwr-Jones, M. (ed.) *British Planning Policy in Transition*, London: UCL Press

Geddes, M. (1995) *Poverty, Excluded Communities and Local Democracy*, London: Commission for Local Democracy

Geddes, P. (1915) *Cities in Evolution*, London: Benn

Gentleman, H. (1993) *Counting Travellers in Scotland*, Edinburgh: Scottish Office

Gentleman, H. and Swift, S. (1971) *Scotland's Travelling People*, Edinburgh: HMSO

Geraghty, P. J. (1992) 'Environmental assessment and the application of an expert systems approach', *Town Planning Review* 63: 123–42

Gestel, T. van and Faludi, A. (2005) 'Towards a European territorial cohesion assessment network: a bright future for ESPON?' *Town Planning Review*, 76(1): 81–92

Gibbs, D. (1994) 'Towards the sustainable city', *Town Planning Review* 65: 99–109

Gibbs, D., Longhurst, J. and Braithwaite, C. (1996) 'Moving towards sustainable development? Integrating economic development and the environment in local authorities', *Journal of Environmental Planning and Management* 39: 317–32

Gibson, J. (1991) 'The integration of pollution control', in Churchill, R., Warren, L. M. and Gibson, J. (eds) *Law, Policy and the Environment*, Oxford: Blackwell

Gibson, M. S. and Langstaff, M. J. (1982) *An Introduction to Urban Renewal*, London: Hutchinson

Giddens, A. (1998) *The Third Way: The Renewal of Social Democracy*, Cambridge: Polity Press

Gilbert, A. and Healey, P. (1985) *The Political Economy of Land: Urban Development in an Oil Economy*, Aldershot: Gower

Gilg, A. W. (ed.) (1983) *Countryside Planning Yearbook 1983*, Norwich: Geo Books

Gilg, A. W. (ed.) (1988) *The International Year Book of Rural Planning*, London: Elsevier

Gilg, A. W. (ed.) (1991) *Progress in Rural Policy and Planning, Volume 1: 1991*, London: Belhaven

Gilg, A. W. (ed.) (1992a) *Progress in Rural Policy and Planning, Volume 2: 1992*, London: Belhaven

Gilg, A. W. (ed.) (1992b) *Restructuring the Countryside: Environmental Policy in Practice*, Aldershot: Avebury

Gilg, A. W. (ed.) (1994) *Progress in Rural Policy and Planning, Volume 4: 1994*, Chichester: Wiley

Gilg, A. W. (ed.) (1995) *Progress in Rural Policy and Planning, Volume 5: 1995*, Chichester: Wiley

Gilg, A. W. (1996) *Countryside Planning: The First Half Century* (second edition) London: Routledge

Gilg, A. W. (1997) *Rural Planning in Practice: The Case of Agricultural Dwellings*, Oxford: Pergamon

Gilg, A. W. (1999) *Perspectives on British Rural Planning Policy, 1994–97*, Aldershot: Ashgate

Gilg, A. W. (2005) *Planning in Britain: Understanding and Evaluating the Post-War System*, London: Sage

Gillett, E. (1983) *Investment in the Environment: Planning and Transport Policies in Scotland*, Aberdeen: Aberdeen University Press

Gillingwater, D. (1992) 'Regional strategy for the East Midlands', in Minay, C. L. W. (ed.) 'Developing regional guidance in England and Wales: a review symposium', *Town Planning Review* 63: 415–34

Gilpin, A. (1995) *Environmental Impact Assessment: Cutting Edge for the Twenty-first Century*, Cambridge: Cambridge University Press

Gilpin, A. (1996) *Dictionary of Environmental and Sustainable Development*, Chichester: Wiley

Gilroy, R. (1999) 'Towards a gentle city?', *Town and Country Planning* 68(11): 334–5

Gilroy, R. with Marvin, S. (1993) *Good Practices in Equal Opportunities*, Aldershot: Avebury

Ginsburg, L. (1956) 'Green belts in the Bible', *Journal of the Town Planning Institute* 42: 129–30

Glaister, S. and Mulley, C. (1983) *Public Control of the British Bus Industry*, Aldershot: Gower

Glaister, S., Burham, J., Stevens, H. and Travis, T. (1998) *Transport Policy in Britain*, London: Macmillan

Glass, R. (1959) 'The evaluation of planning: some sociological considerations', in Faludi, A. (ed.) (1973) *A Reader in Planning Theory*, Oxford: Pergamon

Glass, W. D. (1994) 'Regional rural planning policies', *Town and Country Planning Summer School 1994 Proceedings*: 21–4

Glasson, B. and Booth, P. (1992) 'Negotiation and delay in the development control process', *Town Planning Review* 63: 63–78

Glasson, J. (1992) 'The fall and rise of regional planning in the economically advanced nations', *Urban Studies* 29: 505–31

Glasson, J. (1999) 'The first 10 years of the UK EIA system: strengths, weaknesses, opportunities and threats', *Planning Practice and Research* 14: 363–75

Glasson, J. (2000) *An Introduction to Regional Planning* (second edition), London: UCL Press

Glasson, J., Godfrey, K. and Goodey, B. with Absalom, H. and Borg, J. (1995) *Toward Visitor Impact Management: Visitor Impacts, Carrying Capacity and Management Responses in Europe's Historic Towns and Cities*, Aldershot: Avebury

Glasson, J., Therivel, T. and Chadwick, A. (1998) *Introduction to Environmental Impact Assessment* (second edition), London: UCL Press

GMA Planning, P-E International and Jacques & Lewis (1993) *Integrated Planning and Granting of Permits in the EC* (DoE Planning Research Programme), London: HMSO

Goldmann, K. (2001) *Transforming the European Nation-State*, London: Sage

Goldsmith, F. B. and Warren, S. A. (eds) (1993) *Conservation in Progress*, Chichester: Wiley

Goldsmith, M. (1980) *Politics, Planning and the City*, London: Hutchinson

Goldsmith, M. and Newton, K. (1986) 'Central–local government relations: a bibliographic summary of the ESRC research initiative', *Public Administration* 64: 102–8

Gomez-Ibanez, J. A. and Meyer, J. R. (1993) *Going Private: The International Experience with Transport Privatization*, Washington, DC: Brookings Institution

Goodchild, B. (1997) *Housing and the Urban Environment: A Guide to Housing Design, Renewal and Urban Planning*, Oxford: Blackwell

Goodchild, R. N. and Munton, R. J. C. (1985) *Development and the Landowner: An Analysis of the British Experience*, London: Allen and Unwin

Goodin, R. E. (1992) *Green Political Theory*, Oxford: Polity Press

Goodlad, R., Flint, J., Kearns, A., Keoghan, M., Paddison, R. and Raco, M. (1999) *The Role and Effectiveness of Community Councils with Regard to Community Consultation*, Edinburgh: Central Research Unit, Scottish Office

Goodman, R. (1972) *After the Planners*, Harmondsworth: Penguin

Goodstadt, V. (2003) *President's Report 2003*, London: Royal Town Planning Institute

Goodwin, P. B. (1989) 'The *rule of three*: a possible solution to the political problem of competing objectives for road pricing', *Traffic Engineering and Control* 30: 495–7

Goodwin, P. B. (1992) 'A review of new demand elasticities with special reference to short and long run effects of price changes', *Journal of Transport Economics and Policy* 26: 155–69

Goodwin, P. B. (1995) *The End of Hierarchy? A New Perspective on Managing the Road Network*, London: Council for the Protection of Rural England

Goodwin, P., Hallett, S., Kenny, F. and Stokes, G. (1991) *Transport: The New Realism*, Oxford: Transport Studies Unit, University of Oxford

Gordon, J. E. (ed.) (1994) *Scotland's Soils: Research Issues in Developing a Soil Sustainability Strategy*, Perth: Scottish Natural Heritage

Gore, T. and Nicholson, D. (1991) 'Models of the land development process: a critical review', *Environment and Planning A* 23: 705–30

Gosling, A. (chair) (1968) *Report of the Footpaths Committee*, Gosling Report, London: HMSO

Gourvish, T. (2002) *British Rail 1974–97*, Oxford: Oxford University Press

Gowers, E. (chair) (1950) *Houses of Outstanding Historic and Architectural Interest*, Gowers Report, London: HMSO

Graham, S. and Marvin, S. (1996) *Telecommunications and the City: Electronic Spaces, Urban Places*, London: Routledge

Graham, S. and Marvin, S. (1999) 'Planning cybercities? Integrating telecommunications into urban planning', *Town Planning Review* 70(1): 89–114

Graham, T. (1991) 'The interpretation of planning permissions and a matter of principle', *Journal of Planning and Environment Law* 1991: 104–12

Graham, T. (1993) 'Presumptions', *Journal of Planning and Environment Law* 1993: 423–8

Graham, T. (1996) 'Contaminated land investigations: how will they work under PPG 23?' *Journal of Planning and Environment Law* 1996: 547–53

Grant, J. S. (1987) 'Government agencies and the Highlands since 1945', *Scottish Geographical Magazine* 103: 95–9

Grant, M. (1979) 'Britain's Community Land Act: a post mortem', *Urban Law and Policy* 2: 359–73

Grant, M. (1982) *Urban Planning Law* (supplement 1990), London: Sweet and Maxwell

Grant, M. (1986) 'Planning and land taxation', *Journal of Planning and Environment Law* 1986: 92–106

Grant, M. (1992) 'Planning law and the British planning system', *Town Planning Review* 63: 3–12

Grant, M. (1996) *Permitted Development* (second edition), London: Sweet and Maxwell

Grant, M. (1999a) 'Compensation and betterment', in Cullingworth, J. B. (ed.) *British Planning: 50 Years of Urban and Regional Policy*, London: Athlone

Grant, M. (1999b) 'Planning as a learned profession', paper presented to the Royal Town Planning Institute Council, London, January

Grant, M. (2003) 'Planning gain: why it's such a no-brainer', paper presented to the Joint Planning Law Conference, Oxford, September

Grant, M. (ed.) (1997) *Encyclopedia of Planning Law and Practice* (6 vols), London: Sweet and Maxwell (looseleaf; updated regularly; a supplementary updating *Monthly Bulletin* is issued to subscribers)

Graves, G., Max, R. and Kitson, T. (1996) 'Inquiry procedure: another dose of reform?', *Journal of Planning and Environment Law* 1996: 99–106

Gray, C. (1994) *Government Beyond the Centre: Sub-National Politics in Britain*, London: Macmillan

Gray, D. and Begg, D. (2001) *Delivering Congestion Charging in the UK: What is Required for its Successful Introduction*, Aberdeen: Centre for Transport Policy

Gray, T. S. (1995) *UK Environmental Policy in the 1990s*, London: Macmillan

Grayling, T. and Glaister, S. (2000) *A New Fares Contract for London*, London: Institute for Public Policy Research

Grayson, L. (1990) *Green Belt, Green Fields and the Urban Fringe: The Pressure on Land in the 1980s – A Guide to Sources*, London: London Research Centre

Greater London Authority (GLA) (2001) *Interim Strategic Guidance on Tall Buildings, Strategic Views and the Skyline in London*, London: GLA

Greater London Council (1986) *Race and Planning Guidelines*, London: GLC

Greater London Council (GLC) (1996) *Changing Places*, London: GLC

Greed, C. (1991) *Surveying Sisters: Women in a Traditional Male Profession*, London: Routledge

Greed, C. (1993) *Introducing Town Planning*, Harlow: Longman

Greed, C. (1994) *Women and Planning: Creating Gendered Realities*, London: Routledge

Greed, C. (ed.) (1996a) *Implementing Town Planning: The Role of Town Planning in the Development Process*, Harlow: Longman

Greed, C. (ed.) (1996b) *Investigating Town Planning: Changing Perspectives and Agendas*, Harlow: Longman

Greed, C. (1996c) *Introducing Town Planning*, London: Addison Wesley Longman

Greed, C. (ed.) (1999) *Social Town Planning*, London: Routledge

Greed, C. and Roberts, M. (eds) (1998) *Introducing Urban Design: Interventions and Responses*, Harlow: Addison Wesley Longman

Green, A., Hasluck, C., Owen, D. and Winnett, C. (1993) *Local Unemployment Change in Britain: Leaps and Lags in the Response to National Economic Cycles*, London: Jessica Kingsley

Green, C. (1999) 'Conflicting signals over cluster', *Planning* 27 August: 12

Green, H., Thomas, M., Iles, N. and Down, D. (1996) *Housing in England 1994/95: A Report of the 1994/95 Survey of English Housing Carried Out by the Social Survey Division of ONS on Behalf of the DoE*, London: HMSO

Green, R. E. (ed.) (1991) *Enterprise Zones: New Directions in Economic Development*, Newbury Park, CA: Sage

Green Balance (1992) *Campaigners' Guide to Local Plans*, London: Council for the Protection of Rural England

Green Balance (1999) *Fair Examination? Testing the Legitimacy of Strategic Planning for Housing*, London: Council for the Protection of Rural England

Greenhalgh, L. and Worpole, K. (1995) *Park Life: Urban Parks and Social Renewal*, London: Comedia and Demos

Greenhalgh, L., Worpole, K. and Grove-White, R. (1996) *People, Parks and Cities: A Guide to Current Good Practice in Urban Parks* (DoE), London: HMSO

Greenhalgh, P. and Shaw, K. (2003) 'Regional development agencies and physical regeneration in England: can RDAs deliver the urban renaissance?', *Planning Practice and Research* 18(2–3): 161–78

Greensmith, C. and Haywood, R. (1999) *Rail Freight Growth and the Land Use Planning System*, Sheffield: Centre for Regional, Economic and Social Research, Sheffield Hallam University

Greer, A. (1996) *Rural Politics in Northern Ireland: Policy Networks and Agricultural Development since Partition*, Aldershot: Avebury

Greer, P. (1992) 'The next steps initiative: an examination of the agency framework documents', *Public Administration* 70: 89–98

Gregory, R. and Pearson, J. (1992) 'The parliamentary ombudsman after twenty-five years', *Public Administration* 70: 469–98

Gregory, S. (1998) *Transforming Local Services: Partnership in Action*, York: Joseph Rowntree Foundation

Greg, D. and Begg, D. (2001) 'Delivering Congestion Charging in the UK: What is Required for its Successful Introduction', Policy Paper Series No. 4, Robert Gordon University, Aberdeen: The Centre for Transport Policy

Grieco, M. and Jones, P. M. (1994) 'A change in the policy climate? Current European perspectives on road pricing', *Urban Studies* 31: 1517–32

Griffith, J. A. G. and Street, H. (1964) *A Casebook on Administrative Law*, London: Pitman

Griffiths, D. (1996) *Thatcherism and Territorial Politics: A Welsh Case Study*, Aldershot: Avebury

Griffiths, R. (1993) 'The politics of cultural policy in urban regeneration strategies', *Policy and Politics* 21: 39–46

Griffiths, R. (1995) 'Eurocities', *Planning Practice and Research* 10(2): 215–21

Grove, G. A. (1963) 'Planning and the appellant', *Journal of the Town Planning Institute* 49: 128–33

Groves, P. (1992) 'A chairman's view of a TEC', *Local Government Policy Making* 19: 9–17

Guy, C. (1994) *The Retail Development Process*, London: Routledge

Guy, C. (1998) 'High Street retailing in off-centre retail parks', *Town Planning Review* 69: 291–313

Guy, C. M. (1998) 'Alternative-use valuation, open A1 planning consent, and the development of retail parks', *Environment and Planning A*, 30(1): 37–48

Guy, S. and Marvin, S. (1995) *Planning for Water: Space, Time and the Social Organisation of Natural Resources*, Newcastle upon Tyne: Department of Town and Country Planning, University of Newcastle upon Tyne

GVA Grimley with Environmental Resources Management and University of Westminster (2004) *Reforming Planning Obligations: the Use of Standard Charges*, London: ODPM

Gwilliam, M., Bourne, C., Swain, C. and Prat, A. (1998) *Sustainable Renewal of Suburban Areas*, York: York Publishing Services

Gyford, J. (1990) 'The enabling authority: a third model', *Local Government Studies* 17(1): 1–4

Gyford, J. (1991) *Citizens, Consumers and Councils: Local Government and the Public*, London: Macmillan

Gyford, J., Leach, S. and Game, C. (1989) *The Changing Politics of Local Government*, London: Unwin Hyman

Haar, C. M. (1951) *Land Planning in a Free Society*, Cambridge, MA: Harvard University Press

Hackett, P. (1995) *Conservation and the Consumer: Measuring Environmental Concern*, London: Routledge

Hagman, D. G. and Misczynski, D. J. (1978) *Windfalls for Wipeouts: Land Value Capture and Compensation*, St Paul, MN: West

Hague, C. (1984) *The Development of Planning Thought: A Critical Perspective*, London: Hutchinson

Hague, C. (ed.) (1996) *Planning and Markets*, Aldershot: Avebury

Hague, C. and Jenkins, P. (2004) *Place, Identity, Participation and Planning*, London: Routledge

Haigh, N. (1990) *EEC Environmental Policy* (second edition), Harlow: Longman

Haigh, N. and Irwin, F. (eds) (1990) *Integrated Pollution Control in Europe and North America*, Bonn: Institute for European Environmental Policy; Washington, DC: Conservation Foundation

Hailsworth, A. G. (1988) *The Human Impact of Hypermarkets and Superstores*, Aldershot: Avebury.

Halcrow Fox & Associates and Birkbeck College, University of London (1986) *Investigating Population Change in Small to Medium-Sized Areas*, London: DoE

Halcrow Group (2004) *Unification of Consent Regimes*, London: ODPM

Hale, M. and Lachowicz, M. (1998) *The Environment, Employment and Sustainable Development*, London: Routledge

Hales, R. (2000) 'Land use development planning and the notion of sustainable development: exploring constraint and facilitation within the English planning system', *Journal of Environmental Planning and Management* 43(1): 99–121

Halkier, H., Danson, M. and Damburg, C. (eds) (1998) *Regional Development Agencies in Europe*, London: Jessica Kingsley

Hall, A. C. (1990) 'Generating urban design objectives for local areas: a methodology and case study application to Chelmsford, Essex', *Town Planning Review* 61: 287–309

Hall, A. C. (1996) *Design Control: Towards a New Approach*, Oxford: Heinemann

Hall, D. (1989) 'The case for new settlements', *Town and Country Planning* 58: 111–14

Hall, D. (1991) 'Regional strategies: groping for guidance', *Town and Country Planning* 80(5): 138–40

Hall, D. (1999) 'Practice & process: town expansion – constructive participation (a local visioning conference on the

Stevenage expansion)', *Town and Country Planning*, 68(5): 170–1

Hall, D., Hebbert, M. and Lusser, H. (1993) 'The planning background', in Blowers, A. (ed.) *Planning for a Sustainable Environment: A Report by the Town and Country Planning Association*, London: Earthscan

Hall, P. (1980) *Great Planning Disasters*, London: Weidenfeld and Nicolson

Hall, P. (1981a) *The Enterprise Zone Concept: British Origins, American Adaptations*, Berkeley, CA: Institute of Urban and Regional Development, University of California at Berkeley

Hall, P. (ed.) (1981b) *The Inner City in Context: The Final Report of the Social Science Research Council Inner Cities Working Party*, London: Heinemann (reprinted 1986 by Gower)

Hall, P. (1982) 'Enterprise zones: a justification', *International Journal of Urban and Regional Research* 6: 415–21

Hall, P. (1983) 'The Anglo-American connection: rival rationalities in planning theory and practice, 1955–1980', *Environment and Planning B: Planning and Design* 10: 41–6

Hall, P. (1989) *London 2001*, London: Unwin Hyman

Hall, P. (1991) 'The British enterprise zones', in Green, R. E. (ed.) *Enterprise Zones: New Directions in Economic Development*, Newbury Park, CA: Sage

Hall, P. (1993) 'Planning in the 1990s: an international agenda', *European Planning Studies* 1: 3–12

Hall, P. (1997a) 'Regeneration policies for peripheral housing estates: inward and outward looking approaches', *Urban Studies* 34: 873–90

Hall, P. (1997b) 'The view from London Centre', in Blowers, A. and Evans, B. (eds) *Town Planning into the 21st Century*, London: Routledge

Hall, P. (1999a) 'The regional dimension', in Cullingworth, J. B. (ed.) *British Planning: 50 Years of Urban and Regional Policy*, London: Athlone

Hall, P. (1999b) *Cities in Civilization: Culture, Innovation and Urban Order*, London: Weidenfeld and Nicolson

Hall, P. (1999c) *Sustainable Cities or Town Cramming?*, London: Town and Country Planning Association

Hall, P. (2002a) *Cities of Tomorrow* (third edition), Oxford: Blackwell

Hall, P. (2002b) *Urban and Regional Planning* (fourth edition), London: Routledge

Hall, P. and Hass-Klau, C. (1985) *Can Rail Save the City?*, Aldershot: Gower

Hall, P. and Markussen, A. (eds) (1985) *Silicon Landscapes*, London: Allen and Unwin

Hall, P. and Ward, C. (1998) *Sociable Cities: The Legacy of Ebenezer Howard*, Chichester: Wiley

Hall, P., Gracey, H., Drewett, R. and Thomas, R. (1973) *The Containment of Urban England*, London: Allen and Unwin

Hall, P., Breheny, M., McQuaid, R. and Hart, D. (1987) *Western Sunrise: Genesis and Growth of Britain's Major High Tech Corridor*, London: Unwin Hyman

Hall, S. (1995) 'The SRB: taking stock', *Planning Week* 27 April: 16–17

Hall, S. (2003) 'The "Third Way" revisited: "New Labour", spatial policy and the National Strategy for Neighbourhood Renewal', *Planning Practice and Research* 18(4): 265–78

Hall, S. and Nevin, B. (1999) 'Continuity and change: a review of English regeneration policy in the 1990s', *Regional Studies* 33(4): 447–91

Hall, T. (ed.) (1991) *Planning and Urban Growth in the Nordic Countries*, London: Spon

Hallett, G. (ed.) (1977) *Housing and Land Policies in West Germany and Britain*, London: Macmillan

Hallsworth, A. G. (1992) *The New Geography of Consumer Spending*, London: Pinter

Hallsworth, A. G. (1994) 'Decentralization of retailing in Britain: the breaking of the third wave', *Professional Geographer* 46: 296–307

Halsey, A. J. (ed.) (1972) *Educational Priority*, vol. 1: *EPA Problems and Policies*, London: HMSO

Ham, C. and Hill, M. (1993) *The Policy Process in the Modern Capitalist State* (second edition), London: Harvester Wheatsheaf

Hambleton, R. (1986) *Rethinking Policy Making*, Bristol: School for Advanced Urban Studies, University of Bristol

Hambleton, R. (1990) *Urban Government in the 1990s: Lessons from the USA*, Bristol: School for Advanced Urban Studies, University of Bristol

Hambleton, R. (1991) 'The regeneration of US and British cities', *Local Government Studies* 17(5): 53–69

Hambleton, R. (1993) 'Issues for urban policy in the 1990s', *Town Planning Review* 64: 313–23

Hambleton, R. (1998) *Local Government Political Management Arrangements: An International Perspective*, Edinburgh: Central Research Unit, Scottish Office

Hambleton, R. (2000) 'Modernising political management in local government', *Urban Studies* 37(5–6): 931–50

Hambleton, R. and Hoggett, P. (1990) *Beyond Excellence: Quality Government in the 1990s*, Bristol: School for Advanced Urban Studies, University of Bristol

Hambleton, R. and Sweeting, D. (1999a) 'Change for councils', *Planning* 22 October: 20

Hambleton, R. and Sweeting, D. (1999b) 'Restructuring our decision making', *Planning* 12 November: 16–17

Hambleton, R. and Taylor, R. (eds) (1993) *People in Cities: A Transatlantic Policy Exchange*, Bristol: School for Advanced Urban Studies, University of Bristol

Hambleton, R. and Thomas, H. (eds) (1995) *Urban Policy Evaluation: Challenge and Change*, London: Paul Chapman

Hambleton, R., Stewart, J. and Underwood, J. (1980) *Inner Cities: Management and Resources*, Bristol: School for Advanced Urban Studies, University of Bristol

Hamilton, K. (1999) 'Women and transport: disadvantage and the gender divide', *Town and Country Planning* 68(10): 318–19

Hammersley, R. (1987) 'Plans, policies and the local ombudsman: the Chellaston case', *Journal of Planning and Environment Law*: 101–5

Hamnett, C. and Randolph, B. (1991) *Cities, Housing and Profits: Flat Break-up and the Decline of Private Renting*, London: UCL Press

Hams, T. (2000) 'Local Agenda 21: integration, independence or extinction?', *EG* 6(6): 6–8

Hanley, N., Whitby, M. and Simpson, I. (1999) 'Assessing the success of agri-environmental policy in the UK', *Land Use Policy* 16: 67–80

Hansen, A. (ed.) (1993) *The Mass Media and Environmental Issues*, Chichester: Belhaven

Hanson, J. (1999) *Decoding Homes and Houses*, Cambridge: Cambridge University Press

Harding, A. (1991) 'The rise of urban growth coalitions, UK-style?', *Environment and Planning C: Government and Policy* 9: 295–317

Harding, A., Evans, R., Parkinson, M. and Garside, P. (1996) *Regional Government in Britain: An Economic Solution?* (Joseph Rowntree Foundation), Bristol: Policy Press

Harding, A., Marvin, S. and Sprigings, N. (2004) *Realising the National Economic Potential of Provincial City-regions: The Rationale for and Implications of a 'Northern Way' Growth Strategy*, Salford: SURF Centre.

Hardy, D. (1991a) *From Garden Cities to New Towns: Campaigning for Town and Country Planning, Volume 1, 1899–1946*, London: Spon

Hardy, D. (1991b) *From New Towns to Green Politics: Campaigning for Town and Country Planning, Volume 2, 1947–1990*, London: Spon

Hardy, D. (2000) *Utopian England: Community Experiments 1900–1945*, London: Routledge

Hardy, S., Hart, M., Albrechts, L. and Katos, A. (1995) *An Enlarged Europe: Regions in Competition?*, London: Regional Studies Association and Jessica Kingsley.

Harloe, M. (1995) *The People's Home? Social Rented Housing in Europe*, Oxford: Blackwell

Harman, R. (1995) *New Directions: A Manual of European Best Practice in Transport Planning*, London: Transport 2000

Harrap, P. (1993) *Charging for Road Use Worldwide: A Financial Times Management Report*, London: Financial Times

Harris, N. and Hooper, A. (2004) 'Rediscovering the "spatial" in public policy and planning: an examination of the spatial content of sectoral policy documents', *Planning Theory and Practice* 5(2): 147–69

Harris, N. and Tewdwr-Jones, M. (1995) 'The implications for planning of local government reorganisation in Wales: purpose, process, and practice', *Environment and Planning C: Government and Policy* 13: 47–66

Harris, N. and Thomas, H. (2004) 'Planning for a diverse society? A review of the UK government's planning policy guidance', *Town Planning Review* 75(4): 473–500

Harris, N., Hooper, A. and Bishop, K. (2002) 'Constructing the practice of spatial planning: a national spatial planning framework for Wales', *Environment and Planning C: Government and Policy* 20: 555–72

Harris, R. (1992a) 'The Environmental Protection Act 1990: penalising the polluter', *Journal of Planning and Environment Law* 1992: 515–24

Harris, R. (1992b) 'Integrated pollution control in practice', *Journal of Planning and Environment Law* 1992: 611–23

Harris, R. and Larkham, P. (1999) *Changing Suburbs: Foundation, Form and Function*, London: Spon

Harrison, A. (1992) 'What shall we do with the contaminated site?', *Journal of Planning and Environment Law* 1992: 809–16

Harrison, J. (1994) 'Who is my neighbour?', *Journal of Planning and Environment Law* 1994: 219–23

Harrison, M. (1992) 'A presumption in favour of planning permission?', *Journal of Planning and Environment Law* 1992: 121–9

Harrison, M. L. and Mordey, R. (eds) (1987) *Planning Control: Philosophies, Prospects and Practice*, London: Croom Helm

Harrison, P. (1983) *Inside the Inner City: Life under the Cutting Edge*, Harmondsworth: Penguin

Harrison, R. T. and Hart, M. (eds) (1992) *Spatial Policy in a Divided Nation*, London: Jessica Kingsley

Harrison, T. (1988) *Access to Information in Local Government*, London: Sweet and Maxwell

Harrop, D. O. and Nixon, J. A. (1999) *Environmental Assessment in Practice*, London: Routledge

Hart, D. (2000) 'The impact of the European Convention of Human Rights on planning and environmental law', *Journal of Planning and Environment Law* February: 117–33

Hart, T. (1993) 'Transport investment and disadvantaged regions: UK and European policies since the 1950s', *Urban Studies* 30: 417–36

Hart, T. (1992) 'Transport, the urban pattern and regional change, 1960–2010', *Urban Studies* 29: 483–503

Hart, T., Haughton, G. and Peck, J. (1996) 'Accountability and the non-elected local state: calling training and enterprise councils to local account', *Regional Studies* 30: 429–41

Harte, J. D. C. (1989) 'The scope of protection resulting from the designation of Sites of Special Scientific Interest', *Journal of Environmental Law* 1: 245–54

Harte, J. D. C. (1990) 'The scheduling of ancient monuments and the role of interested members of the public in environmental law', *Journal of Environmental Law* 2: 224–49

Harvey, G. (1997) *The Killing of the Countryside*, London: Cape

Hasegawa, J. (1992) *Replanning the Blitzed City Centre*, Buckingham: Open University Press

Hassan, G. (ed.) (1999) *A Guide to the Scottish Parliament* (Centre for Scottish Public Policy), Edinburgh: The Stationery Office

Hass-Klau, C. (1990) *The Pedestrian and City Traffic*, London: Belhaven

Hass-Klau, C., Nold, I., Bocker, G. and Crampton, G. (1992) *Civilised Streets: A Guide to Traffic Calming*, Brighton: Environmental and Transport Planning

Hastings, A. (1996) 'Unravelling the process of "partnership" in urban regeneration policy', *Urban Studies* 33(2): 253–68

Hastings, A. and McArthur, A. (1995) 'A comparative assessment of government approaches to partnership with the local community', in Hambleton, R. and Thomas, H. (eds) *Urban Policy Evaluation: Challenge and Change*, London: Paul Chapman

Hastings, A., McArthur, A. and McGregor, A. (1996) *Less than Equal? Community Organisations and Estate Regeneration Partnerships*, Bristol: Policy Press

Haughton, G. (1990) 'Targeting jobs to local people: the British urban policy experience', *Urban Studies* 27: 185–98

Haughton, G. and Counsell, D. (2003) *Regions, Spatial Strategies and Sustainable Development*, London: Routledge

Haughton, G. and Hunter, C. (1994) *Sustainable Cities*, London: Jessica Kingsley

Haughton, G. and Lawless, P. (1992) *Policies and Potential: Recasting British Urban and Regional Policies*, London: Regional Studies Association

Haughton, G., Rowe, I. and Hunter, C. (1997) 'The Thames Gateway and the re-emergence of regional strategic planning', *Town Planning Review* 68: 407–22

Hausner, V. A. (1986) *Urban Economic Adjustment and the Future of British Cities: Directions for Urban Policy*, Oxford: Clarendon Press

Hausner, V. A. (ed.) (1987) *Critical Issues in Urban Economic Development* (2 vols), Oxford: Clarendon Press

Hausner, V. A. and Robson, B. (1986) *Changing Cities: An Introduction to the Economic and Social Research Council Inner Cities Research Programme*, Swindon: The Council

Hawes, D. and Perez, B. (eds) (1996) *The Gypsy and the State: The Ethnic Cleansing of British Society* (second edition), Bristol: Policy Press

Hawkins, K. (1984) *Environment and Enforcement Regulation and the Social Definition of Pollution*, Oxford: Oxford University Press

Hayton, K. (1990) *Getting People into Jobs* (DoE), London: HMSO

Hayton, K. (1992) 'Scottish Enterprise: a challenge to local land use planning?', *Town Planning Review* 63: 265–78

Hayton, K. (1996) 'Planning policy in Scotland', in Tewdwr-Jones, M. (ed.) *British Planning Policy in Transition: Planning in the 1990s*, London: UCL Press

Haywood, R. (1999) 'South Yorkshire Supertram: its property impacts and their implications for integrated land use-transportation planning', *Planning Practice and Research* 14: 277–99

Hazell, R. (ed.) (1999) *Constitutional Futures: A History of the Next Ten Years* (The Constitution Unit), Oxford: Oxford University Press

Healey, M. J. and Ibery, B. W. (1985) *The Industrialisation of the Countryside*, Norwich: Geo Books

Healey, P. (1979) *Statutory Local Plans: Their Evolution in Legislation and Administrative Interpretations*, Oxford: Department of Town Planning, Oxford Polytechnic

Healey, P. (1983) *Local Plans in British Land Use Planning*, Oxford: Pergamon

Healey, P. (1985) 'The professionalisation of planning in Britain', *Town Planning Review* 56: 492–507

Healey, P. (1986) 'The role of development plans in the British planning system: an empirical assessment', *Urban Law and Policy* 8: 1–32

Healey, P. (1988) 'The British planning system and managing the urban environment', *Town Planning Review* 59: 397–417

Healey, P. (1989) 'Directions for change in the British planning system', *Town Planning Review* 60: 125–49; 'Comments and response' (by Goodchild, R., Marwick, A., Grant, M., Jones, A., Lyddon, D. and Robinson, D.), *Town Planning Review* 60: 319–32

Healey, P. (1990) 'Places, people and politics: plan-making in the 1990s', *Local Government Policy Making* 17(2): 29–39

Healey, P. (1991a) 'Urban regeneration and the development industry', *Regional Studies* 25: 97–110

Healey, P. (1991b) 'The content of planning education programmes: some comments from recent British experience', *Environment and Planning B: Planning and Design* 18: 177–89

Healey, P. (1991c) 'Models of the development process', *Journal of Property Research* 8: 219–38

Healey, P. (1992a) 'Development plans and markets', *Planning Practice and Research* 7(2): 13–20

Healey, P. (1992b) 'The reorganisation of state and market in planning', *Urban Studies* 29: 411–34

Healey, P. (1992c) 'Planning through debate: the communicative turn in planning theory', *Town Planning Review* 63: 143–62

Healey, P. (1992d) 'An institutional model of the development process', *Journal of Property Research* 9: 33–44

Healey, P. (1993) 'The communicative work of development plans', *Environment and Planning B* 20: 83–104

Healey, P. (1994) 'Development plans: new approaches to making frameworks for land use regulation', *European Planning Studies* 2: 39–57

Healey, P. (1997) *Collaborative Planning*, London: Macmillan

Healey, P. (1998) 'Collaborative planning in a stakeholder society', *Town Planning Review* 69: 1–21

Healey, P. (2004) 'Towards a "social democratic" policy agenda for cities', in Johnstone and Whitehead (eds) *New Horizons in British Urban Policy*, Aldershot: Ashgate

Healey, P. and Barrett, S. M. (1990) 'Structure and agency in land and property development processes', *Urban Studies* 37: 89–104

Healey, P. and Nabarro, R. (1990) *Land and Property Development in a Changing Context*, Aldershot: Gower

Healey, P. and Shaw, T. (1993) 'Planners, plans and sustainable development', *Regional Studies* 27: 769–76

Healey, P., McDougall, G. and Thomas, M. (eds) (1982) *Planning Theory: Prospects for the 1980s*, Oxford: Pergamon

Healey, P., McNamara, P., Elson, M. and Doak, A. (1988) *Land Use Planning and the Mediation of Change*, Cambridge: Cambridge University Press

Healey, P., Purdue, M. and Ennis, F. (1992a) 'Rationales for planning gain', *Policy Studies* 13: 18–30

Healey, P., Davoudi, S., O'Toole, M., Tavsanoglu, S. and Usher, D. (1992b) *Rebuilding the City: Property-Led Urban Regeneration*, London: Spon

Healey, P., Purdue, M. and Ennis, F. (1993) *Gains from Planning: Dealing with the Impacts of Development*, York: Joseph Rowntree Foundation

Healey, P. et al. (eds) (1994) *Trends in Development Plan Making in European Planning*, Newcastle upon Tyne: Department of Town and Country Planning, University of Newcastle upon Tyne

Healey, P., Cameron, S., Davoudi, S., Graham, S. and Madani-Pour, A. (eds) (1995a) *Managing Cities: The New Urban Context*, Chichester: Wiley

Healey, P., Purdue, M. and Ennis, F. (1995b) *Negotiating Development: Rationales and Practice for Development Obligations and Planning Gain*, London: Spon

Healey, P., Khakee, A., Motte, A. and Needham, B. (1997) *Making Strategic Spatial Plans*, London: UCL Press

Hebbert, M. (1992) 'The British garden city: metamorphosis', in Ward, S. (ed.) *The Garden City: Past, Present and Future*, London: Spon

Hebbert, M. (1998) *London: More by Fortune than Design*, Chichester: Wiley

Hedges, B. and Clemens, S. (1994) *Housing Attitudes Survey*, London: DoE

Heim, C. (1990) 'The Treasury as developer-capitalist? British new town building in the 1950s', *Journal of Economic History* 50: 903–24

Hendry, J. (1989) 'The control of development and the origins of planning in Northern Ireland', in Bannon, M. J., Nowlan, K. I., Hendry, J. and Mawhinney, K. (eds) *Planning: The Irish Experience 1920–1988*, Dublin: Wolfhound Press

Hendry, J. F. (1992) 'Plans and planning policy for Belfast: a review article', *Town Planning Review* 63(1): 79–85

Hendry, J. F. (1993) 'Conservation in Northern Ireland', in Royal Town Planning Institute, *The Character of Conservation Areas*, vol. 2: *Supporting Information*, London: Royal Town Planning Institute

Hennessy, P. (1989) *Whitehall*, London: Secker and Warburg (Fontana edition 1990)

Hennessy, P. (1992) *Never Again: Britain 1945–1951*, London: Jonathan Cape (Vintage edition 1993)

Herbert, D. T. (ed.) (1995) *Heritage, Tourism and Society*, London: Mansell

Herbert-Young, N. (1995) 'Reflections on section 54A and plan-led decision-making', *Journal of Planning and Environment Law* 1995: 292–305

Herington, J. (1984) *The Outer City*, London: Harper and Row

Herington, J. (1990) *Beyond Green Belts: Managing Urban Growth in the 21st Century*, London: The Stationery Office (previously published by Jessica Kingsley)

Heriot-Watt University, De Paul University, Studio Cascade and Platt, C. (2003) *Participatory Planning for Sustainable Communities: International Experience in Mediation, Negotiation and Engagement in Making Plans*, London: ODPM

Herrschel, T. and Newman, P. (2002) *Governance of Europe's City Regions*, London: Routledge

Heseltine, M. (1987) *Where There's a Will*, London: Hutchinson

Hewitt, P. (1989) *A Cleaner, Faster London: Road Pricing, Transport Policy and the Environment*, London: Institute for Public Policy Research

Heycock, M., (1991) 'Public policy, need and land use planning', in Nadin, V. and Doak, J. (eds) *Town Planning Responses to City Change*, Aldershot: Gower

Heywood, F. (1990) *Clearance: The View from the Street – A Study of Politics, Land and Housing*, Birmingham: Community Forum

Hibbs, J. (1989) *The History of the British Bus Services*, Newton Abbot: David and Charles

Higgins, J., Deakin, N., Edwards, J. and Wicks, M. (1983) *Government and Urban Poverty: Inside the Policy-Making Process*, Oxford: Oxford University Press

Higgins, M. and Allmendinger, P. (1999) 'The changing nature of public planning practice under the New Right: the legacies and implications of privatisation', *Planning Practice and Research* 14(1): 39–67

Higman, R. (1991) *Local Responses to 1989 Traffic Forecasts*, London: Friends of the Earth

Hill, D. M. (1970) *Participating in Local Affairs*, Harmondsworth: Penguin

Hill, D. M. (1994) *Citizens and Cities: Urban Policy in the 1990s*, Hemel Hempstead: Harvester Wheatsheaf

Hill, D. M. (1999) *Urban Policy and Politics in Britain*, London: Macmillan

Hill, L. (1991) 'Unitary development plans for the West Midlands: first stages in the statutory responses to a changing conurbation', in Nadin, V. and Doak, J. (eds) *Town Planning Responses to City Change*, Aldershot: Gower

Hill, M. (ed.) (1993) *The Policy Process: A Reader*, London: Harvester Wheatsheaf

Hill, M. P. (1987) 'Housing land availability: some observations of the process of housing development in contrasting urban locations', *Land Development Studies* 4: 209–19

Hill, S. and Barlow, J. (1995) 'Single regeneration budget: hope for "those inner cities"?', *Housing Review* 44: 32–5

Hill, S. and Slater, M. (2004) *Planning and Inclusion: A Discussion Paper*, London: RTPI

Hillier Parker and Cardiff University (2004) *Policy Evaluation of the Effectiveness of PPG6*, London: ODPM

Hillier Parker, Dundas and Wilson, Edinburgh School of Planning and Housing (1998) *Review of Development Planning in Scotland* (Central Research Unit, Scottish Office), Edinburgh: The Stationery Office

Hillman, J. (1988) *A New Look for London* (Royal Fine Art Commission), London: HMSO

Hillman, J. (1990) *Planning for Beauty: The Case for Design Guidelines* (Royal Fine Art Commission), London: HMSO

Hillman, J. (1993) *Telelifestyles and the Flexicity: The Impact of the Electronic Home*, Dublin: European Foundation for the Improvement of Living and Working Conditions; Luxembourg: Office for Official Publications of the European Communities

Hillman, M. (1989) 'More daylight, less accidents', *Traffic Engineering and Control* 30: 191–3

Hillman, M. (1992) 'Reconciling transport and environmental policy objectives: the way ahead at the end of the road', *Public Administration* 70: 225–34

Hillman, M. (1993a) *Children's Helmets: The Case For and Against*, London: Policy Studies Institute

Hillman, M. (ed.) (1993b) *Children, Transport and the Quality of Life*, London: Policy Studies Institute

Hillman, M. (1993c) *Time for Change: Setting Clocks Forward by One Hour Throughout the Year*, London: Policy Studies Institute

Hillman, M. and Whalley, A. (1979) *Walking IS Transport*, London: Policy Studies Institute

Hillman, M., Adams, J. and Whitelegg, J. (1991) *One False Move: A Study of Children's Independent Mobility*, London: Policy Studies Institute

Hills, P. J. (1994) 'The car versus mixed use development', in Wood, M. (ed.) *Planning Icons: Myth and Practice* (Planning Law Conference, *Journal of Planning and Environment Law*), London: Sweet and Maxwell

Himsworth, C. M. G. (1994) 'Charging for inspecting the register', *Scottish Planning and Environment Law* 44: 59–60

Hirsch, D. (ed.) (1994) *A Positive Role for Local Government: Lessons for Britain from Other Countries*, London: Local Government Chronicle and Joseph Rowntree Foundation

Ho, S. Y. (2003) *Evaluating British Urban Policy: Ideology, Conflict and Compromise*, Aldershot: Ashgate

Hobbs, P. (1992) 'The economic determinants of post-war British planning', *Progress in Planning* 38(3): 179–300

Hobhouse, A. (chair) (1947) *Report of the National Parks Committee (England and Wales)*, Cmd 7121, Hobhouse Report, London: HMSO

Hobley, B. (1987) 'Rescue archaeology and planning', *The Planner* 73(4): 25–7

Hobson, E. (2004) *Conservation and Planning: Changing Values in Policy and Practice*, London: Spon

Hobson, J., Hockman, S. and Stinchcombe, P. (1996) 'The future of the planning system', in Bean, D. (ed.) *Law Reform for All*, London: Blackstone

Hoch, C. (1994) *What Planners Do: Power, Politics and Persuasion*, Chicago: Planners Press

Hodcroft, D. and Alexander, D. (2004) 'Ecological frameworks in north west England', *Planning Practice and Research* 19(3): 307–20

Hodge, I. (1989) 'Compensation for nature conservation', *Environment and Planning A* 21: 1027–36

Hodge, I. (1991) 'Incentive policies and the rural environment', *Journal of Rural Studies* 7: 373–84

Hodge, I. (1992) 'Supply control and the environment: the case for separate policies', *Farm Management* 8: 65–72

Hodge, I. (1995) *Environmental Economics: Individual Incentives and Public Choices*, London: Macmillan

Hodge, I. (1999) 'Countryside planning: from urban containment to sustainable development', in Cullingworth, J. B. (ed.) *British Planning: 50 Years of Urban and Regional Policy*, London: Athlone

Hodge, I. and Monk, S. (1991) *In Search of a Rural Economy: Patterns and Differentiation in Non-Metropolitan England*, Cambridge: Department of Land Economy

Hogwood, B. W. (1995) *The Integrated Regional Offices and the Single Regeneration Budget*, London: Commission for Local Democracy

Hogwood, B. W. (1996) *Mapping the Regions: Boundaries, Coordination and Government* (Joseph Rowntree Foundation), Bristol: Policy Press

Hogwood, B. W. and Keating, M. (eds) (1982) *Regional Government in England*, Oxford: Clarendon Press

Holden, E. (1977) *The Country Diary of an Edwardian Lady*, London: Michael Joseph

Holford, W. G. (1953) 'Design in city centres', part 3 of MHLG, *Design in Town and Village*, London: HMSO

Hollis, G., Ham, G. and Ambler, M. (eds) (1992) *The Future Role and Structure of Local Government*, Harlow: Longman

Hollox, R. E. and Biart, S. W. (1982) 'Local plan inquiries: a case study', *Journal of Planning and Environment Law* 1982: 17–23

Holmans, A. (1995) *Housing Demand and Need in England 1991 to 2011*, York: Joseph Rowntree Foundation

Holmans, A. and Simpson, M. (1999) *Low Demand: Separating Fact from Fiction*, London: Chartered Institute of Housing

Holmans, A., Morrison, N. and Whitehead, C. (1998) *How Many Homes Will We Need?* London: Shelter

Holstein, T. (2002) 'Moving to a SEA Green Paper', *Town and Country Planning* 71(9): 202–20

Home, R. H. (1982) *Inner City Regeneration*, London: Spon

Home, R. K. (1987) 'Planning decision statistics and the use classes debate', *Journal of Planning and Environment Law* 1987: 167–73

Home, R. K. (1989) *Planning Use Classes: A Guide to the 1987 Order* (second edition), Oxford: BSP Professional

Home, R. K. (1992) 'The evolution of the use classes order', *Town Planning Review* 63: 187–201

Home, R. K. (1993) 'Planning aspects of the government consultation paper on gypsies', *Journal of Planning and Environment Law* 1993: 13–18

Hood, C. (1995) 'Contemporary public management: a new global paradigm', *Public Policy and Administration* 10(2): 104–227

Hooghe, L. and Marks, G. (2001) *Multi-level Governance and European Integration*, Lanham, MD: Rowman and Littlefield

Hooper, A. (1979) 'Land availability', *Journal of Planning and Environment Law* 1979: 752–6

Hooper, A. (1980) 'Land for private housebuilding', *Journal of Planning and Environment Law* 1980: 795–806

Hooper, A. (1982) 'Land availability in South East England', *Journal of Planning and Environment Law* 1982: 555–60

Hooper, A. (1985) 'Land availability studies and private housebuilding', in Barrett, S. and Healey, P. (eds) *Land Policy: Problems and Alternatives*, Aldershot: Avebury

Hooper, A. (1999) *Design for Living: Constructing the Residential Built Environment in the 21st Century*, London: Town and Country Planning Association

Hooper, A., Pinch, P. and Rogers, S. (1988) 'Housing land availability: circular advice, circular arguments and circular methods', *Journal of Planning and Environment Law* 1988: 225–39

Hooper, A., Pinch, P., and Rogers, S. (1989) 'Housing land availability in the South East', in Breheny, M. J. and Congdon, P. (eds) *Growth and Change in a Core Region: The Case of South East England*, London: Pion

Hooper, A., Dunmore, K. and Hughes, M. (1998) *Home Alone: The Housing Preferences of One-Person Households*, Amersham: Housing Research Foundation

Hoskyns, J. (1983) 'Whitehall and Westminster: an outsider's view', *Parliamentary Affairs* 36: 137–47

Hough, B. (1992) 'Standing in planning permission appeals', *Journal of Planning and Environment Law* 1992: 319–29

Hough, B. (1997) 'The erosion of the concept of development: problems in establishing a material change of use', *Journal of Planning and Environment Law* 1997: 895–903

Hough, B. (1998) 'Relevance and reasons in planning matters', *Journal of Planning and Environment Law* July: 625–34

House Builders Federation (HBF) (1987) *Private Housebuilding in the Inner Cities*, London: HBF

Housing Choice (1991) *Planning: A Citizen's Charter*, London: Housing Choice

Howard, E. (1898) *To-morrow: A Peaceful Path to Real Reform*, London: Swan Sonnenschein; republished in facsimile (2003) with a commentary by Hall, P., Hardy, D. and Ward, C.

Howard, E. (1902) *Garden Cities of Tomorrow*, London: Swan Sonnenschein (reprinted by Faber in 1946, with an introduction by F. J. Osborn, and an introductory essay by Lewis Mumford; see also Howard 1898, the original edition

Howard, E. B. and Davies, R. L. (1993) 'The impact of regional out-of-town retail centres: the case of the Metro Centre', *Progress in Planning* 40(2): 89–165

Howe, J. (1996) 'A case of inter-agency relations: regional development in rural Wales', *Planning Practice and Research* 11: 61–72

Hudson, R. and Williams, A. M. (1998) *Divided Europe: Society and Territory*, London: Sage

Hughes, D. (1996) *Environmental Law* (third edition), Oxford: Butterworth

Hughes, D. J. (1995) 'Planning and conservation areas: where do we stand following PPG 15, and whatever happened to Steinberg?', *Journal of Planning and Environment Law* 1995: 679–91

Hughes, J. T. (1991) 'Evaluation of local economic development: a challenge for policy research', *Urban Studies* 28: 909–18

Hughes, M., Clarke, M., Allen, H. and Hall, D. (1998) *The Constitutional Status of Local Government in Other Countries*, Edinburgh: Central Research Unit, Scottish Office

Hull, A. (2000) 'Modernizing democracy: constructing a radical reform of the planning system?' *European Planning Studies*, 8(6): 767–82

Hull, A. (2004) 'A "Winning Wales": an integrated strategy for sustainable development', *Planning Practice and Research* 19(3): 329–44

Hull, A., Healey, P. and Davoudi, S. (1995) *Greening the Red Rose County: Working Towards an Integrated Sub-Regional Strategy*, Newcastle upon Tyne: Department of Town and Country Planning, University of Newcastle upon Tyne

Hull, A. D. (1998) 'Spatial planning: the development plan as a vehicle to unlock development potential?', *Cities* 15(5): 327–35

Hull, A. D. and Vigar, G. (1998) 'The changing role of the development plan in managing spatial change', *Environment and Planning C* 16: 379–94

Humber, J. R. (1980) 'Land availability: another view', *Journal of Planning and Environment Law* 1980: 19–23

Humber, R. (1990) 'Prospects and problems for private housebuilders', *The Planner/Town and Country Planning Summer School 1990 Proceedings* 76(7): 15–19

Hunt, J. (chair) (1969) *The Intermediate Areas*, Hunt Report, London: HMSO

Huppes, G. and Kagan, R. A. (1989) 'Market-oriented regulation of environmental problems in the Netherlands', *Law and Policy* 11: 215–39

Hurrell, A. and Kinsbury, B. (1992) *The International Politics of the Environment: Actors, Interests and Institutions*, Oxford: Clarendon Press

Hutter, B. M. (1989) 'Variations in regulatory enforcement styles', *Law and Policy* 11: 153–74

Hutton, N. (1986) *Lay Participation in a Public Inquiry: A Sociological Case Study*, Aldershot: Gower

Hutton, R. H. (1991) 'Local needs policy initiatives in rural areas: missing the target', *Journal of Planning and Environment Law* 1991: 303–11

Huxley, J. S. (chair) (1947) *Conservation of Nature in England and Wales*, Cmd 7122, Huxley Report, London: HMSO

IAURIF (Institut d'aménagement du territoire et d'urbanisme de la région d'Ile de France) (1991) *La Charte de l'Ile de France*, Paris: IAURIF

IDeA (1998) *Sustainability in Development Control*, Luton: IDeA

Ilbery, B. W. (1990) 'Adoption of the arable set-aside scheme in England', *Geography* 76(1): 69–73

Ilbery, B. W. (1998) *The Geography of Rural Change*, Harlow: Longman

Illich, I. (1971) *Deschooling Society*, New York: Harper and Row

Illich, I. (1977) *Disabling Professions*, London: Boyars

Imrie, R. (1996) *Disability and the City*, London: Paul Chapman

Imrie, R. and Thomas, H. (1999) *British Urban Policy: An Evaluation of the Urban Development Corporations*, London: Paul Chapman

Imrie, R., Thomas, H. and Marshall, T. (1995) 'Business organisations, local dependence and the politics of urban renewal in Britain', *Urban Studies* 32: 31–47

Ince, M. (1984) *Sizewell Report: What Happened at the Inquiry*, London: Pluto

Inland Revenue (1992) *Company Cars: Reform of Income Tax Treatment – A Consultative Document*, London: Inland Revenue

Innes, J. E. (1995) 'Planning theory's emerging paradigm: communicative action and interactive practice', *Journal of Planning Education and Research* 14: 183–90

Innes, J. E. (1998a) 'Challenge and creativity in postmodern planning', *Town Planning Review* 69: v–ix

Innes, J. E. (1998b) 'Information in communicative planning', *Journal of the American Planning Association* 64: 52–63

Insight Social Research (ISR) (1989) *Local Attitudes to Central Advice*, London: ISR

Institute of Logistics and Transport (ILT) (1999) *Sustainable Distribution* (Discussion Paper), London: ILT

International Energy Agency (1989) *Energy and the Environment*, Paris: OECD

Investment Property Databank (1993) *The Investment Performance of Listed Buildings*, London: Royal Institute of Chartered Surveyors

Investment Property Databank (1999) *Investment Performance of Listed Buildings*, Swindon: English Heritage

Jackson, A. R. W. and Jackson, J. M. (1996) *Environmental Science: The Natural Environment and Human Impact*, Harlow: Longman

Jackson, A., Morrison, N. and Royce, C. (1994) *The Supply of Land for Housing: Changing Local Authority Mechanisms*, Cambridge: Department of Land Economy, University of Cambridge

Jackson, T. (1996) *Material Concerns: Pollution, Profit and Quality of Life*, London: Routledge

Jacobs, B. D. (1992) *Fractured Cities: Capitalism, Community and Empowerment in Britain and America*, London: Routledge

Jacobs, B. D. (2003) *Strategy and Partnership in Cities and Regions: Economic Development and Urban Regeneration in Pittsburgh, Birmingham and Rotterdam*, New York: Palgrave

Jacobs, J. (1961) *The Death and Life of Great American Cities*, London: Cape and Penguin

Jacobs, J. (1985) 'Urban development grant', *Policy and Politics* 13: 191–9

Jacobs, M. (1990) *Sustainable Development: Greening the Economy*, London: Fabian Society

Jacobs, M. (1991) *The Green Economy: Environment, Sustainable Development, and the Politics of the Future*, London: Pluto

Jacobs, M. (1999) *Environmental Modernisation: the New Labour Agenda*, London: Fabian Society

James, S. (1990) 'A streamlined city: the broken pattern of London government', *Public Administration* 68: 493–504

James, S. (1997) *British Government: A Reader in Policy Making*, London: Routledge

James, S. (1999) *British Cabinet Government* (second edition), London: Routledge

Jansen, A. J. and Hetsen, H. (1991) 'Agricultural development and spatial organization in Europe', *Journal of Rural Studies* 7: 143–51

Jarvis, R. (1996) 'Structure planning policy and strategic planning guidance in Wales', in Tewdwr-Jones, M. (ed.) *British Planning Policy in Transition: Planning in the 1990s*, London: UCL Press

Jenkins, K., Oates, G. and Stott, A. (1985) *Making Things Happen: A Report on the Implementation of Government Efficiency Scrutinies – Report to the Prime Minister*, London: HMSO

Jenkins, S. (1995) *Accountable to None: The Tory Nationalisation of Britain*, London: Hamish Hamilton (Penguin edition 1996)

Jenkins, T. N. (1990) *Future Harvest: The Economics of Farming and the Environment – Proposals for Action*, London: Council for the Protection of Rural England and WWF

Jenks, M., Burton, E. and Williams, K. (eds) (1996) *The Compact City: A Sustainable Urban Form?*, London: Spon

Jensen, O. B. and Richardson, T. (2003) *Making European Space: Mobility, Power and Territorial Identity*, London: Routledge

Jewell, D. and Raemaekers, J. (1999) 'Local authority archaeological services in Scotland two years after local government reorganisation', *Journal of Environmental Planning and Management* 42: 375–88

Jewell, T. (1995) 'Planning regulation and environmental consciousness: some lessons from minerals?', *Journal of Planning and Environment Law* 1995: 482–98

JMP Consultants (1995a) *Travel to Food Super-stores*, London: JMP Consultants

JMP Consultants (1995b) *PPG 13: A Guide to Better Practice: Reducing the Need to Travel through Land Use and Transport Planning*, London: HMSO

John, P. (1993) *Local Government in Northern Ireland*, York: Joseph Rowntree Foundation

Johnson, D., Martin, S., Pearce, G. and Simmons, S. (1992) *The Strategic Approach to Derelict Land Reclamation* (DoE), London: HMSO

Johnson, D. A. (1995) *Planning the Great Metropolis: The 1929 Regional Plan of New York and its Environs*, London: Spon

Johnson, P. (ed.) (1994) *20th Century Britain: Economic, Social and Cultural Change*, London: Methuen

Johnson, S. P. and Corcelle, G. (1989) *The Environmental Policy of the European Communities*, London: Graham and Trotman

Johnson, T. and Hart, C. (2005) *The Barker Review of Housing and the Planning Gain Supplement*, London: RICS

Johnson, W. C. (1984) 'Citizen participation in local planning: a comparison of US and British experience', *Environment and Planning C: Government and Policy* 2: 1–14

Johnston, B. (1999a) 'New powers and policies for improved performance', *Planning* 24 September: 16–17

Johnston, B. (1999b) 'Know your rights: councils should not panic about the new human rights legislation', *Planning* 19 November: 17

Johnston, B. F. (1995) 'Commission makes case for coherence', *Planning* 20 October: 2830

Johnston, R. J. and Gardiner, V. (eds) (1991) *The Changing Geography of the UK*, London: Routledge

Johnstone, C. and Whitehead, M. (eds) (2004) *New Horizons in British Urban Policy: Perspectives on New Labour's Urban Renaissance*, Aldershot: Ashgate

Johnstone, R. J. (1991) *Environmental Problems: Nature, Economy and State*, Chichester: Belhaven

Jones Land LaSalle (2004) *Land and Property Value Study: Assessing the Property Values Attributable to the Jubilee Line Extension*, London: Transport for London

Jones, A. (1996) 'Local planning policy: the Newbury approach', in Tewdwr-Jones, M. (ed.) *British Planning Policy in Transition: Planning in the 1990s*, London: UCL Press

Jones, B. and Kavanagh, D. (1998) *British Politics Today* (sixth edition), Manchester: Manchester University Press

Jones, C. (1996) 'Property-led local economic development policies from advance factory to English Partnerships and strategic property investment?', *Regional Studies* 30: 200–6

Jones, C. and Wood, C. (1995) 'The impact of environmental assessment on public inquiry decisions', *Journal of Planning and Environment Law* 1995: 890–904

Jones, C., Wood, C. and Dipper, B. (1998) 'Environmental assessment in the UK planning process: a review of practice', *Town Planning Review* 69(3): 315–39

Jones, G. and Stewart, J. (1983) *The Case for Local Government*, London: Allen and Unwin

Jones, M. (1996) 'TEC policy failure: evidence from the baseline follow-up studies', *Regional Studies* 30: 509–32

Jones, M. R. (1995) 'Training and enterprise councils: a continued search for local flexibility', *Regional Studies* 29: 577–80

Jones, M. and Ward, K. (2004) 'Neo-liberalism, crisis and the city: the political economy of Labour's urban policy', in Johnstone and Whitehead (eds) (2004)

Jones, P. (1990) *Traffic Quotes: Public Perception of Traffic Regulation in Urban Areas: Report of a Research Study* (Department of Transport), London: HMSO

Jones, P. (1998) 'The Groundwork network', *Geography* 83: 189–93

Jones, P. (1999) 'Groundwork: changing places and agendas', *Town and Country Planning* 68: 315–17

Jones, P. M. (1991a) 'Gaining public support for road pricing through a package approach', *Traffic Engineering and Control* 32: 194–6

Jones, P. M. (1991b) 'UK public attitudes to urban traffic problems and possible countermeasures: a poll of polls', *Environment and Planning C: Government and Policy* 9: 245–56

Jones, R. (1982) *Town and Country Choice: A Critical Analysis of Britain's Planning System*, London: Adam Smith Institute

Jong, M. de, Lalenis, K. and Mamadouh, V. (2002) *The Theory and Practice of Institutional Transplantation: Experiences with the Transfer of Policy Institutions*, Dordrecht: Kluwer

Jonsson, C., Tagil, S. and Tornqvist, G. (2000) *Organizing European Space*, London: Sage

Joseph Rowntree Foundation (JRF) (1994) *Inquiry into Planning for Housing*, York: JRF

Joseph Rowntree Foundation (2003) *Britain's Housing in 2022*, York: JRF

Journal of Planning and Environment Law (1995) 'Are off-site effects material to material changes of use?' (and case of *Forest of Dean District Council v Secretary of State for the Environment and Howells*), *Journal of Planning and Environment Law* 1995: 889 and 937–43

Jowell, J. (1975) 'Development control [review article on the Dobry Report]', *Political Quarterly* 46: 340–4

Jowell, J. (1977a) 'Bargaining in development control', *Journal of Planning and Environment Law* 1977: 414–33

Jowell, J. (1977b) 'The limits of law in urban planning', *Current Legal Problems 1977* 30: 63–83

Jowell, J. and Grant, M. (1983) 'Guidelines for planning gain?', *Journal of Planning and Environment Law* 1983: 427–31

Jowell, J. and Millichap, D. (1987) 'Enforcement: the weakest link in the planning chain', in Harrison, M. L. and Mordey, R. (eds) *Planning Control: Philosophies, Prospects and Practice*, London: Croom Helm

Jowell, J. and Oliver, D. (1994) *The Changing Constitution*, Oxford: Clarendon Press

JURUE: ECOTEC Research and Consultancy Ltd (1987) *Greening City Sites*, London: HMSO

JURUE: ECOTEC Research and Consultancy Ltd (1988) *Improving Urban Areas*, London: HMSO

Kain, J. F. and Beesley, M. E. (1965) 'Forecasting car ownership and use', *Urban Studies* 2: 163–85

Karn, V. and Lucas, J. with Steele, A., Stefak, P. and Todd, S. (1996) *Home-Owners and Clearance: An Evaluation of Rebuilding Grants* (DoE), London: HMSO

Karppi, I. (2000) *Future Challenges and Institutional Preconditions for Regional Policy: Four Scenario Reports*, Stockholm: Nordregio

Kay, G. (1998) 'The right to roam: a restless ghost', *Town and Country Planning* 67: 255–9

Kay, J. H. (1998) *Asphalt Nation: How the Automobile Took Over America and How We Can Take it Back*, Berkeley, CA: University of California Press

Kay, P. and Evans, P. (1992) *Where Motor Car is Master*, London: Council for the Protection of Rural England

Keating, M. and Jones, B. (1985) *Regions in the European Community*, Oxford: Oxford University Press

Keeble, D., Tyler, P., Broom, G. and Lewis, J. (1992) *Business Success in the Countryside: The Performance of Rural Enterprise* (DoE), London: HMSO

Keeble, L. (1969) *Principles and Practice of Town and Country Planning*, London: Estates Gazette

Keene, Hon. Justice (1999) 'Recent trends in judicial control', *Journal of Planning and Environment Law* January: 30–7

Keith, M. and Rogers, A. (eds) (1991) *Hollow Promises: Rhetoric and Reality in the Inner City*, London: Mansell

Kellett, J. (1990) 'The environmental impact of wind energy developments', *Town Planning Review* 61: 139–55

Kelly, A. and Marvin, S. (1995) 'Demand-side management in the electricity sector: implications for town planning in the UK', *Land Use Policy* 12: 205–21

Kelly, M. and Gilg, A. (2000) 'The analysis of development control decision making: a cautionary tale from recent work on aggregate demand', *Planning Practice and Research* 15(4): 335–42

Kelly, M., Selman, P. and Gilg, A. (2004) 'Taking sustainability forward: relating practice to policy in a changing legislative environment', *Town Planning Review* 75(3): 309–35

Kemeny, J. (1994) *From Public Housing to the Social Market: Rental Policy Strategies in Comparative Perspective*, London: Routledge

Kemp, R. (1980) 'Planning, legitimation and the development of nuclear energy: a critical theoretic analysis of the Windscale inquiry', *International Journal of Urban and Regional Research* 4: 350–71

Kennedy, L. (2004) *Remaking Birmingham: The Visual Culture of Urban Regeneration*, London: Routledge

Kenny, M. and Meadowcroft, J. (eds) (1999) *Planning Sustainability*, London: Routledge

Kent County Council (1995) *A Vision for EuroRegion: Towards a Policy Framework – The First Step*, Maidstone: Kent County Council

Kenyon, R. C. (1991) 'Environmental assessment: an overview on behalf of the RICS', *Journal of Planning and Environment Law* 1991: 419–22

Keogh, G. and Evans, A. W. (1992) 'The private and social costs of planning delay', *Urban Studies* 29: 687–99

Kerr, D. (1998) 'The Private Finance Initiative and the changing governance of the built environment', *Urban Studies* 35: 2277–301

Khan, M. A. (1995) 'Sustainable development: the key concepts, issues and implications', *Sustainable Development* 3: 63–9

Kidd, S. and Kumar, A. (1993) 'Development planning in the English metropolitan counties: a comparison of performance under two planning systems', *Regional Studies* 27: 65–73

King, A. D. (ed.) (1980) *Buildings and Society: Essays on the Social Development of the Built Environment*, London: Routledge

King, A. D. (1990) *Global Cities: Post-imperialism and the Internationalisation of London*, London: Routledge

King, J. (1990) *Regional Selective Assistance 1980–84: An Evaluation by DTI, IDS, and WOID*, London: HMSO

Kingdom, J. E. (1991) *Local Government and Politics in Britain*, London: Philip Allan

Kingdon, J. W. (1984) *Agendas, Alternatives, and Public Policies*, New York: HarperCollins

Kinloch, I. (2005) 'What is wrong with temporary stop notices', *Journal of Planning and Environment Law* January: 7–11

Kintrea, K. and Morgan, J. (2005) *Evaluation of English Housing Policy 1975–2000, Theme 3: Housing Quality and Neighbourhood Quality*, London: ODPM

Kirby, A. (1985) 'Nine fallacies of local economic change', *Urban Affairs Quarterly* 21: 207–20

Kirby, P. (2004) *Best Practice Note on the Use of Conditions*, Planning Officers' Society website: www. planningofficers. org.uk (accessed November 2004)

Kirk, G. (1980) *Urban Planning in a Capitalist Society*, London: Croom Helm

Kirkby, J., O'Keefe, P. and Timberlake, L. (1995) *The Earthscan Reader in Sustainable Development*, London: Earthscan

Kirkwood, G. and Edwards, M. (1993) 'Affordable housing policy: desirable but unlawful?', *Journal of Planning and Environment Law* 1993: 317–24

Kirwan, R. (1989) 'Finance for urban public infrastructure', *Urban Studies* 26: 285–300

Kitchen, T. (1996) 'A future for strategic planning policy: a Manchester perspective', in Tewdwr-Jones, M. (1999) *British Planning Policy in Transition: Planning in the 1990s*, London: UCL Press

Kitchen, T. (1997) *People, Politics, Policies and Plans: The City Planning Process in Contemporary Britain*, London: Paul Chapman

Kitchen, T. (1999a) 'Consultation on government policy initiatives: the case of regional planning guidance', *Planning Practice and Research* 14(1): 5–16

Kitchen, T. (1999b) 'The structure and organisation of the planning service in English local government', *Planning Practice and Research* 14: 313–27

Kitson, T. (1999) 'The European Convention on Human Rights and local plans', *Journal of Planning and Environment Law* April: 321–3

Kivell, P. (1987) 'Derelict land in England: policy responses to a continuing problem', *Regional Studies* 21: 265–9

Kivell, P. (1993) *Land and the City: Patterns and Processes of Urban Change*, London: Routledge

Klein, R. (1974) 'The case for elitism: public opinion and public policy', *Political Quarterly* 45: 406–17

Klosterman, R. E. (1985) 'Arguments for and against planning', *Town Planning Review* 56: 5–20

Knox, P. L. (1994) *Urbanization: An Introduction to Urban Geography*, Englewood Cliffs, NJ: Prentice Hall

Knox, P. L. and Cullen, J. L. (1981) 'Planners as urban managers: an exploration of the attitudes and self-image of senior British planners', *Environment and Planning A* 13: 885–98

Knox, P. L. and Taylor, P. J. (eds) (1995) *World Cities in a World-System*, Cambridge: Cambridge University Press

Koresawa, A. and Konvitz, J. (2001) 'Towards new roles for spatial planning', in OECD, *Towards a New Role for Spatial Planning*, Paris: OECD

Koslowski, J. and Hill, G. (1993) *Towards Planning for Sustainable Development*, Aldershot: Avebury

Kostof, S. (1991) *The City Shaped: Urban Patterns and Meaning through History*, London: Thames and Hudson

Kostof, S. (1992) *The City Assembled: The Elements of Urban Form through History*, London: Thames and Hudson

KPMG Consulting (1999a) *City Challenge: Final National Evaluation*, London: DETR

KPMG Consulting (1999b) *Fiscal Incentives for Urban Housing: Exploring the Options*, London: DETR

Kraan, D. J. and Veld, R. J. (eds) (1991) *Environmental Protection: Public or Private Choice?*, Dordrecht: Kluwer

Krämer, L. (1991) 'The implementation of Community environmental directives within member states: some implications of the direct effect doctrine', *Journal of Environmental Law* 3: 39–56

Krämer, L. (1992) *Focus on European Environmental Law*, London: Sweet and Maxwell

Krämer, L. (1999) *EC Environmental Law* (fourth edition), London: Sweet and Maxwell

Kreukels, A. and Salet, W. (2002) *Metropolitan Governance and Spatial Planning: Comparative Case Studies of European City Regions*, London: Spon

Kromarek, P. (1986) 'The Single European Act and the environment', *European Environment Review* 1: 10–12

Kunzmann, K. R. and Wegener, M. (1991) 'The pattern of urbanisation in Western Europe', *Ekistics* 350 (September–October): 282–91

Labour Party (1995) *A Choice for England: A Consultation Paper on Labour's Plans for English Regional Government*, London: Labour Party

Lambert, A. J. and Wood, C. M. (1990) 'UK implementation of the European directive on EIA', *Town Planning Review* 61: 247–62

Lambert, C. (2004) 'Local strategic partnerships, community strategies and development planning: negotiating horizontal and vertical integration in an emergent local governance', paper presented at the AESOP Conference, Grenoble, July

Lampe, D. and Kaplan, M. (1999) *Resolving Land-Use conflicts through Mediation: Challenges and Opportunities*, Boston, MA: Lincoln Institute of Land Policy

Land Capability Consultants (1990) *Cost Effective Management of Reclaimed Derelict Sites* (DoE), London: HMSO

Land Use Consultants (1986) *Channel Fixed Link: Environmental*

Appraisal of Alternative Proposals (Department of Transport), London: HMSO

Land Use Consultants (1991) *Permitted Development Rights for Agriculture and Forestry* (DoE), London: HMSO

Land Use Consultants (1993) *Local Moves: The Funding and Formulation of Local Transport Policy*, London: Council for the Protection of Rural England

Land Use Consultants (1995a) *Effectiveness of Planning Policy Guidance Notes*, London: HMSO

Land Use Consultants (1995b) *Planning Controls over Agricultural and Forestry Development and Rural Building Conversions* (DoE), London: HMSO

Land Use Consultants (1996) *Reclamation of Damaged Land for Nature Conservation* (DoE), London: HMSO

Land Use Consultants (1999) *Review of National Planning Policy Guidelines*, Edinburgh: Scottish Office

Land Use Consultants (2002) *Information and Communication Technology in Planning*, London: DETR

Land Use Consultants and Wilbraham and Co. (2003) *Formulation of Guidance on the Use of Local Development Orders*, London: ODPM

Landry, C., Montgomery, J. and Worpole, K. (1989) *The Last Resort: Tourism, Tourist Employment and 'PostTourism' in the South East*, Bournes Green, Gloucester: Comedia

Lane, J. and Vaughan, S. (1992) *An Evaluation of the Impact of PPG 16 on Archaeology and Planning*, London: Pagoda Associates

Lane, P. and Peto, M. (1995) *Blackstone's Guide to the Environment Act 1995*, London: Blackstone

Larkham, P. J. (1990) 'The use and measurement of development pressure', *Town Planning Review* 61: 171–83

Larkham, P. J. (1995) *Patterns in the Designation of Conservation Areas*, Birmingham: School of Planning, University of Central England in Birmingham

Larkham, P. J. (1996) *Conservation and the City*, London: Routledge

Larkham, P. J. (1999a) 'Preservation, conservation and heritage: developing concepts and applications', in Cullingworth, J. B. (ed.) *British Planning: 50 Years of Urban and Regional Policy*, London: Athlone

Larkham, P. J. (1999b) 'Tensions in managing the suburbs: conservation versus change', *Area* 31: 359–71

Larkham, P. J. and Chapman, D. W. (1996) 'Article 4 directions and development control: planning myths, present uses, and future possibilities', *Journal of Environmental Planning and Management* 39: 5–19

Larkham, P. J. and Jones, A. (1993) 'Conservation and conservation areas in the UK: a growing problem', *Planning Practice and Research* 8(2): 19–29

Larkham, P. J. and Lodge, J. (1999) *Do Residents Understand What Conservation Means?*, Working Paper Series no. 71, Birmingham: University of Central England

Larkham, P. J., Jones, A. and Daniels, R. (1992) *The Character of Conservation Areas*, London: Royal Town Planning Institute

Lash, S., Szerszynski, B. and Wynne, B. (1996) *Risk, Environment and Modernity: Towards a New Ecology*, London: Sage

Lasok, D. (1994) *Law and Institutions of the European Communities* (sixth edition), London: Butterworth

Last, K. V. (1999) 'Protection of archaeological sites: an outline of the law', *Scottish Planning and Environment Law* 73: 60–1

Lavers, A. and Webster, B. (1994) 'Participation in the plan-making process: financial interests and professional representation', *Journal of Property Research* 11: 131–44

Law Commission (1993) *Administrative Law: Judicial Review and Statutory Appeals*, Law Commission Consultation Paper 126, London: HMSO

Law, C. M. (1992) 'Urban tourism and its contribution to economic regeneration', *Urban Studies* 29: 599–619

Lawless, P. (1986) *The Evolution of Spatial Policy: A Case Study of Inner Urban Policy in the United Kingdom 1968–1981*, London: Pion

Lawless, P. (1989) *Britain's Inner Cities* (second edition), London: Paul Chapman

Lawless, P. (1991) 'Urban policy in the Thatcher decade: English inner-city policy, 1979–90', *Environment and Planning C: Government and Policy* 9: 15–30

Lawless, P. (1994) 'Partnership in urban regeneration in the UK: the Sheffield central area study', *Urban Studies* 31: 1301–24

Lawless, P. (1996) 'The inner cities: towards a new agenda', *Town Planning Review* 67: 21–43

Lawless, P. (1998) *Unemployment and Social Exclusion: Landscapes of Labour Inequality*, London: Jessica Kingsley

Lawless, P. and Dabinett, G. (1995) 'Urban regeneration and transport investment: a research agenda', *Environment and Planning A* 27: 1029–48

Lawless, P. and Robinson, D. (2000) 'Inclusive regeneration? Integrating social and economic regeneration in English local authorities', *Town Planning Review* 71(3): 289–310

Laws, F. G. (1991) *Guide to the Local Government Ombudsman Service*, Harlow: Longman

Lawton, R. and Pooley, C. G. (1992) *The New Geography of London*, London: Edward Arnold

Laxton (1996) *Laxton's Guide to Single Regeneration Budgets*, Oxford: Heinemann

Layard, A., Batty, S. and Davoudi, S. (2001) *Planning for a Sustainable Future*, London: Spon

Layfield, F. (1992) 'The Environmental Protection Act 1990: the system of integrated pollution control', *Journal of Planning and Environment Law* 1992: 3–13

Layfield, F. (1996) 'The planning inquiry: an inspector's perspective', *Journal of Planning and Environment Law* 1996: 370–6

Leach, N. (1997) *Rethinking Architecture: A Reader in Cultural Theory*, London: Routledge

Leach, R. (1994) 'The missing dimension to the local government review', *Regional Studies* 28(8): 797–802

Leach, S. and Game, C. (1991) 'English metropolitan government since abolition: an evaluation of the abolition of the English metropolitan councils', *Public Administration* 69: 141–70

Leach, S. and Stewart, M. (1992) *Local Government: Its Role and Function*, York: Joseph Rowntree Foundation

Leach, S., Davis, H., Game, C. and Skelcher, C. (1991) *After Abolition: The Operation of the Post-1986 Metropolitan*

Government System, Birmingham: Institute of Local Government Studies, University of Birmingham

Leach, S., Stewart, J., Spencer, K., Walsh, K. and Gibson, J. (1992) *The Heseltine Review of Local Government: A New Vision or Opportunities Missed?*, Birmingham: Institute of Local Government Studies, University of Birmingham

Leach, S., Stewart, J. and Walsh, K. (1994) *The Changing Organisation and Management of Local Government*, London: Macmillan

Leather, P. (1997) *The State of UK Housing: A Factfile on Dwelling Conditions*, Bristol: University of Bristol

Leather, P. and Mackintosh, S. (1998) *Housing Renewal: Policy and Practice*, London: UCL Press

Leather, P. and Morrison, T. (1997) *The State of UK Housing*, York: Policy Press

Lederman, P. B. and Librizzi, W. (1995) 'Brownfields remediation: available technologies', *Journal of Urban Technology* 2: 21–9

Ledgerwood, G. (1985) *Urban Innovation: The Transformation of the London Docklands 1968–84*, Aldershot: Gower

Lee, P. and Murie, A. (1999) *Literature Review of Social Exclusion*, Edinburgh: Central Research Unit, Scottish Office

Lees, A. (1993) *Enquiry into the Planning System in North Cornwall District*, London: HMSO

LeGates, R. (ed.) (1998) *Early Urban Planning 1870–1940* (9 vols), London: Routledge

LeGates, R. and Stout, F. (eds) (1996) *The City Reader*, London: Routledge

Leitch, G. (chair) (1977) *Report of the Advisory Committee on Trunk Road Assessment*, Leitch Report, London: HMSO

Leung, H. L. (1979) *Redistribution of Land Values: A Re-examination of the 1947 Scheme*, Cambridge: Department of Land Economy, University of Cambridge

Leung, H. L. (1987) 'Developer behaviour and development control', *Land Development Studies* 4: 17–34

Lever, W. F. (1991) 'Deindustrialisation and the reality of the post-industrial city', *Urban Studies* 28: 983–99

Lever, W. F. (1992) 'Local authority responses to economic change in West Central Scotland', *Urban Studies* 29: 935–48

Levett, R. (1993) *Agenda 21: A Guide for Local Authorities in the UK*, Luton: Local Government Management Board

Levett, R. (2000) 'What counts for quality of life', *EG* 6(3): 6–8

Levett, R. and Christie, I. (1999) *The Richness of Cities: Urban Policy in a New Landscape*, London: Comedia in association with Demos

Levett, R. and Therivel, R. (2002) *Response to the Planning Green Paper*, London: Levett-Therivel

Levett-Therivel Sustainability Consultants (2004) *Assessment of Progress against the Headline Indicators*, London: Sustainable Development Commission

Levin, P. H. (1979) 'Highway inquiries: a study in government responsiveness', *Public Administration* 57: 21–49

Levin, P. H. (1976) *Government and the Planning Process: An Analysis and Appraisal of Government Decision-making Processes with Special Reference to the Launching of New Towns and Town Development Schemes*, London: Allen and Unwin

Levin, P. H. and Donnison, D. V. (1969) 'People and planning', *Public Administration* 13: 473–9

Levy, F., Meltsner, A. J. and Wildavsky, A. B. (1974) *Urban Outcomes: Schools, Streets, and Libraries*, Berkeley, CA: University of California Press

Levy, J. M. (2000) *Contemporary Urban Planning*, London: Pearson

Lewis, J. and Townsend, A. (1989) *The North–South Divide: Regional Change in Britain in the 1980s*, London: Paul Chapman

Lewis, J. P. (1974) *A Study of the Cambridge Sub-region*, London: HMSO

Lewis, N. (1992) *Inner City Regeneration: The Demise of Regional and Local Government*, Buckingham: Open University Press

Lewis, N. C. (1994) *Road Pricing: Theory and Practice*, London: Thomas Telford

Lewis, R. (1995) 'Contaminated land: the new regime of the Environment Act 1995', *Journal of Planning and Environment Law* 1995: 1087–96

Lewis, R. P. (1992) 'The Environmental Protection Act 1990: waste management in the 1990s: waste regulation and disposal', *Journal of Planning and Environment Law* 1992: 303–12

Lichfield, N. (1992) 'From planning gain to community benefit', *Journal of Planning and Environment Law* 1992: 1103–18

Lichfield, N. (1995) *Community Impact Evaluation: Principles and Practice*, London: UCL Press

Lichfield, N. (2003) *Review of Permitted Development Rights*, London: ODPM

Lichfield, N. and Darin-Drabkin, H. (1980) *Land Policy in Planning*, London: Allen and Unwin

Lindblom, C. E. (1959) 'The science of muddling through', *Public Administration Review* 19(2): 79–88 (reprinted in Faludi, A. (ed.) (1973) *A Reader in Planning Theory*, Oxford: Pergamon)

Liniado, M. (1996) *Car Culture and Countryside Change*, Cirencester: National Trust

Lipman, C. (1999) 'Difficult decisions in a rural balancing act' [N. Ireland], (together with 'How the Province deals with countryside issues') *Planning* 29 October: 10–11

Little, J. (1994a) *Gender, Planning and the Policy Process*, Oxford: Pergamon

Little, J. (1994b) 'Women's initiatives in town planning in England', *Town Planning Review* 65: 261–76

Little, J., Peake, L. and Richardson, P. (eds) (1988) *Women in Cities*, New York: New York University Press

Littlewood, S. and Whitney, D. (1996) 'Re-forming urban regeneration? The practice and potential of English Partnerships', *Local Economy* 11: 39–49

Llewelyn-Davies (1994) *Providing More Homes in Urban Areas*, Bristol: School for Advanced Urban Studies, University of Bristol

Llewelyn-Davies (1996) *The Re-use of Brownfield Land for Housing*, York: Joseph Rowntree Foundation

Llewelyn-Davies (1997) *Sustainable Residential Quality: New Approaches to Urban Living*, London: Llewelyn-Davies Planning

Llewelyn-Davies (1998a) *Planning and Development Briefs: A Guide to Better Practice*, London: DETR

Llewelyn-Davies (1998b) *Urban Living: Perceptions, Realities and Opportunities*, Winchester: Hampshire County Council

Llewellyn-Davies (2000) *Urban Design Compendium*, London: English Partnerships and the Housing Corporation

Llewelyn-Davies and JMP Consultants (1998) *Parking Standards in the South East*, London: DETR

Llewelyn-Davies *et al.* (1994) *The Quality of London's Residential Environment*, London: London Planning Advisory Committee

Llewelyn-Davies, CAG Consultants and GHK Economics (2000) *Millennium Villages and Sustainable Communities*, London: DETR

Lloyd, G. (2004) 'The national planning framework and city regional planning in Scotland', paper presented to the RTPI Conference on Regional Spatial Planning, London, October

Lloyd, H. (1998–9) 'Are current visitor numbers at historic properties sustainable? A case study from Batemans, the home of Rudyard Kipling', *Views* 30 (spring) and The National Trust, www. ntenvironment.com/html/building/papers/visitor1.htm

Lloyd, M. G. (1990) 'Simplified planning zones in Scotland: government failure or the failure of government?', *Planning Outlook* 33: 128–32

Lloyd, M. G. (1992) 'Simplified planning zones, land development, and planning policy in Scotland', *Land Use Policy* 9: 249–58

Lloyd, M. G. (1997) 'Structure and culture: regional planning and institutional innovation in Scotland', in MacDonald, R. and Thomas, H. (eds) *Nationality and Planning in Scotland and Wales*, Cardiff: University of Wales

Lloyd, M. G. and Livingstone, L. H. (1991) 'Marine fish farming in Scotland: proprietorial behaviour and the public interest', *Journal of Rural Studies* 7: 253–63

Lloyd, M. G. and Rowan-Robinson, J. (1992) 'Review of strategic planning guidance in Scotland', *Journal of Environmental Planning and Management* 35: 93–9

Lloyd, R. (2001) *A Practical Guide to Compulsory Purchase in Urban Regeneration*, Manchester: Eversheds

Local Government Association (LGA) (2003) *Planning Reform: A Survey of Local Authorities*, London: LGA

Local Government Association and IDeA (1997) *Best Value and Sustainable Development Guidance*, London: IDeA

Local Government Management Board (LGMB) (1992) *Citizens and Local Democracy: Charting a New Relationship*, Luton: LGMB

Local Government Management Board (1993a) *Local Agenda 21 UK: A Framework for Local Sustainability*, Luton: LGMB

Local Government Management Board (1993b) *Local Agenda 21: Principles and Process – A Step by Step Guide*, Luton: LGMB

Local Government Management Board (1995a) *Indicators for Local Agenda 21*, Luton: LGMB

Local Government Management Board (1995b) *Sustainable Settlements: A Guide for Planners, Designers and Developers* (Barton, H., Davis, G. and Guise, R.), Luton: LGMB

Local Government Management Board and Royal Town Planning Institute (1993c) *Local Government Staffs Survey 1992*, Luton: LGMB

Local Government Management Board and Royal Town Planning Institute (1993d) *Planning Staffs Survey 1992*, Luton: LGMB

Lock, D. (1989) *Riding the Tiger: Planning the South of England*, London: Town and Country Planning Association

Lock, D. (1994) 'Keynote address', in Wood, M. (ed.) *Planning Icons: Myth and Practice* (Planning Law Conference), London: Sweet and Maxwell

Lock, D. (1999) 'The new Catch 22 on need', *Town and Country Planning* 68: 111

Loevinger Rahder, B. and O'Neill, K. (1998) 'Women and planning: education for social change', *Planning Practice and Research* 13(3): 247–65

Loftman, P. and Beazley, M. (1998a) *Race, Equality and Planning*, London: Local Government Association

Loftman, P. and Beazley, M. (1998b) 'Racial equality and the planning agenda', *Town and Country Planning* 67(10): 326–7

Loftman, P. and Nevin, B. (1992) *Urban Regeneration and Social Equity: A Case Study of Birmingham 1986–1992*, Birmingham: Faculty of the Built Environment Research Paper, University of Central England

Loftman, P. and Nevin, B. (1995) 'Prestige projects and urban regeneration in the 1980s and 1990s: a review of the benefits and limitations', *Planning Practice and Research* 10: 299–315

Loftman, P. and Pratt, D. (1994) 'Public participation in the UDP process: the need to improve the UDP public consultation process, *Town and Country Planning*, 63(9): 246

Lomas, J. (1994) 'The role of management agreements in rural environmental conservation', *Land Use Policy* 11: 119–23

London and South East Regional Planning Conference: *see* SERPLAN

London Boroughs Association (LBA) (1990) *Road Pricing for London*, London: LBA

London Boroughs Association (1992) *Out of Order: The 1987 Use Classes Order – Problems and Proposals*, London: LBA

London Boroughs' Disability Resource Team (1991) *Towards Integration: The Participation of Disabled People in Planning*, London: Disability Resource Team

London Economics (1992) *The Potential Role of Market Mechanisms in the Control of Acid Rain* (DoE), London: HMSO

London Planning Advisory Committee (LPAC) (1991) *London: World City Moving into the 21st Century*, London: HMSO

London Planning Advisory Committee (1998a) *Sustainable Residential Quality: New Approaches to Urban Living*, London: LPAC

London Planning Advisory Committee (1998b) *Dwellings over and in Shops in London*, London: LPAC

London Research Centre (LRC) (1991) *Much Ado About Nothing: An Examination of the Potential for the Planning System to Secure Affordable Housing with Special Reference to Planning Agreements*, London: LRC

London Research Centre (1992) *Paris, London: A Comparison of Transport Systems*, London: LRC

London Transport Planning (1997) *Review of Office and Retail*

Parking Standards in London Unitary Development Plans, London: London Transport

London Women and Planning Group (1991) *Shaping our Borough: Women and Unitary Development Plans*, London: Royal Town Planning Institute

Longley, P., Batty, M., Shepherd, J. and Sadler, G. (1992) 'Do green belts change the shape of urban areas? A preliminary analysis of the settlement geography of South East England', *Regional Studies* 26: 437–52

Loughlin, M. (1994) *The Constitutional Status of Local Government*, Research Report no. 3, London: Commission for Local Democracy

Loughlin, M., Gelfand, M. D. and Young, K. (eds) (1985) *Half a Century of Municipal Decline*, London: Allen and Unwin

Lovan, W. R., Murray, M. and Shaffer, R. (2004) *Participatory Governance: Planning, Conflict Mediation and Public Decision Making in Civil Society*, Aldershot: Ashgate

Low, N. (1991) *Planning, Politics and the State: Political Foundations of Planning Thought*, London: Unwin Hyman

Lowe, P. and Flynn, A. (1989) 'Environmental politics and policy in the 1980s', in Mohan, J. (ed.) *The Political Geography of Contemporary Britain*, London: Macmillan

Lowe, P. and Goyder, J. (1983) *Environmental Groups in Politics*, London: Allen and Unwin

Lowe, P., Murdoch, J., Marsden, T., Munton, R. and Flynn, A. (1993) 'Regulating the new rural spaces: the uneven development of land', *Journal of Rural Studies* 9: 205–22

Lowe, R. and Bell, M. (1998) *Towards Sustainable Housing: Building Regulation for the 21st Century*, York: Joseph Rowntree Foundation

Lowe, S. (2004) *Housing Policy Analysis: British Housing in Cultural and Comparative Context*, London: Palgrave Macmillan

Lowndes, V. (1998) *Guidance on Enhancing Public Participation in Local Government*, London: DETR

Lutyens, E. and Abercrombie, P. (1945) *A Plan for the City and County of Kingston upon Hull*, Hull: Brown

Lyddon, D. and Dal Cin, A. (1996) *International Manual of Planning Practice* (third edition), The Hague: International Society of City and Regional Planners

Lyle, J. and Hill, D. (2003) 'Watch this space: an investigation of strategic gap policies in England', *Planning Theory and Practice* 4(2): 165–84

Lynch, P. (1999) 'New Labour and the English regional development agencies: devolution as evolution', *Regional Studies* 33: 73–8

McAllister, A. and McMaster, R. (1994) *Scottish Planning Law: An Introduction*, Edinburgh: Butterworths

McAllister, D. (ed.) (1996) *Partnership in the Regeneration of Urban Scotland* (Scottish Office), Edinburgh: HMSO

McAuslan, J. P. W. B. (1975) *Land, Law and Planning*, London: Weidenfeld and Nicolson

McAuslan, J. P. W. B. (1980) *The Ideologies of Planning Law*, Oxford: Pergamon

McAuslan, J. P. W. B. (1991) 'The role of courts and other judicial type bodies in environmental management', *Journal of Environmental Law* 3: 195–208

McBride, J. (1973) 'The urban aid programme: is it running out of cash?' *Quest*, 16 March

McCaig, E., Henderson, C. and MVA Consultancy (1995) *Sustainable Development: What it Means to the General Public*, Edinburgh: Central Research Unit, Scottish Office

McCarthy, J. (1995) 'The Dundee waterfront: a missed opportunity for planned regeneration', *Land Use Policy* 12: 307–19

McCarthy, J. (1999) 'Urban regeneration in Scotland: an agenda for the Scottish Parliament', *Regional Studies* 33(6): 559–66

McCarthy, J. and Newlands, D. (eds) (1999) *Governing Scotland: Problems and Prospects – The Economic Impact of the Scottish Parliament*, Aldershot: Ashgate

McCarthy, J., Lloyd, G. and Fernie, K. (1999) 'Planning for social inclusion in Scotland', *Town and Country Planning* 68: 310–11

McCarthy, P. and Harrison, T. (1995) *Attitudes to Town and Country Planning* (DoE), London: HMSO

McClintock, H. (ed.) (1992) *The Bicycle and City Traffic*, London: Belhaven

McClintock, H. (2001) 'Practitioners' take-up of professional guidance and research findings: planning for cycling and walking in the UK', *Planning Practice and Research* 16(2): 193–203

McClintock, H. (ed.) (2002) *Planning for Cycling: Principles, Practice and Solutions for Urban Planners*, Cambridge: Woodhead

McConville, J. and Sheldrake, J. (eds) (1995) *Transport in Transition: Aspects of British and European Experience*, Aldershot: Avebury

McCormick, J. (1991) *British Politics and the Environment*, London: Earthscan

McCormick, J. (1995) *The Global Environmental Movement*, Chichester: Wiley

McCrone, D. (1997) *Land, Democracy and Culture in Scotland*, Perth: Rural Forum

McCrone, G. (1991) 'Urban renewal: the Scottish experience', *Urban Studies* 28: 919–38

McCrone, G. and Stephens, M. (1995) *Housing Policy in Britain and Europe*, London: UCL Press

MacDonald, R. and Thomas, H. (eds) (1997) *Nationality and Planning in Scotland and Wales*, Cardiff: University of Wales Press

McDougall, G. (1993) *Planning Theory: Prospects for the 1990s*, Aldershot: Avebury

MacEwen, A. and MacEwen, M. (1982) *National Parks: Conservation or Cosmetics?*, London: Allen and Unwin

MacEwen, A. and MacEwen, M. (1987) *Greenprints for the Countryside? The Story of Britain's National Parks*, London: Allen and Unwin

MacEwen, M. (1973) *Crisis in Architecture*, London: Royal Institute of British Architects

MacEwen, M. and Sinclair, G. (1983) *New Life for the Hills*, London: Council for National Parks

McFarlane, R. (1993a) *Community Involvement in City Challenge: A Good Practice Guide*, London: National Council for Voluntary Organisations

McFarlane, R. (1993b) *Community Involvement in City Challenge: A Policy Report*, London: National Council for Voluntary Organisations

MacFarlane, R. (1998) 'What – or who – is rural Britain?', *Town and Country Planning* 67: 184–8

McGill, G. (1995) *Building on the Past: A Guide to the Archaeology and Development Process*, London: Spon

McGregor, A., Glass, A. Higgins, K., MacDougal, L. and Sutherland, V. (2003) *Developing People, Regenerating Pace: Achieving Greater Integration for Local Area Regeneration*, Bristol: Policy Press

MacGregor, B. and Ross, A. (1995) 'Master or servant? The changing role of the development plan in the British planning system', *Town Planning Review* 66: 41–59

MacGregor, S. and Pimlott, B. (eds) (1990) *Tackling the Inner Cities: The 1980s Revisited, Prospects for the 1990s*, Oxford: Clarendon Press

McHarg, I. L. (1967/1992) *Design with Nature*, New York: Wiley

McIntosh, N. (chair) (1999) *Moving Forward: Local Government and the Scottish Parliament*, McIntosh Report, Edinburgh: Scottish Office

Mackay, D. (1995) *Scotland's Rural Land Use Agencies*, Aberdeen: Scottish Cultural Press

McKay, D. H. and Cox, A. W. (1979) *The Politics of Urban Change*, London: Croom Helm

McKay, S. (2003) 'Sheriffs and outlaws: in pursuit of effective enforcement', *Town Planning Review* 74(4): 423–43

Mackie, P. J. (1980) 'The new grant system for local transport: the first five years', *Public Administration* 59: 187–206

McKinnon, A. (1999) 'First steps: the DETR Sustainable Distribution document', *Town and Country Planning* 68: 257

McKinsey Global Institute (1996) *Driving Productivity and Growth in the UK Economy*, London: McKinsey

Mackintosh, M. (1992) 'Partnership: issues of policy and negotiation', *Local Economy* 3(1): 210–24

Mackintosh, S. and Leather, P. (1992) *Home Improvement under the New Regime*, Bristol: School for Advanced Urban Studies, University of Bristol

McLaren, D., Bullock, S. and Yousef, N. (1998) *Tomorrow's World: Britain's Share in a Sustainable Future*, London: Earthscan

Maclennan, D. (1986) *The Demand for Housing: Economic Perspectives and Planning Practices*, Edinburgh: Scottish Development Department

Maclennan, D. (1989) 'Housing in Scotland 1977– 1987', in Smith, M. E. H. (ed.) *Guide to Housing: Main Changes in Housing Law* (third edition), London: Housing Centre

McLoughlin, J. B. (1969) *Urban and Regional Planning: A Systems Analysis*, Oxford: Pergamon

McMaster, R. F. (1995) *Results of Royal Town Planning Institute Survey of Local Planning Authorities' Views on Local Government Reorganisation: A Mixed Picture*, Edinburgh: Royal Town Planning Institute in Scotland

Macnaghten, P., Grove-White, R., Jacobs, M. and Wynne, B. (1995) *Public Perceptions and Sustainability in Lancashire*, Report by the Centre for the Study of Environmental Change, Lancaster University, Preston: Lancashire County Council

Macpherson, W. (1999) *The Stephen Lawrence Inquiry*, Macpherson Report, London: The Stationery Office

McQuail, P. (1994) *Origins of the DoE*, London: DoE

McQuail, P. (1995) *A View from the Bridge*, London: DoE

Macrory, R. (1992) *Campaigners' Guide to Using EC Environmental Law*, London: Council for the Protection of Rural England

McSheffrey, G. (2000) *Planning Derry: Planning and Politics in Northern Ireland*, Liverpool University Press

Madanipour, A. (1996) *Design of Urban Space: An Inquiry into a Social-Spatial Process*, Chichester: Wiley

Madden, M. (1987) 'Planning and ethnic minorities; an elusive literature', *Planning Practice and Research* 2 (March): 29–32

Maddison, D., Pearce, D., Johansson, O., Calthorpe, E., Litman, T. and Verhoef, E. (1996) *The True Costs of Road Transport*, London: Earthscan

Maginn, P. (2004) *Urban Regeneration Community Power and the (In) Significance of 'Race'*, Aldershot: Ashgate

Malbert, B. (1998) *Urban Planning Participation: Linking Practice and Theory*, Gothenburg: Department of Urban Design and Planning, University of Technology

Maloney, W. A. and Richardson, J. J. (1995) *Managing Policy Change in Britain: The Politics of Water*, Edinburgh: Edinburgh University Press

Malpass, P. (1994) 'Policy making and local governance: how Bristol failed to secure City Challenge funding (twice)', *Policy and Politics* 22: 301–12

Malpass, P. (1996) 'The unravelling of housing policy in Britain', *Housing Studies* 11(3): 459–70

Malpass, P. and Means, R. (eds) (1993) *Implementing Housing Policy*, Buckingham: Open University Press

Mandelker, D. R. (1962) *Green Belts and Urban Growth*, Madison, WI: University of Wisconsin Press

Mandelson, P. and Liddle, R. (1996) *The Blair Revolution*, London: Faber

Manley, J. (1987) 'Archaeology and planning: a Welsh perspective', *Journal of Planning and Environment Law* 1987: 466–84 and 552–63

Manley, S. and Guise, R. (1998) 'Conservation in the built environment', in Greed, C. and Roberts, M. (eds) *Introducing Urban Design: Interventions and Responses*, London: Addison Wesley Longman

Manning, P. K. (1989) 'Managing risk: managing uncertainty in the British nuclear installations inspectorate', *Law and Policy* 11: 350–69

Manns, S. and Wood, C. (2001) 'Giving locals a chance to speak', *Planning*, 5 May: 15

Manns, S. and Wood, C. (2002) 'Public representations at development control planning committees', *Public Policy and Administration* 17(1): 39–51

Marquand, D. (1998) *Must Labour Win?*, Ninth ESRC Annual Lecture, Swindon: Economic and Social Research Council

Marriott, O. (1967) *The Property Boom*, London: Hamilton

Marris, P. (1982) *Community Planning and Conceptions of Change*, London: Routledge and Kegan Paul

Marsden, T. (1999) *Rural Sustainability*, London: Town and Country Planning Association

Marsden, T., Murdoch, J., Lowe, P., Munton, R. J. C. and Flynn, A. (1993) *Constructing the Countryside: An Approach to Rural Development*, London: UCL Press

Marshall, J. N. and Wood, P. A. (1995) *Services and Space: Key Aspects of Urban and Regional Development*, Harlow: Longman

Marshall, R. (1988) 'Agricultural policy development in Britain', *Town Planning Review* 59: 419–35

Marshall, S. (1999) 'Restraining mobility while maintaining accessibility: an impression of the city of sustainable growth', *Built Environment* 25: 168–79

Marshall, T. (1991) *Regional Planning in England and Germany*, Oxford: School of Planning, Oxford Polytechnic

Marshall, T. (1995) *Clearing an Industrial Area: Collingdon Road and the Cardiff Bay Development Corporation 1988–1994*, Oxford: School of Planning, Oxford Polytechnic

Marshall, T. (2004) 'English regional planning: recent progress and current government proposals', *Planning Practice and Research* 18(1): 81–94

Marte, L. (1994) *Ecology and Society: An Introduction*, London: Policy Press

Martin, D. and Robert, J. (2002) 'Influencing the development of European spatial planning', in Faludi, A. (ed.) (2002) *European Spatial Planning*, Cambridge, MA: Lincoln Institute of Land Policy

Martin, L. R. G. (1989) 'The important published literature of British planners', *Town Planning Review* 60: 441–57

Martin, S. (1989) 'New jobs in the inner city: the employment impacts of projects assisted under the urban development grant programme', *Urban Studies* 26: 627–38

Martin, S. (1995) 'Partnerships for local environment action: observations of the first two years of *Rural Action for the Environment*', *Journal of Environmental Planning and Management* 38: 149–65

Martin, S. and Pearce, G. (1992) 'The internationalisation of local authority economic development strategies: Birmingham in the 1980s', *Regional Studies* 26: 499–509

Martin, S. J., Ticker, M. J. and Bovaird, A. G. (1990) 'Rural development programmes in theory and practice', *Regional Studies* 24: 268–76

Marvin, S. and Guy, S. (1997) 'Infrastructure provision, development processes and the co-production of environmental value', *Urban Studies* 34: 2023–36

Marvin, S. and Slater, S. (1996) *'Holes in the Road': Roads and Utilities in the 1990s*, Newcastle upon Tyne: Department of Town and Country Planning, University of Newcastle

Marwick, A. (1970) *Britain in the Century of Total War: War, Peace and Social Change 1900–1967*, Harmondsworth: Penguin

Mascarucci, R. (ed.) *Vision*, Rome: Meltemi editore

Masser, I. (1983) *Evaluating Urban Planning Efforts*, Aldershot: Gower

Matarosso, F. (1995) *Spirit of Place: Redundant Churches as Urban Resources*, Bournes Green, Gloucester: Comedia

Mather, A. S. (1991) 'The changing role of planning in rural land use: the example of afforestation in Scotland', *Journal of Rural Studies* 7: 299–309

Mather, A. S. (1999) 'Reducing the need to travel in the City of Bristol: promoting bus use through complementary measures', *Built Environment* 25: 94–105

Mather, G. (1988) *Paying for Planning*, London: Institute of Economic Affairs

Maunder, W. J. (1994) *Dictionary of Global Climate Change* (second edition), London: UCL Press

Maurici, J. (2004) 'Gypsy planning challenges in the High Court', *Journal of Planning and Environment Law* December: 1654–70

Mawson, J. (ed.) (1995) *The Single Regeneration Budget: The Stocktake*, Birmingham: Centre for Urban and Regional Studies, University of Birmingham

Mawson, J. (1996) 'The re-emergence of the regional agenda in the English regions: new patterns of urban and regional governance', *Local Economy* 10: 300–26

May, A. D. (1986) 'Traffic restraint: a review of the alternatives', *Transportation Research A* 20A: 109–21

Meadows, D. H., Meadows, D. L. and Randers, J. (1992) *Beyond the Limits: Global Collapse or a Sustainable Future*, London: Earthscan

Meadows, P. (ed.) (1996) *Work Out – or Work In? Contributions to the Debate on the Future of Work*, York: Joseph Rowntree Foundation

Meller, H. (1997) *Towns, Plans and Society in Modern Britain*, Cambridge: Cambridge University Press

Meny, Y., Muller, P. and Quermonne, J-L. (1996) *Adjusting to Europe: The Impact of the European Union on National Institutions and Policies*, London: Routledge

Meyerson, G. and Rydin, Y. (1996) 'Sustainable development: the implications of the global debate for land use planning', in Buckingham-Hatfield and Evans (eds): 19–34

Meyerson, M. (1956) 'Building the middle-range bridge for comprehensive planning', *Journal of the American Institute of Planners* 22(2): 58–64 (reprinted in Faludi, A. (ed.) (1973) *A Reader in Planning Theory*, Oxford: Pergamon)

Meynell, A. (1959) 'Location of industry', *Public Administration* 37: 9

Midgely, J. (2000) 'Exploring alternative methodologies to establish the effects of land area designations in development control decisions', *Planning Practice and Research* 15(4): 319–34

Midwinter, A. (1995) *Local Government in Scotland: Reform or Decline?*, London: Macmillan

Millar, S. (1999) 'Beauty spot bypass proving failure', *Guardian* 12 July: 3

Miller, C. (2000) *Planning and Pollution Revisited: A Background Paper for the Royal Commission on Environmental Pollution Review of Environmental Planning*, London: Royal Commission on Environmental Pollution

Miller, N. (2002) *Mapping the City*, London: Continuum

Millichap, D. (1995a) 'Law, myth and community: a reinterpretation of planning's justification and rationale', *Planning Perspectives* 10: 279–93

Millichap, D. (1995b) *The Effective Enforcement of Planning Controls* (second edition), London: Butterworths

Mills, E. S. and Hamilton, B. W. (1994) *Urban Economics* (fifth edition), New York: HarperCollins

Milner Holland, E. (chair) (1965) *Report of the Committee on Housing in Greater London*, Cmnd 2605, Milner Holland Report, London: HMSO

Milton, K. (1991) 'Interpreting environmental policy: a social scientific approach', in Churchill, R., Warren, L. M. and Gibson, J. (eds) *Law, Policy and the Environment*, Oxford: Blackwell

Milton Keynes Development Corporation (1992) *The Milton Keynes Planning Manual*, Chesterton Consulting: distributed by Keyne Creations, Milton Keynes

Minay, C. L. W. (ed.) (1992) 'Developing regional planning guidance in England and Wales: a review symposium', *Town Planning Review* 63: 415–34

Minhinnick, R. (1993) *A Postcard Home*, Llandysul, Dyfed: Gomer Press

Mishan, E. J. (1976) *Elements of Cost-Benefit Analysis*, London: Allen and Unwin

Mitchell, C. (chair) (1991) *Railway Noise and the Insulation of Dwellings: Report of the Committee to Recommend a National Noise Insulation Standard for New Railway Lines*, Mitchell Report, London: HMSO

Mitchell, C. (1995) *Renewable Energy in the UK: Financing Options for the Future*, London: Council for the Protection of Rural England

Mogridge, M. J. H. (1990) *Travel in Towns*, London: Macmillan

Mohan, J. (ed.) (1989) *The Political Geography of Contemporary Britain*, London: Macmillan

Mole, D. (1996) 'Planning gain after the *Tesco* case', *Journal of Planning and Environment Law* 1996: 183–93

Möltke, K. von (1988) 'The *Vorsorgerprinzip* in West German environmental policy' (printed as an appendix to the Twelfth Report of the Royal Commission on Environmental Pollution, 1991), *Best Practicable Environmental Option*, London: HMSO

Monk, S., Pearce, B. J. and Whitehead, C. M. E. (1996) 'Land-use planning, land supply, and house prices', *Environment and Planning A* 28: 495–511

Monk, S., Crook, T., Lister, D., Rowley, S., Short, C. and Whitehead, C. (2005) *Land and Finance for Affordable Housing: The Complementary Roles of Social Housing Grant and the Provision of Affordable Housing through the Planning System*, York: Joseph Rowntree Foundation

Montanari, A. and Williams, A. M. (1995) *European Tourism: Regions, Spaces and Restructuring*, Chichester: Wiley

Montgomery, J. (2003) 'Cultural quarters as mechanisms for urban regeneration Part 1: conceptualising cultural quarters', *Planning Practice and Research* 18(4): 293–306

Montgomery, J. (2004) 'Cultural quarters as mechanisms for urban regeneration Part 2', *Planning Practice and Research* 19(1): 3–31

Montgomery, J. and Thornley, A. (eds) (1990) *Radical Planning Initiatives: New Directions for Urban Planning in the 1990s*, Aldershot: Gower

Moore, B. and Townroe, P. (1990) *Urban Labour Markets: Reviews of Urban Research* (DoE), London: HMSO

Moore, B., Rhodes, J. and Tyler, P. (1986) *The Effects of Government Regional Economic Policy* (DTI), London: HMSO

Moore, R. (ed.) (1995) *Soil Diversity: A Literature Review*, Perth: Scottish Natural Heritage

Moore, T. and Thorsnes, P. (1994) *The Transportation/ Land Use Connection: A Framework for Practical Policy*, Chicago, IL: American Planning Association

Moore, V. (2000, 2002) *A Practical Approach to Planning Law* (seventh and eighth editions), London: Blackstone Press

Morgan, K. (1994) *The Fallible Servant: Making Sense of the Welsh Development Agency*, Cardiff: Department of City and Regional Planning, University of Cardiff

Morgan, K. (1995) 'Reviving the valleys? Urban renewal and governance structures in Wales', in Hambleton, R. and Thomas, H. (eds) *Urban Policy Evaluation: Challenge and Change*, London: Paul Chapman

Morgan, K. and Henderson, D. (1997) 'The fallible servant: evaluating the Welsh Development Agency', in MacDonald, R. and Thomas, H. (eds) *Nationality and Planning in Scotland and Wales*, Cardiff: University of Wales

Morgan, P. and Nott, S. (1995) *Development Control: Law, Policy and Practice* (second edition), London: Butterworths

Morphet, J. (1993a) *Greening Local Authorities: Sustainable Development at the Local Level*, Harlow: Longman

Morphet, J. (1993b) *Towards Sustainability: A Guide for Local Authorities*, Luton: Local Government Management Board

Morphet, J. and Hams, T. (1994) 'Responding to Rio: the local authority approach', *Journal of Environmental Planning and Management* 37: 479–86

Morris, A. E. J. (1994) *History of Urban Form: Before the Industrial Revolution* (third edition), Harlow: Longman

Morris, P. and Therivel, R. (eds) (1995) *Methods of Environmental Impact Assessment*, London: UCL Press

Morris, R. (1998) 'Gypsies and the planning system', *Journal of Planning and Environment Law* July: 635–43

Morrison, N. (2003) 'Assessing the need for key-worker housing: a case study of Cambridge', *Town Planning Review* 74(3): 281–300

Morrison, N. and Pearce, B. (2000) 'Developing indicators for evaluating the effectiveness of the UK land use planning system', *Town Planning Review* 71(2): 191–211

Morton, D. (1991) 'Conservation areas – has saturation point been reached?', *The Planner* 7(17): 5–8

Moseley, M. (1997) 'Parish appraisals as a tool of rural community development: an assessment of the British experience', *Planning Practice and Research* 12(3): 197–212

Moseley, M. J. (2002) 'Bottom-up "village action plans": some recent experience in rural England', *Planning Practice and Research* 17(4): 387–405

Motavalli, J. (2001) *Breaking Gridlock: Moving toward Transportation that Works*, San Francisco, CA: Sierra Club Books

Moussis, N. (1999) *Access to European Union: Law, Economics, Policies* (ninth edition), Genval, Belgium: Euroconfidential

Mowat, C. L. (1955) *Britain between the Wars 1918–1940*, London: Methuen

Moxen, J., McCulloch, A., Williams, D. and Baxter, S. (1995) *Accessing Environmental Information in Scotland* (Central Research Unit, Scottish Office), Edinburgh: HMSO

Muchnick, D. M. (1970) *Urban Renewal in Liverpool*, Occasional Papers on Social Administration 33, London: Bell

Murdoch, J. (2003) *The Differentiated Countryside*, London: Routledge

Murdoch, J. and Marsden, T. (1994) *Reconstructing Rurality: The Changing Countryside in an Urban Context*, London: UCL Press

Murray, M. (1991) *The Politics and Pragmatism of Urban Containment: Belfast since 1940*, Aldershot: Avebury

Murray, M. R. and Greer, J. V. (1993) *Rural Planning and Development in Northern Ireland*, Aldershot: Avebury

Murtagh, B. (2001) 'City visioning and the turn to the community: the case of Derry/Londonderry', *Planning Practice and Research* 16(1): 9–19

MVA Consultancy (1995) *The London Congestion Charging Research Programme: Principal Findings* (Government Office for London), London: HMSO

Myerson, G. and Rydin, Y. (1996) 'Sustainable development: the implications of the global debate for land use planning', in Buckingham-Hatfield, S. and Evans, B. (eds) *Environmental Planning and Sustainability*, Chichester: Wiley

Mynors, C. (1992) *Planning Control and the Display of Advertisements*, London: Sweet and Maxwell

Mynors, C. (1993) 'The extent of listing', *Journal of Planning and Environment Law* 1993: 99–111

Mynors, C. (1998) *Listed Buildings, Conservation Areas and Monuments*, London: Sweet and Maxwell

Nadin, V. (1992a) 'Consultation by consultants', *Town and Country Planning* 61: 272–4

Nadin, V. (1992b) 'Local planning: progress and prospects', *Planning Practice and Research* 7: 27–32

Nadin, V. (1998a) 'Planning and the UK presidency', *Town and Country Planning* 67(2): 60–3

Nadin, V. (1998b) 'UK consults on the ESDP', *Town and Country Planning* 67(2): 67

Nadin, V. (1999) 'British planning in its European context', in Cullingworth, B. (1999) *British Planning: 50 Years of Urban and Regional Policy*, London: Athlone

Nadin, V. (2001) 'Sustainability from a national spatial planning perspective', in Konvitz, J. and Koresawa, A. (eds) *Towards a New Role for Spatial Planning*, Paris: OECD

Nadin, V. (2002) 'Visions and visioning in European spatial planning', in Faludi, A. (ed.) *European Spatial Planning*, Harvard, MA: The Lincoln Institute

Nadin, V. (2004) 'The value of transnational co-operation on spatial planning: experience from the Spatial Vision for North-west Europe', in Mascarucci, R. (ed.) *Vision*, Rome: Meltemi editore

Nadin, V. and Brown, C. (1999) *The Management of the Interreg I and II Transnational Spatial Planning Initiatives*, Report for Kent County Council and DG Regio European Commission, Faculty of the Built Environment, Bristol: University of the West of England

Nadin, V. and Daniels, R. (1992) 'Consultants and development plans', *The Planner* 78(15): 10–12

Nadin, V. and Doak, J. (eds) (1991) *Town Planning Responses to City Change*, Aldershot: Gower

Nadin, V. and Dühr, S. (2005) 'Some help with Euro-planning jargon', *Town and Country Planning* March: 82

Nadin, V. and Jones, S. (1990) 'A profile of the profession', *The Planner* 76(3): 13–24

Nadin, V. and Seaton, K. (eds) (2000) *Subsidiarity and Proportionality in Spatial Planning Activities in the European Union: Supplementary Volume*, Faculty of the Built Environment Working Paper 60, Bristol: University of the West of England

Nadin, V. and Shaw, D. (1998a) 'Promoting transnational planning', *Town and Country Planning* 67(2): 70–2

Nadin, V. and Shaw, D. (1998b) 'Transnational spatial planning in Europe: the role of Interreg IIc in the UK', *Regional Studies* 32(3): 281–99

Nadin, V. and Shaw, D. (1999) *Subsidiarity and Proportionality in Spatial Planning Activities in the European Union*, London: DETR

Nadin, V. and Wood, C. (1988) 'More questions than answers . . .', *Planning* 762, August: 6–7

Nadin, V., Cooper, S., Shaw, D., Westlake, T. and Hawkes, P. (1997) *The EU Compendium of Spatial Planning Systems and Policies*, Luxembourg: Office for the Official Publications of the European Communities

Nathan, M., Roberts, P., Ward, M. and Garside, R. (1999) *Strategies for Success? A First Assessment of Regional Development Agencies' Draft Regional Economic Strategies*, Manchester: Centre for Local Government Economic Strategies

National Committee for Commonwealth Immigrants (NCCI) (1967) *Areas of Special Housing Need*, London: NCCI

National Farmers' Union (NFU) (1994) *Real Choices: Report by the Long Term Strategy Group*, London: NFU

National Farmers' Union (1995a) *Taking Real Choices Forward*, London: NFU

National Farmers' Union (1995b) *The Rural White Paper: Submission of the National Farmers' Union of England and Wales*, London: NFU

National Housing Federation (1999) *Closing Doors: Examining the Lack of Housing Choices in London*, London: National Housing Federation

National Housing Forum (1996) *More than Somewhere to Live*, London: National Housing Forum

National Retail Planning Forum (NRPF) (1999) *A Bibliography of Retail Planning*, London: NRPF

National Society for Clean Air and Environmental Protection (NSCA) (1995) *1995 Pollution Handbook*, Brighton: NSCA

National Society of Allotment and Leisure Gardeners (NSALG) and Anglia Polytechnic University (1997) *English Allotments Survey*, Corby, Northants: NSALG

Neill, W. (2003) *Urban Planning and Cultural Identity*, London: Routledge

Neill, W. J. V., Fitzsimons, D. S. and Murtagh, B. (1995) *Reimaging the Pariah City: Urban Development in Belfast and Detroit*, Aldershot: Avebury

Nelson, B. F. and Stubb, A. C-G. (2003) *The European Union: Readings on the Theory and Practice of European Integration* (third edition), Basingstoke: Lynne Riener

Nettlefold, J. S. (1914) *Practical Town Planning*, London: St Catherine's Press

Nevin, B. and Murie, A. (1997) *Beyond a Halfway Housing Policy: Local Strategies for Regeneration*, London: Institute for Public Policy Research

Nevin, B. and Shiner, P. (1995) 'Community regeneration and empowerment: a new approach to partnership', *Local Economy* 9: 308–22

Nevin, M. and Abbie, L. (1993) 'What price roads? Practical issues in the introduction of road-user charges in historic cities in the UK', *Transport Policy* 1: 68–73

Newberry, D. (1990) 'Pricing and congestion: economic principles relevant to pricing roads', *Oxford Review of Economic Policy* 5(2): 22–38

Newberry, D. (1995) 'Royal Commission Report on Transport and the Environment: an economic critique', *Economic Journal* 105(432): 1258–72

Newby, H. (1985) *Green and Pleasant Land: Social Change in Rural England* (second edition), London: Wildwood House

Newby, H. (1986) 'Locality and rurality: the restructuring of rural social relations', *Regional Studies* 20: 209–16

Newby, H. (1990) 'Ecology, amenity, and society', *Town Planning Review* 61: 3–20

Newby, H. (1991) *The Future of Rural Society* (unpublished mimeo), Swindon: Economic and Social Research Council

Newman, P. (1995) 'The politics of urban redevelopment in London and Paris', *Planning Practice and Research* 10: 15–23

Newman, P. (2000) 'Changing patterns of regional government in the EU', *Urban Studies* 36(5–6): 895–908

Newman, P. and Herrschel, T. (2001) *City Regions, Policy and Planning in Europe*, London: Routledge

Newman, P. and Thornley, A. (1996) *Urban Planning in Europe: International Competition, National Systems and Planning Projects*, London: Routledge

Newman, P. W. G. and Kenworthy, J. R. (1989) *Cities and Automobile Dependence: A Sourcebook*, Aldershot: Gower

Newman, P. W. G. and Kenworthy, J. R. (1996) 'The land use-transport connection', *Land Use Policy* 13: 1–22

Newson, M. (ed.) (1992) *Managing the Human Impact on the Natural Environment: Patterns and Processes*, Chichester: Belhaven

Nicol, C. and Hooper, A. (1999) 'Contemporary change and the housebuilding industry', *Housing Studies* 14: 57–76

Nijkamp, P. (1993) 'Towards a network of regions: the United States of Europe', *European Planning Studies* 1: 149–68

Niner, P. (1999) *Insights into Low Demand for Housing*, Foundations Report 739, York: Joseph Rowntree Foundation

Niner, P. (2002) *The Provision and Condition of Local Authority Gypsy and Traveller Sites in the English Countryside*, London: ODPM

Niner, P. (2004) 'Accommodating nomadism: an examination of accommodation options for Gypsies and travellers in England', *Housing Studies* 19(2): 141–59

Noble, B. (1999) 'Walking and cycling in Great Britain', in *Transport Trends 1999* (DETR) London: The Stationery Office

Noise Review Working Party (1990) *Report of the Noise Review Working Party*, London: HMSO

Nolan, Lord (chair) (1995) *First Report of the Committee on Standards in Public Life*, Cm 2850, Nolan Report 1, London: HMSO

Nolan, Lord (chair) (1997) *Third Report of the Committee on Standards in Public Life: Standards of Conduct in Local Government*, Cm 3702, Nolan Report 3, London: The Stationery Office

Nordregio (2000) *Study Programme on European Spatial Planning: Final Report, March 2000*, Stockholm: Nordregio (the full report is included and available on CDRom)

Northcott, J. (1991) *Britain in Europe in 2010*, London: Policy Studies Institute

Norton, D. M. (1993) 'Conservation areas in an era of plan led planning', *Journal of Planning and Environment Law* 1993: 211–13

Nuffield Foundation (1986) *Town and Country Planning*, Nuffield Report, London: Nuffield Foundation

Nugent, N. (1999) *The Government and Politics of the European Community* (fourth edition), London: Macmillan

Oatley, N. (1995) 'Competitive urban policy and the regeneration game', *Town Planning Review* 66: 1–14

Oatley, N. (ed.) (1998) *Cities, Economic Competition and Urban Policy*, London: Paul Chapman

Oatley, N. (2000) 'New Labour's approach to age-old problems: renewing and revitalising poor neighbourhoods – the national strategy for neighbourhood renewal', *Local Economy* 15(2): 86–97

Oatley, N. and Lambert, C. (1995) 'Evaluating competitive urban policy: the City Challenge initiative', in Hambleton, R. and Thomas, H. (eds) *Urban Policy Evaluation: Challenge and Change*, London: Paul Chapman

Oc, T. and Tiesdell, S. (1991) 'The London Docklands Development Corporation 1981–1991: a perspective on the management of urban regeneration', *Town Planning Review* 62: 311–30

Oc, T. and Tiesdell, S. (1998) 'City centre management and safer city centres: approaches in Coventry and Nottingham', *Cities* 15: 85–103

Oc, T., Tiesdell, S. and Moynihan, D. (1997) 'The death and life of City Challenge: the potential for lasting impacts in a limited life urban regeneration initiative', *Planning Practice and Research* 12(4): 367–81

Ofori, S. (1994) 'Urban policy and environmental regeneration in two Scottish peripheral estates', *Environmentalist* 14: 283–96

Ogden, P. (1992) *Update: London Docklands – The Challenge of Development*, Cambridge: Cambridge University Press

Ogilvie, W. (1997) *Birthright in Land*, London: Othila

Olerup, B., Persson, O. and Snickars, F. (eds) (2002) *Reshaping Regional Planning: A Northern Perspective*, Aldershot: Ashgate

Oliver, D. and Waite, A. (1989) 'Controlling neighbourhood noise: a new approach', *Journal of Environment Law* 1: 173–91

O'Neill, J. (1999) 'Moving and shaking: inquiries and hearings', *Planning Inspectorate Journal*, 16 (summer): 18–20

Oregon Chapter of the American Planning Association (1993) *A Guide to Community Visioning*, Chicago, IL: Planners Press

Organisation for Economic Cooperation and Development (OECD): *see* appendix on *Official Publications*

O'Riordan, T. (1985) 'What does sustainability really mean? Theory and development of concepts of sustainability', in *Sustainable Development in an Industrial Economy, Proceedings of a Conference*, Cambridge: UK Centre for Economic and Environmental Development, Queen's College, Cambridge

O'Riordan, T. (1992) 'The environment', in Cloke, P. (ed.) *Policy and Change in Thatcher's Britain*, Oxford: Pergamon

O'Riordan, T. (1997) *Ecotaxation*, London: Routledge

O'Riordan, T. (1999) *Planning for Sustainable Development*, London: Town and Country Planning Association

O'Riordan, T. and Cameron, J. (eds) (1994) *Interpreting the Precautionary Principle*, London: Earthscan

O'Riordan, T. and D'Arge, R. C. (1979) *Progress in Resource Management and Environmental Planning*, New York: Wiley

O'Riordan, T. and Voisey, H. (1997) 'The political economy of sustainable development', *Environmental Politics* 6: 1–23

O'Riordan, T. and Weale, A. (1989) 'Administrative reorganization and policy change: the case of Her Majesty's Inspectorate of Pollution', *Public Administration* 67: 277–94

Osborn, F. J. (1969) *Green Belt Cities*, London: Evelyn, Adams and Mackay

Osborne, D. and Gaebler, T. (1992) *Reinventing Government: How the Entrepreneurial Spirit is Transforming the Public Sector*, New York: Addison-Wesley

Osborne, S. P. and Tricker, M. (2000) 'Village appraisals: a tool for sustainable community development in the UK', *Local Economy* February: 347–58

Osmond, J. (1977) *Creative Conflict: The Politics of Welsh Devolution*, London: Routledge

Osmotherly, E. B. C. (1995) *Guide to the Local Ombudsman Service* (looseleaf, with updating service), London: Pitman

O'Toole, M. (1996) *Regulation Theory and the New British State: The Impact of Locality on the Urban Development Corporation 1987–1994*, Aldershot: Avebury

Ove Arup and University of Reading (1997) *Planning Policy Guidance on Transport (PPG13): Implementation 1994–1996*, London: HMSO

Ove Arup and Partners, Regional Forecasts and Oxford Economic Forecasting (2005) *Regional Futures: England's Regions in 2030*, English Regions Network

Ove Arup Economics and Planning (1998) *Control of Outdoor Advertisements: Fly-Posting* (DETR), London: The Stationery Office

Ove Arup Economics and Planning (1999) *Implementation of PPG13*, London: Ove Arup

Owen, S. (1991) *Planning Settlements Naturally*, Chichester: Packard

Owen, S. (1995a) 'Local distinctiveness in villages', *Town Planning Review* 66: 143–61

Owen, S. (1995b) 'Land, limits and sustainability: a conceptual framework and some dilemmas for the planning system', *Transactions of the Institute of British Geographers* 19: 439–56

Owen, S. (1996) 'Sustainability and rural settlement planning', *Planning Practice and Research* 11: 37–47

Owen, S. (2002a) 'From village design statements to parish plans: some pointers towards community decision making in the planning system in England', *Planning Practice and Research* 17(1): 81–9

Owen, S. (2002b) 'Locality and community: towards a vehicle for community-based decision making in rural localities in England', *Town Planning Review* 73(1): 41–62

Owen, S. and Moseley, M. (2003) 'Putting parish plans in their place: relationships between community-based initiatives and development planning in English villages', *Town Planning Review* 74(4): 445–72

Owens, S. (1984) 'Energy and spatial structure: a rural example', *Environment and Planning A* 16: 1319–37

Owens, S. (1985) 'Potential energy planning conflicts in the UK', *Energy Policy* 13: 546–58

Owens, S. (1986a) 'Strategic planning and energy conservation', *Town Planning Review* 57: 69–86

Owens, S. (1986b) *Energy, Planning and Urban Form*, London: Pion

Owens, S. (1989) 'Integrated pollution control in the United Kingdom: prospects and problems', *Environment and Planning C: Government and Policy* 7: 81–91

Owens, S. (1990a) 'Land use for energy efficiency', in Cullingworth, J. B. (ed.) *Energy, Land, and Public Policy*, New Brunswick, NJ: Transaction

Owens, S. (1990b) 'The unified pollution inspectorate and best practicable environmental option in the United Kingdom', in Haigh, N. and Irwin, F. (eds) *Integrated Pollution Control in Europe and North America*, Bonn: Institute for European Environmental Policy; Washington, DC: Conservation Foundation

Owens, S. (1991) *Energy Conscious Planning*, London: Council for the Protection of Rural England

Owens, S. (1994a) 'Can land use planning produce the ecological city?', *Town and Country Planning* 63: 170–3

Owens, S. (1994b) 'Land, limits and sustainability: a conceptual framework and some dilemmas for the planning system', *Transactions of the Institute of British Geographers* NS 19: 439–56

Owens, S. (1998) 'Better for everyone?', *Town and Country Planning* 67: 329–31

Owens, S. and Cope, D. (1992) *Land Use Planning Policy and Climate Change* (DoE), London: HMSO

Owens, S. and Cowell, R. (1996) *Rocks and Hard Places: Mineral Resource Planning and Sustainability*, London: Council for the Protection of Rural England

Owens, S. and Cowell, R. (2002) *Land and Limits Interpreting Sustainability in the Planning Process*, London: Routledge

Oxford Economic Forecasting (2003) *The Economic Effects of Transport Delays on the City of London*, Oxford: Oxford Economic Forecasting

Oxford Retail Group (1995) *The Implementation of PPG 6*, Oxford: Oxford Institute of Retail Management, Templeton College

PA Cambridge Economic Consultants (1987) *An Evaluation of the Enterprise Zone Experiment* (DoE), London: HMSO

PA Cambridge Economic Consultants (1990a) *An Evaluation of Garden Festivals* (DoE), London: HMSO

PA Cambridge Economic Consultants (1990b) *Indicators of Comparative Regional-Local Economic Performance and Prospects* (DoE), London: HMSO

PA Cambridge Economic Consultants (1992) *An Evaluation of the Belfast Action Team Initiative*, Belfast: DoENI

PA Cambridge Economic Consultants (1995) *Final Evaluation of Enterprise Zones* (DoE), London: HMSO

PA Consulting Group (1999) *Interim Evaluation of English Partnerships: Final Report*, London: DETR

Pacione, M. (1990) 'Development pressure in the metropolitan fringe', *Land Development Studies* 7: 69–82

Pacione, M. (1991) 'Development pressure and the production of the built environment in the urban fringe', *Scottish Geographical Magazine* 107: 162–9

Pacione, M. (1995) *Glasgow: The Socio-spatial Development of the City*, Chichester: Wiley

Pacione, M. (1997) 'The urban challenge: how to bridge the great divide', in Pacione, M. (ed.) *Britain's Cities: Geographies of Division in Urban Britain*, London: Routledge

Pack, C. and Glyptis, S. (1989) *Developing Sport and Leisure* (DoE), London: HMSO

Paddison, R., Money, J. and Lever, B. (eds) (1993) *International Perspectives in Urban Studies*, London: Jessica Kingsley

Paddison, R., Money, J. and Lever, B. (eds) (1994) *International Perspectives in Urban Studies 2*, London: Jessica Kingsley

Paddison, R., Money, J. and Lever, B. (eds) (1995) *International Perspectives in Urban Studies 3*, London: Jessica Kingsley.

Paddison, R., Money, J. and Lever, B. (eds) (1996) *International Perspectives in Urban Studies 4*, London: Jessica Kingsley

Page, S. (1995) *Urban Tourism*, London: Routledge

Pahl, R. E. (1970) *Whose City? and Other Essays in Sociology and Planning*, Harlow: Longmans

Painter, J. (1992) 'The culture of competition', *Public Policy and Administration* 7: 58–68

Pantazis, C. and Gordon, D. (eds) (2000) *Tackling Inequalities: Where Are We Now and What Can Be Done?*, Bristol: Policy Press

Parfect, M. and Power, G. (1997) *Planning for Urban Quality: Urban Design in Towns and Cities*, London: Routledge

Paris, C. (ed.) (1982) *Critical Readings in Planning Theory*, Oxford: Pergamon

Park, P. (2000) 'An evaluation of the Landfill Tax two years on', *Journal of Planning and Environment Law* January: 3–13

Parker, G. (2001) 'Planning and rights: some repercussions of the Human Rights Act 1998 for the UK', *Planning Practice and Research* 16(1): 5–8

Parker, H. R. (1954) 'The financial aspects of planning legislation', *Economic Journal* 64: 82–6

Parkhurst, G. (1995) 'Park and ride: could it lead to an increase in car traffic?', *Transport Policy* 2: 15–23

Parkinson, M. (1985) *Liverpool on the Brink*, Hermitage, Berkshire: Policy Journals Publishers

Parkinson, M. (1989) 'The Thatcher Government's urban policy 1979–1989', *Town Planning Review* 60: 421–40

Parkinson, M. (1998) *Combating Social Exclusion: Lessons from Area-Based Programmes in Europe*, Bristol: Policy Press

Parkinson, M. and Boddy, M. (2004) 'Introduction', in Boddy, M. and Parkinson, M. (eds) *City Matters: Competitiveness, Cohesion and Urban Governance*, Bristol: The Policy Press

Parkinson, M., Hutchins, M., Simmie, J., Clark, G. and Verdonk, H. (2004) *Competitive European Cities: Where do the Core Cities Stand?*, London: ODPM

Parkinson, M., Hutchins, M., Champion, T., Coombes, M., Dorling, D., Parks, A., Simmie, J. and Turok, I. (2005) *State of the Cities: A Progress Report to the Delivering Sustainable Communities Summit*, London: ODPM

Parkinson, M. and Robson, B. (2000) *Urban Regeneration Companies: A Process Evaluation*, London: DETR [available at http://www.urcs-online.co.uk/, accessed 10 March 2005]

Parry, E. (2005) 'Streamlining purchase law', *Planning 1602*, 14 January: 15

Parry, M. and Duncan, R. (eds) (1995) *The Economic Implications of Climate Change in Britain*, London: Earthscan

Parsons, M. L. (1995) *Global Warming: The Truth behind the Myth*, New York: Insight and Plenum

Patten, C. (1989) 'Planning and local choice', *Municipal Journal* 13 October: 18–19

Pattison, G. (2001) 'The role of participants in producing regional planning guidance in England', *Planning Practice and Research* 16(3): 349–60

Paxton, A. (1994) *The Food Miles Report: The Dangers of Long Distance Food Transport*, London: SAFE Alliance

Peake, S. (1994) *Transport in Transition: Lessons from the History of Energy*, London: Royal Institute of International Affairs and Earthscan

Peake, S. and Hope, C. (1994) 'Sustainable mobility in context: three transport scenarios for the UK', *Transport Policy* 1: 195–207

Pearce, B. (1992) 'The effectiveness of the British land use planning system', *Town Planning Review* 63: 13–28

Pearce, B. (2000) 'Mediating a consensus', *Town and Country Planning* 69(9): 246–7

Pearce, D. (ed.) (1991) *Blueprint 2: Greening the World Economy*, London: Earthscan

Pearce, D. (ed.) (1993) *Blueprint 3: Measuring Sustainable Development*, London: Earthscan

Pearce, D. and Turner, R. K. (1992) 'Packaging waste and the polluter pays principle: a taxation solution', *Journal of Environmental Planning and Management* 35: 5–15

Pearce, D., Edwards, L. and Beuret, G. (1979) *Decision-Making for Energy Futures: A Case Study of the Windscale Inquiry*, London: Macmillan

Pearce, D., Markandya, A. and Barbier, E. B. (1989) *Blueprint for a Green Economy: Report for the UK DoE*, London: Earthscan

Pearce, G., Hems, L. and Hennessy, B. (1990) *The Conservation Areas of England*, London: Historic Buildings and Monuments Commission (English Heritage)

Pearlman, J. J. (1995) 'Modification orders made under Wildlife and Countryside Act 1981: an update', *Journal of Planning and Environment Law* 1995: 1106–13

Peck, J. and Ward, K. (eds) (2002) *City of Revolution: Restructuring Manchester*, Manchester: Manchester University Press

Pendlebury, J. (1997) 'The statutory protection of historic parks and gardens: an exploration and analysis of structure, decoration and character', *Journal of Urban Design* 2(3): 241–58

Pendlebury, J. (1999) 'The place of historic parks and gardens in the English planning system: towards statutory controls', *Town Planning Review* 70(4): 479–500

Pendlebury, J. and Townshend, T. (1999) 'The conservation of historic areas and participation', *Journal of Architectural Conservation* 2(5): 72–87

Penn, C. N. (1995) *Noise Control: The Law and its Enforcement* (second edition), Crayford: Shaw

Pennington, M. (1996) *Conservation and the Countryside: By Quango or Market?*, London: Institute of Economic Affairs

Pepper, D. (1993) *Eco-Socialism: From Deep Ecology to Social Justice*, London: Routledge

Pepper, D. (1996) *Modern Environmentalism: An Introduction*, London: Routledge

Percy-Smith, J. (1994) *Submissions to the Commission on Aspects of Local Democracy*, London: Commission on Local Democracy

Perry, M. (1999) 'Clusters last stand', *Planning Practice and Research* 14: 149–52

Perveen, F. (ed.) (1994) *Urban Environment: An Annotated Bibliography*, Manchester: British Council

Peter Scott Planning Services (1998) *Access to the Countryside in Selected European Countries: A Review of Access Rights, Legislation and Associated Arrangements in Denmark, Germany, Norway and Sweden*, Perth: Scottish Natural Heritage

Pezzey, J. (1989) *Economic Analysis of Sustainable Growth and Sustainable Development*, Washington, DC: World Bank

Pezzey, J. (1992) 'Sustainability', *Environmental Values* 4: 321–62

Pharoah, T. (1993) 'Traffic calming in west Europe', *Planning Practice and Research* 8: 20–28

Pharoah, T. (1996) 'Reducing the need to travel: a new planning objective in the UK?', *Land Use Policy* 13: 23–36

Pharoah, T. and Apel, D. (1995) *Transport Concepts in European Countries*, Aldershot: Avebury

Phelps, R. (1995) 'Structure plans: the conduct and conventions of examinations in public', *Journal of Planning and Environment Law* 1995: 95–101

Piatt, A. (1997) 'Public concern: a material consideration?', *Journal of Planning and Environment Law* 1997: 397–400

Pickard, R. (1996) *Conservation in the Built Environment*, Harlow: Addison Wesley Longman

Pickard, R. (2000) *Policy and Law in Heritage Conservation*, London: Spon

Pickering, K. T. and Owen, L. A. (1997) *An Introduction to Global Environmental Issues*, London: Routledge

Pickvance, C. (1982) 'Physical planning and market forces in urban development', in Paris, C. (ed.) *Critical Readings in Planning Theory*, Oxford: Pergamon

PIEDA (1990) *Five Year Review of the Bolton, Middlesbrough and Nottingham Programme Authorities* (DoE), London: HMSO

PIEDA (1992) *Evaluating the Effectiveness of Land Use Planning* (DoE), London: HMSO

PIEDA (1995) *Involving Communities in Urban and Rural Regeneration* (DoE), London: DoE

Plan Local (1992) *The Character of Conservation Areas*, London: Royal Town Planning Institute

Plank, D. (ed.) (1998) *Joined up Thinking: A Directory of Good Practice for Local Authority Empty Property Strategies*, London: Empty Homes Agency

Planning Aid for London (1986) *Planning for Women*, London: Planning Aid for London

Planning Aid for London (1995) *Publicity for Planning Applications*, London: Planning Aid for London

Planning and Environment Law Reform Working Group (1999) 'Planning obligations', *Journal of Planning and Environment Law* February: 113–33

Planning Exchange (1989) *Evaluation of the Use and Effectiveness of Planning Publications*, Edinburgh: Scottish Development Department

Planning Inspectorate (PI) (1996) *Development Plan Inquiries: Guidance for Local Authorities*, Bristol: PI

Planning Inspectorate (annual) *Annual Report* (previously published by the PI; now The Stationery Office)

Planning Officers' Society (POS) (1997) *Better Local Plans: A Guide to Writing Effective Policies*, London: POS

Planning Officers' Society (1998) *Public Involvement in the Development Control Process: A Good Practice Guide*, London: Local Government Association

Planning Officers' Society (Best Value Group) (1999) 'Getting started on Best Value', Barnsley: Barnsley Metropolitan Borough Planning Department

Planning Officers' Society (2000) *Review of the Planning Enforcement System in England: Comments of the Planning Officers' Society*, London: POS

Planning Officers' Society (2004) *Policies for Spatial Plans*, Consultation Draft, London: POS

Plowden, M. (chair) (1967) *Children and their Primary Schools*, Plowden Report, London: HMSO

Plowden, S. and Buchan, K. (1995) *A New Framework for Freight Transport*, London: Civic Trust

Plowden, S. and Hillman, M. (1984) *Danger on the Road: The Needless Scourge – A Study of Obstacles to Progress in Road Safety*, London: Policy Studies Institute

Plowden, S. and Hillman, M. (1996) *Speed Control and Transport Policy*, London: Policy Studies Institute

Plowden, W. (1971) *The Motor Car and Politics 1896–1970*, London: The Bodley Head

Plowden, W. (1994) *Ministers and Mandarins*, London: Institute for Public Policy Research

Plummer, J. and Zipfel, T. (1998) *Regeneration and Employment: A New Agenda for TECs, Communities and Partnerships*, York: Policy Press

Pollock, A. (1999) 'New directions for strategic planning in Northern Ireland', *Town and Country Planning* 68: 337–9

Postle, M. (1993) *Development of Environmental Economics for the NRA* (National Rivers Authority), London: HMSO

Potter, S. (1992) 'New town legacies', *Town and Country Planning* 61: 298–302

Potter, S. (1999) 'Tax and green transport plans', *Town and Country Planning* 68: 48–9

Power, A. (1999) *Estates on the Edge*, London: Macmillan

Power, A. (2004) *Sustainable Communities and Sustainable Development: A Review of the Sustainable Communities Plan*, London: Sustainable Development Commission

Power, A. and Mumford, K. (1999) *The Slow Death of Great Cities? Urban Abandonment or Urban Renaissance?*, York: York Publishing Services

Power, A. and Tunstall, R. (1997) *Dangerous Disorder, Riots and Violent Disturbances in 13 Areas of Britain, 1991–92*, York: Joseph Rowntree Foundation

Poxon, J. (2000) 'Solving the development plan puzzle in Britain: learning lessons from history', *Planning Perspectives* 15(1): 73–90

Pressman, J. L. and Wildavsky, A. B. (1984) *Implementation: How Great Expectations in Washington are Dashed in Oakland; or Why It's Amazing that Federal Programs Work At All, This Being a Saga of the Economic Development Administration as Told by Two Sympathetic Observers who Seek to Build Morals on a Foundation of Ruined Hopes* (third edition), Berkeley, CA: University of California Press

Pretty, J. N. (1995) *Regenerating Agriculture: Policies and Practice for Sustainability and Self-Reliance*, London: Earthscan

Pretty, J. N. (1999) *The Living Land: Agriculture, Food and Community Regeneration in the 21st Century*, London: Earthscan

Price Waterhouse (1993) *Evaluation of Urban Development Grant, Urban Regeneration Grant, and City Grant* (DoE), London: HMSO

Priemus, H. (1999) 'Coordination in planning public transport: evidence from the Netherlands', *Planning Practice and Research* 14(1): 97–106

Priemus, H. and Konings, R. (2000) 'Public transport in urbanised regions: the missing link in the pursuit of the economic vitality of cities', *Planning Practice and Research* 15(3): 233–45

Prior, A. and Allmendinger, P. (1999) 'Certainty or flexibility? The options for neighbour notification in development control in the UK', *Planning Practice and Research* 14: 185–98

Property Advisory Group (1980) *Structure and Activity of the Development Industry*, London: HMSO

Property Advisory Group (1981) *Planning Gain*, London: HMSO

Property Advisory Group (1983) *The Climate for Public and Private Partnerships in Property Development*, London: HMSO

Property Advisory Group (1985) *Report on Town and Country Planning (Use Classes) Order 1972*, London: DoE

Public Sector Management Research Centre (1992) *Parish Councils in England*, London: HMSO

Public Sector Management Research Unit (1988a) *An Evaluation of the Urban Development Grant Programme* (DoE), London: HMSO

Public Sector Management Research Unit (1988b) *Improving Inner City Shopping Centres: An Evaluation of Urban Programme Funded Schemes in the West Midlands* (DoE), London: HMSO

Pugh, C. (ed.) (1996) *Sustainability, the Environment and Urbanization*, London: Earthscan

Pugh-Smith, J. (1992) 'The local authority as a regulator of pollution in the 1990s', *Journal of Planning and Environment Law* 1992: 103–9

Pugh-Smith, J. and Samuels, J. (1993) 'PPG 16: two years on', *Journal of Planning and Environment Law* 1993: 203–10

Pugh-Smith, J. and Samuels, J. (1996a) *Archaeology in Law*, London: Sweet and Maxwell

Pugh-Smith, J. and Samuels, J. (1996b) 'Archaeology and planning: recent trends and potential conflicts', *Journal of Planning and Environment Law* 1996: 707–24

Punter, J. V. (1985) *Office Development in the Borough of Reading 1951–1984: A Case Study of the Role of Aesthetic Control*, Reading: Department of Land Management, University of Reading

Punter, J. V. (1986–7) 'A history of aesthetic control: the control of the external appearance of development in England and Wales' (parts 1 and 2), *Town Planning Review* 57: 351–81 and 58: 29–62

Punter, J. V. (1990) *Design Control in Bristol, 1940–1990: The Impact of Planning in the Design of Office Development in the City Centre*, Bristol: Redcliffe

Punter, J. V. (1992) 'Design control and the regeneration of docklands: the example of Bristol', *Journal of Property Research* 9: 49–78

Punter, J. V. (ed.) (1994) 'Design control in Europe', special issue of *Built Environment* 20(2)

Punter, J. V. (1999) 'Design', in Cullingworth, B. (ed.) *British Planning: 50 Years of Urban and Regional Policy*, London: Athlone

Punter, J. V. and Carmona, M. (1997) *The Design Dimension of Planning: Theory, Content and Best Practice for Design Policies*, London: Spon

Punter, J. V., Carmona, M. C. and Platts, A. (1994) 'The design content of development plans', *Planning Practice and Research* 9: 199–220

Purdom, C. B. (1913) *The Garden City: A Study in the Development of a Modern New Town*, Letchworth: Temple Press

Purdom, C. B. (1925) *The Building of Satellite Towns*, London: Dent

Purdue, M. (1986) 'The flexibility of north American zoning as an instrument of land use planning', *Journal of Planning and Environment Law* 1986: 84–91

Purdue, M. (1989) 'Material considerations: an ever expanding concept?', *Journal of Planning and Environment Law* 1989: 156–61

Purdue, M. (1991) 'Green belts and the presumption in favour of development', *Journal of Environmental Law* 3: 93–121

Purdue, M. (1995) 'When a regulation becomes a taking of land: a look at two recent decisions of the United States Supreme Court', *Journal of Planning and Environment Law* 1995: 279–91

Purdue, M. (1999) 'The changing role of the courts in planning', in Cullingworth, J. B. (ed.) *British Planning: 50 Years of Urban and Regional Policy*, London: Athlone

Purdue, M. and Kemp, R. (1985) 'A case for funding objectors at public inquiries? A comparison of the position in Canada as opposed to the United Kingdom', *Journal of Planning and Environment Law* 1985: 675–85

Purdue, M., Healey, P. and Ennis, F. (1992) 'Planning gain and the grant of planning permission: is the United States' test of the *rational nexus* the appropriate solution?', *Journal of Planning and Environment Law* 1992: 1012–24

Quinn, M. J. (1996) 'Central government planning policy', in Tewdwr-Jones, M. (ed.) *British Planning Policy in Transition: Planning in the 1990s*, London: UCL Press

Rabe, B. G. (1994) *Beyond Nimby: Hazardous Waste Siting in Canada and the United States*, Washington, DC: Brookings Institution

Rackham, O. (1994) *The Illustrated History of the Countryside*, London: Weidenfeld and Nicolson (paperback edition Phoenix Illustrated 1997)

Raco, M., Turok, I. and Kintrea, K. (2003) 'Local development companies and the regeneration of Britain's cities', *Environment and Planning* C 21(2): 277–304

Radcliffe, J. (1991) *The Reorganisation of British Central Government*, Aldershot: Dartmouth

Raemaekers, J. (1995) 'Scots have way to go on strategic waste planning', *Planning* 1106 (17 February): 24–5

Raemaekers, J., Prior, A. and Boyack, S. (1994) *Planning Guidance for Scotland: A Review of the Emerging New Scottish National Planning Policy Guidelines*, Edinburgh: Royal Town Planning Institute in Scotland

Ramsay, J. D. (chair) (1945) *Report of the Scottish National Parks Survey Committee*, Cmd 6631, Ramsay Report 1, Edinburgh: HMSO

Ramsay, J. D. (chair) (1947) *National Parks and the Conservation of Nature in Scotland*, Cmd 7235, Ramsay Report 2, Edinburgh: HMSO

Ranson, S., Jones, G. and Walsh, K. (eds) (1985), *Between Centre and Locality: The Politics of Public Policy*, London: Allen and Unwin

Rao, N. (1990) *The Changing Face of Housing Authorities*, London: Policy Studies Institute

Ratcliffe, D. A. (1994) *Conservation in Europe: Will Britain Make the Grade? The Status of Nature Resources in Britain and the Implementation of the EC Habitats and Species Directive*, London: Friends of the Earth

Ratcliffe, J., Stubbs, M. and Shepherd, M. (2004) *Urban Planning and Real Estate Development*, London: Spon

Ratcliffe, P. (1998) 'Planning for diversity and change: implications of polyethnic society', *Planning Practice and Research* 13(4): 359–69

Ravaioli, C. (1995) *Economists and the Environment*, London: ZED Books

Ravenscroft, N. (2000) 'The vitality and viability of town centres', *Urban Studies* 37(13): 2533–49

Ravetz, A. (1980) *Remaking Cities*, London: Croom Helm

Ravetz, A. (1986) *The Government of Space: Town Planning in Modern Society*, London: Faber

Ravetz, A. (2000) *City Region 2020: Integrated Planning for a Sustainable Environment*, London: Earthscan

Ravetz, A. with Turkington, R. (1995) *The Place of Home: English Domestic Environments 1914–2000*, London: Spon

Rawcliffe, P. (1995) 'Making inroads: transport policy and the British environmental movement', *Environment* 37(3): 6–20, 29–36

Read, L. and Wood, M. (1994) 'Policy, law and practice', in

Wood, M. (ed.) *Planning Icons: Myth and Practice* (Planning Law Conference, *Journal of Planning and Environment Law*), London: Sweet and Maxwell

Reade, E. J. (1982) 'If planning isn't everything . . .', *Town Planning Review* 53: 65–78

Reade, E. J. (1983) 'If planning is anything, maybe it can be identified', *Urban Studies* 20: 159–71

Reade, E. J. (1985) 'Planning's usurpation of political choice', *Town and Country Planning* 54: 184–6

Reade, E. J. (1987) *British Town and Country Planning*, Milton Keynes: Open University Press

Reade, E. J. (1992) 'The little world of Upper Bangor', part 1: 'How many conservation areas are slums?'; part 2: 'Professionally prestigious projects or routine public administration?'; part 3: 'What is planning for anyway?', *Town and Country Planning* 61: 11–12, 25–7 and 44–7

Redclift, M. (1987) *Sustainable Development: Exploring the Contradictions*, London: Methuen

Redclift, M. and Sage, M. (1994) *Strategies for Sustainable Development*, Chichester: Wiley

Redman, M. (1990) 'Archaeology and development', *Journal of Planning and Environment Law* 1990: 87–98

Redman, M. (1991) 'Planning gain and obligations', *Journal of Planning and Environment Law* 1991: 203–18

Redman, M. (1999) 'Compulsory purchase, compensation and human rights', *Journal of Planning and Environment Law* August: 315–26

Redundant Churches Fund (1990) *Churches in Retirement: A Gazetteer*, London: HMSO

Reed, M. (1990) *The Landscape of Britain*, London: Routledge

Reed, R. (ed.) (1999) *Glasgow: The Forming of a City* (revised paperback edition), Edinburgh: Edinburgh University Press

Rees, W. E. (1990) 'Sustainable development as capitalism with a green face: a review article' [Review of Pearce, D., Markandya, A. and Barbier, E. B. (1989) *Blueprint for a Green Economy: Report for the UK DoE*, London: Earthscan], *Town Planning Review* 61: 91–4

Reeve, A. (1998) 'Review article: recent literature in urban design', *Planning Perspectives* 13: 411–16

Reeves, D. (1995) 'Developing effective public consultation: a review of Sheffield's UDP process', *Planning Practice and Research* 10: 199–213

Reeves, D. (2005) *Planning for Diversity: Policy and Planning in a World of Difference*, London: Routledge

Reeves, D. and Burley, K. (2002) 'Public inquiries and development plans in England: the role of Planning Aid', *Planning Practice and Research* 17(4): 407–36

Regional Policy Commission (1996) *Renewing the Regions: Strategies for Regional Economic Development*, Sheffield: Sheffield Hallam University

Regional Studies Association (1990) *Beyond Green Belts*, London: Jessica Kingsley

Reid, A. and Murgatroyd, L. (1999) 'Variations in travelling time: the social context of travel', in DETR *Travel Trends 1999*: 3–13

Reid, C. (1994) *Nature Conservation Law*, Edinburgh: W. Green

Reid, D. (1995) *Sustainable Development: An Introductory Guide*, London: Earthscan

Reith, Lord (chair) (1946) *Interim Report of the New Towns Committee*, Cmd 6759; *Second Interim Report*, Cmd 6794; *Final Report*, Cmd 6876, Reith Reports, London: HMSO

Rendel Geotechnics (1993) *Coastal Planning and Management: A Review* (DoE), London: HMSO

Rendel Geotechnics (1995) *Coastal Planning and Management: A Review of Earth Science Information Needs* (DoE), London: HMSO

Renton, J. (1992) 'The ombudsman and planning', *Scottish Planning Law and Practice Conference 1992*, Glasgow: Planning Exchange

Rhodes, J., Tyler, P., Brennan, A., Stevens, S., Warnock, C. and Otero-Garcia, M. (2002) *Lessons and Evaluation Evidence from Ten Single Regeneration Budget Case Studies: Mid-term Report*, London: ODPM

Rice, D. (1998) 'National Parks for Scotland: a major step forward?', *Town and Country Planning* 67: 159–61

Richards, F. and Satsangi, M. (2004) 'Importing a policy problem? Affordable housing in Britain's national parks', *Planning Practice and Research* 19(3): 251–66

Richardson, G., Ogus, A. and Burrows, A. (1982) *Policing Pollution: A Study of Regulation and Enforcement*, Oxford: Oxford University Press

Richardson, H. W. and Bae, C-H. C. (eds) (2004) *Urban Sprawl in Western Europe and the United States*, Aldershot: Ashgate

Richardson, T. and Jenson, O. B. (1999) *North*, 6: 5 [www.nordregio.se, accessed 10 July 2002]

Ridley, F. F. (1986) 'Liverpool is different: political style in context', *Political Quarterly* 57: 125–36

Ridley, F. F. and Doig, A. (eds) (1995) *Sleaze: Politicians, Private Interests and Public Reaction*, Oxford: Oxford University Press

Ridley, N. (1991) *The Local Right*, London: Centre for Policy Studies

Rietveld, P. and Wissen, L. van (1991) 'Transport policies and the environment: regulation and taxation', in Kraan, D. J. and Veld, R. J. (eds) *Environmental Protection: Public or Private Choice?*, Dordrecht: Kluwer

Rittel, H. W. J. and Webber, M. M. (1973) 'Dilemmas in a general theory of planning', *Policy Sciences* 4: 155–69

Rivlin, A. M. (1971) *Systematic Thinking for Social Action*, Washington, DC: Brookings Institution

Roberts, J., Cleary, J., Hamilton, K. and Hanna, J. (1992) *Travel Sickness: The Need for a Sustainable Transport Policy for Britain*, London: Lawrence and Wishart

Roberts, M. (1974) *An Introduction to Town Planning Techniques*, London: Hutchinson

Roberts, P. (1991) 'Environmental priorities and the challenges of environmental management', *Town Planning Review* 62: 447–69

Roberts, P. (1995) *Environmentally Sustainable Business: A Local and Regional Perspective*, London: Paul Chapman

Roberts, P. (1996) 'Regional planning guidance in England and Wales: back to the future?', *Town Planning Review* 67: 97–109

Roberts, P. (1999) *Urban Regeneration: A Handbook*, London: Sage

Roberts, P. and Lloyd, G. (1998) *Developing Regional Potential: Monitoring the Performance of the Regional Development Agencies*, Dundee: Centre for Planning Research, University of Dundee

Roberts, P. and Lloyd, G. (1999) 'Institutional aspects of regional planning, management and development: models and lessons from the English experience', *Environment and Planning B: Planning and Design* 26: 517–31

Roberts, P., Thomas, K. and Williams, G. (eds) (1999) *Metropolitan Planning in Britain*, London: Jessica Kingsley

Roberts, T. (1998) 'The statutory system of town planning in the UK: a call for detailed reform', *Town Planning Review* 69(1): iii–vii

Robertson, D. J. (ed.) (1966) *The Lothians Regional Survey and Plan, Volume 1 – Economic and Social Aspects*, Edinburgh: HMSO

Robertson, G. (1989) *Freedom, the Individual and the Law* (sixth edition), Harmondsworth: Penguin

Robins, N. (1991) *A European Environment Charter*, London: Fabian Society

Robinson, C. (2001) 'Section 215: effective or neglected? Let's talk dirty', paper presented to the Planning Summer School, University of Exeter, September

Robinson, G. M. (1994) 'The greening of agricultural policy: Scotland's Environmentally Sensitive Areas', *Journal of Environmental Planning and Management* 37: 215–25

Robinson, M. (1992) *The Greening of British Party Politics*, Manchester: Manchester University Press

Robinson, N. (2002) *Transport and the EU*, London: Continuum

Robinson, P. C. (1990) 'Tree preservation orders: felling a dangerous tree', *Journal of Planning and Environment Law* 1990: 720–3

Robson, B. (1988) *Those Inner Cities: Reconciling the Social and Economic Aims of Urban Policy*, Oxford: Clarendon Press

Robson, B. (1999) 'Vision and reality: urban social policy', in Cullingworth, J. B. (ed.) *British Planning: 50 Years of Urban and Regional Policy*, London: Athlone

Robson, B. and Tomlinson, R. (1998) *Updating and Revising the Index of Local Deprivation*, London: DETR

Robson, B., Bradford, M. G., Deas, I., Hall, E., Harrison, E., Parkinson, M., Evans, R., Garside, P. and Harding, A. (1994) *Assessing the Impact of Urban Policy* (DoE), London: HMSO

Robson, B., Parkinson, M., Boddy, M. and Maclennan, D. (2000) *The State of English Cities*, London: DETR

Roche, F. L. (1986) 'New communities for a new generation', *Town and Country Planning* 55: 312–13

Rodriguez-Bachiller, A., Thomas, M. and Walker, S. (1992) 'The English planning lottery: some insights from a more regulated system', *Town Planning Review* 63: 387–402

Rogers, A. (1985) 'Local claims on rural housing', *Town Planning Review* 56: 367–80

Rogers, A. (1999) *The Most Revolutionary Measure: A History of the Rural Development Commission*, Cheltenham: Countryside Agency

Rogers, C. and Stewart, M. (1998) *Sustainable Regeneration: Good Practice Guide*, London: DETR

Rogers, Lord (1999) *Towards an Urban Renaissance: Final Report of the Urban Task Force Chaired by Lord Rogers of Riverside*, London: Spon (for other publications, see Urban Task Force)

Roo, G. de (2003) *Environmental Planning in the Netherlands: Too Good to be True – From Command and Control Planning to Shared Governance*, Aldershot: Ashgate

Roome, N. J. (1986) 'New directions in rural policy: recent administrative changes in response to conflict between rural policies', *Town Planning Review* 57: 253–63

Rosamond, B. (2000) *Theories of European Integration*, London: Macmillan

Rose, C. (1990) *The Dirty Man of Europe: The Great British Pollution Scandal*, London: Simon and Schuster

Rosenbloom, S. (1992) 'Why working families need a car', in Wachs, M. and Crawford, M. (eds) *The Car and the City*, Ann Arbor, MI: University of Michigan Press

Ross, A., Rowan-Robinson, J. and Walton, W. (1995) 'Sustainable development in Scotland: the role of Scottish Natural Heritage', *Land Use Policy* 12: 237–52

Ross, M. (1991) *Planning and the Heritage* (first edition), London: Spon

Ross, M. (1996) *Planning and the Heritage* (second edition), London: Spon

Ross Silcock Ltd and Social Research Associates (1999) *Community Impact of Traffic Calming Schemes*, Edinburgh: Central Research Unit, Scottish Executive

Roth, G. (1996) *Roads in a Market Economy*, Aldershot: Avebury

Rowan-Robinson, J. (1998) *Quality Assessment in Development Control* (Central Research Unit, Scottish Office), Edinburgh: The Stationery Office

Rowan-Robinson, J. and Durman, R. (1992a) *Section 50 Agreements*, Edinburgh: Central Research Unit, Scottish Office

Rowan-Robinson, J. and Durman, R. (1992b) 'Conditions or agreements', *Journal of Planning and Environment Law* 1992: 1003–11

Rowan-Robinson, J. and Durman, R. (1992c) 'Planning agreements and the spirit of enterprise', *Scottish Geographical Magazine* 108: 157–63

Rowan-Robinson, J. and Durman, R. (1993) 'Planning policy and planning agreements', *Land Use Policy* 10: 197–204

Rowan-Robinson, J. and Lloyd, M. G. (1988) *Land Development and the Infrastructure Lottery*, Edinburgh: T. & T. Clark

Rowan-Robinson, J. and Lloyd, M. G. (1991) 'National planning guidelines: a strategic opportunity wasting', *Planning Practice and Research* 6(3): 16–19

Rowan-Robinson, J. and Ross, A. (1994) 'Enforcement of environmental regulation in Britain: strengthening the link', *Journal of Planning and Environment Law* 1994: 200–18

Rowan-Robinson, J. and Young, E. (1987) 'Enforcement: the weakest link in the Scottish planning control system', *Urban Law and Policy* 8: 255–88

Rowan-Robinson, J. and Young, E. (1989) *Planning by Agreement in Scotland*, Glasgow: The Planning Exchange and Edinburgh: Green

Rowan-Robinson, J., Ross, A. and Walton, W. (1995) 'Sustainable development and the development control process', *Town Planning Review* 66: 269–86

Rowan-Robinson, J., Ross, A., Walton, W. and Rothnie, J. (1996) 'Public access to environmental information: a means to what end?', *Journal of Environmental Law* 8: 19–42

Rowell, T. A. (1991) *SSSIs: A Health Check*, London: Wildlife Link

Rowthorn, R. E. (1999) *The Political Economy of Full Employment in Modern Britain* (Kalecki Memorial Lecture 1999) (mimeo), Cambridge: Faculty of Economics and Politics, University of Cambridge

Royal Commission on Environmental Pollution: *see* appendix on *Official Publications*

Royal Fine Art Commission (1997) *Design Quality and the PFI*, London: Thomas Telford

Royal Institute of British Architects (RIBA) (1995) *Quality in Town and Country: A Response to the Secretary of State by a Special Working Party*, London: RIBA

Royal Institution of Chartered Surveyors (RICS) (1991) *Britain's Environmental Strategy: A Response by the RICS to the White Paper 'This Common Inheritance'*, London: RICS

Royal Institution of Chartered Surveyors (1994) *The Effects of Lasting Peace on Property and Construction in Northern Ireland*, Belfast: RICS Northern Ireland Branch

Royal Institution of Chartered Surveyors (1995a) *The Private Finance Initiative: The Essential Guide*, London: RICS

Royal Institution of Chartered Surveyors (1995b) *Considering Private Finance for Public Sector Works: How They Do It Over There*, London: RICS

Royal Society for the Protection of Birds: *see* appendix on *Official Publications*

Royal Town Planning Institute (RTPI) (1976) *Planning and the Future Discussion Paper*, London: RTPI

Royal Town Planning Institute (1987) *Report and Recommendations of the Working Party on Women in Planning*, London: RTPI

Royal Town Planning Institute (1988a) *Access for Disabled People* (Planning Advice Note), London: RTPI

Royal Town Planning Institute (1988b) *Access Policies in Local Plans* (Planning Advice Note), London: RTPI

Royal Town Planning Institute (1988c) *Managing Equality: The Role of Senior Planners*, London: RTPI

Royal Town Planning Institute (1988d) *Planning for Choice and Opportunity*, London: RTPI

Royal Town Planning Institute (1992) *Planning Policy and Social Housing*, London: RTPI

Royal Town Planning Institute (1993) *Ethnic Minorities and the Planning System*, London: RTPI

Royal Town Planning Institute (1995) *Planning for Women* (Practice Advice Note 12), London: RTPI

Royal Town Planning Institute (1997a) *Regional Planning in England*, London: RTPI

Royal Town Planning Institute (1997b) *Slimmer and Swifter: A Critical Examination of District Wide Local Plans and UDPs* (Tewdwr-Jones, M. and Crow, S.), London: RTPI

Royal Town Planning Institute (1997c) *The Role of Elected Members in Plan Making and Development Control* (Darke, R. and Manson, R.), London: RTPI

Royal Town Planning Institute (1999a) *Planning for Biodiversity*, London: RTPI

Royal Town Planning Institute (1999b) *Radical Review of the Development Plan System in England: Process and Procedures*, London: RTPI

Royal Town Planning Institute (2000) *A New Vision for Planning: Delivering Sustainable Communities, Settlements and Places – Mediating Space, Creating Place*, London: RTPI

Royal Town Planning Institute (2001a) *Fitness for Purpose: Quality in Development Plans – A Guide to Good Practice*, London: RTPI

Royal Town Planning Institute (2001b) *Fitness for Purpose: Report of Research on the Quality of Development Plans*, London: RTPI

Royal Town Planning Institute (2001c) *Members' Survey 2001: How Can the RTPI Deliver its New Vision for Planning? The Need for Action*, London: RTPI

Royal Town Planning Institute (2001d) 'RTPI Statement of Intent', *Planning*, 1418 (11 May)

Royal Town Planning Institute (2004a) *A Manifesto for Planning*, London: RTPI

Royal Town Planning Institute (2004b) *Policy Statement on Initial Planning Education*, London: RTPI

Royal Town Planning Institute and Commission for Racial Equality (1983) *Planning for a Multi-Racial Britain*, London: RTPI

Rudlin, D. (1998) *Tomorrow: A Peaceful Path to Urban Reform – The Feasibility of Accommodating 75 percent of New Homes in Urban Areas*, London: Friends of the Earth

Rural Voice (1990) *Employment of the Land*, Cirencester: Rural Voice

Rush, M. (1990) *Parliament and Pressure Politics*, Oxford: Clarendon Press

Russell, H., Dawson, J., Garside, P. and Parkinson, M. (1996) *City Challenge: Interim National Evaluation*, London: HMSO

Russell, S. (2000) 'Environmental appraisal of development plans', *Town Planning Review* 71(2): 529–46

Russell Barber, W. (2000) *Regional Government in England: A Preliminary Review of Literature and Research Findings*, London: DETR

Rutherford, L. A. and Peart, J. D. (1989) 'Opencast guidance: opportunities for green policies', *Journal of Planning and Environment Law* 1989: 402–10

Ryan, J. C. (1991) 'Impact fees: a new funding source for local growth', *Journal of Planning Literature* 5: 401–7

Ryder, A. A. (1987) 'The Dounreay inquiry: public participation in practice', *Scottish Geographical Magazine* 103: 54–7

Rydin, Y. (1984) 'The struggle for housing land: a case of confused interests', *Policy and Politics* 12: 431–46

Rydin, Y. (1986) *Housing Land Policy*, Aldershot: Gower

Rydin, Y. (1992) 'Environmental dimensions of residential development and the implications for local planning practice', *Journal of Environmental Planning and Management* 35: 43–61

Rydin, Y. (1993) *The British Planning System: An Introduction*, London: Macmillan

Rydin, Y. (1998a) 'Land use planning and environmental capacity: reassessing the use of regulatory policy tools to achieve sustainable development', *Journal of Environmental Planning and Management* 41: 749–65

Rydin, Y. (1998b) 'The enabling local state and urban development resources, rhetoric and planning in east London', *Urban Studies* 35: 175–91

Rydin, Y. (1998c) *Urban and Environmental Planning*, London: Macmillan

Rydin, Y. (1999) 'Public participation in planning', in Cullingworth, J. B. (ed.) *British Planning: 50 Years of Urban and Regional Policy*, London: Athlone

Rydin, Y., Home, R. and Taylor, K. (1990) *Making the Most of the Planning Appeals System: Report to the Association of District Councils*, London: Association of District Councils

Safdie, M. and Kohn, W. (1998) *The City after the Automobile: An Architect's Vision*, Boulder, CO: Westview Press

Salet, W. and Faludi, A. (2000) *Revival of Strategic Spatial Planning*, Amsterdam: Edita KNAW

Salter, J. R. (1992a) 'Environmental assessment: the challenge from Brussels', *Journal of Planning and Environment Law* 1992: 14–20

Salter, J. R. (1992b) 'Environmental assessment: the need for transparency', *Journal of Planning and Environment Law* 1992: 214–21

Salter, J. R. (1992c) 'Environmental assessment: the question of implementation', *Journal of Planning and Environment Law* 1992: 313–18

Salter, M. and Newman, P. (1992) 'Minding their own business in the planning department', *Municipal Journal* 50 (11–17 December): 28–9

Samuels, A. (2004) 'Waste', *Journal of Planning and Environment Law* 1471, Nov: 1465–71

Sandbach, F. (1980) *Environment, Ideology and Public Policy*, Oxford: Blackwell

Sanders, A-M. and Rothnie, J. (1996) 'Planning registers: their role in promoting public participation', *Journal of Planning and Environment Law* 1996: 539–46

Sandford, Lord (chair) (1974) *Report of the National Park Policies Review Committee*, Sandford Report, London: HMSO

Sanoff, H. (1999) *Community Participation Methods in Design and Planning*, London: Wiley

Sargent, J. W. (1993) *The Noise Climate Around Our Homes*, London: Building Research Establishment

Sassen, S. (1995) *The Global City: New York, London, Tokyo*, Princeton, NJ: Princeton University Press

Saunders, P. (1986) *Social Theory and the Urban Question* (second edition), London: Routledge

Save Britain's Heritage (1998) *Catalytic Conversion: Revive Historic Buildings to Regenerate Communities*, London: SAVE

Scanlon, K., Edge, A. and Willmott, T. (1994) *The Economics of Listed Buildings*, Cambridge: Department of Land Economy, University of Cambridge

Scargill, D. I. and Scargill, K. E. (1994) *Containing the City: The Role of Oxford's Green Belt*, Oxford: School of Geography, University of Oxford

Scarman, Lord (chair) (1981) *The Brixton Disorders 10–12 April 1981*, Cmnd 8427, Scarman Report, London: HMSO

Schackleton, J. R. (1992) *Training Too Much? A Sceptical Look at the Economics of Skill Provision in the UK*, London: Institute of Economic Affairs

Schaffer, F. (1970) *The New Town Story*, London: MacGibbon and Kee

Schlesinger, A. (1999) 'New deal for communities: one year on', *Town and Country Planning* 68: 345–7

Schmidt-Eichstaedt, G. (1996) *Land Use Planning and Building Permission in the European Union*, Cologne: Deutscher Gemeindeverlag and Verlag W. Kölhammer

Schmitter, P. (1996) 'Some alternative futures for the European Polity and their implications for European public policy', in Meny, Y., Muller, P. and Quermonne, J-L. (eds) *Adjusting to Europe*, London: Routledge

Schofield, J. (1987) *Cost-Benefit Analysis in Urban and Regional Planning*, London: Allen and Unwin

Scholefield, G. P. (1990) 'Transport and society: the Rees Jeffreys discussion papers', *Town Planning Review* 61: 487–93

Schon, D. A. (1971) *Beyond the Stable State*, London: Temple Smith

School of Town and Regional Planning, University of Dundee (1997) *Public Access to Planning Information*, Edinburgh: Scottish Office

Schubert, D. and Sutcliffe, A. (1996) 'The "Haussmannization" of London? The planning and construction of Kingsway-Aldwych, 1889–1935', *Planning Perspectives* 11: 115–44

Schuster, G. (chair) (1950) *Report of the Committee on the Qualifications of Planners*, Cmd 8059, Schuster Report, London: HMSO

Scott, A. (1999) 'Whose futures? A comparative study of Local Agenda 21 in mid Wales', *Planning Practice and Research* 14(4): 401–21

Scott, A. and Bullen, A. (2004) 'Special landscape areas: landscape conservation or confusion in the town and country planning system?', *Town Planning Review* 75(2): 205–30

Scott, A. J. and Roweis, S. T. (1977) 'Urban planning in theory and practice: a reappraisal', *Environment and Planning A* 9: 1097–119

Scott, L. F. (chair) (1942) *Report of the Committee on Land Utilisation in Rural Areas*, Cmd 6378, Scott Report, London: HMSO

Scott, M. (2003) 'Area-based partnerships and engaging the community sector: lessons from rural development practice in Northern Ireland', *Planning Practice and Research* 18(4): 279–92

Scott, P. (1996) *The Property Masters: A History of the British Commercial Property Sector*, London: Spon

Scottish Biodiversity Group (1988a) *Biodiversity in Scotland: The Way Forward*, Edinburgh: Scottish Biodiversity Group

Scottish Biodiversity Group (1988b) *Local Biodiversity Action Plans: A Manual*, Edinburgh: Scottish Biodiversity Group

Scottish Homes (1991) *Planning Agreements and Low Cost Housing in Scotland's Rural Areas*, Edinburgh: Scottish Homes

Scrase, T. (1991) 'Archaeology and planning: a case for full integration', *Journal of Planning and Environment Law* 1991: 1103–12

Scrase, T. (1999) 'The judicial review of local planning authority decisions: taking stock', *Journal of Planning and Environment Law* August: 679–90

Scrase, T. (2000) 'Listed building controls: shifting ground on fixtures and fittings?', *Journal of Planning and Environment Law* March: 235–45

Seaton, K. and Nadin, V. (2000) *A Comparison of Environmental Planning Systems Legislation in Selected Countries*, Background Paper for the Royal Commission on Environmental Pollution, Occasional Paper 8, Bristol: Faculty of the Built Environment, University of the West of England

Seaton, K. and Nadin, V. (2002) *An Update on the EU Compendium of Spatial Planning Systems and Policies*, London: DETR

Seebohm, Lord (chair) (1968) *Report of the Committee on Local Authority and Allied Personal Social Services*, Cmnd 3703, Seebohm Report, London: HMSO

Segal Quince Wicksteed (1992) *Evaluation of Regional Enterprise Grants: Third Stage*, London: Department of Trade and Industry

Segal Quince Wicksteed (1996) *The Impact of Tourism on Rural Settlements*, Salisbury: Rural Development Commission

Self, P. (1982) *Planning the Urban Region: A Comparative Study of Policies and Organisations*, London: Allen and Unwin

Self, P. (1993) *Government by the Market? The Politics of Public Choice*, London: Macmillan

Sellgren, J. (1990) 'Development control data for planning research: the use of aggregated development control records', *Environment and Planning B: Planning and Design* 17: 23–7

Selman, P. (1988) *Countryside Planning in Practice: The Scottish Experience*, Stirling: Stirling University Press

Selman, P. (1995a) 'Local sustainability: can the planning system help get us from here to there?', *Town Planning Review* 66: 287–302

Selman, P. (1995b) 'Theories for rural-environmental planning', *Planning Practice and Research* 10: 5–13

Selman, P. (1996) *Local Sustainability: Managing and Planning Ecologically Sound Places*, London: Paul Chapman

Selman, P. (1998) 'Local Agenda 21: substance or spin?', *Journal of Environmental Planning and Management* 41: 533–53

Selman, P. (1999) *Environmental Planning: The Conservation and Development of Biophysical Resources* (second edition), London: Paul Chapman

Selman, P. and Wragg, A. (1999) 'Local sustainability planning: from interest-driven networks to vision-driven super-networks?', *Planning Practice and Research* 14(3): 329–40

SERPLAN (1992) *SERPLAN: Thirty Years of Regional Planning 1962–1992*, London: London and South East Regional Planning Conference

Sharland, J. (2000) 'Listed buildings: the need for a new approach', *Journal of Planning and Environment Law* November: 1093–110

Sharman, F. A. (1985) 'Public attendance at planning inquiries', *Journal of Planning and Environment Law* 1985: 152–8

Sharp, E. (1969) *The Ministry of Housing and Local Government*, London: Allen and Unwin

Sharp, T. (1940) *Town Planning*, Harmondsworth: Penguin

Sharp, T. (1947) *Exeter Phoenix*, London: Architectural Press

Shaw, D. and Sykes, O. (2003) 'Investigating the application of the European Spatial Development Perspective (ESDP) to regional planning in the United Kingdom', *Town Planning Review* 74(1): 31–50

Shaw, D., Nadin, V. and Westlake, T. (1995) 'The compendium of spatial planning systems and policies', *European Planning Studies* 3: 390–5

Shaw, D., Nadin, V. and Westlake, T. (1996) 'Towards a supranational spatial development perspective: experience in Europe', *Journal of Planning Education and Research* 15: 135–42

Shaw, D., Roberts, P. and Walsh J. (eds) (2000) *Regional Planning and Development in Europe*, Aldershot: Ashgate

Shaw, J. (1998) 'Who's afraid of the double whammy?', *Town and Country Planning* 67(9): 306–7

Shaw, K. (1995) 'Assessing the performance of urban development corporations: how reliable are the official government output measures?', *Planning Practice and Research* 10: 287–97

Shaw, K. and Robinson, F. (1999) 'Learning from experience? Reflections on two decades of British urban policy', *Town Planning Review* 70(1): 49–63

Shaw, T. (1992) 'Regional planning guidance for the North East: advice to the secretary of state for the environment', in Minay, C. L. W. (ed.) 'Developing regional guidance in England and Wales: a review symposium', *Town Planning Review* 63: 415–34

Sheaf, P. (chair) (1972) *Report of the Working Party on Local Authority / Private Enterprise Partnership Schemes*, Sheaf Report, London: HMSO

Sheail, J. (1976) *Nature in Trust: The History of Nature Conservation in Britain*, Glasgow: Blackie

Sheail, J. (1983) 'Deserts of the moon: the Mineral Workings Act and the restoration of ironstone workings in Northamptonshire', *Town Planning Review* 54: 405–24

Sheail, J. (1992) 'The amenity clause: an insight into half a century of environmental protection in the United Kingdom', *Transactions of the Institute of British Geographers* NS 17: 152–65

Sheail, J. (1995) 'John Dower, national parks, and town and country planning in Britain', *Planning Perspectives* 10: 1–16

Shelbourn, C. (1996) 'Protecting the familiar and cherished scene', *Journal of Planning and Environment Law* 1996: 463–9

Shelton, A. (1991) 'The well informed optimist's view', in Nadin, V. and Doak, J. (eds) *Town Planning Responses to City Change*, Aldershot: Gower

Shepherd, J. and Abakuks, A. (1992) *The National Survey of Vacant Land in Urban Areas of England 1990* (DoE), London: HMSO

Shepley, C. (1999) 'Decision-making and the role of the Inspectorate', *Journal of Planning and Environment Law* May: 403–7

Shere, M. E. (1995) 'The myth of meaningful environmental risk assessment', *Harvard Environmental Law Review* 19: 409–92

Sherlock, H. (1991) *Cities are Good for Us*, London: Paladin

Shipley, R. and Newkirk, R. (1999) 'Vision and visioning in planning: what do these terms really mean?', *Environment and Planning B* 26(4): 573–92

Shipley, R., Fick, R., Hall, B. and Earley, R. (2004) 'Evaluating municipal visioning', *Planning Practice and Research*, 19(2): 195–210

Shiva, V. (1992) 'Recovering the real meaning of sustainability', in Cooper, D. E. and Palmer, J. A. (eds) *The Environment in Question*, London: Routledge

Shoard, M. (1980) *The Theft of the Countryside*, London: Maurice Temple Smith

Shoard, M. (1987) *This Land is our Land*, London: Paladin (updated edition, Gaia 1997)

Shoard, M. (1999) *A Right to Roam*, Oxford: Oxford University Press

Short, J. R., Fleming, S. and Witt, S. (1986) *Housebuilding, Planning and Community Action*, London: Routledge and Kegan Paul

Short, M., Jones, C., Carter, J., Baker, M. and Wood, C. (2004) 'Current practice in the strategic environmental assessment of development plans in England', *Regional Studies* 38(2): 177–90

Shorten, J. and Daniels, I. (2000) *Rural Diversification in Farm Buildings: A Report for the Planning Officers's Society*, Faculty of the Built Environment Working Paper, Bristol: University of the West of England

Shorten, J. and Daniels, I. (2001) *Farm Diversification and the Planning System, Final Report for the Planning Officers' Society*, Faculty of the Built Environment Occasional Paper, Bristol: University of the West of England

Shucksmith, M. (1983) 'Second homes: a framework for policy', *Town Planning Review* 54: 174–93

Shucksmith, M. (1988) 'Policy aspects of housebuilding on farmland in Britain', *Land Development Studies* 5: 129–38

Shucksmith, M. and Watkins, L. (1991) 'Housebuilding on farmland: the distributional effects in rural areas', *Journal of Rural Studies* 7: 153–68

Shucksmith, M., Henderson, M., Raybould, S., Coombes, M. and Wong, C. (1995) *A Classification of Rural Housing Markets in England* (DoE), London: HMSO

Sidaway, R. (1994) *Recreation and the Natural Heritage: A Research Review*, Perth: Scottish Natural Heritage

Sillince, J. (1986) *A Theory of Planning*, Aldershot: Gower

Sim, P. A. (1994) 'Mixed use development', in Wood, M. (ed.) *Planning Icons: Myth and Practice* (Planning Law Conference, *Journal of Planning and Environment Law*), London: Sweet and Maxwell

Simmie, J. (1981) *Power, Property and Corporatism*, London: Macmillan

Simmie, J. (1993) *Planning at the Crossroads*, London: UCL Press

Simmie, J. (ed.) (1994) *Planning London*, London: UCL Press

Simmie, J. and French, S. (1989) 'Corporatism, participation and planning: the case of London', *Progress in Planning* 31(1): 1–57

Simmie, J. and King, R. (eds) (1990) *The State in Action: Public Policy and Politics*, London: Pinter

Simmonds, D. (1990) *The Impact of the Channel Tunnel on the Regions*, London: Royal Town Planning Institute

Simmonds, R. and Hack, G. (eds) (2000) *Global City Regions: Their Emerging Forms*, London: Spon

Simmons, I. G. (1993) *Interpreting Nature: Cultural Constructions of the Environment*, London: Routledge

Simpson, I. (1987) 'Planning gain: an aid to positive planning', in Harrison, M. L. and Mordey, R. (eds) *Planning Control: Philosophies, Prospects and Practice*, London: Croom Helm

Sinclair, G. (1992) *The Lost Land: Land Use Change in England 1945–1990*, London: Council for the Protection of Rural England

Sinfield, A. (1973) 'Poverty rediscovered', in Cullingworth, J. B. (ed.) *Problems of an Urban Society*, vol. 3: *Planning for Change*, London: Allen and Unwin

Skeffington, A. (chair) (1969) *Report of the Committee in Public Participation in Planning*, Skeffington Report, London: HMSO

Skelcher, C. (1985) 'Transportation', in Ranson, S., Jones, G. and Walsh, K. (eds) *Between Centre and Locality: The Politics of Public Policy*, London: Allen and Unwin

Skelcher, C., McCabe, A., Lowndes, V. and Nanton, P. (1996) *Community Networks in Urban Regeneration: It All Depends Who You Know*, Bristol: Policy Press

Slater, S., Marvin, S. and Newson, M. (1994) 'Land use planning and the water sector', *Town Planning Review* 65: 375–97

Smallbone, D. (1991) 'Partnership in economic development: the case of UK local enterprise agencies', *Policy Studies Review* 10: 87–98

Smart, G. and Anderson, M. (1990) *Planning and Management of Areas of Outstanding Natural Beauty*, Cheltenham: Countryside Commission

Smeed, J. (1964) *Road Pricing: The Economic and Technical Possibilities*, London: Smeed Report, HMSO

Smith, A. G., Williams, G. and Houlder, M. (1986) 'Community influence on local planning policy', *Progress in Planning* 25: 1–82

Smith, David L. (1974) *Amenity and Urban Planning*, London: Crosby, Lockwood Staples

Smith, Denis (ed.) (1993) *Business and the Environment: Implications of the New Environmentalism*, London: Paul Chapman

Smith, G. (1994) 'Vitality and viability of town centres', in Wood, M. (ed.) *Planning Icons: Myth and Practice* (Planning Law Conference, *Journal of Planning and Environment Law*), London: Sweet and Maxwell

Smith, N. (1996) *The New Urban Frontier: Gentrification and the Revanchist City*, London: Routledge

Smith, R. and Wannop, U. (1985) *Strategic Planning in Action: The Impact of the Clyde Regional Plan 1946–82*, Aldershot: Gower

Smith & Williamson Management Consultants (2002) *Planning Green Paper Report: Processing and Analysis of Responses*, London: DTLR

Smith Morris, E. (1997) *British Town Planning and Urban Design: Principles and Policies*, Edinburgh: Addison Wesley Longman

Solesbury, W. (1976) 'The environmental agenda: an illustration of how situations may become political issues and issues may demand responses from government: or how they may not', *Public Administration* 54: 379–97

Solesbury, W. (1981) 'Strategic planning: metaphor or method?', *Policy and Politics* 9: 419–37

Solesbury, W. (1986) 'The dilemmas of inner city policy', *Public Administration* 64: 389–400

Solesbury, W. (1993) 'Reframing urban policy', *Policy and Politics* 21: 31–8

Solesbury, W. (1998) *Good Connections: Helping People to Communicate in Cities* [working paper of the Comedia/Demos project on *The Richness of Cities* – see Worple and Greenhalgh 1999]

Solesbury, W. (1999) 'Get connected', *Town and Country Planning* 68: 52–3

Solop, F. I. (2001) 'Survey research and "visioning" in Flagstaff, Arizona', *Planning Practice and Research*, 16(1): 51–8

Sorafu, F. J. (1957) 'The public interest reconsidered', *Journal of Politics* 19: 616–39

Sorensen, A. D. and Day, R. A. (1981) 'Libertarian planning', *Town Planning Review* 52: 390–402

Southgate, M. (1994) 'The added value test of merger' [of English Nature and the Countryside Commission], *Town and Country Planning* 63: 134–5

Southgate, M. (1995) 'Nature conservation and planning', *Report for the Natural and Built Environment Professions*, issue 7 (August): 34–7

Sparks, L. (1987) 'Retailing in enterprise zones: the example of Swansea', *Regional Studies* 21: 37–42

Sparks, L. (1998) *Town Centre Uses in Scotland*, Edinburgh: The Stationery Office

Spawforth, P. (1995) 'Vox pop', *Planning Week* 3(27): 16–17

Speer, R., Dade, M. and Cheal, M. (2003) *How to Get Planning Permission*, London: Dent and Stonepound Books

Spencer, S. and Bynoe, I. (1998) *A Human Rights Commission: The Options for Britain and Northern Ireland*, London: Institute for Public Policy Research

Standing Advisory Committee on Trunk Road Assessment (Department of Transport): *see* appendix on *Official Publications* (Transport)

Starkie, D. N. M. (1982) *The Motorway Age: Road and Traffic Policies in Postwar Britain*, Oxford: Pergamon

Stead, D. and Nadin, V. (1999) 'Environmental resources and energy in the UK: the potential role of a national spatial planning framework', *Town Planning Review* 70(3): 339–62

Stead, D. and Nadin, V. (2000) *Urban and Rural Interdependencies in the West of England*, Faculty of the Built Environment Working Paper, Bristol: University of the West of England

Stead, D., Geerlings, H. and Meijers, E. (eds) (2004) *Policy Integration in Practice: The Integration of Land Use Planning, Transport and Environmental Policy Making in Denmark, England and Germany*, Delft: DUP Science

Steel, J., Nadin, V., Daniels, R. and Westlake, T. (1995) *The Efficiency and Effectiveness of Local Plan Inquiries*, London: HMSO

Steele, J. (1995) *Public Access to Information: An Evaluation of the Local Government (Access to Information) Act 1985*, London: Policy Studies Institute

Steer Davies Gleave (1992) *Financing Public Transport: How Does Britain Compare?*, London: Transport 2000

Steer Davies Gleave (1994) *Promoting Rail Investment*, London: Transport 2000

Steer Davies Gleave (1995) *Alternatives to Traffic Growth: The Role of Public Transport and the Future of Freight*, London: Transport 2000

Stein, J. M. (ed.) (1995) *Classic Readings in Urban Planning*, New York: McGraw-Hill

Stevens, R. (chair) (1976) *Planning Control over Mineral Working*, Stevens Report, London: HMSO

Stevens, M. and Whitehead, C. (2005) *Lessons from the Past, Challenges for the Future for Housing Policy: An Evaluation of English Housing Policy 1975–2000*, London: ODPM

Stevenson, R. (2002) *Getting 'Under the Skin' of Community Planning*, Report to the Community Planning Task Force, Edinburgh: Scottish Executive

Stewart, J. D. and Stoker, G. (eds) (1991) *The Future of Local Government*, London: Macmillan

Stewart, J. D. and Stoker, G. (eds) (1995) *Local Government in the 1990s*, London: Macmillan

Stewart, J. D., Walsh, K. and Prior, D. (1995) *Citizenship: Rights, Community, and Participation*, London: Pitman

Stewart, J. (2003) *Modernising Local Government: An Assessment of Labour's Reform Programme*, Basingstoke: Palgrave Macmillan

Stewart, M. (1994) 'Value for money in urban public expenditure', *Public Money and Management* 14(4): 55–61

Stewart, M. (2002) 'Compliance and collaboration in urban governance', in Cars, G., Healey, A., Madanipour, A. and de Magalhaes, C. (eds) *Urban Governance, Institutional Capacity and Social Milieux*, Aldershot: Ashgate

Stewart, M. and Taylor, M. (1995) *Empowerment and Estate Regeneration*, Bristol: Policy Press.

Stoker, G. (1991) *The Politics of Local Government*, London: Macmillan

Stoker, G. (ed.) (1999) *The New Management of British Local Governance*, London: Macmillan

Stoker, G. and Wilson, D. (eds) (2004) *Bristol Local Government into the 21st Century*, Basingstoke: Palgrave Macmillan

Stokes, G., Goodwin, P. and Kenny, F. (1992) *Trends in Transport and the Countryside*, Cheltenham: Countryside Commission

Stone, P. A. (1973) *The Structure, Size and Costs of Urban Settlements* (National Institute of Economic and Social Research), Cambridge: Cambridge University Press

Storey, D. J. (1990) 'Evaluation of policies and measures to create local employment', *Urban Studies* 27: 669–84

Strange, I. (1999) 'Urban sustainability, globalisation and the pursuit of the heritage aesthetic', *Planning Practice and Research* 14: 301–11

Strange, I. and Whitney, D. (2003) 'The changing roles and purposes of heritage conservation', *Planning Practice and Research* 18(2–3): 219–30

Strathclyde Regional Council (1995) *Sustainability Indicators*, Glasgow: The Council

Stubbs, M. (1994) 'Planning appeals by informal hearing: an appraisal of the views of consultants', *Journal of Planning and Environment Law* 1994: 710–14

Stubbs, M. (1999) 'Informality and fairness: unlikely partners in the planning appeal?', *Journal of Planning and Environment Law* February: 106–12

Stubbs, M. (2000) 'Informality and the planning appeal by hearing method: an appraisal of user satisfaction', *Town Planning Review* 71(3): 245–67

Stubbs, M. (2004) 'Heritage-sustainability: developing a methodology for the sustainable appraisal of the historic environment', *Planning Practice and Research* 19(3): 285–307

Suddards, R. W. and Hargreaves, J. M. (1996) *Listed Buildings* (third edition), London: Sweet and Maxwell

Suddards, R. W. and Morton, D. M. (1991) 'The character of conservation areas', *Journal of Planning and Environment Law* 1991: 1011–13

Sullivan, H. and Skelcher, C. (2002) *Working across Boundaries: Collaboration in Public Services*, Basingstoke: Palgrave Macmillan

Susskind, L. and Consensus Building Institute (1999) *Using Assisted Negotiation to Settle Land Use Disputes*, Boston, MA: Lincoln Institute of Land Policy

Sustainable Development Commission (2004) *Shows Promise But Must Try Harder*, London: Sustainable Development Commission

Sustrans (1996) *The National Cycle Network: Guidelines and Practical Details*, Bristol: Sustrans

Sutcliffe, A. (ed.) (1981a) *British Town Planning: The Formative Years*, Leicester: Leicester University Press

Sutcliffe, A. (1981b) *Towards the Planned City: Germany, Britain, the United States and France 1780–1914*, Oxford: Blackwell

Sutcliffe, A. (ed.) (1984) *Metropolis 1890–1940*, London: Mansell

Symes, M. and Steel, M. (2003) 'Lessons from America: the role of business improvement districts as an agent of urban regeneration', *Town Planning Review* 74(3): 301–13

System Three (1999) *Research on Walking*, Edinburgh: Central Research Unit, Scottish Office

Tait, M. and Campbell, H. (2000) 'The politics of communication between planning officers and politicians: the exercise of power through discourse', *Environment and Planning A* 32: 489–506

Tant, A. P. (1990) 'The campaign for freedom of information: a participatory challenge to elitist British government', *Public Administration* 68: 477–91

Tarling, R., Hirst, A., Rowland, B., Rhodes, J. and Tyler, P. (1999) *An Evaluation of the New Life for Urban Scotland Initiative*, Edinburgh: Development Department, Scottish Executive

Tate, J. (1994) 'Sustainability: a case of back to basics?', *Planning Practice and Research* 9: 367–79

Taussik, J. (1992) 'Pre-application enquiries', *Journal of Planning and Environment Law* 1992: 414–19

Taussik, J. and Smalley, J. (1998) 'Partnerships in the 1990s: Derby's successful City Challenge bid', *Planning Practice and Research* 13(3): 283–97

Taylor, A. (1984) 'The planning implications of new technology in retailing and distribution', *Town Planning Review* 55: 161–76

Taylor, A. (1992) *Choosing our Future: A Practical Politics of the Environment*, London: Routledge

Taylor, M. (1995) *Unleashing the Potential: Bringing Residents to the Centre of Regeneration*, York: Joseph Rowntree Foundation

Taylor, N. (1992) 'Professional ethics in town planning', *Town Planning Review* 63(3): 227–41

Taylor, N. (1998) *Urban Planning Theory since 1945*, London: Sage

Taylor, N. (1999) 'The elements of townscape and the art of urban design', *Journal of Urban Design* 4(2): 195–210

Teague, P. (1994) 'Governance structures and economic performance: the case of Northern Ireland', *International Journal of Urban and Regional Research* 18: 275–92

Telling, A. E. and Duxbury, R. M. C. (2002) *Planning Law and Procedure* (twelfth edition), London: Butterworths

TEST (1984) *The Company Car Factor*, London: TEST

TEST (1991) *Wrong Side of the Tracks? Impacts of Road and Rail Transport on the Environment: A Basis for Discussion*, London: TEST

TEST (1992) *An Environmental Approach to Transport and Planning in Cardiff*, London: TEST

Tewdwr-Jones, M. (1994a) 'Policy implications of the plan-led system', *Journal of Planning and Environment Law* 1994: 584–93

Tewdwr-Jones, M. (1994b) 'The development plan in policy implementation', *Environment and Planning C: Government and Policy* 12: 145–63

Tewdwr-Jones, M. (ed.) (1996) *British Planning Policy in Transition: Planning in the 1990s*, London: UCL Press

Tewdwr-Jones, M. (1997) 'Green belt or green wedges for Wales? A flexible approach to planning in the urban periphery', *Regional Studies* 31: 73–7

Tewdwr-Jones, M. (1998) 'Planning modernised?', *Journal of Planning and Environment Law* June: 519–28

Tewdwr-Jones, M. (2002) *The Planning Polity: Planning, Government and the Policy Process*, London: Routledge

Tewdwr-Jones, M. and Allmendinger, P. (1998) 'Deconstructing communicative rationality: a critique of Habermasian collaborative rationality', *Environment and Planning A* 30: 1975–89

Tewdwr-Jones, M. and Crow, S. (1997) *Slimmer and Swifter: A Critical Examination of District Wide Local Plans and UDPs*, London: Royal Town Planning Institute

Tewdwr-Jones, M. and Williams, R. H. (2001) *The European Dimension of British Planning*, London: Spon

Thake, S. and Stauerbach, R. (1993) *Investing in People: Rescuing Communities from the Margin*, York: Joseph Rowntree Foundation

Thatcher, M. (1993) *The Downing Street Years*, London: HarperCollins

The National Forest Company (2004) *The National Forest: The Strategy*, Swadlincote: National Forest Company

Therivel, R. (1995) 'Environmental appraisal of development plans: current status', *Planning Practice and Research* 10: 223–34

Therivel, R. and Partidario, M. R. (1996) *The Practice of Strategic Environmental Assessment*, London: Earthscan

Thirsk, J. (1997) *Alternative Agriculture: A History from the Black Death to the Present Day*, Oxford: Oxford University Press

Thomas, D. (1995) *Community Development at Work: A Case of Obscurity in Accomplishment*, London: Community Development Foundation

Thomas, H. (1992) 'Disability, politics and the built environment', *Planning Practice and Research* 7: 22–6

Thomas, H. (1996) 'Public participation in planning', in Tewdwr-Jones, M. (ed.) *British Planning Policy in Transition: Planning in the 1990s*, London: UCL Press

Thomas, H. (1997) 'Ethnic minorities and the planning system: a study revisited', *Town Planning Review* 68(2): 195–211

Thomas, H. (1999) 'Set the standard: how local authorities are handling the introduction of Best Value', *Planning* 17 September: 17

Thomas, H. (2000) *Race and Planning: The UK Experience*, London: UCL Press

Thomas, H. (2004) 'British planning and the promotion of race equality: the Welsh experience of Race Equality Schemes', *Planning Practice and Research* 19(1): 33–47

Thomas, H. and Healey, P. (1991) *Dilemmas of Planning Practice: Ethics, Legitimacy, and the Validation of Knowledge*, Aldershot: Avebury

Thomas, H. and Krishnarayan, V. (1993) 'Race, equality and planning', *The Planner* 79(3): 17–20

Thomas, H. and Krishnarayan, V. (1994a) 'Race, disadvantage and policy processes in British planning', *Environment and Planning A* 26: 1891–910

Thomas, H. and Krishnarayan, V. (eds) (1994b) *Race, Equality and Planning: Policies and Procedures*, Aldershot: Avebury

Thomas, H. and Lo Piccolo, F. (2000) 'Best value, planning and race equality', *Planning Practice and Research* 15(1): 79–94

Thomas, K. (1990a) *Planning for Shops*, London: Estates Gazette

Thomas, K. (1990b) 'Planning students and careers: the enrolment of planning students and career destinations of new planning graduates 1977– 1989', *The Planner*, 76(36): 13–16

Thomas, K. (1997) *Development Control: Principles and Practice*, London: UCL Press

Thomas, R. (1996) 'The economics of the new towns revisited', *Town and Country Planning* 65: 305–8

Thomas, S. and Watkin, T. G. (1995) 'Oh, noisy bells, be dumb: church bells, statutory nuisance and ecclesiastical duties', *Journal of Planning and Environment Law* 1995: 1097–105

Thompson, H. (1992) 'Contaminated land: the implications for property transactions and the property market', *National Westminster Bank Quarterly Review* November: 20–33

Thompson, P. B. (1995) *The Spirit of the Soil: Agriculture and Environmental Ethics*, London: Routledge

Thomson, J. M. (1969) *Motorways in London*, London: Duckworth

Thornley, A. (1993) *Urban Planning under Thatcherism: The Challenge of the Market* (second edition), London: Routledge

Thornley, A. and Newman, P. (1996) *Replanning European Cities*, London: Routledge

Tiesdell, S. A., Oc, T. and Heath, T. (1996) *Revitalising Historic Urban Quarters*, Oxford: Butterworth-Heinemann

Tietenberg, T. (1990) 'Economic instruments for environmental regulation', *Oxford Review of Economic Policy* 6: 17–33

Till, J. E. (1995) 'Building credibility in public studies', *American Scientist* 83: 468–73

Tillotson, J. (1996) *European Community Law: Text, Cases and Materials* (second edition), London: Cavendish

Tillotson, J. (2000) *European Union Law*, London: Cavendish

Tillotson, J. (2003) *Texts, Cases and Materials on European Union Law*, London: Cavendish

Tinch, R. (1996) *The Valuation of Environmental Externalities* (DoE), London: HMSO

Tindall, F. (1998) *Memoirs and Confession of a County Planning Officer*, Ford, Midlothian: Pantile Press

Titmuss, R. M. (1958) 'War and social policy', in his *Essays on 'The Welfare State'*, London: Allen and Unwin

Todorovic, J. and Wellington, S. (2000) *Living in Urban England: Attitudes and Aspirations*, London: DETR

Tolba, M. (1992) *Saving our Planet: Challenges and Hopes*, London: Chapman and Hall

Tolley, R. (1990a) *Calming Traffic in Residential Areas*, Tregaron, Wales: Brefi Press

Tolley, R. (ed.) (1990b) *The Greening of Urban Transport: Planning for Walking and Cycling in Western Cities*, London: Belhaven

Tomaney, J. and Mitchell, M. (1999) *Empowering the English Regions*, London: Charter 88

Towers, G. (1995) *Building Democracy: Community Architecture in the Inner Cities*, London: UCL Press

Town and Country Planning Association (TCPA) (1989) *Bridging the North–South Divide*, London: TCPA

Town and Country Planning Association (1992) *New Settlements: Planning Policy Guidance* (spoof PPG), London: TCPA

Town and Country Planning Association (1993) *Strategic Planning for Regional Development*, London: TCPA

Town and Country Planning Association (1996) *The People: Where Will They Go?* (edited by Breheny, M. and Hall, P.), London: TCPA

Town and Country Planning Association (1999a) 'Draft PPG3: housing' (summary of Town and Country Planning Association response), *Town and Country Planning* 68: 244–5

Town and Country Planning Association (1999b) *The People: Where Will They Work – Report of Town and Country Planning Association Research into the Changing Geography of Employment* (edited by M. J. Breheny), London: TCPA

Town and Country Planning Association (1999c) *Your Place and Mine: Reinventing Planning*, London: TCPA

Town and Country Planning Association (2000) *Reinventing Planning: The Report of the TCPA Inquiry into the Future of Planning*, London: TCPA

Town and Country Planning Association (2002) *Policy Statement: Green Belts*, London: TCPA

Townroe, P. and Martin, R. (1992) *Regional Development in the 1990s: The British Isles in Transition*, London: Jessica Kingsley

Townsend, A. (1993) 'The urban–rural cycle in the Thatcher growth years', *Transactions of the Institute of British Geographers* 18: 207–21

Townsend, A. (2002) 'Public speaking rights, members and officers in a planning committee', *Planning Practice and Research* 17(1): 59–68

Townsend, P. (1976) 'Area deprivation policies', *New Statesman* 6 August: 168–71

Townshend, T. and Pendlebury, J. (1999) 'Public participation in the conservation of historic areas: case studies from north-east England', *Journal of Urban Design* 4(3): 313–31

Transport 2000 (1995) *Moving Together: Policies to Cut Car Commuting*, London: Transport 2000 Trust

Transport 2000 (1997) *Changing Journeys to Work: An Employer's Guide to Green Commuter Plans* (supported by London First), London: Transport 2000

Transport Planning Research Group (2004) *Developing a Methodology to Capture Land Value Uplift around Transport Facilities*, Edinburgh: Scottish Executive

Travers, T., Jones, G., Hebbert, M. and Burnham, J. (1991) *The Government of London*, York: Joseph Rowntree Foundation

Travers, T., Biggs, S. and Jones, G. (1995) *Joint Working between Local Authorities: Experience for the Metropolitan Areas*, London: Local Government Chronicle and Joseph Rowntree Foundation

Treisman, M. (1995) *Traffic and Planning in Oxford: Commercial Development and the Environment*, Oxford: School of Planning, Oxford Brookes University

Trench, S. and Oc, T. (1990) *Current Issues in Planning*, Aldershot: Avebury

Trench, S. and Oc, T. (1995) *Current Issues in Planning, Volume 2*, Aldershot: Avebury

Trimbos, J. (1997) 'Planning in Northern Ireland', *Journal of Planning and Environment Law* 1997: 904–7

Tromans, S. (1991) 'Roads to prosperity or roads to ruin? Transport and the environment in England and Wales', *Journal of Environmental Law* 3: 1–37

Tromans, S. and Clarkson, M. (1991) 'The Environmental Protection Act 1990: its relevance to planning controls', *Journal of Planning and Environment Law* 1991: 507–15

Tromans, S. and Grant, M. (1997) *Encyclopedia of Environmental Law*, London: Sweet and Maxwell

Tromans, S. and Turrall-Clarke, R. (1994 with 1996 supplement) *Contaminated Land*, London: Sweet and Maxwell

Tromans, S. and Turrall-Clarke, R. (1999) *Contaminated Land: The New Regime*, London: Sweet and Maxwell

Tromans, S., Grant, M. and Nash, M. (eds) (1997) *Encyclopaedia of Environmental Law*, London: Sweet and Maxwell (looseleaf; updated regularly)

Trott, T. (2002) *Best Practice in Regeneration: Because It Works*, Bristol: Policy Press

Truelove, P. (1992) *Decision Making in Transport Planning*, Harlow: Longman

Truelove, P. (1999) 'Transport planning', in Cullingworth, J. B. (ed.) *British Planning: 50 Years of Urban and Regional Policy*, London: Athlone

Tubbs, C. (1994) 'The New Forest: one step backwards', *Ecos* 15(2): 37–42

Tugnutt, A. (1991) 'Design: the wider aspects of townscapes', *Town and Country Planning Summer School 1991 Proceedings*: 19–22

Tunnel, C., White, N. and Hyams, K. (Arup) (2004) *Investigating the Increasing Volume of Planning Appeals in England*, London: ODPM

Turner, T. (1992) 'Open space planning in London: from standards per 1000 to green strategy', *Town Planning Review* 63: 365–86

Turner, T. (1996) *City as Landscape: A Post-Postmodern View of Design and Planning*, London: Spon

Turok, I. (1988) 'The limits of financial assistance: an evaluation of local authority aid to industry', *Local Economy* 2: 286–97

Turok, I. (1989) 'Evaluation and understanding in local economic policy', *Urban Studies* 26: 587–606

Turok, I. (1990a) 'Public investment and privatisation in the new towns: a financial assessment of Bracknell', *Environment and Planning A* 22: 1323–36

Turok, I. (1990b) *Targeting Urban Employment Initiatives* (DoE), London: HMSO

Turok, I. (1992) 'Property-led urban regeneration: panacea or placebo?', *Environment and Planning A* 24: 361–79

Turok, I. (2004) 'Scottish urban policy: continuity, change and uncertainty post-devolution', in Johnstone and Whitehead (eds) (2004)

Turok, I. and Edge, N. (1999) *The Jobs Gap in Britain's Cities: Employment Loss and Labour Market Consequences* (Joseph Rowntree Foundation), Bristol: Policy Press

Turok, I. and Shutt, J. (eds) (1994) 'Urban policy into the 21st century', special issue of *Local Economy* 9(3): 211–304

Turok, I. and Wannop, U. (1990) *Targeting Urban Employment Initiatives* (DoE), London: HMSO

Turok, I. and Webster, D. (1998) 'The New Deal: jeopardised by the geography of unemployment?', *Local Economy* 12: 309–28

Tyldesley, D. (1999) *Planning for Biodiversity*, London: Royal Town Planning Institute

Tyldesley, David & Associates (1996) *Wildlife Impact: The Treatment of Nature Conservation in Environmental Assessment*, Sandy, Beds: Royal Society for the Protection of Birds

Tyler, P., Rhodes, J., Lawless, P. and Dabinett, G. (2001) *A Review of the Evidence Base for Regeneration Policy and Practice*, London: DETR

Tym, R. H. (1994) 'Planning in a rapidly changing economy', in Wood, M. (ed.) *Planning Icons: Myth and Practice* (Planning Law Conference, *Journal of Planning and Environment Law*), London: Sweet and Maxwell

Tym, Roger and Partners (1982–4) *Monitoring Enterprise Zones: Year One Report* (1982); *Year Two Report* (1983); *Year Three Report* (1984a), London: Roger Tym and Partners

Tym, Roger and Partners (1984b) *Land Supply for Housing in Urban Areas*, London: Housing Research Foundation

Tym, Roger and Partners (1988) *An Evaluation of the Stockbridge Village Trust Initiative* (DoE), London: HMSO

Tym, Roger and Partners (1989a) *The Effect on Small Firms of Refusal of Planning Permission* (DoE), London: HMSO

Tym, Roger and Partners (1989b) *The Incidence and Effects of Planning Conditions* (DoE), London: HMSO

Tym, Roger and Partners (1990) *Development Control Performance*, London: National Planning Forum

Tym, Roger and Partners (1991) *Housing Land Availability* (DoE), London: HMSO

Tym, Roger and Partners (1995a) *The Use of Article 4 Directions* (DoE), London: HMSO

Tym, Roger and Partners (1995b) *Review of the Implementation of PPG 16, Archaeology and Planning*, London: English Heritage

Tym, Roger and Partners (1996) *Enterprise Zones Monitoring 1994–95*, London: HMSO

Tym, Roger and Partners (1998) *Urban Development Corporations: Performance and Good Practice*, London: DETR

Tym, Roger and Partners (2004) *Research on Land Use Change Statistics: Measuring Densities of Dwellings and Proportions on Brownfield Land*, London: ODPM

Tym, Roger and Partners (in association with Land Use Consultants) (1987) *Evaluation of Derelict Land Grant Schemes* (DoE), London: HMSO

Tym, Roger and Partners (in association with Oscar Faber TPA) (1996) *The Gyle Impact Study* [retail impact], Edinburgh: HMSO

Tyme, J. (1978) *Motorways versus Democracy*, London: Macmillan

UK Round Table on Sustainable Development (UKRTSD) (1997) *Housing and Urban Capacity*, UKRTSD, London: DETR

UK Round Table on Sustainable Development (1998) *Aspects of Sustainable Agriculture and Rural Policy*, UKRTSD, London: DETR

UK Round Table on Sustainable Development (2000) *Fifth Annual Report*, London: HMSO

Uman, M. F. (ed.) (1993) *Keeping Pace with Science and Engineering: Case Studies in Environmental Regulation*, Washington, DC: National Academy Press

Underwood, J. (1981a) 'Development control: a case study of discretion in action', in Barrett, S. and Fudge, C. (eds) *Policy and Action: Essays on the Implementation of Public Policy*, London: Methuen

Underwood, J. (1981b) 'Development control: a review of research and current issues', *Progress in Planning* 16(3): 175–242

United Nations (1992) *Environmental Accounting: Current Issues, Abstracts and Bibliography*, New York: Department of Economic and Social Development, United Nations

University of Edinburgh (1999) *Summary of Devolved Parliaments in the European Union*, Edinburgh: The Stationery Office

University of Liverpool (Environmental Advisory Unit) (1986) *Transforming our Waste Land: The Way Forward* (DoE), London: HMSO

University of Liverpool, University of Newcastle and University of Warwick (2002) *The Development of a Town and City Indicators Database*, London: ODPM

University of the West of England and Roger Tym and Partners (2004) *Dynamic Smaller Towns in Wales; Final Report to the National Assembly for Wales*, Cardiff: National Assembly for Wales

Unwin, R. (1909) *Town Planning in Practice: An Introduction to the Art of Designing Cities and Suburbs*, London: T. Fisher Unwin

Upton, W. (1999) 'The European Convention on Human Rights and environmental law', *Journal of Planning and Environment Law* April: 315–20

Upton, W. (2005) 'Planning reform: the requirement to replace supplementary planning guidance with supplementary planning documents', *Journal of Planning and Environment Law* January: 34–40

Upton, W. and Harwood, R. (1996) 'The stunning powers of environmental inspectors', *Journal of Planning and Environment Law* 1996: 623–32

Urban Initiatives (1995) *Hertfordshire Dwelling Provision through Planning Regeneration*, Hertford: Hertfordshire County Environment Department

Urban Task Force (1999a) *But Would You Live There? Shaping Attitudes to Urban Living*, London: DETR

Urban Task Force (1999b) *Fiscal Incentives for Urban Housing: Exploring the Options* (KPMG), London: DETR

Urban Task Force (1999c) *Mock Planning Policy Guidance Note on Urban Development* (Tony Burton), London: DETR

Urban Task Force (1999d) *Regional Centres for Urban Development: A Feasibility Study* (Price Waterhouse), London: DETR

Urban Task Force (1999e) *The Future Role of Planning Agreements in Facilitating Urban Regeneration* (Leslie Punter), London: DETR

Urban Task Force (1999f) *Towards an Urban Renaissance: Final Report of the Urban Task Force*, London: Spon

Urban Task Force (2005) *Towards a Strong Urban Renaissance: An Independent Report by Members of the Urban Task Force chaired by Lord Rogers*, Urban Task Force (available at www.urban taskforce.org)

Urban Villages Group (1992) *Urban Villages: A Concept for Creating Mixed-use Urban Developments on a Sustainable Scale*, London: Urban Villages Group

Urban Villages Group (1993) *Economics of Urban Villages*, London: Urban Villages Forum

URBED (1994) *Vital and Viable Town Centres: Meeting the Challenge* (DoE), London: The Stationery Office

URBED (2002) *Towns and Cities: Partners in Urban Renaissance*, London: ODPM

Uthwatt, A. A. (chair) (1942) *Final Report of the Expert Committee on Compensation and Betterment*, Cmd 6386, Uthwatt Report, London: HMSO

Vale, B. (1995) *Prefabs: A History of the UK Temporary Housing Programme*, London: Spon

Varner, G. E. (1987) 'Do species have standing?', *Environmental Ethics* 9: 57–72

Veld, R. J. (1991) 'Road pricing: a logical failure', in Kraan, D. J. and Veld, R. J. (eds) *Environmental Protection: Public or Private Choice?*, Dordrecht: Kluwer

Vergara, C. J. (1995) *The New American Ghetto*, New Brunswick, NJ: Rutgers University Press

Verhoff, E. (1996) *The Economics of Regulating Road Transport*, Cheltenham: Edward Elgar

Verney, R. (chair) (1976) *The Way Ahead: Report of the Advisory Committee on Aggregates*, Verney Report, London: HMSO

Vickerman, R. W. (1991) *Infrastructure and Regional Development*, London: Pion

Vigar, G. and Healey, P. (1999) 'Territorial integration and "plan-led" planning', *Planning Practice and Research* 14(2): 153–69

Vigar, G., Healey, P., Hull, A. and Davoudi, S. (2000) *Planning, Governance and Spatial Strategy in Britain: An Institutionalist Analysis*, London: Macmillan

Vilagrasa, J. and Larkham, P. J. (1995) 'Post-war redevelopment and conservation in Britain: ideal and reality in the historic core of Worcester', *Planning Perspectives* 10: 149–72

Viner, D. and Hulme, M. (1994) *The Climate Impacts LINK Project*, Norwich: Climatic Research Unit, University of East Anglia

Vogel, D. (1986) *National Styles of Regulation: Environmental Policy in Great Britain and the United States*, Ithaca, NY: Cornell University Press

Vogel, D. (1995) 'The making of EC environmental policy', in Ugur, M. (ed.) *Policy Issues in the European Union*, Dartford: Greenwich University Press

VROM (Dutch National Ministry for Spatial Planning, Environment and Housing) (2001) *The Fifth Spatial Planning Report*, The Hague: VROM

Wachs, M. and Crawford, M. (1992) *The Car and the City*, Ann Arbor, MI: University of Michigan Press

Wachs, M. (ed.) (1985) *Ethics in Planning*, New Brunswick, NJ: Center for Urban Policy Research

Wade, E. C. S. and Bradley, A. W. (1993) *Constitutional and Administrative Law* (eleventh edition by Bradley, A. W. and Ewing, K. D.), London: Longman

Wägenbaur, R. (1991) 'The European Community's policy on implementation of environmental directives', *Fordham International Law Journal* 14: 455–77

Wagstaff, P. (1999) *Regionalism in the European Union*, London: Intellect Books

Waite, R. (1995) *Household Waste Recycling*, London: Earthscan

Wakeford, R. (1990) *American Development Control: Parallels and Paradoxes from an English Perspective*, London: HMSO

Wakeford, R. (1993) 'Planning policy guidance: what's the use?', *Housing and Planning Review* April–May: 14–18

Waldegrave, W., Byng, J., Paterson, T. and Pye, G. (1986) *Distant Views of William Waldegrave's Speech*, London: Centre for Policy Studies

Walker, G. (2001) 'Factory outlet centres: passing phenomenon or here to stay?', paper presented to the RTPI Planning and Retail Development Conference, Bournemouth, March

Walker, M. and Reynard, M. V. (1990) *Costs in Planning Proceedings*, London: Longman

Walker, P. (2001) 'What sustainability can learn from Delia Smith: Perry Walker outlines four new participation recipes', *Town and Country Planning* 70(5): 150

Walmsley, D. A. and Perrett, K. E. (1992) *The Effects of Rapid Transport on Public Transport and Urban Development*, Transport Research Laboratory, State of the Art Review 6, London: HMSO

Wannop, U. (1990) 'The Glasgow Eastern Area Renewal (GEAR) project: a perspective on the management of urban regeneration', *Town Planning Review* 61: 455–74

Wannop, U. (1994) *Regional Planning and Governance in Britain in the 1990s*, Strathclyde Papers on Planning no. 27, Glasgow: Centre for Planning, University of Strathclyde

Wannop, U. (1995) *The Regional Imperative: Regional Planning and Governance in Britain, Europe and the United States*, London: Regional Studies Association and Jessica Kingsley

Wannop, U. (1999) 'New towns', in Cullingworth, J. B. (ed.)

British Planning: 50 Years of Urban and Regional Policy, London: Athlone

Warburton, D. (2000) 'Comment [on Blair and environment]', *EG* 6(10): 1–2

Warburton, D. (2002) 'Participation: delivering a fundamental change', *Town and Country Planning* 71(3): 82–4

Ward, C. and Hardy, D. (1990) *Goodnight Campers! The History of the British Holiday Camp*, London: Spon

Ward, S. V. (1988) *The Geography of Interwar Britain: The State and Uneven Development*, London: Routledge

Ward, S. V. (ed.) (1992) *The Garden City: Past, Present and Future*, London: Spon

Ward, S. V. (1998) *Selling Places: The Marketing and Promotion of Towns and Cities*, London: Spon

Ward, S. V. (1999) 'Public–private partnerships', in Cullingworth, J. B. (ed.) *British Planning: 50 Years of Urban and Regional Policy*, London: Athlone

Ward, S. V. (2004) *Planning and Urban Change* (second edition), London: Spon

Wardroper, J. (1981) *Juggernaut*, London: Temple Smith

Warren, H. and Davidge, W. R. (eds) (1930) *Decentralisation of Population and Industry: A New Principle in Town Planning*, London: King

Warwick Business School (1999) *Improving Local Public Services: Interim Evaluation of the Best Value Pilot Programme*, London: DETR

Waste, R. J. (1998) *Independent Cities: Rethinking Urban Policy*, Oxford: Oxford University Press

Waters, G. R. (1994) 'Government policies for the countryside', *Land Use Policy* 11: 88–93

Wates, N. (1977) *The Battle for Tolmers Square*, London: Routledge & Kegan Paul

Wates, N. (2000) *The Community Planning Handbook: How People Can Shape their Cities, Towns and Villages in any Part of the World*, London: Urban Design Group, South Bank University

Watt, P. (1992) 'Publicity for planning applications', *Scottish Planning Law and Practice* 38: 15–16.

Weale, A. (1992) *The New Politics of Pollution*, Manchester: Manchester University Press

Weale, A., O'Riordan, T. and Kramme, L. (1991) *Controlling Pollution in the Round: Change and Choice in Environmental Regulation in Britain and West Germany*, London: Anglo-German Foundation for the Study of Industrial Society

Weber, M. M. (1968–9) 'Planning in an environment of change', *Town Planning Review* 39: 179–95 and 277–95 (reprinted in Cullingworth, J. B. (ed.) (1973) *Problems of an Urban Society*, vol. 3: *Planning for Change*, London: Allen and Unwin)

Webster, B. and Lavers, A. (1991) 'The effectiveness of public local inquiries as a vehicle for public participation in the plan making process: a case study of the Barnet unitary development plan inquiry', *Journal of Planning and Environment Law* 1991: 803–13

Webster, C. and Lai, W-C. (2004) *Property Rights, Planning and Markets: Managing Spontaneous Cities*, London: Edward Elgar

Weir, S. (1995) 'Quangos: questions of democratic accountability', *Parliamentary Affairs* 48: 306–22

Welbank, M., Davies, N. and Haywood, I. (2000) *Mediation in the Planning System*, London: DETR

Welbank, M., Davies, N. and Haywood, I. (2002) *Further Research into Mediation in the Planning System*, London: ODPM

Welford, R. (1995) *Environmental Strategy and Sustainable Development: The Corporate Challenge for the 21st Century*, London: Routledge

Wells, H. G. (1905) *A Modern Utopia*, London: Collins

Wenban-Smith, A. (1998) *Planning Gains: Negotiating with Planning Authorities*, London: Estates Gazette

Wenban-Smith, A. (1999) *Plan, Monitor and Manage: Making it Work*, London: Council for the Protection of Rural England

Wenban-Smith, A. (2002) 'A better future for development plans: making "Plan, Monitor and Manage" work', *Planning Theory and Practice* 3(1): 33–52

Wenban-Smith, A. (2004) *The Missing Links: A Commentary on the Interim Report of Kate Barker's Review of Housing Supply*, London: CPRE

Wenban-Smith, A. and Pearce, B. (1998) *Planning Gains: Negotiating with Planning Authorities* (second edition), London: Estates Gazette

Westmacott, R. and Worthington, T. (1997) *Agricultural Landscapes: A Third Look*, Cheltenham: Countryside Commission

Weston, J. (2000) 'Reviewing environmental statements: new demands for the UK's EIA procedures', *Planning Practice and Research* 15(1–2): 135–42

Westwood, S. and Williams, J. (1996) *Imagining Cities: Scripts, Signs and Memories*, London: Routledge

Whatmore, K. and Boucher, S. (1993) 'Bargaining with nature: the discourse and practice of environmental planning gains', *Transactions of the Institute of British Geographers* 18: 166–78

Whatmore, S., Munton, R. and Marsden, T. (1990) 'The rural restructuring process: emerging divisions of agricultural property rights', *Regional Studies* 24: 235–45

Whitbread, M. and Marsay, A. (1992) *Coastal Superquarries to Supply South-East England Aggregate Requirements* (DoE), London: HMSO

Whitbread, M., Mayne, D. and Wickens, D. (1991) *Tackling Vacant Land: An Evaluation of Policy Instruments for Tackling Land Vacancy* (DoE), London: HMSO

White, P. (1986) 'Land availability, land banking and the price of land for housing: a review of recent debates', *Land Development Studies* 3: 101–11

White, P. (1995) *Public Transport: Its Planning, Management and Operation*, London: UCL Press

White, S., Alden, J. and Tewdwr-Jones, M. (2002) *Peripherality and Spatial Planning in Europe*, London: Routledge

Whitehand, J. W. R. (1989) 'Development pressure, development control, and suburban townscape change', *Town Planning Review* 60: 403–20

Whitehand, J. W. R. and Larkham, P. J. (1991) 'Suburban cramming and development control', *Journal of Property Research* 8: 147–59

Whitehand, J. W. R. and Larkham, P. J. (1992) *Urban Landscapes: International Perspectives*, London: Routledge

Whitehead, A. (1999) 'From regional development to regional devolution', in Dungey, J. and Newman, I. (eds) *The New Regional Agenda*, London: Local Government Information Unit

Whitehead, C. (1999) *Housing Need in the South East*, Cambridge: Department of Land Economy, University of Cambridge

Whitehead, C., Holmans, A., Marshall, D., Royce, R. and Gordon, I. (1999) *Housing Needs in the South East*, Cambridge: Property Research Unit, Department of Land Economy, University of Cambridge

Whitelegg, J. (1993) *Transport for a Sustainable Future: The Case for Europe*, London: Belhaven

Whitelegg, J. (1994) *Roads, Jobs and the Economy*, London: Greenpeace

Whitelegg, J. (1997) *Critical Mass: Transport, Environment and Society in the Twenty-first Century*, London: Pluto

Whitney, D. and Haughton, G. (1990) 'Structures for development partnerships in the 1990s: prac-tice in West Yorkshire', *The Planner* 76(21): 15–19

Widdicombe, D. (chair) (1986, 1988) *The Conduct of Local Authority Business: Report*, Cmnd 9797, Widdicombe Report; *Research Volumes*, Cmnd 9798, 9799, 9800 and 9801, London: HMSO (1986); *Government Response to the Report*, Cm 433, London: HMSO (1988)

Wightman, A. (1996) *Who Owns Scotland?*, Edinburgh: Canongate

Wightman, A. (1999) *Scotland: Land and Power*, Edinburgh: Luath Press in association with Democratic Left Scotland.

Wilcox, S. (1995) *Housing Finance Review 1995/96*, York: Joseph Rowntree Foundation

Wilcox, S. (2002) *UK Housing Review: 2002/2003*, Coventry: Joseph Rowntree Foundation, Chartered Institute of Housing and Council of Mortgage Lenders

Wilcox, S. (2004) *UK Housing Review: 2004/2005*, Coventry: Chartered Institute of Housing and Council of Mortgage Lenders

Wildavsky, A. B. (1973) 'If planning is everything, maybe it's nothing', *Policy Sciences* 4: 127–53

Wildavsky, A. B. (1987) *Speaking Truth to Power: The Art and Craft of Policy Analysis* (second edition), London: Transaction

Wilding, P. (1982) *Professional Power and Social Welfare*, London: Routledge and Kegan Paul

Wilding, R. (1990) *The Care of Redundant Churches: A Review of the Operation and Financing of the Redundant Churches Fund*, London: HMSO

Wilding, S. and Raemaekers, J. (2000) 'Environmental compensation for greenfield development: is the devil in the detail?', *Planning Practice and Research* 15(3): 211–32

Wilhelm, J. (1996) *Fax Messages from the Future*, London: Earthscan

Wilkes, S. and Peter, N. (1995) 'Think globally, act locally: implementing Agenda 21 in Britain', *Policy Studies* 16: 37–44

Wilkinson, D. (1992) 'Maastricht and the environment: the implications for the EC's environmental policy of the Treaty on European Union', *Journal of Environmental Law* 4: 221–39

Wilkinson, D. and Appelbee, E. (1999) *Implementing Holistic Government: Joined-up Action on the Ground*, Bristol: Policy Press

Wilkinson, D. and Waterton, J. (1991) *Public Attitudes to the Environment in Scotland*, Edinburgh: Scottish Office

Wilkinson, D., Bishop, K. and Tewdwr-Jones, M. (1998) *The Impact of the EU on the UK Planning System*, London: DETR

Williams, G., Bell, P. and Russell, L. (1991) *Evaluating the Low Cost Rural Housing Initiative* (DoE), London: HMSO

Williams, G., Strange, I., Bintley, M. and Bristow, R. (1992) *Metropolitan Planning in the 1990s: The Role of Unitary Development Plans*, Manchester: Department of Planning and Landscape, University of Manchester

Williams, H. (1999) 'A need to clarify need', *Town and Country Planning* 68: 112

Williams, K. (1999) 'Urban intensification policies in England: problems and contradictions', *Land Use Policy* 16: 167–78

Williams, K., Burton, E. and Jenks, M. (2000) *Achieving Sustainable Urban Form*, London: E&FN Spon

Williams, N., Shucksmith, M., Edmond, H. and Gemmell, A. (1998) *Scottish Rural Life Update: A Revised Socio-Economic Profile of Rural Scotland* (Central Research Unit, Scottish Office), Edinburgh: The Stationery Office

Williams, P. (ed.) (1997) *Directions in Housing Policy: Towards Sustainable Housing Policies for the UK*, London: Paul Chapman

Williams, R. and Birch, N. (1994) 'The longer-term impli-cations of national traffic forecasts and international network plans for local roads policy: the case of Oxfordshire', *Transport Policy* 1: 95–99

Williams, R. H. (1996) *European Union Spatial Policy and Planning*, London: Paul Chapman

Williams, R. H. (2000) 'Constructing the European Spatial Development Perspective: for whom?', *European Planning Studies* 8(3): 357–65

Williams, R. H. and Wood, B. D. (1994) *Urban Land and Property Markets in the UK*, London: UCL Press

Willis, K. G. (1995) 'Judging development control decisions', *Urban Studies* 32: 1065–79

Willis, K. G. and Garrod, G. (1993) *The Value of Waterside Properties*, Newcastle upon Tyne: Countryside Change Unit, University of Newcastle upon Tyne

Willis, K. G., Turner, R. K. and Bateman, J. (eds) (2001) *Urban Planning and Management*, Cheltenham: Edward Elgar

Willmott, P. (1994) *Urban Trends 2: A Decade in Britain's Deprived Urban Areas*, London: Policy Studies Institute.

Wilson, Alan (chair) (1963) *Committee on the Problems of Noise: Final Report*, Cmnd 2056, Wilson Report, London: HMSO

Wilson, Andrew and Charlton, K. (1997) *Making Partnerships Work: A Practical Guide for the Public, Private, Voluntary and Community Sectors*, York: Joseph Rowntree Foundation

Wilson, Arton (chair) (1959) *Caravans as Homes*, Cmnd 872, Arton Wilson Report, London: HMSO

Wilson, D. and Game, C. (1998) *Local Government in the United Kingdom* (second edition), London: Macmillan

Wilson, E. (1993) *Strategic Environmental Assessment*, London: Earthscan

Wilson, G. K. (1983) 'Planning lessons from the ports', *Public Administration* 61: 265–81

Wilson, K. (1999) 'Urban intensification policies in England: problems and contradictions', *Land Use Policy* 16: 167–78

Winpenny, J. T. (1991) *Values for the Environment*, London: HMSO

Winter, M. (1996) *Rural Politics: Policies for Agriculture, Forestry and the Environment*, London: Routledge

Winter, P. (1994) 'Planning and sustainability: an examination of the role of the planning system as an instrument for the delivery of sustainable development', *Journal of Planning and Environment Law* 1994: 883–900

Winter, P. (1998) *Legal and Liability Issues Preventing Urban Housing Development*, London: Town and Country Planning Association

Wolf, W. (1996) *Car Mania: A Critical History of Transport*, London: Pluto

Wolman, H. and Goldsmith, M. (1992) *Urban Politics and Policy: A Comparative Approach*, Oxford: Blackwell

Wolman, H. L., Cookford, C. and Hill, E. (1994) 'Evaluating the success of urban success stories', *Urban Studies* 31: 835–50

Woltjer, J. (2002) 'The "public support machine": notions of the function of participatory planning by Dutch infrastructure planners', *Planning Practice and Research* 17(4): 437–53

Womack, J. P. (1994) 'The real EV [electric vehicle] challenge: reinventing an industry', *Transport Policy* 1: 266–70

Wong, C. (2002) *Development of Town and City Indicators: Review, Interpretation and Analysis*, London: DTLR

Wong, C., Ravetz, J. and Turner, J. (2000) *The UK Spatial Planning Framework: A Discussion*, London: Royal Town Planning Institute

Wood, C. (1989) *Planning Pollution Prevention*, London: Heinemann

Wood, C. (1991) 'Urban renewal: the British experience', in Alterman, R. and Cars, G. (eds) *Neighbourhood Regeneration: An International Evaluation*, London: Mansell

Wood, C. (1994) 'Local urban regeneration initiatives: Birmingham Heartlands', *Cities* 11 (1): 48–58

Wood, C. (1995) *Environmental Impact Assessment: A Comparative Review*, Harlow: Longman

Wood, C. (1996) 'Private sector housing renewal in the UK: progress under the "new regime" in Birmingham', paper presented to Second ACSP-AESOP Joint International Congress, Toronto, Canada, July

Wood, C. (1999) 'Environmental planning', in Cullingworth, J. B. (ed.) *British Planning: 50 Years of Urban and Regional Policy*, London: Athlone

Wood, C. (2000) 'Ten years on: an empirical analysis of UK environmental statement submissions since the implementation of the Directive 85/337/EEU', *Journal of Environmental Planning and Management* 43(5): 721–47

Wood, C. and Bellinger, C. (1999) Directory of Environmental Impact Statements, Working Paper 179, Oxford Brookes School of Planning

Wood, C. and Jones, C. (1991) *Monitoring Environmental Assessment and Planning* (DoE), London: HMSO

Wood, C. and Jones, C. (1992) 'The impact of environmental assessment on local planning authorities', *Journal of Environmental Planning and Management* 35: 115–28

Wood, M. (ed.) (1994) *Planning Icons: Myth and Practice* (Planning Law Conference, *Journal of Planning and Environment Law*), London: Sweet and Maxwell

Wood, M. (1996) 'Local plans and UDPs: is there a better way?', *Journal of Planning and Environment Law* 1996: 807–15

Wood, R., Handley, J. and Bell, P. (1998) 'The character of countryside recreation and leisure appeals: an analysis of planning inspectorate records', *Journal of Planning and Environment Law* November: 1007–27

Wood, W. (1949) *Planning and the Law*, London: Marshall

Woodward, G. D. and Larkham, P. J. (1994) *Ideal and Reality in Planning: Post-war Development in Worcester*, Birmingham: School of Planning, University of Central England

Woolmar, C. (1997) *Unlocking the Gridlock: The Key to a New Transport Policy*, London: Friends of the Earth

World Commission on Environment and Development: see Brundtland (1987)

World Health Organisation (WHO) (1997) *City Planning for Health and Sustainable Development*, Copenhagen: WHO

Worpole, K. (ed.) (1999) *Richer Futures: Fashioning a New Politics*, London: Earthscan

Worpole, K. and Greenhalgh, L. (1999) *The Richness of Cities*, London: Comedia in association with Demos (Leicester: Econ Distribution)

Worskett, R. (1969) *The Character of Towns*, London: Architectural Press

Wraith, R. E. and Hutchesson, P. G. (1973) *Administrative Tribunals*, London: Allen and Unwin

Wraith, R. E. and Lamb, G. B. (1971) *Public Inquiries as an Instrument of Government*, London: Allen and Unwin

Wrigley, N. (1998) 'Understanding store development programmes in post-property-crisis UK food retailing', *Environment and Planning A* 30: 15–35

Wrigley, N. and Lowe, M. (eds) (1996) *Retailing, Consumption and Capital*, Harlow: Longman

Wye College (1991) *Rural Society: Issues for the Nineties*, Wye, Kent: Wye College

Yardley, D. (1995) *Introduction to Constitutional and Administrative Law*, London: Butterworths

Yelling, J. A. (1992) *Slums and Redevelopment: Policy and Practice in England 1918–1945*, London: UCL Press

Yelling, J. (1999) 'The development of residential urban renewal policies in England: planning for modernization in the 1960s', *Planning Perspectives* 14: 1–18

Yeomans, D. (1994) 'Rehabilitation and historic preservation: a comparison of American and British approaches', *Town Planning Review* 65: 159–78

Yiftachel, O. (1989) 'Towards a new typology of urban planning theories', *Environment and Planning B: Planning and Design* 16: 23–39

Young, E. (1991) *Scottish Planning Appeals*, Edinburgh: Green and Sweet and Maxwell

Young, E. and Rowan-Robinson, J. (1985) *Scottish Planning Law and Practice*, Glasgow: Hodge

Young, K. (1984) 'Metropolitan government: the development of the concept of reality', in Leach, S. (ed.) *The Future of*

Metropolitan Government, Birmingham: Institute of Local Government Studies, University of Birmingham

Young, K. (1986) 'Metropolis, R.I.P.', *Political Quarterly* 57: 36–46

Young, K. (1990) *The Politics of Local Government since Widdicombe*, York: Joseph Rowntree Foundation

Young, K. and Rao, N. (1997) *Local Government since 1945*, Oxford: Blackwell

Young, K., Gosschalk, B., and Hatter, W. (1996) *In Search of Community Identity* (Joseph Rowntree Foundation), York: York Publishing Services

Zetter, J. (2001) *Evaluation of Interreg IIC Projects in the United Kingdom*, London: DTLR

Zetter, J. (2002) 'Spatial planning in the European Union: a vision of 2010', in Faludi, A. (ed.) *European Spatial Planning*, Cambridge, MA: Lincoln Institute of Land Policy

Zonneveld, W. (2005) 'Expansive spatial planning: the new European transnational spatial visions', *European Planning Studies* 13(1): 1–18

Zonneveld, W. and Faludi, A. (1996) 'Cohesion versus competitive position: a new direction for European spatial planning', in Buunk, W. *et al.* (eds) *International Planning: A Dutch Perspective*, Amsterdam Study Centre for the Metropolitan Environment, University of Amsterdam

Zucker, P. C. (1999) *What Your Planning Professors Forgot to Tell You*, Chicago, IL: Planners Press

Official publications

A note on official publications

The student of town and country planning now has a very rich library of official publications to consult. This continues to grow at a rapid rate, and any list quickly becomes out of date. This appendix is a selective one. The first part lists the main sources of planning policy guidance. Other official documents are then presented by topic area related to the main chapters of the book.

The Stationery Office (TSO) is the main publisher of official documents (and is the largest UK publisher). Her Majesty's Stationery Office (HMSO) publishes legislation, Command Papers and statutory instruments and controls Crown copyright. An increasing number of documents are now published by government departments and agencies. Some of these have the advantage of being free of charge, but there can be difficulties in determining the publisher, the cost, and basic details such as the date of publication; names of authors are often omitted from them, especially the web-based versions.

A few words of explanation about the mysteries of government publications may be helpful. Command Papers (which include White Papers) are 'presented to Parliament by Command of Her Majesty'. Command Papers are numbered sequentially and have a short prefix which varies, but which (so far) is always an abbreviation of the word 'command'. The earliest papers (1870–99) were prefixed C, but this was changed to Cd in 1900, to Cmd in 1919, Cmnd in 1956 and Cm, the current prefix, in 1986. Authors and publishers alike have difficulty in getting the prefix correct every time.

Though still used to denote a statement of government policy, the term 'White Paper' now has no precise meaning. Colourful and graphic presentation has become common, and some White Papers bear a strong resemblance to company reports. Until relatively recently, it was reasonable to assume that, if a White Paper was a policy document, it represented the government's view, or its fairly firm proposals. This distinguished it from a 'Green Paper', which was of the nature of a preliminary draft White Paper and offered to guide public debate. The distinction between white and green is now blurred and it is not uncommon to hear that a White Paper has a 'green tinge' (i.e. some of the proposals are still open for discussion). White Papers also review existing policy and present examples of good practice: it can be difficult to sort out what is new and what already is in place (which may be no accident). Similarly, Green Papers may have a 'white tinge' (i.e. certain issues have been firmly decided and are not open for further debate). Of course, circumstances change, and so do the minds of governments. (Thus the 1989 White Paper on *The Future of Development Plans* announced the abolition of structure plans, but it was later decided to retain them.) Green Papers have now largely been superseded by consultation papers which, ironically, are often not published by TSO, but are 'available' from the department concerned. In recent years, there have been large numbers of these. Those concerned with planning are frequently printed in the monthly update to the *Planning Encyclopedia* and are listed on the relevant government website.

Another recent innovation is the publication of annual reports of departments. Again these are

Command Papers (with a range of colours). But they are both more and less than annual reports. They are more in the sense that they present details of recent and planned expenditure on the services which are covered by the budget of the department(s) concerned. In this, they are parts of *The Government's Expenditure Plans* (which is their subtitle). Recent annual reports go further than saying what policies and programmes are in place and what is spent on them and include much 'monitoring' of the achievement of targets and public service agreements. Annual reports are very useful documents, but becoming less useful as more emphasis

is given to 'selling' the work of the department complete with vignettes of policy impacts (inappropriately described as case studies). Executive agencies and other non-departmental bodies also produce their own annual reports.

In the following lists, the publisher is TSO (for the respective countries) unless otherwise indicated. Names in parentheses relate to the author or agency concerned; names without parentheses relate to the publisher. Although we have attempted to be consistent on these matters, it is feared that we have not always succeeded.

Planning policy for England (E)

Planning policy statements and guidance notes (E)

Planning policy guidance notes (PPGs) are being reviewed and replaced by planning policy statements (PPSs). All are available at www.odpm.gov.uk under 'Planning'.

PPS 1   *Delivering Sustainable Development*, 2004
PPG 2   *Green Belts*, 1995
PPG 3   *Housing*, 2000
PPG 3   Update to PPG3: *Housing Planning for Sustainable Communities in Rural Areas*, 2005
PPG 3   Update to PPG3: *Housing: Supporting the Delivery of New Housing*, 2005
PPG 4   *Industrial and Commercial Development and Small Firms*, 1992
PPG 5   *Simplified Planning Zones*, 1992
PPS 6   *Planning for Town Centres*, 1993, revised 2005
PPS 7   *Sustainable Development in Rural Areas*, 2004
PPG 8   *Telecommunications*, 1992, revised 2001
PPS 9   *Biodiversity and Geological Conservation*, 1994, revised 2005
PPS 10   *Planning for Sustainable Waste Management*, 1999, revised 2005
PPS 11   *Regional Spatial Strategies*, 2004
PPS 12   *Local Development Frameworks*, 2004
PPG 13   *Transport*, 1994, revised consultation draft 1999
PPG 14   *Development on Unstable Land*, 1990
PPG 15   *Planning and the Historic Environment*, 1994
PPG 16   *Archaeology and Planning*, 1990
PPG 17   *Planning for Open Space, Sport and Recreation*, 1991, revised 2002
PPG 18   *Enforcing Planning Control*, 1991
PPG 19   *Outdoor Advertising Control*, 1992
PPG 20   *Coastal Planning*, 1992
PPG 21   *Tourism*, 1992
PPS 22   *Renewable Energy*, 2004
PPS 23   *Planning and Pollution Control*, 2004

PPS 23   *Annex 1: Pollution Control, Air and Water Quality*, 2004
PPS 23   *Annex 2: Development on Land Affected by Contamination*, 2004
PPG 24   *Planning and Noise*, 1994, revised version in preparation
PPG 25   *Development and Flood Risk*, 2000

Regional spatial strategies and regional planning guidance notes (E)

The preparation of statutory regional spatial strategies was well underway in 2005. They will replace the regional planning guidance notes, which currently provide statements of regional planning policy. In the mean time, regional planning guidance notes were renamed regional spatial strategies in the 2004 Act reforms, although many documents are still available with the old names. The RPG names are used in this list. If the proposals for submission of new RSS in 2005 are met it is likely to lead to final adoption of the strategies in 2007.

The London Plan: Spatial Development Strategy for Greater London, 2004, replaced RPG3 *Strategic Planning Guidance for London Planning Authorities*, 1996
RPG/RSS 1   *Regional Planning Guidance for the North East*, 2002; the *Regional Spatial Plan for the North East* was planned to be submitted in 2005
RPG/RSS 6   *Regional Planning Guidance for East Anglia*, 1991; *The East of England Plan* (RSS) will replace RPG 6 and RPG 9 where they cover parts of the East of England; the East of England region was created in 2000 and is covered by parts of RPG for East Anglia and the South East
RSS 8   *Regional Spatial Strategy for the East Midlands*, 2005 replaced RPG for the East Midlands, 1994
RPG/RSS 9   *Regional Guidance for the South East*, 1994; *The South East Plan: Regional Spatial Strategy for the South East* is planned to be submitted in 2005
RPG/RSS 10   *Regional Planning Guidance for the South West*, 2001; *The Regional Spatial Strategy for the South West* planned submission end 2005
RPG/RSS 11   *Regional Spatial Strategy for the West Midlands Region*, 2004; this is being revised in a phased approach with a sub-regional strategy for the Black Country to be submitted in 2006 followed by two more submissions on priority issues in 2007 and 2008
RPG/RSS 12   *Regional Planning Guidance for Yorkshire and Humberside, 1996*, and *Selective Review* 2004; the *Regional Spatial Strategy for Yorkshire and Humber* is planned for submission in 2005
RPG/RSS 13   *Regional Planning Guidance for the North West*, 1996, partial revision and regional spatial strategy expected 2005

Minerals policy statements (E)

MPS 2   *Controlling and Mitigating the Environmental Effects of Mineral Extraction in England*, 2005; supersedes MPG 11 *Environmental Impacts and Mineral Working*, 1993

Minerals policy guidance notes (E)

MPG 1   *General Considerations and the Development Plan System*, 1996
MPG 2   *Applications, Permissions and Conditions*, 1998
MPG 3   *Coal Mining and Colliery Spoil Disposal*, 1999
MPG 4   *Revocation, Modification, Discontinuance, Prohibition and Suspension Orders*, 1997
MPG 5   *Stability in Surface Mineral Workings and Tips*, 2000

MPG 6   *Guidelines for Aggregates Provision in England*, 1994
MPG 7   *The Reclamation of Mineral Workings*, 1996
MPG 8   *Planning and Compensation Act 1991: Interim Development Order Permissions (IDOS) – Statutory Provisions and Proceedings*, 1991
MPG 9   *Planning and Compensation Act 1991: Interim Development Orders*, 1992
MPG 10   *Provision of Raw Material for the Cement Industry*, 1991
MPG 12   *Treatment of Disused Mine Openings and Availability of Information on Mined Ground*, 1994
MPG 13   *Guidelines for Peat Provision in England (including the Place of Alternative Materials)*, 1995
MPG 14   *Environment Act 1995: Review of Mineral Planning Permissions*, 1995
MPG 15   *Provision of Silica Sand in England*, 1996
MPG 17   *Oil, Gas and Coalbed Methane* (revised version in preparation)

Marine minerals guidance notes (E)

MMG 1   *Extraction by Dredging from the English Seabed*

Planning policy for Northern Ireland (NI)

This is published by the Northern Ireland Planning Service, some in association with the Department for Regional Development NI. All are available at: www.planningni.gov.uk/.

Regional Planning Policy (NI)

1993   *A Planning Strategy for Northern Ireland* (Belfast: HMSO)
1997   *Shaping our Future: Towards a Strategy for the Region: Draft Regional Strategy Framework for Northern Ireland* (Belfast: DoENI)
1998   *Shaping our Future: Draft Regional Strategy for Northern Ireland* (Belfast: DoENI)

Planning policy statements (NI)

PPS 1   *General Planning Principles*, 1998
PPS 2   *Planning and Nature Conservation*, 1997
PPS 3   *Access Movement and Parking* (revised), 2005
PPS 4   *Industrial Development*, 1997, draft revision 2003
PPS 5   *Retailing and Town Centres*, 1996
PPS 6   *Planning, Archaeology and Built Heritage*, 1999, draft addendum 2004
PPS 7   *Quality Residential Environments*, 2001
PPS 8   *Open Space, Sport, Recreation, Leisure and Community Facilities*, 2004
PPS 9   *The Enforcement of Planning Control*, 2000
PPS 10   *Telecommunications*, 2002
PPS 11   *Planning and Waste Management*, 2002
PPS 12   *Housing in Settlements*, draft 2002, DRD
PPS 13   *Transportation and Land Use*, 2005, DRD
PPS 14   *Sustainable Development in the Countryside*, issues paper, 2004, DRD

PPS 15   *Planning and Flood Risk*, draft 2004
PPS 17   *Control of Outdoor Advertisements*, draft 2004

Development control advice notes (NI)

DCAN 1   *Amusement Centres*
DCAN 2   *Multiple Occupancy*
DCAN 3   *Bookmaking Offices*
DCAN 4   *Hot Food Bars*
DCAN 5   *Taxi Offices*
DCAN 6   *Restaurants and Cafes*
DCAN 7   *Public Houses*
DCAN 8   *Small Unit Housing in Residential Areas*
DCAN 9   *Residential and Nursing Homes*
DCAN 10   *Environmental Impact Assessment*
DCAN 11   *Access for People with Disabilities*
DCAN 12   *Hazardous Substances*
DCAN 13   *Crèches, Day Nurseries and Pre-School Playgroups*
DCAN 14   *Telecommunications Prior Approval Procedures*
DCAN 15   *Vehicular Access Standards*

Design guidance (NI)

A Design Guide for Rural Northern Ireland, 1994
Creating Places: Achieving Quality in Residential Layouts, 2000
Improving the Quality of Housing Layouts in Northern Ireland
Trees and Development: A Guide to Best Practice

Planning policy for Scotland (S)

This is published by the Scottish Executive and available at www.scotland.gov.uk/topics/planning-building.

National planning policy (S)

National Planning Framework for Scotland, 2004

Scottish planning policy and national planning policy guidelines

The national planning policy guidelines are being reviewed and replaced by Scottish Planning Policy. They are available at www.scotland.gov.uk/Topics/ Planning-Building/Planning.

SPP 1   *The Planning System*, 2002
SPP 2   *Economic Development*, 2002
SPP 3   *Planning for Housing*, 2003

NPPG 4   *Land for Mineral Working*, 1994
NPPG 5   *Archaeology and Planning*, 1994
NPPG 6   *Renewable Energy Developments*, 1994
SPP 7   *Planning and Flooding*, 1995, revised 2004
NPPG 8   *Town Centres and Retailing*, 1996, revised 1998
NPPG 9   *The Provision of Roadside Facilities on Motorways and Other Trunk Roads in Scotland*, 1996
NPPG 10   *Planning and Waste Management*, 1996
NPPG 11   *Sport, Physical Recreation and Open Space*, 1996
NPPG 12   *Skiing Developments*, 1997
NPPG 13   *Coastal Planning*, 1997
NPPG 14   *Natural Heritage*, 1999
SPP 15   *Planning for Rural Development*, 1999, revised 2005
SPP 16   *Opencast Coal*, 1999, amended 2001, revised 2005
NPPG 17   *Transport and Planning*, 1999
SPP 17   *Transport and Planning: Maximum Parking Standards, Addendum to NPPG 17*, 2003
NPPG 18   *Planning and the Historic Environment*, 1999
NPPG 19   *Radio Telecommunications*, 2001
SPP 20   *The Role of Architecture and Design in Scotland*, 2005

Scottish planning advice notes

PAN 33   *Development of Contaminated Land*, 1988
PAN 36   *Siting and Design of New Housing in the Countryside*, 1991
PAN 37   *Structure Planning*, 1992, revised 1996
PAN 38   *Housing Land (formally Housing Land Requirements)*, 1993, revised 1996, 2003
PAN 39   *Farm and Forestry Buildings*, 1993
PAN 40   *Development Control*, 1993, revised 2001
PAN 41   *Development Plan Departures*, 1994, revised 1997
PAN 42   *Archaeology: The Planning Process and Scheduled Monuments Procedures*, 1994
PAN 43   *Golf Courses and Associated Developments*, 1994
PAN 44   *Fitting New Housing Development into the Landscape*, 1994
PAN 45   *Renewable Energy Technologies*, 1994, revised 2002
PAN 46   *Planning for Crime Prevention*, 1994
PAN 47   *Community Councils and Planning*, 1996
PAN 48   *Planning Application Forms*, 1996
PAN 49   *Local Planning*, 1996
PAN 50   *Controlling the Environmental Effects of Surface Mineral Workings*, 1996
PAN 50A   *The Control of Noise at Surface Mineral Workings*, 1996
PAN 50B   *The Control of Dust at Surface Mineral Workings*, 1998
PAN 50C   *The Control of Traffic at Surface Mineral Workings*, 1998
PAN 50D   *Control of Blasting at Surface Mineral Workings*, 1999
PAN 51   *Planning and Environmental Protection*, 1997
PAN 52   *Planning and Small Towns*, 1997
PAN 53   *Classifying the Coast for Planning*, 1998
PAN 54   *Planning Enforcement*, 1999

PAN 55   *The Private Finance Initiative and Planning*, 1999
PAN 56   *Planning and Noise*, 1999
PAN 57   *Transport and Planning*, 1999
PAN 58   *Environmental Impact Assessment*, 1999
PAN 59   *Improving Town Centres*, 1999
PAN 60   *Planning for Natural Heritage*, 2000
PAN 61   *Planning and Sustainable Urban Drainage Systems*, 2001
PAN 62   *Radio Telecommunications*, 2001
PAN 63   *Waste Management Planning*, 2002
PAN 64   *Reclamation of Surface Mineral Workings*, 2002
PAN 65   *Planning and Open Space*, 2003
PAN 66   *Best Practice in Handling Planning Applications Affecting Trunk Roads*, 2003
PAN 67   *Housing Quality*, 2003
PAN 68   *Design Statements*, 2003
PAN 69   *Planning and Building Standards Advice on Flooding*, 2004
PAN 70   *Electronic Planning Service Delivery*, 2004
PAN 71   *Conservation Area Management*, 2004
PAN 72   *Housing in the Countryside*, 2005
PAN 73   *Rural Diversification*, 2005
PAN 74   *Affordable Housing*, 2005
PAN 75   *Planning for Transport*, 2005
PAN 76   *New Residential Streets*, 2005

See also:
Designing Places: A Policy Statement for Scotland, 2001
The Scottish Executive Development Department publishes a regular Planning Bulletin.

Planning policy for Wales (W)

This is published by the Welsh Assembly for Wales (NAW) and available at www.wales.gov.uk/subi planning/.

National spatial plan (W)

2000   *Wales Spatial Planning Framework Key Challenges for Wales*
2005   *People, Places, Futures: The Wales Spatial Plan*

National planning policy statements (W)

2000   *Minerals Planning Policy Wales*
2002   *Planning Policy Wales*

Technical advice notes (W)

TAN (W) 1   *Joint Housing Land Availability Studies*, 1997
TAN (W) 2   *Planning and Affordable Housing*, 1996

TAN (W) 3   *Simplified Planning Zones*, 1996
TAN (W) 4   *Retailing and Town Centres*, 1996
TAN (W) 5   *Nature Conservation and Planning*, 1996
TAN (W) 6   *Agriculture and Rural Development*, 2000
TAN (W) 7   *Outdoor Advertisement Control*, 1996
TAN (W) 8   *Renewable Energy*, 1996
TAN (W) 9   *Enforcement of Planning Control*, 1997
TAN (W) 10   *Tree Preservation Orders*, 1997
TAN (W) 11   *Noise*, 1997
TAN (W) 12   *Design*, 1997, revised 2002
TAN (W) 13   *Tourism*, 1997
TAN (W) 14   *Coastal Planning*, 1998
TAN (W) 15   *Development and Flood Risk*, 1998, revised 2004
TAN (W) 16   *Sport and Recreation*, 1998
TAN (W) 17   *Environmental Assessment*, 1998
TAN (W) 18   *Transport*, 1998
TAN (W) 19   *Telecommunications*, 1998, revised 2002
TAN (W) 20   *The Welsh Language: Unitary Development Plans and Development Control*, 2000
TAN (W) 21   *Waste*, 2001

Minerals technical advice notes (W)

MTAN 1   *Aggregates*

European Union (EU) planning policy and studies

Most of these reports can be found on the European Union's website http://europa.int.eu (note there is no 'www'). Some further guidance on EU official documents is given in Chapters 3 and 4. Most EU official documents are published by the Office of the Official Publications of the European Communities, Luxembourg (OOPEC), http://publications.eu.int, and some by the Directorates General.

Reports and papers (EU)

1990   *Green Paper for the Urban Environment*
1991   *Europe 2000: Outlook for the Development of the Community's Territory*
1992   *Fifth Action Programme on the Environment: 1992–2000 – Towards Sustainability*
1992   *The Future Development of the Common Transport Policy*
1992   *The Impact of Transport on the Environment: A Community Strategy for Mobility*
1993   *Growth, Competitiveness and Employment*
1994   *Europe 2000+: Cooperation for European Territorial Development*
1995   *The Citizen's Network: Fulfilling the Potential of Public Passenger Transport in Europe*
1995   *The Trans-European Transport Network*
1995   *Towards Fair and Efficient Policy in Transport*
1996   *European Sustainable Cities Report, EU Expert Group on the Urban Environment*
1997   *Agenda 2000*

1997   *Towards an Urban Agenda in the European Union*
1998   *Community Policies and Spatial Planning*
1999   *Sustainable Urban Development in the EU: A Framework for Action*
1999   *The European Spatial Development Perspective* (Committee on Spatial Development)
2000   *Sixth Action Programme on the Environment* (DG Environment)
2000   *Sixth Periodic Report on the Social and Economic Situation of the Regions in the European Union* (DG Regio)
2000   *TERRA: An Experimental Laboratory in Spatial Planning*
2001   *A Sustainable Europe for a Better World: A European Union Strategy for Sustainable Development*
2001   *European Governance: A White Paper*, COM(2001)
2001   *Second Report on Social and Economic Cohesion* (DG Regio)
2002   *Environment 2010: Our Future, Our Choice* (DG Environment)
2004   *Interim Territorial Cohesion Report* (DG Regio)
2004   *Proposal for a Council Regulation laying down general provisions for the European Regional Development Fund, the European Social Fund and the Cohesion Fund*
2004   *Proposal for a Directive of the European Parliament and of the Council Establishing an Infrastructure for Spatial Information in the Community (INSPIRE)*, COM(2004) 516 final
2004   *Third Report on Economic and Social Cohesion: A New Partnership for Cohesion, Convergence, Competitiveness and Cooperation* (DG Regio)
2004   *Towards a Thematic Strategy on the Urban Environment*, COM(2004) 60
2004   *Working Together for Growth and Jobs: A New Start for the Lisbon Strategy*
2005   *Review of the EU Sustainable Development Strategy: Initial Stocktaking and Future Orientations* (DG Environment)

DG Regional Policy studies (EU)

1992   *Urbanisation and the Function of Cities in the European Community*
1993   *Study of the Prospects in the Atlantic Regions*
1994   *The Prospective Development of the Central and Capital Cities and Regions*
1994   *The Prospective Development of the Northern Seaboard*
1994   *The Regional Impact of the Channel Tunnel throughout the Community*
1997   *The EU Compendium of Spatial Planning Systems and Policies*
1998   *Economic and Social Cohesion in the European Union: The Impact of Member States' Own Policies*
1999   *Inclusive Cities: Building Local Capacity for Development*
1999   *Spatial Perspectives for the Enlargement of the European Union*
2001   *Spatial Impacts of Community Policies and Costs of Non-coordination*
2002   *Creating Smart Systems: A Guide to Cluster Strategies in Less Favoured Regions*
2002   *The Economic Impact of Objective 1 Interventions for the Period 2000–2006*

Agencies of planning

Central and regional government and agencies

1970   *The Reorganisation of Central Government*, Cmnd 4506
1979   *Central Government Controls over Local Authorities*, Cmnd 7634
1984   *Progress in Financial Management in Government Departments*, Cmnd 9297

1988   *Civil Service Management Reform: The Next Steps*, Cm 524

1990   *Report of the Noise Review Working Party (Batho Report)*

1991   *Improving Management in Government: The Next Steps Agencies: Review 1991*, Cm 1730

1994   *Origins of the Department of the Environment* (P. McQuail) (Department of the Environment)

1995   *Review of the Department of the Environment* (Department of the Environment)

1995   *View from the Bridge* (P. McQuail) (Department of the Environment)

1996   *Next Steps Agencies in Government, Review 1995*, Cm 3164

1997   *A Voice for Wales: The Government's Proposals for a Welsh Assembly*, Cm 3718

1997   *Scotland's Parliament*, Cm 3658

1998   *Building Partnerships for Prosperity* (DETR)

1998   *Guidance to the Regional Development Authorities on Rural Policy* (DETR)

1998   *The Future of Regional Guidance*, consultation paper (DETR)

1999   *Guidance to the Regional Development Authorities on Sustainable Development* (DETR)

1999   *Local Plans and Unitary Development Plans: A Guide to Procedures* (DETR)

1999   *Modernising Government*, Cm 4310

1999   *Quality of Life Counts: DETR Indicators for the UK Sustainable Development Strategy* (DETR)

1999   *Regional Chambers* (DETR)

1999   *Regional Development Agencies, HC Environment, Transport and Regional Affairs Committee, Tenth Report, Session 1998–99*, HC 232

1999   *Regional Planning Consultation Draft*, PPG 11

1999   *Regional Strategies* (DETR)

1999   *Rural Economies*, Cabinet Office Performance and Innovation Unit

1999   *Structure Plans: A Guide to Procedures* (DETR)

1999   *Supplementary Guidance to Regional Development Authorities* (DETR)

2000   *Regional Government in England: A Preliminary Review of the Literature and Research Findings* (DETR)

2000   *Reinforcing Standards: Review of the First Report of the Committee on Standards in Public Life*

2003   *Urban Regeneration Companies: A Consultation Paper – Challenging Practice, Testing Innovation* (Scottish Executive)

2004   *Sixth Annual Report of the Planning Audit Unit, 2003* (Scottish Executive)

2004   *Urban Regeneration Companies: Guidance and Qualification Criteria* (ODPM)

2004   *Urban Regeneration Companies Policy Stocktake: Final Report* (ODPM)

Local government

1979   *Central Government Controls over Local Authorities*, Cmnd 7634

1979   *Organic Change in Local Government*, Cmnd 7457

1981   *Committee of Inquiry into Local Government in Scotland (Stodart Report)*, Cmnd 8115

1983   *Rates: Proposals for Rate Limitation and Reform of the Rating System*, Cmnd 9008

1983   *Streamlining the Cities: Government Proposals for Reorganising Local Government in Greater London and the Metropolitan Counties*, Cmnd 9063

1984   *Report of the Committee on Inquiry into the Functions and Powers of the Islands Councils of Scotland (Montgomery Report)*, Cmnd 9216

1986   *Paying for Local Government*, Cmnd 9714

1986   *The Conduct of Local Authority Business: Report of the Committee of Inquiry (Widdicombe Report)*, Cmnd 9797

1992   *Functions of Local Authorities in England (Local Government Review)*

1992 *The Role of Community and Town Councils in Wales*, consultation paper (Welsh Office)

1993 *Local Government in Wales: A Charter for the Future*, Cm 2155

1997 *New Leadership for London: The Government's Proposals for a Greater London Authority*, consultation paper, Cm 3724

1997 *Standards of Conduct in Local Government, Third Report of the Committee on Standards in Public Life (Nolan Reports)*, Cm 3702

1998 *A Mayor and Assembly for London*, Cm 3897

1998 *Guidance on Enhancing Public Participation in Local Government* (DETR)

1998 *Modern Local Government: In Touch with the People*, Cm 4014

1998 *Modernising Local Government: Local Democracy and Community Leadership* (DETR)

1998 *New Ethical Framework for Local Government in Scotland*, consultation paper (Scottish Office)

1999 *Local Government Political Management Arrangements – An International Perspective* (R. Hambleton) (Scottish Office)

1999 *Local Leadership, Local Choice*, Cm 4298

1999 *Report of the Commission on Local Government and the Scottish Parliament (McIntosh Report)* (Scottish Office)

1999 *Report of the Community Planning Working Group* (Scottish Office)

2000 *Preparing Community Strategies: Draft Guidance to Local Authorities*

2004 *The Future of Local Government: Developing a 10 Year Strategy* (ODPM)

The planning policy framework and development control

1942 *Report of the Committee on Land Utilisation in Rural Areas (Scott Report)*, Cmd 6378

1942 *Report of the Expert Committee on Compensation and Betterment (Uthwatt Report)*, Cmd 6386

1944 *The Control of Land Use*, Cmd 6537

1947 *Town and Country Planning Bill 1947: Explanatory Memorandum*, Cmd 7006

1950 *Report of the Committee on the Qualifications of Planners (Schuster Report)*, Cmd 8059

1951 *Town and Country Planning 1943–1951: Progress Report*, Cmd 8204

1965 *The Future of Development Plans: Report of the Planning Advisory Group*

1967 *Town and Country Planning*, Cmnd 3333

1977 *Memorandum on Structure and Local Plans*, DoE Circular 55/77

1978 *Planning Procedures: The Government's Response to the Eighth Report from the Expenditure Committee, Session 1976–77*, Cmnd 7056

1984 *Enforcement of Planning Control in Scotland (J. Rowan-Robinson, E. Young and I. McLarty)* (Scottish Office)

1984 *Memorandum on Structure Plans*, DoE Circular 22/84

1985 *Lifting the Burden*, Cmnd 9571

1986 *Planning Appeals, Call-in and Major Public Inquiries: The Government's Response to the Fifth Report from the Environment Committee*, Cm 43

1989 *The Future of Development Plans*, Cm 569

1990 *Mineral Policies in Development Plans* (Arup Economic Consultants)

1991 *Enforcing Planning Control*, PPG18

1991 *Examination of the Effects of the Use Classes Order 1987 and the General Development Order 1988* (Wootton Jeffreys Consultants and Bernard Thorpe)

1991 *Permitted Development Rights for Agriculture and Forestry* (Land Use Consultants)

1992 *Building in Quality: A Study of Development Control* (Audit Commission)

1992 *Development Plans: A Good Practice Guide*

1992 *Evaluating the Effectiveness of Land Use Planning* (Roger Tym and Partners *et al.*)

1992 *Industrial and Commercial Development and Small Firms*, PPG 4

1992 *Outdoor Advertising Control*, PPG19

1992 *Publicity for Planning Applications*, DoE Circular 15/92

1992 *Simplified Planning Zones*, PPG5

1992 *Use of Planning Agreements* (Grimley J. R. Eve *et al.*)

1993 *Integrated Planning and Granting of Permits in the EC* (GMA Planning, P-E International, and Jacques & Lewis)

1994 *Costs of Determining Planning Applications and the Development Control Service* (Price Waterhouse)

1994 *Guidance Notes for Local Planning Authorities on the Methods of Protecting the Water Environment through Development Plans* (Environment Agency)

1994 *Improving the Local Plan Process*, consultation paper (DoE)

1994 *Mediation: Benefits and Practice* (DoE)

1994 *Planning Out Crime*, DoE Circular 5/94

1994 *Review of the Use Classes Order* (Scottish Office)

1995 *Effectiveness of Planning Policy Guidance Notes* (Land Use Consultants, Department of the Environment)

1995 *Efficiency and Effectiveness of Local Plan Inquiries*

1995 *Evaluation of Planning Enforcement Provisions* (Arup Economic Planning and Linklaters and Paines)

1995 *Planning Controls over Agricultural Land and Rural Building Conversions* (Land Use Consultants)

1995 *Planning Controls over Demolition*, DoE Circular 10/95

1995 *Quality in Town and Country: Urban Design Guidelines*

1995 *Review of Neighbour Notification* (Scottish Office)

1995 *Review of the Town and Country Planning System in Scotland: The Way Ahead* (Scottish Office)

1995 *The Use of Conditions in Planning Permissions*, DoE Circular 11/95

1996 *Planning System in Northern Ireland, HC Northern Ireland Affairs Committee, First Report, Session 1995–96*, HC 53

1996 *Speeding Up the Delivery of Local Plans and UDPs: Report of the Review* (DoE)

1996 *The Planning System in Northern Ireland, HC Select Committee on Northern Ireland Affairs* (HMSO)

1997 *Efficiency and effectiveness of Local Plan Inquiries* (DoE)

1997 *Enforcing Planning Control: A Good Practice Guide for Local Planning Authorities*

1997 *General Policy and Principles*, PPG1

1997 *Review of Development Planning in Scotland* (SDD Research Findings no. 50)

1997 *Speeding Up the Delivery of Local Plans and UDPs*, consultation paper (DoE)

1998 *Future of Regional Planning Guidance* (DETR)

1998 *Impact of the EU on the UK Planning System* (D. Wilkinson, K. Bishop and M. Tewdwr-Jones)

1998 *Modernising Planning: A Policy Statement* (DETR)

1998 *Our Competitive Future: Building the Knowledge Driven Economy*, Cm 4176

1998 *Prevention of Dereliction through the Planning System*, DETR Circular 2/98

1999 *Approach to Future Land Use Planning Policy* (Welsh Assembly)

1999 *Biotechnology Clusters: Report of a Team led by Lord Sainsbury, Minister of Science* (DTE)

1999 *Classes Order: Consultation on Possible Changes to the Use Classes Order and Temporary Uses Provisions* (ODPM)

1999 *Costs in the Planning Service* (Scottish Executive Central Research Unit)

1999 *Development Control and Development Plan Preparation: Local Authority Concerns and Current Government Action* (DETR)

1999   *Development Plans*, PPG12
1999   *Examination of the Operation and Effectiveness of the Structure Planning Process* (M. Baker and P. Robert) (DETR)
1999   *Land Use Planning under a Scottish Parliament and Overview of Responses to Consultation*
1999   *Modernising Planning: A Progress Report* (DETR)
1999   *Modernising Planning: Streamlining the Processing of Major Projects through the Planning System*, consultation paper (DETR)
1999   *Planning Concordat between the Local Government Association and the DETR* (DETR)
1999   *Planning for Telecommunications*, DETR Circular 1999/4
1999   *Planning Policy Guidance Note 11: Regional Planning – Public Consultation Draft* (DETR)
1999   *Proposals for Amendments to Planning Legislation in Northern Ireland*, consultation paper
1999   *Review of National Planning Policy Guidelines* (Land Use Consultants, Scottish Office)
1999   *Town Centres and Retail Developments: Consultation Draft*, PPG6
2000   *By Design: Urban Design in the Planning System: Towards Better Practice*
2000   *Development of Planning Policy in Wales: Report of the Land Use Planning Forum*
2000   *European Spatial Planning and Urban–Rural Relationships: the UK Dimension*
2000   *Planning Inspectorate Executive Agency Annual Report 1999–2000*
2000   *Quinquennial Review of the Planning Inspectorate* (PINS)
2000   *Survey of Urban Design Skills in Local Government*
2000   *The Control of Fly-posting: A Good Practice Guide*
2000   *Training for Urban Design*
2001   *Green Paper Planning: Delivering a Fundamental Change* (DTLR)
2001   *New Parliamentary Procedures for Processing Major Infrastructure Projects* (ODPM)
2001   *Possible Changes to the Use Classes Order and Temporary Uses Provisions*, consultation paper (DTLR)
2001   *Review of Strategic Planning* (Scottish Executive)
2001   *Shaping Our Future: Regional Development Strategy for Northern Ireland 2025* (DoE NI)
2001   *Shaping Our Future: The Family of Settlements Report* (DRD, NI)
2002   *Code of Practice on the Dissemination of Information (Major Infrastructure Projects)* (ODPM)
2002   *Enforcement Appeal Procedures*, ODPM Circular 2/2002
2002   *National Spatial Strategy for Ireland 2002–20*, (Republic of Ireland, Department of Environment and Local Government)
2002   *Planning: Delivering for Wales* (National Assembly for Wales)
2002   *Planning Inquiries into Major Infrastructure Projects: Procedures*, DTLR Circular 2/2002
2002   *The Town and Country Planning (Residential Density) (London and South East England) Direction*, DTLR Circular 1/2002
2003   *Amendments to the GDPO and Listed Building Regulations*, ODPM Circular 8/2003
2003   *Modernising Planning Processes: A Review of the Use Classes Order* (Planning Service Northern Ireland)
2003 and 2004   *Implications for Development Plans in Wales of the Planning and Compulsory Purchase Act* (National Assembly for Wales)
2004   *Making Development Plans Deliver*, consultation paper (Scottish Executive)
2004   *People, Places, Futures, The Wales Spatial Plan* (National Assembly for Wales)
2004   *The National Planning Framework for Scotland* (Scottish Executive)

Land

1940 *Report of the Royal Commission on the Distribution of the Industrial Population (Barlow Report)*, Cmd 6153

1942 *Report of the Committee on Land Utilisation in Rural Areas (Scott Report)*, Cmd 6378

1942 *Report of the Expert Committee on Compensation and Betterment (Uthwatt Report)*, Cmd 6386

1944 *The Control of Land Use*, Cmd 6537

1946 *Reports of the New Towns Committee: Interim Report*, Cmd 6759; *Second Interim Report*, Cmd 6794; *Final Report*, Cmd 6876 *(Reith Report)*

1947 *Town and Country Planning Bill 1947: Explanatory Memorandum*, Cmd 7006

1951 *Town and Country Planning 1943–1951: Progress Report*, Cmd 8204

1952 *Town and Country Planning Act 1947: Amendment of Financial Provisions*, Cmd 8699

1955 *Green Belts*, MHLG Circular 42/55

1957 *Green Belts*, MHLG Circular 50/57

1958 *Town and Country Planning Bill: Explanatory Memorandum*, Cmnd 562

1963 *London – Employment: Housing: Land*, Cmnd 1952

1965 *The Land Commission*, Cmnd 2771

1969 *Modifications in Betterment Levy*, Cmnd 4001

1972 *Development and Compensation: Putting People First*, Cmnd 5124

1972 *Local Authority/Private Enterprise Partnership Schemes*

1974 *Land*, Cmnd 5730

1975 *Development Land Tax*, Cmnd 6195

1977 *Statement on the Non-statutory Inquiry by the Baroness Sharp into the Continued Use of Dartmoor for Military Training*, Cmnd 6837

1981 *Planning Gain* (Property Advisory Group)

1984 *Green Belts*, DoE Circular 14/84

1984 *Land for Housing*, DoE Circular 15/84

1985 *Land Use Planning and the Housing Market* (Coopers and Lybrand)

1987 *Evaluation of Derelict Land Schemes* (Roger Tym and Partners and Land Use Consultants)

1987 *Land Supply and House Prices in Scotland* (PIEDA) (Scottish Office)

1987 *Land Use Planning and Indicators of Housing Demand* (Coopers and Lybrand)

1988 *Urban Land Markets in the UK* (R. N. Chubb)

1988 *Vacant Urban Land: A Literature Review* (G. E. Cameron, S. Monk and B. Pearce) (DoE)

1990 *Contaminated Land, HC Environment Committee, First Report, Session 1989–90*, HC 170 [Government Response Cm 1161]

1990 *Development on Unstable Land*, PPG14

1990 *Mineral Policies in Development Plans* (Arup Economic Consultants)

1990 *Rates of Urbanization in England 1981–2001* (P. R. Bibby and J. W. Shepherd)

1991 *Housing Land Availability* (Roger Tym and Partners)

1991 *Housing Land Availability: The Analysis of PS3 Statistics on Land with Outstanding Planning Permission* (P. Bibby and J. Shepherd)

1991 *Planning and Affordable Housing*, DoE Circular 7/91

1991 *Tackling Vacant Land: An Evaluation of Policy Instruments for Tackling Urban Land Vacancy* (M. Whitbread, D. Mayne, and D. Wickens, Arup Economic Consultants)

1992 *National Survey of Vacant Land in Urban Areas of England 1990* (J. Shepherd and A. Abakuks)

1992   *Relationship between House Prices and Land Supply* (Gerald Eve and Department of Land Economy, University of Cambridge)

1992   *Strategic Approach to Derelict Land Reclamation* (Public Sector Management Research Centre, Aston University)

1992   *Use of Planning Agreements* (Grimley J. R. Eve *et al.*)

1993   *Alternative Development Patterns: New Settlements*

1993   *Effectiveness of Green Belts* (M. Elson, S. Walker and R. Macdonald)

1994   *Assessment of the Effectiveness of Derelict Land Grant in Reclaiming Land for Development*

1994   *Feasibility Study for Deriving Information about Land Use Stock* (A. R. Harrison) (Bristol: Department of Geography, University of Bristol, for DoE)

1994   *Shopping Centres and their Future, HC Environment Committee, Session 1993–94*, HC 359 [Government Response Cm 2767, 1995]

1995   *Derelict Land Prevention and the Planning System*

1995   *Derelict Land Survey (Scotland)* (Scottish Office)

1995   *Green Belts* PPG 2

1995   *Reducing the Need to Travel though Land Use and Transport Planning* (DoE)

1995   *Survey of Derelict Land in England 1993*

1996   *Household Growth: Where Shall We Live?*, Cm 3471

1996   *Minerals Planning Policy and Supply Practices in Europe*

1996   *Survey of Mineral Workings in England*

1997   *Enforcing Planning Control: Legislative Provisions and Procedural Requirements*, DoE Circular 1997/10

1997   *Operation of Compulsory Purchase Orders* (City University Business School)

1997   *Planning Obligations*, DoE Circular 1991/1

1997   *Shopping Centres HC Environment Committee Fourth Report, Session 1996–97*, HC 210 [Government Response, Cm 3729]

1998   *Impact of Large Foodstores on Market Towns and District Centres* (CB Hillier Parker and Savell Bird Avon)

1998   *Our Competitive Future: Building the Knowledge Driven Economy*, Cm 4176

1998   *Planning for the Communities of the Future*, Cm 3885

1998   *Planning for Sustainable Development*

1998   *Prevention of Dereliction through the Planning System*, DETR Circular 2/98

1999   *Biotechnology Clusters: Report of a Team led by Lord Sainsbury, Minister for Science* (DTI)

1999   *Contaminated Land – Draft Circular* (DETR)

1999   *Green Belt Statistics: England 1997*, DETR Information Bulletin 1183

1999   *Housing*, Consultation Draft PPG3

1999   *Land Reform: Proposals for Legislation* SE 1999/1 (Scottish Executive)

1999   *Strategic Gap and Green Wedge Policies in Structure Plans* (DETR)

1999   *Toward an Urban Renaissance (Report of the Urban Task Force)* (London: Spon)

2000   *Derelict Land and Section 215 Powers* (DETR)

2000   *Scottish Vacant and Derelict Land Survey* (Scottish Executive)

2000   *Town and Country Planning (Residential Development on Greenfield Land, England)*, Direction 2000 (DETR)

2001   *Compulsory Purchase and Compensation: The Government's Proposals for Change* (DTLR)

2001   *Reforming Planning Obligations*, consultation paper (DTLR)

2002   *Compulsory Purchase Powers, Procedures and Compensation: The Way Forward* (ODPM)

2002   *Planning Obligations: Delivering a Fundamental Change* (ODPM)

2003   *Compulsory Purchase Orders*, ODPM Circular 2/03, superseded

2003   *The Relationship between Community Strategies and Local Development Frameworks* (ODPM)

2004   *Compulsory Purchase and Compensation Compulsory Purchase Procedure*

2004   *Compulsory Purchase and the Crichel Down Rules*, ODPM Circular 6/04

2004   *Contributing to Sustainable Communities: A New Approach to Planning Obligations, Statement by the ODPM*

2004   *Draft Revised Circular on Planning Obligations: Consultation Document* (ODPM)

2004   *Previously-developed Land that may be Available for Development in 2003* (ODPM)

2004   *The Barker Review of Housing Supply* (HM Treasury)

2005   *Sustainable Communities: People, Places and Prosperity: A Five Year Plan* (ODPM)

The environment

Annual Digest of Environmental Protection and Water Statistics (DETR)

1970   *The Protection of the Environment: The Fight Against Pollution*, Cmnd 4373

1977   *Nuclear Power and the Environment*, Cmnd 6820

1986   *Transforming our Waste Land* (University of Liverpool, Environmental Advisory Unit)

1990   *This Common Inheritance: Britain's Environmental Strategy*, Cm 1200

1991   *Policy Appraisal and the Environment: A Guide for Government Departments*

1991   *Public Attitudes to the Environment in Scotland* (D. Wilkinson and J. Waterton) (Scottish Office)

1992   *Coastal Planning*, PPG 20

1992   *Economic Instruments and Recovery of Resources from Waste* (Environmental Resources Ltd)

1992   *Land Use Planning Policy and Climate Change* (S. Owens and D. Cope)

1992   *Planning, Pollution and Waste Management* (Environmental Resources Ltd and Oxford Polytechnic School of Planning)

1992   *Potential Role of Market Mechanisms in the Control of Acid Rain* (London Economics)

1993   *Eco-management and Audit Scheme for UK Local Government* (DoE)

1993   *Environmental Appraisal of Development Plans: A Good Practice Guide*

1993   *Review of Environmental Expenditure* (ECOTEC)

1994   *Biodiversity: The UK Action Plan*, Cm 2428

1994   *Climate Change: The UK Programme*, Cm 2427

1994   *Contaminated Land and the Water Environment* (Environment Agency)

1994   *Guidance Notes for Local Planning Authorities on the Methods of Protecting the Water Environment through Development Plans* (Environment Agency)

1994   *Managing Demolition and Construction Waste* (Howard Humphreys & Partners)

1994   *Ozone: Report of the Expert Panel on Air Quality Standards*

1994   *Planning and Noise*, PPG 24 (revision in preparation)

1994   *Planning and Pollution Control*, PPG23 (revision in preparation)

1994   *Protecting and Managing Sites of Special Scientific Interest in England* (National Audit Office)

1994   *Sustainable Development: The UK Strategy*, Cm 2426

1995   *Air Quality: Meeting the Challenge: The Government's Strategic Policies for Air Quality Management*

1995   *Contaminants Entering the Sea* (Environment Agency)

1995   *Environmental Agenda for Wales* (Welsh Office)

1995   *Environmental Facts: A Guide to Using Public Registers of Environmental Information* (DoE)

1995   *Guide to Risk Assessment and Risk Management for Environmental Protection*

1995   *Making Waste Work: A Strategy for Sustainable Waste Management in England and Wales*, Cm 3040

1995 *Ozone Layer* (DoE)

1995 *Preparation of Environmental Statements for Planning Projects that Require Environmental Assessment: A Good Practice Guide*

1995 *Risk Assessment and Risk Management for Environmental Protection*

1995 *Sustainable Development, HL Select Committee on Sustainable Development, Report, Session 1994–95*, HL 72 [Government Response Cm 3018]

1995 *Sustainable Development: What it Means to the General Public* (E. McCaig, C. Henderson and MVA Consultancy) (Scottish Office)

1996 *Indicators of Sustainable Development for the UK*

1996 *Integrated Pollution Control: A Practical Guide*

1996 *Scottish Agriculture, HC Scottish Affairs Committee Inquiry into the Future for Scottish Agriculture, Session 1995–96*, HC 629 [Reply, Cm 3548, 1997]

1996 *This Common Inheritance: 1996 UK Annual Report*, Cm 3188

1996 *Waste Management: The Duty of Care – A Code of Practice*

1997 *Air Quality and Land Use Planning*

1997 *Environmentally Sensitive Areas and Other Schemes under the Agri-environmental Regulation, HC Agriculture Committee, Second Report, Session 1996–97*, HC 45 [Government Response, Cm 3707]

1997 *Mitigating Measures in Environmental Statements* (DETR)

1997 *UK National Air Quality Strategy*, Cm 3587

1998 *Biodiversity in Scotland: The Way Forward* (Scottish Biodiversity Group)

1998 *Local Biodiversity Action Plans: A Manual* (Scottish Biodiversity Group)

1998 *Sustainable Development: Opportunities for Change – Consultation Paper on a Revised UK Strategy* (DETR)

1998 *Sustainable Local Communities for the 21st Century: Why and How to Prepare an Effective LA21 Strategy*

1999 *A Better Quality of Life: A Strategy for Sustainable Development in the UK*, Cm 4345

1999 *Air Quality Strategy for England, Scotland, Wales and Northern Ireland*, Cm 4548 (SE 2000/3; NIA 7) (DETR, SE, NAW and DoENI)

1999 *Lessons from the European Commission's Demonstration Programme on Integrated Coastal Zone Management* (Luxembourg: OOPEC)

1999 *National Waste Strategy* (Scottish Executive)

1999 *Operation of the Landfill Tax, HC Environment, Transport and Regional Affairs Committee, Thirteenth Report, Session 1989–99*, HC 150

1999 *Planning and Waste Management*, PPG10

1999 *Reducing the Environmental Impact of Consumer Products, HC Environment, Transport and Regional Affairs Committee, Eleventh Report, Session 1889–99*, HC 149

1999 *State of the Environment of England and Wales: Coasts* (Environment Agency)

1999 *Towards a European Integrated Coastal Zone Management (ICZM) Strategy: General Principles and Policy Options* (Luxembourg: OOPEC)

2000 *Action for Scotland's Biodiversity* (Scottish Biodiversity Group, Scottish Executive)

2000 *Northern Ireland Waste Management Strategy*

2000 *Strategic Environmental Assessment of Plans and Programmes: Proceedings of the Intergovernmental Policy Forum Glasgow*

2000 *The Greening Government Initiative: First Annual Report from the Green Ministers Committee*

2002 *Directing the Flow: Priorities for Future Water Policy* (DEFRA)

2003 *Air Quality Strategy Addendum* (DEFRA)

2003 *Our Energy, Our Future: Creating a Low Carbon Economy White Paper* (ODPM)

2004   *Achieving a Better Quality of Life: Review of Progress towards Sustainable Development, Government Annual Report 2003*

2004   *IPPC: A Practical Guide* (DEFRA)

2004   *Planning and Climate Change: A Guide to Better Practice*

2005   *Securing the Future: Delivering UK Sustainability Strategy* (DEFRA)

Royal Commission on Environmental Pollution

1st Report: *Report of the Royal Commission on Environmental Pollution*, Cmnd 4585, 1972

2nd Report   *Three Issues in Industrial Pollution*, Cmnd 4894, 1972

3rd Report: *Pollution in Some British Estuaries and Coastal Waters*, Cmnd 5054, 1972

4th Report: *Pollution Control – Progress and Problems*, Cmnd 5780, 1974

5th Report: *Air Pollution Control – An Integrated Approach*, Cmnd 6371, 1976

6th Report: *Nuclear Power and the Environment*, Cmnd 6618, 1976

7th Report: *Agriculture and Pollution*, Cmnd 7644, 1979

8th Report: *Oil Pollution of the Sea*, Cmnd 8358, 1981

9th Report: *Lead in the Environment*, Cmnd 8852, 1983

10th Report: *Tackling Pollution – Experience and Prospects*, Cmnd 9149, 1984

11th Report: *Managing Waste: The Duty of Care*, Cmnd 9675, 1985

12th Report: *Best Practicable Environmental Option*, Cm 310, 1988

13th Report: *The Release of Genetically Engineered Organisms to the Environment*, Cm 720, 1989

14th Report: *GENHAZ: A System for the Critical Appraisal of Proposals to Release Genetically Modified Organisms into the Environment*, Cm 1557, 1991

15th Report: *Emissions from Heavy Duty Diesel Vehicles*, Cm 1631, 1991

16th Report: *Freshwater Quality*, Cm 1966, 1992

17th Report: *Incineration of Waste*, Cm 2181, 1993

18th Report: *Transport and the Environment*, Cm 2674, 1994

19th Report: *Sustainable Use of Soil*, Cm 3165, 1996

20th Report: *Transport and the Environment: Developments since 1994*, Cm 3752, 1997

21st Report: *Setting Environmental Standards*, Cm 4053, 1998

22nd Report: *Energy: The Changing Climate*, 2000

23rd Report: *Environmental Planning*, 2002

24th Report: *Chemicals in Products*, 2003

25th Report: *Turning the Tide: Addressing the Impact of Fisheries on the Marine Environment*, 2004

Heritage planning

1950   *Houses of Outstanding Historic and Architectural Interest (Gowers Report)*

1979   *A National Heritage Fund*, Cmnd 7428

1987   *Historic Buildings and Conservation Areas: Policy and Procedures*, DoE Circular 8/87

1990   *Archaeology and Planning*, PPG16

1990   *Conservation Areas of England* (EH)

1992   *Buildings at Risk: A Sample Survey* (EH)

1992   *Development Plan Policies for Archaeology* (EH)

1992   *Managing England's Heritage: Setting our Priorities for the 1990s* (EH)

1993   *Conservation Area Management* (EH)

1993   *Conservation Issues in Strategic Plans* (EH)

1993   *Finding and Minding: A Report on the Archaeological Work of the Department of the Environment Northern Ireland* (DoENI)

1994   *Ecclesiastical Exemption: What it is and How it Works* (EH)

1994   *Our Heritage: Preserving it, Prospering from it*, HC National Heritage Committee, Third Report, Session 1993–94, HC 139

1994   *Planning and the Historic Environment*, PPG15

1995   *Conservation in London: A Study of Strategic Planning Policy in London* (EH)

1995   *In the Public Interest: London's Civic Architecture at Risk* (EH)

1995   *Local Government Reorganisation: Guidance to Local Authorities on Conservation of the Historic Environment* (EH)

1996   *Conservation Areas in Strategic Plans* (EH)

1996   *Conservation Issues in Local Plans* (EH)

1996   *Protecting our Heritage: A Consultation Paper on the Built Environment of England and Wales (DNH)*

1996   *Something Worth Keeping: Post-war Architecture in England* (EH)

1997   *Planning and the Historic Environment: Notifications and Directions by the Secretary of State*, DETR Circular 14/97

1998   *Conservation-led Regeneration: The Work of English Heritage, English Heritage Enabling Development and the Conservation of Historic Assets* (EH)

1998   *Monuments on Record: Celebrating 90 Years of the Royal Commissions on Historical Monuments* (CD-Rom Royal Commission on Historical Monuments in England)

1998   *Planning and the Historic Environment* (Scottish Office)

1998   *Preservation Orders: Draft Regulations*, consultation paper

1998   *Tourism: Towards Sustainability*

1999   *Conservation Plans in Action* (EH)

1999   *Enabling Development and the Conservation of Heritage Areas* (EH)

1999   *Heritage Dividend: Measuring the Results of English Heritage Regeneratio* (EH)

1999   *Historic Core Zones Project, English Historic Towns Forum*

1999   *Historic Royal Palaces: Annual Report and Accounts 1998–99*, HC 598, 1998–9

1999   *Historic Scotland: Annual Report and Accounts 1998–99*, and *Corporate Plan 1999–2000*, SE 1999/7 (also HC 639, Session 1998–9)

1999   *Investment Performance of Listed Office Buildings* (Investment Property Databank) (EH and RICS)

1999   *Protecting and Conserving the Built Heritage in Wales* (National Audit Office Wales)

1999   *Royal Parks Agency: Annual Report and Accounts 1998–99*, HC 600 1998–9

1999   *Tomorrow's Tourism: A Growth Industry for the New Millennium*

2000   *British Government Panel on Sustainable Development: Sixth Report*

2000   *Consultation Paper on the Impact of the Shimuzu Judgement*

2000   *Contemporary Issues in Heritage and Environment Interpretation* (DETR)

2000   *Environment and Heritage Service Annual Report (NI) 1999–2000*, HCP 655, 1998–9

2000   *Indicators of Sustainable Development* (UK Round Table on Sustainable Development)

2000   *New Strategy for Scottish Tourism*

2000   *Not Too Difficult: Economic Instruments to Promote Sustainable Development Within a Modernised Economy* (UK Round Table on Sustainable Development)

2000   *Power of Place*

2000   *Seaside 2000*, consultation paper
2000   *The Stonehenge World Heritage Site Management Plan*
2000   *World Heritage Sites: The Tentative List of the United Kingdom of Great Britain and Northern Ireland*
2001   *Enabling Development and the Conservation of Heritage Assets: Policy Statement and Practical Guide to Assessment* (EH)
2001   *The Historic Environment: A Force for our Future* (DCMS)
2002   *Lottery Funding: The First Seven Years* (DCMS)
2002   *Passed to the Future: Historic Scotland's Policy for the Sustainable Management of the Historic Environment* (Historic Scotland)
2002   *State of the Historic Environment 2002* (EH)
2003   *Heritage Counts: The State of England's Historic Environment* (EH)
2004   *People and Places: Social Inclusion Policy for the Built and Historic Environment* (DCMS)
2004   *Review of Heritage Protection: The Way Forward* (DCMS)

The countryside

Publications of the Countryside Commission (CC), the Rural Development Commission (RDC) and their successor the Countryside Agency (CA) are obtainable from Countryside Agency Postal Sales, PO Box 124, Walgrave, Northampton NN6 9TL.

1942   *Report of the Committee on Land Utilisation in Rural Areas (Scott Report)*, Cmd 6378
1947   *Report of the National Parks Committee (England and Wales) (Hobhouse Report)*, Cmd 7121
1947   *Report of the Special Committee on Footpaths and Access to the Countryside (Hobhouse Subcommittee Report)*, Cmd 7207
1966   *Leisure in the Countryside*, Cmnd 2928
1975   *Food from our Own Resources*, Cmnd 6020
1975   *Sport and Recreation*, Cmnd 6200
1979   *Farming and the Nation*, Cmnd 7458
1983   *New Look at the Northern Ireland Countryside* (Jean Balfour)
1986   *Review of Forestry Commission Objectives and Achievements* (National Audit Office)
1987   *Annual Review of Agriculture 1987*, Cm 67
1990   *Dynamics of the Rural Economy* (ECOTEC, DoE)
1991   *Fit for the Future: Report of the National Parks Review Panel* (CC)
1991   *Permitted Development Rights for Agriculture and Forestry* (Land Use Consultants)
1992   *An Agenda for Investment in Scotland's Natural Heritage* (SNH)
1992   *Business Success in the Countryside: The Performance of Rural Enterprise* (D. Keeble *et al.*)
1992   *Coastal Zone Protection and Planning, HC Environment Committee, Second Report, Session 1991–92*, HC 17 [Government Response, Cm 2011]
1992   *Environmentally Sensitive Areas Wales: Aspects of Designation* (G. O. Hughes and A. M. Sherwood) (Welsh Office)
1992   *Indicative Forestry Strategies*, DoE Circular 29/92
1992   *Protected Landscapes in the United Kingdom* (CC)
1992   *Rural Framework* (Scottish Office)
1992   *Scottish Rural Life: A Socio-economic Profile of Rural Scotland* (Scottish Office)

1992 *Tir Cymen: A Farmland Stewardship Scheme* (CCW)

1993 *Coastal Planning and Management: A Review* (Rendel Geotechnics) (TSO)

1993 *Conserving England's Marine Heritage – A Strategy* (EN)

1993 *Countryside Survey 1990: Main Report* (DoE)

1993 *Development Below Low Water Mark*, discussion paper (DoE)

1993 *English Rural Communities* (Rogers, A.) (RDC)

1993 *Forestry and the Environment, HC Environment Committee, First Report, Session 1992–93*, HC 257

1993 *Managing the Coast*, discussion paper (DoE)

1993 *Role of the Countryside Commission in the Town and Country Planning System* (CC)

1993 *Rural Development and Statutory Planning* (ARUP Economics and Planning) (RDC)

1993 *The Economy and Rural England* (R. Tarling, J. Rhodes, J. North and G. Broom) (RDC)

1993 *Waterway Environment and Development Plans British Waterways*

1993 *Welsh Estuaries Review* (CCW)

1994 *Countryside Stewardship: An Outline* (CC)

1994 *Forestry and Woodlands: HC Select Committee on Welsh Affairs, First Report, Session 1993–94*, HC 35 [Government Response Cm 2645]

1994 *Lifestyles in Rural England* (P. Cloke, P. Milbourne and C. Thomas) (RDC)

1994 *National Forest: The Strategy* (CC)

1994 *Nature Conservation*, PPG9

1994 *Nature Conservation in Environmental Assessment* (EN)

1994 *Our Forests: The Way Ahead*, Cm 2644

1994 *Sustainable Forestry: The UK Programme*, Cm 2429

1995 *Countryside Planning File* (CC)

1995 *Countryside Stewardship Handbook* (CC)

1995 *Environmental Impact of Leisure Activities, HC Environment Committee, Fourth Report, Session 1994–95*, HC 246 [Government Response, HC 761]

1995 *Guide to Measures Available to Control the Recreational Use of Water* (Cobham Resources Consultants) (Scottish Office)

1995 *Planning for Rural Diversification* (M. Elson, R. Macdonald, R. Steenberg and G. Brown)

1995 *Planning for Rural Diversification: A Good Practice Guide* (M. Elson, C. Steenberg and J. Wilkinson)

1995 *Policy Guidelines for the Coast* (DoE)

1995 *Programme for Partnership: Announcement of the Outcome of the Scottish Office Review of Urban Regeneration Policy* (Scottish Office)

1995 *Protecting and Managing Sites of Special Scientific Interest in England, HC Committee of Public Accounts Eleventh Report, Session 1994–95*, HC 252

1995 *Rebuilding the English Countryside: Habitat Fragmentation and Wildlife Corridors in Practical Conservation* (EN)

1995 *Rural England: A Nation Committed to a Living Countryside*, Cm 3016

1995 *Rural Scotland: People, Prosperity and Partnership*, Cm 3041

1995 *Sustainable Rural Tourism* (CC)

1995 *Tourism in the Countryside* (RDC)

1996 *Impact of Tourism on Rural Settlements* (RDC)

1996 *Partnership in the Regeneration of Urban Scotland* (D. McAllister) (Scottish Office)

1996 *People, Parks and Cities: A Guide to Current Good Practice in Urban Parks* (DoE)

1996 *Rural England: The Rural White Paper, HC Environment Committee, Third Report, Session 1995–96*, HC 163

1996   *The Countryside: Environmental Quality and Economic Development* (DoE)
1996   *Working Countryside for Wales*, Cm 3180
1997   *Countryside, Environmental Quality and Economic and Social Development*, PPG 7
1997   *Disadvantage in Rural Areas* (RDC)
1997   *Economic Impact of Recreation and Tourism in the English Countryside* (RDC)
1997   *Hedgerows Regulations 1997: A Guide to the Law and Good Practice* (DoE)
1997   *Protecting Environmentally Sensitive Areas* (National Audit Office)
1997   *Rural Traffic: Getting it Right* (CC)
1997   *Survey of Rural Services* (RDC)
1997   *Towards a Development Strategy for Rural Scotland*, discussion paper (Scottish Office)
1998   *Access to the Countryside for Open-air Recreation* (SNH)
1998   *Access to the Open Countryside in England and Wales*, consultation paper (DETR)
1998   *A Home in the Country? Affordable Housing in Rural England* (RDC)
1998   *Guidance to the Regional Development Agencies on Rural Policy* (DETR)
1998   *Household Growth in Rural Areas* (RDC)
1998   *Investing in Quality: Improving the Design of New Housing in the Scottish Countryside*, consultation paper
       (Scottish Office)
1998   *National Scenic Areas*, consultation paper (SNH)
1998   *Natural Heritage Designations in Scotland* (SNH)
1998   *Protection of Field Boundaries* (HC Environment, Transport and Regional Affairs Committee, HC 969,
       Session 1997/98) [Government Response 1999]
1998   *Review of the Hedgerows Regulations 1997* (DETR)
1998   *Rural Development and Land Use Policies* (RDC)
1998   *Rural Disadvantage: Understanding the Processes* (RDC)
1998   *Sites of Special Scientific Interest: Better Protection and Management*, consultation paper (DETR)
1998   *Towards a Development Strategy for Rural Scotland* (Scottish Office)
1999   *Countryside Recreation – Enjoying the Countryside* (CC)
1999   *Farming and Rural Conservation Agency Annual Report 1998–99*, HC 765
1999   *National Forest – Guide for Developers and Planners* (National Forest Company)
1999   *Our Plan for the Future 2000–2004* (British Waterways)
1999   *Planning for the Quality of Life in Rural England* (CC)
1999   *Rural Economies* (Cabinet Office Performance and Innovation Unit)
1999   *Rural Wales: A Statement by the Rural Partnership* (Welsh Assembly)
1999   *State of the Environment of England and Wales: Coasts* (Environment Agency)
1999   *Town and Country Parks Twentieth Report of the Environment, Transport and Regional Affairs Committee,
       Session 1998–99*, HC 477 [Government Response, Cm 4550, 2000]
1999   *Unlocking the Potential: A New Future for British Waterways* (DETR)
2000   *National Parks (Scotland) Act – Explanatory Notes*
2000   *Sites of Special Scientific Interest: Encouraging Positive Partnerships*, public consultation paper on code of
       guidance
2000   *The Rural Development Plan for Wales 2000–06* (National Assembly for Wales)
2002   *A Winning Wales: The National Economic Development Strategy of the Welsh Assembly Government* (National
       Assembly for Wales)
2002   *Plan for Wales 2001* (National Assembly for Wales)
2002   *Review of National Park Authorities* (DEFRA)

2002   *Strategy for Sustainable Farming and Food: Facing the Future* (DEFRA)
2003   *The Haskins Report: Rural Delivery Review*
2004   *Delivering the Essentials of Life: DEFRA's Five Year Strategy*
2004   *Review of the 2000 Rural White Paper*
2004   *Social and Economic Diversity Change and Diversity in Rural England* (DEFRA)
2004   *The Rural Strategy* (DEFRA)

Housing and urban policy

1965   *Report of the Committee on Housing in Greater London (Milner Holland Report)*, Cmnd 2605
1967   *Areas of Special Housing Need, National Committee for Commonwealth Immigrants*
1968   *Older Houses into New Homes*, Cmnd 3602
1968   *The Older Houses in Scotland: A Plan for Action*, Cmnd 3598
1971   *Fair Deal for Housing*, Cmnd 4728
1973   *Better Homes: The Next Priorities*, Cmnd 5339
1973   *Homes for People: Scottish Housing Policy in the 1970s*, Cmnd 5272
1973   *Towards Better Homes: Proposals for Dealing with Scotland's Older Housing*, Cmnd 5338
1973   *Widening the Choice: The Next Steps in Housing*, Cmnd 5280
1977   *Policy for the Inner Cities*, Cmnd 6845
1981   *Priority Estates Project 1981: Improving Problem Council Estates*
1985   *Development in the Countryside and Green Belts*, SDD Circular 24/85 (Scottish Office)
1985   *Home Improvement: A New Approach*, Cmnd 9513
1985   *Home Improvement in Scotland: A New Approach*, Cmnd 9677
1985   *Land Use Planning and the Housing Market* (Coopers and Lybrand)
1985   *Urban Programme, National Audit Office (1984–85)*, HC 513
1986   *Demand for Housing: Economic Perspectives and Planning Practices* (D. MacLennan) (Scottish Office)
1986   *Evaluation of Environmental Projects Funded under the Urban Programme* (JURUE)
1986   *Evaluation of Industrial and Commercial Improvement Areas* (JURUE)
1986   *Urban Programme and the Young Unemployed* (C. Whitting)
1987   *Land Use Planning and Indicators of Housing Demand* (Coopers and Lybrand)
1987   *Review of Data Sources for Urban Policy* (ECOTEC)
1988   *Action for Cities* (Cabinet Office)
1988   *Evaluation of the Urban Development Grant* Programme (Public Sector Management Research Unit, Aston University)
1988   *Improving Inner City Shopping Centres: An Evaluation of Urban Programme Funded Schemes in the West Midlands* (Public Sector Management Research Centre, Aston University
1988   *New Life for Urban Scotland* (Scottish Office)
1988   *Socio-demographic Change and the Inner City* (M. Boddy *et al.*)
1988   *Stockbridge Village Trust: Building a New Community* (Roger Tym and Partners)
1988   *Urban Development Corporations, National Audit Office (1987–88)*, HC 492
1990   *Area Renewal, Unfitness, Slum Clearance and Enforcement Action*, DoE Circular 6/90
1990   *Employment in the 1990s*, Cm 540
1990   *Evaluation of Garden Festivals* (PA Cambridge Economic Consultants)
1990   *Green Paper on the Urban Environment* (CEC)

1990   *Patterns and Processes of Urban Change in the UK* (T. Fielding and S. Holford)

1990   *Regenerating the Inner Cities, National Audit Office (1989–90)*, HC 169

1990   *Targeting Urban Employment Initiatives* (I. Turok and U. Wannop)

1990   *Tourism and the Inner City: An Evaluation of the Impact of Grant Assisted Tourism Projects* (Polytechnic of Central London School of Planning *et al.*)

1990   *Urban Labour Markets: Reviews of Urban Research* (B. Moore and P. Townroe)

1990   *Urban Scotland into the 90s: New Life: Two Years On* (Scottish Office)

1990   *US Experience in Evaluating Urban Regeneration* (T. Barnekov, D. Hart and W. Benfer)

1991   *Planning and Affordable Housing*, DoE Circular 7/91

1992   *Developing Indicators to Assess the Potential for Urban Regeneration* (M. Coombes, S. Raybould and C. Wong)

1992   *Estate Action: New Life for Local Authority Estates: Guidelines for Local Authorities* (DoE)

1992   *Relationship between House Prices and Land Supply* (Gerald Eve and Department of Land Economy, University of Cambridge)

1992   *Scottish House Condition Survey 1991*

1993   *Evaluation of Urban Grant, Urban Regeneration Grant, and City Grant* (Price Waterhouse)

1993   *Future of Private Housing Renewal Programmes*, consultation paper (DoE)

1993   *Progress in Partnership* (Scottish Office)

1993   *Rural Housing, HC Welsh Affairs Committee, Third Report, Session 1992–93*, HC 621 [Government Response Cmnd 2375]

1994   *Assessing the Impact of Urban Policy* (B. Robson *et al.*) (DETR)

1994   *Employment Department Baseline Follow-up Studies* (Coopers & Lybrand)

1994   *Planning for Affordable Housing* (J. Barlow, R. Cocks and M. Parker)

1995   *Final Evaluation of Enterprise Zones* (PA Cambridge Consultants)

1995   *Impact of Environmental Improvements on Urban Regeneration* (PIEDA)

1995   *Involving Communities in Urban and Regional Regeneration* (DoE)

1995   *Programme for Partnership: Announcement of the Outcome of the Scottish Office Review of Urban Regeneration Policy* (Scottish Office)

1995   *Projections of Households in England to 2016*

1995   *Single Regeneration Budget, HC Environment Committee, First Report, Session 1995–96*, HC 26

1996   *City Challenge: Interim National Evaluation*

1996   *Evaluation of Six Early Estate Action Schemes* (Capita Management Consultancy) (DoE)

1996   *Government Response to the Environment Committee Report into the Single Regeneration Budget*, Cm 3178

1996   *Home-owners and Clearance: An Evaluation of Rebuilding Grants* (V. Karn, J. Lucas *et al.*)

1996   *Household Growth: Where Shall We Live?*, Cm 3471

1996   *Housing in England 1994/95: A Report of the 1994/95 Survey of English Housing* (H. Green *et al.*)

1996   *Housing Need, HC Environment Committee, Second Report, Session 1995–96*, HC 22

1996   *Partnership in the Regeneration of Urban Scotland* (D. McAllister) (Scottish Office)

1996   *People, Parks and Cities* (L. Greenhalgh, K. Worpole and R. Grove-White)

1996   *Planning and Affordable Housing*, DoE Circular 13/96

1996   *Private Sector Renewal: A Strategic Approach*, DoE Circular 17/96

1997   *Economic Determinants of Household Formation: Where Will They Go?* (G. Bramley, M. Munro and S. Lancaster) (DETR)

1997   *Housing Action Trusts: Evaluation of Baseline Condition* (Capita Management Consultants) (DETR)

1997   *Neighbourhood Renewal Assessment and Renewal Areas* (Austin Mayhead and Co) (DETR)

1997   *Planning Obligations*, DoE Circular 1/97

1997   *Towards an Urban Agenda in the European Union* (CEC)

1998   *Bringing Britain Together: A National Strategy for Neighbourhood Renewal (Social Exclusion Unit)*, Cm 4045

1998   *Community-Based Regeneration Initiatives: A Working Paper* (DETR)

1998   *English House Condition Survey 1996*

1998   *Evaluation of the Single Regeneration Challenge Fund Budget: A Partnership for Regeneration – An Interim Evaluation* (A. Brennan, J. Rhodes and P. Tyler) (DETR)

1998   *Household Growth in Rural Areas: The Household Projections and Policy Implications* (RDC)

1998   *Housing, HC Select Committee on the Environment, Transport and Regional Affairs, Tenth Report, Session 1997–98*, [Government Response, Cm 4080]

1998   *Impact of Urban Development Corporations in Leeds, Bristol and Central Manchester* (Centre for Urban Policy Studies, University of Manchester) (DETR)

1998   *New Deal for Communities: Phase I Proposals* (DETR)

1998   *Planning and Affordable Housing*, DETR Circular 6/98

1998   *Planning for the Communities of the Future*, Cm 3885

1998   *Regenerating London Docklands* (Cambridge Policy Consultants *et al.*) (DETR)

1998   *Regeneration Programmes – The Way Forward*, discussion paper (DETR)

1998   *Sustainable Urban Development in the European Union: A Framework for Action* (CEC)

1998   *Urban Development Corporations: Performance and Good Practice* (Roger Tym and Partners) (DETR)

1999   *City Challenge: Final National Evaluation* (KPMG Consulting) (DETR)

1999   *City-wide Urban Regeneration* (M. Carley and K. Kirk) (Scottish Executive Central Research Unit)

1999   *English Partnerships: Assisting Local Regeneration* (National Audit Office)

1999   *Evaluation of the New Life for Urban Scotland Initiative* (R. Tarling *et al.*) (Scottish Executive)

1999   *Housing: PPG3, HC Environment, Transport and Regional Affairs Committee, Seventeenth Report, Session 1998–99*, HC 490

1999   *Interim Evaluation of English Partnerships: Final Report* (PA Consulting Group) (DETR)

1999   *Literature Review of Social Exclusion* (P. Lee and A. Murie) (Scottish Office)

1999   *Local Evaluation for Regeneration Partnerships: Good Practice Guide* (DETR)

1999   *New Deal for Communities: Learning Lessons: Pathfinders' Experiences on NDC Phase I* (DETR)

1999   *Opportunity for All: Tackling Poverty and Social Exclusion*, Cm 4445 (Department of Social Security)

1999   *Projections of Households in England to 2021* (DETR)

1999   *Running and Sustaining Renewal Areas: Good Practice Guide* (DETR)

1999   *Sustainable Urban Development in The European Union: A Framework for Action* (Luxembourg: OOPEC)

1999   *Toward an Urban Renaissance (Report of the Urban Task Force)* (London: Spon)

2000   *City Challenge: Final National Evaluation*

2000   *Conversion and Redevelopment: Process and Potential*

2000   *Coordination of Area-based Initiatives*, research working paper (DETR)

2000   *Good Practice Guide on Monitoring the Provision of Housing through the Planning System* (DETR)

2000   *Local Strategic Partnerships*, consultation document

2000   *Millennium Villages and Sustainable Communities: Final Report*

2000   *Monitoring Provision of Housing through the Planning System: Towards Better Practice*

2000   *New Deal for Communities: Delivering Change (and Other Guidance)*

2000   *Quality and Choice: A Decent Home for All: the Way Forward for Housing* (DETR)

2000   *Town and Country Planning (Residential Development on Greenfield Land, England), Direction 2000* (DETR)

2000   *Urban Regeneration Companies: A Process Evaluation* (M. Parkinson and B. Robson) (DETR)

2001 *A New Commitment to Neighbourhood Renewal* (Cabinet Office)
2001 *A Smart, Successful Scotland* (Scottish Executive)
2001 *Our Towns and Cities: The Future – An Implementation Plan* (ODPM)
2002 *Better Communities in Scotland: Closing the Gap* (Scottish Executive)
2002 *Building Better Cities* (Scottish Executive)
2002 *Review of Scotland's Cities: The Analysis* (Scottish Executive)
2002 *Urban Policy Evaluation Strategy, Consultation* (ODPM)
2002 *What Works? Reviewing the Evidence Base for Neighbourhood Renewal* (ODPM)
2003 *Assessing the Impact of Spatial Interventions; Regeneration, Renewal and Regional Development, Guidance* (ODPM)
2003 *English House Condition Survey* (ODPM)
2003 *Modernising Scotland's Social Housing*, consultation paper
2003 *Neighbourhood Renewal Strategy: People and Place* (DSD, NI)
2003 *Scottish House Condition Survey 2002* (Communities Scotland)
2003 *Sustainable Communities: Building for the Future*
2004 *Decent Homes, House of Commons ODPM: Housing, Planning, Local Government and the Regions Committee Fifth Report*
2004 *Housing Health and Safety Rating System Guidance (Version 2)* (ODPM)
2004 *Housing in England 2001/2: A Report on the 2001/02 Survey of English Housing* (ODPM)
2004 *Making it Happen: Urban Renaissance and Prosperity in our Core Cities: A Tale of Eight Cities* (ODPM)
2004 *Property Advisory Group Annual Report 2002–03* (ODPM)
2004 *Sustainable Communities: An Urban Development Corporation for West Northamptonshire: A Decision Document* (ODPM)
2005 *English House Condition Survey: Key Findings for 2003* (ODPM)
2005 *Improving the Prospects of People Living in Areas of Multiple Deprivation in England* (Cabinet Office)
2005 *Sustainable Communities: Homes for All* (ODPM)

Transport

1963 *Traffic in Towns (Buchanan Report)*
1964 *Road Pricing: The Economic and Technical Possibilities (Smeed Report)*
1966 *Transport Policy*, Cmnd 3057
1967 *Public Transport and Traffi*, Cmnd 3481
1968 *Transport in London*, Cmnd 3686
1969 *Roads for the Future: A New Inter-Urban Plan*
1972 *New Roads in Towns*
1973 *Report of the Urban Motorways Project Team to the Urban Motorways Committee*
1976 *Transport Policy: A Consultative Document*
1977 *Transport Policy*, Cmnd 6836
1978 *Policy for Roads, England 1978*, Cmnd 7132
1978 *Report on the Review of Highway Inquiry Procedures*, Cmnd 7133
1980 *Policy for Roads, England 1980*, Cmnd 7908
1981 *Lorries, People and the Environment*, Cmnd 8439
1982 *Policy for Roads, England 1981*, Cmnd 8496
1982 *Public Transport Subsidy in Cities*, Cmnd 8735

1983   *Policy for Roads in England*, Cmnd 9059

1983   *Public Transport in London*, Cmnd 9004

1983   *Roads in Scotland: Report for 1982*, Cmnd 9010

1984   *Buses*, Cmnd 9300

1985   *Airports Policy*, Cmnd 9542

1986   *The Channel Fixed Link*, Cmnd 9735

1987   *Policy for Roads in England 1987*, Cm 125

1989   *National Road Traffic Forecasts (Great Britain) 1989*

1989   *New Roads by New Means – Bringing in Private Finance*, Cm 698

1989   *Roads for Prosperity*, Cm 693

1990   *Road Pricing for London* (London Boroughs Association)

1990   *Traffic Quotes: Public Perception of Traffic Regulation in Urban Areas* (P. Jones)

1990   *Trunk Roads, England: Into the 1990s* (DoT)

1991   *Cycling: HC Transport Committee, Minutes of Evidence, 8 May 1991, Session 1990–91*, HC 423

1991   *Railway Noise and the Insulation of Dwellings: Report of the Committee to Recommend a National Noise Insulation Standard for New Railway Lines (Mitchell Report)*

1991   *Urban Public Transport: The Light Rail Option, HC Transport Committee Fourth Report, Session 1990–91*, HC 14

1992   *Developers' Contributions to Highway Works: Consultation Document* (DoT)

1992   *Developers' Contributions to Highway Works: Efficiency Scrutiny Report and the Department's Response* (DoT)

1992   *Role of Investment Appraisal in Road and Rail Transport* (DoT)

1992   *Traffic in London: Traffic Management and Parking Guidance*, DoT Local Authority Circular 5/92

1993   *Paying for Better Motorways*, Cm 2200

1994   *Charging for the Use of Motorways, HC Transport Committee, Fifth Report, Session 1993–94*, HC 376

1994   *Impact of Transport Policies in Five Cities* (M. Dasgupta *et al.*) (Transport Research Laboratory)

1994   *Trunk Roads, England: 1994 Review* (DoT)

1995   *Better Places through Bypasses: Report of the Bypass Demonstration Project*

1995   *Guidance on Decriminalised Parking Enforcement outside London*, DoT Local Authority Circular 1/95

1995   *Guidance on Induced Traffic*, DoT Guidance Note 1/95, (DoT)

1995   *London Congestion Charging Research Programme: Principal Findings* (MVA Consultancy)

1995   *Managing the Trunk Road Programme* (DoT)

1995   *PPG13: A Guide to Better Practice: Reducing the Need to Travel through Land Use and Transport Planning* (JMP Consultants)

1995   *Urban Road Pricing, HC Transport Committee, Third Report, Session 1994–95*, HC 104 [Government Response, Cm 3019]

1996   *Cycling in Great Britain (Transport Statistics Report)* (DoT)

1996   *National Cycling Strategy* (DoT)

1996   *Transport Strategy for London*

1996   *Transport: The Way Forward*, Cm 3234

1997   *Developing a Strategy for Walking* (DETR)

1997   *Keeping Buses Moving: A Guide to Traffic Management to Assist Buses in Urban Areas*, DoT Local Transport Note 1997/1

1997   *Keeping Scotland Moving: A Scottish Transport Green Paper*, Cm 3565

1997   *National Road Traffic Forecasts (Great Britain)* (DETR)

1997   *Regulation of Heavy Lorries* (National Audit Office)

1998   *Breaking the Logjam: The Government's Consultation Paper on Fighting Traffic Congestion and Pollution through Road User and Workplace Parking Charges* (DETR)

1998   *Guidance on Local Transport Plans* (DETR)

1998   *Moving Forward: Northern Ireland Transport Policy Statement* (DoENI)

1998   *New Deal for Transport: Better for Everyone*, Cm 3950

1998   *New Deal for Trunk Roads in England* (DETR)

1998   *Parking Standards in the South East* (Llewelyn Davies and JMP Consultants) (DETR)

1998   *Role of Technology in Implementing an Integrated Transport Policy* (Office of Science and Technology DTI)

1998   *Transporting Wales into the Future* (Welsh Office)

1998   *Travel Choices for Scotland: The Scottish Integrated Transport White Paper*, Cm 4010

1999   *Benefits of Green Transport Plans* (DETR)

1999   *Community Impact of Traffic Calming Schemes in Scotland* (Scottish Executive Central Research Unit)

1999   *From Workhorse to Thoroughbred: A Better Role for Bus Travel* (DETR)

1999   *Integrated Transport White Paper: Report of the HC Environment, Transport and Regional Affairs Committee, Session 1998–99*, HC 32 [Government Response Third Special Report, Session 1998–99, HC 708]

1999   *Local Transport Strategies: Preliminary Guidance* (SDD)

1999   *Research on Walking (System Three)* (Scottish Executive Central Research Unit)

1999   *Revision of Planning Policy Guidance Note 13, Transport: Public Consultation Draft*, PPG13

1999   *School Travel: Strategies and Plans. A Best Practice Guide for Local Authorities* (DETR)

1999   *Sustainable Distribution: A Strategy, DETR1999 School Travel: Strategies and Plans: A Best Practice Guide for Local Authorities* (DETR)

1999   *Tackling Congestion: The Scottish Executive's Consultation Paper on Fighting Traffic Congestion and Pollution through Road User and Workplace Parking Charges* (Scottish Executive)

1999   *Transport and the Economy* (SACTRA)

1999   *Transport Statement for Northern Ireland*

1999   *Transport: The Way Forward*, Cm 3234

1999   *Travel Choices for Scotland: Strategic Roads Review*

2000   *Commission for Integrated Transport First Annual Report*

2000   *Encouraging Walking: Advice to Local Authorities* (DETR)

2000   *Good Practice Guide for the Development of Local Transport Plans* (DETR)

2000   *Guidance on Full Local Transport Plans* (DETR)

2000   *Guidance on the New Approach to Appraisal and Understanding the New Approach to Appraisal, 2000*

2000   *Road Charging Options for London* (DETR)

2000   *Transport 2010: The 10 Year Plan* (DETR)

2000   *Waterways for Tomorrow*

2001   *Transport 2010: Meeting the Local Transport Challenge* (DETR)

2003   *Managing our Roads* (DfT)

2004   *The Future for Transport*, White Paper (DfT)

2004   *Walking and Cycling Action Plan*

2004   *Walking and Cycling: Success Stories* (DfT)

Standing Advisory Committee on Trunk Road Assessment

1977   *Trunk Road Assessment Report*

1979   *Trunk Road Proposals: A Comprehensive Framework for Appraisal*

1986   *Urban Road Appraisal*
1992   *Assessing the Environmental Impact of Road Schemes*
1994   *Trunk Roads and the Generation of Traffic*
1999   *Transport and the Economy*

Planning, the profession and the public

1950   *Report of the Committee on the Qualifications of Planners (Schuster Report)*, Cmnd 8204
1957   *Report of the Committee on Administrative Tribunals and Enquiries (Franks Report)*, Cm 218
1969   *Information and the Public Interest*, Cmnd 4089
1969   *People and Planning: Report of the Committee on Public Participation in Planning (Skeffington Report)*
1978   *Planning Procedures: The Government's Response to the Eighth Report from the Expenditure Committee, Session 1976–77*, Cmnd 7056
1978   *Report on the Review of Highway Inquiry Procedures*, Cmnd 7133
1986   *Planning Appeals, Call-in and Major Public Inquiries: The Government's Response to the Fifth Report from the Environment Committee*, Cm 43
1991   *The Citizen's Charter: Raising the Standard*, Cm 1599
1992   *Green Rights and Responsibilities: A Citizen's Guide to the Environment* (DoE)
1992   *The Citizen's Charter: First Report*, Cm 2101
1994   *Mediation: Benefits and Practice* (DoE)
1995   *Accessing Environmental Information in Scotland* (J. Moxen *et al.*) (Scottish Office)
1995   *Attitudes to Town and Country Planning* (P. McCarthy and T. Harrison)
1995   *Community Involvement in Planning and Development Processes*
1995   *Environmental Education; Is The Message Getting Through?* (MORI) (Scottish Office)
1995   *Environmental Facts: A Guide to Using Public Registers of Environmental Information* (DoE)
1997   *Involving Communities in Urban and Rural Regeneration* (revised) (ODPM)
1997   *Rights Brought Home: The Human Rights Bill*, Cm 3782
1997   *Your Right to Know: The Government's Proposals for a Freedom of Information Act*, Cm 3818
1998   *Guidance on Enhancing Public Participation in Local Government* (V. Lowndes) (DETR)
1998   *Mediation in the Planning Process*
1999   *An Open Scotland: Freedom of Information – A Consultation*, SE/1999/51
1999   *Freedom of Information: Consultation on Draft Legislation*, Cm 4355
1999   *Freedom of Information Draft Bill, HC Select Committee on Public Administration, Third Report, Session 1998–99*, HC 570
1999   *Improving Enforcement Appeal Procedures*, consultation paper (DETR)
1999   *Role and Effectiveness of Community Councils with Regard to Community Consultation* (R. Goodlad) (Scottish Executive Central Research Unit)
2001   *Local Strategic Partnerships: Government Guidance* (ODPM)
2001   *Preparing Community Strategies: Government Advice to Local Authorities*
2002   *Developing Effective Relationships between Community Strategies and Local Development Frameworks* (ODPM)
2002   *Further Research into Mediation in the Planning System* (ODPM)
2002   *Information Communications Technology in Planning* (ODPM)
2002   *Participatory Planning for Sustainable Communities* (ODPM)
2002   *Reforming Public Services: Principles into Practice* (ODPM)

2003 *Equality and Diversity in Local Government in England: A Literature Review* (ODPM)
2003 *Searching for Solid Foundations: Community Involvement and Urban Policy* (ODPM)
2004 *A Draft Practical Guide to Strategic Environmental Assessment Directive* (ODPM)
2004 *Community Involvement in Planning: The Government's Objectives* (ODPM)
2004 *Creating Sustainable Communities: Consultation on PPS 1* (ODPM)
2004 *Investigating the Increasing Volume of Planning Appeals in England* (ODPM)
2004 *LSP Evaluation and Action Research Programme: Case-Studies Interim Report: A Baseline of Practice* (ODPM)
2004 *Review of the Planning Inspectorate*
2004 *Review of the Publicity Requirements for Planning Applications* (ODPM)
2005 *Citizen Engagement and Public Services: Why Neighbourhoods Matter* (Home Office)

Index of Statutes

Index of Authors

General Index